When Professor Richards' classic book first appeared in 1952, the tropical rain forests could be thought of as stretching endlessly over vast areas of the humid tropics, but now their nearly complete destruction is not far from realisation. Over the years, ecological ideas have changed in many respects. The old notion of the stability of rain forest over long periods of time has been replaced by a dynamic concept of rain forests as kaleidoscopic mosaics continually reacting to climatic changes and human pressures.

The enormous growth of interest in tropical forests has led to an explosion of new data and ideas. This new and completely rewritten version provides a wide-ranging view of the field by one of the major contributors to our current understanding of rain-forest ecology. New chapters have been contributed on climate and microclimate by R.P.D. Walsh, and on soils of the humid tropics by I.C. Baillie, and there is an appendix on numerical methods in rain forest by P. Greig-Smith.

The new book will stand as a record of what the rain forest was like in the twentieth century.

The tropical rain forest
an ecological study

The tropical rain forest
an ecological study

Second edition

P.W. Richards

Emeritus Professor of Botany, University of Wales, Bangor

with contributions by
R.P.D. Walsh
Department of Geography, University of Wales, Swansea
I.C. Baillie
Environmental Consultant, Bedford
and P. Greig-Smith
Emeritus Professor of Botany, University of Wales, Bangor

CAMBRIDGE
UNIVERSITY PRESS

PUBLISHED BY THE PRESS SYNDICATE OF THE UNIVERSITY OF CAMBRIDGE
The Pitt Building, Trumpington Street, Cambridge CB2 1RP, United Kingdom

CAMBRIDGE UNIVERSITY PRESS

The Edinburgh Building, Cambridge CB2 2RU, UK http://www.cup.cam.ac.uk
40 West 20th Street, New York, NY 10011-4211, USA http://www.cup.org
10 Stamford Road, Oakleigh, Melbourne 3166, Australia

First published 1952
Second edition first published 1996
Reprinted 1998

Printed in the United Kingdom at the University Press, Cambridge

Typeset in Linotronic Aldus 9.5/12 pt [wv]

A catalogue record for this book is available from the British Library

Library of Congress Cataloguing in Publication Data

Richards, Paul W. (Paul Westmacott)
The tropical rain forest : an ecological study / P.W. Richards;
with contributions by R.P.D. Walsh, I.C. Baillie, and P. Greig-
Smith. – 2nd ed.
p.   cm.
Includes bibliographical references and index.
ISBN 0 521 42054 7. – ISBN 0 521 42194 2 (pbk.)
1. Rain forest ecology. I. Title.
QK938.R34R53 1996
581.5′2642′0913–dc20   93-49019 CIP

ISBN 0 521 42054 7 hardback
ISBN 0 521 42194 2 paperback

To the memory of my friends
Carl Schroeter (1855–1939) whose *Pflanzenleben der
Alpen* inspired me to write this book
and
Agnes Arber (1879–1960) to whose encouragement
and constructive criticism it owes so much

Contents

VI Human impacts and the tropical rain forest

Preface to the second edition

The publication of the first edition of this book in 1952 was followed by reprints in 1957, 1964, 1966, 1972 and 1976 and a paperback edition in 1979. All of them included small corrections and additions, but none was fully revised. The present edition is virtually a new book as it has been completely rewritten; very little of the original text remains. The short section of Chapter 4 on the flowers and fruits of rain-forest trees has been expanded into a new chapter and the chapters on climate, microclimates and soil conditions have been replaced by entirely new chapters, the first two by Dr R.P.D. Walsh, the third by Dr Ian C. Baillie. Two appendices, the first on tree species recognition in the field, and the second (by Professor P. Greig-Smith) on quantitative methods in rain forest ecology, have been added.

Since *The tropical rain forest* first appeared, knowledge of rain-forest ecology has grown enormously. Interest in tropical forests has also spread among the scientific and wider public throughout the world. The term 'tropical rain forest' itself, which was once familiar only to professional botanists and foresters, is now seen in newspapers and popular magazines and heard almost daily on radio and television. At the same time the scientific literature on the ecology of rain forests has increased so rapidly that it is hardly possible any more for a single individual to keep abreast of it.

But even more important than the transformation in human awareness and attitudes to tropical rain forests is the change in the forests themselves. Up to the middle of this century they could still be thought of as stretching endlessly over vast areas of the humid tropics, but now their nearly complete destruction, foretold in the first edition of this book, is not far from realisation. Only relatively small areas remain where the natural forest has not been replaced by farms, plantations and secondary vegetation. Though the forest at Moraballi Creek, Guyana, which formed much of the subject matter of Edition 1, still stood in 1979 more or less as it was in 1929, many of the other areas referred to have been altered so much as to be no longer recognizable. Fortunately, owing to the efforts of some enlightened governments, the World Wide Fund for Nature, and other organizations, a few rain-forest areas are being conserved, but there is an urgent need for much more to be done.

The pace of change in both rain-forest ecology and the forests themselves has made the writing of this new *Tropical rain forest* a long and difficult task. The ever-expanding bulk of literature has made it impossible to cover the field to the extent which was possible in 1952. Like its predecessor, the present book does not aim to be encyclopaedic. What has been put in and what left out reflects the author's experience and particular interests. Some branches of rain-forest ecology have been dealt with in cavalier fashion, for example productivity and biomass, not because they have been judged unimportant, but because the author feels ill-equipped to discuss them. Some subjects, for example mangrove ecology, are now very adequately dealt with in other books and do not need to be treated at length here. On conservation, vitally important though it is, only a short Postscript (Chapter 19) has been given; a full discussion would have greatly increased the size of the book. It has been impossible to deal adequately with the role of animals in rain-forest ecosystems; the book has had to be written (as one of its critics has said) from 'a botanist's point of view'.

Since 1952 ecological ideas, including the author's,

have changed in many respects. The old notion of the stability of rain forests over long periods of time has been replaced by a dynamic concept of rain forests as kaleidoscopic mosaics continually reacting to climatic changes and human pressures. The rather rigid concept of rain-forest structure and stratification of Edition 1 has become less formal.

This book will, I hope, be regarded as a record of what the tropical rain forest was like in the twentieth century. Revising a book originally published forty years ago is like renovating an old building; the facade may retain something of its old appearance, but it is hoped that the interior has been sufficiently modernized to be still serviceable.

P.W. Richards
Cambridge

Preface to the first edition

The scope of ecology is not easy to define—it has even been said that the only definition of ecology is that it is the subject-matter of the *Journal of Ecology*. In writing a book about the ecology of the Tropical Rain forest I have therefore had to decide for myself what was and what was not relevant to my theme; in this I have been influenced, no doubt, by my own particular whims and prejudices.

Because ecology is a synthetic science, embracing or touching many other disciplines, it has been my ambition to interest many who are neither botanists nor foresters—zoologists, geographers, in fact anyone who is concerned with the rain forest as a plant community or an environment. I have dealt scarcely at all with the economic aspects of my subject; my aim has been to provide a basis for future work, whether 'pure' or 'applied'. Because I hope the book will be of use to those not trained as professional botanists, I have tried to make the text as self-explanatory as possible and to avoid unnecessary technical terms.

No general account of the Tropical Rain forest has been written since A.F.W. Schimper published his great *Plant Geography* (1898; English edition, 1903), which has since been revised and expanded by Prof. F.C. von Faber (1935). My main qualification for such a formidable task is first-hand experience of rain-forest vegetation adding up to nearly two years. As this experience, though short, was very intensive, and as I had the unusual good fortune to visit each of the three chief tropical regions—South America, Africa and Malaysia—within a space of seven years, this qualification is perhaps not as painfully inadequate as it appears. The great development of interest in tropical vegetation during the last fifteen years has given rise to a voluminous and very scattered literature. In writing the book I have endeavoured to make full use of this, but much has had to be deliberately neglected and still more has probably been unintentionally overlooked. In a work of this kind it is inevitable that many statements will prove to be wrong, and in some places the facts may prove to have been misinterpreted. These shortcomings may not matter if the book stimulates further work. In my travels I have been impressed by the large amount of valuable ecological information which exists unpublished in the minds and notebooks of foresters and buried in departmental reports; I hope the publication of this book may coax some of these data from their hiding places. Much valuable information has been obtained from letters from various friends; the source of such data is indicated in the text by the name of my correspondent in brackets without date. In every chapter I have tried to point out the chief gaps in present knowledge and to suggest lines for future work. No better prospect for my work can be wished than that it may soon become out of date.

Many ecologists would agree that their science is not yet ripe for a rigid theoretical framework, but since a theoretical background of some kind is necessary, the general principles of the Anglo-American school of ecologists have been followed. The absence of a chapter on biotic factors is due, not to a failure to realize their importance, but to the lack of a sufficient body of suitable data.

Part of the matter in the book has appeared in a series of papers published from 1933 onwards. As might be expected, I have since modified some of the views and interpretations given in those papers.

With regard to the nomenclature of species, it is obvious that in a work of this kind, in which names are quoted from papers and books dealing with the flora or vegetation of many different countries, the author cannot answer for the correctness of every name used, though the nomenclature has been checked as far as time and opportunity have allowed. I am much indebted to various members of the staff of the Kew Herbarium for helping me in this part of the work. Where information has been taken from published books or papers the names given here are not always those used in the original, but some synonyms will be found in the Index of Plant Names. In a few instances names of plants in the text have been placed in inverted commas; this indicates that the validity of the name or the correct citation is doubtful.

I could not have written a general account of the tropical rain forest without the help of many kind friends. Though it is impossible to acknowledge individually the help of all who have provided data, references to literature, or who have assisted in other ways, a word of special thanks is due to Dr Agnes Arber, F.R.S., who has given me much valuable advice on matters of presentation and has read and criticized a large part of the manuscript. Also to Sir Edward Salisbury, C.B.E., F.R.S., whose help in planning the book was invaluable; it was also a suggestion of his that gave rise to the 'profile-diagram' technique which has proved such a useful tool in the study of tropical vegetation. Special thanks for help of various kinds are also due to Dr J.S. Beard, Prof. H.G. Champion, Dr E.M. Chenery, Mr E.J.H. Corner, Mr T.A.W. Davis, Dr G.C. Evans, Mr P.J. Greenway, Prof. F. Hardy, the late Mr A.P.D. Jones, Mr R.W.J. Keay, Prof. J. Lebrun, Prof. G. Manley, Mr R. Ross, Dr C.G.G.J. van Steenis, Mr C. Swabey, Prof. J.S. Turner, Prof. T.G. Tutin, and Dr Frans Verdoorn. A word of gratitude is also due to the librarians of several libraries who have assisted me in searching for literature, especially the Librarian of the Imperial Forestry Institute, Oxford. For permission to reproduce figures and photographs, I have to thank Dr J.R. Baker, Dr J.S. Beard, Mr W.J. Eggeling, Dr G.C. Evans, Dr E.W. Jones, Prof. F.W. Went, the Director of the Musée Royale d'Histoire Naturelle de Belgique, the Editor of the *Bulletin du Jardin Botanique de Buitenzorg*, the Forestry Department, Malayan Union and the Editor of the *Journal of Ecology*. Lastly, I am indebted to my wife for much help, especially in preparing the indexes.

P.W. Richards
Botany School, Cambridge
August 1948

Acknowledgements

In addition to the three contributors named on the title page, I am grateful to a great many friends and colleagues all over the world who have helped me with information, sent me reprints of their papers and allowed me to reproduce their figures or photographs. I am especially indebted to Dr Paul Adam, Professor P.S. Ashton, Professor D.A. Janzen, Dr Michael Lock, Professor G.T. Prance, Professor Richard E. Schultes, Professor Len Webb, Dr T.C. Whitmore and Dr R.J. Whittaker, who have criticized parts of the book and helped me in other ways. Dr Sean Edwards has given me much valuable advice on photography as well as drawing several of the figures. A special word of thanks is due to Mrs Wendy Whitmore for the immense trouble she has taken in typing the lists of references and plant names. Roy Perry has given me invaluable help with proofreading. Lastly I owe very much to the support and patience of my wife who, as well as compiling the lists of references and plant names, has endured all the deprivations and inconveniences that writing the book has involved.

I thank the Royal Society and the Tansley Fund of the New Phytologist Trust for contributing to the cost of typing the manuscript.

Note on geographical names

The geographical names used in this book are, with a few exceptions, those in *The Times Atlas of the World*, seventh comprehensive edition (Times Books, London, 1985). 'Tropical America' is used for Central and South America. 'Malesia' is used for the Malay Peninsula and Malay Archipelago, including New Guinea and the Philippines, as in *Flora Malesiana* (Nijhoff, The Hague, 1948–). To avoid tedious repetition, 'Malaya' is used for the Malay Peninsula (Peninsular Malaysia plus Singapore).

Most studies of vegetation have been carried out in Europe, and I am of the opinion that owing to a paucity of material these investigations have begun with an inverted viewpoint. When studying the manifold types of vegetation, comparing them and relating them to each other, one ought logically to start with the richest and to derive from it the less complicated, impoverished types which have arisen from it by selection. The richest type of vegetation in number of species, volume and density, is found in the tropics. It is not the impoverished anthropogenic vegetation of Europe which should be the starting-point of one's investigations.

C.G.G.J. VAN STEENIS (1937, transl.)

Chapter 1

Introduction

ITS LANDS ARE high and there are in it very many sierras and very lofty mountains, beyond comparison with the island of Teneriffe. All are most beautiful, of a thousand shapes, and all are accessible and filled with trees of a thousand kinds and tall, and they seem to touch the sky. And I am told that they never lose their foliage, as I can understand, for I saw them as green and as lovely as they are in Spain in May and some of them were flowering, some bearing fruit and some in another stage, according to their nature.

In these words from Christopher Columbus's account (Jane 1930) of the island of Española, written in 1493, we have a glimpse of a tropical landscape and perhaps the earliest description in Western literature of the kind of vegetation with which this book is concerned. Evergreen tropical forest, which in Columbus's time covered most of the West Indian islands, is today rapidly decreasing in area, but not long ago stretched over almost all of the equatorial lowlands of South America, Africa and southeastern Asia. It is the characteristic vegetation of the humid tropics and occupies (or formerly occupied) all land surfaces with a sufficiently hot climate, and a sufficiently heavy and well-distributed rainfall, except for the small areas where the ground is too swampy, is otherwise unsuitable for trees to grow, or where, as on young volcanic lava, there has not been time for it to develop. In ecological terms, evergreen forest is the climax, or potential natural vegetation, of humid equatorial climates.

In scientific literature this evergreen forest is most often referred to as tropical rain forest (*tropische Regenwald*), a term coined for it by A.F.W. Schimper in his classical *Plant Geography* (1898, 1903); the Latin

Fig. 1.1 Tropical rain forest, 800 m above sea-level, Arfak Mountains, West Irian (New Guinea) (*ca.* 1930). From van Straelen (1933, pl. XXXVIII).

equivalent *pluviisylva* is also sometimes used. Humboldt, one of the first adequately to describe the great South American rain forest, referred to it as the *Hylaea* (from ῦλη, a forest), and German writers often use the same word for the evergreen tropical forest of Africa. In German the word *Urwald*, is also frequently used, meaning original or primitive forest, a term which has no necessary tropical implication, but which custom has attached to the tropical rain forest rather than to other types of primitive forest. To the Englishman in the tropics the rain forest was usually the 'bush' or the 'jungle', to Australians 'brush' or 'scrub'.

The name 'rain forest' is commonly given, not only to the evergreen forest of moist tropical lowlands – the plant formation (or better, the formation-type) dealt with in this book – but also to the somewhat less luxuriant broad-leaved evergreen forest found on tropical mountains, and to the evergreen forests of oceanic subtropical and temperate climates, in southern China, southern Chile, South Africa, New Zealand and eastern extra-tropical Australia. These other formation-types will here be called montane rain forest, subtropical and temperate rain forest[1], respectively. In this book 'rain forest' without qualification means tropical rain forest.

What is a tropical rain forest? Schimper (1903, p. 260) gave this brief diagnosis: 'Evergreen, hygrophilous in character, at least 30 m. high, but usually much taller, rich in thick-stemmed lianes and in woody as well as herbaceous epiphytes'. This definition fits the concept of tropical rain forest as most writers since Schimper have used it and is the one which will be adopted here, but some modern workers wish to use the term in a less comprehensive sense or to abandon it altogether. In this narrower sense the term rain forest would be reserved for the almost completely non-seasonal forest of tropical climates with the most evenly distributed rainfall; it would not be applied to the more seasonal types of evergreen forest found in areas with a distinct annual dry season. To this question we shall return in Chapters 16 and 17, where an attempt will be made to make clearer the differences between tropical rain forest on the one hand and, on the other, montane rain forest, subtropical and temperate rain forest, and the various more or less deciduous tropical forest formations classified by Schimper (1903) as monsoon forest, savanna forest and thorn forest.

A short definition, even if it could be precise, would give but a poor idea of what a tropical rain forest is, especially to those who are familiar only with the vegetation of temperate countries such as Europe and North America. What are its chief characteristics? What kind of an impression does it make on an observer?

Most people who have no first-hand experience of tropical vegetation derive their ideas about it from books of travel, and these unfortunately are often biased or exaggerated, even when not positively inaccurate. Errors arise because travellers have been too quick to generalize from a single area, or because they have travelled mainly by river and have been misled into supposing that the interior of the forest is like the riverside fringe. The main source of inaccuracy, however, is that tropical vegetation has a fatal tendency to produce rhetorical exuberance in those who describe it. Few writers on the rain forest seem able to resist the temptation of the 'purple passage', and in the rush of superlatives they are apt to describe things they never saw or to misrepresent what was really there. In attempting to paint an accurate picture of the rain forest it will be necessary to point out and correct some of these common errors.

One of the outstanding features of the tropical rain forest, indeed, of all the vegetation of the humid tropics,[2] is that the overwhelming majority of the plants are woody and are of the dimensions of trees. Not only do trees form the dominants of the rain-forest community, but most of the climbing plants and some of the epiphytes are also woody. The undergrowth largely consists of woody plants: seedling and sapling trees, shrubs (some of which have a single mainstem and may be termed treelets, see p. 18), and young woody climbers. The only herbaceous plants are some of the epiphytes and a proportion, often relatively small, of the undergrowth. Families of plants which in temperate regions are represented exclusively by low-growing herbs here assume the size and woodiness of trees. Spruce (1908, vol. 1, p. 256) says of the Amazon forests:

Nearly every natural order of plants has here trees among its representatives. Here are grasses (bamboos) of 40, 60, or more feet in height, sometimes growing erect, sometimes tangled in thorny thickets, through which an elephant could not penetrate. Vervains form spreading trees with digitate leaves like the horse-chestnut. Milkworts, stout woody twiners ascending to the tops of the highest trees, and ornamenting them with festoons of fragrant flowers not their own. Instead of your periwinkles we have here handsome trees exuding a milk which is sometimes salutiferous, at others a most deadly poison, and bearing fruits of corresponding qualities. Violets of the size of apple trees. Daisies (or what might seem daisies) borne on trees like alders.

---

[1] The name 'rain forest' is sometimes given, but with little justification, to temperate forests which are wet and evergreen but have almost nothing else in common with the rain forests of the tropics, e.g. the 'Olympic rain forest' on the Pacific coast of North America.

[2] For a definition of this term (with maps), see Fosberg *et al.* (1961).

Fig. 1.2 Profile diagrams. A Tropical rain forest, Trois Sauts, French Guiana. B Oak–maple (*Quercus rubra* – *Acer rubrum*) forest, Tom Swamp Tract, Harvard Forest, Massachusetts. After Oldeman in Hallé *et al.* (1978). Both diagrams show plots 30 × 20 m, drawn to the same scale (note man in each). Mature trees in outline; immature trees densely stippled. Boulders (in B) cross-hatched. Smaller undergrowth omitted.

In the rain-forest region woody plants not only form the larger proportion of the mature forest vegetation, but also play the chief part in its successional stages, as in the colonization of bare rock or soil, and in the building up of vegetation in lakes and swamps.

The trees in rain forests are extremely numerous in species and range from treelets of one or two metres to giants over fifty or even eighty metres high. As will be shown in Chapter 2, trees of more or less similar height often form (or seem to form) layers (also called storeys or strata). For this reason there is generally what Humboldt called 'a forest above a forest'.

The dimensions reached by the largest individual rain-forest trees have sometimes been much exaggerated. Data on the dimensions of unusually large rain-forest and other trees are given in Table 1.1.[3] The average height of the taller trees in a rain forest is usually not more than 46–55 m (150–180 ft). In the Amazon forest trees taller than 55 m are uncommon, although occasional individuals possibly reach over 60 m (Pires & Prance 1977). Trees of 60 m are uncommon in tropical Africa and it seems that it is only in Malesia that rain-forest trees of 60 m and over are commonly found. On the Mahakam river (E. Kalimantan) the tallest trees usually exceed 70 m in height (T. Kira). In New Guinea the giant conifers *Araucaria cunninghamii* and *A. hunsteinii* grow to over 80 m (Gray 1975), but they are mainly species of upland rather than lowland rain forest (see p. 424). In Europe and North America broad-leaved trees commonly attain heights of 30 m and under very favourable conditions may reach 46 m or more. Thus,

---

[3] It is interesting that in Ghana, Hall & Swaine (1976) found that in the wet evergreen forest type the trees tended to be less tall than in the moist evergreen and moist semideciduous types, which are found in climates with a slightly more marked annual dry season (see p. 410).

Table 1.1. *Dimensions of giant forest trees*

Metres	Species	Locality	Source
Height			
	(a) Tropical		
89[a]	*Araucaria hunsteinii*	New Guinea	S. Cavanaugh (Gray 1975)
86	*Koompassia excelsa*	Sabah (Borneo)	P. F. Cockburn
75	*Shorea superba*	Sabah (Borneo)	Ashton (1982, Fig. 9, p. 256)
71	*Eucalyptus deglupta*	New Britain	Lane-Poole (1925, p. 214)
70	*Agathis alba*	Sulawesi (Celebes)	van der Koppel (1926, p. 529)
66	*Dinizia excelsa*	Amazonia	Pires & Prance (1977, p. 189)
63	*Couratari pulchra*	Amazonia	Pires & Prance (1977, p. 189)
62	*Vochysia maxima*	Amazonia	Pires & Prance (1977, p. 189)
61	*Piptadeniastrum africanum*	Ghana	Taylor (1960, p. 233)
60	*Entandrophragma cylindricum*	Ghana	Taylor (1960, p. 191)
57	*Tieghemella heckelii*	Sierra Leone	Richards (1971, unpubl.)
	(b) Temperate		
115	*Sequoia sempervirens*	California	Becking (1967)
107	*Eucalyptus regnans*	Victoria (Australia)	Tiemann (1935, p. 904)
75	*Agathis australis*	New Zealand	van der Koppel (1926, p. 530)
49	*Ulmus* sp. (probably *U. americana*)	Tennessee	Anon. (1971)
46	*Fagus sylvatica*	France	Elwes & Henry (1906–7, 1, p. 13)
Girth			
	(a) Tropical		
17	*Entandrophragma cylindricum*	Nigeria	Unwin (Kennedy 1936, p. 176)
14	*Entandrophragma* sp. cf. *angolense*	Nigeria	Richards 1946 (unpubl.)
14	*Bertholletia excelsa*	Amazonia	Pires & Prance (1977, p. 189)
12[b]	*Neobalanocarpus heimii*	Malay Peninsula	Foxworthy (1927, p. 53)
ca. 12[c]	*Dryobalanops sumatrana*	Malay Peninsula	Foxworthy (1927, p. 45)
	(b) Temperate		
23[c]	*Sequoiadendron giganteum* ('General Sherman')	California	Tiemann (1935, p. 909)
23[c]	*Agathis australis*	New Zealand	Tiemann (1935, p. 911)
17	*Tilia* sp.	Europe	Kannegiesser (Büsgen & Münch 1929, p. 38)
15	*Castanea sativa*	Europe	Kannegiesser (Büsgen & Münch 1929, p. 38)

[a] This appears to be the tallest tree ever recorded in the tropics (Gray 1975).
[b] 'This was the largest tree of which we have any record in the Peninsula' (Foxworthy 1927).
[c] Calculated from diameter.

although rain-forest trees are usually taller than those in temperate forests, they never reach the gigantic dimensions of the Californian redwoods or the huge gums (*Eucalyptus*) of Australia. In girth and diameter the largest rain-forest trees also fall far short of those of higher latitudes. The largest recorded girth measurement appears to be 17 m (56 ft), but trees of girth greater than a metre are uncommon, and rain-forest trees in general are more often remarkable for the slenderness than for the large girth of their trunks.

Although the largest individual trees are not as large as those in some other kinds of forest, the mass of trunks, branches, twigs and leaves from the tops of the tallest trees to the ground is enormous. The total quantity of living plant material and wood, including roots, on an area of primary tropical rain forest is generally very large, but varies considerably. This quantity, the biomass (or more correctly the phytomass, since animal matter is not included) is very difficult to measure accurately. Reliable estimates of the above-ground phytomass of primary lowland rain forests are mostly in the range 300–500 t ha^{-1}, although in the very tall forests of Malaya and Borneo even larger figures have been reported. There are very few estimates of the root phytomass, which seems to be about 10% of that above ground.

The phytomass is dependent mainly on the net annual increment of organic matter (primary productivity),

Fig. 1.3 Canopy of lowland rain forest, Moraballi Creek, Guyana, seen from 34 m above ground level (1929).

Fig. 1.5 Canopy of lowland rain forest, Korup National Park, Cameroon (*ca.* 1990). Photo. P.E. Parker (courtesy of World Wide Fund for Nature).

Fig. 1.4 Undergrowth of Mora (*Mora excelsa*) forest, Moraballi Creek, Guyana (1929). The buttressed tree is *M.excelsa*, belonging to the A storey. The small tree on the right is *Duguetia* sp. and shows the habit characteristic of C storey trees.

which is not as large as might be expected from the input of solar energy which the forests receive. The best estimates of the productivity of tropical rain forests are in the range 1000–3500 g m^{-2} a^{-1} (mean 2200 g m^{-2} a^{-1}), compared with 600–2500 g m^{-2} a^{-1} in temperate forests (Whittaker & Likens 1975). It is noteworthy that the higher productivity of rain forests is mainly due to their high rate of leaf production; their wood productivity is not significantly greater than that of temperate forests (Jordan 1985). The very rapid growth and high productivity of young secondary forest (pioneer) trees is discussed in Chapter 18. A useful discussion of rain-forest biomass and productivity is given by Whitmore (1984a).

In European or North American forests the large trees belong to a few, sometimes only one, species; even in the richest temperate forests such as the 'Mixed Mesophytic forest' of China and southeastern North America the number is only 20–30 (p. 293). Tropical rain forests are very much richer: on a single hectare of primary forest there are seldom fewer than 40 (and often over 100) tree species over 10 cm in diameter. Wallace (1878, p. 65) says:

If the traveller notices a particular species and wishes to find more like it, he may often turn his eyes in vain in every direction. Trees of varied forms, dimensions and colours are around him, but he rarely sees any one of them repeated. Time after time he goes towards a tree which looks like the one he seeks, but a closer examination proves it to be distinct. He may at length, perhaps, meet with a second specimen half a mile off, or may fail altogether, till on another occasion he stumbles on one by accident.

The richness of the tree flora is indeed one of the most important characteristics of the rain forest and on this many of its other features are directly dependent. Trees of different species are most commonly found mixed in fairly even proportions; more rarely, one or two species are much more abundant than the rest.

The rain forest is thus usually a community with a large number of co-dominant species, but sometimes there are only one or two dominants. Rain-forest communities with numerous dominants are here termed mixed forests; those with a single strongly dominant species, single-dominant forests (see Chapters 11–12).

Though so numerous in species, rain-forest trees are on the whole remarkably uniform in their general appearance and physiognomy. The trunks are as a rule straight and slender and do not branch till near the top. The base is very commonly provided with plank buttresses, flange-like outgrowths which are a highly characteristic feature of rain-forest trees and are very little developed in other formation-types. The bark is often relatively thin and smooth without deep fissures or conspicuous lenticels. In the majority of the mature trees, as well as the saplings and shrubs, the leaves are large, leathery, and dark green with entire or nearly entire margins; they often resemble the leaves of the cherry laurel (*Prunus laurocerasus*) of gardens in size, shape and texture. When the leaves are compound, as in leguminous trees, each leaflet tends to approximate in size and shape to the undivided leaves of the other trees (although there are exceptions). So uniform is the foliage that the non-botanical observer might be excused for supposing that the forest was predominantly composed of species of laurel. The similarity and sombre colouring of the majority of the leaves are mainly responsible for the monotonous appearance of the forest. Large and strikingly coloured flowers are not very common; many of the trees and shrubs have inconspicuous, often greenish or whitish, flowers.

Fig. 1.6 Undergrowth of lowland rain forest, Okomu Forest Reserve, Nigeria (1935). This forest resembles primary forest, but is probably secondary forest several centuries old (p. 470). The large tree is *Entandrophragma angolense* and shows the characteristic surface roots.

Uniform in general appearance as are most of the woody species, some diverge widely from the normal type in appearance, notably the palms and the species of *Dracaena* and *Pandanus*, but such bizarre and unfamiliar-looking plants are usually far less conspicuous in typical rain forest than many descriptions would suggest. They are seldom common enough to affect greatly the physiognomy of the community as a whole and quite often they are entirely absent.

The undergrowth of the rain forest consists of shrubs, herbaceous plants and vast numbers of sapling and seedling trees. Travel books often give a misleading impression of the density of tropical forests. Mature (primary) rain forest is usually not difficult to penetrate. On river banks and in natural gaps or clearings, where much light reaches the ground, there is a dense growth which is indeed often quite impenetrable, but in the interior of old undisturbed forest, if one's hands are free to bend back a twig here and there, it is not difficult to walk in any direction, though sometimes a detour has to be made to avoid a fallen log or a mass of lianes which have slithered down from above. When in the forest it is usual to carry a cutlass or parang, but as much in order to mark a path to return by if necessary as to hack one's way. It is the slippery clay soil and the abundance of fallen logs and branches which make progress in the forest slow and laborious, rather than the thickness of the vegetation. Photographs give an exaggerated notion of the density of the undergrowth; it is usually possible to see another person at least 20 m away. The forest interior is gloomy, but when the sun is shining the floor is dappled with sun-flecks. The herbaceous ground flora is sparse; the number of species of ground herbs (per unit area) is usually less than in temperate broad-leaved woodland and the number of individuals fewer. Owing to the rapid rate at which organic matter decomposes at high temperatures, the ground is only thinly covered with dead leaves and there are often patches of nearly bare soil. Tropical rain forests with unimpeded drainage in which the ground is covered with 'age-long accumulations of rotting vegetation' are figments of the imagination, though in swamps peat may be formed (pp. 365–71).

Besides trees, shrubs and ground herbs, rain forest is composed of climbing plants of all shapes and sizes, and of epiphytes growing on the trunks, branches and even on the living leaves of the trees and shrubs. The abundance of climbers is one of the most characteristic features of rain-forest vegetation. The majority of these climbers are woody (lianes) and often have stems of great length and thickness; stems as thick as a man's thigh are not uncommon. Some lianes cling closely to the trees that support them, but most ascend to the forest canopy like cables or hang down in loops or festoons. The number of species of climbing plants is enormous, and there is great variety of form and structure among them.

The epiphytic vegetation, as well as including algae, mosses, liverworts and lichens, as in temperate forests, consists of large numbers of orchids and other flowering plants and many ferns. The tree-top epiphytes (which are often not visible from the ground) include shrubby as well as herbaceous species, also the semi-parasitic Loranthaceae and Viscaceae (mistletoes) and the curious 'stranglers' (such as some species of *Ficus*), which begin life as epiphytes and often develop afterwards into independent trees rooted in the ground. In no other plant community, except some types of montane and temperate rain forest, are epiphytes more abundant and luxuriant.

The tropical rain forest is thus extremely complex in structure and is built up of plants, mostly woody, but of the most varied life-forms, some of them quite unlike any found in temperate vegetation. Yet although it is in fact the most elaborate of all plant communities in structure and the richest in species, the first impression it makes is of sombreness and monotony. Though this is mainly due to the overwhelming predominance of woody plants and to the uniformity of foliage which prevails among most of the life-forms, it is increased by the absence of marked seasonal changes. In the rain forest there is no winter or spring, only a perpetual midsummer; the aspect of the vegetation is much the same at any time of year. There are, it is true, seasons of maximum flowering at which more species are in flower than at other times, and also seasons of maximum production of young leaves, but, for the most part, plant growth and reproduction are continuous and some flowers can be found at any time. This is a direct consequence of the climate, which is characterized by relatively slight seasonal variation, as well as by high temperature and humidity. No cold season or severe drought interrupts plant activities.

1.1 The biological spectrum of the rain forest

One of the most significant differences between tropical rain forest and temperate forests lies in the relative proportions in which plants of different life-forms are represented. This is strikingly shown by comparing biological spectra of rain-forest and temperate floras, based on either a physiognomic life-form classification such as Du Rietz's (1931) 'Main life-form system' or some modification of the well known epharmonic life-form system of Raunkiaer (1934) (Fig. 1.7). The validity of applying the latter to tropical plants has sometimes been questioned, e.g. by Aubréville (1963), but as Lebrun (1966) showed, the objections are not entirely justified and there is no doubt that Raunkiaer life-form spectra

have value both for comparing different tropical plant formations (see Chapters 16 and 17) and for demonstrating relationships between vegetation and climate in different climatic zones.

In the tropics it is not always easy to assign a plant to a particular life-form. Some species vary in life-form with environmental conditions, e.g. the African *Haumania liebrechtsiana* (Marantaceae) which can be either a rhizome geophyte or a climbing phanerophyte (Germain & Evrard 1956). Two versions of Raunkiaer's system modified for use in the tropics which have been widely adopted are those of Lebrun (1947) and Ellenberg & Mueller-Dombois (1967).

Table 1.2 and Fig. 1.7 give a life-form spectrum based on a collection of flowering plants made in 1929 in the rain forest at Moraballi Creek, Guyana. This is of particular interest as the whole collection came from within a radius of about 8 km in an area entirely forested except for some very small clearings. The forest was partly primary (but not all of one type; see Chapter 11) and partly secondary, but most of the collection was from primary forest. The list includes over 400 species but is far from complete: some constituents of the flora (tall trees, lianes and epiphytes) are certainly less fully represented than others (herbaceous and smaller woody plants).

A number of biological spectra for both large and small areas of rain forest have been published, e.g. for Amazonia (Cain *et al.* 1956), Zaïre (Germain 1957, Evrard 1968), Sarawak (Brünig 1968, 1970) and Queensland (Cromer & Pryor 1942). Spectra from large areas usually include plants from swamps, clearings and other non-forest habitats, some of which belong to life-forms absent or rare in undisturbed forests. Although the published spectra are not all strictly comparable with one another and some are incomplete, certain general conclusions are evident.

The most striking feature in all rain-forest biological spectra is the great preponderance of woody phanerophytes, which include numerous climbers as well as trees and shrubs. If figures were available for numbers of individuals as well as numbers of species, the dominance of phanerophytes would be overwhelming. The proportion of climbing plants varies: in number of species they sometimes equal the erect phanerophytes (Evrard 1968, p. 14). In some types of periodically flooded forest practically the whole flora, apart from epiphytes, consists of erect and climbing woody phanerophytes.

Another very significant feature of rain forest is that the lower layers are composed of plants of life-forms different from those forming the undergrowth of tem-

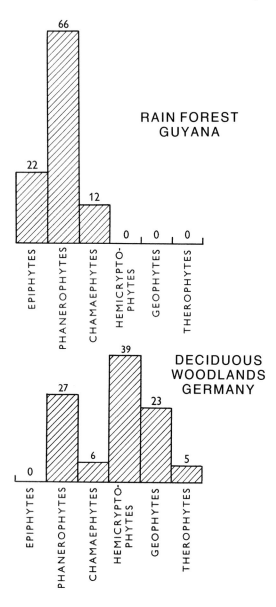

Fig. 1.7 Raunkiaer biological spectrum of lowland rain-forest flora of Moraballi Creek, Guyana, compared with deciduous woodlands, Germany. The columns show the number of species of each life-form as a percentage of the total flora. The spectrum for Moraballi Creek is approximate only and was calculated as explained at the foot of Table 1.2. The spectrum for the German woodlands is the average for six associations in northwest Germany (Tüxen 1929).

Table 1.2. *Life-forms of flowering plants in the tropical rain forest at Moraballi Creek, Guyana*

	No. of species	Life-forms according to Raunkiaer (1934)	Life-forms according to Du Rietz (1931)
Trees:			
Dicotyledons	163	Mega- and Mesophanerophytes	Tall, High and Low trees
Tall palms	3		
	166		
Shrubs and ground herbs:			
Green herbs and 'shrubs'	80	Micro- and Nanophanerophytes, Chamaephytes	Dwarf and Pigmy trees, Ctonophytic shrubs (?), Herbaceous plants
Saprophytes	13	—	—
Parasites	1	—	
Adventives (in clearings)	8	Therophytes	Herbaceous plants
	102		
Lianes and other climbers:	83	Meso- and Megaphanerophytes	Woody and Herbaceous lianes
Epiphytes:			
Shrubby (including hemiepiphytes)	*ca.* 10	Epiphytes	Epiphytoidic shrubs, Epiphytoidic Dwarf shrubs, etc.
Herbaceous	*ca.* 81	Epiphytes	Epiphytoidic Herbaceous plants
Semiparasitic (Loranthaceae)	7	Epiphytes	Parasitic Dwarf shrubs
	98		

The data in this table give no exact indication of the relative numbers of phanerophytes and chamaephytes. If, however, we estimate the chamaephytes as half the 'green herbs and shrubs', plus the parasites and saprophytes (certainly an overestimate), and take the total flora, omitting the adventives, as 441, the Raunkiaer biological spectrum would be:

Stem succulents	Epiphytes	Phanero-phytes	Chamae-phytes	Hemicrypto-phytes	Geophytes	Helo- and Hydrophytes	Therophytes
0	22	66	12	0	0	0	0
		78					

Source: From the data of N.Y. Sandwith.

perate forests. As has been mentioned, the smallest erect phanerophytes are mostly treelets rather than typical shrubs. Chamaephytes and geophytes, if present at all, occur in insignificant numbers. Hemicryptophytes, which in temperate forests are by far the most numerous life-form in the ground flora, are absent or very rare in tropical rain forests: most of the ground herbs are herbaceous phanerophytes or chamaephytes. Therophytes such as annual grasses are entirely absent except in gaps and along paths. The number of species of epiphytes, which in rain forests include both woody and herbaceous plants of various growth habits, often exceeds that of ground herbs.

It will be seen that the biological spectra of tropical rain forest closely reflect the continually favourable, non-seasonal climate. This is shown by the absence or scarcity of life-forms in which the perennating buds are protected from cold or desiccation (hemicryptophytes,

geophytes) and of plants adapted to survive unfavourable periods as seeds (therophytes). There is a corresponding abundance of evergreen phanerophytes and epiphytes, the life-forms least suited to resisting cold and drought. It may be noted also that the buds of rain-forest phanerophytes tend to be larger and relatively fewer than in temperate trees (p. 91). An extreme example of lack of adaptation to climatic hazards is found in palms. The 'cabbage' of a palm is the solitary perennating bud; there are no reserve buds, so if the apex is injured the whole shoot dies. Though some palms branch from the base, others never have more than a single stem. 'Like a foolhardy gambler', says Schroeter (quoted by Rikli 1943, p. 30, transl.), 'the palm stakes all on a single card'. Some other tropical plants such as pandans (*Pandanus* spp.) and the unbranched 'tuft trees' (*Schopfbäumchen*) (p. 72), a peculiar type of dicotyledonous phanerophyte very characteristic of rain forest, resemble small palms in their life-form. The dominance of phanerophytes in most stages of successions leading to rain forest (Chapters 13, 14 and 18) is good evidence that their abundance is in fact an expression of the climate.

1.2 Recent distribution

The area of the tropical rain-forest forms a belt round the whole Earth, bisected somewhat unequally by the equator, so that rather more of it lies in the northern than in the southern hemisphere (Fig. 7.1). Owing to the presence of mountain ranges and plateaux, and to the uneven distribution of the controlling factors (which in turn depends on the distribution of land and sea), the belt is very irregular in outline. Its northern and southern boundaries do not coincide exactly with any latitudinal limits: in some places they do not reach the Tropics of Cancer and Capricorn, in others they extend somewhat beyond them.

The tropical rain-forest formation-type (or pan-climax) comprises three floristically different formations occupying the three main regions of the world with a humid tropical climate: the American rain forest in South and Central America, the African rain forest in tropical Africa, and the Indo-Malayan rain forest, which is found from India to southern China and New Guinea. A fourth formation, much smaller than the others, is found in northeastern Australia. Though often regarded as an extension of the Indo-Malayan rain forest, it is probably better regarded as distinct and autochthonous, as Webb *et al.* (1986) have shown. Most of the plant species, and many of the families and genera, are confined to one of these formations and not shared with the others. In spite of the dissimilarity of the flora, the structure and physiognomy of the climax communities, as well as the successional stages in their development, are in many respects remarkably alike in all three formations. The climax vegetation of each, as will be seen later, also varies in a parallel manner with local differences of climate and soil, as do its responses to human and natural disturbances.

Until comparatively recently, the detailed distribution of the tropical rain forest formations was very inadequately mapped. During the past fifty years it has become much better known, but there are still parts of the world where boundaries can be only vaguely drawn. Vegetation maps of several large parts of the humid tropics have been published. Some of the most important are Hueck & Seibert's (1981) map of South America (scale 1:8 million), the Projeto RADAM maps of Brazil

(Projecto RADAM 1973–75, Projeto RADAMBRASIL 1975–83), the AETFAT map of Africa south of the Tropic of Cancer (White 1981, 1983) (1:10 million), van Steenis' (1958a) map of Malesia (1:5 million) and Whitmore's (1984b) of Malaya (1:5 million). Guillaumet & Adjanohoun (1971) have a map of the vegetation of Côte d'Ivoire (1:500 000). Hou (1979) has published a vegetation map of China (1:4 million). Champion & Seth (1968a) give a map of the forests of India and Sri Lanka (1:19 million) and Gaussen, Legris and others (see Gaussen 1959, Legris & Viart 1959) have produced a series of regional vegetation and bioclimatic maps of India, Sri Lanka and Indo-China on a scale of 1:1 million.

Mapping tropical vegetation long depended mainly on ground surveys. Recent improvements in knowledge of the extent and detailed distribution of the tropical rain forest are largely due to the increasing availability of air photography and of remote sensing techniques such as landsat. An advantage of using radar for remote sensing and mapping of vegetation is that the wavelengths used (*ca.* 1 cm) are able to penetrate the cloud cover which in the humid tropics is present for a large proportion of the daylight hours. Another advantage is that radar images lend themselves well to electronic data processing, thus eliminating the subjective element always involved to some extent in interpreting conventional air photographs.

The most extensive surveys of tropical rain forest by means of radar so far are the Projeto RADAM (1973–75) and Projeto RADAMBRASIL (1975–83) studies of Amazonia referred to above. The rate at which the world's forest resources are dwindling has made the accurate mapping and evaluation of the remaining tropical rain forests increasingly urgent. The methods available and future research needs were reviewed by UNESCO (1978).

The distribution of the three tropical rain forest formations is shown on a small scale in Fig 7.1 and will now be briefly described. Further discussion of the boundaries and the climatic and other factors determining them will be found in later chapters.

The American rain forest. By far the largest mass of rain forest in the world is in South America, mainly in the Amazon and upper Orinoco basins. This forest stretches from the lower slopes of the Andes east to the Guianas and south to about 15°S in western Brazil and northern Bolivia. In Amazonia the area of terra firme forest (i.e. forest not permanently or seasonally flooded) was estimated in 1973 as 3 303 000 km^2 (Pires, quoted by Goodland & Irwin 1975, p. 103). Apart from man-made clearings, the Guiana – Rio Branco savannas and the montane vegetation of the Guiana highlands, the Amazon forest is broken only by relatively small

A

B

Fig. 1.8 Primary Amazonian rain forest, *ca.* 11–12°S, Mato Grosso, Brazil. (1968). A Rio Liberdade. B Rio Suiá Missu. In 1968 the forest in both showed no signs of recent disturbance. Since then, following the opening of the Transamazon Highway, an airstrip has been built in A and forest clearings have been made.

and scattered 'cerrados' (savanna woodlands) and 'campos limpos' (treeless grasslands) (see Chapter 16).

Along the east coast of Brazil there is a narrow strip of tropical rain forest, the 'Atlantic forest', which is separated from the Amazonian 'hylaea' by a wide corridor of cerrado and dry deciduous forest. This extends southwards from about 7°S near Recife to about 28°S in Rio Grande do Sul, though there is a gap where the land is lower between about 20 and 23°S. The Atlantic forest thus extends well beyond the southern tropic, but it retains much of its tropical character throughout. A large part of this forest, especially in the north and the coastal plain, has been long since cleared and cultivated.

West of the Andes a broad band of rain forest extends northwards from about 0°23'S in Ecuador through Colombia to Panama. This area, especially the Chocó region in Colombia, is one of the wettest rain forest regions in the world (p. 180).

The distribution and regional variations of the South American rain forest are described more fully by Hueck (1966) and by Daly & Prance (1988). A detailed account of the eastern and southern boundaries of the Amazon forest is given by Castro Soares (1953).

From Panama, tropical rain forest extends northwards, though not quite continuously, to southern Mexico, reaching its limit at about 19°N in Vera Cruz. In the Antilles a considerable amount of rain forest formerly existed, especially in the larger islands, Jamaica, Hispaniola, Puerto Rico, Dominica, Martinique and Trinidad. In these islands, most of the rain forest that still remains is montane rather than lowland in type.

The African rain forest. In Africa the tropical rain forest is much less extensive than in tropical America. The total area in Zaïre, which contains about 60% of the African evergreen tropical forest, has been recently estimated to be about 500 000 km² (Catinot in UNESCO 1978, p. 22): this is less than one sixth of the size of the Amazon forest. Much of what is often mapped as rain forest is in fact semideciduous forest. The destruction of the forest by people began earlier and until recently has been much greater in South America, but it also seems clear that even before the extensive clearances of the past 100 years the area must have been much smaller. For a description of the effects of exploitation and other human pressures on rain forests in West Africa, see Martin (1991).

From the large mass of rain forest near the equator in the Zaïre (Congo) basin, the forest continues westwards into Gabon and Cameroon. From there a narrow belt follows the coast of the Gulf of Guinea through Nigeria to Ghana and beyond, finally ending in Guinea at about 10°N. This western extension of the rain forest is interrupted from western Nigeria to a little west of the Volta in Ghana by a stretch mainly of savanna about 300 km wide, the Dahomey Gap (p. 405). South of Zaïre the African rain forest extends into Angola to about 9°S. Towards its southern limit the rain forest is increasingly confined to river valleys.

In East Africa the area of continuous forest reaches its eastern limit at Bwamba in western Uganda (*ca.* 30°E) (Hamilton 1974). East of the Western Rift Valley, forest similar to tropical rain forest is absent except for outliers of various sizes, e.g. Budongo forest and fragments near Lake Victoria in Uganda, a relic near Kakamega (34°47′E) in western Kenya and some small areas in northwestern Tanzania (Langdale-Brown *et al.* 1964, Hamilton 1974). The isolated wet evergreen forests on mountains further east in Tanzania (Uluguru etc.), although floristically similar to rain forests further west, are lower montane rather than lowland in type (p. 445). In the coastal region of Kenya there are relics of forest in which some of the genera and species strongly suggest a connection with the central African rain forest in some past climatic period.

Tropical rain forest is probably the natural climax at low altitudes in Fernando Po and the other islands in the Gulf of Guinea; relics of rain forest of a peculiar insular type are found in Mauritius and Réunion in the Indian Ocean (p. 448). A broad band of rain forest, much of which has now been destroyed, extends through the wetter eastern part of Madagascar to *ca.* 25°S (Guillaumet 1984).

The Indo-Malayan and Australian rain forest. In the eastern tropics the Indo-Malayan rain forest is found from southern India and Sri Lanka to Thailand and southern China, extending southwards into Indo-China and through the Malay Peninsula to the islands of the Malay Archipelago, the Philippines, New Guinea, the western Pacific islands and Australia. The largest forest areas are in the Malay Peninsula, Sumatera and Borneo: here the Indo-Malayan rain forest reaches its greatest luxuriance and floristic wealth.

In India, rain forest relics are found in the Western and Eastern Ghats (Pascal 1984) and, more extensively, in the eastern Himalayas, the Khasia hills and Assam. In Burma, Thailand, Cambodia and Vietnam, rain forest is found only locally and the total area is not very large, most of the forests being semideciduous or deciduous. In China, tropical rain forest reaches its northern limit in Hainan, at Lan Hsu island at the southern tip of Taiwan, and on the mainland at about 28°N (Hou 1979; see also Kira 1991). In Yunnan Province there are small areas of

tropical rain forest and much larger areas of 'monsoon' (semievergreen) forest (Li & Walker 1986).

In Java, except for a small area in the west, and in the islands to the east of it (Bali, Timor, etc.) the dry season is too severe for lowland rain forest to develop extensively, although it may occur very locally as a postclimax in favourable situations, but in the lowlands of New Guinea tropical rain forest is the climatic climax almost everywhere except in the small seasonally dry region in the south-east near Port Moresby.

In Australia, tropical rain forest forms a narrow band in Queensland: it rarely extends more than 160 km from the coast. This evergreen forest continues southwards into New South Wales and there is a gradual transition to subtropical and temperate rain forest (Chapter 17).

Rain forest is also found in some of the islands of the western Pacific (Solomons, New Hebrides, Fiji, Samoa, New Caledonia, etc.). In the Hawaiian group (*ca.* 20–30°N) there is a peculiar type of lowland wet evergreen forest dominated by *Metrosideros* spp. (lehua), which little resembles normal tropical rain forest.

Potential and actual areas. The total area actually occupied by 'wet evergreen forest' in the mid-twentieth century has been variously estimated as 850 million hectares (FAO 1963, 1976) and 280 million hectares (Persson 1974). The total area present in 1990 has been estimated at 718 Mha (FAO 1993). These estimates are subject to such large unavoidable errors that they should be accepted only with reservations (Chapter 19). It must also be realised that the actual forest area is much less than the potential area (see Figs. 19.1 and 19.2). The distribution of the tropical rain forest described in the preceding section is the area within which it seems to be the climatic climax or potential natural vegetation at elevations of up to about 500–1000 m. Within the boundaries of this area, as mentioned earlier, there are some habitats where rain forest has not yet developed, or apparently cannot develop. Locally such habitats may be fairly extensive, but under the conditions which existed over a large part of the humid tropics until about 200 years ago, or even later, so little forest had been cleared or seriously disturbed by humans that the actual area fell not much short of the potential area.

At the present time the situation is entirely different: in all but the most remote and thinly populated regions, vast areas of rain forest have been cleared. Farmland, plantations and cattle ranches have appeared and are expanding rapidly. In regions where there is a dense human population, the natural vegetation has been so completely replaced by cultivation and anthropogenic plant communities that it may be difficult or impossible to reconstruct with certainty the natural climax communities. So fast are these changes taking place that within a few decades climax tropical rain forest may have disappeared almost everywhere except in the most inaccessible sites or where it has been deliberately conserved.

Besides forest land which has been converted to agriculture or some other forms of land-use, in some countries there are enormous areas which were formerly cultivated but have been abandoned to the natural regrowth of vegetation. If there is little further disturbance, the vegetation on such land soon takes on the aspect of a forest but it remains recognizable as secondary forest for many years. Old secondary forest can eventually become indistinguishable from true primary forest but only if left undisturbed for a period which is probably as long as 200 years or more. In some parts of the tropics it can be seen that as a result of population movements forest has recolonized land which was formerly cultivated (Chapter 18).

It often happens that young secondary forest is cleared and used again for agriculture after a few years. When this is often repeated and fires are frequent, the soil becomes degraded and what was once primary forest becomes changed into secondary vegetation, usually savanna or grassland (section 16.4). Anthropogenic savannas within the potential area of the tropical rain forest are termed 'derived savannas'.

In many parts of the tropics, especially in Africa, derived savannas and other biotic climaxes now occupy enormous areas. The replacement of forest by savanna has often been attributed to a progressive change of climate and much has been written about the supposed 'desiccation of Africa' (Stebbing 1937 and others). In West Africa, it has been alleged, the climate has become drier in recent centuries, leading to a southward advance of the Sahara which is still continuing at the present time. This is supposed to be causing a corresponding southward retreat of the 'closed forest', together with the Guinea savanna zone and other formations of the rain forest – desert ecotone.

Undoubtedly the savannas in West Africa appear to have increased in area in recent years and the forest has decreased, but when the evidence has been critically examined the conclusion has always been that what is happening is replacement of forest or savanna woodland by secondary vegetation such as derived savannas, which are a result of over-cultivation, over-grazing, excessive burning and other forms of land abuse (Aubréville 1938, 1947). There is no firm evidence that boundaries such

as those between the West African 'closed forest' and the Guinea savannas have changed significantly during the past two or three centuries. A.P.D. Jones (1947) pointed out that the description of northern Nigeria, as it was in 1822–24, by the travellers Denham and Clapperton indicates conditions very like those of 1947 and offers no support to the idea of a rapid deterioration of climate and vegetation. Fluctuations of climate, such as the run of dry years in West Africa in the 1960s and 1970s, are probably within the normal range of climatic variation and do not necessarily indicate a long-term trend.

Changes in climate and vegetation in recent centuries should not be confused with those measured in thousands rather than hundreds of years. The evidence for these secular changes in the climates of the humid tropics is discussed in section 7.9; their probable effects on the distribution of rain forests in West Africa are considered next.

1.3 The tropical rain forest during the Tertiary and Quaternary periods

Fossil evidence shows that there were forests apparently very like modern tropical rain forests at the beginning of the Tertiary period some seventy million years ago; as will be seen later, they probably came into existence even earlier, in the Cretaceous. Since they first appeared their boundaries seem to have been constantly shifting in response to changes of climate and sea-level and to movements of the Earth's crust such as mountain building and continental drift. These past changes are not merely of historical interest but probably have an important bearing on many features of the rain forest as it now exists, in particular on variations in its species richness and its dynamics.

The evidence for these changes is of many different kinds, coming from biogeography (especially relict species and disjunct distributions), archaeology, geology (geomorphology, freshwater and marine sediments and fossil soils) and, above all, from the fossil remains of animals and plants, especially pollen.

Unfortunately the difficulty of identifying pollen from species-rich and often inadequately investigated tropical floras is only one of the formidable problems which pollen analysis faces in the tropics. Another is that most tropical trees, unlike the conifers and catkin-bearing trees of the temperate zone, are not wind-pollinated and produce relatively little pollen: abundant species may thus be poorly represented in the pollen record. The tropical pollen profiles so far studied come mainly from lakes and bogs in mountainous regions (northern Andes, East Africa, New Guinea) or from coastal deposits. Much has been learnt from these profiles about climatic changes in the tropics generally, and about movements of montane vegetation zones upwards or downwards caused by them, but much less about the history of the lowland rain forest. In the tropical lowlands, especially in the interior of large expanses of forest such as those in Amazonia, sites suitable for pollen analysis are unfortunately rare. What is at present known about the history of the tropical rain forest has been discussed in detail by Flenley (1979) and a general review of tropical palaeogeography and palaeoclimatology was published by UNESCO (1978) in which full references can be found. Here only a few important conclusions relevant to later chapters of this book can be mentioned.

Tertiary period. It is remarkable that in the early Tertiary period forests apparently very similar to the tropical rain forests of today were found not only near the equator, e.g. in the Caribbean region and in Borneo (Germeraad *et al.* 1968), but also far outside the tropics, even allowing for changes of latitude of parts of the Earth's surface since the Tertiary due to plate movements. Two examples of Tertiary rain forests at high latitudes may be briefly mentioned.

Wolfe (1972; see also Upchurch & Wolfe 1987) in his work on fossil floras in Alaska shows that in the Paleocene and early Ravenian (late Middle Eocene) a type of forest which he calls 'para-tropical rain forest' existed at what is now more than 60°N. In both its floristic composition and its physiognomy this closely resembles a Recent Indo-Malayan lowland rain forest. Tree genera identified include *Alangium, Macaranga, Gluta, Parashorea* and *Stemonurus*, all of which are well represented in Malesia today: out of thirty-six genera present in the early Ravenian, thirty-two are now found living in Malesia and seven are restricted to it. Many of the plants belong to families which today are entirely tropical. Even more significant than the floristic resemblances to a modern Malesian forest is the similarity in physiognomy. Leaf anatomy shows that the forest was overwhelmingly evergreen and 65% of the leaves, most of which belonged to the mesophyll size-class, had entire margins. These are characteristic tropical rain-forest features (Chapter 4). Drip-tips were also common. It is estimated that about a quarter of the species were lianes.

Even before the end of the Eocene, the climate of Alaska seems to have become cooler. Deciduous species began to appear and by the Oligocene a temperate broad-leaved forest with a smaller proportion of entire-margined leaves had replaced the 'para-tropical rain forest'. In the Miocene and Pliocene, coniferous forests had supplanted the broad-leaved trees.

Another example of an early Tertiary fossil flora which seems to indicate the presence of forest very like tropical rain forest far to the north of its present limits is that of the London Clay (Lower Eocene) and other deposits in southeastern England at about 51°N (Reid & Chandler 1933, Collinson 1983, Collinson & Hooker 1987). Like the Eocene floras of Alaska, this consisted mainly of woody plants with strong Indo-Malayan affinities. Some of the families represented are now exclusively tropical and many others, e.g. Lauraceae, Meliaceae, Sterculiaceae and Palmae, are mainly tropical. No Dipterocarpaceae have been found. Only a few genera, e.g. *Cornus*, seem to indicate a non-tropical climate. As the remains are mainly seeds, fruits and wood fragments, little can be said about the physiognomy of the community from which they came, but the floristic composition strongly suggests some type of tropical rain forest. The abundance of fruits of the palm *Nypa burtini*, which closely resemble those of the Recent estuarine *Nypa fruticans* of Malesia (pp. 372ff), and the presence of two other mangroves, *Palaeobruguiera* and *Ceriops*, shows that there were extensive mangrove swamps.

Reid & Chandler (1933) argued cogently that the London Clay fossils were the debris of a large river flowing into a sea occupying what is now the London Basin and communicating with the huge Tethys Sea, which stretched from southern Europe to India and North Africa: they supposed that they were the remains of plants which grew not far from their place of deposition. This view was contested by van Steenis (1962a, b) who believed that they (and other remains of tropical plants in the Eocene of western Europe) were drift carried by the Tethys from much further south. Van Steenis' arguments cannot be fully discussed here, but it is hard to reconcile this view with what is now known about the London Clay flora and fauna (Collinson & Hooker 1987). The flora seems to have grown in the catchment area of the large river in which its remains were deposited. At least its tropical components probably grew in a frost-free climate with a mean temperature of perhaps 25 °C. Independent evidence from oxygen isotope analysis indicates that the mean minimum annual temperature of the southern North Sea at a period slightly later than the deposition of the London Clay was between 20 and 27 °C (Buchardt 1978, quoted by Collinson 1983). Growth rings in some of the larger wood fragments suggest that the climate was slightly seasonal.

Later European fossil floras show that as the Tertiary progressed the climate cooled; it is generally accepted that by the Pliocene it was no warmer in temperate regions than at present. We can thus probably assume that when the successive glacial and interglacial periods of the Pleistocene began, the tropical rain forest had retreated in both the New World and the Old to something more like its modern boundaries.

Quaternary period. At one time it was thought that the tropical rain forest was relatively little affected by the glaciations and other climatic changes that had such a profound effect on ecosystems outside the tropics from the end of the Tertiary onwards. During the past fifty years, evidence has been coming forward from many sources showing that this cannot be true. In the Pleistocene, and to a somewhat smaller degree during the Holocene, large parts of the tropical zone experienced alternating phases of wet (pluvial) and dry (interpluvial) climate. During the pluvials, the rainfall was as great as or greater than that in recent centuries. Both the lowland and montane rain forests expanded their area and it was probably in a late Pleistocene or Holocene pluvial period that the two West African forest blocks (p. 405) were connected across what is now the Dahomey Gap. In some (perhaps the same) pluvial episode, the African rain forest spread much further east than it does now: the various isolated patches of rain forest and forest containing species related to rain-forest species, east of the Rift valley (referred to earlier), are probably relics from this period. In tropical America and probably also in the Indo-Malayan region, similar expansions of the rain forest probably took place in pluvial periods.

During interpluvials the rainfall was much lower and presumably more seasonal. Not only must the total area of all three rain-forest formations have been much smaller than in recent times, but large expanses such as the Amazon forest became fragmented into separate blocks occupying refuges where climate and soil were favourable, divided by wide corridors of savanna and seasonal forest.

In the Pleistocene and Holocene, considerable temperature changes accompanied the variations in rainfall. During pluvials, the temperature was relatively high; during interpluvials it was lower. Thus, about 6000 yr BP, according to van Geel & van der Hammen (1973), the vegetation zones in the northern Andes were several hundred metres higher than at present and the mean temperature in the lowlands of South America was probably higher than it is now. On the other hand, evidence from pollen analysis indicates that during the dry inter-

pluvial period at about 12 000 yr BP the mean temperature in this area was about 2.5 degrees lower than at present (UNESCO 1978, p. 80). This is in fair agreement with calculations of surface sea temperature in the Caribbean based on foraminifera and oxygen isotope ratios in marine sediments (Shackleton et al. 1981).

For many of these climatic changes during the Quaternary there are reliable radiocarbon dates, though much more information on local stratigraphy is needed before it will be possible to follow them in detail or to be sure how far the various pluvial and interpluvial episodes were synchronous in different parts of the tropics. But it is already well established that the major interpluvials in the tropics were more or less contemporary, not with the interglacials, as was formerly supposed, but with the glaciations and later relatively cold periods in Eurasia and North America. It has been shown from both radiocarbon-dated pollen profiles and archaeological evidence that the long dry period in many parts of the tropics culminating at about 20 000 yr BP was contemporary with the Würm glaciation in northern latitudes. The dry periods coincident with the Pleistocene glaciations probably affected the tropical rain forest not only through their effect on rainfall and temperature but also through the eustatic fall of sea-level amounting in the glacial maxima to over 100 m. This probably caused the climate of areas far from the coast to become drier and more seasonal. In all parts of the tropics the past 10 000 years seem to have been moister than the dry, cool period which preceded them and has allowed the rain forests to spread out everywhere from their restricted refuges.

Not enough is yet known about the Quaternary climatic changes to trace their effects on the tropical rain forest in more than general terms. The evidence for the fragmentation of the American and African rain forests (Figs. 1.9 and 1.10) is at present largely biogeographical but is supported by evidence from geomorphology and lake sediments. The view that the Amazon forest was formerly less continuous than now was first put forward by Haffer (1969) who suggested that the amount of diversity and local endemism in the bird fauna of tropical America could be understood only if the forest had once consisted of a number of widely separated refuges (Fig. 1.9): he suggested that the most marked fragmentation had taken place during the dry periods of the Pleistocene and that during a dry post-Pleistocene episode the forests of the upper and lower Amazon had been divided by a corridor of grassy savanna running more or less north and south through the Obidos–Santarem region. Similar conclusions have been drawn by other zoologists from the study of South American butterflies, lizards and other animals. Meggers (1987) pointed out that the two most recent of the postulated dry periods were subsequent to the entry of humans into South America; she found evidence in ethnography, archaeology and linguistics consistent with the hypothesis that the Amerindian peoples of lowland South America had experienced periods of forest recession in the past.

Knowledge of the flora and plant communities of Amazonia (Chapters 11–12) is still too incomplete to provide conclusive evidence for the theory of forest fragmentation put forward by the zoologists, but Prance (1987), especially from a detailed examination of distribution patterns of trees in the families Caryocaraceae, Chrysobalanaceae, Dichapetalaceae and Lecythidaceae, found much to support the idea of forest fragmentation in the late Pleistocene. Direct evidence from pollen

analysis within the lowland forest is still rather scanty (Absy 1985). Some authorities familiar with Amazonia, e.g. Sioli (1983), Connor (1986), reject or are critical of the 'refuge theory' and its implications. A critical account of the botanical and zoological evidence is given in a volume edited by Whitmore & Prance (1987).

In tropical Africa it is clear that dry periods during the Quaternary led to drastic reductions in the forest area and to its restriction to widely separated refuges, but the size and location of these is still uncertain (Fig. 1.10). In Africa, as in South America, there was a lowering of temperature in the late Pleistocene. On the East African mountains the glaciers advanced and there are indications that the vegetation zones were lowered, The chronology of these climatic changes has been firmly established from lake levels, sediments and pollen analysis in Lake Victoria and other East African lakes (Livingstone 1976, Kendall 1969). A long dry period ended about 12 500–10 000 yr BP. Hamilton (1982) has given a general account of the environmental history of East Africa in the Quaternary.

In West Africa, concentrations of endemic plants and animals and the disjunct distribution patterns of many species of forest plant (pp. 311–13) suggest the existence of Pleistocene forest refuges in the region of Liberia and in the Cameroun – Equatorial Guinea area (Fig. 1.10), although, as Hall & Swaine (1981) pointed out, these are both high rainfall regions and the biogeographical facts could be explained by present-day conditions without recourse to historical factors. Data from pollen analysis from two lakes in the lowland forest zone are now becoming available. From Lake Bosumtwi in Ghana cores rich in pollen and other plant remains have provided a continuous record of the vegetation from 27 500 yr BP to recent times. From Lake Barombi-Mba in western Cameroun, data from 25 000 BP have been obtained.

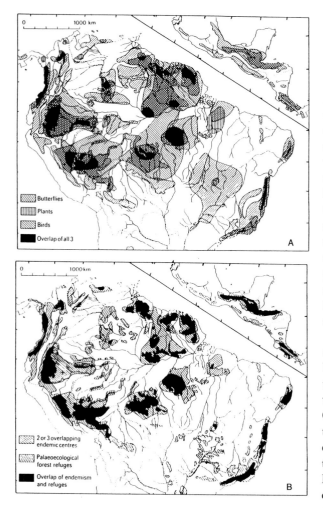

The preliminary results of Livingstone (1976), Talbot et al. (1984) and Maley & Livingstone (1983) from Lake Bosumtwi show that the surrounding vegetation, which today is semideciduous forest, derived savanna and cultivation, was grassland with scattered trees in the Late Quaternary. It is remarkable that pollen of the montane tree *Olea capensis*, which is now found in West Africa only above about 1200 m, is abundant from the base of the core up to *ca.* 9000 yr BP. Cuticle fragments of Pooideae, grasses now found on Cameroun Mountain only above 2000 m, are also plentiful in these layers. At *ca.* 9000 yr BP, closed semideciduous forest suddenly returned and the montane plants disappeared. At Barombi Lake (Maley 1987) *Olea capensis* was also abundant in the lower part of the profile but disappeared

Fig. 1.9 Pleistocene refuges in tropical America. A Centres of plant and animal endemism. B Overlap of A with position of palaeoecological refuges deduced from soils, geomorphology and other evidence. After Whitmore & Prance (1987).

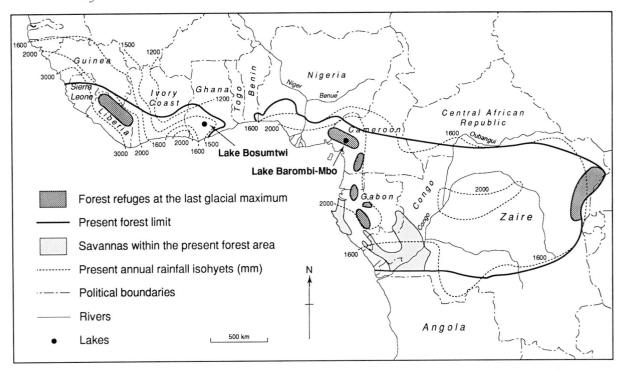

Fig. 1.10 Refuges for lowland tropical rain forest in tropical Africa during the last glacial maximum. After Maley (1987), modified.

soon after 15 000 yr BP, earlier than at Bosumtwi. Grasses were abundant from *ca.* 20 000–15 000 yr BP, but the quantity of tree pollen suggested that forest was present at the base of the profile.

The data from both lakes show that there was a period of dry climate and a decrease of temperature probably of about 5–8 degrees during the Late Pleistocene. Maley (1987) suggests that the latter may have been caused by persistent cloud due to upwelling of cold Atlantic water in the Gulf of Guinea (see also Maley 1990).

The history of the central mass of forest in Zaïre is almost unknown, but the presence of dune sands in the western part of the existing forest (de Plöey 1965) is evidence of a dry climatic period from *ca.* 30 000–50 000 to 10 000 yr BP or later, when the forest must have been less extensive than in recent times.

The relative species poverty of some parts of the African rain forest, e.g. southwestern Nigeria, the distribution patterns of plants and animals in tropical Africa generally (Hamilton 1976) and facts such as the existence of fossil termitaria of savanna species within the present forest boundaries, suggest that much of the African forest is not very old and has spread relatively recently from the refuges to which it was confined during the dry periods of the late Pleistocene.

The tropical rain forest of Malesia, like that of Africa and America, has been profoundly affected by past climatic changes. During the glacial maxima of the Pleistocene the eustatic lowering of sea-level was about 200 m, uniting the Malay Peninsula, Sumatera, Borneo and Java into a large continental land mass, Sundaland. Part of this probably had, as now, an ever-wet climate, but pollen analyses from Peninsular Malaysia and southern Borneo (Morley & Flenley 1987) suggest that in some parts of Sundaland the climate was seasonally dry, the vegetation was perhaps semideciduous or deciduous forest or savanna, forming a corridor for migrations across the equator of plants adapted to annual dry seasons (p. 18). The northern half of Borneo, and perhaps other smaller areas, were refuges where the rain forest survived. Pollen profiles from northwestern Borneo (Germeraad *et al.* 1968, Muller 1972) have no long periods in the Quaternary when grasses were abundant and show that in that region rain forest maintained itself throughout the Quaternary.

Evidence from pollen analysis (Lezine 1988) shows that in the coastal region of Senegal mesophilous forest floristically resembling the semideciduous forest of today (p. 409) advanced *ca.* 2 degrees north of its present limit during a pluvial episode of the early Holocene (*ca.* 9000–7500 yr BP), replacing dry grassy 'Sudanian' vegetation.

To the south of Sundaland, in New Guinea and north-eastern Australia, there is much evidence of past climatic changes. In New Guinea it is the effects of lower temperatures in the Pleistocene, rather than dryness, which are most apparent. On the high mountains, where now only a few small ice caps persist, there were extensive glaciers from before 22 000 to 14 000 yr BP, even on mountains such as Mt Wilhelm (4510 m) which have no ice now. The snow line was at about 3500–3600 m and on Mt Wilhelm the vegetation zones were depressed by 700–1200 m before 10 000 yr BP (Hope 1976). Earlier, at *ca.* 30 000 yr BP, the upper limit of forest on the New Guinea mountains was as low as 1500 m. The area of lowland rain forest within the modern coastline was much reduced: Walker & Chen (1987) estimate that in central New Guinea lowland rain forests occupied only 75% of their present area and 60% of their present altitudinal range. To what extent they covered the continental shelf, then exposed by the low-ered sea-level but now submerged, is unknown.

In Australia, where rain forests had covered much of the continent during the early Tertiary, their area became much more restricted by the end of the Quaternary. In Queensland the history of tropical rain forest has been traced by pollen analysis far back into the Pleistocene, particularly on the Atherton Plateau, which is close to the western edge of the present forest (Walker 1986, Walker & Chen 1987; see also references in Walker & Singh 1981). From *ca.* 120 000 yr BP to 80 000 yr BP there is evidence of a sequence of rain-forest types comparable to those of today, except that they were richer in conifers (*Araucaria, Dacrydium, Podocarpus*). After *ca.* 30 000 yr BP, rain forest was replaced by grassy sclerophyll woodlands, probably similar to some of the existing *Eucalyptus* woodlands. Then over a period lasting from *ca.* 10 000 to *ca.* 5000 yr BP, the rain forests returned; these had a smaller proportion of conifers than those of the Pleistocene and were more similar to the modern rain forests.

The gradually emerging picture of the tropical rain forest during the Quaternary is thus one not of stability, but of continual adjustment to changes in the environment. The nature of these changes will no doubt become clearer as research progresses, but they certainly involved both rainfall and temperature. At present it seems that, in most parts of the tropics, changes in rainfall were more important than variations in temperature, though this was evidently not true everywhere. There may be some tropical areas where the rain forest has survived *in situ* since the early Tertiary, a period

of many millions of years, but if so most of them could hardly have been more than relatively small refuges. Outside the refuges the forest has contracted and expanded many times.

No doubt some components of the rain forest could respond to environmental changes more easily than others and it is likely that there are relict species and even relict communities which were left behind when the forest retreated as well as similar relics of savanna and seasonal forest biota left behind in periods of forest expansion (p. 414).

1.4 The origin of the tropical rain forest and the evolution of angiosperms

It is generally believed that the angiosperms arose in the tropics, probably some 136 million years ago at the beginning of the Cretaceous period (Takhtajan 1969, Hughes 1976, Doyle 1978). Takhtajan goes further and maintains that the cradle of the flowering plants was in the Indo-Malayan region 'between Assam and Fiji', and in montane forests rather than in the lowlands. Whether the first appearance of the angiosperms was in the Cretaceous or, as some believe, much earlier, it is clear from fossil evidence that in the Cretaceous they began a phase of rapid evolution and adaptive radiation (Crane 1987). Hughes (1976) plausibly associates this with a general rise of world temperatures which culminated in the 'Radmax' of Maestrichtian (late Cretaceous) times.

By the late Cretaceous large dicotyledonous trees had certainly evolved[4]. Both Stebbins (1974) and Doyle (1978) believe that the earliest angiosperms were comparatively small shrubs or 'treelets', which were probably adapted to semiarid and disturbed habitats. Doyle (1978) makes the interesting suggestion that the first forest angiosperms were restricted to river margins, gaps and undergrowth: only later did they become able to compete with the Mesozoic conifers and replace them as dominants of tall forests. If this were so, it was perhaps quite late in the Cretaceous, about 65 million years ago, that forests recognizably like the tropical rain forests of today first came into existence. It is interesting to speculate whether, before the broad-leaved dicotyledonous trees became completely dominant, there may not have been a phase when humid tropical forests were mixtures of conifers and dicotyledonous trees. In most modern tropical rain forests conifers are entirely absent but it is tempting to suggest that exceptions such as the heath forests of Borneo with their populations of

---

[4] Corner's durian theory of angiosperm evolution is referred to in Chapters 4 and 5.

Agathis and the forests of Queensland and New Guinea with their gigantic araucarias may be survivors of these mixed angiosperm–conifer forests of the late Mesozoic.

Modern humid tropical forests, especially lower montane rain forests and the lowland forest of certain areas such as Queensland, are richer than other plant formations in 'archaic angiosperms'. These are groups such as some of the Magnoliales and Laurales with many supposedly primitive characters, e.g. free carpels, various features of wood anatomy, presence of leucoanthocyanins. As Stebbins (1974) says, 'archaic taxa' do not form a very large percentage of the total rain-forest flora but their relative abundance is one of the principal arguments used by Takhtajan and others to support their views on where the angiosperms originated.

Sporne (1970, 1973, 1980) used thirty-nine primitive characters from which an 'advancement index' can be calculated: this allows quantitative comparisons to be made between different plant formations. The average advancement indices of the dicotyledonous families found in tropical rain-forest areas in Guyana, Uganda and Sarawak are, respectively, 45.1, 47.2 and 47.5. The differences between these figures and 56.6, the average for the whole world, are statistically highly significant. If Sporne's interpretation is correct, there is no doubt that the tropical rain forest has an unusually high percentage of archaic taxa, but Wood (1970) has suggested that the characters used in calculating the advancement index are adaptations to the rain-forest environment rather than primitive features. However, there does not seem to be any reason why a character conferring a high degree of adaptation to a rain-forest environment could not also be primitive.

The presence of apparently primitive angiosperms in the lowland and montane rain forests of the tropics raises a further question: did they evolve there, as Stebbins and others believe, or has the continuously favourable environment of these forests merely provided a refuge where they have been able to survive? An argument used by Takhtajan in favour of Southeast Asia as the original home of the angiosperms is that archaic taxa are much more numerous there than in tropical America or Africa. But, as Hughes (1976) points out, this may be because in the early Cretaceous when the angiosperms were rapidly evolving, what are now the rain-forest regions of America and Africa were outside the tropics and were only later carried to their present latitudes by plate movements of the earth's crust.

The view that the vegetation of the temperate regions was derived by a process of selection from that of the tropics is said to have been first proposed by the Russian plant geographer Krassnov in 1894 (Takhtajan 1969): since then it has been put forward by Bews (1927), Aubréville (1969) and many others. Converging evidence of many kinds indicates that temperate floras have directly or indirectly a tropical origin, of which many temperate plants still show traces in their phenology (Diels 1918) and organization. Axelrod (1959, 1966) believes that the migration of the angiosperms from the tropical zone into higher latitudes began in the latter part of the Cretaceous period and continued throughout the Tertiary[5]. Like that of the temperate regions, much of the flora of the tropical savannas and dry forests seems to be derived from that of the tropical rain forest. Even if the very early angiosperms inhabited semiarid habitats, they seem to have left few direct descendants in the seasonal tropics.

[5] See Upchurch & Wolfe (1987).

Part I

Structure and physiognomy

Man hat eine Pflanzengesellschaft noch nicht 'verstanden', wenn man weiss, unter welchen Bedingungen sie vorkommt. Viel wichtiger ist, zunächst zu ergründen, wie ein Pflanzenverein aufgebaut, wie seine Struktur ist. Die Vegetationskunde ist eine Formenlehre der Pflanzengesellschaft.

A plant community is not 'understood' if it is known merely under what conditions it is found. It is much more important first to discover how it is built up, what is its structure. The science of vegetation is the study of the morphology of plant communities.

H. Meusel (1935, p. 269), transl.

Chapter 2

Structure of primary rain forest

2.1 The synusiae

LIKE THE INDIVIDUALS in a human society, the plants and animals in complex natural ecosystems form classes or groups, the members of each group having a similar function in the community as a whole and a similar relationship to their physical and biotic environment. The inviduals in the classes of human societies are bound together by physiological and economic links, ecological groups of plants in an ecosystem by their common dependence on solar energy.

The analogues of human social groups in a plant community are called synusiae, a term originally introduced by Gams (1918) but often used vaguely and in different senses (see Lebrun 1961): here it is applied to a group of plants of similar life-form which play a similar part in the community to which they belong. A synusia is thus an aggregation of species or individuals making similar demands on a similar habitat. The species belonging to the same synusia are mostly unrelated taxonomically but are to a large extent ecologically equivalent and so, to use the term of Whittaker *et al.* (1973), they occupy the same ecotope[1]. In all complex plant communities the synusiae are numerous and have a characteristic arrangement in space.

In a temperate broad-leaved forest such as a European oakwood or North American beech–maple forest, the component synusiae can be easily recognized and have a relatively simple spatial arrangement. Commonly the woody plants form two layers (strata), one of tall trees and another of shrubs and smaller trees, while below these there are one or more layers of herbs and undershrubs (the ground or 'field' layers), and sometimes also a layer of mosses and liverworts close to the ground. Besides these layers, each of which can be regarded as a separate synusia, there are groups of plants dependent on the trees and shrubs for mechanical support, the climbers (vines), and epiphytes, the latter in most temperate climates consisting entirely of nonvascular plants. There are also synusiae of saprophytes and parasites which are only indirectly dependent on sunlight for energy; these include a few flowering plants as well as fungi and other micro-organisms.

In tropical rain forests, which are much richer in species than temperate forests, the synusiae are more numerous and less clearly defined; their spatial arrangement is also much less obvious. At first sight a tropical forest appears to be a bewildering chaos of vegetation: in Junghuhn's (1852) phrase, nature seems to show here a *horror vacui* and to be anxious to fill every available space with stems and leaves. Closer examination shows, nevertheless, that in a rain forest, as in other plant communities, the plants can be grouped into a limited number of synusiae with a discernible, though complicated, three-dimensional distribution. The latter is repeated as a pattern in rain forests throughout the tropics, so the structure of climax rain forest is essentially similar in South America, Africa and Southeast Asia. Species of corresponding synusiae in widely separated geographical regions, though taxonomically different, are alike in life-form and physiognomy. As will be seen later, there are local variations on this pattern, but they are relatively unimportant.

[1] This term implies a habitat as well as a role in an ecosystem. Since the term 'niche' cannot be easily applied to plants (Richards 1969), 'ecotope' has much to recommend it.

Table 2.1. *Synusiae of the tropical rain forest*

A Autotrophic plants (with chlorophyll)
1. Mechanically independent plants
(a) Trees and shrubs or treelets
(b) Herbs
2. Mechanically dependent plants
(a) Climbers
(b) Epiphytes (including semi-parasitic epiphytes)
(c) Hemi-epiphytes (including 'stranglers')
B Heterotrophic plants (without chlorophyll)
1. Saprophytes
2. Parasites

The extent to which the synusiae are divided up and the way in which they are classified is to some extent arbitrary. The following scheme (which does not include micro-organisms) is based on the plant's means of obtaining energy and has proved simple and easy to apply (Table 2.1).

The autotrophic (photosynthetic) plants fall into two groups according to their method of competing for sunlight. The mechanically independent (or self-supporting) plants (A 1) are able to reach the light without assistance from other plants; the mechanically dependent plants, A 2 (the guilds or *Genossenschaften* of Schimper) cannot do so: they require a supporting plant (phorophyte) which is usually a tree or shrub, but sometimes a liane. The independent plants, as we have seen, vary in size from giant trees to tiny herbs. They are usually regarded as forming several strata or storeys: the nature and number of these will be discussed later. These plants form the 'framework' of the forest and largely determine the microclimates within it (Chapter 8).

The dependent plants also have a characteristic three-dimensional spatial arrangement but this depends mainly on the branching and foliage (architecture) of the trees and shrubs. The most important group is the climbers (lianes) (A 2a) which are rooted in the ground and need support for their long flexible stems. They contribute a considerable part of the total leaf-area of the forest (p. 46) and play a significant part in its dynamics.

Another group of dependent plants, the epiphytes (A 2b) are mostly short-stemmed, but as the majority of them are intolerant of deep shade and are not normally able to grow on the ground, they are found attached to the stems, branches and twigs of trees. Semi-parasitic epiphytes, represented chiefly by the mistletoes (Loranthaceae and Viscaceae). Unlike true epiphytes, they are attached to their phorophytes by haustoria through which they absorb water and dissolved substances; they are thus partial parasites. True epi-

phytes, though sometimes spoken of as parasites, depend on phorophytes for mechanical support only, although their supply of mineral nutrients may include small quantities of ions leached out of the surrounding foliage by rain.

The remaining group of dependent autotrophic plants, the hemi-epiphytes and stranglers (A 2c) occupy an intermediate position: they begin life as epiphytes and later send roots down to the ground. Some of them, for example the strangling figs (*Ficus* spp.), become very large and may eventually kill and supplant the phorophyte on which their seeds germinated. Stranglers are thus at first dependent and later may become independent. They cannot be sharply separated from other hemi-epiphytes such as some Araceae (p. 134), which send down roots which reach the soil but do not usually become self-supporting. Hemi-epiphytes are a life-form which hardly exists outside tropical forests.

Lastly there are the synusiae of saprophytic and parasitic flowering plants (B 1 and B 2), which together form the class of heterotrophic plants. They are not necessarily unresponsive to light, but obtain organic food from other living or dead plants and are not directly dependent on light as a source of energy.

The subdivisions of these various groups will be further discussed in Chapter 6, where some of their characteristics will be considered.

As far as is known, there is no climax community to which the term tropical rain forest can be properly applied in which all the plant groups in Table 2.1 are not present. There is also no region of the rain forest which possesses local synusiae not found elsewhere. The variations in structure depend partly on differences in the relative importance of the various synusiae. Thus in some rain-forest communities, epiphytes are more abundant (in species and individuals) than in others. Similarly, some rain forests have more climbing plants than others do.

The contributions of the various synusiae to the structure and functioning of the rain-forest ecosystem as a whole is very unequal, whether measured in number of species and individuals or as fractions of the total biomass. This is well illustrated in an analysis by Klinge *et al.* (1975) of the structure of a 0.5 ha plot of mixed lowland rain forest near Manaus in Amazonia (Table 2.2). The phytomass was separated into roots (about 25.8%) and aerial phytomass. The latter was divided into trees and other independent plants, lianes, epiphytes and parasites. The last three groups formed only 6.3% of the aerial phytomass, the rest consisting of dicotyledonous trees and shrubs, palms and herbaceous plants. Among the trees it is striking that the very tall trees

Table 2.2. *Phytomass on a 0.2 ha plot, near Walter Egler Forest Reserve, Manaus, Amazonia*

	Fresh mass (t ha^{-1})	% aerial phytomass
Independent plants		
Dicotyledonous trees >1.5 m high		
Leaves	14.1	1.9
Branches and twigs	202.2	27.5
Stems	465.5	63.3
>1.5 m high		
Leaves	0.34	0.5
Stems	2.1	0.3
Plants <1.5 m high		
Leaves	0.6	0.1
Branches and twigs	0.2	N^a
Stems	0.6	0.1
Total independent plants	697.7	
Dependent plants		
Lianes	46.0	6.3
Epiphytes	0.1	N^a
Parasites	0.1	N^a
Total dependent plants and parasites	46.2	6.3
Total aerial phytomass	743.9	100.0
Root mass		
Fine roots	49.0	—
Other roots	206.0	—
Total roots	255.0	—
Total phytomass	989.9	—

^{a} N, negligible.

Source: From Klinge *et al.* (1975).

of the A and B strata (pp. 30ff.) make by far the biggest contribution to the phytomass (over 85%). The rain forest, although composed of many varied synusiae, is thus overwhelmingly dominated by large trees, though as will be seen later (p. 38) small trees and shrubs form a large proportion of the species.

A further point, of great importance to the functioning of the rain forest as an ecosystem, is that, in the forest studied, wood represented some 97.4% of the phytomass and leaves of dicotyledonous trees and palms only 2.5%. It is not surprising that Klinge *et al.* (1975) found that the biomass of wood-eating animals (mainly termites and beetle larvae) was large relative to that of consumers of leaves, flowers and fruits.

These results refer to a single small sample of Amazonian forest only, but are probably fairly typical of mature rain forests generally. Although the abundance

of lianes, epiphytes and parasites varies in different types of forest, they seem always to form a comparatively small part of the whole biomass. The proportion of tall trees also varies, but in general they are probably by far the largest contributors to the total biomass.

2.2 The forest mosaic

Primary rain forest is never homogeneous in structure, even where there has been no felling or other artificial disturbances. Stretches of high forest are interrupted by gaps and here and there by 'climber tangles' or thickets of immature trees. The gaps vary in size and frequency. In a lowland forest in Malaya, which showed no signs of human disturbance, Poore (1968) found that gaps up to 600 m^2 in area occupied 9.9% of the total area. Most gaps are due to the death of large trees from old age or disease, others to damage caused by wind, lightning, landslips or by large animals such as elephants or wild pigs. In all mature forests the populations of most tree species consist of young, mature and senescent (over-mature) individuals, the trees of the future, the present and the past, as Hallé *et al.* (1978) call them. These age-groups are not evenly distributed and cause patchiness in forest structure.

The natural forest is thus always an irregular mosaic of developmental stages, which can be called the mature, gap and building phases (Watt 1947) (Figs. 3.1–3.3). The mature phase (high forest, the 'homoeostatic forest' of Hallé *et al.* 1978) is dominated by mature and senescent trees, while the gaps usually contain few or no living trees larger than seedlings and young saplings. The building phase represents the stage of recovery as a gap is gradually filled by young, vigorously growing trees. The three phases represent stages in a continuing process and the area occupied by each is constantly changing. Regeneration processes are discussed more fully in the next chapter. It should be noted that what is often dealt with in descriptions of tropical forest structure is only the mature phase, disregarding the gap or building phases.

2.3 Density and dispersion of trees

The density (number per unit area) of the trees in primary tropical rain forests varies within wide limits and depends on many factors. It can be dealt with only briefly here (see data in Table 4.4 of Golley *et al.* 1975 and UNESCO 1978). In mature forest with few gaps on more or less level, free-draining lowland sites the number of trees per hectare with a diameter at breast height (d.b.h.) greater than or equal to 10 cm (girth ≥ 31 cm) is usually about 300–700 (10 cm d.b.h. is an arbitrary limit roughly corresponding to trees about 15–26 m high).

The factors controlling tree density in rain forests are complex and not well understood. Apart from the effects of natural and anthropogenic disturbances, they certainly include drainage and other soil conditions, as is well shown in the catenas of forest types in Guyana (pp. 302–4). In general heath forests and other forests on porous sandy soils have high tree densities and those on swampy sites with impeded drainage have, as might be expected, low densities. In hilly country the density of the trees is often strikingly greater on ridge tops than on the slopes, as Wyatt-Smith (1960) found in Malaya and Ashton (1964) in Brunei.

Like density, the diameter-class distribution of the trees is very variable, some mature forests having relatively large numbers of trees of 40–60 cm d.b.h. or more and others comparatively few. It also seems that there is little correlation between the number of trees of very large diameter and the total number above some fairly small arbitrary lower diameter limit such as ≥10 cm. A vast amount of data on diameter (girth) class distributions in tropical forests can be found in the literature and in the files of Forestry Departments: a useful short review of this information is given by UNESCO (1978) but much of it is still unanalysed. It should be noted that though the diameter (or girth) class distribution of individual species often departs widely from the mean, the diameter class distribution of whole stands follows an exponential model fairly closely, but only for stems of about 20 cm d.b.h. or more (UNESCO 1978). Thus, if the trees in a stand of rain forest are grouped into 10 cm intervals from 20 cm d.b.h. upwards, the number of trees in each interval is generally about twice that in the next larger class though the ratio increases slightly from each class to the next higher one. Below about 20 cm d.b.h. the number of individuals is usually considerably greater than would be expected from this rule. Since there is some correlation, but not a close one, between diameter and height (p. 71), the diameter-class distribution reflects the height distribution to some extent and gives some indication of the vertical structure of the forest.

The dispersion as well as the density and diameter distribution of the trees contributes to the structural pattern characteristic of rain forests. In stands of supposedly uniform floristic composition the dispersion of the trees, taking all species together, is usually random, as Poore (1968) showed for 100 m² plot sizes in lowland forest in Malaysia. But this is not always so. Golley *et al.* (1975) found in Panama that in 'tropical moist forest', 'Pre-montane Wet forest' and in riverain and mangrove forest, tests of goodness of fit to a Poisson distribution showed that the trees were not randomly dispersed. It has also been shown repeatedly (see Chapters 3 and 12) that individual tree species, especially those with inefficient methods of seed dispersal, often show markedly clumped (contagious) dispersal patterns. Even where the larger trees are randomly distributed the smaller species, which can only grow in the spaces left free by the large individuals, are likely to be contagiously distributed. Since many large rain-forest trees (over about 30 cm d.b.h.), are buttressed and most of the smaller trees unbuttressed (p. 79), the area of the forest floor effectively occupied by the former is considerably larger than the basal area calculated from their diameters.

2.4 Stratification

The idea that in tropical forests the crowns of the trees form several superposed strata or storeys (the words layer, tier and canopy are also used) has been current in the literature for a very long time; it may have originated in von Humboldt's (1808) description of the South American hylaea as 'a forest above a forest'. The term stratification as applied to rain forests has been variously interpreted; its meaning is often misunderstood. Sometimes it is stated categorically that there are three tree strata (according to a few authorities, more than three). Brown (1919, pp. 31–2), for example, describes the stratification of a Philippine dipterocarp forest in these words: 'The trees are arranged in three rather definite stories. The first, or dominant story forms a complete canopy; under this there is another story of large trees, which also form a complete canopy. Still lower there is a story of small scattered trees'. Some writers give the impression that the strata of the rain forest are as well defined and as easy to recognize as in an English 'coppice-with-standards', but Brown continues: 'The presence of the three stories of different trees is not evident on casual observation, for the composition of all stories is very complex and few of the trees present any striking peculiarities, while smaller trees of a higher story always occur in a lower story and between the different stories'.

There are also authors who state, or imply, that any grouping of the trees according to their height is arbitrary and that 'strata' have no objective reality. This is the opinion of Mildbraed (1922, pp. 103–4), who with special reference to the forest of southern Cameroon, says explicitly:

It is often stated that the rain forest is built up of several tiers or stories. These terms may easily give a wrong idea to anyone who does not know the facts at first hand. What is meant is merely that the woody plants can be grouped into three, four, five, or perhaps more, height classes, according to taste. The space can indeed be thought of as consisting of height intervals of 5, 10 or 20 metres, and species are found which normally reach these height intervals when fully grown. As, however, trees of all intermediate heights are present and the mixture of species is so great, these hypothetical height intervals never really appear as stories. It is truer to say that the whole space is more or less densely filled with greenery. (Transl.)

In recent years the concept of stratification has been much debated, some authors having views similar to Brown's and others agreeing with Mildbraed. Recent studies of rain-forest structure show that both views are over-simplifications. It is also now evident that, even though it can still be said that all tropical rain forests have a basically similar structure, there is considerable variation among them. This is partly caused by the differences between the phases of development already mentioned, but even if only the mature or 'high forest' phase is considered, structure can vary considerably. Forest structure also depends greatly on species composition: single-dominant forests in which one tree species forms a large percentage of the whole stand (Figs. 2.8B, 2.9, 2.10) differ in structure from the more widespread mixed types (Figs. 2.3, 2.4 etc.).

The meaning of the term stratum needs some clarification. The space from ground level to the tops of the tallest trees is never uniformly filled: there are always more leaves and branches at some levels than at others. The term stratum can be applied to a layer of tree crowns between certain limits of height (or with a certain 'mid-crown height'). Sometimes a discontinuity can be seen above such a layer, and in some cases also below it, so that its limits are clearly defined (Figs. 2.8B) but more often there is no discontinuity and the limits are vague or quite arbitrary (Figs 2.3, 2.8A). In this book the word stratum, storey or layer is used for a group or set of tree crowns between certain heights, whether or not it is clearly separated above and below. The layers of shrubs and other small plants below the trees can also be termed strata. As Hallé et al. (1978, p. 333) say, strata are a useful concept for analysing forest structure and whether or not they 'exist' (i.e. can be objectively demonstrated) is more or less irrelevant.

Grubb et al. (1963) emphasize the distinction between stratification of individual trees and stratification of species. Here we are mainly concerned with the stratification of individuals. The vertical distribution of the average heights attained by different species at maturity is difficult to study because of lack of data and it would be hard to determine whether it is continuous or discontinuous, although it is apparent that mature trees of certain groups or families, e.g. Dipterocarpaceae, Caesalpinioideae, are found mainly in the upper strata of the forest and that others, e.g. many Annonaceae, Diospyros and Eugenia spp., are mostly small trees belonging to lower storeys.

The word 'canopy' is often used in descriptions of forest structure but unfortunately in several different senses. Sometimes it means 'the sum total of the crowns of the trees of all heights' (Grubb et al. 1963, p. 580) and sometimes, as mentioned above, it is regarded as equivalent to stratum (as when rain forest is described as a 'multi-canopy forest'). In zoological and in popular usage the canopy generally means the upper, well-illuminated levels of the forest, i.e. the euphotic layer (p. 48). If the term is to be used at all, it is perhaps best to define it as a more or less continuous layer of tree crowns forming the 'roof' of the forest. This surface, as will be seen later, may be formed by the crowns of the highest tree stratum or, more often, by the highest and the next lower stratum together. Where the canopy is a layer of close-packed crowns, neighbouring trees show 'crown-shyness', i.e. they are usually separated by a narrow gap (Ng 1977b, Whitmore 1984a). The cause of this is not clear, but it seems to depend on the reactions of the apical buds.

'Emergent' is another ill-defined term. Sometimes it means no more than the taller trees of the forest, but in this book it is used for trees rising 'head and shoulders' above their neighbours so that most of their foliage is fully exposed to sunlight.

Useful discussions of stratification in tropical forests have been given by Grubb et al. (1963) and by Hallé et al. (1978, Chapter 5). An extensive review of the subject has been published by Rollet (1974).

2.4.1 Methods of studying stratification

The differing opinions on the stratification of tropical rain forest are no doubt mainly due to the difficulty of seeing the whole height of the forest from ground level. On the forest floor the observer is, as Mildbraed says, a prisoner, and visibility is restricted both horizontally and vertically. The mass of leaves and branches above allows only occasional glimpses of the crowns of the taller trees, although here and there gaps made by fallen trees will reveal rather more. A river, road or clearing

Fig. 2.1 Views of canopy of mixed dipterocarp forest, Pasoh Forest Reserve (Selangor), Malaya, from *ca.* 25 m on a ladder. (A, March 1974; B, August 1974.) A shows a deciduous tree bare of leaves.

may make it possible to see the whole profile of the forest but such cross-sections are often misleading, because when more light is able to reach the lower levels of the forest, a vigorous outburst of growth takes place in the undergrowth and climbing plants grow down from above and up from the ground until the exposed surface becomes covered with a curtain of foliage which conceals the natural spacing of the tree crowns. The structure of mature forest on river banks or on the edge of a clearing, unless very recently made, is always different from that in the interior (p. 43).

Much can be learned about forest structure by climbing trees. In recent years permanent ladders (Fig. 2.1) and other tree-climbing devices have been set up in various rain-forest localities, usually with the assistance of skilled local tree climbers (Mitchell 1982, 1986). In some places observation posts in tall trees have been established and used for long periods. At Bukit Lanjan in Peninsular Malaysia a walkway, part of which was

over 30 m above ground level, was maintained for some years (Muul & Lim 1970) (Fig. 2.2). In most cases the chief object of these contrivances has been to make observations on mammals, birds and insects, or on microclimates, but they have also been used to study the flowering, fruiting and pollination of tall forest trees (Chapter 9). Often only a relatively small part of the forest canopy can be seen from a tree-top observation post or ladder but useful impressions of the spatial arrangement of the tree crowns and the part played by lianes in the forest structure can be obtained.

Another valuable source of information about the upper strata of the forest is low-level aerial photography. Its use for studying forest structure (as distinct from making forest inventories and the like in economic forestry) has been developed especially by Holdridge in Costa Rica (Holdridge *et al.* 1971). From photographs at a scale of 1:5000 it was possible to draw 'canopy maps' with contours at 5 ft (1.5 m) intervals: these show

Fig. 2.2 Walkway in canopy of mixed deciduous forest, Bukit Lanjan (Selangor), Malaya (March 1974).

25 ft (7.6 m) wide, preferably not less than 200 ft (60 m) long, and then remove all undergrowth and small trees of less than an arbitrary minimum height of 15 ft (5 m). The positions of the remaining trees were then mapped and their diameters recorded. The total height, height to lowest (large) branch, lower limit of crown and width of crown of each tree were then measured. At first it was considered necessary to fell all the trees to obtain these measurements, but later it was found that in many types of forest if a few trees were cut down, sufficiently accurate measurements of the heights of the other taller trees could be made with an altimeter: the heights of the smaller trees can then be estimated with the help of a graduated pole. The trees were all identified as far as possible and herbarium material of each species was collected for reference.

Since profile diagrams were first introduced, various modifications of the original method have been suggested. Widths of 10 or 20 m, instead of the original 7.6 m, and various lengths, are sometimes used. The tree dimensions measured are not always the same and various minimum heights or diameters have been adopted. The style of representation varies from highly diagrammatic to attempts at 'realism'. Because of the many methods and styles, the profile diagrams of different authors are not always readily comparable.

When the information needed is available, it is useful to indicate mature, immature and over-mature trees by different kinds of outline or shading as is done by Oldeman (Figs. 1.2A, 2.13) (Oldeman 1974, Hallé *et al.* 1978). By distinguishing the mature trees from those that are young or senescent in this way, the forest structure becomes clearer. Holdridge's 'idealized profiles' (Holdridge *et al.* 1971, p. 24) have a similar aim: a profile for an area of 10 m × 100 m is 'reconstructed' from sketches of actual trees using only mature individuals of the more abundant species. Some authors give a ground plan of the tree bases or a crown projection diagram with each profile. Aubréville (1965) and others have attempted to draw three-dimensional block diagrams to illustrate forest structure, but to do this accurately is very time-consuming.

Robbins (1959) and UNESCO (1978) have given critical appraisals of the profile diagram method and have suggested that there should be some standardization of the measurements that should be made and of the conventions used in the diagrams.

There is no doubt that profile diagrams are useful for portraying certain general features of forest structure that are not easily appreciated from photographs and quantitative data alone. They are particularly valuable in comparing forest types differing in structure. But the

very clearly the spatial relations of the tree crowns and how the two upper strata of the forest combine to form a canopy. Such canopy maps proved very useful in showing the structural differences between lowland rain forest and other types of forest.

Profile diagrams. Because direct observation of the profile of tall tropical forests is so difficult, it is often useful to construct profile diagrams to scale from measurements of the trees on narrow sample strips of forest. This technique, first applied in the forest of Guyana (Davis & Richards 1933–34) (Fig. 2.3) has since been widely used in many parts of the world. Though somewhat laborious, it has proved a valuable method of recording and comparing the structure of tropical forest communities.

Various modifications of the profile diagram method have been suggested. As originally carried out, the procedure was to mark out a rectangular strip of forest

Fig. 2.3 Profile diagram of mixed forest, Moraballi Creek, Guyana. From Davis & Richards (1933–34, p. 368). The diagram represents a strip of forest 135 ft (41 m) long and 25 ft (7.6 m) wide. Only trees over 15 ft (4.6 m) high are shown.

limitations of the method need to be understood and its value should not be over-rated. In the first place the areas sampled are necessarily very small and they are usually selected on a subjective rather than systematic or random basis. It is seldom practicable to draw a large number of diagrams of the same type of forest, but since forest structure is so variable it is possible to draw very different-looking diagrams of forest on closely adjoining sites (Fig. 2.4). It is necessary to be very cautious in drawing quantitative conclusions from even a large number of profiles. As it is hardly feasible to show lianes accurately in profile diagrams, they are at best incomplete representations of the forest structure.

2.4.2. Structure of mixed rain forests

Figs. 2.3–2.6 (also Figs. 2.8A and 2.14) are profile diagrams of mixed rain forests in the mature phase, as defined above. Fig. 2.3 shows mixed forest in Guyana

and Fig. 2.5 mixed dipterocarp forest in Sarawak (Borneo) (without single-species dominance but with family dominance of the Dipterocarpaceae, p. 311): these are both examples of primary lowland forest in ever-wet climates. Fig. 2.6 represents a sample of evergreen seasonal forest, a type of rain forest found in climates with a distinct dry season (pp. 36–7). It was originally regarded as primary forest (Richards 1939), but was more probably old secondary forest which had not been disturbed for some time. It is assumed to have a structure approximating to that of the original primary forest of the region.

The following descriptions of these profiles, based on notes taken in the field, supplement the diagrams. The strata are referred to as A, B, C, D and E (from above downwards).

Mixed forest at Moraballi Creek, Guyana (Fig. 2.3). This forest can be described as having three strata of

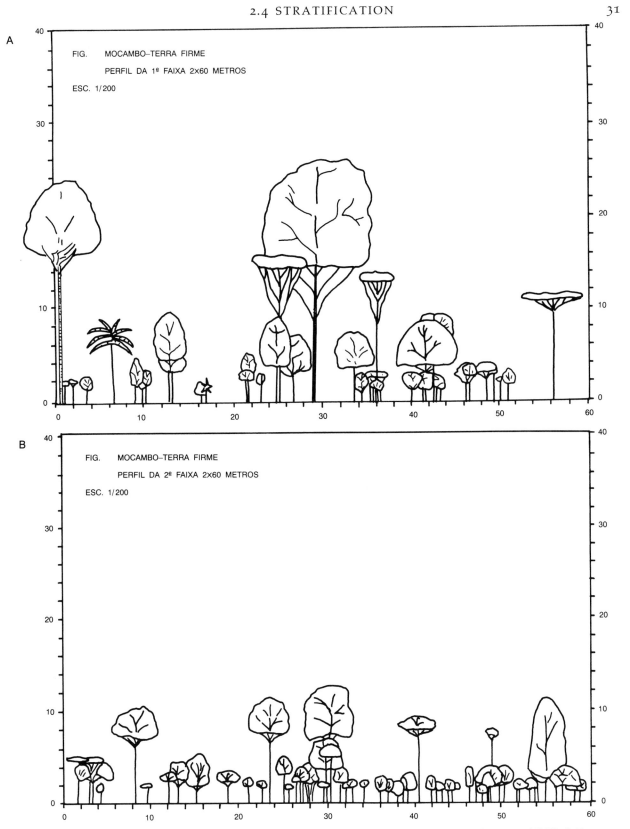

A

FIG. MOCAMBO–TERRA FIRME

PERFIL DA 1ª FAIXA 2x60 METROS

ESC. 1/200

B

FIG. MOCAMBO–TERRA FIRME

PERFIL DA 2ª FAIXA 2x60 METROS

ESC. 1/200

Fig. 2.4 Profile diagrams of adjacent strips of primary forest, Mocambo (Pará), Brazil. After Pires & Moraes (1966). A–E are successive strips 2 m wide. In F they are combined to form a plot 10 m × 60 m.

Fig. 2.4 (*cont.*). For legend see previous page.

Fig. 2.4 (*cont.*).

trees, though their boundaries are very indefinite. Only the lowest (C) is laterally continuous, but it should be remembered that all trees less than 15 ft (5 m) high have been omitted: the C layer merges below into the stratum of treelets and saplings (D). The two upper tree strata are more or less open and not laterally continuous: they are not demarcated by any clear horizontal boundary above or below, though towards the left side of the diagram the B stratum becomes somewhat more distinct. In the original description of this profile (Davis & Richards 1933–34, pp. 362–72) the A and B strata were regarded as a single layer of somewhat irregular profile. Most of the gaps in the highest (A) stratum are closed below by crowns in stratum B, so A and B together form a fairly complete canopy of foliage.

The height of the trees in stratum A on the diagram is about 35 m, but outside the sample strip they are somewhat higher (to about 42 m). Those of the B stratum are about 20 m high, while stratum C includes trees from the arbitrary lower limit of 15 ft (4.6 m) to about 15 m; the average height is about 14 m. In the profile strip (135 ft = 41 m long) there are sixty-six trees 15 ft high; seven of these can be reckoned as belonging to the A stratum, twelve to the B and the rest to the C stratum.

The crowns of the A storey trees are only here and there in contact laterally, but this stratum is more closed than it appears to be in the diagram because trees whose crowns overlap the sample strip but whose bases are outside it are not shown. The crowns are mostly wider than deep and tend to be umbrella-shaped. The trees of this storey belong to many species and families (Lecythidaceae, Lauraceae, Araliaceae, etc.).

Stratum B is more continuous but has occasional gaps. Like A it is composed of many species belonging to numerous families: many of these are not represented in A, though a considerable proportion of the trees are immature individuals of A storey species.

In stratum C there are very few gaps and the density of foliage is probably greater than at any other level of the forest, higher or lower. More than half the total number of trees are young individuals of species which may eventually reach the higher strata. The remainder are species peculiar to stratum C, mostly belonging to families scarcely represented in the A and B strata (especially Annonaceae and Violaceae). Both the young A and B trees and the true C species usually have long tapering crowns, several times deeper than wide.

Below the three storeys represented in the profile diagram there are two other strata consisting chiefly or partly of woody plants; both are ill-defined. The upper of these (D), the average height of which is about 1 m, consists of young trees, small palms, tall herbs (Marantaceae etc.) and large ferns, as well as mature woody plants of shrub or treelet form (pp. 74–5). The lowest stratum is the ground or field layer (E). This consists chiefly of seedlings of trees and lianes; herbaceous plants (dicotyledons, monocotyledons, ferns and *Selaginella*) form only an insignificant proportion of the total number of individuals in this layer. Like the D stratum, this layer is usually discontinuous, the plants being very scattered, except in openings or here and there where a social species forms a patch of closed vegetation.

There is no moss layer on the forest floor. Except for patches of mosses such as *Fissidens* spp. on disturbed soil (by overturned trees, armadillo holes, etc.), bryophytes are confined to the surface of living or dead trees (see p. 150).

The forest stratification was also examined by climbing a tall tree in forest similar to, but not actually adjoining, the profile strip. The following notes were made on the spot at a height of 110 ft (33.6 m).

There is no flat-topped canopy; there are two more or less clear [upper] layers, but they are both discontinuous, so the general effect is very uneven. Any two tall trees may be separated from one another by one or more lower trees. The lower trees do not grow under, and are not much overshadowed by the higher. Practically all the lower trees of the canopy are covered and bound together with lianes. [These do not usually reach the trees of the highest stratum]. Apart from the ordinary upper canopy [i.e. stratum A] trees, there are rare ones which tower far above all others. Two such 'outstanding' (emergent) trees were seen whose whole crowns were well clear of all surrounding trees.

A view from this observation point is shown in Fig. 1.3; one of these 'outstanding trees' is visible on the horizon. No 'outstanding trees' are shown in the profile diagram. Such trees are probably not more frequent than 1 per square kilometre on average, and their height is about 40–45 m. They belong to species rarely found in stratum A, e.g. *Hymenaea courbaril, Peltogyne pubescens.*

Mixed dipterocarp forest, Gunung Dulit, Sarawak (Fig. 2.5). A full description of this profile was given by Richards (1936). This forest also can be regarded as three-storeyed, but instead of a clearly defined C stratum and two higher strata arbitrarily separated from each other, as in the last example, the highest (A) stratum, though discontinuous laterally, is fairly well separated vertically from the B and C storeys. If the average heights of the trees in the A, B and C strata are taken

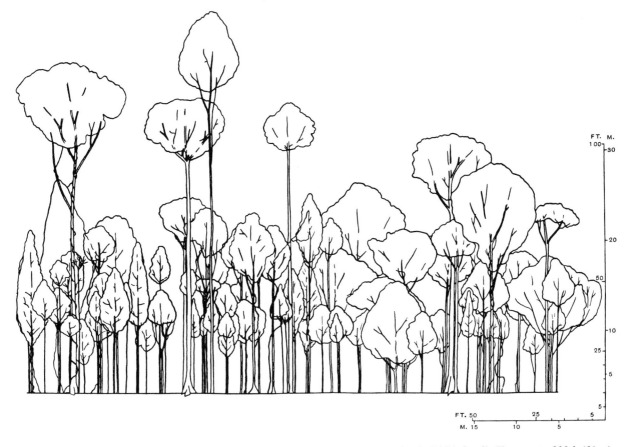

Fig. 2.5 Profile diagram of mixed dipterocarp forest, G. Dulit, Sarawak. From Richards (1936, fig. 2). The strip is 200 ft (61 m) long and 25 ft (7.6 m) wide.

as 35, 18 and 8 m, respectively, there are seven trees in A and eighty-six in B and C taken together. These numbers cannot be compared directly with those of the Guyana profile because the sample strip was 200 ft (61 m) long instead of 135 ft (41 m) and only trees of height 25 ft (7.6 m) or more were measured.

The crowns of the A storey trees are so large and well spaced, and so well raised above the very dense lower storeys, that when the canopy is looked at from above from the crest of the Dulit escarpment or from the air (Fig. 11.10) the individual trees can be distinguished from a long distance. They are about 6 m deep on the average and tend to be wider than deep. The majority of the trees in this storey belong to the Dipterocarpaceae.

The B and C strata are also laterally continuous, each crown being more or less in contact with several others. The individual crowns are about 4.5–6 m deep and are mostly at least twice as deep as wide. Some young dipterocarps are present but the majority of the trees in these strata belong to other families. It is interesting

that, as in Guyana, Annonaceae are numerous among the smaller (C storey) trees. The lack of a clear division between the B and C strata is due partly to the large number of young trees and partly to the varying heights reached by the smaller species at maturity.

Below the C tree stratum the vegetation becomes much less dense. No very clear stratification is evident in the smaller undergrowth, but it is convenient to divide it into a shrub stratum averaging 4 m high, consisting of 'shrubs', palms and young trees, and a ground layer up to 1–2 m high, consisting of tree seedlings and herbs, including ferns, *Selaginella* and rarely one or two species of very small palms. The number of herbaceous plant species is considerable, but the tree seedlings outnumber them in individuals; on ten quadrats 1 m × 1 m the total number of tree seedlings and other woody plants was 184 and of herbaceous plants 135 (shoots). The density of the ground layer is uneven. Over large areas it is represented only by widely scattered plants; here and there it is fairly dense, though seldom as dense as in a deciduous wood in Europe.

Mixed (evergreen seasonal) forest, Omo Forest Reserve, Nigeria (Fig. 2.6).   The structure of this was described in detail by Richards (1939). Only the lowest (C) storey of trees up to about 15 m high is continuous. Above this there are trees of various heights, the tallest of which is 46 m high. The crowns of these taller trees are occasionally in lateral contact but there is no closed canopy above stratum C. Observations in the surrounding area showed that these taller trees could be regarded as an A storey from about 37 to 46 m (average about 42 m) and a B storey from about 15 to 37 m (average about 27 m) but only an arbitrary division could be made between them. The A stratum is represented on the profile by a single individual.

The A stratum thus consists of very large trees scattered through the forest (though much more densely than the 'outstanding trees' in Guyana). Their crowns are umbrella-shaped, extremely heavy, and 25 m or more wide. They are rarely in lateral contact and are

mostly raised well above those of stratum B. This storey consists of comparatively few species, chiefly *Lophira alata* and *Erythrophloeum ivorense* in the area studied.

Stratum B is also very open and the crowns are only occasionally in lateral contact. In the diagram the gap in this layer beneath the crown of the large A stratum tree is noteworthy. The crowns of the B storey trees are smaller and narrower than those of the A storey and are usually under 10 m wide. The species are numerous and belong to a wide range of families.

Stratum C is very dense and almost without gaps. The crowns are packed closely together and are usually tightly bound together by lianes. The majority of the trees in this layer are species which never reach a higher storey; they belong to various families, but there is a tendency for a single species (the actual species varying from place to place) to be locally dominant. The remaining trees in the stratum are chiefly young B species; young A species seem to be strikingly rare. The C

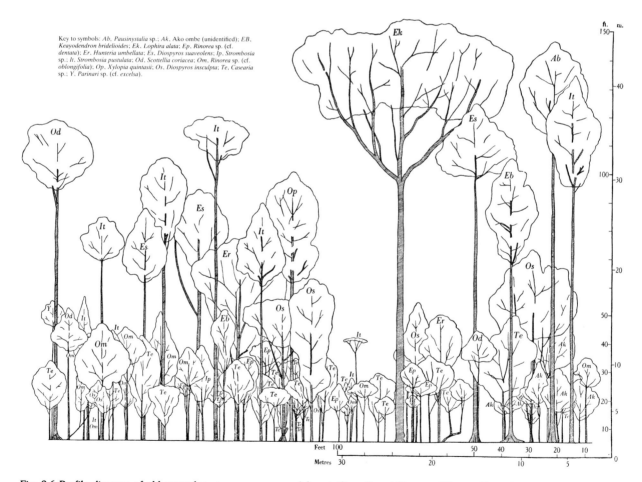

Key to symbols: *Ab*, *Pausinystalia* sp.; *Ak*, Ako ombe (unidentified); *EB*, *Keayodendron bridelioides*; *Ek*, *Lophira alata*; *Ep*, *Rinorea* sp. (cf. *dentata*); *Er*, *Hunteria umbellata*; *Es*, *Diospyros suaveolens*; *Ip*, *Strombosia* sp.; *It*, *Strombosia pustulata*; *Od*, *Scottellia coriacea*; *Om*, *Rinorea* sp. (cf. *oblongifolia*); *Op*, *Xylopia quintasii*; *Os*, *Diospyros insculpta*; *Te*, *Casearia* sp.; *Y*, *Parinari* sp. (cf. *excelsa*).

Fig. 2.6 Profile diagram of old secondary evergreen seasonal forest, Omo Forest Reserve, Nigeria. From Richards (1939, Fig. 4). The strip is 200 ft (61 m) long and 25 ft (7.6 m) wide. Only trees 15 ft (4.6 m) and over are shown.

Fig. 2.7 Canopy of old secondary forest, Omo Forest Reserve, Nigeria, from 24 m above ground (1935). In the foreground, note C storey trees bound together in a dense mass by lianes; in the background are trees of A and B storeys.

species, like those of the corresponding layer in Guyana and Sarawak, mostly have narrow conical crowns, but old individuals sometimes have much wider and heavier crowns.

The D stratum is very indefinite. It consists largely of young trees belonging to strata B and C, so there is no clear division between it and the lowest tree storey. Individuals of species properly belonging to stratum D, most of which are treelets rather than true shrubs, are few. The density of this stratum is very variable; in undisturbed forest it is never so dense as to make progress difficult, and in some places both shrubs and ground layers are almost wanting.

The lowest stratum of the forest is the ground layer (E), which consists of plants varying from a few centimetres to 1 m or more high, including tree and liane seedlings, dicotyledonous and monocotyledonous herbs, and ferns, the first generally predominating. This stratum is even more unevenly developed than D. Large stretches of the forest floor may be almost completely bare, but in places, especially in openings, the ground

may be concealed by a dense growth of herbaceous plants and tree seedlings. There are no mosses on the ground.

A view from a tree, 24 m above ground, in forest similar to the sample strip is shown in Fig. 2.7. Like the diagram, this shows the dense mass of small trees with interwoven crowns about 9–12 m high. The taller trees rising above this compact layer do not form a closed canopy at any level, so that above 11 m it is possible to see clearly for some distance in any direction. When climbing this tree one seemed to emerge into full daylight as soon as the dense C stratum was passed. The view of the stratification obtained from this tree is thus very similar to that shown by the profile diagram.

The openness of the two upper strata is a remarkable feature of this profile. Very discontinuous A and B strata are a general feature of mature mixed forest in West Africa, most of which is of the evergreen seasonal type (Chapter 16), e.g. Okomu Forest Reserve, Nigeria (Jones 1955, Fig. 2) and Southern Bakundu Forest Reserve, Cameroon (Richards 1963a). Aubréville (1933) described the 'closed forest' of Côte d'Ivoire as a mass of vegetation 20–30 m high, dominated by scattered taller trees, somewhat like a European *tallis-sous-futaie* (coppice-with-standards). Whether the open structure of the upper strata should be attributed to the relatively severe dry season or is the result of past disturbance (almost universal in West African forests) is not clear. The rather similar structure of rain forests in relatively seasonal climates in South America, e.g. the undisturbed terra firme forest of the Serra do Navio in eastern Amazonia (Rodrigues 1963) and the *Carapa–Eschweilera* forest in Trinidad (Fig. 2.8A) perhaps supports the first suggestion.

Conclusions. The three examples of mixed forest described above are similar in some features of their stratification but show differences that may be significant. From these data and other descriptions of mixed forest, the following conclusions seem justified.

(i) For descriptive purposes, mixed rain forest can be regarded as having five strata of independent plants. The tree strata A, B and C are never very clearly definable. Sometimes, as in the Guyana profile, the lowland forest of Ecuador (Grubb *et al.* 1963) and other examples, there is no clear vertical discontinuity of structure at any level and the strata can be delimited only arbitrarily. In the Sarawak profile and in some, but perhaps not all, mixed dipterocarp forests in Malesia, the A stratum is fairly clearly separated from the B stratum. In West Africa the A and B strata are usually fairly sharply separated from the C stratum but not from each other. In all mixed forests the D and E strata of herbaceous and small woody plants are only very vaguely definable.

(ii) The height of each stratum varies, but not within very wide limits. Thus the height of the A stratum is about 30 m or more in the Guyana forest described above, about 35 m in the Borneo example and about 42 m in the Nigerian profile. Similarly, the height of the B stratum is about 20, 18 and 27 m, respectively, and that of the C stratum 14, 8 and 10 m, respectively, in the three examples.

(iii) In mixed forests the A stratum is always more or less discontinuous: the tree crowns are mostly well separated, giving the canopy a characteristic 'humped' and uneven appearance when seen from the air. The B stratum may be continuous or more or less discontinuous. The C stratum is always more or less continuous and often appears to be the densest layer of the forest. Lianes rarely reach the A stratum but often play a large part in binding together the trees in the B and C strata.

(iv) Each stratum has a different and characteristic floristic composition, but in all the strata, except A and B, young trees of species that reach higher strata when mature form a large proportion of the total number of individuals.

(v) In each stratum the trees have a characteristic shape of crown. The ratio of depth to width of crown tends to decrease with the height of the tree, so that in the A storey the crowns are mostly wide, sometimes umbrella-shaped; in the middle levels the majority are about as deep as wide or somewhat deeper, while in the C storey most are conical and tapering. This is doubtless an aspect of what Horn (1971) calls their 'adaptive geometry'. In Chapter 4 it will be shown that the size and shape of the leaves, and other physiognomic features of the trees, also vary with height (or stratum).

The reason why the tree crowns are rarely clearly stratified in mixed rain forests is no doubt chiefly because they are composed of very large numbers of tree species each differing in growth potential and in their reactions to light and other factors. Similarly, the relatively simple structure of broad-leaved temperate forests is a result of their comparative poverty in species. The sharply defined strata of British broad-leaved woodlands are no doubt in part caused by past management but as Salisbury (1925) pointed out, they are mainly due to the very small number of species present in any one stand. Dawkins (1966) believes that because of the growth-pattern of tropical trees clearly defined stratification is unlikely to be found in species-rich natural forests with a random species distribution. It is, however, likely to be found where a single species forms a large proportion of the stand, as in what are here called single-dominant forests.

2.4.3 Structure of single-dominant forests

Climax communities in which a single tree species forms a large proportion, sometimes eighty per cent or more, of the stand, occur in all the main geographical divisions of the rain forest. The floristic composition of these communities is dealt with in Chapters 11 and 12. Some cover hundreds of square kilometres, while others are found locally in mixed forest as patches of only a few hectares. Single-dominant forests differ in structure from mixed forests and, as expected, their stratification is much more distinct. The A storey is usually more or less continuous, so they tend to have a smooth even canopy which makes them readily recognizable in aerial photographs.

An example of a forest showing these characteristics is the Mora association of Trinidad, dominated by *Mora excelsa*, a leguminous tree which may reach a height of 58 m (Beard 1946a). Profile diagrams of this community and of the *Carapa–Eschweilera* forest (Fig. 2.8), a mixed evergreen seasonal association also found in the lowland of Trinidad, show that in the former the tall *Mora* trees form an almost completely continuous level canopy, strikingly different from the uneven surface of the latter. 'The individual crowns of the moras are shaped in conformity with the adjacent ones, fitting together into a most striking mosaic. . . . Viewed from the air the canopy has the same undulating but continuous character as the waves of the sea' (Beard 1946a, p. 173). *M. excelsa* commonly forms 85–95% of all trees ≥30 cm diameter. Below the A stratum there are B and C storeys at about 12–25 m and 3–9 m, respectively, but neither forms a continuous layer and there is no clear boundary between them. In the D and E layers seedlings and saplings of *M. excelsa* are dominant to the exclusion of most other plants. Probably because of the large amount of light intercepted by the A stratum, the two lower tree storeys are very thinly stocked.

The *Mora excelsa* forest of Guyana has a continuous and undulating canopy like that of Trinidad when seen from the air but profile diagrams are not available. Lindeman & Moolenaar's (1959) profile of mora forest on a levee in swampy land in Surinam in which *M. excelsa* formed 43% of trees ≥2 m high is less tall and does not have such a continuous A storey.

Other types of single-dominant forest in the American tropics which seem to be similar to Mora forest in structure are the *Mora megistosperma* association found

Key to symbols: *A.*, *Pera arborea*; *B.*, *Clathrotropis brachypetala*; *B.c.*, *Amaioua corymbosa*; *B.ch.*, *Diospyros ierensis*; *B.l.*, *Ryania speciosa*; *B.m.*, *Pentaclethra macroloba*; *B.p.*, *Swartzia pinnata*; *B.t.*, *Rudgea freemani*; *C.*, *Carapa guianensis*; *Ca.*, *Virola surinamensis*; *Cb.*, *Guarea glabra*; *Coc.*, *Maximiliana elegans*; *Cp.*, *Brownea latifolia*; *G.*, *Eschweilera subglandulosa*; *Ga.*, *Esenbeckia pilocarpoides*; *In.*, *Protium guianense*; *L.s.*, *Ocotea wachenheimii*; *La.*, *Cordia* sp.; *M.*, *Mora excelsa*; *Ma.*, *Sterculia caribaea*; *Mc.*, *Euterpe langloisii*; *Mk.*, *Sapium aucuparium*; *Mi.*, *Miconia* sp.; *N.*, *Calliandra guildingii*; *P.*, *Tabebuia serratifolia*; *Pd.*, *Inga* sp.; *S.*, *Rheedia* sp.; *W.*, *Warszewiczia coccinea*; *W.c.*, *Tovomita eggersii*; *W.co.*, Rubiaceae sp.; *W.o.*, *Terminalia amazonia*; *Y.o.*, *Buchenavia capitata*.

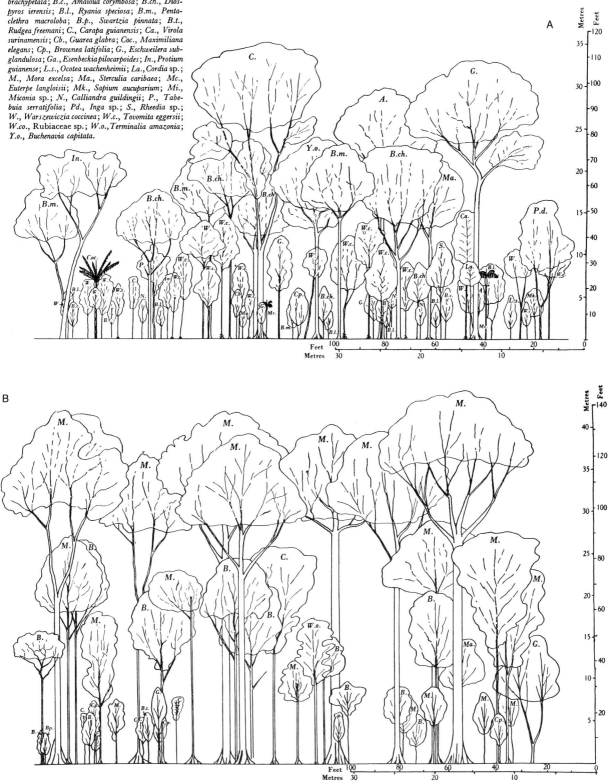

Fig. 2.8 Profile diagrams of rain forest in Trinidad. A Crappo–Guatecare forest (*Carapa–Eschweilera* association). B Mora forest (*Mora excelsa* association), Mayaro District. From Beard (1946a). Each strip is 200 ft (61 m) long and 25 ft (7.6 m) wide.

on the Pacific coast mainly from Costa Rica southwards and the *Prioria copaifera* (cativo) association found on river alluvium in the same region (Chapter 14).

A type of single-dominant forest quite different from those just mentioned is the wallaba forest of Guyana (Fig. 2.9). This association is characteristic of 'white sands' (podzols) and is dominated by the leguminous tree *Eperua falcata*; two other species of *Eperua* are also abundant (pp. 300–2). The diagram shows that in this community also the A stratum is more even in height than in mixed forest and is almost continuous. Below this there is a sharply defined layer of small trees mostly 8–15 m high, corresponding to the C stratum of mixed forest, but a layer at an intermediate height corresponding to the B stratum is scarcely recognizable. The crowns are in fact mainly in two layers. Beneath the C layer there is a rather dense layer of treelets and a rather sparse ground layer of herbs and seedlings. Large lianes are scarce and do not make an important contribution to the forest structure. Because of its uniform and continuous A storey, wallaba forest has an easily recognized even-textured appearance in aerial photographs (Welch *et al.* 1972).

The caatinga forests on white sands in Amazonia (pp. 322–4) are floristically similar to the Guyana wallaba forest and the 'high caatingas' are also similar to it in structure, for example, the forest on the Rio Negro dominated by *Eperua purpurea* (Fig. 12.6) described by

Aubréville (1961) and Rodrigues (1961a,b). In the 'low caatingas', which are open woodlands rather than high forests, the trees are only about 15–17 m high, e.g. the 'campina' on the Rio Uaupés, dominated by *Aldina* spp. (Aubréville 1961, Takeuchi 1961a).

In the African rain forest the most widespread single-dominant association is the *Gilbertiodendron dewevrei* forest which covers hundreds of square kilometres in the eastern Congo (Zaïre) basin (Fig. 2.10). The structure of this unusual community has been described by Louis (1947a) and others (pp. 331–2). The dominant, a leguminous tree growing to a height of 35–40 m, may form more than 90% of trees ≥40 cm d.b.h.; usually only an occasional individual of some other species equals or surpasses *G. dewevrei* in height but on some sites *Julbernardia seretii* forms a small proportion of the stand. The very dense and continuous A storey casts a deep shade and in its young stages *G. dewevrei* is remarkably shade-tolerant. The rather sparse C storey consists of saplings of the dominant and a few other small woody plants. Between the A and C storeys there are only scattered trees, mostly immature *G. dewevrei*. Lianes are few and do not usually reach the crowns of the A storey trees.

A less extensively developed African single-dominant community is the *Brachystegia laurentii* association, which forms patches some hectares in extent in the central Congo basin (region of Yangambi) (Germain &

Key to symbols: Am, Amaioua guianensis; As, Aspidosperma excelsum; B, Byrsonima sp.; C, Cassia pteridophylla; Ca, Catostemma sp.; D, Duguetia neglecta; E, Eperua falcata (soft wallaba); Eg, Eperua grandiflora (ituri wallaba); Ec, Ecclinusa psilophylla; Em, Emmotum fagifolium; Es, Eschweilera sp.; L, Licania heteromorpha; M, Matayba inelegans; Ma, Marlierea schomburgkiana; Ms, Matayba sp.; O, Ocotea sp.; Or, Ormosia coutinhoi; P, Pouteria sp.; Sw, Swartzia sp.; T, Tovomita cephalostigma; U, Unidentified.

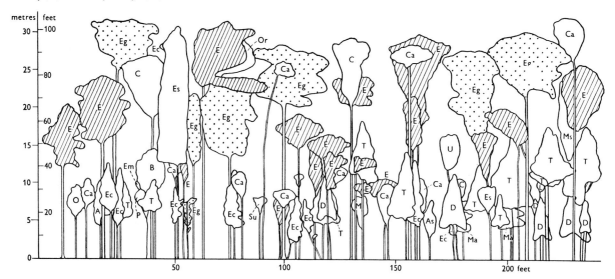

Fig. 2.9 Profile diagram of Wallaba forest (*Eperua* association), Barabara Creek, Mazaruni River, Guyana. After T.A.W. Davis (unpublished).

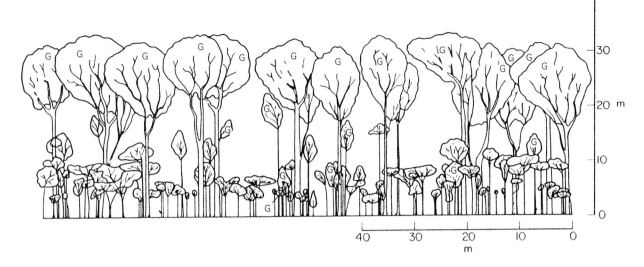

Fig. 2.10 Profile diagram of *Gilbertiodendron dewevrei* association, Ngula, Zaïre. From Louis (1947a, p. 913). The strip is 100 m × 10 m. Woody plants 4–8 m high shown only on a strip 5 m wide. Plants under 4 m omitted.

Evrard 1956) (pp. 332–4). The structure (Fig. 12.10) is somewhat different from that of the *Gilbertiodendron* forest, probably because the dominant is a somewhat less shade-tolerant species. The A storey is formed almost exclusively of the dominant, which grows to a height of 45–50 m and has an extremely broad and heavy crown (30–40 m wide). Although this stratum forms a level canopy, its cover is estimated at not more than 65% because there are some gaps between neighbouring crowns and because the leaves of *B. laurentii* are crowded at the tips of the branches. There are fairly distinct B and C strata. The former, with an average cover of *ca.* 50%, consists of trees about 25 m high, mostly young *B. laurentii*, which tend to be grouped underneath openings in the A storey. The rather thinly stocked C stratum is mainly a mixture of immature individuals of A and B storey species. Below this is a C storey 3–4 m high (cover *ca.* 55%) and a ground layer of herbs and seedlings.

Another African single-dominant association, the *Cynometra alexandri* forest of Uganda and the eastern Congo basin, is a type of semideciduous rather than rain forest (p. 34): in structure it is like the Nigerian mixed forest described above except that the upper storeys consist mainly of one species (Eggeling 1947).

In the Malesian rain forest single-dominant communities are common in peat swamps and other swampy habitats, but on well-drained sites climax single-dominant associations are rare. One of the few examples is the ironwood (*Eusideroxylon zwageri*) forest in Borneo and formerly in Sumatera (p. 330). This has a

very unusual structure; the B stratum to which the dominant belongs is extremely dense and there is almost no A stratum. Gresser (1919) describes the crowns of the ironwood trees as forming a flat compact 'roof' supported by trunks 15–20 m high. Only rarely is this canopy pierced by an emergent *Koompassia*, *Shorea* or *Intsia* rising far above the general level. The undergrowth is very open but young individuals of the dominant are extremely abundant. The now almost entirely destroyed *Eusideroxylon* forests of Sumatera, of which Franken & Roos (1981) have given profile diagrams (see Chapter 12, p. 330), seem to have had a rather different structure from those in Borneo.

The heath (kerangas) forests found in Borneo, Sumatera and less extensively in other parts of Southeast Asia, mainly on sandy podzolic soils, are similar in many respects to the wallaba forests and Amazonian caatinga of South America, but single species are usually less strongly dominant in them. In structure they differ considerably from the mixed dipterocarp forests found on kaolisols in the same region. The heath forests of Sarawak (pp. 325–6) vary considerably in composition with differences in soil and other site factors, so it is difficult to generalize about their structure. Most types of heath forest resemble wallaba forest in having a very dense C storey of shrubs and saplings and a rather compact, even canopy: in such types of heath forest the individual tree crowns are hardly noticeable in high-altitude aerial photographs (Fig. 11.10), in contrast to huge 'cauliflower-like' crowns of the large dipterocarps which are so characteristic of mixed dipterocarp forest.

Fig 2.11 Profile diagram of heath forest, Badas Forest Reserve, Brunei (1959). The strip is 150 ft × 25 ft (46 m × 7.6 m). All trees over 15 ft (4.6 m) shown. Compare with profile of mixed dipterocarp forest in same region (Fig. 2.5, p. 35).

Key:

Agathis borneensis 8, 18, 32, 33, 53, 62, 77, 80, 85, 93, 95
Annonaceae 7
Burseraceae 12
Calophyllum sp. 45, 55, 92
Canthium umbelligerum 46, 52
Cotylelobium burckii 9, 44, 70
Dialium indum 13, 35, 59, 88
Diospyros sp. (dead) 31
Dipterocarpus borneensis 2, 6
Elaeocarpus sp. 63, 66, 71
Eugenia bankensis 1, 4, 5, 11, 14, 16, 17, 19, 21, 24, 25, 26, 43, 54, 56, 60, 72, 82
Eugenia sp. 15, 28, 34, 36, 37, 39, 40, 42, 47, 56a, 65, 68, 79, 81, 83, 84, 86, 90, 97
Euphorbiaceae 29

Ficus sp. 57
Kokoona ovatolanceolata 50
Koompassia malaccensis 22, 41, 49
Lauraceae 3, 30, 38, 48, 51, 73, 87
Memecylon sp. 74
Nephelium maingayi 67, 78
Palaquium sp. 20, 64, 69, 89
Parinari sp. 23, 61, 96
Quercus sp. 75, 76
Sapotaceae 58
cf. *Stemonurus* sp. 10
Tristaniopsis sp. 91
Vatica mangachapoi 27

Field identifications by P.S. Ashton and B.E. Smythies.

In other types of heath forest the canopy is much less level, for example those in which the conifer *Agathis borneensis* is abundant. In these, crowns of the latter species stand out as emergents well above the rest of the canopy. According to Brünig (1974), in a majority of the heath forest stands sampled the strata were distinct but in others 'they cannot be recognized visually'. In stands dominated by single species or by 'ecological species groups' the strata were often well separated vertically by gaps.

2.4.4 Structure of mixed forests with tall emergents

In Guyana, as mentioned on p. 34, occasional trees of certain very tall species occur as emergent trees overtopping the forest canopy. *Dinizia excelsa* is similarly found as widely scattered emergents in evergreen seasonal forest in Amazonia (Baur 1964) and *Koompassia excelsa* occurs similarly in rain forest in Malesia. However, types of forest also exist in which emergents are much more numerous and form a discontinuous upper storey overtopping a dense canopy formed by a mixture of species, so that there indeed appears to be 'a forest above a forest'.

The most striking examples of such forests are the *Araucaria* forests in New Guinea, in which giant trees of *Araucaria hunsteinii* and *A. cunninghamii*, two species with slightly different ecological preferences, form an 'over-storey' above a layer consisting mainly of a mixture of many dicotyledonous species. In a stand of this kind described by Paijmans (1970), *A. hunsteinii* forms a pure but very open layer at 55–67 m above a fairly compact mixed layer at 20–30 m: occasional individuals belonging to the latter may grow to over 46 m but never become as tall as the *Araucaria*. Though *A. hunsteinii* is in fact represented in all height classes, the layer of giant emergents appears 'as if added to' a mixed broad-leaved stand. The now largely destroyed *Araucaria angustifolia* forests of subtropical Brazil are similar in structure to the *Araucaria* forests of New Guinea but the emergents are much less tall (Hueck 1966).

2.4.5 Structure of forests on slopes and river banks

So far only the structure of mature forests on more or less level ground has been considered. On steep slopes the crowns of the taller trees, instead of forming a level or undulating more or less continuous canopy, are arranged in a step-like fashion or tend to be asymmetric and to overlap like tiles. This type of structure, due to the combined effect of gravity and the one-sided incidence of light, has been discussed by Hallé *et al.* (1978) who suggest that it partly explains the greater floristic richness of forest on slopes which has been noted in French Guiana and elsewhere. The imbricate arrangement of the crowns allows more light to penetrate between them than when trees are growing on a horizontal surface, so that shade-intolerant smaller species can more easily find sites where they are not overtopped. On some types of rock, conditions on steep slopes for

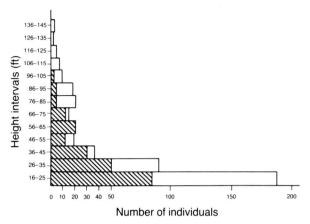

Fig 2.12 Total heights, and heights of lowest leaves, of trees on clear-felled plots, Moraballi Creek, Guyana. From Davis & Richards (1933–4, p. 367). The open bars represent distribution among height intervals, the hatched bars the distribution of the heights of lowest leaves for all trees over 15 ft (4.6 m) on a total area of 400 ft × 50 ft (122 m × 15.3 m).

shade-intolerant species are also improved by landslips, which create gaps and lead to a very irregular forest structure.

Another type of forest structure different from that usually represented in profile diagrams is found on the banks of rivers wide enough to admit sunlight to ground level. The trees develop very one-sided branching systems and their trunks tend to lean outwards (see Hallé *et al.* 1978, Fig. 106D). As mentioned earlier, lianes often form a blanket over most of the exposed edge of the forest. A somewhat similar structure may develop on the margins of clearings and large natural gaps.

2.5 Quantitative aspects of stratification

Profile diagrams, though valuable in giving a general picture of the stratification, cannot provide the quantitative information needed for understanding the functioning of the rain forest as an ecosystem. The stratification is ecologically important because it determines the microclimates within the forest: the leaves, trunks and branches intercept sunlight and other forms of radiation and act as barriers impeding air movements in all directions (Chapter 8). To measure these effects data on phytomass density and leaf areas from ground level upwards are required, but the difficulties in obtaining such data are considerable. Some useful work has been done, but so far only on mixed forests and in a few localities, so any conclusions must be provisional.

Table 2.3. *Aerial phytomass of strata on a 0.2 ha plot, near Walter Egler Forest Reserve, Manaus, Amazonia*

Stratum or storey	Mean crown height (m)	Number of individuals per ha		Basal area (m² ha⁻¹)	Mean crown area (ha ha⁻¹)	Fresh mass (t ha⁻¹)				
		Dicotyledons	Palms			Leaves	Branches and twigs	Stems	Total	Percentage of total
A	23.7–35.4	50	0	7.1	0.5	2.3	48.7	139.2	190.2	27.6
B	16.7	315	0	14.6	1.3	7.1	123.1	269.3	399.5	58.0
C1	25.9	760	15	5	0.9	3.9	26.1	47.3	77.3	11.2
C2	8.4–14.5	2765	155	2	0.8	2.0	3.6	10.0	15.6	2.3
D	3.6–5.9 1.7–3.0	5265	805	1	0.5	2.2	0.7	1.8	4.7	0.7
E	0.1–1.0	83 650		1	3	0.6	0.2	0.6	1.4	0.2
Total		93 780		30.7	7	18.1	202.4	468.2	688.7	100
Percentage of total						2.6		29.4	68	

Sources: From Klinge (1973) and Klinge *et al.* (1975).

2.5.1 Vertical distribution of tree heights and crowns

Attempts have been made to study the stratification by drawing curves or histograms of the numbers of trees in arbitrarily defined height-classes. Davis & Richards (1933–34, p. 367) plotted the heights of all trees of height 15 ft (4.6 m) or over on clear-felled plots of mixed forest in Guyana measuring 400 ft × 50 ft (122 m × 15.3 m, *ca.* 0.16 ha) (Fig. 2.12). This gave a continuous curve with a small peak in the 76–85 ft (23–27 m) height-class, corresponding approximately to the B storey (compare the profile diagram, Fig. 2.3). Most other published data on height-class distributions refer to undefined areas or deal with rather small numbers of trees.

One reason why curves of total tree heights give comparatively little information about stratification is because of the variation in crown depth and in the ratio of crown depth to total height ('free trunk height' in the terminology of Hallé *et al.* 1978). Ogawa *et al.* (1965) attempted to overcome this disadvantage by drawing 'crown depth diagrams', taking the difference between total height and height to the lowest living branch as a measure of crown depth. Paijmans (1970) gives histograms for mid-crown height and curves for total crown width by tree height classes in sample plots of mixed forest in hilly areas in New Guinea. These show not very marked peaks at about 12 and 20 m.

Measurements of the vertical distribution of phytomass in a 0.2 ha plot in the central Amazonian forest (Klinge 1973, Klinge & Rodrigues 1974, Klinge *et al.* 1975) (Table 2.3) indicate that the most densely filled layer of the forest is the B storey, which accounts for more than half the total above-ground phytomass. This layer is arbitrarily defined as that with mean crown heights of 16.7 to 25.9 m, i.e. somewhat lower than the B storey in the Guyana profile (Fig. 2.3) and in the height-class curve in Fig. 2.12.

2.5.2 Leaf-area profiles

The vertical distribution of leaf-area in a tall tropical forest is for practical reasons difficult to determine. Some information has been obtained by photography using hemispherical ('fish-eye') lenses (cf. Fig. 8.9) and by other optical methods (Odum *et al.* 1963, Johnson & Atwood 1970) but more reliable results have been obtained by stratified sampling. It is of interest to consider leaf-area relative to ground area (leaf-area index) and in relation to volume occupied (leaf-area density).

In the forest in Southwestern Cambodia, Kira *et al.* (1969) felled a sample of 100 trees to determine the leaf-area index. For the whole stand the leaf-area index was estimated to be about 7.4 m^2 m^{-2}, which is similar to the values found for lower montane rain forest in Puerto Rico (Odum 1970a) and for temperate broad-leaved evergreen forest in Japan, though higher than Müller & Nielsen's (1965) value (*ca.* 2.2) for plants above a height of 15 m in rain forest in Côte d'Ivoire and considerably lower than both the dry season and the wet season values found by Golley *et al.* (1975) for 'tropical moist forest' in Panama. What is of greater interest in the present context is the large differences in the leaf-area index between the strata as shown by the following figures.

	Leaf-area index
'first layer' (45–30 m)	2.0
'second and third layers' (30–15 and 15–5 m)	3.8
'ground layer' (below 5 m)	1.6
total	7.4

It is evident that in this type of forest the largest proportion of the foliage is borne by trees and lianes of the B storey. It is, however, also important to note that Kira *et al.* (1969) found from calculations of photosynthetic rates that the very large (A storey) trees at 40–45 m, the crowns of which are almost fully exposed to the light, contributed more than half the total photosynthetic productivity of the stand, though they had only about 20% of the leaf-area.

In a later study of a lowland mixed dipterocarp forest at Pasoh in Malaysia, Kira and his associates (Kira 1978) determined the leaf-area index and leaf-area density on a clear-felled strip 100 m × 20 m. The total leaf-area index was 8.0 ha ha^{-1} but leaves formed only 1.7% of the above-ground phytomass and the average leaf-area density was only 0.14 m^2 m^{-3}, the lowest yet recorded for a forest stand. This result, at first sight surprising, is accounted for by the way in which the leaves are distributed. The profile of leaf-area density and leaf-biomass density (Fig. 2.13) shows two maxima, one at about 20–35 m, which in this very tall forest represents the top of the B storey, and another much larger one in the layer of shrubs, small palms, saplings, seedlings and herbaceous plants below 5 m. The abrupt decrease of leaf-area density above 35 m corresponds to the transition from the relatively compact B storey to the relatively isolated crowns of the large A storey trees. Below 20 m leaf-area density falls off gradually but according to Kira (1978, p. 566) the very marked minimum in this diagram at 5–10 m is probably exaggerated compared with the Pasoh forest generally. About a third of

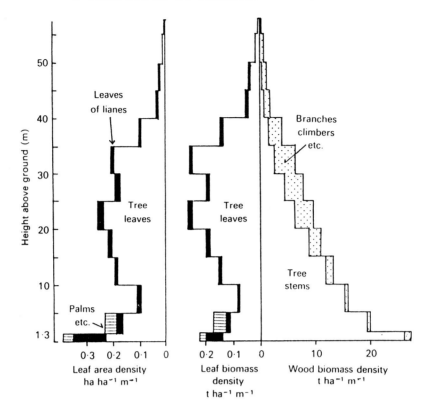

Fig 2.13 Profile structure of mixed dipterocarp forest, Pasoh Forest Reserve, Malaysia, based on a clear-felled strip 200 m × 20 m. From Kira (1978, Fig. 24.4.)

the leaf mass in the stratum of trees over 4.5 cm d.b.h. was contributed by lianes. These plants reduce the leaf-area of the trees to which they are attached, but the loss is about equivalent to the leaf-area of the lianes.

If, as seems likely, the data for Pasoh forest are typical of mixed rain forest in the *Indo-Malayan* region, it appears that the B storey is normally the densest stratum of the forest and carries the largest proportion of the total leaf-area. The A storey in forests like Pasoh has a much lower leaf density than the B stratum owing to the wide spacing of the large tree crowns. It is the high photosynthetic activity of these crowns, due to their full exposure to light, which causes the A storey to be the most productive layer of the forest and to be, quite literally, the dominant stratum.

As far as can be judged from the data of Klinge (1973) and Klinge *et al.* (1975), mentioned above, the leaf-area profile in mixed rain forests in Amazonia seems to be much like that just described for Pasoh forest (Table 2.3). The B stratum here accounts for 18.5% of the leaf-area, compared with about 7% in the more open A stratum. The largest contribution to the total leaf-area comes from the E stratum, which carries only

a small fraction of the total leaf mass. It cannot yet be said whether the vertical distribution of foliage in this Amazonian plot is typical of tropical American mixed rain forests or whether, as the Guyana profile (Fig. 2.3) seems to suggest, the C stratum is sometimes denser than the B stratum (though perhaps its total leaf area is not larger).

2.6 Ecological significance of stratification

Rain forests are obviously not unorganized aggregations of trees of different heights and stem diameters, but, as has been shown in this chapter, only single-dominant forests are stratified in the literal sense in which the term is often used. Mixed rain forests, which form by far the largest part of the rain forests of the lowland tropics, are not stratified in this way. For descriptive purposes, however, it is convenient to regard them as having three tree storeys, A, B and C, and two lower storeys of smaller, mainly woody, plants. Although there is sometimes a fairly clear boundary between the A and B storeys, or between the B and C storeys,

Fig 2.14 Profile diagram of mixed rain forest, Rivière Sinnamary, French Guiana. After Oldeman (1974, Fig. 39.50). The dotted line shows the inversion surface. Crowns with thick outlines are mature trees ('trees of the present'); stippled crowns are immature trees ('trees of the future'). The tree with dotted outline is rooted outside the plot.

objectively definable boundaries between all three tree storeys are not generally found.

Although a division of the forest into storeys is useful for some purposes, it is becoming increasingly clear that ecologically the most important horizontal boundary is that separating the lower layer of closely packed tree crowns, which are often interwoven with lianes, from the much more open layer above where the crowns are broader and more widely spaced. This boundary is often in the B storey, as in Figs. 2.5 and 2.6, but sometimes at the upper limit of the C storey as in most West African rain forests (Fig. 2.6). It corresponds closely to what Oldeman (1974) and Hallé *et al.* (1978) call the 'morphological inversion' surface. This is an imaginary plane joining the heights of lowest branching of the A storey trees or the B storey trees where the B storey has an open structure (see Fig. 2.14). The layers above this plane (often called the canopy, as noted earlier) are termed here the euphotic zone and those below it ('the undergrowth') the oligophotic zone (Richards 1983).

In the euphotic zone the tree crowns are only a little, if at all, shaded by their neighbours and their leaves are almost fully exposed to the sun. They also experience the whole range of temperature and humidity of the local microclimate; air movement round them is little impeded. In the oligophotic zone illumination is much reduced and the available light is mostly transmitted through leaves or reflected, although moving sun-flecks penetrate to ground level (Chapter 8). Both temperature and humidity are much more constant than in the euphotic zone and air movements are slight. All small plants, including the seedlings and juvenile stages of the trees and lianes, must be adapted to these conditions unless they can establish themselves in gaps.

Conditions in the oligophotic zone contrast sharply with those in the euphotic zone. The latter, because it receives more energy, is much the more productive of the two zones. Leaves, flowers and fruits provide food for an abundant and diverse population of birds, mammals, insects and other animals and the amount of space unoccupied by stems and leaves allows them to move freely by flying, gliding, jumping and other methods. Below the inversion surface flowers and fruits are comparatively scarce and the most abundant animal foods are wood and decaying plant materials. The litter of leaves, flowers and fruits dropping from the euphotic zone supplies a considerable part of the diet of the consumers and decomposers living at and near ground level. It may not be too fanciful to compare the euphotic zone of the rain forest to the well-illuminated plankton-inhabited photic zone of the ocean, which supplies food and energy to the dark or poorly illuminated water below it.

It is of course a simplification to divide the forest merely into two zones. Within each there is a variety of micro-environments available for organisms with sometimes widely differing requirements. A striking example of a special environment within the euphotic zone is that provided by the epiphytes on tall trees (pp. 147ff.). Similarly, a variety of different types of environment exists at various levels in the oligophotic zone.

An ecologically important feature of rain-forest structure is the degree of 'roughness' of the canopy surface. Earlier in this chapter it was seen that in mixed rain forests the canopy is generally uneven because the crowns of the taller trees form a more or less discontinuous A stratum; the smaller trees of the B and C strata fill the intervening spaces except where there are gaps down to ground level. In single-dominant forests the canopy is usually much smoother, with the crowns of the tall trees often packed together to form a comparatively level surface. Lianes, which usually seem to be scarcer in single-dominant forests than in mixed forests, are an important feature of canopy structure.

Brünig (1970, 1974; see also Givnish 1984) has called attention to the probable ecological significance of differences in canopy structure. The degree of canopy roughness affects the relative amount of solar radiation falling on the tree crowns and, by its influence on the turbulence of the surrounding air, it affects heat losses and evapotranspiration. Brünig calculated a measure of aerodynamic canopy roughness for the canopies of various forest types along gradients in Sarawak and found that canopy roughness tended to decrease with increasing liability of the forest to drought.

Canopy roughness is also important as a factor affecting the wind resistance of forest stands and individual trees (pp. 194ff.). It may well be true, as Whitmore (1974) suggests, that in parts of the tropics particularly subject to destructive cyclones, such as the Solomon Islands, there may be selection in favour of trees of even height with smooth canopies.

Chapter 3

Regeneration

THE PROCESSES INVOLVED in the natural regeneration of rain forests ('sylvigenesis' of Hallé *et al.* 1978) are complex. Although these processes are of obvious importance for conservation and practical forestry, they are by no means fully understood. Much of what has been written about 'natural regeneration' refers only to the reproduction of comparatively few economically valuable tree species after removal of timber or under silvicultural management. Before regeneration under these artificial conditions can be scientifically controlled, more needs to be known about the dynamics of the whole ecosystem under natural conditions.

As was mentioned in the last chapter, the regeneration processes operate unevenly, so that extensive areas of rain forest, even if undisturbed by logging or other human activities, are always patchworks or 'mosaics' of gap and building phases as well as stands of mature high forest. The gaps are formed by the death of old trees, windfalls and other natural causes. When large trees die they are replaced by others, often of different species. Trees of rain forests, like those of other kinds of forest, vary widely in their light requirements: some are light-demanding (shade-intolerant) and can develop to maturity only in gaps or openings; others are shade-tolerant and their seedlings are able to survive and reach maturity sometimes even under an unbroken canopy. Many (but not all) the trees of the A and B storeys are light-demanders but those of the lower layers are mostly able to establish themselves in shade. The phases are continually changing, so the floristic composition of a hectare of forest varies over a relatively short period of years, although that of larger areas remains more or less constant.

As Whitmore (1982) has pointed out, the regeneration processes are similar in essentials in tropical and temperate forests, but in the former they are more complex and the patterns of phases more diverse because of their much greater richness in tree species.

3.1 Life-span of individual trees

Determination of tree age. A major difficulty in studying the regeneration of tropical rain forests is that the ages of evergreen tropical trees cannot be estimated by counting growth rings in their trunks. The activity of the cambium is linked to leaf flushes (pp. 242–3) or may be more or less continuous. In some trees leaf flushes and cambial growth take place only in one sector of the trunk at a time. In trees of tropical climates with a well marked dry season, especially deciduous species, clearly recognizable growth rings are commoner; in a few of these, such as the Central American *Cordia alliodora*, growth rings have been shown to be more or less annual (Tschinkel 1966). In the wood of trees in the seasonally flooded várzea and igapó forests of Amazonia, regular annual growth rings are found (Worbes 1986). In general, however, growth rings are an unreliable means of determining the age of tropical trees (see Ogden 1981).

The age of many palms that produce a definite number of leaves each year can be estimated relatively easily (Uhl & Dransfield 1987). For this reason they have been used for demographic studies of their populations (Piñero *et al.* 1977, Sarukhán 1978).

Other methods are those based on historical data,

radio-carbon dating of heartwood or growth measure-ments. Historical evidence sometimes gives indications of the age of trees, as in an example mentioned below, where land now occupied by forest is known to have been farmed in past centuries (see Chapter 18). Occasionally the age of individual trees (but usually not trees growing under forest conditions) can be estimated from historical records, e.g. the Cotton tree (*Ceiba pentandra*) in Freetown, Sierra Leone, beneath which slaves used to be liberated: it must already have been a sizeable tree in the late eighteenth century.

Radio-carbon dating has so far been little used for determining the ages of tropical trees. Its limits of accuracy are so wide (at best about ±70 years) that it can be useful only for dating the largest and oldest individuals. An example from lowland rain forest is a very large 'overmature' specimen of the dipterocarp *Shorea curtisii* in Malaya, the age of which was found to be 800 ± 70 years (Burgess, quoted by Whitmore 1984a, p. 93). The age of a recently felled *Nothofagus pullei* of 1.8 m d.b.h. in lower montane rain forest in the New Guinea highlands was determined as 550 ± 85 years (Walker 1966, p. 514). A tree of *Dacryodes excelsa* in Dominica (West Indies), also in forest of lower mon-tane type, which from the photograph seems to be of about 1.0–1.5 m d.b.h., was dated at 370 ± 150 years BP (Odum 1970b, p. H 21). Radio-carbon dates for large rain-forest trees on the Atherton Plateau, Queens-land, gave ages of 630 ± 70 and 620 ± 100 years for two dicotyledonous species and 1060 ± 65 years for the conifer *Agathis microstachys* (Nicholson, quoted by Ogden 1981).

The most commonly used method of estimating the age of tropical trees is by extrapolation from growth measurements over a period of years. This involves several sources of error which cannot be entirely elimin-ated. The most serious of these is that the growth rate of a tree changes considerably during the course of its life and is much affected by the surrounding conditions, particularly by the amount of light the tree receives. In the case of a tree in natural forest, which may live for several hundred years, neighbouring trees may fall down or young trees grow up so that conditions during the period of measurement may differ from those earlier in its life in ways which are impossible to estimate. Very accurate measurements of the girth of tropical trees themselves pose technical problems (UNESCO 1978, pp. 201–5).

Nicholson (1965, see also Whitmore 1984a, pp. 110–11) estimated the age of nineteen tree species in a 4.5 acre (1.8 ha) plot of virgin rain forest in the Sepilok Reserve, Sabah. Some of the species were shade-

intolerant and some shade-tolerant (pp. 60ff.) and they ranged from giant A storey ('emergent') to small C storey ('understorey') species. From measurements of girth increments over a four-year period, Nicholson calculated the minimum and mean age of each species at maximum size and at 1.2 m girth. The minimum age was based on the increments of the fastest-growing individual in each 1.2 m girth-class, the mean age on those of all individuals. The calculations showed that at maximum size the minimum age of the largest species, *Dipterocarpus acutangulus*, was 400 years and the mean age 570 years; at 1.2 m girth its minimum age was 120 years and mean age 260 years. The other A storey and the B storey species had minimum ages at maximum size of 60–310 years and at 1.2 m girth of 60–180 years; their mean ages were 110–720 years at maximum size and 110–450 years at 1.2 m girth. Although the figures suggest that, as might be expected, the A storey trees tend to live longer than those of the B storey, it is remarkable that some of the quite small C storey species may reach ages equalling or exceeding those of the taller trees: thus the minimum age of *Gluta* sp. at 1.2 m girth (the maximum size) was 270 years. The very high estimated mean age for this species (720 years) may, however, be due only to the large proportion of very slow-growing individuals not destined to survive to maturity.

D. & M. Lieberman (1987) obtained similar results at La Selva, Costa Rica. The oldest tree in a sample of 12.4 ha was estimated to be 440 years old. Their calcu-lations, unlike those of Nicholson (1965) indicated that the small C storey trees were relatively short-lived.

Similar calculations were made by E.W. Jones (1956, p. 99) for various species of large trees in a sample plot of evergreen seasonal forest in the Okomu Forest Reserve, Nigeria, which was probably secondary (300–400 years old) rather than virgin forest (p. 470). Using records of girth increments over five years, Jones esti-mated the age at which each species might be expected to reach a girth of 10.5 ft (3.2 m). For the relatively slow-growing species *Guarea cedrata*, *G. thompsonii* and *Lophira alata* this was 270, 250 and 250 years, respectively, if based on the five fastest-growing individ-uals in each girth class, and 315, 345 and 255 years if based, like Nicholson's 'mean age', on all the trees in each class. Trees of greater girth than 10.5 ft were present but, because the growth rate of large trees decreases as they approach senescence, estimates of the age at death from growth rates are probably very unre-liable. Jones (1956) suggests that the largest trees of *Lophira*, many of which were obviously dying, could not be less than 144 years old and might be as much as 336. Some species were much faster-growing than the three mentioned; for example, *Triplochiton scleroxylon* would probably reach 10.5 ft girth in 50 years or less.

Heinsdijk (1965) from growth measurements of 120 000 trees in Amazonia calculated that in theory their ages might range from 27 to 418 years.

It is hard to draw firm conclusions from the data summarized above, but the evidence suggests that the age of mature A storey rain-forest trees is usually in the range of about 100–400 years. Some species evidently have longer life-spans than others and it is not unlikely that exceptionally large old trees of the very slow-growing and extremely hardwooded Malayan tree *Neobalanocarpus heimii* may be at least 1400 years old, as Cousens (1965) suggests. Ages exceeding 1000 years were also suggested by Pires & Prance (1977) for an individual of *Bertholletia excelsa* with a girth of 14 m and for other exceptionally large trees in Amazonia.

Although the relatively slow-growing dominant trees of primary rain forests may live for several hundred years, it is certain that the early seral (pioneer) and other secondary forest species which colonize gaps and often persist into the building phase are much shorter-lived, as well as faster-growing, than typical mature forest species (see Chapter 18). It seems that among tropical trees life-span and growth rate tend to vary inversely, although as Dawkins (1965) pointed out, some species such as *Ceiba pentandra*, which grow rapidly when young, also live to a great age.

3.2 Developmental phases

The view that primary tropical rain forests consist of ever-changing developmental phases was first suggested by Aubréville (1938) (see pp. 67–8). Later it was formulated somewhat differently by Whitmore (1978, 1982, 1984a), who applied the 'pattern and process' concept of Watt (1947) to the mixed dipterocarp forests of the Far East, adopting the terms 'mature', 'gap' and 'building' which Watt had given to the phases in the beech (*Fagus sylvatica*) woodlands of southern England. Whitmore illustrated his account by a 'canopy map' of a 2 ha plot of forest in Malaysia (Fig. 3.1) and by a profile diagram. A similar map of lowland forest in Sulawesi is reproduced in Fig. 3.2.

The phases are arbitrarily divided stages in the developmental cycle found in all tropical rain forests. They have ill-defined boundaries and differ in structure and floristic composition. The mature phase has a closed canopy formed by the A and B storey trees, as described in the previous chapter. When these trees become senile

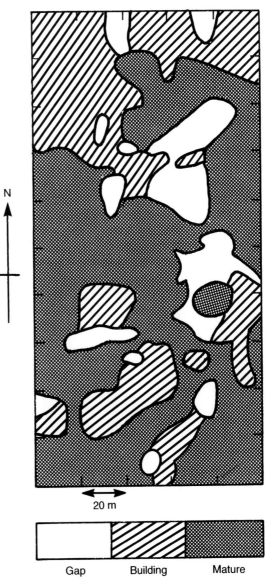

20 m

Gap   Building   Mature

Fig 3.1 Developmental phases of lowland mixed dipterocarp forest, Sungei Menyala, Malaya (1971). From Whitmore (1984a, Fig. 2.3). The larger gaps were formed by multiple windfalls enlarging smaller gaps resulting from the falling of single trees. The large area of the building phase at the north end of the plot represents regrowth after partial clearance 54 years earlier.

they shed large limbs and eventually die or fall down, forming gaps and damaging the smaller trees and other plants beneath them. The gaps are quickly filled with a mass of herbs, climbers and young trees. Some of these plants originate from roots and stumps or from a seedling bank of suppressed seedlings and saplings (pp. 64–5); others arise from seeds brought by the seed rain or dormant in the soil (pp. 115–16). As the tree seedlings grow into saplings and later into poles, the gap passes into the building phase. Many years later the mature phase is re-established.

The cycle of vegetative phases is accompanied by changes in microclimates (Chapter 8). If the gaps are large enough, sunlight can reach ground level and the temperature and humidity regimes approach those of open clearings. These changes allow the germination and establishment of shade-intolerant (light-demanding) tree species; these often include rapidly growing early seral (pioneer) species characteristic of young secondary forests (Chapter 18). As time goes on most of the relatively short-lived light-demanding trees are replaced by more slow-growing shade-tolerant species characteristic of the mature phase.

Torquebiau (1986) has described the mosaic patterns of dipterocarp forests in Sumatera (Fig. 3.3) and Kalimantan. Following Oldeman (1983), he regards the mosaic as composed even-aged 'eco-units' of varying sizes, originating from treefall gaps or other disturbances. Each eco-unit passes through a series of stages: (i) reorganizing, (ii) aggrading, (iii) steady state and (iv) degrading. The first two correspond to Whitmore's gap and building phases, (iii) and (iv) to the mature phase. Torquebiau found that each stage was characterized by the architecture of its 'canopy' (i.e. A and B storey) trees (see pp. 30ff.). In stages (i) and (ii) they are juvenile, with single-leader dominance and self-pruning branches. In the later stages the dominance of the single leader is lost and they develop permanent branches and a clear bole, the latter forms an increasing proportion of their total height as they grow taller. In this way they adjust themselves to the change from the environment of stages (i) and (ii) to the more exposed conditions of the canopy.

By using these architectural characters of the trees, the eco-units could be accurately mapped and by planimetry the area of forest in each stage could be measured. In the plots studied, steady-state units covered over 50% of the area, reorganizing units less than 5% and aggrading units 9.8–16.6%.

Formation of natural gaps. The commonest natural cause of forest gaps is the falling of large trees. A single tree may make a gap of several hundred square metres and a single falling branch may make a sizable gap. Often groups of two or more trees fall at the same time. Openings a hectare or more in extent are usually the result of catastrophic winds such as cyclones and line-squalls. Occasionally, very extensive areas of forest

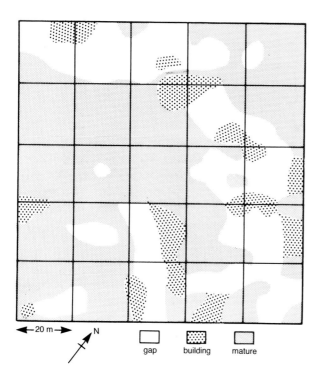

Fig 3.2 Developmental phases in lowland rain forest, Toraut, Sulawesi. From Whitmore & Sidiyasa (1986, Fig. 2).

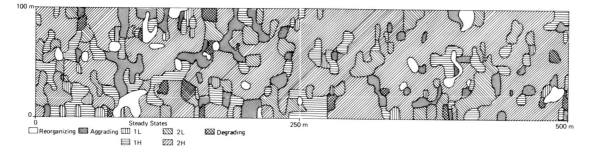

Fig 3.3 Mosaic of developmental phases in lowland rain forest, Batang Hari Jambi, Sumatra. From Torquebiau (1986, Fig. 5). Reorganizing, aggrading and degrading eco-units are shown; IL, 2L, 1H and 2H refer to steady states with trees of different heights.

are destroyed by wind. In 1880 an exceptional storm, followed by fire, devastated many hundreds of square kilometres of forest in Kelantan (northern Malaya); the results are still evident today (Whitmore 1984a). In some parts of the tropics such as the Caribbean and Solomon islands hurricanes or cyclones occur so frequently that very little forest remains undisturbed for even a century.

Dead and dying trees, both standing and fallen, can be seen everywhere in undisturbed rain forests (see data in Jones 1956, Poore 1968, Hallé *et al.* 1978, D. & M. Lieberman 1987, etc.). Wind often gives rain-forest trees their *coup de grâce*, as Pires found at Mocambo (p. 59), (Moraes & Pires 1967), but the crash of a falling tree is not uncommon even in calm weather (Jones 1956, p. 95). By no means all trees that fall are dead or obviously diseased. Trees are more liable to be overturned if the soil does not provide a firm anchorage for their roots and for this reason treefalls are particularly common in wet weather, as Hallé *et al.* (1978) note in Surinam. The roots of a large overturned tree often throw up a mound of soil a metre or more high: on some soils a pattern of mounds and hollows formed in this way over the years is a common feature of the forest floor.

Other important causes of gap formation are landslips and lightning. Landslips are common in forests on steep slopes, especially on certain types of rock (see Guariguata 1990). In some countries, e.g. Panama and New Guinea, they are triggered by earthquakes. In northern Papua New Guinea two earthquakes were together responsible for the denudation of 130 km^2 of forest (Garwood *et al.* 1979). Lightning is a frequent cause of death among tall rain-forest trees (Brünig & Huang 1989). In Sarawak Brünig (1964) found that in the *Shorea albida* peat swamps (p. 371) lightning might kill or badly damage single trees or large groups and might affect the undergrowth to a varying extent. Lightning damage is also common in the Solomon Islands; sometimes, though rarely, the rain forest is dry enough for fire to spread from the original strike into the surrounding area (Whitmore 1984a).

Gaps, and even quite large openings, caused by the death of trees defoliated by the caterpillars of a hymantrid moth, occur in the peat swamps of Sarawak. Damage of this kind does not seem to have been noticed in mixed dipterocarp forests or in rain forests elsewhere.

The size and shape of gaps caused by fallen trees depends on the weight of their crowns, how many large branches have been previously lost, and the density of the surrounding canopy. When a small tree falls it may do little damage and the increased light may stimulate one-sided growth of the neighbouring crowns like that on trees by a river or road (p. 43), thus quickly closing the gap. On the other hand, the crown of a large emergent tree is often 15–18 m or more across; the fall of such a tree, or even of one or two of its branches, may destroy or seriously damage a large number of its neighbours. Jones (1956, p. 95) records that in Nigeria a single large tree of *Parinari excelsa* destroyed most of the trees in a 100 ft × 100 ft (30.6 m × 30.6 m) plot, an area larger than any of the natural gaps found by Poore (1968) in the Jengka forest in Malaya. A group of neighbouring trees is sometimes so tightly bound together by lianes that they fall together. Once a gap has been formed it often tends to enlarge; it has often been reported that small gaps can grow into extensive windthrows. In Africa and Malaya where there are elephants their browsing may cause gaps to enlarge (Fig. 18.13). Lightning gaps in the *Shorea albida* forest in Sarawak, after the trees had become dead and defoliated, are often the origin of windthrows in the form of narrow swaths extending sometimes a very long way into the surrounding forest. One such swath was no less than 8 km long (Palmer, quoted by Whitmore 1984a, p. 82). In mixed rain forests, because of their more complex structure, there may be less tendency for gaps to extend in this way. Hubbell & Foster (1986) have given a detailed analysis of gap formation in the forest of Barro Colorado Island, Panama.

Seeding and seedling establishment. The populations of seedlings[1] of trees and other woody plants on the forest floor are often large (Figs. 3.4, 3.5). The seedlings are usually unevenly distributed and their abundance varies from one time of year to another and from year to year. In a species-rich mixed rain forest the species composition of the seedling population may reflect fairly closely that of the tree storeys. Thus in the evergreen seasonal forest of Trinidad Beard (1946b, p. 63) found that the species commonest in the higher storeys were mostly also the commonest as seedlings. This is also true of the mixed forest of Guyana (see p. 54) and Table 3:1; however, it is not always so. Sometimes species abundant as seedlings are relatively rare as adult trees. In the Mapane forest in Surinam Schulz (1960) found that seedlings of *Vochysia guianensis* and *V. tomentosa* were extremely numerous but mature trees were scarce. In wallaba forest in Guyana the palm *Jessenia bataua* is sometimes abundant as seedlings though

---

[1] In this chapter, as in most rain-forest literature, 'seedling' is used in a wide sense to include established plants which are too small to be called saplings (Fig. 3.4) although they may be several months or even years old.

Table 3.1. *Size-class representation of dominant species on sample plots 122 m × 122 m of four associations at Moraballi Creek, Guyana*

Association	Average no. of seedlings under 2 m high per m²	No. of young trees under 10 cm diam. and over 4.6 m high	No. of trees 10 cm diameter and over: Diameter-class (cm)				
			10–20	21–30	31–40	41–60	61 and over
Mora excelsa (Mora forest)	4.7	120[a]	28	19	17	24	21
M. gonggrijpii (Morabukea forest)	11.7	318	39	17	9	38	16
Ocotea rodiei (Greenheart forest)	['Seedlings fairly abundant']	16[a]	2	5	10	38	18
Eperua falcata (Wallaba forest)	1.8	80[a]	47	43	37	64	3

[a] On these plots the trees of this size-class were counted on two strips together equal to one-eighth of the total area and the figure obtained multiplied by 8.

Source: After Davis & Richards (1933–4).

adult individuals are rare or absent. More frequently it has been found that seedlings of certain tree species, usually large emergents, are absent or very rare so that it might be supposed that these species were disappearing from the community. The significance of this at first sight puzzling state of affairs is discussed below.

In single-dominant rain forests, such as various consociations in Guyana (Chapter 11), seedlings and other young stages of the dominant species are almost always plentiful (Table 3.1). Dense thickets of established seedlings and young saplings round the base of the parent trees are particularly characteristic of the *Mora excelsa* and *M. gonggrijpii* consociations. In the *Brachystegia laurentii*, *Gilbertiodendron dewevrei* and other single-dominant communities in Africa there is also a great abundance of young regeneration (pp. 331–4).

The abundance of tree seedlings in a given place depends on the availability of viable seeds which drop from the canopy, are carried from elsewhere in the 'seed rain' or are dormant in the soil. Different species of trees differ very much in the quantity and frequency of seed production. The phenology of flowering and fruiting in rain-forest trees is dealt with in Chapter 9 and the dispersal, viability and germination of seeds in Chapter 5. The following note by T.A.W. Davis (unpublished) gives some idea of the many factors affecting seed availability in the Guyana forest.

Seed production is usually seasonal in the canopy, but in the undergrowth flowers and fruit are often borne almost continually in small quantities or at irregular but frequent intervals. The majority of those which form the main canopy (A and B strata) however, flower either annually, biennially or even thrice a year, though seed is not always set, or if set, ripened: they probably produce good seed in fair quantity at least once in three years. The commonest species generally seed frequently; some, e.g. most species of *Eschweilera*, *Licania alba* and *Ocotea rodiaei*, flower once or twice every year and usually ripen some fruit, though they may bear well only every second or third year, others, e.g. *Mora excelsa* and *M. gonggrijpii*, flower at intervals of roughly eighteen months or two years and normally have a good crop of seed each time they flower . . . Flowering by no means always results in the ripening of even a little seed. Sometimes no fruit sets, presumably because the weather is unfavourable for fertilization. This seems to happen especially to species with a very short flowering season . . . When seed is set wastage is, as a rule, not serious from such causes as weather and insects, sufficient being produced to allow for the loss of a considerable proportion of the crop. A number of trees, however, suffer heavy loss through parrots which feed principally on unripe fruits and are particularly fond of leguminous seeds. A flock of parakeets (*Brotogeris chrysopterus*) was seen feeding in a *Bombax surinamense* daily for weeks together on its soft and juicy unripe seeds: it is probably not a rare occurrence for certain trees to be stripped in this way of practically the whole of their crop.

A large proportion of seeds reaching the ground are destroyed by insects or small mammals or become infected by fungi. Their chances of survival may be increased if they become pressed into the soil (Schulz 1960, p. 226, Soepadmo 1972). In spite of enormous

Fig 3.4 Established seedlings at base of old tree of *Dicymbe altsoni*, Bartica–Potaro road, Guyana (1979).

Fig 3.5 Carpet of seedlings of *Shorea multiflora* (Dipterocarpaceae) nine months after flowering, lowland dipterocarp forest. Andulau Forest Reserve, Brunei (August 1959). (Photo. P.S. Ashton.)

losses a heavy seed fall may result in an abundant crop of seedlings. Ng (1978) notes that in Malaya seedlings of rapidly germinating species carpet the ground after a seed fall (Fig. 3.5); those that germinate over a longer period do not do this, but some seedlings are present at all times. In the Malayan dipterocarp forests enormous numbers of seedlings die in their first year and many in the first few weeks or months. Fox (1973) records that only 14% of seedlings of *Dipterocarpus caudatus* in Sabah survived more than one month. Among the light-demanding *Shorea* spp. only a very small proportion of the seedlings survives longer than two years. Since seed in these trees is only produced at intervals of some years (p. 252), 'we must visualize successive waves of short-lived seedlings, of which every one is doomed unless there happens to be a suitable gap an

eventuality that may occur only once in a good many years, and not, as in a managed forest, at regular intervals as the parent trees reach economic maturity' (Watson 1937, p. 147). For the more shade-tolerant species gaps are less essential and the average life of seedlings is much longer but, as with the light-demanding species, mortality is very heavy. Because of the variations in the timing and quantity of seed falls and the heavy losses before and after seed germination, the number of seedlings on the forest floor fluctuates considerably from year to year as is shown by the figures of Fox (1972, 1973, 1976) for seedlings of various dipterocarps in the rain forest of Sabah.

Seedling mortality is due to various causes. The most important are probably starvation from lack of light and destruction by predators. Root competition certainly affects the growth rate of seedlings and probably their survival. Fox (1973) found that on a trenched plot in Sabah the growth of seedlings was twice as fast as on an adjacent untrenched plot. Weather conditions, particularly drought, are occasionally fatal to seedlings. In one locality in Guyana a whole thicket of *Mora gonggrijpii* seedlings died during the drought of 1926 (T.A.W. Davis). A study of the abundance and survival of dipterocarp seedlings in the lowland rain forest of Malaya (Burgess, quoted by Whitmore 1984a) showed the importance of soil and other site conditions for seedling survival and establishment. On soils derived from granite, establishment seemed to be on the whole more successful than on shale; slope and other features of the site also seemed to be important. Nicholson's interesting observation (quoted by Fox 1972) that in Sabah seedlings of *Parashorea tomentella* failed to establish themselves successfully when grown on non-forest soils suggests that mycorrhizal fungi may be necessary for seedling establishment, at least for some species.

Large numbers of young tree seedlings are killed by insect predators that eat their cotyledons, and by various small mammals. Burgess (1975) found that in Malaya the majority of the seedlings of *Shorea curtisii* were destroyed by ants: of those that escaped most were later eaten by other animals.

The importance of predators that destroy the seeds before they are shed or the young seedlings has been much stressed by Janzen (1970) who considers that they are the main cause of the widely scattered distribution of many tree species which is so characteristic of most rain-forest communities. Janzen believes that many of the predators, particularly seed-eating insects, are host-specific, attacking only a single species or group of species. A tree species may survive by means of 'escape mechanisms' such as satiation of predators by very abundant seed production and mast fruiting at long and irregular intervals, or by defences such as toxic secondary compounds in the seeds or leaves. But in any case the chances of survival are likely to be greater for an isolated individual among trees of other species than for several individuals of the same species growing close together. Selection may therefore favour widely dispersed populations.

Single-dominant forests, and the occurrence in both mixed and single-dominant forests of trees such as the *Mora* species mentioned above (and see Fig. 3.4), in which the seedlings are aggregated round the parent trees, seem to conflict with Janzen's theory. In these cases it seems possible that the seedlings are better protected from predators than those of species in which they are more widely dispersed. Nearly all the dominants of single-dominant consociations, it should be noted, belong to the Leguminosae, a family in which secondary compounds (possibly with a protective function) are common (Chapter 12).

Forget's (1989) study of regeneration in wallaba (*Eperua falcata*), a single-dominant over large areas in the Guianas (pp. 300–2) is interesting in this connection. The average mass of the seeds is 7.4 g; Forget found that 60% fell within 10 m of the parent tree. No seeds were attacked by insects, but large numbers were eaten or damaged by peccaries, rodents and other animals. Predation was, however, not distance-dependent and seedling mortality was low. Saplings were abundant, reaching their greatest density at 10–15 m from the nearest parent tree; this is about half the average distance between neighbouring trees. In *E. grandiflora* (Fig. 3.6 A, B) which is often associated with *E. falcata*, the seeds are even larger (mean fresh mass 47.6 g) and very few are carried far from where they fall. A large proportion are destroyed by insects and fungi, but some develop into juveniles, which form a thicket round the parent tree (Forget 1992). (See p. 114.)

The building phase. If a gap is left to itself, the fallen trees and debris soon decay and are replaced by young

Fig 3.6 A Germinating seed of *Eperua grandiflora*. B Established seedlings of *Eperua* sp., in clearing, near Kwapau, Mazaruni river, Guyana (1979).

vegetation. The latter may consist mainly of herbaceous plants and young trees but sometimes, especially if it is large, a gap becomes a dense tangle of lianes through which young trees have difficulty in growing (Rollet 1974). In Southeast Asia, particularly where there is seasonal drought, forest gaps often become filled with thickets of bamboos, up to 10 m high, which exclude most other plants for a long time. Gap vegetation grows much faster than the surrounding forest, responding to the temporarily more favourable conditions. The microclimate of gaps is considered in Chapter 8. It will be seen, as well as the higher average light intensity, there is more air movement, more variable humidity and a greater range of temperature in a gap than in the mature phase. It is also reasonable to assume that the wider the gap, the greater the differences from conditions in undisturbed forest. Root competition must also for a while be less severe and the surface of the ground is often broken and uneven. Nutrients released by the decay of the fallen trees are added to those already in the soil (Chapter 10).

The trees and other plants filling gaps come from more than one source. Often, many are already present as a 'seedling bank' of young individuals that were undamaged when the gap was formed. Others are sprouts from the stumps of trees that were injured but not killed when they fell. Seedlings may also grow from seeds dropped from the surrounding trees or carried from further away by animals or the wind. Seeds of most primary rain-forest trees have a very short period of viability (pp. 114ff.) and generally cannot accumulate in the soil to form a 'seed bank'. In forests which have never been cleared or cultivated, the viable seeds that are found buried in the soil, are chiefly those of weeds and secondary forest trees (pp. 115–16).

The species composition of the building phase thus depends mainly on the species already present on the site and on seed sources that happen to be available when the gap is formed. Herwitz (1981a, p. 71) found that in gaps in primary forest in Corcovado National Park, Costa Rica, 'incoming and ungerminated propagules contribute very little to the regeneration process after a tree has fallen'. This conclusion may not apply in other places and different forest types: if it is generally true it would emphasize the importance to young trees of the ability to survive long suppression periods as seedlings or saplings. It does not apply to large. A storey species, such as *Hymenaea courbaril*, which cannot establish themselves in shade and are rarely found as seedlings except in gaps.

The young trees that grow up in gaps include relatively shade-tolerant as well as relatively intolerant (light-demanding) species and some of intermediate characteristics. The growth rate of all the young trees that are present when a gap is formed increases as a result of the improved illumination, but the intolerant species generally respond more quickly than the others. Among the intolerant species, especially in gaps where the soil has been disturbed, are often very fast-growing 'pioneer' (early seral) trees, for example *Cecropia* spp. and *Ochroma lagopus* in tropical America, *Musanga cecropioides* and *Zanthoxylum macrophyllum* in Africa and in the Indo-Malayan region many species of *Macaranga*, *Mallotus* and other genera. These pioneers have only a temporary foothold: after a few years they disappear and do not occupy a permanent place in the mature phase of the forest. In gaps they are present only in small numbers; in extensive clearings they become dominants over large areas. As well as trees, herbs and climbers, both herbaceous and woody, are abundant in gaps. The changes in the vegetation of a forest gap are to some extent a small-scale telescoped version of the secondary succession in a large clearing, but the two successions are not exactly similar (see Chapter 18).

The increased exposure to sunlight when a gap is formed is favourable for the growth of many species of trees, but on some it has adverse effects. For instance, seedlings of *Ocotea rodiei*, a common large tree of the Guyana rain forests, make little growth in large clearings (T.A.W. Davis) and saplings under 2 m high may die if suddenly exposed to full light (Fanshawe, quoted by Baur 1964, p. 142). Similar effects have often been reported for rain-forest trees in other parts of the tropics, e.g. Dipterocarpaceae in Malesia. The physiological causes of these effects are not clear; they seem to result from excessive heating and water stress due to exposure to direct sunlight.

Both the increase in illumination and the changes in other microclimatic factors when a gap is formed vary with its size (and to some extent its shape). It would therefore be expected that there would be a relation between the successional changes in gaps and their size. The evidence on this is somewhat contradictory. Blanford (1929) found that in Malaya the dominants regenerated best in gaps not more than 20 ft (6 m) wide: in larger gaps young regeneration was lacking at the centre. Similarly, in upland (probably lower montane) forest on Gunung Gede in Java, Kramer (1933) showed experimentally that existing regeneration of primary forest trees survived and made good growth in artificial clearings of not more than 1000 m² but in clearings of 2000–3000 m² the young primary forest trees were suppressed by the luxuriant growth of invading pioneers. In the lowland forest of the Douala–Edea Reserve (Cameroon) in gaps less than 20 m wide, the trees regenerate mainly from a bank of suppressed seedlings that were present before the gap was formed, but in larger gaps they regenerate mainly from seeds (D.W. Thomas, unpublished).

Whitmore (1978), Hartshorn (1978), Denslow (1980) and others emphasize the importance of gap size. However, Herwitz (1981a) found that in the Corcovado National Park, Costa Rica, there was no correlation between the floristic composition of the vegetation and gap size in gaps from 100 to 1200 m² in area, most species being randomly distributed among different sized gaps. Pioneer species common in large clearings, such as *Cecropia obtusifolia* and *Ochroma lagopus*, occurred only occasionally and in small numbers in the forest gaps. A lack of relation between gap size and flora is to be expected if the latter develops mainly from preexisting seedlings and saplings. It may be that there is a critical size, possibly depending on local conditions, above which the contribution of incoming tree seeds to the gap vegetation becomes larger and the latter tends to be different from that of smaller gaps. Uhl *et al.* (1988b) have given an account of gap colonization at S. Carlos (Amazonia). They also found that treefall gaps benefit chiefly established seedlings and saplings, colonists being of minor importance.

It is evident that gaps due to the death of old trees and other causes are necessary for the reproduction of the more light-demanding species, but there are also relatively shade-tolerant species which are not able to regenerate successfully without them. For example, among the rain-forest species of Guyana some very large super-emergent and A storey species such as *Hymenaea* spp. seem to be entirely dependent on gaps: young individuals are rarely if ever found in the shade of the mature phase. Other large trees, e.g. *Ocotea rodiei*, need at least a small gap to reach the A storey. Others, again, such as *Licania alba*, have a wide tolerance and can regenerate under an unbroken canopy as well as in gaps. In Amazonia the majority of the large emergent trees reproduce only in gaps, but there are some exceptions, e.g. *Manilkara huberi* (Pires & Prance 1977). Whitmore (1984a) has discussed the degree of dependence on gaps of intolerant and tolerant species in the forests of South-east Asia. Similar variations in species behaviour are found in all rain-forest communities.

Denslow (1980), Pickett (1983) and others have suggested that there is 'gap partitioning' among the regenerating species, each being adapted to a particular range of gap sizes, and that this may contribute to the high species diversity of rain-forest communities.

To test this hypothesis Brown & Whitmore (1992), working in the Danum Valley (Sabah), compared the responses of seedlings of three dipterocarp species known to differ in shade tolerance, *Hopea nervosa*, *Parashorea malaanonan* and *Shorea johorensis*, to artificial gaps ranging from 10 m² to 1500 m² (with two closed canopy plots as controls). Seedlings of these species, probably at least four years old, were thinly but patchily distributed over the whole area and the plots were placed where seedling density was relatively high. After the gaps were made the number and size of the seedlings was monitored for forty months. The results showed, as expected, significant differences in seedling survival between the gap and control plots. Height growth of all the species was greater in the gaps. Although height growth increased with gap size, the most important determinant of seedling growth and survival was seedling size at the time of gap creation, regardless of species. *Hopea nervosa*, known to be the most shade-tolerant of the three, was the most favoured by the gaps because its seedlings were the largest under closed canopy conditions. The other species, which were regarded as less shade-tolerant, benefited less because under the canopy most of their seedlings were small. There was no evidence that any species was better adapted to a particular range of gap sizes than the others. It was clear that the ability of seedlings to persist for a long time under a closed canopy was of great importance.

Turnover rates. If a stand of climax rain forest of more than a certain minimum area is in equilibrium, then the number of trees as well as the total phytomass should remain more or less constant over a period of time, measured perhaps in centuries or thousands of years, the death of old trees being balanced by the germination, establishment and growth of younger ones. This is generally assumed to be true; direct evidence is available from a few sample areas (see below).

Of more importance than the maximum life-span of individual tree species is the turnover rate of whole forest communities, i.e. the rate at which trees die and are replaced. In various parts of the tropics sample plots have been monitored over periods of some years and records kept of tree mortality and recruitment, so that turnover rates can be estimated. A good example is a 2 ha plot of mixed 'terra firme' forest in the Mocambo Reserve near Belém in Amazonia (Pires & Prance 1977). In 1956 the plot contained 1144 trees of over 30 cm girth, with a total basal area of 55.4 m². By 1971, 868 of these trees had survived, 276 (with a basal area of 128.6 m²) had died and 112 young trees had reached the minimum girth limit, so that the total number of living trees had fallen to 970, a loss of 23.9% over the 15 year period or an average annual loss of 1.6%. Since 15 years is a short time relative to the probable life-span of most of the trees and there is much variation in mortality from year to year, the decrease in the total number of trees is unlikely to indicate a progressive trend.

The total basal area of trees ⩾30 cm girth, which is an approximate measure of the above-ground phytomass, remained almost constant. In 1971 it was 55.5 m², an increase of less than 0.2%, which is probably not significant.

In an earlier report (1967) Moraes & Pires analysed the dead trees in a 5.5 ha plot in the same forest. In 1967 there were 246 standing and fallen dead trees which had not decomposed beyond recognition. The percentage mortality varied very little between girth classes. Many of the standing dead trees, which, as might be expected, included more individuals of smaller diameter than the fallen trees, had mostly died from indeterminable causes, but there was much evidence that windthrow was the commonest immediate cause of death. 37.4% of the fallen trees had exposed roots, indicating that they had probably been overthrown when still alive. The distribution of the dead trees over the whole 5.5 ha plot was markedly uneven. Pires attributed the clumping of dead trees in certain quadrats to wind-storms but it could perhaps be partly due to the death of even-aged groups of trees that had grown up in gaps.

Records comparable to those for Mocambo have been kept since 1949 (for some plots since 1947) for a number of sample plots of primary and secondary rain forest in Malaya. A report on the results up to 1959 by Wyatt-Smith (1966) is summarized by Whitmore (1984a).[2] Of particular interest is a sample plot at Sungei Menyala (Negri Sembilan) in a relic of lowland mixed forest rich in *Dipterocarpus* spp. and *Shorea* spp. of the 'red meran-ti' group, a type of rain forest formerly widespread in Malaya. A small part of the plot is secondary forest on a clearing made in 1917 but the remainder (1.6 ha) is primary forest in the mature phase; two small natural gaps developed in this about 1951.

In 1947 the total number of trees ⩾30 cm girth in the primary part of the plot was 1075. By 1959 the number had fallen slightly to 988, 233 of the trees present in 1947 having died and 146 young trees having reached the lower girth limit of 30 cm. These figures are remarkably similar (allowing for the slightly smaller area) to those for the Brazilian forest. The mortality over the twelve-year period was 20.1%; calculated as a

yearly average it was 1.6%, exactly the same as that at Mocambo. In both places the mortality was fairly evenly spread between the girth-classes. At Sungei Men-yala the basal area per hectare was 32.4 m² in 1947 and 32.0 m³ in 1959, so it can be concluded that the phyto-mass, as at Mocambo, remained approximately constant.

Some changes in species composition took place during the period of observation. Out of the 237 species of trees ⩾30 cm girth recorded in 1947, eighteen had disap-peared by 1959. All but three of these were present only as single individuals; eighteen new species had appeared.

Changes in the population of saplings (defined as plants not less than 1.5 m high and less than 30 cm girth), recorded on a subsample strip of 0.08 ha, throw an interesting light on the dynamics of the forest. It is noteworthy that species which when mature belong to different storeys were very unequally represented: 'emergent' (A storey), 'main canopy' (B storey) and 'understorey' (C storey) species formed 6, 22 and 72%, respectively, of the saplings.

For mixed dipterocarp forest in the Jengka Reserve in Malaya, Poore (1968) calculated the mean period between the formation of successive gaps on the same spot. The percentage area occupied by gaps was known and, depending on whether the gaps were assumed to take twenty, twenty-five or thirty years to fill, the intervals between gaps would be 250, 312 or 375 years.

This is a much longer time than the turnover period estimated by Hartshorn (1978) for the permanent study plots at La Selva, Costa Rica. Here, from measurements of the total gap area and an estimate of the frequency with which gaps were formed, he found the mean time needed for the whole area to be covered by gaps on four different sites (one an alluvial swamp forest) to be 118 ± twenty-seven years. Later work has shown that Neotropical rain forests are more 'dynamic' than those of the Old World.

It has been assumed in the past that mature tropical rain forests, unless disturbed by hurricanes, logging or other external causes, are in equilibrium, their turnover rate remaining more or less constant. Phillips & Gentry studied data from twenty-one samples of rain forest with no history of logging and disturbance, in which the turnover had been monitored two or more times at intervals of at least ten years. They claimed that in both the Neotropics and the Palaeotropics there was evidence of a significant increase of turnover rate during the 1950s, which accelerated for 1980 and might have been due to environmental causes. However, Phillips & Hall (unpublished) subsequently found that the supposed increase was an artefact due to the statistical method

---

[2] Results up to 1981 (thirty-four years) have since been published (Manokaran & Kochummen 1987).

used. Changes in the turnover rate with time need further study.

3.3 Growth and development

Size-class representation. Inventories of natural populations of tropical trees by girth- or diameter-classes are very numerous; only certain aspects of the subject can be dealt with here. A large amount of information, much of it from Venezuelan Guiana, has been collected and critically discussed by Rollet (1974). There is also a shorter general review of diameter distributions for individual species in UNESCO (1978). Jones' papers on the Okomu forest in Nigeria (1950, 1956) give a useful discussion of the problems of interpreting size-class distributions, which because they depend on two variables, growth and mortality, cannot be readily converted into age-class distributions, as pointed out earlier. In tropical trees, especially large emergents, the ratio of girth to height tends to increase with age, so girth-classes are not exactly equivalent to height-classes.

In natural rain forests there are many more young individuals than mature trees in the majority of species and in the population as a whole (the 'total structure' of Rollet) the relation of girth-classes to number of individuals follows an exponential model fairly closely, but only for stems of about ⩾20 d.b.h. (Rollet 1974). If the trees are grouped into 10 cm girth-classes, the number in each class is usually approximately twice that in the next larger class, although the ratio increases slightly from each class to the next higher one. Below about ⩾20 cm d.b.h. the number of individuals is generally greater than would be expected from this rule. These relations appear to hold fairly generally throughout the tropics for stands of primary mixed rain forest of a few hectares: above this area, according to Rollet (1974, p. 142) the numbers fit the exponential model less closely. There are, however, considerable divergences from the 'normal' pattern of size-class distribution in forests of different history and seral status as well as in different tree species.

In mature mixed rain forests there are striking differences between the size-class distribution of shade-intolerant (light-demanding) and shade-tolerant tree species. These are well illustrated by Pires & Prance's data (1977) for the height and girth of various species on a 6000 m × 10 m transect in primary forest on the Jari river in Amazonia (Fig. 3.7).

In the mature phase of the forest, young individuals of intolerant species are scarce or quite lacking, a remarkable example being *Bertholletia excelsa* (Brazil nut), in which most of the trees on the transect are 50–60 m high. *Dinizia excelsa*, which is often found in the Amazon forest as extremely large super-emergents (p. 43), has a similar deficiency of smaller sizes. A somewhat less extreme size distribution is found in *Geissospermum sericeum*. On the other hand in shade-tolerant species, e.g. *Eschweilera* spp. *Protium tenuifolium*, young trees of all heights are plentiful and the size–height/girth curve is close to the usual logarithmic form. The intolerant species in which young individuals are lacking on the transect are tall A storey trees, whereas the shade-tolerant species are mainly trees of low or medium stature that belong to the B or C storeys when fully grown. However, there are exceptions: *Manilkara huberi* (Fig. 3.3c), as mentioned above is an example of a very tall emergent species which is shade-tolerant; small individuals of this species are numerous. The Jari river data do not include early seral species such as *Cecropia* and *Vismia* spp., which may have occurred transiently in gaps: these are very intolerant of shade and form only single-generation populations (Chapter 18).

Differences between the size-class representation of shade-tolerant and intolerant species are well known in other parts of tropical America and the Palaeotropics. Schulz (1960) found that in the rain forest of Surinam the great majority of the tree species had a good representation of the smaller diameter-classes, the curve of size-class distribution following the normal pattern, but in a few light-demanders, e.g. *Goupia glabra*, *Ocotea rubra* and *Quassia amara*, there was usually a marked deficiency in the small and/or middle size-classes. A detailed study of regeneration in *Goupia glabra*, a strongly light-demanding species which seemed able to regenerate only in fairly large openings, showed that there were considerable variations in the size-class structure of the population in different sites and localities. As there was little information on the past history of the stands, it was impossible to be sure of the cause of these variations: most probably they were related to the incidence of natural or artificial disturbances in the past. Rollet's (1974) voluminous data on regeneration in the rain forests of Venezuelan Guiana can be only very briefly summarized here: they show that there is a great variety of size-class structures among different species. 'Shade-tolerant' and 'light-demanding' are only relative terms; between the extremes there are many intermediates. There are consequently many patterns of size-class distribution, each representing different regeneration strategies. Rollet recognizes seven types of diameter-class distribution ⩾10 cm d.b.h.: (1) species with stems of 10–20 cm only; (2) those with a 'bell-

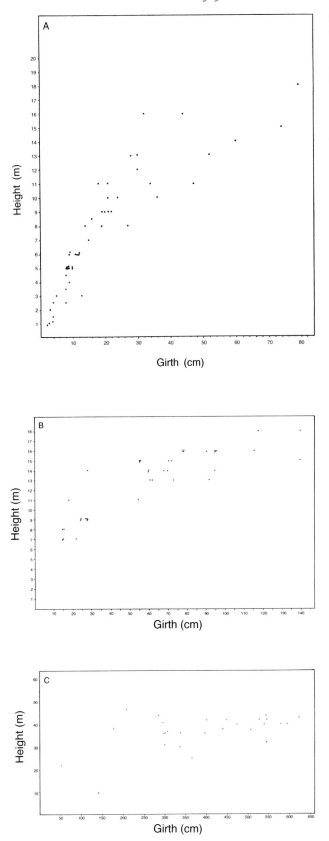

shaped' size-class distribution 'more or less truncated to the left' (i.e. with deficiencies in the smaller classes); (3) those with an 'erratic' distribution; (4) those with an exponential distribution giving a curve in the form of a tall L (i.e. with very large numbers in the smaller sizes); (5) those with a normal exponential distribution; (6) those with an exponential distribution giving 'an L-shaped curve with a short foot'; (7) those with a bimodal distribution. Examples of each of these types of distribution are given in Table 3.2.

In his 155.5 ha sample Rollet found that the great majority of the tree species belonged to Types 4 and 5, which had an abundant representation of the smaller diameter-classes. They have an equilibrium population structure and can regenerate in moderate shade. On the other hand, as individuals, Types 1, 2, 3, 6 and 7 form more than half the stand. Early seral trees such as *Cecropia* and *Didymocarpus morototoni* and other light-demanding trees belong to Types 2, 3 and 6.

Rollet (1974) regards the relatively small proportion of light-demanding species in the Venezuelan forests as evidence that these forests are primitive and have had little natural or other disturbance. He terms the ratio of the number of light-demanding species (types 2, 3 and 7) to the total population (⩾10 cm d.b.h.) the 'coefficient of equilibrium'. He believes that this can be used as a measure of the 'state of equilibrium' (or degree of disturbance) of a stand when comparing forests in different sites and regions.

In the Old World, as in tropical America, populations of rain-forest trees consist of species with different degrees of shade-tolerance, the more intolerant often having a size-class distribution noticeably different from that characteristic of shade-tolerant species. The intolerant (light-demanding) species, as in America, are mainly the tall emergents and the seral species found temporarily in gaps.

In Africa, particularly in West Africa, observers have often commented on the apparent scarcity of young individuals of the large dominant trees in the evergreen and semideciduous forests. The apparent lack of adequate regeneration has given rise to various speculations and seems at one time to have been regarded as evidence

Fig 3.7 Height and girth of trees, Jari river, Pará, Brazil. After Pires & Prance (1977, figs. 18, 23 and 24). A *Protium tenuifolium*, B *Geissospermum sericeum*, C *Bertholletia excelsa*. All individuals over 15 cm girth shown. A is a shade-tolerant species with many young individuals. B is a light-demanding species with no very small individuals in mature forest. C is one of the largest emergents, which is represented only by very large trees.

Table 3.2. *Examples of types of size-class distribution of tree species in inventory of 155.5 ha of lowland rain forest, Venezuelan Guiana*

Diameter-class	Species	Number of individuals in class (cm)									
		10–19	20–29	30–39	40–49	50–59	60–69	70–79	80–89	90–99	≥100
1	*Lacistema aggregatum*	10	—	—	—	—	—	—	—	—	—
2	*Inga alba*	84	90	105	73	27	20	3	5	2	—
3	*Tabebuia serratifolia*	3	1	3	3	2	1	1	—	3	1
4	*Trichilia schomburgkii*	1225	140	10	1	—	—	—	—	—	—
5	*Licania densiflora*	1632	796	450	237	106	33	4	—	—	—
6	*Sclerolobium* sp.	49	38	23	24	14	10	9	4	1	2
7	*Erisma uncinatum*	149	61	36	42	37	52	27	30	24	29

Source: After Rollet (1974, table 100, p. 134).

for the then current view that the African rain forest was not in equilibrium with the present climate. In the forest of Côte d'Ivoire, Aubréville (1938) found that on some sample plots young individuals of species common in the 'dominant' (A and B storeys), such as *Piptadenias-trum africanum*, *Canarium schweinfurthii* and *Parkia bicolor*, were very poorly represented, sometimes quite absent, in the young regeneration; according to the local people these trees 'ne font jamais des petits'. He came to the conclusion, as will be seen later, that in any small area the species composition of the forest must be constantly changing.

Although in West Africa the scarcity of seedlings and other young individuals of the dominant tree species is often striking when compared with tropical America (see Schnell 1965), it may have been exaggerated, perhaps because many of the reports come from foresters who have naturally concentrated on the large economic species, many of which are undoubtedly shade-intolerant when young.[3] In Ghana, Hall & Swaine (1976) found that in samples of moist evergreen forest the numbers of young individuals of most tree species were quite large and probably adequate for replacement.

Regeneration in the Okomu Reserve (Nigeria), an old secondary forest with a long history of disturbance, was studied in detail by Jones (1956) (Fig. 3.8). The condition of the forest at the time is described on pp. 470–2.

Seedlings of most of the woody species were probably present but those of some species were extremely local in distribution, tending to be found in patches, sometimes near parent trees. In the B and C storey species, which formed the great majority of the tree population, young individuals of all sizes were numerous and the size-classes had a 'normal' logarithmic representation.

[3] It should be noted that of at least one economic species, *Pericopsis elata*, remarkably few seedlings are found even in gaps (Taylor 1960).

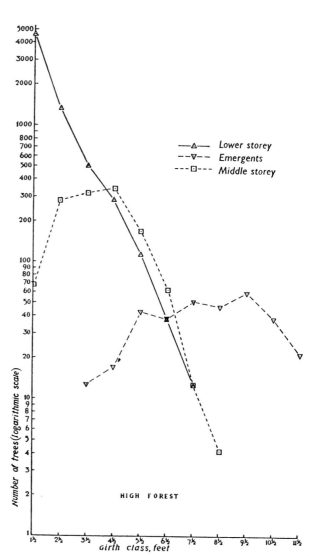

Fig 3.8 Frequency distribution of girths of trees in 'High forest', Okomu Forest Reserve, Nigeria. From Jones (1956, Fig. 3). The figures are from 47 plots and have been adjusted to give the numbers of trees per 200 plots (18.6 ha).

However, in the emergent (A storey) trees the situation was very different (Fig. 3.8). Three types of size-class distribution were found among these: (a) species of which young individuals were almost completely absent in the lower storeys, e.g. *Alstonia boonei*, *Canarium schweinfurthii*; (b) species that were fairly abundant as seedlings and saplings but which were almost unrepresented in the middle size-classes, e.g. *Guarea thompsonii*, *Lophira alata*, (c) species with a 'normal' size-class distribution like that of the B and C storey species, e.g. *Guarea cedrata*, *Piptadeniastrum africanum*. On seven sample plots with a total area of 18.6 ha, 21% of the stems and 24% of the species belonged to type (a), the corresponding figures for type (b) being 45 and 38%. The remaining 32% of stems and 38% of species belonged to type (c). Species with an 'abnormal' size-class distribution thus formed over 60% of both individuals and species among the A storey trees.

The type (a) species are strongly light-demanding and can probably establish themselves only in large gaps, type (b) seem to be somewhat less intolerant. One explanation of the deficiency of middle sizes in these species is that growth through this stage is very rapid, another, which Jones (1956) thinks more likely, is that it is only intermittently, perhaps when canopy gaps are available, that small saplings grow up into the middle size-classes. Type (c) are the most shade-tolerant of the A storey trees: one of the latter, *Lovoa trichilioides*, is said to be the most shade-tolerant of the Nigerian Meliaceae. Even so, the species of type (c) seem unable to reach mature size without the help of at least a small gap. The A storey is thus dominated by species that are light-demanding in various degrees. The Okomu forest in 1947–48, when Jones' work was done, was a patchwork of 'broken forest' and 'scrub' due to windthrows and past felling, as well as high forest, and it is evident that light-demanding trees may have been able to grow and reproduce successfully along with a large number of more shade-tolerant species.

Contrasting with West African mixed forests, in which light-demanding species always seem to play a considerable part, are the consociations in Zaïre dominated by *Gilbertiodendron dewevrei* and *Brachystegia laurentii* and referred to earlier (pp. 40–1). In the former the dominant and to a lesser degree the most important subordinate species, *Julbernardia seretii*, are shade-tolerant and seem to have the size-class distribution characteristic of such species (Gérard 1960). Experiments showed that young plants of *Gilbertiodendron*, though able to grow within shade of the parent tree, grew better in slightly improved illumination but not if exposed to full daylight. Apart from rare individuals of *Alstonia boonei* and secondary forest species found in clearings

(presumably also in large natural gaps) other large tree species are almost entirely absent. *Brachystegia laurentii* is also shade-tolerant: it has a typical logarithmic size-class distribution (Germain & Evrard 1956, Fig. 5).

In Malesia, studies of size-class distributions have been mainly concentrated on mixed dipterocarp forests. In these, members of the Dipterocarpaceae usually form more than half of all trees ⩾40 cm d.b.h. and thus dominate the A storey. The pattern of regeneration tends to depend on the reproductive characteristics of this family, which, as previously mentioned, include abundant seeding at irregular and often very long intervals, and seeds of short viability (Chapters 5 and 7). Established seedlings in the forest undergrowth often have the ability to survive long periods of suppression but require the increased light from at least a small gap to grow up into saplings, poles and eventually mature A or B storey trees.

Typically the size-class distribution of dipterocarps is of the 'reversed J' shape, with an abundance of established seedlings, poles and mature trees but relatively few small saplings (Fig. 3.4). This distribution is like that of Jones' (1956) type (b) species in Nigeria. Some dipterocarps, e.g. *Shorea maxwelliana*, are more shade-tolerant than others, e.g. *S. leprosula*, but if generalization is possible, it seems that in general Malesian dipterocarps are neither very shade-tolerant nor very light-demanding compared with large rain-forest trees in other parts of the tropics. Much further information on the regeneration of dipterocarps is given by Wyatt-Smith (1963) and Whitmore (1984a).

Many of the large non-dipterocarps in the mixed dipterocarp forests seem to be more or less light-demanding, but emergents with size-class distributions like that of the Amazonian *Bertholletia* and some of the African emergent trees do not seem to occur: the most extreme light-demanders are apparently early seral species such as *Macaranga* and *Mallotus* spp., which occur transiently in gaps. If this is really so, these forests must have a very small coefficient of equilibrium (*sensu* Rollet), perhaps indicating that they have been less subject to natural disturbances from wind and other causes than most of the primary forests in tropical America and Africa.

Brünig (1974) gives a short account of regeneration in the kerangas (heath) forests of Sarawak and Brunei, which are found usually on sandy podzolic soils and differ from the mixed dipterocarp forests in structure and floristic composition Chapter 12). The undergrowth of the kerangas is better illuminated and the small tree and shrub layers (C and D storeys) denser. Seedlings and saplings of most of the taller species are plentiful but they vary in abundance with soil conditions and micro-relief, perhaps because these factors affect water stress in dry periods. Diameter-class distributions for various species are similar to those found in mixed dipterocarp forest but sometimes show deficiencies in certain middle sizes so that there are two or even three peaks. Thus *Agathis borneensis*, one of the most abundant A storey species, had a good representation of all size-classes on one sample plot but in other stands showed a deficiency between 10 and 30 cm d.b.h. *Dryobalanops fusca*, a dipterocarp, and *Casuarina nobilis*, a common B storey tree, also showed bimodal distributions.

The regeneration of the rain forest on Kolombangara (Solomon Islands) has been studied in detail by Whitmore (1974, 1989), who has monitored the changes in the floristic composition and population structure of sample plots in six floristically different types of forest over a period of twenty-one years. In contrast to the relative freedom from major natural disturbances in Borneo and Malaya, the forest here is liable to be frequently devastated by cyclones, so that areas of 'broken forest' and large gaps are common. A study of the regeneration strategies of twelve tree species in the island of Kolombangara showed that they could be divided into four groups: (1) species whose seedlings can establish themselves and grow to maturity in the shade of the 'high forest' (mature) phase; (2) those that can establish themselves under high forest but appear to benefit from gaps; (3) those that can establish themselves under high forest but definitely require gaps for further growth; and (4) those that establish themselves mainly in gaps and grow to maturity only in gaps (these are the pioneer (early seral) species). The size-class distribution of the more light-demanding species such as *Endospermum medullosum* (group 4) shows the usual deficiency in the smaller size-classes and is markedly different from that of shade-tolerant species such as *Parinari papuana* subsp. *salomonensis*.

Growth rates. The various patterns of size-class distribution found in rain-forest trees can be fully understood only when something is known of the rates of growth and mortality at each stage of their development. Data on growth rates are plentiful in the forestry literature (see review in UNESCO 1978) but many of the measurements have been made in plantations, where light and other conditions are different from those in a natural forest and tree growth is usually much faster (see N.E. Johnson 1976, Whitmore 1984a). Observations are commonly made over periods of a few years, a very short time compared with the probable life-span of the trees, and more information is available about girth (or

diameter) increments than about growth in height, which is more relevant to the processes of forest regeneration. The data show that there is much variation between individual trees as well as between species and at present only a few generalizations are possible. The very fast growth of early seral trees is dealt with in Ch. 18.

The growth projections of D. & M. Lieberman (1987), based on d.b.h. measurements of forty tree species $\geqslant$10 cm d.b.h. on a 12.4 ha plot over thirteen years, showed a wide range of growth rates. The shade-tolerant 'canopy' (A storey) trees grew faster than the B storey species. The most abundant species in the 'canopy' group, *Pentaclethra macroloba*, was projected to reach maximum diameter (140 cm) (from 10 cm) in 145–300 years, but there were wide differences between the growth rates of species in this group. Light-demanding trees such as *Goethalsia meiantha* could reach maximum size in forty to sixty years. The slowest-growing species were those in the C storey: many individuals of these showed zero growth in the thirteen-year period (cf. pp. 60ff.).

In most young A and B storey trees, which start life as seedlings in the shade, after the first few foliage leaves have been produced, growth is very slow for a long time: measurements show that they often cease to grow at all for many years. It is not until the young tree is 'released' by the formation of a gap in its neighbourhood, or if it eventually succeeds in growing tall enough for its crown to be no longer overshadowed by its neighbours, that the growth rate begins to accelerate. Growth in height may then be relatively fast for many years. Growth in girth also increases when the illumination improves. In the later stages of the tree's life (at least in large A and B storey trees) growth in height begins to fall off before growth in diameter so that, as mentioned earlier, the ratio between the two changes. In growth measurements of a large number of trees in the lower montane forest of Puerto Rico, Wadsworth (1947) found that the average annual growth increment of trees with crowns fully exposed to light was about three times that of trees with shaded crowns. Similar differences are found in lowland rain forest.

Ability to survive a suppression period in which growth is very slow or nil seems to be a feature of the development of many, perhaps most, rain-forest trees except the extreme light-demanders, which can only establish themselves in large gaps. It is not known how long young trees can endure suppression. In the Mocambo forest in Amazonia, Pires & Prance (1977) recorded zero growth over a period of fifteen years in some species. Large numbers of young trees having

Table 3.3. *Growth in diameter of three species of tree in the dipterocarp forest at ca. 100 m, Mount Makiling, Philippine Islands*

Species	Diameter-class (cm)	No. of individuals measured[a]	Mean annual diameter increment (cm)	Mean no. of years in diameter-class
Parashorea malaanonan	0–5	4	0.07	71
(in forest)	5–10	10	0.27	19
	10–20	21	0.38	26
	20–30	19	0.49	20
	30–40	12	0.74	13.5
	40–50	28	0.82	12
	50–60	7	0.94	11
	60–70	10	0.75	13
	70–80	1	0.84	12
P. malaanonan (in open)	0–5	13	0.42	12
	5–10	27	0.55	9.1
	10–15	7	0.73	6.8
Diplodiscus paniculatus	0–10	14	0.15	66.7
	10–20	20	0.31	32.3
	20–30	8	0.44	22.7
Dillenia philippinensis	0–10	1	0.46	22
	10–20	1	0.55	18
	20–30	6	0.39	26
	30–40	—	0.24	42

[a] Not all trees were measured every year throughout the four-year period.

Source: After Brown (1919).

successfully survived the dangerous early seedling stage, probably die in suppression periods (see, for instance, Wyatt-Smith's data (1958) for three Malayan species), but saplings of many rain-forest species seem to retain for a remarkably long time the ability to respond by rapid growth as soon as conditions become more favourable. For shade-tolerant species a 'bank' of suppression seedlings and saplings thus plays a similar part to a seed bank of buried seeds for light-demanding trees. Ability to endure a long suppression period is a specialized characteristic, which is no doubt an important part of the biological equipment of many rain-forest trees. The large heavy seeds that are common among tall-growing species (p. 111) must confer a considerable advantage in enabling the seedlings to grow to a size at which they have some chance of surviving a long period of suppression. Herbert (1929) plausibly suggested that one reason why so few species of *Eucalyptus* have been able to invade rain forests is that they lack the ability to survive suppression.

The incidence of a suppression period and the great differences in the growth rate at different stages of development are well illustrated by the early work of Brown & Matthews (1914) and Brown (1919) on the growth of trees in the dipterocarp forests of the Philippines. Their results, although referring to a single area and to very few species, are probably of wide application to rain-forest trees in general.

Careful measurements were made over a period of four years of the diameter of a large number of individual trees of three species in the evergreen forest at *ca.* 100 m above sea-level on Mt Makiling. One of these species, *Parashorea malaanonan*, a typical dipterocarp, was the dominant species in the A stratum; the others were *Diplodiscus paniculatus*, the commonest tree in the B stratum, and *Dillenia philippinensis*, another common B species. At the same time measurements were made of a group of *Parashorea* trees in an open situation at the edge of the forest. The diameters were measured once a year to the nearest 0.5 mm. The individuals of each species were grouped into diameter-classes, and for each diameter-class the average annual increment of diameter was determined; from this the average number of years spent in each diameter-class was calculated. Brown considered that the individual variation in the growth rate of trees in the 0–5 cm diameter-class was so great that there was a large error in estimating the age of trees in that class, but there was probably not a large error in the estimates of the ages of trees in the larger classes.

The results are summarized in Table 3.3 and those for *Parashorea* and *Diplodiscus* are shown in Fig. 3.9. The higher points on the curve for *Parashorea* in the open have been calculated on the assumption that trees of large diameter grow equally fast in the forest and in the open.

The data for *Parashorea* show the effect of the suppression period in a very striking way. The trees in the forest grow exceedingly slowly until they reach 5 cm diameter, then much faster. They take nearly as long to grow to 5 cm diameter as from 5 to 40 cm. The trees in the open, on the other hand, grow quite rapidly even when small, and their growth rate is approximately constant throughout their development. The slow growth of the forest trees under 5 cm, for which competition for light must be largely responsible, is clearly due to their unfavourable environment. Trees of the 5–10 cm diameter-class are already 'high in the forest', while those over 20 cm diameter are in the highest storey and living under conditions which, at least as far as illumination is concerned, approximate to those in the open.

The figures for *Diplodiscus* show a growth rate slightly slower than that of *Parashorea*, but here, too, there is a marked suppression period. The calculated growth rates for the smaller diameter-classes appear to be higher than those for *Parashorea*, because no trees of

Diplodiscus less than 5 cm in diameter were measured; if measurements of such trees had been included *Diplodiscus* would certainly have proved slower growing than *Parashorea* at all stages of its development. The figures for *Dillenia philippinensis* are based on very few measurements, but they also indicate a slower growth rate than that of *Parashorea*. Brown & Matthews (1914), as mentioned earlier, concluded that B storey trees in general grew slower than the A storey dipterocarps.

In the open even very small *Parashorea* trees grow rapidly, showing that the slow growth of young trees of this species in the forest is not due to inherent characteristics but is imposed on them by external conditions. The growth rate of a tree at all stages of its development is no doubt determined by two groups of factors, environmental and genetic. The latter differ between individuals as well as from species to species: light-demanding species, given sufficient illumination, usually grow faster than shade-tolerant species.

The wide variation in growth rate between species and individuals of the same species is also clearly shown by the data of Nicholson (1965; and in Whitmore 1984a, p. 111) for a 4.5 acre (1.82 ha) sample plot of primary mixed dipterocarp forest in the Sepilok Reserve in Sabah. In this there were 194 species of trees ≥ 12 in (30 cm) girth. The girth increments of 1140 individuals were recorded over a four year period. The mean girth increment of all the trees was only 0.19 in (0.5 cm) per year, equivalent to about 12 ft^2 $acre^{-1}$ a^{-1} (0.22 m^2 ha^{-1} a^{-1}) basal area, but this low figure is partly accounted for by the very small growth increments of the smaller trees: the average girth increment rises with girth up to 7 ft (1.22 m) and above this it is unrelated to girth.

The differences in growth rate between individuals even in the same girth-class were so great that they largely, but not entirely, masked inter-specific differences. It was found that the growth rate of an individual was in general correlated with the size of its crown and for the majority of trees the annual girth increment could almost be predicted from the appearance of the tree. However, in about 15% of the trees the increment was larger or smaller than expected. A relation between girth increment and crown size is obviously to be expected but other factors, one of which may be the age of the tree, seem to affect the growth rate.

Trees with larger crowns than others of the same species were mostly near gaps and the size of the crown was probably a consequence of increased illumination. This was true of many of the 'understorey' (B and C storey) trees as well as of those of the top layer. While some understorey species, e.g. *Eugenia* spp. and *Symplocos*, seemed able to take advantage of increased light,

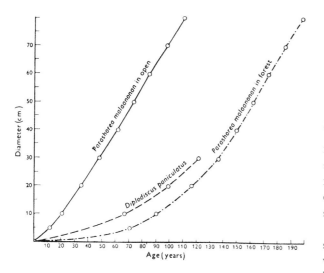

Fig 3.9 Growth rate of *Parashorea malaanonan*, an emergent (A storey) dipterocarp, and *Diplodiscus paniculatus*, a B storey species, at 300 m, Mt Makiling, Philippine Islands. After Brown (1919, Fig. 2).

others, e.g. *Madhuca kingiana* and *Gluta* sp., seemed
unable to respond so markedly.

The overriding importance of competition for light in
controlling the growth rates of young trees, and the
regeneration processes generally, is obvious, but the
nature of the physiological differences between species
underlying the differences in their responses are much
less clear.

Mortality. The heavy mortality among trees at the
seedling and suppression stage has already been discussed. It is sometimes supposed that after a seedling has
become established and entered a period of comparatively
rapid growth, there is a high probability that it will
survive to become a mature tree. In fact this does not
seem to be true: in natural forests the mortality rate
is quite large in all size-classes, as can be seen in the
data of Nicholson (1965) for Sabah, of Wyatt-Smith
(1958) for Malaya, of Milton *et al.* (1994) for Barro
Colorado Island, and of Moraes & Pires (1967) for
Amazonia. In a study of tree mortality over a period
of 13 years at La Selva, Costa Rica, Lieberman *et al.*
(1985c) estimated that the mortality rate of trees
⩾10 cm d.b.h. was independent of size.

Mervart (1972, 1974) has analysed in some detail the
growth and mortality of trees in natural 'high forest'
(evergreen seasonal and semideciduous types) in Nigeria.
The average mortality over a five year period in a
sample of 6224 trees was 3.95%; as this is a small
fraction of the whole population, the error is probably
large. No very substantial differences were found in
mortality between size-classes or between tree species
with different rates of growth, though Mervart considers
that the 'small pole' stage (trees of *ca.* ⩽30 cm girth)
is the most critical for survival after the seedling stage.
Among trees of the same size-class the individuals with
the smallest growth increments are less likely to survive
than those that grow faster. There is in fact no stage
at which a tree's future is assured; and those that lag
behind in upward growth have a less than average
chance of survival.

3.4 Dynamic patterns

Aubréville's 'Mosaic theory'. The development of the
modern concept of rain forests as dynamic systems of
ever-changing mature, gap and building phases owes
much to Aubréville's (1938) hypothesis, which was
referred to in the first edition of this book (Edn 1) as his
'Mosaic' or 'Cyclic' theory of regeneration. According to
his view, which was based on experience in the 'forêt

dense' (evergreen seasonal and semideciduous forest) of
Côte d'Ivoire, the particular combination of species that
forms the dominants of a given small area of mixed
tropical forest is constant neither in space nor in time.
The composition of the 'dominant' (A) storey varies
from place to place (Chapter 12) and, according to
Aubréville, at the same place over a long period of
years: the combination of species at a given place and
time is succeeded, not by a similar combination but by
a different one. No combination of species can be in
permanent equilibrium with the environment.

To borrow a comparison from algebra, one could say that
the conditions of equilibrium between the determining factors of the habitat and the numerous independent variables
constituted by the characteristic species of the community
form a number of equations much fewer than the number
of independent variables. All kinds of solutions are thus
possible satisfying the equations of equilibrium.
(Aubréville, 1938, p. 140, transl.)

On this hypothesis an extensive stand of mixed forest
is a mosaic, the pieces being different combinations of
tree species which replace one another in a more or less
cyclic fashion.

Aubréville does not specify the size of the units in
the mosaic and gives little indication of the extent of
the observations on which he bases his conclusions. To
illustrate his theory he quoted a sample plot of 1.4 ha
in 'untouched primitive forest' in the Massa Mé Reserve
(Côte d'Ivoire): on this 74 species ⩾0.1 m d.b.h. were
recorded[4]. The commonest large species was *Piptadeniastrum africanum*, of which there were 7 trees ⩾0.5 m
d.b.h. but only two of 0.1–0.5 m and no young plants
of ⩾0.1 m d.b.h. Several other very large species had
a similar size-class distribution and it seemed justifiable
to conclude that when the existing generation of 'dominants' died a new combination of species would take
the place of the present one. As has been noted earlier,
similar size-class distributions are not uncommon in
African forests and can also be found in other parts of
the tropics. Such communities have a large proportion
of shade-intolerant species and have a low coefficient of
equilibrium in Rollet's sense (pp. 60ff).

Aubréville's theory has been much discussed and has
contributed to the replacement of older static views of
rain-forest communities by the more dynamic concept
now widely accepted. Also, as Schulz (1960) pointed out,

---

[4] Hall & Swaine (1976, p. 935) and Swaine & Hall (1988, p. 255) have
pointed out that in Edn. 1, pp. 50–1 and Table 6, Aubréville's data
for the Massa Mé plot are reproduced incorrectly. The blanks in
column 1 (young plants) in the Table should indicate 'not recorded'
and not 'absent', except for seven selected species (*Khaya ivorensis*
etc.).

Aubréville's recognition of the part played by chance in rain-forest regeneration is important. But even at the time when it was first put forward, it was clear that Aubréville's conception of a climax community could not be applied to all mixed rain-forest associations. In the mixed forest of Guyana, for example, the commonest species in the larger diameter-classes are also well represented in the smaller, although their relative abundance in the two is not quite the same (Table 3.2). The original data (Davis & Richards 1933–34, Table IV) also show that no species capable of reaching large dimensions is well represented in the smaller diameter-classes but not in the larger. In a mixed forest such as this it seems likely that the species composition of the higher storeys will remain about the same for an indefinite time. Schulz's (1960) work on the regeneration of rain forests in Surinam leads to a similar conclusion and it probably also applies to most primary mixed dipterocarp forests in Malaya and Borneo.

Hall & Swaine (1976) found that in the moist evergreen forest of Ghana (which is very similar to the *forêt dense* of Côte d'Ivoire) seedlings of most tree species were fairly abundant under 'closed canopy' (i.e. in the mature phase of the forest) with the exception of long-lived, strongly light-demanding species such as *Lophira alata* and *Nauclea diderichii*. An indication that the existing composition of the 'canopy' (A and B storeys) was likely to be maintained without much change was provided by a 320 m² sample strip in which there were six species reaching a height of 30 m or more. All but one of these were represented in the regeneration on the same strip. If it is assumed that each young tree, irrespective of species, has an equal chance of replacing the six existing canopy trees, the probability of reaching the canopy can be calculated for each species. It was found that, of the four species with a better than even chance of reaching the canopy, three were already represented there. Hall & Swaine (1976, p. 937) conclude that it is probable that the 'canopy composition changes stochastically, gaps being filled in a more or less random manner by the same or by other available species' rather than 'kaleidoscopically' as postulated by the Mosaic Theory. A fuller discussion of Aubréville's theory in relation to the Ghana forests is given in a later paper (Swaine & Hall 1988).

Aubréville himself concedes that stable edaphic climax communities such as a mature mangrove or the association of *Cynometra megalophylla*, *Pterocarpus santalinoides* etc., characteristic of river banks in the Côte d'Ivoire (p. 361), cannot be ever-varying in composition. In these associations one combination of dominants, consisting of species with special soil requirements or tolerances, seems able to maintain itself permanently. The same is certainly true of most single-dominant climax rain-forest communities such as the *Eperua* association of Guyana and the *Gilbertiodendron dewevrei* association of Africa. In these, as already mentioned, the younger size-classes of the dominant species are always well represented and there is no reason to believe that the dominant will be supplanted by a different species as long as the environment remains unchanged.

It is evident that scarcity or apparent absence of regeneration in certain species does not necessarily imply impending disappearance from the stand. It has to be remembered that most rain-forest trees are probably long-lived and very few young individuals are required to replace the old trees when they die. Secondly, regeneration may be successful only on certain sites, as with strongly light-demanding species, which require relatively large openings. It is possible also, although it has not been definitely demonstrated, that some species regenerate intermittently, perhaps only at intervals of many years. In a few cases the absence of young plants may indeed indicate that the species is in the process of extinction. This is probably true of the endemic tree *Canarium paniculatum* in the 'upland climax forest' of Mauritius (Vaughan & Wiehe 1941), but here the conditions are not natural as the forest has long been reduced to mere vestiges and has been altered in other ways by human interference.

The role of chance in regeneration. Modern work suggests that regeneration is a less predictable and more irregular process than Aubréville's Mosaic Theory implied. Tropical rain forests, especially mixed forests, are populations of numerous species differing in their reproductive strategies and in their responses to light and other environmental conditions. Further complications are introduced by the intermittent seeding of many species, the varying efficiency of seed dispersal, and other factors. In such an exceedingly complex situation, it is not surprising that regeneration sometimes tends to be patchy and uneven. It is doubtful whether a cyclically changing pattern of regeneration such as that postulated by Aubréville (1938) ever actually exists. More obvious, and probably more important, is the irregular pattern of gap, building and mature phases which is found in all rain forests, the relative extent of the three phases varying with the amount and frequency of external disturbances. Since the formation of a gap is largely a chance event, and there is also an element of chance in determining which species succeed in colonizing a gap, regeneration could be regarded, as Hall & Swaine (1976) suggested, as a stochastic process.

However, the fact that the distribution of most tree species in rain-forest stands is usually not random and that minor variations in the species composition of mixed associations can often be correlated with small differences in soil, topography and other habitat features (Chapter 12), seems to indicate that the regeneration pattern is determined only in part by chance.

Webb *et al.* (1972) have discussed the part played by determinate and indeterminate factors in the regeneration of Australian rain forests. They conclude that the factors influencing regeneration, for example the particular assemblage of species colonizing a gap, are neither completely predictable nor completely unpredictable and are best regarded as probabilistic.

Different species of trees, even if taxonomically closely related, may differ in characteristics affecting their ability to regenerate under different climatic and microclimatic conditions as well as on different sites within the forest. Grubb (1977a) has suggested that these differences in the equipment of species for the 'regeneration niche' may explain, at least in part, the puzzling coexistence of numerous congeneric and apparently ecologically equivalent tree species in rain forests and thus contribute to the great species richness that is so typical of them.

Chapter 4

Trees and shrubs: I. Vegetative features

ANYONE SEEING AN evergreen tropical forest for the first time cannot fail to be struck by the extraordinary prevalence of certain physiognomic characteristics in the vegetation, some of which are seldom or never met with in other plant formations. These characteristics recur throughout the rain forest of the American and Old World tropics; in fact, they are typical of rain forest wherever it exists and are not peculiar to any geographical region. As was pointed out in the introduction, this uniformity in the aspect of the rain-forest flora contrasts sharply with its taxonomic diversity.

A newcomer, for instance, would very soon notice the wing-like expansions at the base of a large proportion of the bigger tree trunks. Closer acquaintance with the forest shows that these buttressed trees do not belong to any one family or group of families; buttressing is a characteristic of species of the most diverse systematic position. Again, many rain-forest trees, especially the smaller ones, bear their flowers directly on the trunk or larger branches. This unusual habit (dealt with in the next chapter) is called cauliflory and is highly characteristic of tropical rain forests; like buttressing, it occurs in species belonging to many quite unrelated families.

Besides such striking features as buttressing and cauliflory, various other peculiar morphological characters are widespread among rain-forest trees and, taken together, give them an easily recognizable physiognomy, which is shared to some extent by the trees of the most nearly related formation-types, the montane and subtropical rain forest. This is true not only of the trees, but in a smaller degree also of the other components of the community, the shrubs, ground herbs, epiphytes and climbers, all of which have their own special fea-

tures, many of them no less remarkable than those of the trees.

Since the characters that give the tropical rain forest its distinctive physiognomy are found throughout the tropics and are largely independent of the taxonomic position of the species, it must be assumed that they depend in some way on the environment and have evolved as a result of natural selection.

It is important to realise that, although most rain-forest trees have features in common, each stratum of the forest has a somewhat different physiognomy. For example, the type of leaf characteristic of the A storey (emergent) trees differs in some respects from that of the trees and shrubs in the lower storeys. Buttresses are found mainly on upper and middle (A and B) storey trees and are generally absent on those of the lowest (C and D) layers. Physiognomic characters thus depend considerably on the stratum to which the tree belongs. In large emergent trees they change markedly during the plant's development.

There are also important differences in vegetative characteristics, as well as in reproductive and physiological features, between the long-lived, relatively slow-growing tree species that dominate the mature phase of primary forest and the short-lived, fast-growing trees of secondary forest and gaps in the building phase of primary forest (see section 18.2).

4.1 Tree architecture

Some of the most important characters of rain-forest trees are those of their growth habit or architecture.

[70]

The latter term is used here to include the height : diameter and crown width : diameter ratios of the trees, as well as their branching systems and the 'architectural models' (Hallé *et al.* 1978) to which they belong. All these features are more easily seen in trees on the edges of clearings and isolated individuals than in the interior of the forest where the form of the whole tree is often hard to determine. It must, however, be remembered that the form of a tree, though in part genetically determined, also depends on the conditions under which it has developed: trees that have grown up more or less in the open may be very different from individuals of the same species in natural high forest.

In tall trees in mature forest the 'free bole' forms at least half the total height, but in trees of the same species that have grown up in large gaps or open conditions, the lower branches are not shed and the bole is proportionately shorter (see Torquebiau 1986). Under forest conditions most of the trees in all storeys have straight or only slightly sinuous trunks but usually a surprising number have been damaged by competition with their neighbours or by falling trunks and branches, resulting in 'bayonet-joints' or other deformities.

Shape and dimensions. The maximum height and girth reached by exceptionally large rain-forest trees were briefly discussed in Chapter 1. Rain-forest trees often give the impression of being more slender than temperate trees: this is particularly striking in trees of the A and B storeys, which sometimes have an almost etiolated appearance. Measurements confirm this impression. Kira (1978; see also Kira & Ogawa 1971) found that in Southeast Asia there is a south–north trend in the diameter : height ratio. At Sebulu, East Borneo (*ca.* 0°N) a tree of 1 m d.b.h. may be expected to be about 61 m high; at Pasoh, Malaya (*ca.* 2°44′N) it would be 47 m; at Sakaerat, central Thailand (*ca.* 15°N) 39 m; while at Ping Kong in northern Thailand (*ca.* 19°N) it would be only 32 m. Along this gradient the forest changes from typical lowland rain forest to evergreen seasonal forest and then deciduous forest; Kira believes that the changing relation between tree diameter and height depends on the decreasing amount and increasing seasonality of the rainfall. When individual species of rain-forest trees are compared with allied species found in other types of vegetation, their straightness and thinness are often very apparent. The rain-forest species of *Eucalyptus*, for instance, differ from the rest of the genus in having more 'columnar' trunks (Herbert 1929). Similarly, in many genera there is a striking contrast in the ratio of diameter to height and

the straightness of the trunk between rain-forest and savanna species (pp. 412).

The impression of great height and slenderness made by rain-forest trees depends partly on the relatively large proportion of the total height taken up by the unbranched bole. In mature trees growing under closed forest conditions the lowest branches tend to be near the top. This depends of course on the competition of neighbouring trees and the low light intensity in the undergrowth. In gaps and on river banks and in open situations, trees of the same species usually branch nearly to the base.

The shape of the crown varies in different species and in the same species in the course of its development. As mentioned in Chapter 2, the prevailing crown shape is different in different strata, ranging from the wide, often more or less umbrella-shaped, crowns of large emergents such as the South American *Peltogyne pubescens* and the African *Lophira alata*, which may be 30 m or more across, to the more or less isodiametric crowns common in the B layer and the narrow tapering crowns characteristic of most C storey trees (see Figs. 2.3, 2.5). In large trees the crown becomes wider relative to its depth (height) with age. The striking difference between the shape of the crown in young and old trees of the same species is particularly well shown in some Malayan dipterocarps, e.g. *Dryobalanops sumatrana*. Brünig (1974) has classified the crown shapes of emergent trees in the lowland forests of Sarawak and Brunei into seven types (disc, umbrella, composite sphere, etc.) which he regards as different ways of adapting to the heat load and liability to water stress.

In forest trees in general there is a relation between crown width and trunk diameter which is more or less independent of the age of the tree and the site characteristics. This has considerable practical importance, particularly in forest surveys where, if the relation between the crown diameter and the diameter (or basal area) of the emergents is known, it is possible to estimate the timber volume per unit area from aerial photographs. Dawkins (1963) investigated the ratio of crown diameter to trunk diameter in seventeen tropical economic species growing in plantations and found that for all but the largest and smallest trees the relationship was:

Crown diameter = $a + (b \times$ bole diameter).

The constants a and b are both specific; a is related to the light tolerance of the species and is high in tolerant and low in intolerant species. The relation is effectively a straight line but would probably become sigmoid if extended downwards to zero and upwards to include large senile trees.

Branching. Tropical trees and shrubs differ from those of temperate forests in their degree of branching. The delicate tracery of branches characteristic of most European and North American trees is rare in the tropics where many species are relatively little branched and the twigs and branches are thicker and have fewer buds. Wiesner (1895) found that branching of the fifth to eighth degree is usual in Europe but in tropical rainforest trees the highest degree of branching is generally the second, third or at most the fourth. In Java, however, Koorders (Schimper 1935, p. 462) found that higher degrees of branching than the fifth occur, though rarely.

Sometimes branching is completely or almost absent, as in tree-ferns, all rain-forest palms, some Pandanaceae and a considerable number of small dicotyledonous trees and 'treelets' of many different families, e.g. species of *Agrostistachys*, *Clavija* and *Pycnocoma* as well as the extraordinary West African *Cylicomorpha solmsii* and *Phyllobotryon soyauxianum* (Fig. 1.1). These monocaulous trees, the 'Schopfbäume' (tuft trees) of German writers, the pachycauls of Corner (see below), are mostly only a few metres high, but *Hortia regia* (Guyana) reaches 34 m (Sandwith 1931). The leaves are generally clustered at the top of the stem and very large (over 2 m long in some Meliaceae and even longer in some palms). Some large rain-forest trees that are normally branched when adult remain unbranched with the leaves crowded near the apex until they are 7 m high or more, e.g. *Lophira alata* (Africa) and *Sterculia pruriens* and *S. rugosa* (tropical America).

In some trees epicormic (coppice) shoots are produced at the base when the trunk has died or been injured. If the tree continues to grow, only one of these usually develops into a new trunk but in a few species self-coppicing regularly takes place and leads to the formation of a cluster of trunks of about equal height, somewhat like a multi-stemmed palm such as *Euterpe*. Remarkable examples are in two leguminous trees, *Dicymbe altsoni* and *D. corymbosa*. Both species are rain-forest dominants over large areas in the interior of Guyana; *D. corymbosa* is also found in the Rio Negro region of Amazonia (p. 321, Fig. 4.1). The primary trunk, which reaches a height of 30 m or more and a diameter of over half a metre, becomes surrounded by numerous coppice shoots at its base. When the main trunk dies, up to about ten coppice shoots take its place and grow to similar dimensions. In time coppice shoots may grow up round the secondary main trunks, forming tertiary main trunks, and so on. The result of this extraordinary method of growth is that most individual trees have the form of a clump of trunks of various sizes, hence the vernacular name, clump wallaba. On a sample plot

in the Bartica-Potaro region 800 m² in area there were twenty-six individuals and sixty-one trunks of 2 in (5 cm) d.b.h. (T.G. Tutin).

A somewhat similar 'self-pollarding' habit is found in *Trigonobalanus verticillata* (Fagaceae), a tree of lower montane forest in Malaysia and Southeast Asia (Soepadmo 1972, Corner 1990) and in *Castanopsis acuminatissima* in Java (C.G.G.J. van Steenis). It is interesting to speculate on the ecological significance of the self-coppicing habit, which can be regarded as a kind of vegetative reproduction (Chapter 5). Coppice shoots from stumps and fallen logs, which may grow into mature trees, are formed by some rain-forest trees, particularly in heath forest and some types of freshwater swamp.

Interest in the branching systems of tropical trees has been stimulated by the Durian Theory of angiosperm evolution of Corner (1949, 1953, 1954a, b, etc.). Corner emphasizes the correlation existing between the thickness of axes and the weight of the fruits, flowers and leaves they have to support. He expresses this relation in two complementary 'rules' of tree construction: (1) axial conformity (in a given species, the more massive the axis, the larger and more complicated its appendages); and (2) diminution and ramification (the greater the degree of branching, the thinner the branches and the smaller their appendages). Much-branched trees with slender twigs Corner terms leptocaulous; these he considers to be evolutionarily advanced. Those that have relatively few and thick branches he calls pachycaulous and regards as retaining features of the primitive angiosperms[1]. There are of course all gradations between leptocauls and pachycauls. A large majority of rain-forest trees, including almost all the emergents, are leptocaulous, but as mentioned above their twigs tend to be thicker than those of temperate trees. The non-climbing palms and most of the other monocaulous trees and treelets are pachycaulous, although most of them have thinner stems and narrower apical meristems than more extreme pachycauls such as *Cycas* and the pawpaw, *Carica papaya*.

Tree models. A comprehensive classification of the growth forms of tropical trees is provided by the system of architectural tree models of Hallé *et al.* (1978). Only a brief outline need be given here.

The models, of which twenty-one are recognized at present, are first subdivided into two groups, one unbranched with a single apical meristem (monoaxial) and the other branched with more than one apical meri-

---

[1] Compare Doyle's views on primitive angiosperms (p. 18).

Fig. 4.1 *Dicymbe altsoni* (A) and *D. corymbosa* (B) (clump wallabas), Bartica–Potaro road, Guyana (1979). In both species mature trees become multi-stemmed by self-coppicing.

stem (polyaxial). The models in the second group are then subdivided into those in which the axes are orthotropic and all alike, and those in which they are differentiated into trunk and branches. In the latter the axes are either all orthotropic, or some are orthotropic and some plagiotropic, or, as in many tropical and temperate trees, the axes are mixed, changing during development from orthotropic to plagiotropic or vice versa. The complete system is set out with examples in Fig. 4.2. The models are named after eminent botanists in order to avoid cumbersome terminology and were not originally assigned numbers in order to avoid implications of linear descent between successive models. (The numbers used here and in Fig. 4.2 are purely for convenience of reference.)

Though originally based on studies of tropical trees in Côte d'Ivoire and French Guiana, Hallé & Oldeman's system can be applied to all tropical and temperate trees and can probably be extended to include lianes and non-woody plants, although among these latter there may exist models not found among trees (Chapter 6).

All twenty-one models exist in the tropics, but not all are found among rain-forest trees. The fact that the trees of temperate forests belong to a much smaller number of models (the great majority probably to not more than three or four) seems to support the view that temperate tree floras are derived from those of the tropics (pp. 18–19).

Some models are not found in mature primary rain forests and some are not represented by large trees. Thus, Tomlinson's model (3), examples of which are various palms such as *Euterpe* spp. and the banana-like *Phenakospermum guyanensis*, does not include any woody dicotyledons. Schoute's model (4), which includes the estuarine *Nypa fruticans* and a few other palms, is not represented in mature forests on terra firme. Model 6 (McClure's) includes bamboos, some of which occur in rain-forest gaps, but no dicotyledonous trees. In Stone's model (17) there are only Pandanaceae and small herbaceous plants. Some models, notably Holttum's (1) and Corner's (2), which together include all the monocaulous trees discussed on p. 74–5, Mangenot's (21), and several others, seem to be found only among the smaller trees and shrubs of the lower (C, D, rarely B) storeys and never among emergent trees. Trees of Troll's model (23) are much the most numerous group: they probably include the majority of tall rain-forest trees. Hallé *et al.* (1978) estimate that 20–30% of trees worldwide conform more or less closely to it. Among Malayan dipterocarps Massart, Roux and Rauh, but not Troll, models are found (Hallé & Ng 1981; but see also Whitmore 1984a, p. 23).

The value of Hallé & Oldeman's system for plant morphology is obvious: each architectural model implies a different strategy for absorbing and utilizing light-energy. It is also important for understanding the ecology of tropical forests, but the ecological significance of the models requires further investigation. Little is known at present about the roles of the various models in the ecosystem or the numerical frequency of different models in different types of forest as species and individuals. It is likely that the spectrum of models differs in forests on different sites and soils, for example in forests on kaolisols (latosols) and those on podzols. It would

also be useful to know to what extent it varies in different climates and altitudinal zones. General impressions suggest that the proportion of different models changes during succession. This has been confirmed by Shukla & Ramakrishnan (1986) in India. In general it seems that rain-forest communities that provide habitats for relatively short-stemmed plants with high light requirements, such as swamp forests, forests on steep slopes and early seral communities, have more different models than undisturbed climax rain forest in its mature phase. Great species richness and variety of tree models do not seem to be necessarily associated.

Shrubs and treelets. Besides young individuals of trees and lianes belonging to the higher strata, the D storey of rain forests includes woody and herbaceous plants of various growth-forms. Some are monocotyledons such as palms, bamboos, *Dracaena* spp. and Scitamineae, which grow to 8 m or more and may reach the C storey but are not woody in the anatomical sense. Others are woody dicotyledons belonging to many families and showing considerable diversity in branching habit. The majority of the latter have a distinct main axis. These

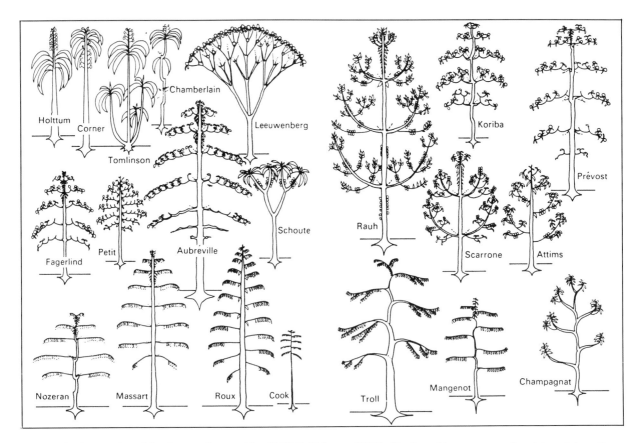

Fig. 4.2 Architectural models of tropical trees. After Hallé & Oldeman (1970). The branching system and position of the inflorescences are shown diagrammatically in one plane.

resemble trees in miniature and are sometimes termed treelets ('dwarf trees' and 'pigmy trees' of Du Rietz 1931). They can be assigned to at least three or four of the architectural models of Hallé *et al.* (1978). Treelets that are entirely unbranched with leaves crowded at the top, e.g. the West African *Angylocalyx oligophyllus* and *Sphenocentrum jollyanum* differ from the monocaulous trees mentioned earlier only in their smaller size. Others, probably all of Leeuwenberg's model, are multi-stemmed and are true shrubs, e.g. some *Psychotria* spp. and other Rubiaceae, although it has been asserted (Müller & Nielsen 1965, p. 87) that shrubs ('buissons') do not exist in tropical rain forests. Treelets are certainly commoner than 'true shrubs', but plants such as the Rubiaceae just mentioned have more than one main axis and differ in habit from the shrubs of temperate woodlands only in being less branched. The distinction between treelets and 'true shrubs' is in fact arbitrary (Opler *et al.* 1980b) and in this book 'shrub' is used for both.

Small palms are often abundant in the D storey. In tropical America they are represented by numerous species of *Bactris*, *Geonoma* and other genera, while in Malesia species of *Eugeissona*, *Licuala*, *Pinanga*, etc. are common and sometimes dominant in the D layer. In West Africa there are very few species of small palms but *Sclerosperma mannii* and *Podococcus barteri* are locally common. Bamboos and tall Scitamineae are mostly light-demanding plants and are abundant only in openings. Tree ferns, though often conspicuous in montane forests, are comparatively rare in lowland forests and when present are usually restricted to gullies and other damp sheltered situations.

A growth-form very unlike that of most small rain-forest trees is found in the West African *Anthonotha macrophylla* (Jones 1955, 1956) and *Scaphopetalum amoenum* (Jeník 1969, Hall & Swaine 1976). Both species are said sometimes to develop into erect trees *ca.* 10 m high (Hutchinson & Dalziel 1954–72), but after reaching a height of a few metres the stem usually bends over and roots at the apex; numerous upright branches then grow out from the original base and the arching stem (Fig. 4.3). The process can be repeated indefinitely, so that by vegetative reproduction dense stands of up to 0.25 ha may be formed. These thickets exclude other species and hinder tree regeneration. In the Okomu Reserve (Nigeria) Jones (1956) found that *Anthonotha* thickets often developed in old secondary forest after windfalls (Fig. 4.3). *Scaphopetalum*, which is found only from Ghana westwards, is probably also dependent on disturbance, and both species belong to the gap phase rather than to the high forest.

Fig. 4.3 *Anthonotha macrophylla*, Okomu Forest Reserve, Nigeria (1948). This tree forms dense thickets in forest gaps. The stems arch over and root at the tip.

4.2 Bark

Apart from the very thorough work of Whitmore (1962a, b) on bark structure in the Dipterocarpaceae of Southeast Asia, the bark of rain-forest trees has been rather little studied. Taxonomic descriptions of species rarely refer to bark characters though as will be seen later (Appendix 1) they are often of great value in identifying trees in the field. It is, however, mainly the internal characteristics of the bark, revealed by a 'slash', which are useful in species recognition; roughness, colour and other surface features are seldom specific, varying with the age and environment of the tree.

In spite of the great differences between genera, species and even individuals, the bark of rain-forest trees tends to have common characteristics. It is often said to be thinner, smoother and lighter-coloured than that of temperate trees but there are many exceptions and quantitative data for a valid comparison are not available (Grubb *et al.* 1963). It is certainly true that the bark of many rain-forest trees, even some large emergents, is only a few millimetres thick. In extreme cases (mostly secondary forest species such as various Euphorbiaceae, Sterculiaceae and some *Ficus* spp.), so little cork is formed that the chlorophyllous layer underlying the phellogen is visible. Trees with very thick rugged bark like that of a European oak or pine are usually uncommon.

The smoothness that is such a common feature of the bark, especially in the smaller rain-forest trees, is no doubt a consequence of its thinness. The phellogen is often superficial and continuous; the bark therefore does not tend to fissure and is often shed in very small flakes or granules, which are barely noticeable.

Flaking and fissured barks are, however, not rare. The bark of some African species of *Ochna* and the characteristically bright red bark of the African *Distemonanthus benthamianus* is shed in large flakes like that of a plane (*Platanus*). That of the Malesian *Pouteria obovata* comes off in longitudinal flakes; in *Tristaniopsis* spp., which are common in certain types of rain forest in Malesia, flakes several decimetres long are formed, which remain attached at the base of the tree. Furrowed and fissured bark is commoner than flaking bark; it occurs in many large Malayan dipterocarps and in the huge African tree *Baillonella toxisperma*. In the remarkable Malesian tree *Pertusadina eurhyncha* the bark has deep longitudinal perforations, so that it forms a lattice somewhat resembling an old strangling fig.

The frequency of smooth thin bark among rain-forest trees contrasts with the prevalence of thick, corky, often deeply fissured bark among savanna trees. The difference is well illustrated by comparing *Lophira alata*, a tall tree of the African rain forest, with the very closely related *L. lanceolata*: the bark of the former is thin and only slightly roughened, whereas in the latter it is very thick and fissured. A similar comparison could be made between many related forest and savanna species in South America as well as in Africa. It is usually believed that the thick bark of savanna trees is one of the adaptive characters that enable them to tolerate fires (which are not a normal feature of the rain-forest environment). It should be noted, however, that Whitmore (1962a, b) found no systematic differences in bark characteristics between Dipterocarpaceae of the ever-wet Malayan forests and those of dry deciduous and fireclimax forests in Burma, Thailand and other seasonally dry parts of Southeast Asia.

Thorny trunks, a characteristic of trees of tropical forests in dry climates (Chapter 16), are occasionally found in some rain-forest trees. Examples are *Hura crepitans* and *Zanthoxylum* spp. in tropical America and *Zanthoxylum* spp. and *Hylodendron gabunense* in Africa. Many forest palms have thorny stems. In some palms and dicotyledonous trees the thorns are modified roots (Jeník & Harris 1969, Jeník 1978). In the small West African tree *Citropsis articulata*, and in *Balanites wilsoniana*, especially when young, the twigs and smaller branches are armed with sharp thorns (cf. p. 414).

4.3 Root systems

The root systems of tropical trees are very varied and are often differentiated into subsystems differing in morphology and functions. In many species in addition to roots in the soil there are adventitious roots growing out of the trunk or branches, stilt-roots and other types of aerial roots. The buttresses at the base of the trunk, which are so common in large rain-forest trees, are also part of the root system, as they are mainly outgrowths of superficial lateral roots, although the cambium of the stem also contributes to them. In many tropical trees of poorly drained soils and of freshwater and mangrove swamps, specialized aerial 'breathing roots' (pneumorhizae) of various kinds (knee roots, peg roots, etc.) form part of the root system (pp. 89–90). All these above-ground root structures are conspicuous physiognomic features of tropical forests. The underground parts of the root systems of tropical trees, on the other hand, have received much less attention. The chief reason for this is of course that underground tree roots can be studied only by laborious excavations, or in wind-thrown trees, or root systems exposed by soil pits and road works. Data from overturned trees may not be very reliable, as such trees may have untypically shallow roots, and in road cuttings important parts of the root system may be destroyed. The work of Jeník and his collaborators on the roots of West African trees, as well as the increasing interest in the part played by roots in nutrient cycles, has stimulated interest in the root systems of rain-forest trees. A certain amount of information, including some developmental and experimental data, is available for a few species of economic importance such as cacao, rubber and oil palm (references in Jeník 1971a and Hallé *et al.* 1978, pp. 71–3) and for some timber trees, but it is still no exaggeration to say that 'for the majority of tropical trees the root system is entirely unknown' (Jeník 1978, p. 323).

Types of root system. An attempt to classify the root systems of Malayan trees on the same lines as those of European trees was made by Wilkinson (1939). He recognized four types, of which he gives numerous examples: (1) lateral roots well developed, tap-root not persistent; (2) lateral well developed, tap-root persistent but not well developed; (3) laterals and tap-root both well developed; (4) laterals weak or absent, tap-root well developed. A similar classification for Javanese trees had previously been made by Coster (1932a, 1933). A much more comprehensive, but probably not complete, classification has been proposed by Jeník (1978) who recognizes twenty-five models of root systems (Fig. 4.4). It should be noted that only a proportion (about half) of Jeník's dicotyledonous models are found in trees of primary terra firme rain forest; the remainder are found in secondary forest, freshwater swamp forest or mangroves.

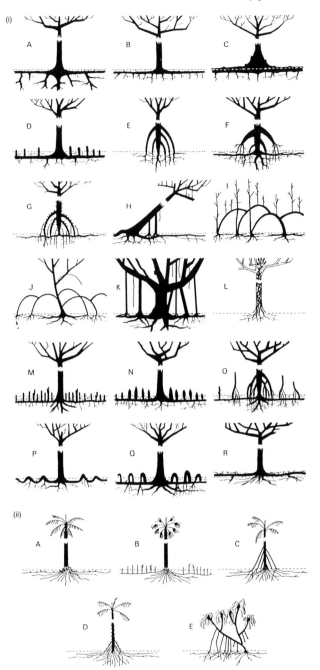

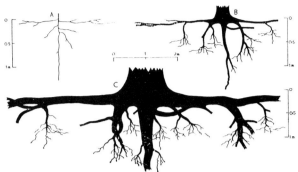

Fig. 4.5 Schematic drawings of root system of *Milicia excelsa*, in Ghana. From Mensah & Jeník (1968, Fig. 2). A Seedling; B pole tree; C mature tree.

Fig. 4.4 Types of root system in tropical trees. After Jeník (1978, Figs. 14.1 and 14.2). (i) Dicotyledonous trees: A *Milicia excelsa*, B *Cariniana pyriformis*, C *Piptadeniastrum africanum*, D *Xylocarpus mekongensis*, E *Uapaca guineensis*, F *Heritiera utilis*, G *Bridelia micrantha*, H *Protomegabaria stapfiana*, I *Scaphopetalum amoenum*, J *Rhizophora mangle*, K *Ficus benjamina*, L *Ficus leprieurii*, M *Avicennia germinans*, N *Sonneratia alba*, O *Xylopia staudtii*, P *Bruguiera gymnorrhiza*, Q *Mitragyna stipulosa*, R *Alstonia boonei*. (ii) Monocotyledonous trees: A *Elaeis guineensis*, B *Mauritia flexuosa*, C *Socratea* exorrhiza, D *Crysophila aculeata*, E *Pandanus candelabrum*.

The component roots of the underground subsystems are, as in temperate trees, of more than one kind. It is useful to distinguish first between skeletal roots and distal roots. The former consist largely of secondary xylem (except in palms and other monocotyledonous trees) and are usually much thicker than the distal roots to which the absorbing rootlets are attached. According to Jeník (1971a) the differences among the distal roots between 'leading roots' (macrorhizae) and smaller branch roots (microrhizae) are less marked among tropical than among temperate trees. The smaller roots are sometimes differentiated in various ways: in the African swamp tree *Aeschynomene elaphroxylon* there are brachyrhizae of limited growth which differ in their anatomy from the macrorhizae of unlimited growth (Jeník & Kubikova 1969). Similar differentiation exists in other swamp trees.

As will be clear from Wilkinson's classification of root systems, the skeletal part of root systems in rain-forest trees may include a well marked tap-root growing more or less vertically down from the base of the trunk, as in *Milicia excelsa* (Mensah & Jeník 1968) (Fig. 4.5), but often there is no recognizable tap-root in the mature root system. The growth of the tap-root in the seedling may be overtaken by that of the laterals, or the tap-root may atrophy. In young individuals of the large African tree *Cynometra alexandri* the tap-root atrophies *pari passu* with the growth of the buttresses (Lebrun 1936c) (Fig. 4.9). There is some evidence that among African rain-forest trees in general large (emergent) trees tend to have shallowly hemispherical or saucer-shaped root systems, often lacking a tap-root, while in the smaller (B and C storey) trees the root system has the shape of a narrow inverted cone (J. Louis). Beneath large horizontally running skeletal roots there are usually numerous small descending 'sinkers'.

Fig. 4.7 *Aspidosperma excelsum*, Mile 76, Bartica–Potaro road, Guyana (1979). Fluted trunks are characteristic of several species in this genus.

Fig. 4.6 Two types of root system in tropical trees. (Photos P.S. Ashton.) A *Alstonia angustiloba*, Negri Sembilan. With tap root. B *Shorea falcifera*, Kuantan. No tap root.

4.4 Buttresses

Plank buttresses or 'tabular roots' are more or less flat triangular plates of wood subtended by the angle between a tree-trunk and a lateral root running at or a little below the surface of the soil (Fig. 4.8). They are produced by epinastic secondary growth along the upper side of the root, the centre of growth always remaining at the base (Fig. 4.10). Stilt-roots (pp. 84–5) are also modifications of lateral roots but are formed from descending adventitious roots springing from the main axis above ground level, while buttresses are developed on roots arising at the base of the tree. Buttresses and stilt-roots grade into one another and if a stilt-root undergoes thickening mainly in a vertical plane it becomes flattened and a 'flying buttress' is formed. Sometimes the soil beneath a buttressed lateral root is eroded away so that a space is formed beneath it and the buttress comes to resemble a flattened stilt-root. Some trees develop both typical buttresses and stilt-roots, e.g. the African *Grewia coriacea* and some Malesian *Hopea* spp. Young & Perkocha (1994) have shown that in large trees with asymmetric crowns in Panama and Costa Rica buttressing is significantly greater on

Fig. 4.8 Buttressed trees. A *Mora excelsa*, Moraballi Creek, Guyana (1929). B Unidentified tree, Okomu Forest Reserve, Nigeria (1948).

the side of the tree opposite to the heavy part of the crown.

The trunks of rain-forest trees are often ribbed or flanged along their whole length. In some species the trunk may be elaborately stellate in section, e.g. *Aspidosperma excelsum* (Guyana) (Fig. 4.7). Such ribs or flanges are not usually included in the term 'buttress', which is generally applied to outgrowths extending only up the lower part of the trunk. Thick local expansions of the base of the trunk (*empattements* or *accottements* of French writers) such as are common in temperate trees are also found in some tropical trees and are particularly characteristic of certain swamp species, e.g. the mangrove *Pelliciera rhizophorae* (Fig. 15.6).

It is difficult to give exact figures for the relative frequency of buttresses among trees of different diameter- or height-classes, but observations in all parts of the tropics show that the largest proportion of buttressed trees is found among the emergent trees (A storey)

over about 30 m high. In the B storey buttresses are also very common but there are more unbuttressed species and individuals. Among the smallest trees the buttressed habit is rare except in young individuals that may later belong to the higher storeys. There is thus an obvious relation, but not a very close one, between the size of the tree and the presence of buttresses. Since the width, and presumably the mass, of the crown is a function of tree height, there must also be a correlation between the presence of buttresses and crown mass.

In individual trees the number of buttresses increases with the diameter and age of the tree and may be ten or more. In very large trees the buttresses may extend up the trunk for at least 9 m, and outwards for a similar distance. The buttresses of the huge Malayan tree *Koompassia excelsa* are sometimes used for making large dining tables.

The number, size and profile of buttresses also vary between species. Anybody experienced in identifying tropical trees in the field knows that buttress characteristics often provide useful diagnostic characters. Thus, in Guyana, trees of *Mora gonggrijpii* generally have smaller buttresses than comparable individuals of *M. excelsa* (Fig. 4.11).

Chipp (1922) suggested that the ratio of height to base (or angle of inclination of the outer edge) of the buttresses was a particularly useful diagnostic character. This is undoubtedly true of some species, but it must be remembered, however, that in most, perhaps all, buttressed trees the base : height ratio of the buttresses increases with the size and age of the tree. Francis (1924), found in Australia that in trees of *Argyrodendron trifoliolatum* of about 10, 21 and 61 cm diameter above the buttresses this ratio was 41 : 100, 45 : 100 and 98 : 100, respectively.

Buttresses are found in trees of many unrelated families but among those represented in tropical rain forests some have a much greater tendency to produce buttresses than others. The Dipterocarpaceae, Leguminosae and Sterculiaceae, for instance, include many species that normally have large buttresses, whereas in the Annonaceae and Fagaceae (the Malayan *Lithocarpus bennettii* is an exception, according to Corner) buttresses are usually absent. They are also absent, at least in the Malayan species, in the Myristicaceae (E.J.H. Corner). There are great differences between genera in one family: the Bombacaceae includes trees with enormous buttresses such as *Bombax* spp. and *Ceiba pentandra* and also *Catostemma*, a genus with several species in the Guyana rain forest which have no buttresses at all. Similarly, although the majority of large rain-forest Leguminosae are more or less strongly buttressed, in the Malesian genus *Sindora* (which includes large trees) and some others, buttresses are never found (E.J.H. Corner). In the Dipterocarpaceae, the genera *Dipterocarpus*, *Hopea* and *Shorea* include both buttressed and unbuttressed species, as is also true in Malaya of the large genera *Diospyros* (Ebenaceae) and *Eugenia* (Myrtaceae) (E.J.H. Corner). There is some evidence that species with thin bark tend to be more strongly buttressed than those with thick bark (Smith 1979).

Within one species buttressing appears to be a fairly fixed character, but some species are plastic in this respect. Ghesquière (1925) says that buttresses may or may not be present in the African *Petersianthus macrocarpus* and *Tetrapleura tetraptera*. *Eschweilera sagotiana* in Guyana is usually almost unbuttressed but occasionally an individual has one or more large buttresses (T.A.W. Davis). *Ceiba pentandra* is always buttressed in rain forest but has a savanna form (possibly a distinct ecotype) which is unbuttressed (Baker 1965).

Petch (1930) found that a seedling from an unbuttressed tree of *Delonix regia* became buttressed itself.

From what has been said, it is evident that buttressing is a character that is partly genotypically and partly phenotypically determined.

Structure and development. As already mentioned, buttresses are formed mainly by epinastic growth on lateral roots. The stem cambium may also contribute to them and the proportion of stem tissue to root tissue in buttresses probably increases as they become older, as Francis (1951) found in *Argyrodendron trifoliolatum*. When clear growth-rings are visible in cross-section, it can be seen that the early growth of a buttressed root is more or less symmetrical but after a time an excessive amount of secondary xylem is formed on the upper side (Fig. 4.10). The roots that form buttresses thus begin with a normal structure. This point is of some importance in considering the processes that may be involved in buttress formation.

The stage of growth at which buttresses appear varies in different species and individuals: often they begin to develop when the tree is quite young and increase in number and size as it grows (Fig. 4.9). In some Queensland species buttresses are already formed when the tree is only 1/17–1/9 of its maximum stem diameter and 2/9–1/4 of its maximum height (Francis 1924). In plantations in Nigeria MacGregor (1934) found that *Terminalia superba* began to produce buttresses when 8 m high and less than five years old and under garden conditions in Ceylon (Sri Lanka) Petch (1930) observed buttresses on trees of *Delonix regia* that were two to three years old and less than 12 ft (2.7 m) high.

It should be noted that in trees with well-developed buttresses the maximum girth is usually just above the buttresses. Usually, but not always, the trunk tapers gradually downwards from this point. In a large *Sloanea woollsii*, in Queensland, for example, Francis (1924) found that the diameter of the trunk at the top of the buttresses was 60.4 cm but at the surface of the soil it was only 22.6 cm. Young buttressed trees less than 20 cm in diameter did not show a downward taper; it became pronounced only when the buttresses grew large. The growth mechanism causing the downward taper is not entirely clear (see Kira in Jeník 1978, p. 343).

In a few buttressed species, e.g. *Canarium commune* and *Ceiba pentandra*, small wing-like expansions similar to the buttresses on the roots are found in the angles between the branches and the main trunk.

Buttressing and environment. Buttressed trees are highly characteristic of tropical rain forests but they

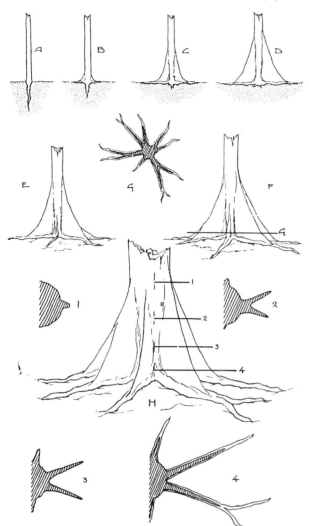

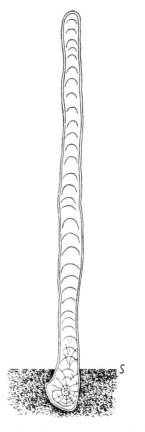

Fig. 4.10 Vertical section through a buttress of *Argyrodendron trifoliolatum*. After Francis (1951). S, soil level.

Fig. 4.9 Development of buttresses in *Cynometra alexandri* in Zaïre. After Lebrun (1936c). A–D Stages in development in young trees. E–F Branching of buttresses. G Transverse section of tree at ground level (roots continuing buttresses unshaded). H Base of mature tree (1–4, successive sections from above downwards).

are not entirely restricted to them. The distribution of buttressing shows that the climatic conditions under which it is most strongly developed are high rainfall, not necessarily throughout the year, combined with high mean temperature.

Within the tropics, buttressing is common only at fairly low altitudes and where there is a considerable annual rainfall. It is found in all types of tropical rain forest, including evergreen seasonal forests (p. 394), and also occurs in semideciduous and deciduous forests with a dry season, such as the 'monsoon forests' of India and Burma. In the West African semideciduous forests

with an annual rainfall of about 1200–1500 mm, buttressed trees are as frequent, or nearly so, as in the more constantly humid evergreen rain forests. In tropical regions with a long dry season, buttressed trees are usually found only in gallery (fringing) forests by rivers, although in Sri Lanka trees with large buttresses, e.g. *Bombax malabaricum*, occur in the open patanas (savannas) as well as in forests (Petch 1930). Porter (1971) reported buttressing in *Bursera graveolens* growing in very arid conditions in the Galapagos Islands. In the Mato Grosso (Brazil), with a dry season of about five months, there are no buttressed trees in the cerrado (savanna) but a few of the larger species in the permanently moist swampy gallery forests such as *Pouteria* sp., *Qualea wittrockii* and *Symphonia globulifera* are buttressed (Ratter *et al.* 1973, p. 460).

On tropical mountains buttressing is found in the lower montane forests but is usually absent in the upper montane rain forests at higher altitudes. In the 'mossy forest' on Gunung Dulit in Sarawak, which covers the top of the mountain down to about 970–1100 m, buttressed trees are quite absent. Shreve (1914a) also found none in the montane rain forest on the Blue Mountains in Jamaica. In Australia where rain forest extends from the tropical into the temperate zone (pp. 449–53), the number of buttressed species diminishes with increasing latitude. In Queensland buttressed trees are abundant in the rain forest wherever the annual rainfall exceeds about 60 in (1524 mm) (Francis 1924) but further south in the subtropical rain forest of northern New South Wales (32°S) buttressed trees are few (Fraser & Vickery 1938). Apparently even in the same species the production of buttresses decreases southwards; *Dendrocnide excelsa* is buttressed in northern New South Wales but unbuttressed further south (N.A. Burges).

The rarity of buttressed trees in the broadleaved forests of the north temperate zone is a remarkable fact which has to be taken into account in trying to assess the functions and adaptive value of buttresses. Some European and North American trees occasionally develop buttresses, e.g. *Populus nigra* var. *italica* (Senn 1923), *Fagus sylvatica* and others, but the buttresses are never as large or as thin as those of tropical trees. There are probably no north temperate tree species that are regularly and distinctly buttressed.

Buttressing varies with local site conditions as well as with climate, particularly drainage and soil texture. This is well shown in the catena of rain-forest types and soils at Moraballi Creek, Guyana (Chapter 11). There the size of the buttresses in the most abundant species (Fig. 4.11) and the percentage of buttressed species are clearly correlated with soil and other factors. In the mora forest on the silt of the flood-plain the dominant species, *Mora excelsa*, has larger buttresses than the most abundant species of the other forest types, and the percentage of species with buttresses is larger. On the other hand in the wallaba forest on white sand (podzol) on the ridges and plateaux the dominant *Eperua falcata* is almost unbuttressed and very few species of the subordinate species are strongly buttressed. A similar relation between buttressing and soil type also was found in Sarawak (Richards 1936) (Chapter 11) where the proportion of buttressed species was greater and buttresses were larger in the mixed dipterocarp forest on heavy loamy and clayey soils than in the heath forest on white sandy podzols.

It has often been noted that buttressed as well as stilt-rooted trees are more frequent on swampy ground than in rain forest on freely draining sites. In Malaya the large leguminous tree *Koompassia excelsa* has larger buttresses in swamp forests than when growing on hillsides and the swamp tree *Pometia alnifolia* has bigger buttresses than *P. pinnata*, which grows on firm soil on stream banks and on hillsides (E.J.H. Corner).

The relation between buttressing and soil conditions probably depends on the fact that buttresses develop only in trees with large, more or less horizontal lateral roots

(species of Wilkinson's types (1), (2) and (3); see p. 76 above, and of Jeník's models, especially A–D). It is likely that the soil factor with which buttressing is most closely correlated is not texture as such, but aeration and effective depth. In temperate forest soils, descending roots encounter increasing resistance to penetration and, as oxygen tension decreases with soil depth, less favourable conditions for respiration. On the other hand, availability of water and mineral nutrients may become more favourable. The depths at which roots are actually found represent a balance between various factors. In hot humid climates oxygen concentrations in all soils tend to be low (p. 275); those that are poorly drained or heavy in texture are even less well aerated than those that are porous and freely draining. Also, in at least some rain-forest soils, available nutrients tend to be more abundant close to the surface litter than deeper down (Chapter 10). It is probable, therefore, that superficial horizontally running roots, and consequently buttress formation, will be commoner on poorly drained or shallow soils than on those that are porous and well drained. For similar reasons buttresses are likely to be better developed on shallow than on easily penetrable deeper soils.

Trees are also less likely to have persistent tap-roots on soils which are shallow or impervious than on those which are more permeable. It is not surprising, therefore, that many but not all, buttressed trees lack a well developed tap-root. This has been observed by Spruce (1908, vol. 1, p. 21) in Amazonia and by many others. It is also supported by the author's observations in Borneo, Nigeria and elsewhere. Even enormous buttressed individuals of *Ceiba pentandra* lack a tap-root, the whole root system consisting of lateral roots with small 'sinkers' penetrating not more than about 0.5 m into the soil. Exceptions are found. Foxworthy (quoted by Chipp 1922) records a buttressed tree of *Dracontomelon dao* in the Philippines with a tap-root. Francis (1924, p. 32) saw a specimen of *Argyrodendron trifoliolatum* in Queensland that had a tap root although it had six buttresses: the trunk of this tree was only 9.6 cm in diameter and in an older individual the tap-root would perhaps not have been recognizable. In Malaya *Eugenia grandis* and *Scaphium affine* have moderate-sized buttresses yet have tap-roots (E.J.H. Corner).

Trees with and without tap-roots are sometimes found close together, presumably because their root systems react differently to similar soil conditions. Thus, in Zaïre *Julbernardia seretii*, a strongly buttressed species in which the root system consists mainly of shallow lateral roots, is often associated with *Gilbertiodendron dewevrei*, an unbuttressed species with a strong tap-root giving off large laterals at various depths (Louis & Fouarge 1949, Gérard 1960).

Mora excelsa (*Mora excelsa* association)

Mora gonggrijpii (*Mora gonggrijpii* association)

Eschweilera sagotiana (Mixed forest association)

Ocotea rodiei (*Ocotea rodiei* association)

Eperua falcata (*Eperua falcata* association)

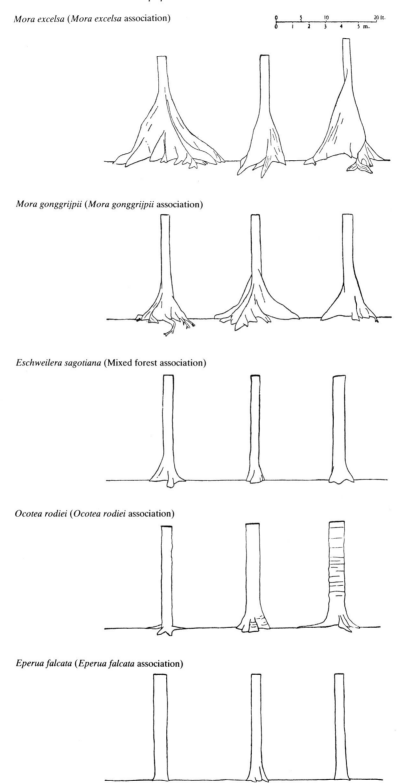

Fig. 4.11 Buttresses of typical mature individuals of the most abundant tree species in five forest types, Moraballi Creek, Guyana. After Davis & Richards (1933–34, Fig.2).

A

B

Fig. 4.12 Stilt-rooted trees. A *Tovomita* sp., Moraballi Creek, Guyana (1929). B *Musanga cecropioides*, Omo Forest Reserve, Nigeria (1935).

Although buttressed trees usually have superficial root systems, the converse is not necessarily true. Some rain-forest trees have huge roots running horizontally for long distances above the surface of the ground, but such trees are not always buttressed. A tree of *Entandrophragma angolense* in the Okomu Reserve, Nigeria (Fig. 1.6) had thick surface roots extending over 20 m from the trunk and was almost unbuttressed. Jeník (1971b) studied similar above-ground lateral roots in this species in Ghana and concluded that they had been pulled out of the ground when the tree was rocked by violent winds. Similar large surface roots are a feature of the South American *Caryocar nuciferum*, a tree that also has moderate-sized buttresses.

4.5 Stilt-roots

Aerial adventitious roots are common in tropical trees; they usually spring from the lower part of the trunk, but sometimes from the branches, e.g. *Ficus* spp. In some trees the aerial roots may dangle freely in the air or remain quite short, but in many species the aerial roots from near the base of the trunk bend downwards, often in a graceful curve, and enter the soil so that the tree appears to be standing on stilts (Fig. 4.12). The origin of such roots is generally within 1–2 m of the ground, but sometimes as high as 5 m. As they grow downwards they may fork and occasionally anastomose; when they reach the soil they branch freely and give rise to smaller roots and rootlets. In the mangrove *Rhizophora mangle* the aerial roots do not branch above ground unless they have become anchored in the substratum, or have been injured (Gill & Tomlinson 1969). In some stilt-rooted species, as the tree grows older, the stilt-roots become as thick or thicker than the part of the trunk to which they are attached. Sometimes, from the point of attachment of the highest stilt-roots downwards, the trunk becomes difficult to recognize because it either ceases to grow in diameter or rots away.

Stilt-roots may be round in section, as in *Uapaca* spp. and the non-epiphytic species of *Clusia*, or more commonly, as in *Casuarina nobilis*, *Musanga* and *Virola surinamensis*, they become laterally flattened. In a few swamp-forest species both pneumorhizae and stilt-roots are present. Gill & Tomlinson (1969, 1971a) have made detailed studies of the anatomy and development of the aerial roots in *Rhizophora mangle*; Jeník (1973) gives a general discussion of stilt-roots.

In rain forests stilt-roots are found in a species of at least nineteen dicotyledonous families. In monocotyledons they occur in some *Dracaena* spp. and many palms (Dransfield 1978) and Pandanaceae. The stilts (often called prop-roots) of monocotyledons lack secondary thickening of the type found in dicotyledons. Some ground herbs, some Acanthaceae, Malayan Zingiberaceae and *Mapania* spp. have small stilt-roots. Stilt-rooting, like buttressing, is evidently determined partly genetically and partly phenotypically. In many species stilts are invariably present, but some trees are plastic in this respect. Thus the West African *Anthocleista nobilis* is usually stilt-rooted in swampy places but often not so on drier ground. Evrard (1968) gives other examples among swamp forest species in the Central Congo Basin in Zaïre. According to Kunkel (1965a) in Liberia *Uapaca togoensis* is not stilt-rooted in open grassland but always has stilts when growing in bush islands on savanna or in the dense vegetation at the forest margin. In Malaya, also, certain species that are normally stilt-rooted when in swamp forest do not produce stilts on well-drained sites, e.g. *Ploiarium alternifolium* and species of *Myristica* and *Macaranga* (E.J.H. Corner).

The majority of stilt-rooted trees are small to medium-sized (mostly less than 20 cm d.b.h.), but a few, such as the African *Heritiera utilis*, are taller and belong to the B or even the A storey. In the dipterocarp genus *Hopea*, some of the very large species such as *H. mengerawan* are often, but not constantly, stilt-rooted and have many buttresses as well (Symington 1943).

Stilt-rooting is sometimes supposed to be found mainly in mangroves and freshwater swamp forest trees, but in fact it is common in primary and secondary forests that are not swampy, although rarely abundant enough to be a striking feature. *Rhizophora* is the only genus of mangroves with well-developed stilt-roots, although they are sometimes found in *Avicennia* and other genera. At Moraballi Creek, Guyana, no difference was found in the abundance of stilt-rooted trees in five forest types on soils varying from waterlogged silt liable to flooding to well-drained sand and lateritic clay. In southern Nigeria stilt-rooted trees formed 7.3% of all trees ≥10 cm d.b.h. on a swamp-forest plot with soft, yielding soil, 1.3% in a similar plot on free-draining clay, and 1.2% on a plot on sandy soil (Richards 1939). In Malaya, stilt-rooted species are abundant, especially in freshwater swamp forests. In Singapore and South Johore, Corner (1978) found seventy-five species with stilt-roots, but he recorded only three in forests on well-drained soil in some 'montane' and 'lower submontane' forests.

The stilt-rooting habit, like buttressing, is rare outside the tropics. In temperate regions swamp and riverside species of *Alnus* and *Salix* often form abundant aerial roots up to the level reached by floods; these occasionally develop into short stilt roots. According to Jeník (1973) in Europe the development of aerial roots into stilts is checked by cold or dry periods which kill them before they have grown to any size. Drought may well also be the reason why stilt-roots are rather uncommon in the seasonally dry types of tropical forest. It seems plausible to suggest that the frequency of stilt-roots in trees of humid tropical environments, and their relative scarcity in non-tropical climates, may be due to negative selection against them (Jeník 1973).

4.6 Functional significance of buttresses and stilt-roots

Buttresses and stilt-roots differ in their morphology and development but their chief function seems to be similar. There is good evidence for believing that both are of adaptive value by increasing the resistance of the tree to stresses due to gravity and strong winds. Buttresses act as stays and props ('cables de résistance') supporting the trunk: they also reduce strains on the more superficial roots which could break them or, as Jeník (pp. 76–7 above) showed, may actually pull them out of the ground, when the tree is rocked by violent winds. Stilt-roots, which, as mentioned above, are found mainly in the smaller trees, have a 'snow-shoe effect' and act much like buttresses in helping to keep the trunk upright. As overturning by the wind or by failure of the root system under the stress of gravity is probably the commonest cause of death in mature rain-forest trees on both level ground and steep slopes, it is probable *a priori* that adaptations increasing the mechanical stability of the tree will have survival value.

The forces acting on buttressed trees have been analysed by Smith (1972) and by Henwood (1973) whose structural models are shown in Fig. 4.13. Several simplifying assumptions are made: that the trunk is perfectly rigid, that deflection or rotation of the tree will be only through small angles, and that under lateral loading the root platform will rotate about a point approximately under the edge of the compressed side of the trunk. This model shows that the mechanical advantages of buttressing will be greater the shallower the root system, and also when, as is sometimes the case, the crown of the tree is asymmetric and unbalanced. A large buttressed tree is thus in effect comparable to a tall building supported on a raft which may be 20 m or more in diameter.

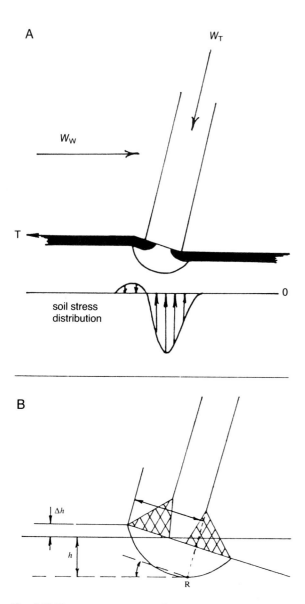

Fig. 4.13 Forces acting on an unbuttressed and a buttressed tree. After Henwood (1973). A Forces acting on an unbuttressed tree. B Dynamics of stresses in a buttressed tree. In both A and B there is lateral loading due to wind from the left (W_W). T, tension, Δh, increment; R (in B), point of rotation.

Jeník (1971b) attempted to estimate the forces acting on the large surface roots of *Entandrophragma angolense* mentioned earlier (p. 84). In a tree 40 m high, with four lateral roots in the form of a cross, the tension tending to pull a single root out of the ground on the windward side would be 20 t at a wind speed of 80 km h^{-1}; on the lee side the roots would be pushed downwards, but owing to the resistance of the soil, the movement would be slight.

Evidence for the view that the function of buttresses is primarily mechanical is of several kinds. As it has often been pointed out, the trees with the largest buttresses are the emergents with heavy crowns fully exposed to the wind. In smaller trees the crowns are lighter and more sheltered by their neighbours and their buttresses, if present, are smaller and fewer. Some large trees, as mentioned above, begin to develop buttresses early in life when the crown is still small. If buttressing is partly a genotypically determined character, this is not surprising and is not an argument against buttresses being mechanically important later in the life of the tree.

Data on the orientation of buttresses in relation to the prevailing wind direction, slope and other factors have been collected by various authors. Senn (1923), after his interest in the problem had been aroused by a visit to Java, made a detailed study of the buttresses on the Lombardy poplar (*Populus nigra* var. *italica*) in Switzerland. In this species the buttresses are smaller and thicker than in most tropical trees, but are otherwise similar: there seems no reason why the causes of the phenomenon should be different in the two cases. Senn examined the buttresses in a large number of trees in various localities. He could find no relation between their orientation and the distribution of water or nutrients in the soil. There was, on the other hand, an apparent relation to unilateral heat radiation, buttress formation being greater near heat-reflecting surfaces such as walls. Much more striking, however, was the correlation between the orientation of the buttresses and the direction of the prevailing wind, which is remarkably constant in many Swiss valleys. Out of 461 trees examined, 443 had more or less distinct buttresses and 407 (91.9% of the total) had buttresses chiefly on the side of the tree *facing* the prevailing wind. Senn concluded that buttress formation is due to stimulation of the cambium on the upper side of the proximal end of the lateral roots and the adjoining part of the trunk by tensions set up by the wind.

Among the exceptions were eight trees which had buttresses on the *leeward* side and also had a steep bank of rock abutting on that side. Senn suggests that when the wind blows from the prevailing direction, the lower part of the trunk tends to be bent slightly, owing to the resistance of the bank or rock; it therefore acts as a lever pulling on the roots on the leeward side and pushing on the roots on the windward side. Thus the stresses are the opposite of those in a tree on level ground without obstructions, and hence the orientation of the buttresses is the opposite to the usual one. Three more exceptions are ascribed to the influence of unilateral heat radiation. In any case, as Senn points out, a mathematically exact correlation with wind direction cannot be expected, because buttresses are formed on pre-existing roots, the orientation of which is not influenced by the wind.

Senn also found that buttressing was most marked where the lower part of the trunk is shaded, that is, where it is surrounded by damp air. This might help to explain its frequency in tropical rain forests.

In tropical trees a relation between the orientation of the buttresses and the direction of the prevailing wind is seldom obvious. The only species for which much statistically significant information is available is *Ceiba pentandra*. Twenty-eight trees were studied in Cuba by Navez (1930) who found that there was a significant tendency for the buttresses to be on the side of the tree facing the prevailing wind. In Zaïre a similar result was obtained by van den Brande (1936, quoted by Lebrun 1936c). More recently Baker (1973) examined two sets of *Ceiba* trees at Achimota (Ghana), where the wind direction is remarkably constant throughout the year. Set A consisted of a row of twenty-two fifteen-year-old trees planted as a wind break along the edge of a field, set B of forty-three trees of various sizes and ages growing in various places but under similar conditions. In set B, in which the trees were mostly older and larger than in set A, there was a highly significant preponderance of buttresses on the windward side of the trees. In the young trees of set A there was more variation, although again most of the buttresses were on the windward side. Evidence that more buttresses are formed on the windward than the leeward side of trees was also found for two other species, *Bombax buonopozense* and *Hildegardia barteri*.

Observations on leaning trees and trees on slopes support the view that buttress development is related to tensile stresses: in the former the buttresses tend to be on the upper side of the trunk and in the latter on the upslope side. In *Cynometra alexandri* in Zaïre Lebrun (1936c) found that in each diameter-class, buttresses were more numerous in trees on slopes than in those on level ground. Stresses due to running water seem to act in a somewhat similar way to those due to

wind and gravity; Evrard (1968) observed that in swamp forests liable to flooding in the Central Zaïre (Congo) basin (pp. 361ff.) the bases of trunks of *Entandrophragma palustre* and *Guibourtia demeusei* develop 'accotements' (broad thickenings) at the base on the upstream side and thin buttresses on the downstream side.

Although facts such as those set out above seem to indicate that the primary function of buttresses is to give mechanical support, other evidence suggests that their mechanical importance may not be as great as is generally assumed. Buttresses are sometimes absent where one might expect them to be most needed. Thus in the Guyana wallaba forest and the heath forest of Borneo, both of which grow on loose sandy soils, buttressing is less common than in forests on firm clays and loams (p. 82 above). On a mountain in the Philippines, Whitford (1906) found that the buttressed habit disappeared altogether on the ridges where wind exposure is greatest. The data of Moraes & Pires (1967) also cast doubts on the survival value of buttresses. In 5.5 ha of rain forest at Mocambo, Amazonia (p. 59 above), 30% of the trees were buttressed, but among those that died from wind-throw during a ten year period the proportion of buttressed trees was nearly as large (28.6%). Among the twenty-five most abundant tree species the range in mortality (from all causes) was about equal in the buttressed and unbuttressed species. In *Protium cuneatum*, a strongly buttressed species, the mortality over ten years was 18% and wind-throw was responsible for a considerable proportion of the deaths. The mortality in *Eschweilera odora*, which is also buttressed, was only slightly less than in *E. corrugata*, which is not. These figures cannot be critically assessed without information on tree height, size of crown and soil depth, but they lend some support to Moraes & Pires' conclusion that buttresses are not as effective supports as might be expected.

Whether or not the chief value of buttresses is mechanical support, they may be functionally important to the tree in other ways. From observations in Sri Lanka, Petch (1930) concluded that buttress formation had little to do with stresses due to the wind. In *Canarium commune* he could find no correlation between wind direction and the orientation of buttresses and he also noticed that buttress-like wings sometimes developed on horizontal roots at some distance from the trunk where they could be of no value as supports and could not be affected by tensile stresses. Petch emphasized the significance of the lack of a tap-root in most buttressed trees and suggested that in trees in which the root-system consists mainly of superficial lateral roots, the transpiration stream in the trunk carrying 'nutrients' from the soil mainly follows limited paths on the same radii as the lateral roots. His theory assumes that lateral conduction in the trunk is relatively inefficient, and suggests that growth in the sectors of the cambium on the radii carrying the transpiration stream are stimulated to grow at the expense of the intervening sectors, thus leading to the formation of ridges, which in time become buttresses. The frequency of buttressing in the humid tropics, according to Petch, is therefore due to poor soil aeration, which prevents the development of tap-roots. Not all trees are buttressed because in some species the tap-root may be more tolerant of poor aeration than others.

Petch's (1930) suggestion that in many tropical trees there is relatively little lateral movement of water is supported by some anatomical evidence. Chalk & Akpala (1963) found that vessel anastomoses and xylem elements such as fibre tracheids, which might be expected to facilitate lateral conduction, were less common in buttressed than in unbuttressed trees. In the buttressed African tree *Triplochiton scleroxylon* no vessel anastomoses could be found and in another buttressed species none could be demonstrated although from the occurrence and disappearance at a higher level of radial multiples in the pores it was inferred that some anastomoses must exist. On the other hand in two unbuttressed species, *Populus marilandica* and the African savanna tree *Daniellia oliveri*, anastomoses were easy to observe. Chalk & Akpala also found that in 142 tree species from Ghana, belonging to many unrelated families, there were considerable differences in the wood anatomy between species with large buttresses, those with moderate-sized buttresses (or which were only occasionally buttressed) and those without buttresses.

'Reaction wood' (in which gelatinous fibres replace normal wood fibres) is commonly found in the trunks and roots of temperate trees subjected to mechanical stress, but it is not characteristic of the buttresses of tropical trees. Fisher (1982) found that in wood from the buttresses of forty-eight species of tropical trees reaction wood was present in only five, though it was found in the aerial roots of *Ficus* spp and in the stilt roots of *Cecropia* sp.

A great difficulty in finding a convincing 'explanation' of buttressing in tropical trees is to suggest why buttresses are so little developed in temperate and tropical montane climates where they might be expected to confer an equal selective advantage. A widely accepted view is that the prevalence of buttressing in the tropics depends on the tendency of roots in tropical forests to be concentrated in the better aerated upper layers of the soil profile, which are often also the richest in

nutrients. In such conditions root systems tend to be superficial and the need for buttresses to increase the stability of the trunk is consequently greater. This may well be so, but shallow and waterlogged soils are also found in some situations in temperate regions and by itself this hardly seems a satisfying hypothesis.

Smith (1972) has suggested that the decrease of buttressing with increasing latitude and altitude is due to a decreasing selective advantage, or that there may be negative selection against them. By means of a simplified model he calculated the volume and surface area of the lower part of four large buttressed trees in Costa Rica and found that though the amount of wood was about the same as in a cylindrical trunk of the same diameter, the surface area, as might be expected, was considerably greater. He also suggests that the high surface: volume ratio may be a disadvantage to trees exposed to low winter temperatures. It should be pointed out, however, that the risks of damage from freezing due to a high surface: volume ratio apply equally to leptocaulous trees with slender twigs such as *Betula* spp., which do not seem to be at a disadvantage even in climates with very cold winters (Givnish 1978a).

Stilt-roots increase the stability of the tree as buttresses do, but as this is so it seems surprising that most stilt-rooted trees are comparatively small and many grow in the lower storeys of the forest where they are little affected by the wind. Some, like the mangrove *Rhizophora*, are found on soft swampy soils but many grow on firm ground. Corner (1966), who describes stilt-rooting as 'an unexplained deviation from the normal method of trunk making', suggests tentatively that they may be of value in raising the trunk of the tree above flood level although many, probably the majority, of the stilt-rooted species grow on sites not liable to flooding.

Kunkel (1965a) noted that in West Africa stilt-rooted species such as *Musanga* and *Myrianthus arboreus* were particularly numerous in young secondary growth and suggested that stilt-rooting might be useful in competition by enabling the tree to raise itself above the dense tangled undergrowth. This seems plausible but it should be realized that stilt-roots do not actually 'lift up' the tree. They do, however, provide an economical means by which a slender tree can maintain its stability without diverting a large part of its resources into the growth in diameter of its trunk.

Many forest palms have stilt-roots. Schatz *et al.* (1985) show that in Costa Rica two stilt-rooted species, *Iriartea gigantea* and *Socratea durissima*, produce tall stems earlier in development than *Welfia georgii*, which lacks stilt-roots; the latter forms a large underground

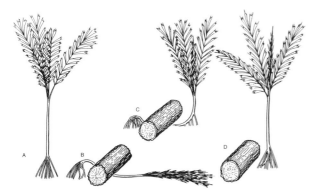

Fig. 4.14 'Walking palm' (*Socratea exorrhiza*), Peruvian Amazon. After Bodley & Benson (1980). A–D show stages in the 'walking' of a young plant flattened by the fall of a tree trunk five years earlier (see below).

axis before the stem starts to elongate above ground. In some species at least, mechanical support may not be the most important function of stilt-roots. Bodley & Benson (1980) found that in the dry forests of eastern Peru the stilt-roots of the palm *Socratea exorrhiza* enabled the entire plant to move when broken or smothered by fallen trunks, branches and leaves so that it literally 'walks away' from such impediments (Fig. 4.14). The palm may thus be able to form a trunk some distance away from where the seed germinated. Seedlings can move short distances in a few months and older juveniles 2 m or more in two or three years. It seems likely that other stilt-rooted palms, and probably also some dicotyledonous trees, may be able to 'walk' in a similar way and thus escape what is one of the major hazards for young trees establishing themselves in forest undergrowth.

4.7 Pneumorhizae

In freshwater and saline swamp forests in the tropics aerial roots emerging from the ground are often a striking feature. They are particularly well developed in mangrove communities, where they are sometimes so numerous as to make walking difficult (Chapter 15). These roots are called pneumorhizae[2] in reference to their presumed function as aerating or 'breathing' roots. In form, size and mode of development they vary greatly. Often they have the appearance of loops or 'knees', as in *Symphonia globulifera*, the African *Mitragyna* spp. (Fig. 4.15), *Bruguiera* and other mangroves,

---

[2] 'Pneumorhiza' seems preferable to 'pneumatophore' because the latter term has been used for a large variety of very diverse organs in both plants and animals (See Jeník 1978).

and in many Indo-Malayan freshwater swamp-forest trees. Peg-roots, another very common type of pneumorhiza, are erect negatively geotropic roots, sometimes resembling cigars or shoots of asparagus, which grow to a height of 20 cm or more above ground or water level. As they are branches of horizontal lateral roots they are often arranged in straight lines radiating from the trunk of the tree. In the swamp forests of Singapore and southern Johore, Corner (1978) found thirty-eight species of eighteen families with pneumorhizae. A general account of the morphology of pneumorhizae is given by W. Troll (1938–43).

In *Xylopia staudtii* of freshwater swamps in Ghana slender vertical branches rise to a height of up to 2 m from the lateral roots: these in turn send down positively geotropic branches to the soil so that the whole structure becomes a 'stilted peg-root' (Jeník 1970a, 1978). Corner (1978, p. 24) describes similar roots in *Elaeocarpus macrocera* and other Malayan species as 'λ-roots'. In some freshwater swamp plants aerenchymatous roots are produced, which float limply in small pools, e.g. *Combretocarpus rotundatus* in peat-swamps in Borneo and the liane *Tetracera alnifolia* in Sierra Leone. The roots of certain trees when growing in waterlogged soil form large numbers of pneumathodes; these are very small (0.5–2 cm) upright aerial branches the apices of which are covered with a mealy mass of aerenchymatous cells, e.g. in the oil palm *Elaeis guineensis* (Yampolsky 1924) and the mangrove *Laguncularia racemosa*. In the latter the pneumathodes are produced in successive short-lived crops (Jeník 1970b). In the mangrove *Rhizophora*, and in some freshwater genera, the aerial roots are abundantly provided with lenticels and do double duty as stilts and pneumorhizae. The African *Alstonia boonei* on waterlogged soils pro-

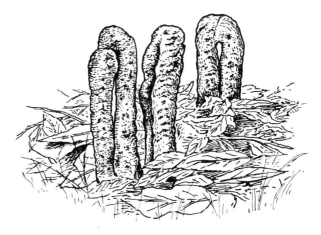

Fig. 4.15 Loop-like pneumorhizae of *Mitragyna ciliata* in swamp forest, Omo Forest Reserve, Nigeria (1935).

duces horizontal roots that grow along the surface; they have many very large lenticels and apparently function as pneumorhizae (Jeník 1967).

Some trees have pneumorhizae of more than one kind, e.g. *Virola albidiflora* which in small pools in the swampy gallery forest of the Mato Grosso (Brazil) has both loops and peg-roots (Ratter *et al.* 1973).

Pneumorhizae are found in dicotyledonous trees, mostly medium- to large-sized, belonging to many unrelated families. Both the loop and the peg types occur in monocotyledons, e.g. *Euterpe, Raphia* and other palms, and *Phenakospermum guianense* (Strelitziaceae) (Hallé *et al.* 1978). In some genera, they are present in some species and absent in others, e.g. *Anthocleista* (Jeník 1971c). It seems certain that ability to produce pneumorhizae is genetically determined but in most cases its expression is probably facultative. Thus in *Anthocleista nobilis* (Jeník 1971c) and *Mitragyna ciliata* in West Africa and *Symphonia globulifera* in Guyana (T.A.W. Davis) peg-roots are abundant on very swampy ground but rare or absent on drier sites. Jeník's (1971c) observation that on an area of well-drained soil in Ghana *Anthocleista nobilis* produced peg-roots only at the bottom of a ditch suggests that pneumorhiza production may be a response to low oxygen tension. Outside the tropics pneumorhizae are occasionally found, e.g. loop-like roots in *Nyssa aquatica* and the well-known 'knees' of the North American conifer *Taxodium distichum*, which, like those of some tropical swamp trees, originate as local thickenings on the upper side of the lateral roots.

Pneumorhizae arise in several different ways; roots that begin as pegs sometimes develop into loops or knees. Except in monocotyledons, most types of pneumorhizae form secondary xylem early in their development and become stiff and woody. McCarthy (1962) has described the development of pneumorhizae in *Mitragyna stipulosa* in Uganda. In this tree large woody knees can develop by localized cambial activity along the upper surface of lateral roots or from the arching of lateral roots above the soil surface. They can also develop from negatively geotropic upright roots, which, after reaching a height of 3–5 cm above the ground, begin to curve downwards and form loops. A similar bending-over of slender upright pneumorhizae to form loops has also been described in the Malayan *Xylopia fusca* (Corner 1978).

The surface of pneumorhizae is usually covered with a corky bark. Where this is exposed to the air it has numerous lenticels, which often become very large and conspicuous (although according to Adamson (1910) the pneumorhizae of the Indian riverside tree *Terminalia*

arjuna lack both periderm and lenticels). The internal structure varies: generally it is soft and spongy, with air spaces forming a large part of the total volume.

Both the structure and the mode of occurrence of pneumorhizae suggest that their primary function is to act as ventilating organs for roots growing in substrata deficient in oxygen. Gas exchange between pneumorhizae and their environment has been studied experimentally only in mangroves (Chapman 1976, Tomlinson 1986). By blowing through pneumorhizae it can readily be shown that air passes through the air spaces and escapes through the lenticels. Scholander *et al.* (1955) were able to show conclusively that the pneumorhizae of *Avicennia germinans* and the stilt-roots of *Rhizophora* enable oxygen to reach roots living in the completely anaerobic mud. A direct gas connection between the active roots and the lenticels was demonstrated. If *Avicennia* is deprived of its pneumorhizae or the lenticels on the stilt-roots of *Rhizophora* are blocked with grease, the oxygen concentration within the buried roots drops in two days from its maximum of 15–18% during low tide to 2% or less. Water does not readily enter intact root systems. There is a diurnal rhythm of gas pressure and oxygen concentration due to the respiration of the internal tissues; on this a tidal rhythm is superimposed. The internal gas pressure is lowest when the lenticels are first uncovered by the falling tide and air is then drawn into the air spaces of the exposed roots. In freshwater swamp trees not affected by tides, the pneumorhizae are not submerged at daily intervals but it is likely that they function in a similar way to those of mangroves.

4.8 Leaves of trees and shrubs

The leaves of rain-forest trees are characteristic, not only in their shape, size and structure, but also in the way their arrangement, their mode of development and their means of protection (or lack of it) in bud. All of these features are doubtless in some way dependent on the environment and have evolved as a result of selective pressures. The majority of the trees, and probably all the shrubs, of the rain forest are evergreen, but some are deciduous. The distinction between the evergreen and deciduous habits is, however, not as clear as in temperate climates, as will be seen in Chapter 9 where leaf fall and renewal are discussed in relation to seasonal climatic changes.

Leaf buds. Many rain-forest trees grow continuously or 'cease and resume growth as internal and external conditions permit' (Zimmerman & Brown 1971, p. 9). Others grow in 'flushes' once a year, or at intervals of less or more than a year (Chapter 9). In those in which growth is more or less continuous, there are no dormant buds, but in those that grow intermittently or are deciduous, dormant buds are usually formed (Koriba 1958). As noted in Chapter 1, they are fewer and larger than in temperate trees and generally much less well protected against desiccation and other hazards.

When there are true dormant buds they may be enclosed by the already expanded leaves, or by hairs, or by mucilaginous or resinous secretions. In a large number of tropical trees there is protection by the bases of petioles, by a pair of stipules (as in many Rubiaceae), or, as in the Moraceae, by a stipular hood, which in *Cecropia* and *Musanga* may be 30 cm or more long. Potter (1891) showed that if the stipular hood that protects the buds of *Artocarpus altilis* was removed the young leaves became dwarfed and deformed. The most specialized type of bud covering, leaves modified as bud scales, is occasionally found even in humid tropical climates, e.g. in the wet zone of Sri Lanka and in such characteristically tropical genera as *Litsea* (Lauraceae) (Holtermann 1907). Resting buds with scales are also found in the genera *Castanopsis* and *Quercus*, which are well represented in lowland and montane tropical forests. Resvoll (1925) found that the buds of five species of *Quercus* in the rain forests of Java had dry suberized bud scales and differed very little in structure from those of the temperate *Q. robur* except that the scales overlapped less closely. Resvoll concluded that the nature of the buds was conditioned by factors common to the genus *Quercus* as a whole.

Raunkiaer (1934) classified phanerophytes into (i) evergreen phanerophytes without bud-covering, (ii) evergreen phanerophytes with bud-covering, and (iii) deciduous phanerophytes. Owing to the variety of bud-coverings and the lack of a clear distinction between evergreen and deciduous trees, this classification is not easy to apply in the tropics, but there is certainly a relation between the bud protection of tropical trees and the humidity of the climate. Lebrun (1936b) found that in the forests of the Congo (Zaïre) basin the proportion of trees with protected buds increased from the humid centre to the periphery where there is a more marked dry season. It is probably generally true in the tropics that bud protection increases as the rainfall becomes more seasonal.

Young leaves and leaf development. One of the most beautiful sights of a tropical forest is a tree covered with newly expanded young leaves. Until one or two weeks after reaching their full size the leaves of many rain-forest trees are quite limp and appear almost lacking in chlorophyll; often they are bright red or crimson, sometimes purple or steel blue and, not infrequently, white or very pale green, contrasting with the dark colour of the mature foliage. In Guyana, species of *Duguetia* (Annonaceae) with a flush of young leaves look as if sprinkled with pieces of white paper. The development of chlorophyll is delayed and the young leaves do not reach their full photosynthetic capacity until some days after they are completely expanded.

The red and purple colouring of young leaves in tropical trees is due to anthocyanin pigments in the cell sap and it is more strongly developed in some species than in others. In cacao (*Theobroma cacao*) it varies in different varieties and clones (Baker & Hardwick 1973). Red colours are found in the young leaves of some palms as well as in dicotyledonous trees. They are less common in herbaceous plants than in trees, but young fern leaves are sometimes red.

Brightly coloured young leaves, though very characteristic of wet evergreen forests, are also common in savannas, e.g. in *Daniellia* and *Isoberlinia* spp. in Africa and many cerrado trees in South America. In temperate plants such colouring is rare, though well shown in a few species, e.g. the cultivated Asiatic *Pieris formosa*.

It is often supposed that the red pigments in the young leaves of tropical trees have an adaptive value. It has been suggested that the pigments might shield the chloroplasts from the ill effects of strong light or from ultraviolet radiation (Bünning 1947), but observations by Lee & Lowry (1980) on six woody species in Malaya show that red young leaves reflect much less ultraviolet than older leaves; they also have a higher content of anthocyanin and total phenols. In *Pometia pinnata* five-day-old leaves have a lower reflectance, especially in the UV-B range, than older leaves. As the leaves mature the development of cuticle and intercellular spaces increases the reflectance at all wavelengths and simultaneously the anthocyanin disappears and the total phenols decrease in amount. Later work on developing leaves of mango (*Mangifera indica*) and cacao (*Theobroma cacao*) (Lee *et al.* 1987) gives no support to the view that anthocyanins give a selective advantage to the tree and suggests that they may be merely by-products in the synthesis of flavonoid compounds in the young leaves.

A common peculiarity of young leaves in rain-forest trees, which is often associated with their unusual colouring, is that for some days after expanding to nearly full size they remain quite limp and hang down as if wilted. Only later, as they become fully green, do they become stiff and assume a more erect position. This feature is common in rain-forest trees in all parts of the tropics. It is particularly striking in some Leguminosae, although it also occurs in many other families. In *Amherstia nobilis*, species of *Brownea* (Fig 4.16), *Saraca* and many other Caesalpinioideae the young twigs as well as the large compound leaves droop when young. In these trees the buds expand very rapidly, often overnight, so that Treub (quoted in Schimper 1935, p. 484) described them as 'pouring out' (ausschütten) their leaves. Funke (1929) found that the young shoots of *Amherstia* took ten to fourteen days to reach their full length: by then the leaves were fully green but still hanging limply. The stiffening and lifting up of the shoots and leaves took about another two weeks. The drooping of the young shoots of *Amherstia* is due to lack of mechanical tissues, not to absence of turgor (Czapek 1909). In *Brownea grandiceps* Büsgen (1903, p. 438) found that there was an increase of dry as well as of fresh mass during leaf expansion, indicating some true growth as well as the enlargement of cells already formed.

The limpness and deficiency in chlorophyll common in the newly expanded leaves of these trees show that in them the sequence of events in leaf development is different from that usually found in temperate plants. In the latter the various processes involved in leaf maturation proceed more or less in parallel and are completed quite soon after the opening of the buds, as Maksymowych (1973) showed in *Xanthium*. But in many tropical trees the lignification of the vascular tissues and the building up of chlorophyll and photosynthetic activity have hardly begun when the leaves have already almost finished their expansion. It can often be observed that the development of chlorophyll in young leaves

spreads from the midrib and side veins over the rest of the lamina: in those with red or purple colouring the anthocyanin disappears simultaneously from the veins outwards.

One rain-forest species in which the physiological changes during leaf maturation have been studied in detail is cacao (*Theobroma cacao*) in which the young leaves are limp and pale green or tinged with red. Baker & Hardwick (1973) found that chlorophyll builds up very slowly during leaf expansion: the linear phase of chlorophyll synthesis does not begin until expansion has been completed. The curves for total chlorophyll and photosynthesis per unit leaf area reach their maxima about fifteen days after the leaf has reached full size. In the cultivated mango (*Mangifera indica*), probably originally a native of monsoon forest rather than true rain forest, the young leaves are reddish brown and hang limply as in cacao. The metabolic and other changes during their development are described by Taylor (1970–71). The young leaves are at first erect and up to day 7 they grow by cell division, after this by cell enlargement. During this stage they become limp and hang down. As Czapek (1909) found in *Amherstia*, this is not because of loss of turgor, but coincides with minimum dry mass per unit area and is due to lack of rigid cell wall material. Photosynthesis is still absent, or at a very low-level, not becoming active until day 28. The anthocyanins reach a maximum about day 14 and disappear by day 28. Kursar & Coley (1992) have given details of the delayed greening and associated biochemical changes in the young leaves of four species of small trees in Barro Colorado Island.

Leaf age. Precise information about the life-span of the leaves of rain-forest trees is scanty. In the lower montane rain forest of El Verde, Puerto Rico, the leaves of various species were found to reach ages of eighteen months to three years or more (Odum 1970b). From what is known about flushing and leaf-fall in lowland rain-forest evergreen trees (Chapter 9), it may be surmised that the life-span of their leaves falls within similar limits. The leaves of small trees and shrubs in the C and D storeys probably live longer than those of the emergent trees. Bentley (1979) tagged the leaves of many 'under-storey' trees at La Selva (Costa Rica) and found that 39% were still attached to the plant after two years. The leaves of small palms seem to be particularly long-lived: those of *Iguanura wallichiana* in Malaya live more than six years (Kiew 1982) and those of *Podococcus barteri* in Cameroon have a 'working life' of about five years (Bullock 1980). Leaves of deciduous and leaf-exchanging trees (see Chapter 9), and probably

Fig. 4.16 Developing leaves of *Brownea* sp., Aburi Botanical Garden, Ghana (1968).

the soft-textured leaves of some early seral trees, probably live for less than a year.

In most tropical trees the leaves cease to grow after expansion is complete. The apex is normally late in becoming mature but in some tropical Meliaceae it remains meristematic indefinitely and the leaves have unlimited growth (Sinia 1938, Skutch 1946). The most remarkable examples are species of *Chisocheton*, small rain-forest trees or treelets with compound leaves 1–2 m long. They have a bud (pseudogemmula) at the apex from which additional leaflets arise and the rachis has seasonal increments of growth. In *C. pentandrus* subsp. *paucijugus* eight pairs of leaflets are formed over eight years, the older ones falling off when they are four to five seasons old (Corner 1964, Fig. 42, p. 127). In two extraordinary New Guinea species, *C. pohlianus* and *C. tenuis*, the leaves are not only 'ever-growing', but bear epiphyllous inflorescences, so that the leaf closely resembles a shoot (Mabberley 1979). The unlimited leaf growth of these Meliaceae is an extreme example of the prolonged leaf development characteristic of many rain-forest trees. Corner (1954a) regards it as a primitive trait that has been lost in most modern angiosperms.

Leaf size and shape. In the tropical rain forest, as in other plant communities, there is a wide range in the shape and dimensions of the leaves, but the dominance of what Warming (1909) termed the 'laurel type' belonging to the mesophyll size-class of Raunkiaer (1934) is very striking. In trees and shrubs of almost every dicotyledonous family the leaves are oblong-lanceolate to elliptical in shape with entire or finely serrate margins; often they have a long and distinct acumen, which is sometimes exaggerated into a pronounced drip-tip. The colour of the leaves when mature is a deep sombre green and the upper surface is glabrous, often shiny; any indumentum is usually confined to the lower surface. The texture is usually somewhat leathery but, although the leaves of rain-forest trees are often classified as sclerophyllous,[3] they are generally larger also thinner and less hard than the typical sclerophylls of Mediterranean climates (Warming's 'myrtle type'). In compound leaves the individual leaflets tend to conform to the size and shape of the prevailing type of simple leaf. Among the few conspicuous exceptions are the bipinnate leaves of *Parkia*, *Pentaclethra* and other Mimosaceae, which are common rain-forest trees especially in Africa and America; in these the individual leaflets are of nanophyll or leptophyll dimensions but so close

together as almost to appear to be a single lamina. The uniformity of the foliage is clearly shown by studying representative samples of leaves (Figs. 4.17 and 4.18).

Analysis of leaf sizes by Raunkiaer's (1934) classification shows that in primary lowland mixed rain forest

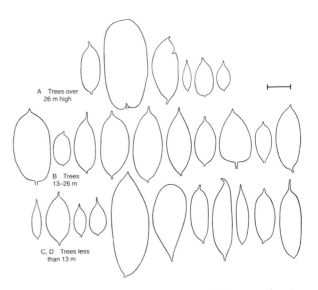

Fig. 4.17 Leaves of trees in primary mixed dipterocarp forest, Gunung Dulit, Sarawak. Each is a typical mature leaf from a different species of tree. Scale bar, 6 cm.

Fig. 4.18 Leaves of rain-forest plants from two localities in southern Nigeria. Each is a typical mature leaf from a different species. Only trees, 'shrubs' and ground herbs are shown. Letters indicate storeys; scale bar, 6 cm.

---

[3] On sclerophylly and the anatomy of the leaves of rain-forest trees, see Grubb (1974).

at least 70% of the tree species have leaves in the mesophyll size-class[4] (area 2025–18 225 mm² (Fig. 4.19). Webb (1959) proposed restricting the mesophyll class to leaves of 4500–18 225 mm², separating the smaller leaves in the range 2025–4500 mm² as notophylls. On this basis in the 'mesophyll vine forest' of Australia (p. 451) (approximately equivalent to lowland tropical rain forest) 50–70% of the species and 60–70% of the individuals are mesophylls and 30–50% of the species and 30–40% of the individuals notophylls. In Ghana, Hall & Swaine (1981) found that in all the 'closed canopy' forest types (p. 410) species with notophylls were slightly more numerous than those with mesophylls (in Webb's sense). On the data at present available it seems that in tropical rain forests it may be about equal but generally the proportion of mesophylls and notophylls varies; in the lowland forest of New Guinea there are five times as many mesophyll as notophyll species (Grubb 1974). The next most frequent size-class is the macrophylls. The percentage of microphylls is small and the nanophyll and leptophyll classes are usually scarce and often quite absent. Plants with very large leaves, such as some palms and Scitamineae, the megaphylls of Raunkiaer's classification, though often regarded as especially characteristic of the humid tropics, are comparatively scarce in the mature phase of lowland high forest but often represented in gaps and secondary communities as well as in swamp forest and lower montane rain forest. Some typical size-class spectra are shown in Fig. 4.19. Further data are given by Brown (1919), Cain et al. (1956), Cain & Castro (1959) and others.

The majority of rain-forest trees and shrubs (apart from palms) have simple leaves; in the figures given above, the leaflets of compound leaves have been taken as units. The percentage of species with compound leaves varies, though not within very wide limits. In the virgin dipterocarp forest at Mt Makiling (Philippines) Brown (1919) found that 19% of the species and 22% of the individual woody plants more than one or two metres high had compound leaves. At Castanhal near Belém in Amazonia 30.7% of the trees in a sample plot had pinnate or otherwise divided leaves. At another locality 46% of the tree species had compound leaves; the percentage increased from twenty-seven among trees 8–30 m high to thirty-seven in those over 30 m (Cain et al. 1956).

The proportion of species with entire-margined leaves is much greater in tropical rain forest than in temperate forests. Bailey & Sinnott (1916) found that in a sample

from Amazonia 90% of the trees and 87% of the 'shrubs' had leaves with entire margins, compared with 10 and 14%, respectively, in broad-leaved forest in the United States. In a large sample of leaves from two rain-forest localities in Nigeria 80% of the species had entire leaves or leaflets, and of those with non-entire margins only a few (4% of the whole sample) were deeply lobed or incised. At Mt Makiling (Brown 1919) 76% of the species and 86% of the individual trees had leaves with entire margins. Similar figures would be obtained for any mature rain-forest community. No wonder that in the Amazon forests Spruce (1908, vol. 1, p. 37) found it a 'rare treat' to see a deeply divided leaf.

It is remarkable how the rain-forest environment seems, as it were, to mould the foliage of all species coming under its influence to a particular form. This is well seen when the leaves of tropical species are compared with those of near relatives in other climates. For example, in the oaks (Quercus), a genus with many species in both tropical Asia and America as well as the north-temperate zone, the leaves vary greatly in size, shape and texture. The lowland rain-forest species have entire, leathery, often longly acuminate, leaves of meso-

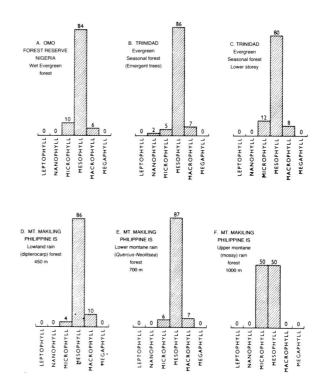

Fig. 4.19 Leaf sizes of rain-forest trees and shrubs according to Raunkiaer's classification. The columns represent the percentage of species in each class. B and C from Beard (1946b); D–F from Brown (1919).

---

[4] A sufficiently accurate estimate of leaf area is given by 2/3 the length of the lamina (excluding the drip-tip, if present) × the width (Cain & Castro 1959).

phyll or notophyll size (Brenner 1902). Another example is *Acer laurinum*, a Malayan rain-forest tree with entire, finely acuminate leaves, very different from those of most temperate species of the genus.

Although the leaves of rain-forest trees are predominantly of the mesophyll and notophyll size-classes, there are differences in leaf shape, particularly in the leaf apex, in different strata. The small trees and shrubs of the C and D storeys nearly all have leaves with a long acumen, which is sometimes prolonged into a distinct drip-tip. In the B storey the leaves have a shorter acumen or none at all, while the leaves of the tall emergent trees of the A storey generally have no acumen or at most a short, ill-defined one; sometimes they are even emarginate. Compound leaves seem to be commoner among the larger than among the smaller trees (see Beard 1946b, Cain *et al.* 1956). Associated with the differences in shape, there is a change from the flexible, relatively thin leaves of the lower-storey trees to the stiffer and thicker foliage characteristic of

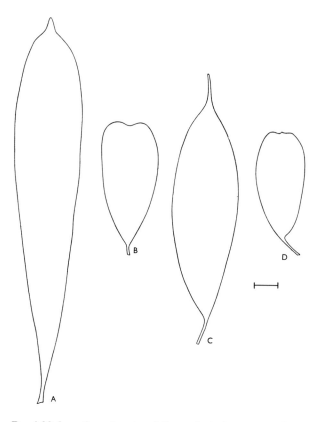

Fig. 4.20 Juvenile and mature foliage of rain-forest trees. A Juvenile leaf of *Lophira alata* from sapling 4 m high. B Leaf from mature tree (over 30 m high) of same species. C Leaf from seedling of *Catostemma fragrans*. D Leaf from mature tree of same species. A and B from Omo Forest Reserve, Nigeria; C and D from Guyana. Scale bar, 3 cm.

the emergents. No quantitative data on leaf texture are available but the change clearly indicates a trend towards a more sclerophyllous type of leaf structure and a difference in the ratio of leaf mass to leaf area (specific leaf area of Evans 1972). Both the shape of the leaf apex and the leaf structure seem to be correlated with the microclimatic differences between the well-illuminated, relatively exposed euphotic zone and the more sheltered and shady undergrowth (Chapter 8).

Differences between the leaves of juvenile and mature individuals, which are very marked in some species (Fig. 4.20), contribute to the differences between the foliage of the upper and lower storeys. The juvenile leaves are often several times larger than the adult leaves and so unlike them in shape and texture that they might easily be supposed to belong to another species, though the differences are not as extreme in trees as in some lianes (pp. 128–9). In many Dipterocarpaceae the young leaves are much larger and more longly acuminate than the adult leaves; thus in *Dryobalanops rappa* seedlings leaves are up to 114 mm long and 31 mm wide with a drip-tip up to 28 mm long, whereas mature leaves are 64–89 mm long and 25–41 mm wide with an acumen of only 15 mm (Meijer & Wood 1964). Juvenile leaves usually have a drip-tip even when the adult type of leaf is obtuse or emarginate (Fig. 4.20). Sometimes the juvenile leaves are compound or divided and have non-entire margins while the mature leaves are simple and entire, e.g. *Artocarpus elasticus*. In *Catostemma commune* (Guyana) the juvenile leaves are palmate with three or four leaflets but the adult leaves have only one.

The size-class spectrum and other leaf characteristics of rain-forest trees vary with climate, soil and other site conditions and can be used in classifying forest types as Webb (1959) showed in Australia and Hall & Swaine (1981) in Ghana (see below). Under similar climatic and soil conditions leaf characteristics probably vary with the stage of succession; general observations suggest that the range of leaf sizes is greater and the uniformity of leaf shapes much less in young secondary forest than in climax high forest (see Janzen 1975, plate 1).

In primary forest macrophyll and megaphyll leaves seem to be commoner in swamp, and other humid sites than in drier habitats. Microphylls and other small leaves, on the other hand, are usually better represented on sandy soils and on exposed ridges. Ashton (1964) compared the leaf-size spectrum on sites differing in soil and topography in lowland mixed dipterocarp forest in Brunei (Borneo). Using an ordination method he found that the total basal area of macrophyll trees decreased from valleys to ridges while that of micro-phylls increased. Mesophylls and notophylls were together much the most abundant size-classes, except on one site.

The foliage of heath forests on white sands (podzols) in Borneo and the wallaba and other similar forests on white sand in South America (pp. 321ff.) is generally smaller and more sclerophyllous than that of mixed forests on kaolisols (latosols).

Brünig (1974) compared leaf-size spectra for sites in various types of heath (kerangas) forest with one for mixed dipterocarp forest estimated from Ashton's data, using a leaf-area index, $I=\frac{1}{2}(G+Sp)$, where G is the percentage of trees by basal area in each leaf-size class and Sp the percentage by species. The comparison is complicated by the occurrence of gregarious species in the heath forest, which tend to overweigh certain classes, e.g. *Agathis borneensis* (microphyll) and *Casuarina nobilis* (leptophyll). The results indicate, however, that in heath forest notophylls tend to be the largest class and mesophylls (*sensu* Webb 1959) are less well represented than in mixed dipterocarp forest. A considerable proportion of microphylls is characteristic of heath forest; the two smallest size-classes, leptophylls and nanophylls, which are usually absent in mixed dipterocarp forest, are often well represented in heath forest. The sclerophylly of heath forests and its relation to nutrient deficiency and water availability in the soil is discussed in Chapter 12.

In their ordination study of 'closed canopy forest' in Ghana, Hall & Swaine (1976, 1981) showed that leaf size was related to climate. As mentioned above, notophylls were the largest class in all forest types, but along a gradient of decreasing environmental moisture microphylls increased from about 10 to 40% of the species. Macrophylls were confined to the wetter forest types. There was also a decrease in the percentage of acuminate leaves from the wetter to the drier types.

The changes in leaf-size spectra and other leaf characteristics with increasing altitude are considered in Chapter 17.

Leaves of rheophytes. In the tropics trees and other plants growing by fast-flowing streams, or in their beds (rheophytes) often have linear to lanceolate leaves, much narrower than those of related species living in the interior of the forest. This is particularly marked in Malesian species of *Eugenia*, *Garcinia*, *Homonoia*, *Ixora*, *Phoebe* and other genera in various unrelated families; in those with compound leaves, such as *Dysoxylum*, the leaflets are narrower than in species from other habitats. In *Dipterocarpus oblongifolius*, which charac-

teristically grows by fast streams, the leaves in the juvenile stage are narrowly lanceolate and very unlike those of other species of the genus (Fig. 4.21B). Some herbaceous plants, e.g. the African orchid *Ancistrorhynchus clandestinus*[5], including ferns, show the same feature. Beccari (1904) first noticed this phenomenon in Sarawak, and termed it stenophyllism. Although rheophytes are particularly numerous and varied in western Malesia, they are also found in other parts of the tropics and in temperate regions. In West Africa examples include *Deinbollia saligna* (Fig. 4.21a), *Eugenia* sp. and *Psychotria* spp., and in tropical America, *Salix humboldtiana* and *Tessaria integrifolia*.

Van Steenis (1981, 1987), in his monograph on rheophytes of the world, discusses the significance of their narrow leaves. He provides experimental evidence supporting his conclusion that they are an adaptation that reduces resistance to fast-flowing water. Other characteristics of rheophytes, such as flexible stems and petioles, have a similar effect.

Drip-tips. The exaggerated acumen or drip-tip characteristic of the leaves of many plants in the undergrowth of rain forests has attracted attention mainly because of its supposed value as an adaptation to damp, rainy conditions. Drip-tips occur in both herbaceous and woody plants and in ferns as well as monocotyledons and dicotyledons. Drip-tips like those of modern plants can also be seen on some fossil leaves of Tertiary age. Outside tropical lowland and lower montane rain forests well-developed drip-tips are rare (see map in Ellenberg 1985). In both montane and seasonally dry tropical forests they are absent or less well developed. In Sri Lanka Holtermann (1907) found that the majority of plants in the wet forests had longly acuminate leaves, but not those of the dry zone and the mountain tops. This difference was shown among species of the same genus and sometimes within a single species; for example, in *Memecylon varians* specimens from wet localities had drip-tips 2–3 cm long, but those from exposed peaks had none. As was mentioned earlier (p. 95), drip-tips are commoner and more distinct in the lower than the upper strata of lowland rain forest; in tall emergent trees they are usually found only on the juvenile leaves. All the evidence points to a close connection between the development of drip-tips and the shaded humid environment of the rain-forest undergrowth.

---

[5] In this species in addition to the form growing by rivers, which has narrow leaves up to 70 cm long, there is a forest form (ecotype?) with shorter leaves (W.W. Sanford).

A

B

Fig. 4.21 Leaves of rheophytes (compare with Figs. 4.17 and 4.18). A *Deinbollia saligna*. River bed, Kwa Falls, near Calabar, Nigeria (1978). B Seedlings of *Dipterocarpus oblongifolius* on river bank, Brunei (P.S. Ashton).

Drip-tips vary in length. In some species they may be 4 cm or more long (Fig. 4.22) but usually they are shorter. In the leaves of dicotyledons generally, the apical part often develops somewhat in advance of the rest of the lamina. In leaves with drip-tips the apical growth continues longer than usual, although not as long as in the 'ever-growing' leaves considered earlier (p. 93). Holtermann (1907) found that in many plants in Sri Lanka a drip-tip was present in the young leaf but it withered away later, the mature leaf becoming obtuse or emarginate at the apex: this appeared to be due to shortage of water, the main vein being the only vascular supply to the leaf apex.

Jungner (1891) was the first to suggest a possible adaptive value of drip-tips for rapidly draining the leaf surface. On Cameroon Mountain in West Africa, where for most of the year rainfall is extremely heavy (p. 446), he was impressed by the frequency of leaves with drip-tips in the forest undergrowth. He claimed that leaves with a drip-tip dried more quickly after rain

and were less often overgrown with epiphyllous algae, bryophytes, etc. (see section 6.4.14, p. 149) than those without one. He thought that the presence of epiphyllae would interfere with carbon assimilation to such an extent as to be a serious handicap to the plant and that, by allowing water to run off quickly, drip-tips would help to keep the leaf surface cleansed from the propagules of epiphyllae as well as from the eggs and larvae of insects and soluble substances that might favour their growth. Later Stahl (1893) showed experimentally in Java that the removal of drip-tips greatly increased the time taken by the leaves to dry, but he was doubtful of any correlation between drip-tips and the presence or absence of epiphyllae. The chief value of the drip-tips, he suggested, was to prevent the lingering of a surface film of water, which would lower the leaf temperature and depress the rate of transpiration. Stahl believed that a low rate of transpiration would be disadvantageous to the plant by checking the uptake of mineral salts.

The views of Jungner and Stahl were later criticized

by Shreve (1914) and others. Working in the under-growth of 'montane' (probably lower montane) rain forest of Jamaica, mainly with herbaceous plants, Shreve was unable to show that the removal of drip-tips had much effect on the rate of drying. He found that epiphyllae were abundant on leaves of many types, including those with drip-tips; their occurrence seemed to depend chiefly on the humidity of the atmosphere surrounding the plant. The lowering of the temperature of the leaves by a surface film of water was too slight to affect their rate of transpiration. A water film, however, decreased the water uptake of the leaves from the stem, probably partly by stopping cuticular transpiration which was 'slightly more than half the water loss in leaves of rain-forest hygrophytes', and partly because of actual absorption of water through the lightly cuticularized epidermis. McLean (1919, p. 31) came to similar conclusions from observations at Rio de Janeiro.

The abundance of epiphyllae in rain-forest undergrowth depends on various factors (p. 152); Shreve's (1914b) observation that their occurrence is not related to the shape of the leaf apex can be easily verified. There is also little evidence that epiphyllae seriously damage the host plant. They begin to colonize leaves only when the latter are several months old, and usually do not cover a large part of the surface until the leaf is senescent and near the end of its working life.

More recent work on drip-tips has dealt mainly with their effect on the rate of drying of leaf surfaces and its supposed influence on leaf functions. Gessner (1956a) and others have further investigated the effect of artificially removing drip-tips and have now attempted to show that they are of adaptive value to plants in very wet environments, but some authors have now rejected all teleological interpretations of their significance.

Seybold (1957) gives a number of reasons why drip-tips probably have little functional importance. For example, some leaves with drip-tips have unwettable

leaf surfaces, some plants have drip-tips on some leaves and not on others, and some plants with long drip-tips, e.g. *Ficus religiosa*, grow in dry environments. He also criticizes earlier experiments comparing the rate of drying of leaves with and without drip-tips and claims that his own experiments showed hardly any difference. Seybold found that water films had very little effect on light absorption by leaves and also points out that because most leaves with drip-tips have stomata only on the lower surface, which is generally unwettable, water films cannot affect carbon dioxide absorption.

Ellenberg (1985) argues against drip-tips giving possible selective advantages, largely because of their presence or absence in different types of vegetation and at different altitudes in South America. He found that, though common in lowland tropical rain forests, they are absent in the temperate rain forest of southern Chile and in 'cloud forests' at high altitudes in the Andes, both of which are continuously wet but have much lower temperatures. Drip-tips are thus frequent only in environments that are hot as well as wet. Ellenberg also criticizes Gessner's (1956a) results. His own experiments show that whether the leaf is upright, horizontal or inclined affects the rate of drainage of water more than whether it has a drip-tip and that when this is taken into account there is very little difference in the rate of drainage between comparable leaves with different shapes of leaf apex.

Ellenberg (1985) concludes that the frequency of drip-tips in hot, wet climates (defined as having more than six 'wet' months and a mean air temperature $>15\ °C$) has no adaptive significance but depends on the conditions under which the young leaves develop. In rain-forest undergrowth most leaves have little bud protection and in the warm moist atmosphere the apex of the leaf usually grows rapidly in length, becoming fully differentiated before the rest of the lamina is fully expanded. The effect of the environment on the leaf apex can be shown experimentally in *Coffea arabica*, in which the leaves have a drip-tip when grown in a hot humid atmosphere but not in cooler and drier conditions. Ellenberg's interpretation would account for the presence of drip-tips in juvenile plants of emergent trees and their absence in mature individuals, as mentioned above.

Leaf joints. Swollen petiolar joints are a very common feature of the petiole in leaves of tropical trees: according to Funke (1929, 1931) they are even more characteristic of tropical leaves than their size, shape and other well-known attributes. Usually the joint is at the base of the petiole, but in some woody species there are one or more joints along its length. In many compound

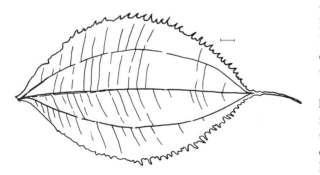

Fig. 4.22 Leaf of small tree, *Miconia schwackei*, to show drip-tip. Ponte Negra, Manaus, Brazil (1968). Scale bar, 1 cm.

leaves each leaflet is jointed to the rachis. Similar joints are found on the leaves of ferns and other herbaceous plants, e.g. those of tall herbs such as *Leea, Strobilanthes* and some Marantaceae. Though common in tropical plants of many families, leaf joints are much less common in temperate plants.

Leaf joints vary in structure. Funke (1929, 1931) describes fifteen different types. The functions of these types have not been adequately studied. In *Albizia* and some other Leguminosae the joints are true pulvini and bring about sleep movements by a turgor mechanism. The great majority seem to be pulvinoids or abscission joints, or to combine both functions (van der Pijl 1952). Pulvinoids remain meristematic and bring about movements of the leaf or leaflet by differential growth. Competition for light is so severe in tropical forests that mechanisms for adjusting leaf positions, especially of long-lived leaves, are no doubt very important, but it is not clear why they are so much less common in temperate forests.

Leaf characteristics in relation to the rain-forest environment. The remarkable uniformity, especially in size, among the leaves of rain-forest trees of the most diverse systematic affinities cannot be due to chance. The fact that the leaflets of large compound leaves such as those of the Leguminosae often mimic simple leaves in shape and dimensions, as well as the slight but distinct differences in leaf characteristics between leaves in different strata, shows clearly that the prevailing uniformity is a response to the environment. Fedorov (1966) suggested that the leaf characteristics of rain-forest trees have no distinct adaptive significance and are the result of genetic drift. This is difficult to disprove but it would seem much more probable that the prevailing uniformity in foliage is adaptive and due to natural selection.

To discuss in any detail why entire-margined mesophyll and motophyll leaves are so strongly predominant among the trees, shrubs and lianes of the rain forest would lead far into developmental physiology and experimental morphology. The crude teleological 'explanations' of the past century are no longer acceptable but there is little as yet to put in their place. It is evident that the leaves of rain-forest plants must be adapted to the humid tropical climate and to the microclimate of the strata in which they develop and function: their size and shape will be governed by a balance between many factors, including the temperature and humidity of the environment and the radiation they absorb, as well as less predictable factors such as resistance to the effects of violent rain, wind and herbivorous animals. These factors are probably not of equal import-

ance and it seems likely *a priori* that those most directly affecting the water balance of the leaf and its photosynthetic functions are the most significant.

Bailey & Sinnott (1916) speculated on why such a large proportion of the leaves in humid tropical climates have entire margins and pointed out that even in the wettest tropical rain forests the leaves of the trees often have a 'xerophilous' (xeromorphic) structure. They attribute this to the frequency of short periods of water stress when the leaves are exposed to strong sunlight (see p. 94). They suggest that during such periods leaves with entire margins would be less liable to damage than those with lobes or teeth. Bews (1927) also believed that the leaves of rain-forest trees were xeromorphic. He related this not to the climate but to the low water conductivity of the wood, which he said was characteristic of tropical and subtropical evergreen trees. Xeromorphic features are much more pronounced in heath forests on white sands than in rain forests on kaolisols. The significance of this is discussed in Chapter 12.

More recent work seems to show that the size as well as the structure of the leaves is primarily determined by the balance of water loss and water uptake. In a discussion of leaf size in relation to climate, Parkhurst & Loucks (1972, p. 534b) conclude that 'natural selection leads to organisms having a combination of form and function optimal for growth and reproduction in the environment in which they live'. Applying this specifically to leaves, they make the assumption that, as a first approximation, the optimal leaf size for a given

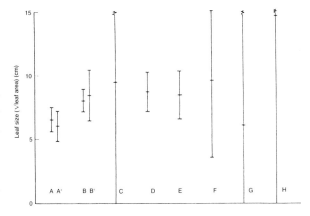

Fig. 4.23 Leaf sizes of plants in various synusiae in Amazonian rain forest (data of Cain *et al.* 1956). From Parkhurst & Loucks (1972). A Trees taller than 38 m; B trees 8–30 m; C woody plants 2–8 m; D woody plants 0.25 – 2 m; E lianes; F chamaephyte herbs; G hemicryptophyte herbs; H geophyte herbs; A' as in A, but leaf widths weighted by species importance; B' as in B, but leaf widths weighted by species importance.

environment is the size giving the maximum efficiency in water use (defined as grams of carbon dioxide assimilated per gram of water lost). By means of a complex calculation involving seven independent variables, four of them climatic (convection coefficient, temperature, relative humidity and absorbed radiation), and three dependent on leaf structure (stomatal resistance, mesophyll resistance and distribution of stomata between the two leaf surfaces) Parkhurst & Loucks developed a model from which several predictions can be made: one is that selection should favour large leaf sizes in environments where temperature is high and radiation low, as in the undergrowth of tropical forests. They compared the predicted leaf sizes for the undergrowth of tropical rain forests with the data of Brown (1919), Cain *et al.* (1956) and others summarized earlier in this chapter. There is good agreement, not only in the average leaf size, but also in the leaf sizes of plants in different height-classes or strata (Fig. 4.23). The model also predicts the decrease in leaf size with altitude which is found when lowland rain forest is compared with montane forest (Chapter 17).

As might be expected, the fit between the predicted and actual sizes is not always satisfactory, indicating that other factors than those included in the calculation may sometimes be important. It is noteworthy that in both tropical rain forests and temperate forests the water-use model does not apply as well to ground herbs as to woody plants. The reason for this discrepancy is not known; there is possibly some physiological difference between herbaceous and woody plants that has been overlooked. Givnish (1978b, 1984) gives a useful discussion of the leaf characteristics of tropical forest trees in relation to environmental factors.

Chapter 5

Trees and shrubs: II. Reproductive biology

TROPICAL RAIN-FOREST TREES belong to a very large number of different families and genera; their flowers and fruits are correspondingly varied and much more diverse than those of forest trees in temperate climates. In their reproductive biology they show traits that are clearly related to the relatively non-seasonal climate in which they live. The connection is largely indirect, depending on the year-round availability of insects, birds and mammals able to act as pollinators and seed dispersers. This chapter is mainly concerned with the breeding systems, pollination and dispersal of rain-forest trees and shrubs, aspects of their reproductive biology that are closely linked to the rain forest's characteristic patterns of structure and species composition. The phenology of flowering and fruiting, which also reflects the permanently favourable climatic environment, is dealt with in Chapter 9. Various aspects of the reproductive ecology of tropical plants have been reviewed in a volume edited by Bawa & Hadley (1990). A modern account of the evolutionary biology of tropical flowers is given by Endress (1994).

5.1 Vegetative reproduction

Most rain-forest trees reproduce mainly or exclusively by seed: in them vegetative reproduction is much less important than in trees of some other tropical ecosystems such as savanna woodlands. When vegetative multiplication occurs, it is usually by means of buds on roots running more or less horizontally near the surface of the ground, or from lateral buds on fallen trunks, which develop into upright stems. Reproduction by the latter method is common in *Dimorphandra conjugata* and the swamp tree *Cyrilla racemiflora* in Guyana (T.A.W. Davis), in the African *Grewia coriacea* and in many other species. Vegetative reproduction is frequent in tree-like monocotyledons such as Pandanaceae and some Scitamineae. In palms that produce buds or stolons at the base of the primary stem, the secondary stems often become independent plants, e.g. in the West African *Podococcus barteri*. (Bullock 1980). Vegetative reproduction is relatively common in trees of heath forests (Chapter 12), secondary forests (Chapter 18) and some types of swamp forest. In all of these its frequency is probably related to difficulties in seedling establishment.

The extraordinary method of vegetative multiplication of the African shrubs *Anthonotha macrophylla* and *Scaphopetalum*, which form dense thickets in gaps and secondary forests, is like that of some lianes and does not occur in large trees of mature forest. The formation of 'clumps' by *Dicymbe* and *Trigonobalanus* (p. 321) is a method of crown enlargement rather than reproduction, as the secondary trunks do not usually become independent trees.

5.2 Reproduction by seed

5.2.1 Breeding systems

In recent years the great interest in the species diversity of rain-forest trees has directed attention to their breeding systems, but information is available for relatively few species and only tentative conclusions are possible. It should be noted that, although data on species diversity in tropical trees are plentiful, very little is known about genetic diversity below the species level. The experience of foresters with trees of economic importance suggests that at least in common species there is much intra-specific genetic diversity between individuals and populations. This has been very little studied experimentally, but Gan *et al.* (1981) have investigated intra-specific diversity of isozymes by electrophoretic analysis in populations of three widely distributed Malayan rain-forest trees, the dipterocarps *Shorea leprosula* and *S. ovalis*, both tall emergents, and *Xerospermum noronhaianum (intermedium)* (Sapindaceae), an 'understorey' (B or C storey) species. Leaves from a number of mature trees and numerous seedlings of *S. leprosula*, a smaller number of seedlings of *S. ovalis* and 68 mature trees of *X. intermedium* were tested for various enzyme systems. In *S. leprosula* a high level of gene polymorphism was indicated and in *X. intermedium*, although the number of individuals and enzyme systems tested was smaller, there also seemed to be considerable polymorphism. These results suggest that both species are outbreeding. *S. leprosula* is known to be self-incompatible and *X. noronhaianum*, the pollination of which has since been carefully investigated by Appanah (1982),[1] is an androdioecious species with delayed self-compatability (p. 109). The authors suggest that in these species outcrossing, possibly combined with short-range pollination and seed dispersal, may lead to physical isolation of populations, a situation that might lead to speciation. In *Shorea ovalis*, which is probably an autotetraploid and is known to be self-compatible, only a low level of isozyme variation was found.

The work summarized above, though concerned with only three species, is of some interest because Fedorov (1966) had earlier suggested that the great species diversity of tropical rain-forest trees was dependent on widespread self-pollination and inbreeding due to the very scattered populations of most species (see Chapter 11). He believed that the great species diversity had resulted mainly from random genetic drift. Ashton (1969, 1977) rejected this view: he believed that the majority of large rain-forest trees were outbreeders and argued that selection and ever-increasing specialization were the processes chiefly responsible for the high species diversity.

Mechanisms, some of them very complex, leading to obligate or facultative cross-pollination abound among rain-forest trees. The commonest obligate mechanisms are dioecy (found in *Cecropia*, *Diospyros* and many other widespread genera) and dichogamy. Ashton (1969) found that in an area of approximately 9 ha of lowland

dipterocarp forest in Sarawak 26% of the tree species were dioecious (compared with 5% in the world flora and about 2% in the British flora); a further 14% had unisexual or dichogamous flowers. Dividing the trees into girth-classes showed that in the lower storeys of the forest, where spatial separation of individuals is lowest, the percentage of dioecious individuals is considerably greater than in the upper storeys. In evergreen seasonal forest in Nigeria 40% of the tree species are dioecious (Bawa & Opler 1975, from the data of Jones 1955). In the semi-evergreen forest of Barro Colorado Island (Panama) only 9% of the whole flora is dioecious, but among the trees and shrubs the percentage is much higher (21%) (Croat 1978, 1979).

Sometimes tree populations include both unisexual and hermaphrodite individuals. In the Malayan tree *Xerospermum noronhaianum*, referred to earlier, Appanah (1982) found that the population in Pasoh forest consisted of male and hermaphrodite individuals. The flowers of the latter last for three days and are at first functionally female. Both types of flower are visited by an array of unspecialized insects, including trigonid bees and butterflies, which can bring about cross-pollination but which also visit many other species. The anthers of the hermaphrodite flowers do not dehisce but on the second or third day their walls break down and self-pollination can occur; experiments showed that this can lead to fruit formation. An interesting feature is that nectar is produced in pulses in both kinds of flower; as the pulses tend to alternate between the male and hermaphrodite flowers, this may promote crossing, but when visitors were excluded by bagging nearly as many fruits were set as when there was open pollination.

Dichogamy is found in a very large number of rain-forest trees as well as in shrubs and lianes. Protandry is common in Leguminosae, Rubiaceae and many other families. A striking example is *Parkia* (Mimosoideae) in which the large globular masses of small bat-pollinated flowers are all in the staminate condition on one night and a night or two later are all pistillate (Baker & Harris 1957). Protogyny seems to be less common, although the peculiar form of protogyny characteristic of the very large and diverse genus *Ficus* is well known (p. 105).

The literature (up to 1981) on incompatibility systems in tropical plants was reviewed by Baker *et al.* (1983). Few species of humid tropical forests have so far been investigated and much further research is required. It may be noted here that some self-incompatible rain-forest species are heteromorphic (heterostyled), but as far as is known at present most of these are shrubs, e.g. some Rubiaceae, or herbs. Homomorphic self-incompatibility is much commoner and has been investi-

---

[1] See also Ha *et al.* (1988a).

gated in some detail in *Theobroma cacao* (cacao). In this species both self-compatible and self-incompatible forms exist in the wild as well as in cultivation (Purseglove 1968). In the rain forests on the eastern slopes of the Andes, the presumed centre of origin of the species, wild populations consist only of self-incompatible forms, but the greater the distance from this centre, the larger the proportion of self-compatible forms. Purseglove (1968) suggests that there is some selection against self-compatibility but it may have an advantage when the species is invading new areas (or grown as a crop plant). Though outbreeding, even if it is likely only between relatively near neighbours, seems to be the general rule among rain-forest trees, exceptions are probably not rare.

Apomixis has long been known in various cultivated tropical trees and is probably widespread among rain-forest species. Ashton (1977) found reason to suspect that apomixis might be widespread among climax forest trees in Malesia. Kaur *et al.* (1978, 1986) demonstrated apomixis in two Malayan Dipterocarpaceae, *Shorea agamii* and *S. ovalis* subsp. *sericea*, and inferred its occurrence in several other members of the family. In *Shorea macroptera*, which is also apomictic (Kaur *et al.* 1978), apomixis may be related to the scarcity of a suitable pollen vector. The mass flowering of various rain-forest dipterocarps takes place in sequence (pp. 252–3) and the pollen of at least some species is carried by thrips, the numbers of which build up rapidly as successive species come into flower (Appanah & Chan 1981). It may be significant that the first of these, *S. macroptera*, is apomictic, as it flowers when the thrips population may still be inadequate for reliable pollination. *Garcinia parvifolia*, a common small tree (C–B storey) at Pasoh, has been shown to be usually apomictic and to produce fruit parthenocarpically like the cultivated *G. mangostana* (Ha *et al.* 1988b). Richards (1990) has recorded facultative agamospermy in several other Malayan species of this genus.

It is remarkable that even putative natural inter-specific hybrids among rain-forest trees have been recorded very rarely. This can hardly be due entirely to lack of careful observations. Ashton (1969) says that the only well-authenticated records of hybrids in the Dipterocarpaceae are from the 'species-poor seasonal and Dry Dipterocarp forests of the Philippines, Burma and Thailand, and not from the rain forests of Borneo and the Malay Peninsula'. He concludes that there must be a strong selection pressure against hybridization and that crossing between different species is likely to be infrequent because they often do not flower simultaneously.

5.2.2. Pollination

In rain-forest trees that are not self-pollinating or apomictic, exchange of genes between individuals and populations is brought about by cross-pollination, usually by animals. Pollination mechanisms and the behaviour of different types of pollinator therefore need to be considered, particularly in relation to two of the rain forest's most important characteristics, its great species diversity and the prevalence of very thinly scattered species populations.

It should be realised, however, that pollination has been carefully studied only in a very small proportion of rain-forest species. Discussions on the pollination ecology of rain-forest plants often depend on assumptions based on syndromes of floral characters and these can be misleading, as van Steenis (1972, 1980) has rightly emphasized. Orchids, for example, have elaborate mechanisms that would seem to ensure cross-pollination, yet many species are regularly self-pollinated, as in numerous tropical genera (J.J. Smith, quoted by van Steenis 1972). Similarly, the flowers of Leguminosae appear to be beautifully adapted for cross-pollination, but in many species selfing takes place before the flower buds open.

A further difficulty is that many flowers attract insects, birds and other animals that are not necessarily effective pollinators: they may collect no pollen or steal it without transferring it to other flowers. An example of this is the durian (*Durio zibethinus*). The flowers open in the late afternoon and then attract insects of many different kinds. Soepadmo & Eow (1976) have found that in Malaya the effective pollinator is a nectarivorous bat, *Eonycteris spelaea*, which visits the flowers at night, together with moths and frugivorous bats. Until the pollination of many more rain-forest trees has been investigated experimentally, important problems of rain-forest ecology, such as the effective size of outbreeding populations, will remain unsolved.

Useful sources of information on the pollination of rain-forest plants are Janzen (1975), Faegri & van der Pijl (1982) and Baker *et al.* (1983).

Wind pollination. Pollination by wind, which in temperate climates is characteristic of many broad-leaved and all coniferous trees, is very uncommon in tropical rain forests. Tropical genera nearly related to wind-pollinated temperate genera are usually animal-pollinated. For example, although the Malesian, like the temperate, *Quercus* species have pendulous male catkins and are wind-pollinated, those of the allied genus *Lithocarpus* have erect male spikes and scented flowers, which attract large numbers of small bees, hoverflies and beetles (Soepadmo 1972).

Well-authenticated examples of wind-pollinated tropical trees and shrubs are in fact very few. In the lowland rain forests of the eastern tropics they probably include the conifers *Agathis, Araucaria, Dacrydium* and *Podocarpus*, but all of these are rather localized in distribution. Among angiosperms wind pollination, in addition to occurring in *Quercus*, is also probable in the Casuarinaceae, bamboos and Pandanaceae and perhaps also occurs in some palms (Uhl & Dransfield 1987), *Engelhardtia* and some Euphorbiaceae. *Trophis involucrata* (Moraceae), a dioecious C storey tree of lowland rain forest in Costa Rica, is almost certainly wind-pollinated (Bawa & Crisp 1980). The anthers dehisce explosively and the flowers lack nectar and other attractants. At La Selva (Costa Rica), male and female plants of this species are randomly distributed; their average distance apart is about 6.6 m. Wind pollination has also been suggested, though not definitely proven, in other Moraceae, e.g. *Cecropia* (Holthuijzen & Boerboom, p. 66, 1982). Wind pollination is not unlikely in this genus and in the related African *Musanga cecropioides*; it may be significant that trees of both genera often grow in nearly pure stands in which transport of pollen by wind might be effective.

The rarity of wind-pollinated plants in tropical rainforests (and closed tropical forests generally) is no doubt due to selection against it. Except where single species are dominant, as in the seral communities of *Cecropia* and *Musanga* just mentioned and in some types of primary forest (Chapter 12), conspecific individuals are widely scattered. Under such conditions pollen transport by air currents, which even in temperate forests involves a large waste of energy, would be extremely inefficient. There is also little air movement (Chapter 8). It is notable that among the herbs of the ground layer, where the turbulence of the air is minimal, animal pollination is the rule, even though many species grow in extensive patches: even some forest grasses are habitually pollinated by insects (p. 125). Although in a closed tropical forest air currents are an unreliable pollen vector, insects and other animals capable of acting as pollinators are abundant at all times of year. Even in tropical forests with severe annual droughts most birds and insects remain active through the dry season (Janzen 1975). Some observations (e.g. Appanah 1982) suggest that, especially in the lower storeys of the forest, competition for pollinators may be important. If this is so, the relatively large number of rain-forest flowers pollinated by bats and night-flying insects is understandable.

Animal pollination has probably prevailed among the angiosperms since the origin of tropical rain forests, although in Mesozoic times wind-pollinated conifers were probably relatively more abundant than now. There is much morphological evidence that wind-pollinated angiosperms are derived from animal-pollinated ancestors (Robertson 1904); according to Cockburn (in Whitmore 1974), the Malesian Fagaceae show 'evolution from insect to wind pollination currently in progress'. All the most primitive living families (Magnoliaceae, Winteraceae, etc.) are insect-pollinated, but mainly by beetles, small Diptera and thrips, rather than large specialized pollinators such as bees and Lepidoptera (Gottsberger 1970, 1974, Thien 1980).

Animal pollination. The pollinators of tropical trees include birds, bats and a few small non-flying mammals, as well as insects. The great variety in taxonomic affinity, size and behaviour of their pollinators distinguishes tropical from temperate trees (in which the vectors are, with very few exceptions, wind or insects). Throughout the rain forest, flower-visiting vertebrates play an important part. In the Neotropics the hummingbirds (Trochilidae) are the chief, but not the only, family of bird pollinators. In the Old World there are no hummingbirds and their place is taken by passerine birds of several families, including the Nectarinidae (sunbirds) and the Zosteropidae (white-eyes) (Fig. 5.1).

Bats also pollinate many rain-forest trees, shrubs and lianes (Fig. 5.2, 5.3). The nectarivorous bats of the Old World all belong to the Megachiroptera, but in the New World, where this suborder is absent, they belong to the Microchiroptera (Baker 1973). Sussman & Raven (1978) believe that there is evidence that the place in

Fig. 5.1 Sunbird (*Nectarinia ventusa*) visiting a Bottlebrush tree (*Callistemon citrinus*) in Naro Moru, Kenya. Photograph by Gregory G. Dimijian, courtesy of Oxford Scientific Films.

the forest ecosystem now occupied by the flower-visiting bats was earlier occupied by small non-flying arboreal mammals. Marsupials, lemurs and even small monkeys still play a role as pollinators of rain-forest trees in some parts of the tropics (Janson *et al.* 1981).

Insects of several different orders act as pollinators for tropical trees and shrubs. The most conspicuous, and certainly the most generally important, are bees, wasps and Lepidoptera, both diurnal and nocturnal, but beetles, Diptera and even such seemingly unlikely insects as thrips are important to some trees. The C storey tree *Theobroma cacao* is pollinated by aphids, thrips, midges and ants (Cuatrecasas 1964).

It is remarkable that, although ants are by far the most abundant of all insects in tropical rain forests, they rarely seem to pollinate flowers except by accident: it is doubtful whether they are the normal pollinators of any species, although they often visit flowers and take nectar from them. It is well known that many flowers pollinated by bees or Lepidoptera have barriers of hairs or other structural features that appear as if designed to exclude ants. Although ants eagerly collect sugar from extra-floral nectaries, they seem in general less attracted by the nectar of flowers or by pollen. This has led to the suggestion that, at least in some plants, the flower nectar may contain substances such as alkaloids or phenolic compounds, which are repellent to ants (Janzen 1977a). H.G. and I. Baker (1981) found that small amounts of alkaloids, phenolics and non-protein amino acids were present in nectar from flowers and extra-floral nectaries of various plants but seemed to have little or no deterrent effect. Experiments in Costa Rica by Guerrant & Fiedler (1981) on the flowers of a number of forest plants showed that the floral nectar of most species was palatable to ants; they concluded that structural barriers and chemical defences in the floral tissues rather than the nectar were probably the effective repellents.

Pollination mechanisms. Different kinds of animals are attracted to flowers by different attractants and seek different rewards. Thus flowers pollinated by hummingbirds differ in colour range and in structure from those attractive to bees because flower-visiting birds and insects differ in their colour vision and foraging behaviour. For many birds and insects nectar is the most important reward, but for others it is pollen or occasionally something else. Some rain-forest trees produce no nectar but their flowers are eagerly sought after by bees collecting pollen, e.g. *Cassia* and various Melastomataceae (Baker *et al.* 1983). In Java *Zosterops* sp. (a white-eye) visits the flowers of *Elaeocarpus ganitrus* to feed on the mites that infest them (Faegri & van der Pijl 1982).

In what may be a considerable number of tropical plants, oil rather than nectar is the reward and the flowers are visited by bees of special oil-collecting taxa, e.g. some thirty-six genera of Malpighiaceae (Vogel 1974) and the tribe Memecyleae of the Melastomataceae. The floral biology of one of the latter, *Mouriri parvifolia*, a small tree common in the rain and semideciduous forests of Panama, has been described in detail by S.L. & M.D. Buchmann (1981). They found that the oil secreted by glands in the flowers contained at least 13 fatty acids as well as amino acids and other substances.

Pollination mechanisms are generally adapted to the habits of the pollinators. For example, hummingbirds when visiting flowers hover and do not require a landing platform, unlike bats and Palaeotropical flower-visiting birds such as the Nectariniidae. Different attractants are needed for diurnal Lepidoptera, night-flying moths and bats. Because of such differences among pollinators and the flowers they visit, syndromes of floral characters can be recognized which are associated with different classes of pollinators. These have been fully described by Faegri & van der Pijl (1982) and others. The degree of adaptation between flowers and their animal pollinators varies greatly: the relatively large flowers of the Annonaceae, which are pollinated mainly by beetles (Gottsberger 1970), are little specialized, while in those of rain-forest trees belonging to more 'advanced' families such as the Lecythidaceae, which are pollinated by large bees and wasps (Prance 1976), floral structure and the behaviour of the pollinators are much more elaborate and are closely interrelated.

Mutual adaptation of plants and their pollinators has led to highly complex patterns of co-adaptation, of which rain-forest plants provide some of the most striking examples. The extremely long corolla tubes of flowers such as *Rothmannia* spp. and other African Rubiaceae, visited by moths with very long proboscides, and the strange pollination 'devices' of some tropical orchids, are well known. The even more extraordinary pollination mechanisms of *Ficus*, a genus which includes many rain-forest trees and stranglers, involve complicated co-adaptations between the structure and development of the inflorescences (syconia) and the life-cycles of the minute fig-wasps (Agaonidae) which pollinate them and on which their reproduction is completely dependent. Fig-wasps are reputed to be very host-specific (Ramirez 1970, Galil & Eisikowitch 1971, Janzen 1977b, 1979), but their specificity does not seem to have been investigated in the very numerous rain-forest *Ficus* species: if the wasps are strictly host-specific the availability of suitable wasps might limit the distribution of some *Ficus* species.

The complexity of many of the pollination mechanisms and the enormous variety of animal pollinators in tropical rain forests have been able to evolve because both flowers and pollinators are available all through the year. The phenology of flowering in rain-forest plants is dealt with in Chapter 9, but here it should be noted that only a small proportion of the trees are 'ever-flowering'; most species flower for limited periods only. But flowers of some kind or another are always available, though more abundantly at some times of year than at others. Flowers with similar pollination syndromes, which may or may not belong to the same genus or family, often flower in succession over many months, e.g. Bignoniaceae in tropical America (Gentry 1974). As a result, the links between birds, bats, insects and flowers are generally to fairly large numbers of species and rather rarely to one or a small number. In Costa Rica, Opler *et al.* (1975) found that some *Cordia*

species are visited by several hundred species of insects. In the terminology of Faegri & van der Pijl (1982) most rain-forest trees are generally polyphilic and their pollinators are polytropic (polylectic).

Flower-visiting animals depend on flowers for their nutrition to varying degrees, and nectar and pollen are produced by flowers in amounts differing greatly from one species to another. Some flowers attract many visitors but have little or no nectar. Others, especially those pollinated by birds and bats, produce it in very large quantities; an inflorescence of *Parkia* or a single flower of *Ochroma lagopus* (balsa) (Fig. 5.2), both bat-pollinated, may produce 15 ml of nectar in one night (Baker *et al.* 1983). Nectar varies in composition as well as in volume; in addition to sugars it contains many other organic substances, including lipids, amino acids and vitamins (Baker 1978, H.G. & I. Baker 1981, Appanah 1982).

There is evidence that the nectar of flowers visited by animals that are dependent mainly on it for food is richer in amino acids than that of flowers visited by animals that obtain much food from other sources. But nectarivorous animals, especially the larger ones, in fact rarely feed entirely on nectar and generally obtain some protein in other ways. Pollen is the chief source of protein for most bees. Flower-visiting birds and bats are usually also insectivorous; hummingbirds catch large numbers of insects, especially the females when feeding the young. Gilbert (1972, 1975) has described a very curious method of obtaining amino acids in the tropical American *Heliconius* butterflies. The females collect pellets of pollen from staminate flowers of Cucurbitaceae (lianes), regurgitate nectar onto them and re-imbibe it when amino acids from the pollen have diffused out.

Foraging strategies of pollinators. Gene interchange between trees of the same or related species depends on the foraging habits, range of movement and vertical distribution in the forest of the pollinators. Janzen (1971, 1975) recognized two contrasting foraging strategies, opportunistic and 'trap-lining'. In the former, nectar or pollen is collected mainly from trees that produce relatively conspicuous flowers (or inflorescences) in very large numbers at one time, while in the latter the flowers visited tend to be smaller, less conspicuous and produced in small numbers over a much longer period. In the opportunistic strategy the pollinator visits flowers that it happens to meet in the course of foraging excursions; in trap-lining it seeks out flowers of particular kinds and this may involve visits over many successive days, implying an ability on the part of the insect or bird to remember complex flight paths. Trap-lining

Fig. 5.2 Bat pollinating *Ochroma lagopus*, in cultivation, Ibadan, Nigeria. (Photo. S.R. Edwards.)

P. discolor

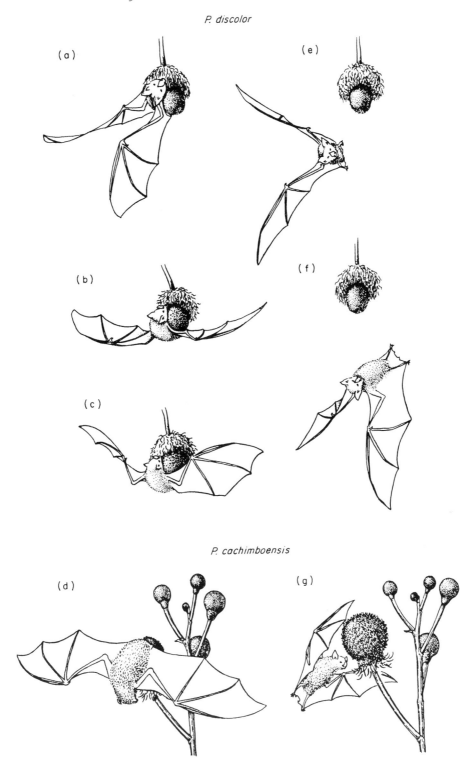

P. cachimboensis

Fig. 5.3 Two species of phyllostomid bats visiting flowers of *Parkia* spp. in Brazil. (From photographs by H.C. Hopkins 1984.)

behaviour was noted by Janzen (1971, 1975) in euglossine bees in Costa Rica, but is probably found in other insects such as hawk moths (Sphingidae), and certainly in hummingbirds and bats. Foraging does not always conform entirely to one pattern or the other; some hummingbirds, for instance, can be both opportunists and trap-liners.

Opportunist pollination is a strategy well adapted to emergent trees and large lianes of the canopy; in some of the former the flowers are made even more conspicuous by the fall of the leaves before flowering, as in *Couratari pulchra* of Guyana, *Bombax* and *Tabebuia*. The very large number of flowers produced in such species are not necessarily followed by an equally heavy crop of seeds, but as Janzen (1971, 1975) points out, this need not be regarded as wasteful if it ensures cross-pollination. Trap-lining pollination seems to be especially characteristic of the lower storeys of the forest where, owing to the dense mass of leaves and branches, even large brightly coloured flowers are less easily visible than in the canopy.

Some trap-lining insects and bats fly very long distances and can therefore be efficient cross-pollinators of tree species growing in widely separated populations. Large solitary bees in Costa Rica are reported as flying as much as 22 km on a foraging route (Janzen 1975, p. 21), and the bats that pollinate the mangrove *Sonneratia* and the durian in Malaysia may visit trees 38 km from their roosting places (Start & Marshall 1976). Even *Trigona erythrogaster*, a quite small bee, is believed on circumstantial evidence to have a flying range of 1100 m in Malaysia, although it is not known whether it flies such long distances in the forest undergrowth (Appanah 1980).

The maximum distance within which cross-pollination is likely to be effective is limited in practice by other factors besides the flying range of the pollinators. Flower visits by hummingbirds are restricted in range because the males are territorial and the females generally forage not far from their nests. Social bees such as *Melipona* also tend to visit flowers only within a certain distance from their nests: for this and other reasons they are not among the most important pollinators of rain-forest plants (Janzen 1975). Large social bees range more widely and probably play a large part as pollinators in the canopy of the rain forest.

Stratification of pollinators. The vertical distribution of pollinators is as important as the distance they can cover. In the mature phase of the forest the pollinators active in the relatively open euphotic zone (A and B storeys) among the emergent trees are to a considerable extent different from those found in the more shady lower strata, although some of them frequent both zones. In gaps and on forest edges bordering rivers and clearings these vertical differences are less marked because the butterflies, bees and other insects, which in closed forest live mainly in the canopy, here often forage down to ground level.

Ashton *et al.* (1977) and Appanah (1980) recognize several 'breeding synusiae' in the primary forest of Malesia, each consisting of taxonomically unrelated trees or other plants with similarities in their breeding systems and adaptations to pollination. Apart from the ground herbs and epiphytes, which form synusiae of their own, there are three breeding synusiae: (1) the trees and large lianes 'of the canopy' (i.e. the euphotic zone); (2) the trees and lianes of the shade (oligophotic) zone; and (3) the trees and other plants of gaps and clearings. The differences between the synusiae are largely due to the fact that flowers are much more easily seen in the canopy than in the less open layers below it. In the lower storeys even quite conspicuous flowers are not visible from a distance and, owing to the stillness of the atmosphere, even powerful scents may not carry far. Below the canopy also, flying is often impeded by the close-packed leaves and twigs. The stratification thus determines the type of lure and flower presentation needed to attract pollinators and the foraging methods of the insects, birds and bats themselves.

In cauliflorous and basiflorous trees (section 5.5) the flowers are borne on the lower part of the trunk or at its base and may be at a much lower level than if they were on the twigs. This may have the advantage that they are more easily accessible to some types of pollinators than they would be if they were in the crown of the tree.

In the Malesian rain forest the pollinators in the canopy include Nectarinidae (sunbirds) and other birds and many large insects. All these are capable of flying long distances and therefore have high energy requirements. Among the insects, large solitary carpenter bees (Xylocopidae) (van der Pijl 1954) and other large bees such as *Apis dorsata* (Appanah 1980) are particularly important. Although some tall trees and lianes have relatively inconspicuous flowers, it is in the two upper strata, where they are easily seen by birds and insects flying above the trees, that most of the large brightly coloured (and often also strongly scented) flowers are found. In most species of these strata flowering is seasonal or intermittent and in young trees it usually does not begin until the crown has reached the canopy. The pollination by thrips of *Shorea* spp., which flower gregariously at intervals of some years, was mentioned earlier (p. 103). In some of the other Dipterocarpaceae, e.g. *Dryobalanops* and *Neobalanocarpus*, abundant flowering occurs much more frequently and the pollinators are probably bees.

In contrast to the canopy species, the trees and lianes of the C and D storeys mostly have inconspicuous flowers: although some Annonaceae and Myristicaceae are scented, many of the smaller trees lack obvious visual or olfactory attractions. Most species produce a few flowers at a time over long periods from an early age; some are ever-flowering. Common pollinators do not usually include birds but there are numerous polytropic flies, beetles and thrips, and also 'imprecise trapliners' such as meliponid bees, Lepidoptera and solitary wasps. Appanah's (1982) study of *Xerospermum noronhaianum*, a typical lower-storey tree of the Malaysian forest, showed that pollination was probably carried out by an array of insects, including trigonid bees, wasps, beetles and butterflies, opportunists that also visit the flowers of many other species.

The breeding synusiae of gaps and seral communities seems to be similar to those of the mature forest canopy, except that many species begin to flower when young and do so intermittently or more or less continuously through the year. The flowers tend to be rather conspicuous and many are brightly coloured. Appanah (1980) believes that some large pollinators with high energy requirements, such as birds and Xylocopidae, use gaps and clearings as a base because flowers are more continuously available, and extend their foraging to the forest canopy only when trees come into flower.

A stratification of pollinators somewhat similar to that described above for Malesia is probably found in the rain forests of tropical America, except that in the latter birds play a larger part. Hummingbirds often visit flowers in the lower storeys as well as in the canopy. In Africa stratification of pollinators is perhaps more similar to that in the forests of Southeast Asia, but little is known about it.

5.3 Seed dispersal

The seeds and fruits of tropical rain-forest trees and shrubs are extremely varied in their methods of dispersal, as well as in size, shape and other characters. Much detailed information on the subject is given in the standard works of Ridley (1930) and van der Pijl (1982), both of which include data gathered during long experience in the eastern tropics, also in Roosmalen's (1985) book on the fruits of woody plants in Guiana. Field observations and experiments on dispersal are available for relatively few rain-forest species, so for most of them the method by which the seeds are dispersed has to be inferred mainly from structural characters.

Fruits and seeds can be classified by their known (or presumed) means of dispersal, or by their morphological and functional adaptations. A useful structural classification is that of Dansereau & Lems (1957). Lebrun (1960a) proposed a simplified version of this for tropical plants, to which Evrard (1968) made some further modifications. The dispersal classes represented among tropical and temperate trees are similar, although there may be some differences in the relative number of species in each.

A very large proportion of rain-forest trees have fruits or seeds attractive to mammals and birds. Roosmalen (1985) estimates that 87–90% of the woody species in the high forest of the Guianas are animal-dispersed. Most of these are eaten and probably dispersed by them, although frugivorous animals are not necessarily efficient dispersers. Occasionally even reptiles such as iguanas act as dispersal agents. Some animal-dispersed plants often have berries, drupes or fleshy fruits of other kinds; sometimes the seeds are embedded in pulp. In other species the seeds rather than the whole fruit may be the chief attraction; often they are rich in oil, starch or sugars and may have coloured arils or similar structures. Fruits attractive to birds tend to be brightly coloured, often with a striking contrast between the seeds and the fruit wall, as in some Leguminosae, Sterculiaceae and Ochnaceae, but in those sought after by fruit bats and other mammals odours seem commonly to be the chief attractants.

Seeds and fruits are the chief, sometimes the only, food of many forest-inhabiting rodents, birds, gorillas (Tutin *et al.* 1991) and monkeys, but other mammals, even carnivores such as civets, mongooses and tigers, are said to eat fruits. A large proportion of rain-forest birds are frugivorous and many, such as parrots, toucans and hornbills are highly specialized for eating palm fruits, nuts and hard seeds. Snow (1981) has shown that fruits such as the berries of the Melastomataceae, which are dispersed by more or less opportunistic frugivores, are small, watery and contain mainly carbohydrates: those dispersed by more specialized frugivores are larger and more nutritious. The forests of tropical Africa are poorer than those of Asia and America in frugivorous birds and also in plant families such as the Burseraceae, Lauraceae and palms whose fruits often form an important part of their diet. According to Janzen (1975) oily fruits tend to be eaten by vertebrates that depend mainly on fruits, but those rich in starch and sugar are preferred by those that also eat leaves and insects. Some frugivores eat a wide variety of fruits and seeds; others are more selective.

Seeds and fruits eaten by animals may be later regurgitated or excreted unharmed, but some animals destroy most of the seeds they eat. In feeding experiments on a tapir in Costa Rica, Janzen (1981) found that all seeds of one species and 78% of another were killed when ingested. Fruits and seeds are often eaten and dispersed by ground-living animals after falling to the forest floor. Rodents, such as agoutis (*Dasyprocta*), make hoards of seeds, some of which may germinate before they are used. In some tropical trees the percentage germination is significantly higher if the seeds have passed through birds, monkeys or bats, e.g. in *Cecropia palmata* (C. & A.M. Hladik 1969, Fleming & Heithaus 1981 and references given there).

Very few plants of primary rain forests, apart from grasses and other weeds that spread along paths, are dispersed by hooked or sticky diaspores attached to animals externally.

In many rain-forest trees, especially light-demanders, the diaspores are light and easily dispersed by air currents. Most of them have winged fruits, as in *Terminalia*, most Dipterocarpaceae and some Leguminosae, such as *Pterocarpus* and *Gossweilerodendron*. More rarely it is the seeds themselves that are adapted for wind-dispersal; they may be plumed (some Apocynaceae), winged (Bignoniaceae, many Meliaceae) or entangled with hairs to make a fluffy mass, as in *Ceiba*, *Bombax* and *Ochroma*. Wind-dispersed fruits and seeds are common among tall emergent trees, but usually absent, as might be expected, among trees of the sheltered lower levels. In a semideciduous forest in Nigeria, Keay (1957) found that the percentages of wind-dispersed species in the 'emergent' (A), 'upper' (B) and 'lower' (C) strata were 56, 25 and 2, respectively. Similar gradients would no doubt be found in primary rain forests, although the total number would probably be smaller. Wind-dispersed trees appear to be commoner in some regions than others; thus they are numerous in Malesia, but quite rare in mature primary forest in Guyana (T.A.W. Davis). They are also commoner in secondary than in primary forests.

The seeds of trees in riparian and periodically flooded rain forests may be dispersed by water; some species, such as *Guibourtia demeusii* of the Congo basin, have quite complex flotation mechanisms (Evrard 1968). But ability to float does not necessarily lead to long-distance transport. Stopp (1956) found that only a small proportion of the seeds in the river drift of the Congo were viable and suggested that flotation was important mainly in enabling the seeds to find favourable sites for germination, which were often quite close to the parent tree. In Amazonia the seeds of various trees and shrubs of the seasonally flooded forests, especially those by the nutrient-poor blackwater rivers, are eaten by fish. Those of many species are swallowed intact and are probably dispersed in this way (Gottsberger 1978, Goulding 1980, Kubitzki & Ziburski 1994). In Malaysia also the seeds of some streamside and freshwater swamp trees are eaten and probably dispersed by fish (Whitmore 1984a).

The seeds or indehiscent fruits of some rain-forest trees are extremely large and heavy. They seem to have no obvious adaptations for dispersal, and are generally classed as barochores (dispersed by gravity), but it is likely that many of them are eaten and transported by terrestrial birds or mammals after falling to the ground. The egg-like fruits of the Borneo ironwood (*Eusideroxylon zwageri*) may indeed be gravity-dispersed, though Meijer (1974) suggests (with some doubt) that they are dispersed by porcupines. They weigh over 200 g and roll downhill; according to Witkamp (1925) this is why the tree is absent on steep slopes.[2] The seeds of all the *Mora* species are very large. The biggest are those of *M. megistosperma* (Fig. 5.4), which are about 18 cm in diameter and weigh about 500–850 g when fresh (D.A. Janzen); they are said to be the largest of all dicotyledonous seeds. As they are flattened and do not roll, they may be transported by water during floods.

'Self-dispersed' seeds, such as those of some Moraceae and the well-known *Hura crepitans*, which shoots its seeds explosively to distances of up to 50 m (Swaine & Beer 1977), may perhaps, like barochores, be transported by animals after being ejected.

Fig. 5.4 Germinating seeds of *Mora megistosperma*. Swamp forest, Osa Peninsula, Costa Rica (1965). Said to be the largest dicotyledonous seed.

---

[2] There are further references to the ecology of this species on p. 330.

The biological requirements for successful seed dispersal give rise to a variety of selection pressures, the relative importance of which differs in different species. For all plants it is an advantage for the seedlings to develop far enough from their parents to avoid competition with them. Janzen (1970) has also argued that in tropical forests the chance of escaping predators is likely to increase with distance from other individuals of the same species and that this favours the widely scattered distribution patterns characteristic of many rain-forest trees. In addition there is also a requirement for seeds to fall in sites where light and soil conditions are suitable for germination and seedling establishment ('safe sites' of Harper 1977).

Seed mass and dispersibility. Salisbury (1942) showed that in British plants seed mass is related to the ecological requirements of the seedlings as well as to the method of dispersal. In species that germinate in deep shade (and also in sites where competition is severe, as in grasslands) the seeds are relatively heavy, because they must carry enough food reserves for the young plant to survive until it has become physiologically independent. Salisbury also pointed out that 'opportunist' (pioneer) trees which colonize open ground, such as ash (*Fraxinus excelsior*) and birch (*Betula*) have small light seeds (<0.1 g), because for such species it is more important for the seeds to be easily dispersed and numerous than for the seedlings to have an abundant food reserve.

In the rain forest the counterparts of Salisbury's shade species are the shade-tolerant trees ('dryads') of the mature phase: these generally have relatively large heavy seeds. The intolerant species, which regenerate only in gaps, and the *r*-selected early seral nomads (pioneers) correspond to the ash and birch in Britain and like them have light (and usually smaller) seeds (see Chapter 18).

The large food reserves in the seeds of typical trees of the mature forest phase enable the seedlings to establish themselves in deep shade and grow quickly to a size large enough to withstand the long period of suppression that usually awaits them. Their large food reserves also enable them to survive other hazards such as heavy losses and defoliation by predators (see Ng 1978, Forget 1992). For tall trees in a dense forest another advantage is that heavy seeds drop to the ground and do not become lodged in the foliage below them: Symington (1943) noted that the large winged fruits of *Dipterocarpus* may remain caught up in palm clumps or crevices of branches for years.

Unlike shade-tolerant trees, most early seral and intolerant species have light seeds, which are usually dispersed by wind, birds and bats. They can thus be quickly transported to newly-formed gaps and clearings. Probably an even greater advantage than their lightness is that they can be produced in very large numbers, often through most of the year. This is possible not only because the seeds are relatively small, but perhaps because, like other *r*-selected species, these trees invest a larger proportion of their resources in reproduction than do primary forest species (see Harper *et al.* 1970).

Shade-tolerant and shade-intolerant species, as pointed out in Chapter 3, are not clear-cut categories: there are intermediates and also species that share some, but not all, of the characters of one group or the other. The Dipterocarpaceae are somewhat anomalous in their method of dispersal. Some of them are shade-intolerant, although many are typical 'dryads', yet have winged fruits, which are wind-dispersed. The fruits are relatively light in *Shorea* and *Hopea*, but in some species of *Dipterocarpus* they weigh up to 30 g. In several genera a few species have wingless fruits. Ashton (1969, p. 182) describes the dispersal of dipterocarps as 'ineffectual', but the spinning flight of their fruits is not ill-adapted to transporting the seeds to gaps fairly near the parent tree (Fig. 5.6).

The biological advantages of dispersal to sites suitable for seedling establishment at a sufficient distance to avoid competition with the parent plant are obvious, but seed transport over hundreds or thousands of kilometres involves enormous wastage and is less certainly advantageous. Yet mechanisms favouring short-range dispersal may also favour long-distance migration. It is therefore not surprising that early seral and gap species are generally more geographically widespread than shade-tolerant dryads. Pioneer trees such as *Trema orientalis* and species of *Macaranga* in the eastern tropics, *Musanga* in Africa, and many species of *Cecropia*, *Inga* and *Vismia* in the Neotropics, have very wide geographical ranges and also effective dispersal by birds or bats. The wide Indo-Malayan distribution of the Bornean *Ficus* species and other Moraceae contrasts with the more localized distribution of the Dipterocarpaceae, many of which are endemic to the island of Borneo or to quite small areas within it (Ashton 1969, 1982).

Seed shadows. Seeds, whether dispersed by wind, animals or other means, are carried to varying distances from the parent plant, and rarely uniformly in all directions. The area within which seed from an individual plant falls is known, somewhat inappropriately, as a seed shadow (Fig. 5.5). The density of seed fall within the shadow decreases with distance from the source. In homogeneous surroundings the relation of density to distance would be a leptokurtic distribution with its peak at the parent plant (Janzen 1970, Fleming & Heithaus 1981), but many factors, such as air turbulence and the vagaries of animal dispersers, complicate this relationship, so that at any particular point the seed density may be much lower or higher than predicted (see Forget 1992). Usually a very large fraction of the total seed falls close to the source. Janzen (1970, p. 514) (Fig. 5.5) says that the total seed shadow of a tree species (its 'population recruitment surface') may be represented as 'a gently undulating surface with tall peaks of various shapes centered on the reproductive adults, and occasional low rises where seed shadows overlap or dispersal agents concentrate owing to habitat heterogeneity. The general height of the entire surface, and the height of the peaks around the parents, will be a function of

the efficiency of the dispersal agents, of the predispersal seed predators, and of the parent's productivity'.

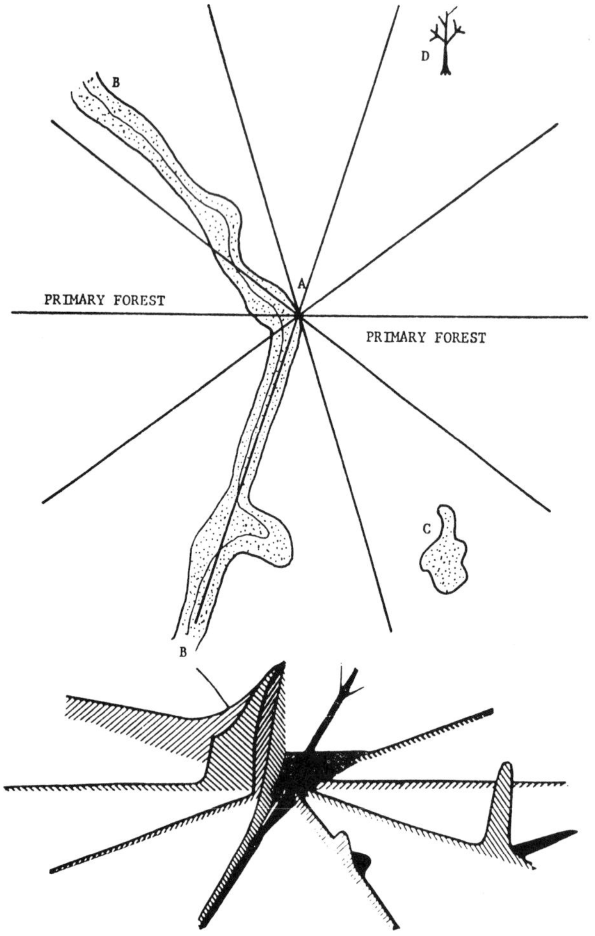

It is evident that the probability of seeds falling far from the parent plants must be considerably greater for the very light fruits of *Koompassia*, the fluffy seed masses of *Ceiba* or seeds eaten by birds and fruit-bats, than for heavy seeds with no apparent means of dispersal. The characteristic 'circles' of seedlings and juveniles surrounding mature trees of the tropical American *Mora gonggrijpii* (and other *Mora* species and *Eperua* species) (p. 114) clearly result from the heaviness of the seeds (p. 110) and the inefficiency of their dispersal. Quantitative data on the shapes and dimensions of the seed shadows are scarce. General observations suggest that, as Harper (1977) found for various temperate trees and herbaceous plants, the dispersal curves (log density/log distance) would be lines differing in slope from species to species, the slope being steeper for those with inefficient mechanisms than for trees well adapted to wind or animal dispersal.

There are a few observations in the literature on the distance wind-dispersed seeds of tropical trees are actually carried. Ridley (1905, 1930) recorded the distance seeds or fruits of various trees were carried in fairly open conditions in the Singapore Botanic Garden. He found that a large proportion of dipterocarp fruits fell within 10 m and rarely as far as 30 m from the parent and concluded that the chance of even a few fruits being transported over very long distances must be very small. In a forest it might be expected that the distances would be less, but in rain forest in Sabah Fox (1972, 1973) found that the fruits of dipterocarps with large wings mostly fell within about 40 m of the source and those with smaller wings within 10 m. Although the probability of such fruits being carried for more than a few metres is small, it may not be negligible: Webber (1934) recorded that the fruits of various dipterocarps and the samaras of *Koompassia* were transported for about 1 km by a tornado in Malaya.

With animal-dispersed seeds the characteristics of the seed shadow will depend on the kinds of animals eating the fruits or seeds and their patterns of behaviour. Many factors are involved, such as the distance the animals fly or move on the ground, also their foraging, feeding and defecating habits, as well as nesting, roosting, food-storing, territorial and migrating behaviour. Birds and bats may carry seeds many kilometres from the parent plant and even terrestrial mammals such as pigs, deer and elephants may deposit them in their faeces far from where they were eaten. Over long distances it is likely that animal dispersal is generally more effective than transport by wind.

Fig. 5.5 Complex seed shadow (hypothetical). After Janzen (1970, Fig. 8).

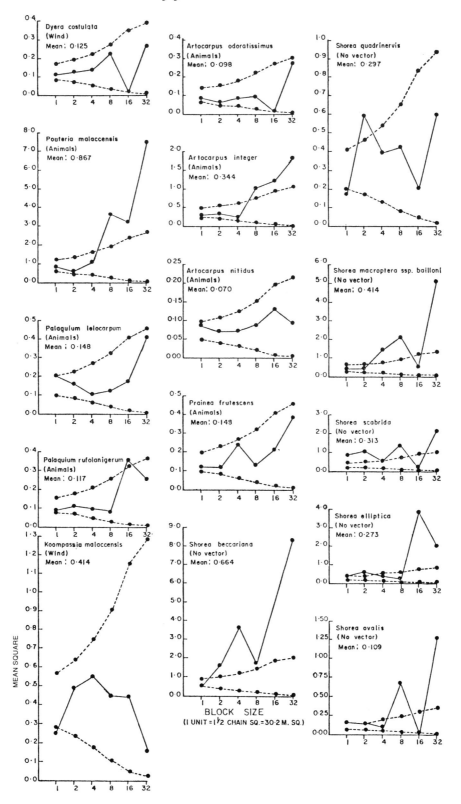

Fig. 5.6 Pattern analysis of wind-, animal- and 'inefficiently' dispersed tree species in mixed dipterocarp forest, Sarawak. From Ashton (1969, Fig. 5). The broken lines represent 5% significance levels.

Some idea of the influence of behavioural factors on animal seed dispersal can be obtained from observations such as those of Howe & Primack (1975) and Howe (1977) on bird dispersal of the tree *Casearia corymbosa* in Costa Rica and of Fleming & Heithaus (1981) and others on bat dispersal.

Howe & Primack (1975) found that a large number of bird species fed on the seeds of *Casearia* and that obligate frugivores such as toucans (*Ramphastos* spp.) were more effective dispersers than opportunists such as flycatchers (*Myiozetetes* spp.) which feed on insects as well as fruits.

Fleming & Heithaus (1981) found that in the 'Premontane moist forest' (a type of seasonally dry forest) of the Santa Rosa National Park, Costa Rica, frugivorous bats produced dense seed shadows after feeding on the fruits of *Cecropia peltata*, *Ficus* spp. and other trees. Most of the seeds were dropped beneath either the trees on which the bats had been feeding or the trees on which they roosted, which were usually less than 50 m away. The bats fed on more than one species, often several during the same night, and so produced multi-species seed shadows.

Dispersal and distribution patterns. As mentioned elsewhere (see Chapters 3 and 11), many rain-forest trees have distribution patterns which are clumped (contagious) to various extents, in contrast to other species which are randomly (and often very thinly) dispersed. Clumping might be expected to have some relation to the efficiency of the dispersal system.

Ashton (1969) found that contagious distribution patterns are common in mixed dipterocarp forest in Borneo, as Poore (1968) had previously found in Malaya. He also showed that clumping is most pronounced in trees of families in which means of dispersal 'do not exist or are unreliable' (Ashton 1969, p. 166). Among the latter he includes the Dipterocarpaceae, the dispersal methods of which were discussed above (p. 111). Species with apparently inefficient dispersal methods were compared with those with efficient wind dispersal, e.g. *Koompassia malaccensis* and some Apocynaceae, or which are animal-dispersed, such as Burseraceae, Sapotaceae and Moraceae. Pattern analysis showed that clumping is much more marked in the first group than in the second (Fig. 5.6).

It is probably true that some dispersal methods produce a more random and others a more clumped (denser) seed shadow, but it does not necessarily follow that the distribution pattern of the adult trees is similar to that of the shadow. Fleming & Heithaus (1981), in their study of bat-dispersal in Costa Rica referred to above,

studied the distribution patterns of both bat-dispersed trees and species not dispersed by them. All the bat-dispersed species had clumped distributions, except one *Ficus* species, which had a random distribution, but most of the species not dispersed by bats were also clumped. The authors concluded that clumping was the result of several processes, among which dispersal of viable seeds is only the first: other factors affecting germination and seedling establishment might also be non-randomly distributed. Single-dominant rain forests, e.g. those formed by *Mora* and *Eperua* spp. in South America (Chapters 11 and 12), can be regarded as extreme cases of clumping, but the relatively inefficient dispersal of their heavy seeds can be only one among a number of factors responsible for the development of such communities. Forget (1989) found that in French Guiana seeds of *Eperua falcata* (wallaba) (Fig. 3.6), which weigh *ca.* 7 g, have a maximum dispersal distance of 30 m, 60% falling within 10 m of the parent. Contrary to expectation (p. 56), many of those germinating close to it grow into saplings.

5.4 Dormancy and germination

A capacity for seed dormancy is almost universal in plants of climates that are seasonally cold or dry (or both). In such environments viable seeds can accumulate in the soil as a seed bank and buried seeds may remain alive for many years. In the relatively non-seasonal climates of the humid tropics, dormancy is much less common. Typical primary rain-forest trees seldom form seed banks and in many species populations of suppressed seedlings take their place (pp. 64–5). As a rule it seems to be chiefly the seeds of early seral (pioneer) species that remain viable for long periods and accumulate as seed banks.

In some tropical plants, such as mangroves, the development of a seedling follows fertilization without interruption. Even under controlled storage conditions seeds of most primary rain-forest trees are 'recalcitrant' and remain alive for only weeks or months rather than years (Longman & Jeník 1987). The difficulty of storing seeds can be a major problem in establishing plantations of economic species. The seeds of *Theobroma cacao*, in origin a C storey rain-forest species, have a maximum viability of three months and are intolerant of low humidity; exposure to temperatures below about 8 °C reduces the percentage germination drastically. Probably for this reason, they usually lose their viability when transported by air at high altitude (Hunter & Burrows, quoted by Cuatrecasas 1964, pp. 429–30). The short viability of the seeds of dipterocarps is probably due to the fact that most species lack endosperm (Ashton 1982). Under natural conditions, exposed to predators and fungal infections, the life of tree seeds is likely to be even shorter; it is not surprising that in many primary rain-forest trees germination is of the simultaneous type of Salisbury (1929), the seeds germinating within a few days after falling to the ground. In such species large, but mainly transient, populations of seedlings appear on the forest floor; they are a characteristic feature after the mass flowerings of the Dipterocarpaceae in Malesia (pp. 252–3).

However, germination in rain-forest trees is by no means always simultaneous or rapid: in some species it takes place continuously or discontinuously for many weeks. Thus in Guyana germination in *Ocotea rodiei* is spread over a prolonged period, although in *Mora excelsa* all the seeds of the same crop germinate together, usually immediately after falling (T.A.W. Davis). Seeds that do not germinate simultaneously must have some inherent or 'opportunistic' dormancy mechanism (see Harper 1977). Hartshorn (1978) says that at La Selva, Costa Rica, *Saccoglottis trichogyna*, a species in which the seeds are dispersed by flotation, is one of the very few primary forest trees with seed dormancy. However, further investigations may show that in the primary rain forests of tropical America, as in other parts of the tropics, some degree of seed dormancy is not rare.

An investigation of germination under nursery conditions of 180 species of Malayan rain-forest trees by Ng (1978)[3] showed that in 65%, including most of the A and B storey species and nearly all the Dipterocarpaceae, the viable seeds had all germinated within twelve weeks (many within one to two weeks) ('rapid germination'). In 7% there was a dormant period; germination did not begin until after at least twelve weeks and took up to 158 weeks to complete ('delayed germination'). The remaining 28% of species were intermediate; germination began in one to eleven weeks, but continued for up to 104 weeks. Ng believes that in some of his delayed and intermediate species differences in dormancy among seeds of the same crop are an important strategy. It reduces the hazards from competition and predation and makes it possible for viable seeds to be always available for animal dispersal or to respond to changes in the environment.

The species in Ng's (1978) list show little correlation between the pattern of germination and seed size or method of dispersal, but in some families, notably the Leguminosae, dormancy is commoner than in others.

Ng points out that in some Malayan leguminous trees, e.g. *Parkia javanica*, the whole seed, including the embryo as well as the testa, is hard and the water content low, while in others e.g. *Parkia speciosa*, *Koompassia* and *Pithecellobium*, the seeds are soft, can be easily cut through with a sharp razor, and have a higher water content. The hard-seeded species generally have intermediate or delayed germination, but the soft-seeded ones have rapid germination.

As might be expected, germination is more seasonal in trees of forests with marked dry seasons than in those of ever-wet rain forests. Aiyar (1932) found that in the forests of the Western Ghats (India), which are of the evergreen seasonal type, tree seeds falling just before or during the monsoon period do not germinate, or if they do, make little progress until the wettest period is past. In *Mesua ferrea*, for example, the seeds fall before the monsoon, in May or June, but usually do not germinate until September, the beginning of the dry season. *Palaquium ellipticum* drops its seeds during the monsoon and they germinate immediately, although the seedlings make little growth until the wet weather is over. Jones (1956, p. 112) found that in the Okomu forest in Nigeria (p. 470) prolonged dormancy is not rare, especially among the tall animal-dispersed trees and perhaps also the understorey species.

At Barro Colorado Island (Panama), Garwood (1983) showed that, among the several hundred species studied, the mean length of dormancy was 2 to 370 days and there were three types of germination behaviour. In 18% of the species the seeds were dispersed in the rainy season and remained dormant until the beginning of the next rainy season, four to five months later. In 42% they were dispersed in the dry season and germinated early in the rainy season one to two months later. In the third group (40%) both dispersal and germination took place in the rainy season; in these species the timing of dispersal, not dormancy, controls the time of germination.

Seed banks. Although a capacity for dormancy is occasionally found in some dryads of the mature rain forest, it is much more characteristic of the early seral trees of secondary communities and the gap phase of primary forests; it is chiefly the seeds of these that accumulate in the soil to form seed banks. Symington (1933) discovered that large numbers of viable seeds were present in the soil of forests in Malaya which as far as was known had never been cultivated or disturbed by people. These included seeds of various early seral trees as well as those of *Imperata* (alang-alang grass) and other herbaceous weeds (pp. 478–80). Since then

---

[3] Additional data in Ng (1980a).

the existence of buried viable seeds in forest soils has been demonstrated in many parts of the humid tropics. The literature on the subject is now extensive. It has been reviewed by Garwood (1989) and can be dealt with here only briefly. The role of seed banks in secondary successions is discussed in Chapter 18.

Seeds become buried in the soil by burrowing rodents, worms and other animals, as well as by physical processes. How long they may remain alive is uncertain. Some are destroyed by fungi and bacteria or eaten by soil animals, but large numbers probably survive for many years. In rain forest in Surinam, Holthuijzen & Boerboom (1982) found that *Cecropia* seeds in earthenware pots sunk in the ground were viable after at least sixty-two months, but in the soil itself they might not have survived so long. The source of buried seeds is not exclusively the vegetation in the immediate neighbourhood: some seed banks contain seeds carried from a distance in the seed rain (Hall & Swaine 1980; see also Swaine & Hall 1983). The building up of seed banks over two years at Gogol in New Guinea has been described by Saulei & Swaine (1988).

The abundance (density) and the number of species of seeds in the soil are very variable, even over short distances. In the topsoil of wet evergreen forest in Ghana, Hall & Swaine (1980, see also Swaine & Hall 1983) found about 163 seeds m^{-2} of twenty-two species. In primary mixed dipterocarp forest at Pasoh (Malaysia), Putz & Appanah (1987) estimated that there were 132 seeds m^{-2} of thirty-one species at 0–10 cm depth; in similar forest in Sabah at 0–15 cm depth, Liew (1973) found sixty seeds m^{-2} of thirty-one species. In secondary and disturbed forests the densities of seeds in the soil may be much greater. In Amazonia Uhl *et al.* (1981) found about 9000 seeds m^{-2}, of which 43% were of *Cecropia* and other pioneer species, in soil of clearings five years after cultivation, compared with 69 seeds m^{-2} in the soil of 'old growth forest'.

The depth to which viable seeds are found varies, probably depending on soil texture and other factors. In Surinam primary forest soils seeds of *Cecropia* were found down to 20 cm depth, but the greatest mean density was at 0–1 cm (Holthuijzen & Boerboom 1982). In secondary forest in Costa Rica considerable numbers of seeds were present at 20–40 cm depth (Young 1985).

When buried seeds are brought to the surface in clearings or natural gaps they are exposed to conditions very different from those in the soil. In many species light is the stimulus triggering germination, but some are temperature regulated. Vazquez-Yanes (1976) and Vazquez-Yanes & Smith (1982) showed that the seeds of the pioneers *Cecropia obtusifolia* and *Piper auritum* have a light-sensitive phytochrome system; light rich in far-red wavelengths such as that at the ground surface in high forest induces dormancy, but white light with a higher red to far-red ratio breaks dormancy. Seeds of the African early seral tree *Musanga*[4] as well as of the longer-lived pioneers *Milicia* spp. and *Nauclea didderrichii* are also light-sensitive (Longman & Jeník 1987), but the germination of the tropical American early seral tree *Ochroma lagopus* is determined by temperature (Vazquez-Yanes 1974). The oil palm *Elaeis guineensis*, which is common in secondary forest in West Africa, also has temperature-regulated germination (Purseglove 1968) and according to Keay (quoted by Baur 1964, p. 136), seedlings do not come up in the shade of mature forest because the temperature is not high enough.

The germination of seeds buried in the soil accounts for the appearance of vast numbers of pioneer tree seedlings two or three weeks after forest is cleared or a treefall gap is formed. The ability of the seeds to remain dormant in the soil is an important part of the biological equipment of most pioneer species and is probably shared by light-demanding primary forest trees which regenerate only in gaps. However, as will be seen in Chapter 18, not all the plants that colonize gaps and clearings originate from buried seeds: some are carried in by the seed rain. At Pasoh, where the population of buried seeds is relatively small, Putz & Appanah (1987) showed that during the first nine months after the formation of a gap buried seeds accounted for about seven times as many seedlings as did seeds from the seed rain.

5.5 Cauliflory

A characteristic of many tropical rain-forest trees, which is rarely found in trees outside the tropics, is cauliflory, the production of flowers on the leafless trunks, rather than on the twigs and smaller branches (Figs. 5.7 A–C). A familiar example is cacao, *Theobroma cacao*, a C storey tree native in the rain forests of tropical America. To a visitor from temperate climates cauliflory seems one of the strangest features of tropical trees. With rare exceptions, European and North American trees bear their flowers on twigs or slender branches of the current or previous years, never on thick branches or the main trunk. In many tropical trees the flowers or inflorescences (and later the fruits) are sessile or on very short leafless stems springing directly from the trunk or larger branches, while the finer branches, and often the whole crown, remains wholly vegetative.

---

[4] This is perhaps why its seeds failed to germinate in the experiment reported by Aubréville (1947) (p. 116).

Fig. 5.7 Cauliflory. A A tall cauliflorous tree, *Theobroma simiarum*, Las Horquetas, Costa Rica (1965). B A small cauliflorous tree, *Napoleonaea imperialis*, Idanre, Nigeria (1948). C A small cauliflorous tree, *Quassia* sp., Woopen Creek, Queensland (1981).

Cauliflory may arise in two ways, but in either case the flowers or inflorescences probably always develop from buds originally subtended by foliage leaves. The commoner way is for the axillary buds on the young shoots to continue producing flowers year after year until they have become thick branches or even the main trunk. The tree may then bear flowers anywhere from its crown to the base of the trunk; in an old tree the buds on the trunk may grow into short shoots with scale leaves and abbreviated internodes. Should accessory buds develop on these shoots, thick burrs may be formed. The second way in which cauliflory may arise is for the flower buds on the primary stems to remain dormant for a long period. By the time they become active they have become buried by secondary tissues and may actually have to push their way through the bark. The developmental aspects of cauliflory have been considered by Thompson (1943, 1946, 1951). Pundir (1981) has described the development of the cauliflorous syconia in *Ficus hispida*, but further research is needed.

Several somewhat different morphological phenomena are grouped together under the general heading of cauliflory. Mildbraed (1922, pp. 115–25), distinguishes between four types.

(1) Simple cauliflory. The flowers are borne on the main stem and branches, as well as on the twigs, because the buds remain active for an indefinite period, as in *Theobroma cacao* and species of *Angylocalyx*, *Diospyros*, *Drypetes* and many other genera.

(2) Ramiflory. Flowers on the larger branches, usually not on the twigs or main trunk; in this case the buds remain active for a limited period only, e.g. *Polyalthia insignis*, *Turraeanthus africanus*.

(3) Trunciflory. Flowers on the main trunk only, as in *Grias cauliflora*, *Theobroma speciosum*, the African *Omphalocarpum procerum* (Fig. 5.7A) and many species of *Cola* and *Diospyros*.

(4) Basiflory. Flowers or inflorescences restricted to the base of the trunk, e.g. *Chytranthus villiger* (Fig. 5.8), *Polyalthia flagellaris*, or to branches without foliage leaves arising near the base, as in the well-known cannon-ball tree, *Couroupita guianensis*. A variant of basiflory is flagelliflory, in which the flowers are on whip-like branches lying on the surface of the ground and arising at the base of the trunk, as in *Caloncoba flagelliflora* (in which they may be as much as 11 m long) (Mildbraed 1922), or below ground level, e.g. in *Duguetia rhizantha* and some species of *Ficus*.

Simple cauliflory is probably the commonest type, but intermediates exist between all four. Different types of cauliflory (as well as production of flowers in normal positions) can sometimes be found in the same genus,

e.g. *Durio* (Whitmore 1972a) or *Theobroma* (Cuatrecasas 1964), and there may be variation between individuals of the same species; in *Diospyros suaveolens* some trees show simple cauliflory and others ramiflory.

Mildbraed (1922) suggested that ramiflory is more primitive than other types of cauliflory and it is certainly tempting to regard ramiflory, simple cauliflory, trunciflory and basiflory (in this order) as evolutionary stages. In this connection the genus *Ficus* is of great interest because it offers examples of a great variety of types of cauliflory. In a detailed study of the Malayan species of the subgenus *Covellia*, Corner (1933) showed that a graded series can be recognized. In *F. lepicarpa* the syconia are borne in pairs in the axils of the leaves; in *F. fistulosa* a few may be found on the leafy shoots, but most are on the older branches in clusters looking like bunches of enormous grapes; in *F. obpyramidata* all the syconia are borne on the older leafless branches,

Fig. 5.8 A basiflorous tree, *Chytranthus villiger*, Atewa, Ghana (1969).

and the main and fruiting branches reach a much greater length; in *F. hispida* most of the syconia are on leafless branches arising from the main stem, but some of these are up to 1 m long, bearing fruit along their whole length and dangling. At the end of the series are the geocarpic figs, e.g. *F. beccarii*; in these the fruiting branches are attached to the trunk not more than 1 m above ground level and grow into branched whip-like strands up to 6 m long and about 2.5 cm thick at the base. Near the growing point they bear caducous stipules, but no foliage leaves. As they elongate, these branches hang down and creep along the soil surface; when they meet an obstruction, such as a bank of earth, they burrow into it like a root. The fruits ripen underground and the seeds are presumably dispersed by soil animals. This series might represent stages in an evolutionary process. A habit similar to that of the geocarpic figs is found in *Polyalthia hypogaea* and *Saurauia callithrix* (Corner 1933). *Jacaratia dolicaula*, which has flowers only in the lower part of the crown, may represent a step towards cauliflory.

Cauliflory is found in trees and lianes of several different architectural models and in many unrelated families, although in some, e.g. Dipterocarpaceae, Lauraceae and Rubiaceae, it is absent or relatively rare. Cauliflorous plants are found in all regions of the tropics and many different tropical plant communities, but they are much more numerous in lowland tropical rain forest than in montane or seasonally dry forests.

It is a striking fact that true cauliflory is very rare outside the tropics, although *Cercis* spp. the Judas trees, and the carob (*Ceratonia siliqua*) of the north temperate zone, are ramiflorous and occasionally simply cauliflorous, as is the Caucasian *Pterocarya fraxinifolia*. *Forsythia* and other north temperate genera are ramiflorous to some extent. In the temperate rain forest of northern New Zealand a species of the Indo-Malayan genus *Dysoxylum*, and occasionally *Fuchsia excorticata* and *Planchonella costata*, are ramiflorous or cauliflorous. A tall liane, *Tecomanthe speciosa*, found on the Three Kings Islands (some 50 km north of New Zealand) is also said to be cauliflorous (Dawson & Sneddon 1969). *Eucalyptus lehmannii* of southern Australia is a surprising example of an extra-tropical cauliflorous tree.

It is noteworthy that most cauliflorous plants, other than woody climbers, are trees of small to moderate height (*ca.* 5–20 m) belonging to the B or C storeys, or treelets or shrubs of the D storey. Among large emergent trees, cauliflory is rare; *Couroupita guianensis* is one of the few examples. The almost complete restriction of cauliflory to the humid tropics, and to lianes and small or medium-sized trees, must clearly be taken

Fig. 5.9 Leaf of *Phyllobotryon soyauxianum*, Oban, Nigeria, a treelet 2–3 m high. (Photo. S.R. Edwards.) The flowers are adnate to the midrib of the leaves and are pollinated by bees. Note tree frog.

into account in attempting to suggest interpretations of the cauliflorous habit.

Cauliflory has received various speculative 'explanations' for which there is little or no experimental evidence. It has been regarded as an adaptation to particular classes of pollinator and has also been interpreted in terms of physiology or phylogeny, but these various 'explanations' are not necessarily mutually exclusive. A well-known theory is that of Wallace (1878, pp. 34–5), who suggested that cauliflory was a device by which the smaller forest trees, excluded from the abundant light and space of the canopy, could display their flowers where they could be easily seen and visited by shade-loving insects. As was emphasized in Chapter 2, the lowest (C) tree storey is often the layer most densely filled with leaves and twigs and the space below the crowns of these small trees is relatively open; thus at this level flowers are more easily accessible to flying animals than if they were in the C storey crowns. Information on the pollinators of cauliflorous plants is somewhat scarce. Many probably require cross-pollination by insects and some are dioecious. It seems that these insect pollinators are more often beetles, flies and small bees rather than Lepidoptera. Some cauliflorous species are pollinated by bats, e.g. *Durio* (van der Pijl 1956, 1969). Another way in which flowers are made accessible to large flying pollinators is penduliflory, a curious condition in which the flowers are borne on slender flexible peduncles which may reach a length of 2 m or more, hanging down from the branches so that they are well clear of the foliage: examples are *Eperua falcata*, *Kigelia pinnata* (sausage tree) and the African shrub *Ochna subcordata*. In considering cauliflory in relation to animals it has to be remembered that cauliflorous plants are also caulicarpic; the position of the fruits must therefore be adapted to its dispersal agent. However well adapted some cauliflorous plants may be to particular kinds of pollinators or seed-dispersers, it certainly seems rather unlikely that cauliflory originally arose as an adaptation to pollination or dispersal.

Haberlandt (1926) and Klebs (1911) approached the problem of cauliflory from a physiological point of view. According to Haberlandt, evergreen trees, which often expand their leaves gradually, rather than in large numbers simultaneously, store in their trunks and large branches only such food materials as may be necessary for the growth of the flowers and fruits. If the latter are borne on the trunk and branches, the food need not be translocated back to the twigs and there is thus a saving of time and energy. The most obvious objection to this theory is that the taller the tree, the more advantageous should cauliflory be, while it has already been seen that cauliflory is found chiefly in small trees. Klebs held that flower formation can only take place when there is a surplus of photosynthates relative to the amount of mineral ions available. In tropical trees, in which assimilation can proceed without interruption all the year round, large quantities of assimilates accumulate in the trunk and branches. As at the same time the leaves keep up a constant drain on the supplies of mineral substances, conditions in the bark are always favourable to flower production from the buds.

Corner (1966) looks on cauliflory as a stage in the evolution of trees from the pachycaulous to the lepto-caulous stage. It allows the advantages of a leptocaulous crown with slender twigs and moderate-sized leaves to be combined with retaining heavy durian-like fruits. This theory deserves more detailed consideration than it has yet received. Most cauliflorous trees have rather heavy fruits, although not many are as large and heavy as the durian. Cauliflory seems to be commoner in the more primitive than in the more advanced families of flowering plants. Corner's view has the advantage that, as very heavy fruits are a feature mainly of tropical trees, it gives a plausible explanation of why cauliflory is rare outside the tropics. Also, since very heavy fruits are commoner in small trees than in tall emergents, it suggests why cauliflory is mainly restricted to lower-storey trees.

Chapter 6

Ground herbs and dependent synusiae

THE HERBACEOUS GROUND flora of the rain forest as well as the climbers, epiphytes, saprophytes and parasites depend on other plants for mechanical support, or for nutrition and live under specialized conditions; many of the species in synusiae are themselves highly specialized in structure and physiology. Climbing plants, especially the woody species (lianes), form a large part of the biomass in most rain forests and compete actively with the trees for light and space, but the rest of these very diverse groups of plants play only minor parts in the structure and economy of the forest ecosystem. They are, nevertheless, of considerable interest to the ecologist. The epiphytes, especially, have been much studied, because of their wealth of curious adaptations to their environment. They are also of great interest because of the remarkable animal communities for which they provide food and shelter.

Fig. 6.1 A ground herb, *Phyllagathis* sp., young lianes and other seedlings on forest floor, Pasoh Forest Reserve, Malaya (1974).

6.1 Ground herbs

The synusia of ground herbs is not synonymous with the 'ground flora' or lowest (E) stratum of the forest, because, as was pointed out in earlier chapters, the majority of the plants in this layer are young trees, shrubs and woody climbers. On the other hand, not all plants that can be classified morphologically as herbs form part of the E layer. Some of them, e.g. the taller Scitamineae, although their stems are not woody, reach a height of 6 m or more and belong to the D, or even to the C, storey. The banana *Musa ingens*, which grows in gaps in the New Guinea lower montane forest, is over 12 m high and is said to be the world's tallest herbaceous plant (Argent 1979).

In lowland rain forests, as was noted in Chapter 2, luxuriant herbaceous vegetation is often found in natural and artificial openings and by paths and rivers where there is good illumination, but in the interior of mature rain forest herbs occur chiefly as scattered individuals and are sometimes almost absent. Social species forming patches of closed vegetation are found, but mainly in gaps. On steep slopes and in upland rain forests, where for one reason or another the canopy is less dense, ground herbs are generally more plentiful.

Occasional non-flowering individuals of the herbaceous species which are abundant in openings are also found in the shade, except in the darkest parts of the forest, but some shade-loving species, e.g. *Trichomanes* spp., and some other ferns, never grow in openings. There are thus two ecological groups among rain-forest herbs, the shade-tolerant and the shade-demanding, but they are not sharply defined. The first reaches its maximum development at almost full daylight, while the latter seems to be intolerant of exposure. The factor chiefly responsible for separating the two groups is not necessarily light as such; the varying needs of species for a saturated atmosphere and perhaps tolerance of root competition may also be involved. The shade-tolerant species are to some extent analogous to the 'wood-marginal' plants of European woodlands and are more numerous than the shade-demanding species.

The synusia of ground herbs includes, in addition to dicotyledons and monocotyledons, many ferns and often species of *Selaginella*, but the total number of species and families represented is generally relatively small, although it varies from one region to another. In the richest areas it is probably never more than a small fraction of the number of trees and shrubs in the same area. For example, in each of five sample plots (1.5 ha) at Moraballi Creek (Guyana) there were fewer than twenty species of herbaceous flowering plants, ferns and *Selaginella*, compared with sixty to ninety-one species of trees ≥10 cm d.b.h. In sample plots in West Africa the number of species of flowering plants was similar and ferns were the most abundant herbaceous plants (Richards 1939, 1963a). In Malesia, however, herbaceous species are somewhat more numerous. In the G. Dulit district in Sarawak the floristic composition of the ground herb synusia varied greatly from place to place, some species occurring on one or two neighbouring ridges and apparently not elsewhere in the district.

Fig. 6.2 *Psychotria* sp., flowering in deep shade, Ankasa Forest Reserve, Ghana (1969).

Even where the number of herbaceous species is relatively large, rather few families are represented. In all lowland rain forests small Rubiaceous herbs are common and grasses are usually present although, in contrast to more seasonal tropical forests and savanna woodlands, they are not numerous in species or individuals. In tropical America other families of herbs that are well represented include the Gesneriaceae, Melastomataceae, Cyperaceae and Marantaceae; in Africa the Acanthaceae, Araceae and Commelinaceae are common. In Malesia a wide range of families, as of species, is found; in Borneo the commonest families are Araceae, Begoniaceae, Cyperaceae, Gesneriaceae, Melastomataceae and Zingiberaceae.

Many, but not all, of the ground ferns are shade-demanders and increase in abundance in particularly damp and shady places, such as stream valleys. These ground ferns belong to many genera; species of *Trichomanes* are often present, though seldom abundant.

In heath forests on white sands in Guyana and Borneo the author found that ground herbs were fewer in species than in mixed forest. Gentry & Emmons (1987) also noted that in samples of Neotropical forests herbs were markedly fewer in species and individuals on white sand sites than on 'wet lateritic soils'.

The ground herbs are varied in morphology. Their shoot architecture has been little studied but several of the models of Hallé *et al.* (1978) are probably represented among them. In life-form they differ considerably from the ground herbs of temperate woodlands. Plants of compact habit, such as rosette plants, are rare and, as might be expected from the uniformity of microclimatic conditions throughout the year, there are very few plants that form resting buds on or just below the soil surface. Many of the herbs have elongated aerial stems which live for several years (Kiew 1986); the majority are probably, in Raunkiaer's terminology, herbaceous phanerophytes, and not, as in temperate deciduous forests, hemicryptophytes or geophytes. Plants with underground rhizomes or stolons are frequent, but the rhizomes are adapted for multiplication and migration rather than perennation. In some ground herbs the plants are linked by aerial shoots, which become prostrate and buried in the soil (Burtt 1977). Examples of social species forming dense patches several square metres in extent are various Melastomataceae in tropical America, the grass *Leptaspis cochleata* (Fig. 6.3) and *Dorstenia* and *Geophila* spp. in Africa, as well as various Melastomataceae and many Zingiberaceae in the Malayan forest. In these, and probably most other ground herbs of the mature phase of the primary forest, reproduction appears to be chiefly vegetative. In some

Fig. 6.3 A forest grass, *Leptaspis cochleata*, Bukit Lanjan, Selangor, Malaya (1974).

Begonia spp. and in the West African *Haemanthus cinnabarinus* reproduction often takes place by buds formed on the leaves. In the Malayan *Phyllagathis rotundifolia* (Fig. 6.1) flowering is infrequent and reproduction is mainly by fragments of leaves and stems (Kiew 1986).

Although plants with underground rhizomes, tubers, corms and bulbs are not uncommon, true geophytes in which the aerial shoots die down for some part of every year are rare, and, as would be expected, seem to be found only (or chiefly) near the climatic limit of the rain forest where there is a regularly recurrent dry season. They are probably absent in the ever-wet rain forest of Borneo and Guyana, but in the evergreen seasonal forest of southern Nigeria there are some Araceae (*Anchomanes* and *Amorphophallus* spp.) that are tuberous geophytes dying down for several months every year, the inflorescence being produced in some

species before, in others with or after, the new leaves. *Cyanastrum cordifolium* (Tecophilaceae), a common herb in the West African forest, was found to behave as a true geophyte in the Omo forest, Nigeria (Richards 1939), perennating by means of its corm, but in the Okomu forest and in Cameroon, perhaps because conditions are moister, the foliage appeared to remain green throughout the dry season. In the rain forest of Malaysia the tuberous *Tacca* spp. retain their leaves throughout the year, but *T. pinnatifida* is deciduous in more open vegetation on sandy soils on the east coast. *Amorphophallus* spp. behave as geophytes in forest on limestone in northern Malaya which has a regular but not severe dry season (R.E. Holttum). In the seasonally flooded forests of Zaïre there are herbaceous plants with bulbs or rhizomes which produce aerial shoots during the dry season only (Lebrun 1936a); these could be regarded as either geophytes or hydrophytes.

The stems of rain-forest herbs tend to be sappy and brittle, no doubt because they have little lignified tissue, but they sometimes become more or less woody when old, as in many Malesian species of *Didymocarpus* and *Sonerila*. The lower part of the stem may be horizontal or inclined and supported by aerial prop-roots, e.g. in the African *Acanthus montanus*, and some Malesian Zingiberaceae. This possibly enables the plant to 'walk away' from debris such as dead leaves and branches falling from above, as in some small palms (p. 89) (Chapter 4).

The leaves of rain-forest herbs are varied, contrasting with the monotonous foliage of the shrubs and young trees. The majority are mesophyll or notophyll in size, and thin and soft in texture, rarely glossy and sclerophyllous. The variation among the leaves of ground herbs is shown more in their margins, surface and colouring than in their size and shape. Leaves larger than mesophyll are rare in ground herbs, but leaves of the smaller size-classes are perhaps commoner than among the trees and shrubs. The broadly elliptical to oblong-lanceolate acuminate shape is as prevalent as in the higher strata, but leaves with non-entire margins are more frequent. The leaves are mostly acuminate and long drip-tips are rarer than among the shrubs and young trees. Kiew (1986) found that the leaves of some Malayan rain-forest herbs expand over a period of about three months and live for about eighteen to thirty-three months, longer than those of most of the trees.

The prevalence of broad leaves among rain-forest plants is strikingly shown in the grasses and Cyperaceae (Fig. 6.5). Most species of these families in the ground flora of mature primary forest are very different in appearance from their relatives in other environments,

Fig. 6.4 A forest grass, *Guaduella humilis*, Oban, Nigeria (1978). (Photo. S.R. Edwards.)

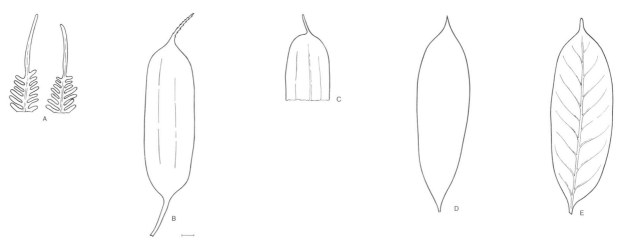

Fig. 6.5 Convergence in leaf form among rain-forest plants of different systematic affinities, all showing drip tips. A Pinnae of *Pteris preussii*, Nigeria. B, C Leaves of two unidentified Cyperaceae and D leaf of a grass from undergrowth of mixed dipterocarp forest, Gunung Dulit, Sarawak. E Leaf of *Goniothalamus malayanus*, a dicotyledonous shrub from same locality as B and C. Scale bar, 2 cm.

and from the weed species (*Panicum, Paspalum, Fimbristylis* etc.) which infiltrate into the forest in openings and along paths. True forest grasses such as *Pariana* and *Pharus* (tropical America) and *Leptaspis cochleata* (Old World tropics) have broad leaves and resemble Scitamineae, or, as in the African *Commelinidium gabunense*, the Commelinaceae, rather than typical grasses. They grow in patches or small loose tufts and are not tussock- or turf-forming like savanna grasses. Unlike all other lowland tropical grasses (including the weeds mentioned above) these forest grasses have C_3 metabolism and lack the 'Kranz anatomy' associated with the C_4 metabolism characteristic of the grasses of savannas and other non-forest habitats in the lowland tropics (Smith & Brown 1973, Livingstone & Clayton 1980, Klink & Joly 1989). Rain-forest Cyperaceae also have foliage very unlike that of non-forest species; some species of *Mapania*, for example, have extraordinarily broad short leaves with a 'pseudopetiole'.

The leaf surface in rain-forest herbs is seldom smooth and glossy as in most of the woody plants; in some the leaves have a peculiar velvety surface, e.g. in various Araceae and Orchidaceae and some dicotyledonous herbs, such as *Neckia* sp. in Borneo: the seedling leaves of some woody plants, e.g. *Cecropia* spp. and a few Rubiaceae, are similar. The velvety surface is due to the cells of the epidermis, which project as papillae. Stahl (1896) supposed that this would cause raindrops to spread out to form a water film and suggested that this would increase the absorption of radiation. In fact the surface of these leaves is usually water-repellent and this may increase their photosynthetic efficiency.

Other features common in the leaves of ground herbs, but rare in other rain-forest plants are iridescence, red pigmentation of the lower surface, and variegation.

Bluish iridescence can be observed in plants of deep shade such as some *Selaginella* spp. and terrestrial species of *Trichomanes* as well as in flowering plants such as some *Mapania* spp., e.g. *M. monostachya* (Borneo). This peculiar feature is found only in plants of deep shade and disappears when they are grown in stronger light. According to Gentner (1909) it is due to granules of cutin in the cell walls. Lee & Lowry (1975) could not confirm this. They found that the iridescence disappeared when the leaves were immersed in water and in *Selaginella willdenowii* evidence from reflectance spectra showed that part of the outer walls of the epidermal cells acted as a quarter-wavelength interference filter. They suggest that this may have adaptive value by increasing the absorption of the photosynthetically active wavelengths of light in the 600–680 nm range.

Red colouring of the lower (abaxial) leaf surface is common in *Begonia* spp., various Commelinaceae and Melastomataceae, and other rain-forest herbs. It is occasionally seen in woody species but usually only in juvenile plants. Lee *et al.* (1979) studied four Malayan herbaceous species with this feature and found that the red pigment was located, not in the epidermis, but in the mesophyll cells immediately beneath the palisade layer. In three of the species red colouring was a polymorphic character present only in some individuals. Early workers such as Stahl (1896) believed that red coloration led to increased heat absorption and was one of several adaptations to increase the transpiration rate, which was supposedly beneficial to plants living in a saturated atmosphere. Lee *et al.* (1979; see also Lee 1986, Lee & Graham 1986) could detect no differences in temperature between leaves with and without red abaxial pigmentation: they suggested that the pigmented layer increased the efficiency of photosynthesis by back-scattering light into the palisade cells, so bringing about greater energy absorption.

A more puzzling feature common in rain-forest herbs is variegation. In many species the leaves are spotted, blotched or striped with white, pale green or various shades of red, e.g. *Begonia* spp., tropical American *Cephaelis* spp., and many species of *Sonerila* in South-east Asia. Variegation is also found among mono-cotyledons; the leaves of the West African *Dracaena phrynoides* are mottled with yellowish green and those of the handsome *D. goldieana* of southeastern Nigeria are transversely striped in dark green and silver grey. Variegation does not seem to occur as a normal feature in any rain-forest trees or shrubs, although frequent in cultivars. Some epiphytes, e.g. South American species of *Peperomia* and *Vriesia*, are also variegated. It is difficult to suggest any possible adaptational significance for variegation; it is perhaps related to some aspect of the plant's biology such as the deterrance of predators or the attraction of pollinators, rather than to its metabolism.

The ability to reproduce vegetatively is common in rain-forest herbs and many probably usually reproduce in this way, but some reproduce by seed. Kiew (1986) found that of the four species of Malayan ground herbs she studied two relied on regular seed production for survival. Another (*Didymocarpus platypus*) rarely set seed, but was estimated to live for twenty years. Most ground herbs appear to be pollinated by insects or birds, but the majority, like the woody plants, have inconspicu-ous flowers.

It is remarkable that at least some rain-forest grasses are probably pollinated by insects and not by wind. At Moraballi Creek (Guyana) the flowers of *Pariana radiciflora* were habitually visited by small bees (*Melipona* spp.) and flies (Phoridae) and it was suggested that insect pollination in this grass might be related to the extreme stillness of the air in the rain-forest undergrowth (Davis & Richards 1933–34, p. 372). In forest near Belém (Amazonia) another species of *Pariana* when in flower was surrounded by veritable clouds of bees and other insects, which seemed to be attracted by

Fig. 6.6 *Pariana* sp. in undergrowth of várzea forest, near Belém, Pará, Brazil. An insect-pollinated grass. The flowers have ten conspicuous stamens and attract numerous insects.

the bright yellow stamens (Fig. 6.6). Soderstrom & Calderón (1971) found that in various parts of the South American rain forest insects (bees, Diptera of several groups and others) visited flowers of *Pariana* and *Olyra*. Entomophily in these forest grasses is perhaps a second-ary adaptation to the windless environment of the forest undergrowth rather than a primitive character.

In many ground herbs the flowers are borne at or just above the ground on stalks which are leafless or bear only scale leaves. In Borneo, species of *Curculigo* and *Forrestia* and many other herbaceous plants bear their inflorescences at the base of the plant; in *Cyrtandra penduliflora* the inflorescence axis lies limply on the surface of the ground. In many Malesian Zingiberaceae the inflorescences, which are more vividly coloured than in most of the ground flora, spring from underground rhizomes often a metre or more from the flowerless leafy shoots. In some members of this family the fruits ripen within pocket-like water-holding bracts and even dehisce when wet. In other species the bracts rot and form a wet mass around the developing fruit (R.E. Holttum). The author has observed this also in an unidentified member of the Rapateaceae in Amazonia.

Many ground herbs have berries or fleshy fruits which may be dispersed by small mammals, pigs or birds. Others are probably ant-dispersed.

6.2 Climbers

The size and abundance of the climbing plants, especially large woody climbers (lianes), are one of the most characteristic features of tropical rain forests; their strange, often fantastic, forms contribute much to its unfamiliar aspect. The ivy, clematis and other climbers of temperate woodlands give but a faint notion of the

diversity of such plants in the tropics. Lianes (often called vines or bush-ropes) are almost always present in both primary and secondary forest as well as being early colonists of gaps and clearings; their stems, sometimes rope- or wire-like, may be as thick as an arm or thigh and vanish into the mass of foliage overhead, hanging down here and there in gigantic loops. They often grow from tree to tree and link the crowns so firmly that sometimes, even if a trunk is cut through at the base, the tree will not fall. Some tropical lianes attain almost incredible lengths. Treub (1883) measured the stem of a rattan (climbing palm) 240 m long in Java; Burkill (1966, p. 1913) records another (*Calamus* sp.) of 169 m from Malaya. Most tropical climbers do not grow to such enormous lengths and probably seldom exceed about 70 m. The stems sometimes become very thick: at Belém in Amazonia Ducke & Black (1953) saw a stem of *Bauhinia siqueiraei* of 60 cm diameter. Lianes

Fig. 6.7 Lianes in old secondary forest, Omo Forest Reserve, Nigeria (1935). The roots of a strangling fig (*Ficus* sp.) are attached to the trunk of the large tree on the right.

are troublesome to study and collect, so in spite of their ecological importance and striking appearance, they remain, as Janzen (1975) said, the most poorly understood of the major synusiae of rain-forest plants.[1]

Although climbers are a fairly well-defined plant form, weak-stemmed trees like the South American *Annona haematantha* and half-climbing shrubs such as the African *Icacina trichantha*, are transitional between woody climbers and fully independent plants. Similarly, no sharp line can be drawn between climbers and epiphytes. Some epiphytes have climbing stems and many root climbers, e.g any Araceae, some species of the tropical American genus *Carludovica*, the Malayan orchid *Dipodium pictum*, and perhaps some Marcgraviaceae, start life rooted in the ground but eventually lose their connection with it and become epiphytes. Some hemiepiphytes (pp. 133ff.) when fully grown closely resemble large woody climbers.

Climbing plants are found in most climates and every type of vegetation where there are trees to support them, but they are more abundant, and far more numerous in species, in the tropical rain forest than in any other plant formation. Schenck (1892–93) estimated that more than 90% of all species of climbing plants occur within the tropics. In the West Indies they form about 8.5% of all flowering plants in contrast to Europe where they are less than 2% (Grisebach, quoted by Schenck 1892–93, p. 57). According to Schenck the South American rain forest is the richest in lianes. Southeast Asia is also very rich: in Malaya 14 out of 48 genera and 67 out of 198 species of the Annonaceae alone are lianes and in Borneo there are nearly 200 species of climbing palms (J. Dransfield). The relative scarcity of climbers in temperate forests is perhaps not easy to explain, but their abundance in the tropics is not surprising in view of the great advantages of the climbing habit in very tall closed plant communities. In a tropical rain forest a liane can reach the well-illuminated canopy with great economy of stem material and then concentrate its resources on the production of leaves and flowers.

6.2.1 Morphology and floristic composition

Climbing plants can be classified according to their means of attachment to their supports into scramblers, twiners, root-climbers and tendril-climbers. Scramblers neither twine nor have specialized sensitive organs of attachment, although many of them have structures that passively assist them in climbing such as hooks, recurved

spines and branches diverging at right angles. Twiners also often lack tendrils, but the tip of the young stem is able to revolve, so that the plant becomes securely wound round its support. Root-climbers attach themselves by aerial roots, which cling to the surface over which the plant grows. Tendril-climbers, the most specialized of the four classes, possess organs of varied morphology which are sensitive to contact with a support to which they fix themselves actively, usually by curling round it. Some climbing plants have more than one means of attachment: species of *Hoya* twine and also bear roots at the nodes, and some Bignoniaceae twine and also have tendrils and adherent roots (Schenck 1892–93). Many species lack special attachment organs in the early stages of development and some lose them when they become mature.

Rain-forest climbers fall into two (not always sharply separated) groups (synusiae): (1) shade-tolerant and (2) light-demanding (skiophytic and photophytic climbers of Grubb *et al.* 1963). Group 1, which includes ferns, Araceae, tropical American Cyclanthaceae, and plants of a few other families, are mainly herbaceous root-climbers. They are found chiefly in the oligophotic zone (C and D storeys), although some grow up trees to 20 m or more. Group 2 are the large, nearly always woody, lianes, which are tolerant of shade only when young and need the sunlight of the forest canopy to become mature. In undisturbed primary forest these large Group 2 lianes are seldom conspicuous, but in natural and artificial gaps and secondary vegetation they are abundant, forming impenetrable tangles through which young trees can grow only with difficulty. Curtains of lianes are also characteristic of river banks, while in logged forests and on farmland lianes often envelop standing dead trees, forming 'climber towers'. In tree plantations lianes are a serious pest.

The synusia of large climbers includes many more species than Group 1. The majority of these are dicotyledons of many families; the most numerous are Annonaceae, Apocynaceae, Bignoniaceae, Celastraceae, Combretaceae, Connaraceae, Convolvulaceae, Dilleniaceae, Leguminosae, Malpighiaceae, Menispermaceae, Passifloraceae and Sapindaceae. The climbing *Bauhinias* and Bignoniaceae, which are so conspicuous in South American rain forests, are absent in Africa.

The climbing palms (rattans or rotans) are the largest group of climbing monocotyledons. Though found in both Old and New World rain forests, they are more abundant in Malesia than anywhere else: there they are represented by over 479 species (J. Dransfield) and play an important part in the physiognomy of the vegetation. Crowns of palms are scattered everywhere in the

---

[1] Putz & Mooney's (1991) *Biology of vines* unfortunately appeared too late to be fully utilised in writing this section.

canopy; they belong not to tree palms but to rattans. In the Queensland rain forest climbing palms ('lawyer vines') are also conspicuous. In West Africa climbing palms are found chiefly in secondary and swamp vegetation and are smaller, less abundant and much fewer in species than in the eastern tropics. In tropical America the only genus of climbing palms is *Desmoncus*.

Other monocotyledonous families represented among the climbers of Group 2 are *Freycinetia* and some species of *Gnetum*.

A mass of lianes brought down by a fallen tree may seem a chaotic tangle of leaves stems and branches, yet lianes, like trees, are built according to recognizable architectural models. These have not yet been fully explored. Hallé *et al.* (1978) found that some lianes can be assigned to models that also include trees, but others do not conform to known tree models. Cremers (1973, 1974) studied the architecture of twenty West African lianes, but much more work is needed. Peñalosa (1984) has shown that some lianes have shoots of several different types (see below).

The stems of tropical climbers, especially the large woody species, show great diversity in their external appearance and internal anatomy, as well as in their attachment organs (Troll 1938–43). They vary much in thickness and may be round, angular or flattened in cross-section; sometimes they are twisted or cable-like, sometimes winged or almost chain-like. Many of the tropical American species of *Bauhinia* (Caesalpinioideae), the 'turtle-step bush ropes' of the Guyana Indians, are ribbon-shaped with undulate edges.

In their internal structure, liane stems show many different kinds of anomalous secondary growth. Although their superficial features and anatomical peculiarities are to some extent characteristic of the family or genus, they are not a reliable guide to the taxonomy (Obaton 1960). According to Schenck's (1892–93) classic study, stem structure in lianes also has little relation to the method of climbing. Anomalous secondary growth is often found only in the elongated climbing stems and not in the short 'sun shoots' of the crown. In the larger stems the xylem commonly consists of a number of partly or completely separated strands embedded in softer tissue. Liane stems thus often resemble a rope internally as well as externally and combine flexibility with great tensile strength; both properties are important in equipping the plant for its special mode of life.

The morphology and anatomy of rattans have peculiar and little-known features (Dransfield 1978, Uhl & Dransfield 1987). In the large Indo-Malayan genus *Korthalsia* the climbing stems branch at the top when they reach the canopy, but in other rattan genera, as in most palms, the aerial stems remain unbranched. Many rattans (including *Korthalsia*) branch at the base.

The subterranean roots of some lianes show anatomical anomalies similar to those in their climbing stems (Schenck 1892–93) and the aerial roots of root climbers have various special features. In the Araceae, for instance, they may be of two kinds, adhering and nutritive. The former grow more or less horizontally away from the light and fix the plant to its support, while the latter are strongly positively geotropic. The internal structure of the two kinds of root is different, the mechanical tissues being better developed in the adhering roots, the conducting in the nutritive roots.

The leaves of most rain-forest climbers are evergreen and similar in size to those of the trees which support them: the majority probably belong to the (mesophyll and notophyll size-classes, but in shape they tend to be somewhat different from the leaves of the trees. Simple elliptical or oblong-lanceolate leaves are relatively uncommon and compound leaves are perhaps more frequent than among the trees. There is a marked tendency in tropical lianes, as in all climbing plants, for simple leaves (and the leaflets of compound leaves) to be broadest at the reniform or cordate base and relatively short. Some rain-forest climbers have peltate leaves in which the main veins usually diverge palmately from the insertion of the petiole. The leaf apex may or may not be acuminate, but drip-tips, if present, are found only in juvenile leaves. The lamina is usually at a wide angle to the petiole.

This combination of characters is so prevalent in the leaves of climbing plants, and in species of so many unrelated families, that it is likely that it has some functional significance. Lindmann (1900) suggested that broad leaves widest at the base have the advantage of providing a large transpiring area, which he considered to be necessary owing to the resistance to water conduction of a very long and relatively slender stem: it also allows the leaf to be easily inserted into a pre-existing leaf mosaic.

Young individuals of lianes live under very different conditions from adult plants, which have most of their foliage in the forest canopy, and they probably pass more quickly than young trees from one environment to the other. It is therefore not surprising that the shade leaves or 'bathyphylls' of many rain-forest climbers are strikingly different in many respects from the sun leaves or 'acrophylls' (Holttum 1938a). Extreme examples of this type of heterophylly are seen in some Araceae (Goebel 1930, pp. 388–91), and some species of *Ficus*, subgenus *Synoecia* (Corner 1939). In the tropical American genus *Marcgravia* the bathyphylls and acrophylls are sometimes so different that they have sometimes been thought to belong to different species or even genera. Young plants of *Marcgravia* are root-climbers. The leaves are small, shortly rectangular in shape, have a velvety surface, and are tightly pressed against the bark of the supporting tree. When the plant reaches the canopy, the stem branches and produces orthotropic shoots without adhering roots. The leaves on these shoots project freely from the stem and are quite different in shape and texture from those on the young plagiotropic shoots; they are elongated, sclerophyllous and have a glossy cuticle. The bathyphylls no doubt help to retain moisture round the aerial roots beneath them; as the author noted in Guyana, they wilt more quickly when picked than the adult leaves.

Another extraordinary heterophyllous liane is *Triphyophyllum peltatum*, an endemic of the West African Western Block forests (Fig. 6.8). The mature climbing stems bear leaves with two apical hooks by which they attach themselves to other plants. On juvenile non-climbing stems there are lanceolate leaves, usually without hooks, and leaves in which there is a continuation of the midrib covered with sticky glandular hairs: these catch, and possibly digest, small insects (Green *et al.* 1979).

Remarkable examples of heterophylly are found in the climbing ferns *Lomogramma* and *Lomariopsis* (Holttum 1938a). The stems adhere to tree trunks by adventitious roots and the acrophylls are almost as different from the bathyphylls as in *Marcgravia*.

6.2.2 Physiology

The movements and responses to stimuli of climbing plants have been much studied, but other aspects of their physiology have been largely neglected until recently. Of particular ecological interest is the water flux through the stems of tropical lianes, which owing to their very slow rates of growth in diameter (Putz 1990), are of small cross-sectional area and, as was seen earlier, are often extremely long. Several authors have suggested that resistances to water conduction may limit the length to which they can grow and perhaps also determine their rather unusual branching systems (see below).

The vessels in liane stems are said to be among the widest and longest in the plant kingdom. They are often over 0.3 mm and sometimes over 0.6 mm wide (Gessner 1956b); some over 7.7 m long have been recorded (Ewers *et al.* 1991), but such very large vessels are accompanied by others of smaller diameter and length. It is evident that the water conducting capacity of liane stems must

Fig. 6.8 *Triphyophyllum peltatum*, a reputedly carnivorous liane (Green *et al.* 1979). Secondary forest, Black Johnson Beach, Sierra Leone 1971. A Juvenile stage (non-climbing) with two types of leaf in one of which the lamina is replaced by sticky stipitate glands which trap insects. B Mature climbing stage in which the leaves have two hooks at the apex.

be greater than that of most woody plants. Berger (1931) measured the conductivity for *Vitis* and some other temperate climbers and found that it was almost 100% of that theoretically possible. Unfortunately there is little comparable information for tropical lianes, but Tammes (1938) found that in *Merremia nymphaeifolia* in Sulawesi the amount of water conducted by pieces of stem 5 m long was only about 66% of that conducted by pieces 1 m long under similar conditions. This suggests that the resistance increases quite rapidly with the length of the stem. Gessner (1956b) concluded that the length of liane stems is limited by their ability to conduct water and suggested that very long lianes can exist only in climates in which the rate of evaporation is never very great.

Coster (1932b) measured the speed of the transpiration stream in liane stems in Java by means of dyes. In various species he found that under sunny conditions the rate was 100–200 cm min^{-1}; in *Aristolochia brasiliensis* it reached 250 cm min^{-1}, the highest rate recorded in any plant.

Negative pressures of several atmospheres develop in the stems of climbers which are transpiring rapidly. Scholander *et al.* (1957); see also Scholander 1958) made manometric measurements of pressures in stems of the liane *Tetracera* sp. at Barro Colorado Island (Panama) and found that large negative pressures develop during the day when transpiration is fast; even in the night when rain was falling the pressure at the base of the stem was negative. Of particular interest because of the ability of tropical lianes to survive and recover from injuries are Scholander's observations on air embolisms in stems that were cut through or wounded. In *Tetracera* he found that embolisms did not extend beyond the severed vessel segments and did not damage the water cohesion in the stem as a whole. Even when the leaves had wilted, they would revive if the cut end of the stem was placed in water. After cutting through the stem the pressure gradient increased; it was estimated that in a liane stem 40 m long it might reach 10 atm or more.

More recent work on water conduction in lianes and its relation to their stem anatomy is discussed by Ewers *et al.* (1991) and Fisher & Ewers (1992), but there is much in the water economy of lianes, particularly the role of positive root pressure, which is still unclear. Further work is needed on a greater variety of species in natural environmental conditions.

6.2.3 Reproduction and development

In general the floral biology of large lianes seems to be similar to that of tall forest trees. Many lianes have conspicuous, brightly coloured, flowers; in tropical America a tree covered with a flowering mass of one of the climbing Bignoniaceae is an unforgettable sight. With the exception of the climbing species of *Gnetum*, which are probably wind-pollinated (Faegri & van der Pijl 1982), large lianes seem to be pollinated by insects, birds, bats or possibly other mammals: the syndromes of floral characters are like those found in trees and other plants with the same type of pollen vector. Some lianes, e.g. various Menispermaceae and West African species of *Pararistolochia* (Fig. 6.9), are cauliflorous, flowering only on the unbranched lower part of the stem.

The seeding of rain-forest climbers is often extremely abundant and their fallen fruits and/or seeds are sometimes conspicuous on the forest floor: those of *Calycobolus africanus* cover the ground in the Okomu Reserve in Nigeria in February. In their fruiting, and probably in their requirements for germination and early development, any large lianes (Group 2) show characteristics similar to those of large shade-intolerant trees, while others (especially herbaceous climbers such as *Merremia* spp. and other invaders of large clearings) are perhaps more like early seral nomad trees in these respects. Dispersal of the larger lianes is usually by wind or animals: many have winged seeds or fruits, e.g. many Bignoniaceae, some Convolvulaceae, *Banisteriopsis* and *Serjania*, but others have berries or fleshy fruits dispersed presumably by birds, monkeys, bats or other

Fig. 6.9 A cauliflorous liane, *Pararistolochia flos-avis*, Omo Forest Reserve, Nigeria (1935).

mammals. *Strychnos* spp. have massive, very heavy fruits, which drop to the ground, the seeds being probably dispersed later by rodents and other ground-living animals. Keay (1957) found that in old secondary forest in Nigeria 48% of the lianes that fruited high up in the canopy were wind-dispersed but none of those fruiting near the ground, most of which were cauliflorous. Gentry (1982a) found that in many forest sites in tropical America the proportion of wind-dispersed species was larger in lianes than in trees.

Viable seeds of some lianes have been found in seed banks, e.g. *Pericampylus incanus* in Malaya (Putz & Appanah 1987); the ability to remain dormant in the soil is probably common among liane seeds. Small juvenile lianes are often plentiful in the shade of mature forest and counts show that they may form a large proportion of the population of small woody plants. In Nigeria small plants of *Calycobolus* and *Neuropeltis* can survive in a suppressed condition for many years (Keay 1957). These seem comparable to the 'seedling banks' of some trees, but may in fact have arisen vegetatively. Janzen (1975) found in Costa Rica that most young plants of lianes were not seedlings but were connected to lateral roots of established plants.

Most lianes can probably reproduce vegetatively as well as reproducing prolifically by seed. They also have a great capacity for recovery when cut down or when the supporting tree falls down; this is partly why they are so troublesome in plantations and managed forests. Schenck (1892–93, p. 259) found that detached portions of liane stems could remain alive for many months. He attributed the ability of lianes to survive injuries to the intimate mixture of parenchyma cells containing food reserves with other xylem elements in their stems.

In the rattan *Korthalsia* the branches at the base serve for vegetative spread; in this way a plant of *K. flagellaris* in a peat swamp forest in Sumatera was able to form a branching system of thirty apices spread over about 2300 m^2 of forest canopy (Dransfield 1978). Many other rattans form clumps and some have stolons which spread along the surface of the ground before turning up to become climbing stems. In some *Plectocomia* spp., in which the shoots are hapaxanthic, small axillary branches from near the base grow into independent plants when the old stem collapses. Observations on vegetative spread in dicotyledonous lianes, many of which may have a capacity for reproduction as great or greater than that of rattans, would be useful. The tropical American climbing palm *Desmoncus isthmius* produces from its base a series of increasingly large stems until a clump is formed (Putz 1983a).

The efficacy of vegetative reproduction in lianes is shown by Peñalosa's (1984) study of the behaviour of *Ipomoea phillomega* and *Marsdenia laxiflora* in rain forest at Las Tuxtlas (Mexico). Both species have non-climbing stolons which arise from near the base of the erect twining stems which grow up into the canopy. In primary forest all the climbing stems were traceable to established plants, but seedlings were occasionally found in clearings. In *I. phillomega* the stolons grow at a mean rate of 13.6 cm d^{-1} and for distance up to 30 m. If their apical buds are destroyed by herbivores, they branch. In well-illuminated sites some of the branches become twining shoots, but sometimes the stolons change their orientation and themselves begin to grow upwards and to twine. The growth pattern of *M. laxiflora* is similar except that the twining stems, which in *I. phillomega* are leafy, are bare in *M. laxiflora*, in which most of the foliage is on short branches in the canopy. In both these lianes the stolons have an exploratory function, enabling the plant to invade gaps and clearings. Clones are formed as some of the twining stems become independent plants (Fig. 6.10).

In open sites with no near supports some young climbers remain for a long while as erect shrubs with no tendency to climb and may flower when only about a metre high, for example *Agelaea trifolia* in clearings in Nigeria; inside the forest it becomes a liane 30 m or more high. Janzen (1975) noted similar behaviour in some lianes in deciduous forest in Costa Rica: after an apparently inactive stage the stem suddenly begins to elongate and grows about 5 cm d^{-1}. Other species had no inactive stage and climbed from the start.

The nature of the support seems of little importance for climbers: it may be a tree or the stem of another climber and root-climbers often attach themselves to rocks. Twiners cannot climb trees of more than a certain diameter.

The attachment of old climbers is often at a considerable height above ground, so that a great length of unsupported stem hangs down from above. In such cases the climber may have made use of a smaller supporting tree which has long since died and disappeared or it may have originally made contact with a young tree and kept pace with its upward growth.

6.2.4 Role in the ecosystem

Climbing plants play an important part in rain-forest ecosystems and make a considerable contribution to their structure and productivity. However, the abundance of climbers varies very much between different forest types and seral stages; quantitative information is scarce. Because of their growth form lianes have much higher

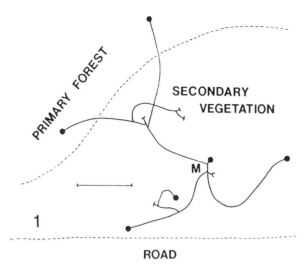

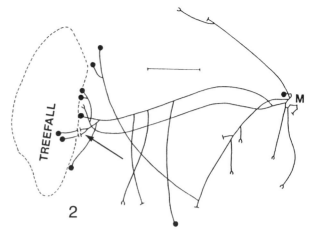

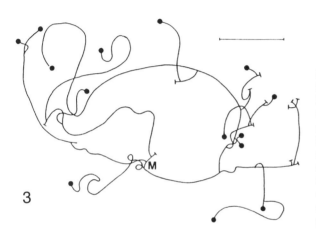

Fig. 6.10 Maps of stolon systems of two species of liane showing vegetative spreading in rain forest, Veracruz, Mexico. After Peñalosa (1984, figs. 8, 9 and 11). (1) and (2) *Ipomoea phillomega*, (3) *Marsdenia laxiflora*. In (1) *I. phillomega* is invading primary forest from adjacent secondary vegetation; in (2) it is invading a tree-fall gap (branches of fallen tree indicated by arrow) from surrounding forest. (3) is a map of a large stolon system of *M. laxiflora* in primary forest. M, 'manifold' (bud) from which stolons and climbing shoots arise. Closed circles show position of liane canopies. Branches with 'Y' endings are actively extending; those with 'T' endings have lost their growing apex.

ratios of leaf area to cross-sectional area of stem than trees and neither stem counts nor basal area measurements are useful indices of their importance in forest communities, which is more appropriately measured by biomass and leaf-area index (l.a.i.).

In a sample of primary mixed forest in Amazonia Klinge & Rodrigues (1974) estimated that the dry mass of lianes was about 46 t ha^{-1} and about 6% of the total above-ground biomass. In a forest on nutrient-poor lateritic soil at San Carlos (Venezuela) Putz (1983b) found the mass of lianes to be only 15.7 t ha^{-1} and *ca.* 4.5% of the total biomass. The average l.a.i. was 1.2 mmm^{-2}, *ca.* 19% of the total. In mixed dipterocarp forest at Pasoh (Malaya) the abundance of lianes is much less than in the previous examples: their dry mass was only 9 t ha^{-1} and the l.a.i. 1.9% of the total

(Kato *et al.* 1978). Figures for other rain-forest sites are given by Hegarty & Caballé (1991) who conclude that lianes seldom contribute more than 5% of the total above-ground biomass of humid tropical forests and that the biomass of their leaves is *ca.* 5–20% of the total, compared with 1–2% for the leaves of trees.

Because lianes are generally light-demanding, their abundance in forest communities seems to depend largely on the amount of natural and anthropogenic disturbance. The great abundance of lianes in the rain forest of Gabon (Hladik 1974) may be due to damage by elephants (Putz 1983b). But other factors are also involved. According to Fox (1969) and Jacobs (1976) many species of liane are unable to invade large bare areas, possibly because in their young stages they require high atmospheric humidity. Gentry (1982a) and

Putz & Chai (1987) suggest that lianes tend to be abundant on nutrient-rich soils, and Janzen (1975) believes that they need a high concentration of soil nutrients to maintain their rapid growth rates. Nutrient deficiency may be the cause of the relative scarcity of lianes on white sand in the Guyana wallaba forest (Davis & Richards 1933–34) and in heath forest in Sarawak (Richards 1936, Putz & Chai 1987)[2].

In Amazonia, many hundreds of square kilometres are covered with 'liana forest', an open community with some large trees and a great abundance of lianes (Pires & Prance 1985). The factors responsible for the occurrence of this forest type are not clear. Large areas of it in the Xingu basin have been used for shifting cultivation in the past and may be late seral stages (Balée & Campbell 1990). In general lianes are less abundant in montane than in lowland rain forests, e.g. in Ecuador (Grubb *et al.* 1963) (see Chapter 17).

Lianes seldom reach the tops of very tall trees, but they help to close the canopy between neighbouring B storey trees, thus increasing the shade below. They also increase the resistance of the stand to wind by binding the tree crowns together, although when windfalls occur or trees are felled they may cause one tree to damage or pull down others when it falls. Lianes are formidable competitors with trees for space and light, and often make their crowns one-sided or misshapen. Twiners may have a marked constricting effect, leaving the trunk of the supporting tree with deep spiral grooves when they are removed. A heavy load of climbers often breaks the branches of a tree and slides to the ground; they may then die or begin to climb up to the canopy once more. In forest at La Selva, Costa Rica, Clark & Clark (1990) found that the load of lianes and hemiepiphytes varied greatly between different species of tree and that, as might be expected, the load had a highly significant negative effect on the diameter growth of the host tree. In large gaps blankets of climbers, which may persist for several years, often make it difficult for tree seedlings and saplings to survive.

Lianes are such formidable competitors with trees that it is not surprising that the latter have evolved defence mechanisms against them. Some trees, according to Janzen (1966) and Putz (1984), have structural characters which make it difficult for climbers to attach themselves, such as flexible trunks and drooping leaves with very smooth surfaces from which tendrils easily slide off. The most remarkable, and the best authenticated, of

these defence mechanisms is the symbiotic relationship between some small trees and ants which was originally suggested by Belt (1874) and later substantiated by Janzen (1966, 1967). *Acacia cornigera* and related 'bullshorn acacias' of second growth in seasonally dry parts of Central America have large paired thorns inhabited by ants of the genus *Pseudomyrmex*. One of the important beneficial results of the partnership for the tree is that the ants bite off the tendrils and growing points of climbers that attempt to attach themselves to their host. The ants also attack other plants in the neighbourhood, so that circular bare areas are formed round acacias inhabited by ants (but not round those which ants have failed to colonize). A similar relationship exists between the small West African rain-forest tree *Barteria fistulosa* and ants of the genus *Pachysima* (Janzen 1972). In the Okomu Reserve in Nigeria, and other places where *Barteria* is common, the circular patches almost free of lianes and tree seedlings beneath these trees are well known and are popularly supposed to be due to a 'ju-ju' (Fig. 18.11).

Because of their ill effects on the growth and regeneration of timber trees, foresters regard lianes as pernicious weeds. Young tree plantations often provide ideal conditions for their growth and the opening of the forest canopy in routine thinning operations also encourages their development. For these reasons regular and repeated 'climber-cutting' is important in forest systems management in the humid tropics (see Putz *et al.* 1987).

The synusia of small climbers, such as climbing ferns, the smaller aroids and *Carludovica* spp., which never grow far above ground level, is much less important in the rain-forest ecosystem than that of the large lianes. They are sometimes a conspicuous feature of riverain and swamp forests and of some of the better-illuminated forest types on well-drained sites, but their contribution to the biomass and productivity of the community is probably small.

6.3 Hemiepiphytes and 'stranglers'

Hemiepiphytes and 'stranglers' are one of the most remarkable ecological groups in the tropical rain forest and belong to a life-form that has no parallel in north temperate forests. Stranglers are plants that begin life as epiphytes and later send down roots to the soil, becoming independent, or almost independent, plants and often killing the trees by which they were originally

[2] Though apparently not at Gunung Mulu (Proctor *et al.* 1983a).

supported (Fig. 6.11). Although no sharp line can be drawn between stranglers and hemiepiphytes, which send down roots to the soil but do not become mechanically self-supporting or kill their supporting trees, the two groups play different roles in the ecosystem. Stranglers form a synusia which stands on the borderline between the dependent and independent plants. Most of them belong to the genera *Ficus* (Moraceae), *Clusia* (Guttiferae) and *Schefflera* (Araliaceae). A somewhat similar habit is seen in tropical American species of *Coussapoa* (Moraceae) and *Posoqueria* (Rubiaceae), some species of *Metrosideros* (Myrtaceae) and probably other genera, but the life-history of these plants has been little studied and there are differences in the way they develop. A general account of hemiepiphytes has been given by Putz & Holbrook (1986).

In the humid tropics of the Old World the strangling figs (*Ficus* spp.) are abundant as species and individuals

Fig. 6.11 Strangling fig (*Ficus* sp.), on tree at Noah Creek, Queensland (1981). The lower part of the crown on the left of the tree is that of the fig.

and often play a considerable part in the physiognomy of the rain forest (Fig. 6.12). The seeds, usually carried by birds, germinate on tall trees, not on the ground, often in a fork between the trunk and a large branch, although some stranglers prefer small branches near the outside of the crown (Todzia 1986). The seedling grows into a stout bush, which sends down positively geotropic aerial roots, some of which are in close contact with the trunk of the host tree, while others descend vertically through the air like perpendicular cables. The roots eventually reach the ground and ramify in the soil; it is likely that in the strangling figs, as in *Ficus benjamina*, the aerial part of the root contracts and becomes taut. The descending roots multiply, branch and anastomose until the trunk of the supporting tree becomes encased in a network of extremely strong, woody meshes; the crown of the fig has meanwhile become large and heavy. The crown of the host tree suffers from competing with the fig for space and light and after a time the host tree may die and rot away, leaving the strangler as a hollow, but quite independent, tree in its place. The exact cause of death in the host is not certain: probably it is due partly to shading and partly, perhaps, to constriction of its bark by the fig and competition with its roots. Some species of *Ficus* are more lethal to their hosts than others (Corner 1988). Strangling figs may become trees of great size, belonging to the highest storey of the forest (see Hallé *et al.* 1978, Fig. 106 C). Michaloud & Michaloud-Pelletier (1987) have shown that in rain forests hemiepiphytic *Ficus* spp. colonize certain 'host' trees, e.g. *Parinari excelsa*, much more frequently than other, often commoner, tree species.

In South American rain forests, in the author's experience, strangling species of *Ficus* are not as abundant as in Asia and Africa, and the epiphytic species that occur there are mostly hemiepiphytes sending roots to the ground rather than true stranglers. By far the most important genus of stranglers in this region is *Clusia*, which is represented by numerous species and is often abundant (Fig. 6.13). The seedlings, like those of the strangling figs, germinate high up on first- or second-storey trees. Aerial roots grow down to the ground and anastomose; short clasping roots are produced which fix the plant firmly to its support. These clusias develop large crowns, but do not form such a stout or close network of roots as the strangling figs. They seldom kill their hosts, although they often greatly deform and stunt their crowns, and they do not usually become independent trees. A detailed account of the life-history of the West Indian *C. rosea* is given by Schimper (1888).

Fig. 6.12 Development of a strangling fig (*Ficus* sp.). After Corner (1988) (re-drawn).

6.4 Epiphytes

Epiphytes grow attached to the trunks and branches of trees and other plants, some even on the surface of living leaves. In closed rain forests the majority of the epiphytes grow high above the ground where relatively strong illumination compensates for lack of soil and the precarious water supply, but some grow on twigs and tree bases at low levels where the light climate is little more favourable than on the ground. The mode of life of epiphytes is highly specialized; epiphytic flowering plants and ferns consequently differ greatly in physiognomy as well as in physiology from the herbaceous ground flora to which indeed most of them are not closely related taxonomically. Rain-forest epiphytes are mainly small plants, although a few grow to several metres high. Their structural and physiological adaptations to the conditions under which they live are extraordinarily varied and they show more diversity in their water economy and patterns of metabolism than any other rain-forest synusiae.

In open situations epiphytes sometimes grow on rocks or even on the ground, but in closed forests they are wholly dependent on supporting plants (phorophytes). It has been generally assumed that the dependence is for mechanical support alone and that, unlike parasites,

Fig. 6.13 Hemiepiphytic *Clusia* sp. with clasping roots attaching it to host tree (*Aldina* sp.), in degraded caatinga, Ponte Negra, Manaus, Brazil (1968).

epiphytes do not require organic food from other plants. As will be seen later, it has been suggested that other epiphytes besides the obviously parasitic mistletoes (Loranthaceae and Viscaceae) may not be completely autotrophic. The assumption that most epiphytes are harmless to the trees on which they grow may also be questionable (pp. 144ff.).

Epiphytes are interesting not only in themselves and because of their relations with other plants; they provide the chief, and in some cases the only, habitat for a rich and varied fauna which plays an important part in the rain-forest ecosystem. Epiphytic flowering plants and ferns provide the chief nesting sites for many species of arboreal ants. Some epiphytes are specialised myrmecophytes in which there is a symbiotic relationship between the ants and the plants with which they live. The masses of humus collected by many epiphytes have a large fauna of small invertebrates. The tropical American Bromeliaceae collect considerable quantities of water as well as organic debris in the 'tanks' (phytotelmata)

formed by their rosettes of overlapping leaves (Fig. 6.14); organisms of many kinds, including even crabs and amphibians, live in these tanks and form miniature ecosystems or 'microcosms'. In forests where bromeliads are abundant their tanks constitute veritable 'aerial marshes' (Picado 1912, 1913). Bromeliads can be important foci of human and animal diseases as the larvae living in them may include those of mosquitoes carrying malaria and yellow fever.

In tropical rain forests epiphytes are not always very abundant, but they are always present; on a given area there are often many more species, and probably a greater biomass, of epiphytes on the trees than of herbaceous plants on the ground. The epiphytic flora includes a wealth of flowering plants and pteridophytes (macroepiphytes), as well as algae, fungi, lichens and bryophytes (microepiphytes), and it is the presence of these macroepiphytes which is particularly characteristic of tropical and other rain forests. Even in the so-called 'Olympic rain forest' of western North America (in reality a very humid coniferous forest), where epiphytes grow luxuriantly, the only vascular species are two pteridophytes.

An abundance of macroepiphytes usually indicates a permanently moist atmosphere and is characteristic of montane and subtropical as well as tropical rain forests. In these other formations the mean temperature is lower and the air may be even more constantly humid. In the mountains of Malesia, New Guinea and northern South America the trees are often almost hidden by masses of epiphytic vegetation, which reaches a degree of luxuriance not seen in lowland rain forests. In Ecuador Grubb *et al.* (1963) found 2555 epiphytes of ninety-one species in a sample plot of montane forest

Fig. 6.14 Epiphytic bromeliads, orchids, bryophytes and lichens on tree in arboretum, Turrialba, Costa Rica (1973).

Fig. 6.15 Epiphytes on branch of *Aldina* sp. in degraded caatinga, Ponte Negra, Manaus, Brazil (1968). Bromeliads, several species of orchids, bryophytes and lichens present.

at 1710 m and only 268 individuals of fifty-seven species in a similar plot of lowland forest at 380 m.

Hosokawa (1943, 1950) showed that within the tropics the 'epiphyte quotient' i.e. the percentage of 'typical' epiphyte species in the total vascular flora of a given area, was a useful measure of the humidity of the climate. He found that in the western Pacific islands this quotient varied from zero on very dry islands to fifteen or more on islands such as Kusaie and Ponape (Carolines) which have a humid equatorial climate.

In tropical rain forests some epiphytes are woody and may become shrubs of some size, e.g. *Coussapoa* spp. (tropical America), *Preusiella* spp. (West Africa) and *Rhododendron* spp. (Malesia). Schimper (1903) regarded the presence of these woody epiphytes as a feature distinguishing tropical rain forest from other plant formations (see p. 389), but they are also not uncommon in tropical montane rain forests, as well as in the temperate rain forests of the southern hemisphere.

6.4.1 Family representation

The epiphytic flowering plants and ferns of the tropical rain forest belong to an enormous number of species and genera, but relatively few families. Benzing (1990, from the data of Kress 1986), estimates that epiphytes (defined as plants that 'anchor on bark too often to be regarded as accidental epiphytes') form about 10% of all vascular plant species, but only a proportion, possibly less than half, occur in tropical rain forest; the rest are found in other types of tropical vegetation or outside the tropics. Forty-three genera include at least a hundred epiphytic species, but these belong to only twelve families.

By far the largest family of epiphytes is the Orchidaceae, which includes hundreds of genera and many thousands of species of 'typical' epiphytes. Like other epiphytes, orchids are much more numerous in Southeast Asia and tropical America than in Africa. The Malay Peninsula alone has about 800 species of orchid, the majority of which are epiphytes (Holttum 1957), about twice as many as in the whole of West Africa. But even in Africa orchids are the largest family of epiphytes: in the Nimba mountains (Liberia) Johansson (1974) found that out of a total of 153 species of macroepiphyte 101 were orchids. The comparative poverty of African rain forests in orchids, as in other plant groups well represented in rain forests elsewhere, is not easy to explain by present climatic conditions, although admittedly most African forests have a more marked dry season than large parts of tropical Asia and America.

Fig. 6.16 Epiphytic orchid on palm trunk, Utria National Park, Chocó Department, Colombia. (Photo. Alan Watson.)

It may be due to periods of dry climate in the past (pp. 292–3).

Many of the families to which epiphytes belong are pantropical, but the Bromeliaceae, one of the most abundant and species-rich families in tropical America, is absent in the Old World except for one species, *Pitcairnia feliciana*, which is restricted to West Africa (Guinea) and occurs, not as an epiphyte, but in rock crevices. The abundance of bromeliads, many of which are tolerant of extreme habitats not colonized by most other macroepiphytes (p. 139), partly accounts for the apparent luxuriance of epiphytic vegetation in the Neotropics. The absence of this family in the rain forests of the Old World could perhaps be due to some kind of 'evolutionary accident'.

The Cactaceae, of which several genera and many species are epiphytes in the rain forests of South America, are represented in the Old World only by *Rhipsalis baccifera*. This occurs in tropical Africa and Sri Lanka as well as in South and Central America. Buxbaum

(1969) suggested that it was introduced into the Old World by English sailors who used it as a substitute for mistletoe (*Viscum album*) in Christmas decorations, but long-distance dispersal by birds also seems possible.

In addition to the three families mentioned above, some eighty-one families of vascular plants include at least one epiphytic species, but only twelve include a hundred or more species of epiphyte (Benzing 1990). Among the latter may be mentioned Ericaceae, Gesneriaceae, Melastomataceae, Piperaceae and Rubiaceae among dicotyledons and Araceae among monocotyledons. Epiphytic pteridophytes include ferns of various families as well as species of *Lycopodium*, *Selaginella* and *Psilotum*. No family consists entirely of epiphytes. It is also of interest to note that many very large pantropical families include no epiphytes, e.g. grasses, Leguminosae and Labiatae.

Fig. 6.17 Epiphytic cactus *Rhipsalis cassytha* on fallen tree, Okomu Forest Reserve, Nigeria (1955). (Photo. D. E. Coombe.)

Fig. 6.18 Epiphytic fern, *Asplenium africanum*, Omo Forest Reserve, Nigeria (1935).

Of the three tropical rain-forest formations, the American is the richest and Africa by far the poorest in epiphytes (species). Madison (1977) estimated that there were 15 510 species of macroepiphytes in the Neotropical region and only 12 560 in the Palaeotropics.

6.4.2 Origin of epiphytes

The relatively small number of families of macroepiphytes raises the question of their origin. Schimper (1888) believed that epiphytes had evolved from terrestrial plants living in wet shady forests. He also suggested that their ancestors were pre-adapted by possessing suitable seed dispersal mechanisms and certain other characters making it possible for them to adopt an epiphytic mode of life. This view is perhaps plausible for the Araceae, many of which live on the ground in shady conditions (Simmonds 1949), but the shade-intolerance of most epiphytes, and their lack of close relatives in the herbaceous ground flora of the forest, would seem to be against it. Pittendrigh (1948) argued in favour of a theory proposed earlier by Tietze (1906) that the epiphytic Bromeliaceae originated from plants growing on rocks or soil in open, arid or even semi-desert conditions. More recently Benzing & Renfrow (1971a,b) and Benzing (1990), in the light of modern work on their physiology (pp. 143–4 below) have suggested that the ancestral habitat of the bromeliads was mesic and infertile rather than arid. It could perhaps have been some type of white sand savanna.

The epiphytic cacti belong to an evolutionarily advanced section of their family (Buxbaum 1969) and have no near terrestrial relatives in the rain forest. They (and probably epiphytes of other families) may have arisen in open, seasonally dry environments.

6.4.3 Epiphytic habitats

Large trees offer a range of habitats for epiphytes, differing in microclimatic and other factors, but all epiphytic habitats differ in important respects from those of terrestrial plants. The fact that the substratum is above ground level and is inclined at angles varying from horizontal on large branches to vertical on the trunk itself creates problems for seed dispersal and seedling establishment. Secondly, the substratum, except on old thick-barked trees, is more or less smooth and may provide an insecure anchorage, although forks and branch scars may offer easy sites for colonization. Epiphytic habitats are necessarily temporary as well as discontinuous. The twigs and branches on which most epiphytes grow are available for a period always shorter than the life-span of the whole tree. Benzing & Davidson (1979) estimate that in Florida surfaces on which epiphytes grow last on average twenty-five to thirty years. Thus epiphytes, like the colonists of forest gaps, are 'biological nomads' (p. 463).

Another important feature of epiphyte habitats is the lack of normal soil. The mineral nutrition of epiphytes has been discussed in detail by Benzing (1990). Here it must suffice to say that their only sources of nutrients are atmospheric precipitation and dust, the small quantities of solutes 'leaking' from their phorophytes, and organic debris brought to them in various ways. Small amounts of humus and other debris are usually present in cracks and hollows of the bark, especially on horizontal branches. Once the seedlings of epiphytes are established, soil-forming materials begin to accumulate round their roots. Many epiphytes, particularly the bracket, nest and tank epiphytes (p. 140), are able to collect considerable quantities of organic materials which their roots are able to exploit. The ants which inhabit the root systems of most macroepiphytes, except the smallest, gather fragments of leaves and flowers, seeds and animal debris from wide areas and play an active part in humus accumulation. Only extreme 'atmospheric' epiphytes (p. 140) depend entirely on the very small quantities of nutrients in dust and precipitation.

Under favourable conditions, as in upland rain forests where the lower average temperature favours the build-up rather than the decomposition of dead organic matter, masses of humus develop under colonies or large individuals of epiphytes; these may be 10 cm or more thick and may amount in total to several tonnes of dry matter per hectare (Klinge 1966, see also Pócs 1980). Such accumulations are sometimes referred to as 'sols suspendus' (Delamare-Deboutteville 1951), but they are comparable to the A horizon rather than to the whole profile of a terrestrial soil. They consist mainly of incompletely decomposed plant material, among which, according to Lötschert (1969), root fragments predominate, but it also includes animal excreta and dust particles; near active volcanos, as in parts of Central America, the latter consist mainly of volcanic ash. In addition to an invertebrate fauna the 'sols suspendus' contain micro-organisms of many kinds. Miehe (1911) found nitrifying and cellulose-decomposing bacteria in humus from the roots of epiphytes in Java and found no evidence to suggest that the microbiological processes taking place were different from those in a terrestrial soil. Schwabe (1962) identified potentially nitrogen-fixing blue-green algae such as *Nostoc* in 'sols suspendus' from Central America. Klinge (1962, 1963, 1966) found that the humus masses on trees in montane forests in El Salvador had similar characteristics to the moder humus of terrestrial soils. They were acid with a very high base-exchange capacity; the phosphorus content was fairly high but combined nitrogen was low and nearly all in the ammonium form. Mycorrhizal and other fungi probably play a large (but inadequately studied) part in epiphyte nutrition (see p. 146).

Even when organic material has accumulated to form a 'sol suspendu', the volume that the roots of epiphytes can exploit is limited and this, combined with free drainage and intermittently high rates of evaporation, renders epiphytes liable to frequent periods of water stress. This is another important difference between epiphytic habitats and that of the rain-forest ground flora. Even in climates where rain falls almost daily, epiphytes may lose large amounts of water between showers, and in most rain forests rainless periods of several days or weeks are not uncommon (Chapter 7). Epiphytes therefore need to be able to absorb water quickly when it is available and to conserve it when it is not. Most epiphytes are drought-enduring and drought-resistant. During the dry year 1923 in north Queensland large numbers of epiphytes died (Herbert 1935).

The humus accumulations on trees are no doubt important to epiphytes as reservoirs of water as well as for the nutrients they contain. Johansson (1974) gives data on the water-holding capacity and retention of water by humus from accumulations under some West African epiphytic ferns. In the lower montane forest of the Uluguru mountains (Tanzania) Pócs (1976a) found that the epiphytes, together with the associated humus masses, intercepted about 8% of the total precipitation. The importance of 'sols suspendus' to epiphytes varies: they vary in humus requirements from pronounced humiphiles at one extreme to humiphobes at the other. The latter actually appear to avoid habitats where humus has accumulated (see Johansson 1974). In Java Went (1940) found that in exposed situations *Pholidota ventricosa* and other species always grew on large masses of humus, but in shady places they could grow with very little humus. It seems possible that the preferences of different epiphytes for sites with or without humus accumulations may depend more on their drought-resistance than on their nutrient requirements.

Branches of tall forest trees and outcropping rocks are similar habitats in some respects and it is not surprising that, as already noted, many epiphytes also grow on rocks where climatic conditions are suitable (see Sanford 1974). In both habitats there is a lack of soil and the water supply tends to be intermittent. 'Atmospheric' epiphytes (p. 139), such as some *Tillandsia* spp., are so well adapted to extreme conditions that in West Indian towns they can be seen growing abundantly on telephone wires and similar substrata.

6.4.4 Morphology

Macroepiphytes have an extraordinary variety of morphological features which can be regarded as adaptations to their special mode of life. These were first described in Schimper's classical *Die epiphytische Vegetation Amerikas* (1888). Full accounts of the subject are available in later literature, e.g. Schimper (1903, 1935), W. Troll (1938), Schnell (1970), Walter (1971), Benzing (1990), and it need be dealt with only briefly here.

Schimper classified the macroepiphytes, by their means of collecting water and nutrients, into four groups: protoepiphytes, nest, bracket and tank epiphytes. In addition he recognized hemiepiphytes (pp. 133–5 above) and the hemiparasitic Loranthaceae and Viscaceae (pp. 154–5). These are still useful categories, although more refined classifications have since been proposed (see Benzing 1990).

Protoepiphytes have no specialized water- or humus-collecting structures and do not send down aerial roots to the ground. This group is the least specialized class and the least protected from the effects of drought and lack of soil. Some have creeping stems and are thus able to exploit a large area, e.g. species of *Peperomia* and many ferns. Most protoepiphytes have xeromorphic features and some have water-storing organs: in species of *Codonanthe*, *Dischidia* and other genera the leaves are succulent and in many orchids there are swollen internodes known as pseudobulbs. In the epiphytic Cactaceae, as in most members of the family, the photosynthetic organs are the green succulent leafless stems. Root tubers are found in Malesian species of *Vaccinium* and *Pachycentria*. Several genera of epiphytic orchids are leafless and have photosynthetic roots (*Taeniophyllum*, some *Bulbophyllum* spp).

In orchids and some aroids and other protoepiphytes the aerial roots are covered with a multi-layered velamen, the cells of which are non-living when mature and have large pores in their walls. Beneath the velamen there is an exodermis with passage cells, and beneath this the green photosynthetic cells of the root cortex. The chief function of the velamen has generally been assumed to be the absorption and storage of water, but Dycus & Knudson (1957) found that water does not easily penetrate into intact velamen-covered roots except where they are in close contact with a substratum; here the velamen may be locally absent or different in structure from elsewhere. They suggested that the chief function of the velamen might be to protect the underlying tissues from overheating and excessive evaporation. Sanford & Adanlawo's (1973) observation that in Nigerian orchids the velamen is better developed, and has more layers, in species from districts with a severe dry season than in those from rain forests seems to give some support to this view.

However, it is also possible that the storage of runoff water in the velamen is important to epiphytes, as it gives time for the very small amounts of dissolved nutrients in the water to be absorbed, a function similar to that postulated for the hyaline cells of the moss *Sphagnum*. Went (1940) found at Cibodas that the water running down a tree is richer in mineral nutrients at the beginning of a shower of rain than later on; if this water can be absorbed by the velamen the nutrients could be slowly extracted from it.

It is interesting to note that velamen is found on the roots of a few terrestrial plants as well as in epiphytes (Schimper 1888, von Goebel 1922). It may have been part of the equipment that enabled certain species to become adapted to epiphytic life.

In nest and bracket epiphytes dead materials accumulate from which the roots can obtain water and nutrients. In nest epiphytes, of which the large fern *Asplenium nidus* of the eastern tropics and many other ferns, aroids and orchids are examples, the roots form a dense interwoven mass resembling a bird's nest. Ants and other small animals usually colonize the humus which collects in this. In bracket epiphytes the leaves, or some of them, are bracket-like and retain debris. In some species, e.g. the Malesian *Dischidia collyris*, the leaves are convex and closely appressed to the bark; the space beneath becomes filled with roots and humus. Ferns of the genus *Drynaria* and many species of the pantropical genus *Platycerium* have leaves of two kinds, one of which (the mantle leaves) is convex and rounded; they become hard and stiff and hold humus efficiently when both living and dead. In the climbing epiphytic cactus *Strophocactus wittii* of central Amazonia the large flattened phylloclades act as brackets.

Ecologically similar in some ways to the bracket epiphytes are the specialized myrmecophytic epiphytes. In the small Malesian climbing species *Dischidia major* the leaves are dimorphic: some leaves are rounded, flat and *ca.* 2 cm in diameter, a few are up to *ca.* 7 cm long, hollow and shaped like an inverted pitcher, and contain debris, ants and other animals and a mass of small adventitious roots (Holttum 1954, Janzen 1974b, Huxley 1980). The ferns of the genus *Solanopteris* of western tropical America have tubers, which are a remarkable parallel to the pitcher leaves of *Dischidia major*. The

Rubiaceous genera *Hydnophytum* and *Myrmecodia* are perhaps the most bizarre in appearance of the myrmecophytic epiphytes (Huxley 1978) (see Fig. 6.19). In these grotesque plants the stems may be several decimetres long and wide and are swollen to form 'tubers' honeycombed with cavities, which are almost always inhabited by ants. The cavities are not produced by the ants and are formed even when they are excluded. In some ferns of the eastern tropics (e.g. *Lecanopteris*) the rhizomes are much enlarged and flattened and also contain ant-inhabited cavities. The plant – ant relationships in epiphytic myrmecophytes are discussed later (p. 144).

The tank or cistern epiphytes belong to one family only, the Bromeliaceae, although some Malesian species of *Pandanus* (B. Stone) are to some extent comparable in structure. Bromeliads are abundant throughout the rain forests of tropical America and are one of their most characteristic features. In tank-forming species the long narrow leaves form a rosette round the short stem and their sheathing bases overlap to form a reservoir which in large plants may hold up to 5 l of water. Dust and debris of all kinds fall into the tanks and, as already mentioned, they often have a large and varied living fauna and flora. Not all bromeliads form tanks: in the so-called atmospheric type, such as certain *Aechmea* and *Tillandsia* species, which grow on exposed twigs and branches, the leaves do not overlap so as to hold water, or, as in the Spanish moss, *Tillandsia usneoides*, the stems are long and slender and the leaves not arranged in rosettes.

In nearly all Bromeliaceae the leaves bear large numbers of peltate trichomes, which are able to absorb water and solutes: water cannot escape through them and they act as one-way valves. These unique organs have often been described (see references in Benzing 1990). The water absorption process is dealt with in detail by Benzing (1990). The trichomes differ somewhat in structure in different members of the family but in all the epiphytic rain-forest bromeliads their function is basically similar. In most tank-forming species the roots are probably able to absorb some water, but in the atmospheric (p. 140) and other extreme xerophytic species the roots, when present, serve as holdfasts only, so that the trichomes are the only absorbing organs. In these latter species the trichomes cover almost the whole leaf surface but in the tank species they are widely scattered and are sometimes confined to the part of the leaf surface lining the tank.

6.4.5 Reproduction

The seeds and seedlings of epiphytes have various characteristics which are clearly related to their special problems of dispersal and establishment. Their propagules are almost always relatively small and suitable for wind or animal dispersal. The spores of ferns and the very small seeds of orchids, *Aeschynanthus*, etc., are well adapted to wind transport; for instance, the seeds of the orchid *Stanhopea oculata* weigh only 3×10^{-6} g (Ulbrich 1928) and those of some *Aeschynanthus* spp. 2×10^{-5} g (Beccari, quoted by von Goebel 1889, p. 153). Seeds with parachute appendages are found in various Asclepiadaceae, Bromeliaceae (subfamily Tillandsioideae) and other epiphytes. Fleshy fruits, often with sticky pulp which helps to attach the seeds to the bark, are found in *Rhipsalis* (Cactaceae) and many Bromeliaceae, as well as in the hemiparasitic Viscaceae and Loranthaceae. Seeds of the last-named are often deposited on telephone wires where the seedlings grow for a time like the bromeliads mentioned earlier.

Propagules attractive to ants are found in many epiphytes, including cacti (Buxbaum 1969, p. 595) and some orchids. In Java the myrmecophyte *Myrmecodia* and various other epiphytes are ant-dispersed (Docters van Leeuwen 1929), although some species may be dispersed by birds as well. The seeds of epiphytes carried by ants usually contain oil drops, whereas those unattractive to ants do not. Thus the ant-dispersed *Dischidia major* has oil in its seeds but *D. lanceolata* has none and is not dispersed by ants. The spores of the myrmecophilous fern *Lecanopteris* sp. are carried by ants, which seem to be attracted by the oil in the thin-walled cells of the sporangium wall. Arboreal ants carry seeds and spores of epiphytes to their nests where many of them germinate, forming 'ant gardens' (p. 144).

The method of dispersal of *Tillandsia usneoides*, the well known Spanish moss of America[3] is similar to that of the dangling mosses of the family Meteoriaceae and lichens of the genus *Usnea*. Beard-like masses of the plant are torn off and distributed by wind or used by birds for making nests; seeds are rarely produced and roots are found only on young seedlings.

Like other biological nomads, epiphytes reproduce very abundantly. They are mostly polycarpic and, with a few exceptions such as *Tillandsia usneoides* which have other means of reproduction, produce their usually small seeds in very large numbers. In his discussion of epiphyte life-strategies Benzing gives some data on their survivorship and seedling mortality. Their mortality is not density-dependent and he concludes that they 'favour a large Malthusian coefficient' (rate of unconstrained population growth) (Benzing 1990, p. 178).

6.4.6 Physiology

Epiphytes are as varied in their physiology as in their morphology. Early work on epiphyte physiology was summarized by Gessner (1956b) and critical surveys have been given more recently by Benzing (1990) and by various authors in Lütge (1989). A few comments may be made here on the water relations and carbon metabolism of epiphytes, two closely connected aspects of their physiology which have an important bearing on their ability to succeed in rather inhospitable habitats. Most of the experimental work has been done on the largest groups, orchids, bromeliads and ferns, but there is no reason to think that the conclusions cannot be extended to other epiphytes.

The water supply available to epiphytes between rainfalls, except for those with tanks or which accumulate very large masses of humus, depends entirely on water from dew and mist condensation or that stored in their own tissues. Water stress in dry weather is much more severe for epiphytes in exposed positions such as the twigs of tall emergent trees than for those in shade or sites where moisture collects, as in branch forks. A few macroepiphytes are poikilohydric and can take up water from saturated air (p. 143 below), but most, including atmospheric bromeliads, can absorb it only in liquid form (Lieske 1915). Rapid and efficient water uptake by roots, or, as in bromeliads, by specialized trichomes, is therefore very important. Many epiphytes lose water slowly during drought periods, but some, like other xerophytes, are capable of transpiring faster than typical mesophytes when water is freely available (Gessner 1956b). There is in fact great variety in transpiration rates and stomatal behaviour among epiphytes, as well as in their ability to absorb and store water. As might be expected, shade epiphytes differ greatly in their water economy from sun epiphytes; some (but not all) epiphytes can be regarded as xerophytes. Even species with overlapping ecological ranges may show considerable differences in their water relations. Thus Spanner (1939) found that the tuberous myrmecophytes *Myrmecodia* and *Hydnophytum*, which are superficially alike and often grow together, differ in stomatal behaviour and transpiration rates. Stomatal regulation is slow in *Hydnophytum* and rapid in *Myrmecodia*. The very thorough

---

[3] This plant is widespread in the southeastern USA and the Caribbean region but in the rain forests of mainland South America seems to be found only in northeastern Brazil (Ducke & Black 1953, p. 9) and in the Guianas.

study by Boyer (1964) of the autecology of the common West African epiphytic ferns *Platycerium angolense* and *P. stemaria*, which often grow close together on the same tree, showed that the former was more drought-resistant than the latter, though in both the prothalli and young sporophytes were less resistant than older plants.

Epiphytes of exposed sites (sun epiphytes), as might be expected, can generally withstand greater water losses than those growing in shade; some of them can lose over 50% of their fresh mass without apparent injury (Schimper 1935, p. 26). The osmotic pressure of the cell sap in the leaves is usually considerable higher in sun epiphytes than shade epiphytes (Senn 1913, Walter 1971).

Walter (1964, 1971) made some observations on the water loss from shade and sun epiphytes from the lower montane rain forest at Amani (Tanzania). The shade epiphyte *Asplenium nidus* when kept dry on a veranda for twenty days lost a large fraction of its water content but did not die; in its natural habitat the water held in the mass of roots and humus forming the 'bird's nest' would probably be sufficient for the plant's survival in normal dry periods. The sun epiphytes examined included orchids with and without pseudobulbs. The former lost water at a steady rate of 2.5% per day for twelve days, but those without pseudobulbs lost a similar amount in nine days and showed symptoms of water stress after ten to twelve days; under forest conditions they would have received at least some water from dew in this period. When wetted, the roots of both types of orchids absorbed many times their own mass of water in a few minutes. The succulent *Rhipsalis baccifera* lost 28% of its fresh mass in twenty-eight days and showed 'visible damage'. Unlike the cacti of arid regions, this epiphyte is evidently unable to withstand long dry periods. Walter concluded that the behaviour of the sun epiphytes at Amani was comparable to that of non-extreme European succulents such as species of *Sedum* and *Sempervivum*.

Some epiphytic ferns, e.g. *Polypodium* spp., and a few other macroepiphytes, resemble bryophytes and lichens in being poikilohydric: their degree of hydration closely follows the humidity of the ambient air. The filmy ferns (Hymenophyllaceae), which are common in most tropical rain forests, are 'semipoikilohydric' (Benzing 1990). Water can be quickly absorbed over the whole surface of their leaves and as rapidly lost again, but the amount of water involved is small. In a dry atmosphere their leaves curl up; they expand again when moist conditions return.

Benzing & Renfrow (1971a) found that the thin leaves of shade-tolerant Bromeliaceae dried out more at low humidities, and recovered more slowly when wetted, than those of 'atmospheric' species which were more xeromorphic and more thickly covered with absorbent trichomes.

The water economy of epiphytes is closely connected with their patterns of carbon assimilation. Many of them have crassulacean acid metabolism (CAM) which enables them to absorb carbon dioxide in the dark and close their stomata during daylight: this requires a smaller expenditure of water per unit of dry mass synthesized. CAM metabolism was first demonstrated in epiphytes from the Atlantic (Subtropical) rain forest of Brazil by Coutinho (1964). Benzing & Renfrow (1971b) later found CAM in atmospheric bromeliads. Apart from some lianes (Adams *et al.* 1988) and weeds of paths and clearings, epiphytes seem to be the only rain-forest components that do not conform to the usual C_3 carbon cycle characteristic of mesophytes.

Although in many epiphytes the carbon metabolism is of the C_3 type, CAM is now known in many ferns and orchids as well as epiphytes of other groups. There is considerable variation in detail in the metabolism of CAM epiphytes. During dry weather when the stomata are closed many CAM epiphytes can recycle CO_2 within the leaves for long periods (Benzing 1990). Some have a capacity for following either C_3 or CAM pathways, in some cases according to conditions and sometimes even in different parts of the plant (Benzing 1990).

Wallace (1981) found that in epiphytes of the subtropical rain forest in Australia there is a clear relation between the type of metabolism and the microclimate in which the plant lives. Many of the sun epiphytes of the canopy were CAM plants, but all the shade-tolerant epiphytes of the undergrowth were C_3 plants. Similarly in Trinidad (Griffiths & Smith 1983) the bromeliads of sunny and exposed sites included both CAM and C_3 species but the shade species were all C_3 plants.

Sun and shade epiphytes, like sun and shade plants of other habitats, differ in their photosynthetic response to light. Benzing & Renfrow (1971a) in an investigation of photosynthesis in excised leaf tissues from 21 species of bromeliads showed that the thick-leaved 'xeric' species behaved as typical sun plants: they had a high compensation point (>70 ft-c, about 42 lux) and their photosynthetic efficiency at low light intensities was low. Most of the more mesic species, on the other hand, showed photosynthesis–light curves typical of shade-adapted plants. It is also interesting that in two xeric species carbon dioxide absorption was much slowed down when the leaves were wet but in two mesic species wetting made no difference (Benzing & Renfrow 1971b).

6.4.7 Epiphytes and ants

The microcosms dependent on epiphytes, especially bromeliads, are too large a subject to deal with here, but a few remarks are needed on the interrelations between epiphytes and ants. These are still inadequately understood and are of several kinds for which van der Pijl (1955) uses terms such as myrmecophily, myrmecotrophy and myrmecodomy. The term 'myrmecophyte' is generally used only when the association is constant and there is some evidence that it is of mutual advantage to both partners. Some of the most remarkable myrmecophytes are epiphytic but many examples of myrmecophytes can be found in other rain-forest synusiae. Epiphytic myrmecophytes are not confined to, or even particularly abundant in, tropical rain forests.

Although almost any well-established epiphyte will be found to have ants living among its roots or leaves, the species of ant vary and in most cases the co-existence of plant and ant seems to be facultative and incidental. In rain forests epiphytes frequently grow on the nests of arboreal ants. The 'ant gardens' described by Spruce (1908) and Ule (1906) in Amazonia consisted of many species of epiphyte growing on the carton nests of *Azteca* spp. and other arboreal ants. Similar ant gardens of *Aeschynanthus* and *Dischidia* spp. and other epiphytes were found in Java by Docters van Leeuwen (1929). The seeds of most of the species growing in the 'garden' are probably carried there by the ants and the plants appear to benefit from the nutrients available in the nests.

In the most specialized ant-inhabited epiphytes, such as the ferns *Solanopteris* and *Lecanopteris*, and the many species of *Hydnophytum* and *Myrmecodia* (Rubiaceae) (Fig. 6.19), the plant has cavities in its rhizome or aerial stem, or between its leaves, which ants use as nesting sites (ant-houses). The two latter genera have been much investigated (Schimper 1935, Spanner 1939, Huxley 1978, 1980). The cavities are not made by ants and it has been shown that the plant can grow well in their absence. It was formerly thought that the main importance of the much swollen stems was as water-storing organs, but it is now clear that the plant can absorb, and probably benefit from, nutrients brought to it by the ants (*Iridomyrmex* spp.) living in the cavities. The system of chambers is complex: some parts are inhabited by the ants and their commensals and used by the former for rearing their broods, other parts are used by the ants for depositing their faeces and other debris. The walls of these latter chambers have wart-like projections. Huxley (1978) showed by means of a radioactive tracer that solutes could be absorbed by these warts.

In the myrmecophytic bromeliads *Tillandsia butzii* and *T. caputmedusae*, the rosette of leaves encloses a space which is not filled with water as in a tank-forming species, but remains dry and provides a nesting place for ants of several species. Benzing (1970) found that this space contained much nitrogenous debris and from which the plant can probably obtain nutrients. Gay (1993) has also demonstrated the absorption of ant-derived nutrients by the lining of the rhizome cavities in *Lecanopteris* spp.

Janzen (1974b) studied the epiphytes growing on small trees in the 'padang' vegetation on white sands (podzols) at Bako in Sarawak (p. 326); they included species such as *Hydnophytum formicarium*, *Myrmecodia tuberosa* and *Dischidia major*, which have ant-inhabited cavities, and others such as *Pachycentria tuberosa*, which do not. He concluded that in the myrmecophytes there was a 'mutualistic' relationship between the plant and the ants. The ants benefit the plant by providing it with nutrients and in the extremely nutrient-poor conditions at Bako this supplementary nutrition probably contributed to their survival. It was doubtful whether the relatively unaggressive *Iridomyrmex* ants that lived in these plants defended them against herbivores as the ants can be seen to do in the African *Barteria* and the tropical American 'bullshorn acacias' (p. 133). The only benefit received by the ants seemed to be relatively dry and permanent nesting sites; they obtained little or no food from the epiphytes. The non-myrmecophytic epiphytes such as *Pachycentria* were regarded by Janzen as parasites because they exploited the supplies of nutrients imported by the ants but provided the latter with nothing in return.

6.4.8 Effects of epiphytes on phorophytes

The relationship between epiphytes and the plants on which they grow is generally assumed to be one-sided: the phorophyte is supposed to provide its epiphytes with mechanical support only and, unlike the hemiparasitic mistletoes, they do not deprive it of organic nutrients. This conclusion seems to be supported by the fact that epiphytes often grow well on dead trees; for example, in 1963 some of the dead trees in Gatun Lake in the Panama Canal, which had presumably died some forty years earlier, bore numerous aroids and orchids. These epiphytes could perhaps have obtained nutrients from the decayed wood of the trees, but could not have been actively parasitic.

A

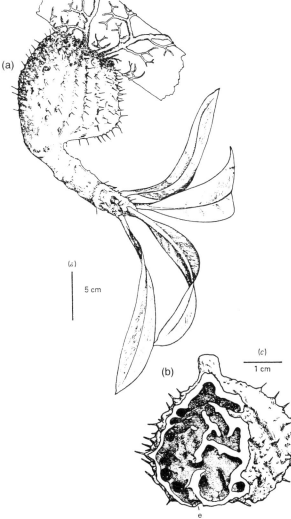

(a)

(ä)

5 cm

(c)
1 cm

(b)

C

B

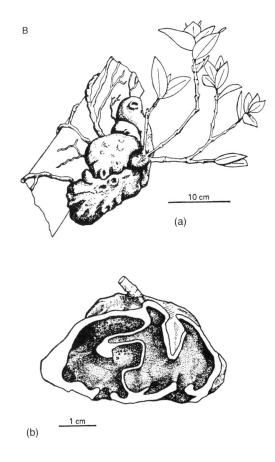

(a)

10 cm

(b)

1 cm

Fig. 6.19 Myrmecophilous epiphytes. From Huxley (1980, figs. 1–3). A *Myrmecodia tuberosa*: (a) young plant; (b) young plant cut open to show ant-inhabited cavities; e, entrance hole. B *Hydnophytum* sp. cf. *papuanum*: (a) young plant, (b) young plant cut open, showing two cavities each with one entrance hole. C *Dischidia major*: 'flask leaf' cut open, showing adventitious roots and leaf apex bent inwards to form inner cavity.

However, some evidence suggests a different view. Trees heavily overgrown with epiphytes often appear sickly and have die-back and other symptoms of disease. This might be because they are old or infected by pathogens and is not necessarily a consequence of the presence of epiphytes, although epiphytic vegetation if very luxuriant can compete with the phorophyte for light and it may, by holding water, encourage the growth of parasitic fungi and bacteria. Epiphytes are often popularly referred to as parasites; planters in the tropics frequently assert that they harm cocoa, coffee and other cultivated trees and take trouble to 'clean' them.

The first serious scientific study of this problem was that of Ruinen (1953) who found that in Java cultivated trees laden with ferns and orchids appeared to be less healthy than those with few or none. She studied the disease symptoms and found that they included reduced leaf area, die-back and abnormally small buds, as well as anatomical features such as the blocking of the wood vessels by tyloses and gum, asymmetric cambial growth necrotic spots in the xylem, and degeneration of the phloem. These pathological developments were always accompanied by the presence of fungal hyphae in the cells of the epidermis, cork and stem cortex.

In *Barringtonia racemosa*, *Randia dumetorum* and other shrubs or small trees growing in the Bogor Botanical Garden an improvement in the number, size and colour of the leaves soon followed if the epiphytes were removed. In another experiment, trees of *Brunfelsia americana* were 'infected' with the fern *Drymoglossum piloselloides*: after a few months the epiphyte was growing well, but the leaf area and general condition of the phorophyte had deteriorated compared with 'uninfected' controls.

Ruinen believed that some species were susceptible to the effects of epiphytes such as *Drymoglossum* and others more or less resistant. She suggested that the pathological symptoms were due, not to the epiphytes directly, but to mycorrhizal fungi associated with them. In her view, the latter live in the roots of the epiphytes in a state of 'controlled parasitism', but tend to invade the tissues of the phorophyte and become actively parasitic on them. In several instances fungal hyphae could be traced from the roots of the epiphyte to the internal tissues of the phorophyte. All epiphytic orchids, as is well known, contain fungal endophytes, and, according to Ruinen and to von Faber (Schimper 1935), many other epiphytes are mycorrhizal.

Ruinen's theory of 'epiphytosis' is certainly plausible, but it may well apply to some epiphytes and not to all. It is also possible that the supposed antibiotic effects of epiphytes occur in some climatic and soil conditions and not in others. Further experimental work is needed and, in particular, it would be desirable to demonstrate by means of radio-tracers the actual transfer of nutrients and other materials from the living cells of the phorophyte to the epiphyte.

The quasi-parasitism of epiphytes remains somewhat controversial. In his most recent assessment of the problem Benzing (1990) points out that most of the symptoms of tree disease attributed to epiphytes are illusory. Epiphytes flourish on dead branches and senile trees mainly because these are old and provide firm, well-illuminated surfaces on which they can grow. They may indeed deprive their phorophytes of nutrients by collecting organic matter as well as by intercepting rainfall and throughfall, immobilising them for long periods. This activity, better described as nutrient piracy than as parasitism, may be of some importance in heath forests where nutrient shortage is perhaps critical.

6.4.9 Abundance and biomass

Epiphytic vegetation varies much in abundance and diversity with climate, maturity of stand and other factors. In primary lowland rain forest, even under the most favourable conditions, usually only a small proportion of the trees bear macroepiphytes. On clear-felled plots of mixed forest at Moraballi Creek, Guyana, on two plots, out of 193 trees 5 m or more high, 30 (16%) had macroepiphytes, and on a third plot, out of 55 trees 14 m high or more, 21 (38%) (Davis & Richards 1933–34, p. 378). In lowland rain forest at Gunung Dulit, Sarawak, 13% of all trees on one plot and 11% on another bore macroepiphytes (Richards 1936, p. 16). Two plots in evergreen seasonal forest in Nigeria gave figures of 15 and 24%, respectively. In the lowland plot in Ecuador of Grubb *et al.* (1963) referred to above, the percentage was 60, although 40% of the individual epiphytes were borne on only three trees.

It is difficult to determine the quantity of epiphytes by counting or weighing except on recently felled trees and few data are available. The figures of Grubb *et al.* (1963) for Ecuador were mentioned on pp. 136–7. In the Nimba Mountains (Liberia) at about 500 m Johansson (1974) counted 1171 individual epiphytes on one tree and 1857 on another.

In a sample area of rain forest in central Amazonia near Manaus, Fittkau & Klinge (1973) estimated the biomass of macroepiphytes (excluding Loranthaceae) as 0.1 Mt ha^{-1}. The very small figure (1.6 kg ha^{-1}) of Golley *et al.* (1975, p. 139) for the biomass of epiphytes in 'Tropical Moist' (evergreen seasonal) forest may be an under-estimate; their figures for riverine and mangrove forest are considerably larger and that for 'Premontane Wet' (lower montane) forest is almost a thousand times as large. Pócs (1980) estimated the dry mass (including microepiphytes) in the submontane (lower montane) forest in the Uluguru Mountains in Tanzania as 2130 kg ha^{-1}.

The largest number of species of macroepiphyte on a single tree noted by the author was about fifteen in Guyana and thirteen in Nigeria, but much larger numbers have been reported, e.g. thirty-seven in Johore (Malaya) (Corner 1978) and forty to forty-five on single trees in Budongo forest, Uganda (Eggeling 1947). In upland forests the number of species as well as the number of individuals may be much larger. In Sumatera Bünning (1947, p. 78) found over fifty species of fern on one tree.

6.4.10 Vertical distribution

The habitats of epiphytes are better illuminated and are exposed to a wider range of microclimates than those of similar plants living on the ground. In closed forest, as will be shown later (Chapter 8), there are microclimatic gradients from soil level, where conditions are extremely constant, to the tops of the emergent trees, where the microclimate approximates to that in the open. The average rate of air movement, as well as the illumination and the maximum temperatures and saturation deficits reached during daylight hours, all increase upwards.

Macroepiphytes may be found at all levels, but individual species differ considerably in their microclimatic requirements, as was demonstrated by the early work of Wiesner (1907, pp. 135–8), Went's (1940) observations at Cibodas (Java), Johansson's (1974) in West Africa and much later work. Some species are restricted to strongly illuminated sites, some to shady sites, while some avoid both strong light and deep shade and yet others have a wide range of tolerance. Such species preferences are partly responsible for the striking differences in the composition of the epiphytic vegetation at different heights above ground and from tree to tree.

In a mature forest stand the climatic gradients vary from place to place because of the unevenness of the tree storeys; there are also inter-specific and individual differences in the density of the tree crowns. As a result the epiphytes are not regularly stratified but form ill-defined communities at varying heights above ground. Microclimatic factors other than those due to the vertical gradients also sometimes affect epiphytes. In swampy areas the epiphytic flora is often different from that in normal forest; thus in Budongo forest (Uganda) *Angraecum infundibulare* and other epiphytes are found only near water (Eggeling 1947).

The height at which epiphytes are found differs in different forest types depending on their structure and its effect on light penetration. In a community where the A and B tree storeys are dense the average vertical range of the epiphytes is high; in the more open forest types it is always lower. This is strikingly shown in the five types of primary forest at Moraballi Creek, Guyana. The average illumination in the undergrowth is highest in the mora and wallaba types and is low in the morabukea, mixed and greenheart types (pp. 299–300). The vertical range of the epiphytes shows a corresponding variation. In the three poorly illuminated forest types few epiphytes are visible from the ground; typical sun epiphytes (see below) such as *Tillandsia* spp., rarely occur below about 18 m from the ground, except for stray individuals which do not flower (lowest record

of a mature plant, 9 m) and shade epiphytes are very scarce. In the lighter types of forest (mora and wallaba) epiphytes descend much lower and flowering individuals of *Tillandsia* spp. could be found at 3 m from the ground (see Fig. 11.5). It was noticed in the darker types of forest that when a plant of a sun epiphyte fell off a tree on to the ground, it nearly always soon died, while in the lighter types it often continued to flourish[4]. In the wallaba forest *Tillandsia* plants were sometimes met with which had apparently established themselves on the ground; when examined these plants were always found to be growing on the remains of a rotten branch. ter Steege & Cornelissen (1989), in a study of epiphytes in greenheart, mixed and wallaba forest near Mabura Hill, Guyana, found that in all three types of forest the number of species was greatest in the 'lower canopy,' i.e. the lower part of the euphotic zone.

Similarly, in Borneo epiphytes grow at lower levels in the well-illuminated heath forest than in the darker mixed dipterocarp forest. In the former they occasionally grow actually on the ground; in one locality the epiphyte *Asplenium nidus* was seen rooted in sandy soil by a road through heath forest (Richards 1936). Polak (1933a) found species of *Bulbophyllum* and other epiphytes growing on the ground in sandy 'scrub' areas (padang) in West Borneo. In all parts of the tropics epiphytes grow at much lower levels on trees by rivers and on isolated trees in gardens and clearings than in the interior of the forest.

The variation in the vertical distribution of epiphytes seems to depend more on light than on the average humidity of the air, which in rain forests varies more or less inversely with the light intensity (Davis & Richards 1933–34, p. 382). Pittendrigh (1948), who studied the vertical and geographical range of Bromeliaceae in the forests and cacao plantations of Trinidad, also concluded that light is the chief controlling factor.

6.4.11 Distribution on individual trees

Microclimatic factors are of course not the only ones to which epiphytic vegetation responds. What epiphytes are present on an individual tree largely depend on its age and species. The effect of age is illustrated by the data of Went (1940) and Yamada (1975–77) on the epiphytes in the lower montane forest at Cibodas in Java. Went observed that the 'epiphytic association' on young trees of *Altingia excelsa*, which are smooth-barked, is different from that on the scaly-barked older

---

[4] It is interesting that Benzing (1978) found that plants of the extreme sun epiphyte *Tillandsia circinata* were soon killed by contact with moist soil.

trees. Yamada found that out of 237 trees examined only 38% of those in the 10 cm d.b.h. class had epiphytes but all those in the 80 cm class had them.

Differences in the abundance and composition of epiphytic vegetation apparently depending on the species of phorophyte can often be observed. Thus on the large bamboo *Dendrocalamus* and other trees with very smooth, hard bark there are seldom any epiphytes other than lichens. Some trees seem to be avoided for less obvious reasons, e.g. *Ceiba* and *Ficus* spp; on the other hand, the abundance of epiphytes on the rain tree *Albizia saman*, compared with other trees commonly planted in tropical towns, is remarkable. There are various references in the literature to 'preferred' phorophyte species. Johansson (1974) gives a list of tree species rich in epiphytes in West Africa, and Wallace (1981) gives similar data for subtropical rain forests in Australia.

It has sometimes been stated that certain epiphytes are confined to a single species of phorophyte, but such claims are not usually supported by statistics. Schimper (1888) noted that in the southern United States *Epidendrum conopseum* was chiefly but not exclusively found on *Magnolia*. In the Philippines the fern *Stenochlaena areolaris* is said to grow only on *Pandanus utilissimus* (Copeland 1907).

At Cibodas (Java) Went (1940) recorded the species of epiphyte on a large variety of trees in the lower montane forest by means of observations with a telescope and binoculars. The abundance of macroepiphytes (on an arbitrary scale) was tabulated in relation to the species of tree, roughness of bark and other factors. Great differences were apparent between the epiphytic 'associations' on different species, although their composition changed with the age of the tree. The differences between the epiphytes on different species of trees were greater than those between individuals of the same species, indeed, in some instances, the assemblage of epiphytes was so distinctive that the tree could be identified by its epiphytes, e.g. various species of *Castanopsis*.

Went considered that the differences between the epiphytic flora on different phorophytes were probably not due only to physical factors such as the roughness, method of peeling or water-holding capacity of the bark, or to the amount of light penetrating the crown of the tree. He concluded that chemical differences in the stemflow water, or in the bark itself, were probably responsible. It was significant that epiphytes growing in large humus accumulations showed no preference for particular tree species.

Evidence from outside the tropical rain forest supports the view that the composition of the stemflow water and the bark of the phorophyte may be of great import-

ance to epiphytes. In the southern United States Schlesinger & Marks (1977) found that the bromeliad *Tillandsia usneoides* was common on *Quercus* spp. but rare on pines; they attributed this to toxins in the bark of the latter. In a 'cloud forest' in Oaxaca (Mexico) Frei & Dodson (1972) found that orchids and other epiphytes were abundant on *Quercus castanea* and *Q. vicentensis*, but rare or absent on other trees such as *Q. magnoliaefolia*. Germination experiments in which powdered bark was added to an agar medium showed that bark from the avoided species was toxic to the seeds and protocorms of orchids, but bark from *Q. castanea* and *Q. vicentensis* had no inhibiting effects. Phenolic substances such as gallic acid derivatives and leucoanthocyanin tannins were found in the bark of the inhibiting species and not in the others; these were perhaps responsible for the effects observed.

Benzing & Renfrow (1974) studied the growth and mineral content of the epiphytes *Encyclia tampense* and *Tillandsia circinata* on normal and dwarfed trees of *Taxodium ascendens* in subtropical Florida and found that the epiphytes were more vigorous and had a higher mineral content on the former: other factors in addition to nutrition might have been involved.

On isolated trees in the tropics, as in other parts of the world, epiphytes are usually not uniformly distributed round the trunks, some species being more abundant on some aspects than on others, depending on exposure to wind, rain and sun. Within the forest, as might be expected, the epiphytic vegetation is little affected by aspect, even on the topmost branches; there is no 'mossy side' to the trunks to help the lost traveller in finding the right direction.

6.4.12 Succession

The first colonists on rain-forest trees are usually algae, lichens and bryophytes; these often provide sites where orchids, ferns and other macroepiphytes can establish themselves. In West Africa (Johansson 1974) and probably elsewhere in the Old World tropics, orchids are commonly the first macroepiphytes, but in the New World bromeliads often establish themselves even earlier than orchids; because of their water-holding ability, they often provide starting points for other epiphytes. Schimper (1888) noted in Dominica that the strangler *Clusia rosea* frequently established itself at the base of large epiphytic bromeliads.

Johansson (1974) says that in the lower montane rain forest of the Nimba mountains (West Africa) moss stages are non-existent below 1000 m, but become increasingly important at higher altitudes. There is much evidence of one species of epiphyte actively competing with others; thus the fern *Drynaria laurentii* often suppresses orchids such as *Bulbophyllum* spp., which are slower-growing. On one tree the changes in the epiphyte community were recorded over two years. Evidence of changes over longer periods was obtained by analysing dead plant fragments in humus accumulations, one of which was over 16 cm thick.

When masses of epiphytes become very large and heavy they may become detached and fall to the ground: the succession can then begin again.

6.4.13 Synecology

The first attempt to group epiphytes into communities was that of Schimper (1888) who called them *étages*, (storeys), a somewhat misleading term because, as we have seen, epiphytes are not regularly stratified, their distribution depending on other factors as well as the change of microclimate with height above ground. Went (1940) refers to 'epiphyte associations' but stresses that the groupings found in the analysis of his data from Cibodas were not associations in the usual phytosociological sense. Hosokawa (1943, 1968) developed an elaborate classification based on what he terms 'aerosynusiae'. Johansson (1974), who regarded the existing terminology applied to epiphytes as confused, termed synecological groups of epiphytes 'epiphyte communities' in the Nimba mountains (West Africa). He was able to recognize seven of these, four of which occurred below 1000 m and three at higher altitudes; each community could be found in 'initial', 'intermediate' and 'final' (i.e. climax) successional stages.

In tropical rain forests the epiphytic vegetation is so fragmented and variable in composition that the value of applying any formal phytosociological classification to it seems questionable. For practical purposes it is probably enough to distinguish two synusiae, shade epiphytes and sun epiphytes (corresponding to the skiophytes and photophytes of Barkman 1958). The former live in complete or partial shade, mainly at low levels in the forest, while the latter are found high above the ground and more or less fully exposed to sunlight. Each of these synusiae can be subdivided if necessary.

Shade epiphytes are found mainly in the C storey of the forest, on the trunks and branches of small trees, both C storey species and young A- and B-storey trees, as well as on the stems of lianes. Macroepiphytes are usually scarce on large trunks. On very large emergent trees with dense foliage, shade epiphytes occasionally reappear above the first fork in the centre of the crown. The shade epiphyte community consists chiefly of ferns,

often including filmy ferns (*Trichomanes* and, less commonly, *Hymenophyllum* spp.). Orchids and other flowering plants are seldom abundant, but in Guyana, for instance, the minute *Peperomia emarginella* and one orchid, *Cheiradenia cuspidata* (always found clinging to large trunks at about 1–2 m above ground) belong to this synusia. In Malaya shade epiphytes, including *Hedychium longicornutum*, *Medinilla* spp., *Lycianthes parasitica* and many small orchids, are sometimes very plentiful, especially on trees of *Saraca* growing by streams (E.J.H. Corner).

The synusia of sun epiphytes is usually much the richer of the two communities in species and individuals. In mature forest it occurs chiefly on the larger branches of A- and B-storey trees. On very large trees sun epiphytes sometimes spread down the trunk below the first fork, probably after first establishing themselves on the branches. This synusia includes numerous species of orchids, and also flowering plants of other families and many ferns.

The most extreme of the sun epiphytes are those found on the outermost branches and twigs of the taller trees. On exceptionally large emergent trees they may be found throughout the greater part of the crown. This group includes few ferns and consists chiefly of flowering plants highly specialized in structure and physiology. At Moraballi Creek, Guyana, for instance, the chief representatives are narrow-leaved species of *Tillandsia* such as *T. bulbosa* and *Aechmea* spp. (Bromeliaceae), *Codonanthe crassifolia* (Gesneriaceae), *Rhipsalis baccifera* (Cactaceae) and certain orchids. Among the extreme sun epiphytes are the 'atmospheric' bromeliads (p. 140) and such orchids as the remarkable Malesian *Thrixspermum arachnites* and the African *Solenangis* spp., which dangle freely in the air from long slender aerial roots attached to twigs.

6.4.14 Microepiphytes

The epiphytic vegetation of the tropical rain forest includes algae, fungi, lichens and bryophytes as well as vascular plants. Some of these microepiphytes live in close association with epiphytic ferns and flowering plants and can be regarded as forming part of the same synusiae; the larger bryophytes, for example are of the same order of size as the epiphytic filmy ferns and compete with them on a more or less equal basis. Microepiphytes, as mentioned earlier, often provide a seed bed in which macroepiphytes can establish themselves. In terms of biomass and productivity, microepiphytes are insignificant in lowland rain forests, but they probably play a not unimportant part in their nutrient economy, as will be seen later, and help in maintaining the very humid uniform microclimate characteristic of the lower strata. A general account of the ecology of tropical forest bryophytes (including epiphyllae) is given by Richards (1984a).

In most lowland rain forests, all the lichens and almost all the mosses, hepatics and algae are epiphytic. As was pointed out in Chapter 2, there are usually no bryophytes on the forest floor, although on very steep slopes and in montane forests some bryophytes may be found on the ground. One reason for their absence from the floor in lowland forests is probably the smothering effect of the blanket of dead leaves; bryophytes are found on tree-trunks down to ground level and are abundant on stones and fallen logs from which the leaves easily slide off. Freshly disturbed soil, on which a thick layer of leaves has not had time to collect, such as termites' nests and earth thrown up by burrowing animals or the roots of overturned trees, is also often colonized by small mosses (especially *Fissidens* spp.) and hepatics.

Besides the communities on stones, rotten logs, and tree-stumps, non-vascular plants are found in the rain forest as shade epiphytes, sun epiphytes and as epiphyllae on living leaves. The shade-epiphyte community occurs on trunks and twigs in the undergrowth and, like vascular shade epiphytes, may reappear in the crowns of large trees with thick foliage. The non-vascular sun epiphytes are found on the branches and smaller twigs of trees in the A and B storeys. Epiphyllae are found mainly in the shady undergrowth and sometimes also in the crowns of the taller trees if their foliage is dense enough to provide the moist shady conditions they require.

Microepiphytes on trunks and branches. The shade community consists chiefly of bryophytes of many families (mosses, Lejeuneaceae, *Plagiochila* spp., etc.); lichens and algae do not play an important part in it. Unlike the corresponding community of flowering plants and ferns, it is much richer in species than the sun community. The twigs and branches of shrubs and small C storey trees, especially in the most humid places, such as near streams, are often so thickly covered with mosses and hepatics that the bark is completely concealed; the trunks of larger trees are seldom covered with a continuous carpet of bryophytes. Besides tufted and carpet-forming bryophytes of growth-forms more or less familiar in temperate floras, a characteristic feature of the shade community is the presence of pendent mosses, mostly belonging to the family Meteoriaceae;

in these the primary stem creeps along a twig and its branches, in some species a metre or more long, hang freely in the air. The occurrence of these pendent mosses, which are found only in tropical and other rain forests, is no doubt related to the constantly high humidity, as the pendent habit offers the maximum exposure to evaporation. The bryophytes of the shade community have a typically hygromorphic structure; their leaves are usually large and many of the species have very large cells with thin walls.

The non-vascular sun epiphytes form a strong contrast to the shade epiphytes and few species are common to both communities. Lichens are abundant, including large foliose forms such as species of *Parmelia*. Mosses, e.g. *Macromitrium* and *Groutiella* spp., and hepatics (*Frullania* spp., Lejeuneaceae) are abundant but the number of species in any one district is usually not large. Most of the bryophytes are of compact creeping or tufted habit and pendent species are absent or rare. Their structure is usually xeromorphic. The leaves are small and closely imbricated when dry, bending back rapidly when moistened; the leaf cells are small and thick-walled. The differences between the shade and sun bryophytes in tropical rain forests are thus similar to those between shade and sun epiphytes in European woodlands (Olsen 1917), but, if anything, more strongly marked.

Epiphyllae.   Epiphytes on living leaves (epiphyllae) are highly characteristic of wet tropical forests and are common in tropical, montane and subtropical rain forests; elsewhere they are found, but only sparingly, in evergreen forests in Japan, in the Macronesian islands and the southeastern United States. Epiphyllae grow mainly on long-lived evergreen leaves, usually only on the upper surface. They require a constantly humid warm atmosphere. In rain forests which are seasonally relatively dry, such as most of those in Africa, they are usually absent, except near streams and in swampy places. Epiphyllous lichens are less exacting in their requirements and able to tolerate somewhat drier conditions.

The epiphyllous flora of tropical rain forests includes algae (e.g. Trentepohliaceae, *Phycopeltis* and Cyanophyta), lichens (*Dictyonema*, *Strigula* spp. and many other genera) and leafy liverworts (chiefly Lejeuneaeae and *Radula* spp.). Epiphyllous mosses, e.g. *Crossomitrium* spp. in tropical America, *Ephemeropsis tjibodensis*, *Taxithelium* spp. and others in Malesia, are rather uncommon. Associated with epiphyllous plants are micro-organisms of many kinds and small animals such

Fig. 6.20 Epiphyllae. A Early stages of leaf colonization by lichens and hepaticae Oban, Nigeria (1978), B Later stages, Noah Creek, Queensland (1981). C Later stages, várzea forest, Belém, Pará, Brazil (1969).

as rotifers, worms, mites and insect larvae; the epiphyllous community is thus one of the many subsystems or 'microcosms' within rain-forest ecosystems.

Epiphyllous species do not often occur also on bark but bryophytes growing on twigs such as mosses of the genus *Floribundaria* sometimes spread on to the leaves, although they may not be capable of establishing themselves there in the first place. In montane rain forests in New Guinea, weevils (*Gymnopholus* spp.) are found in which the elytra are always overgrown with lichens and bryophytes, all of which probably also occur on leaves (Gressit *et al.* 1965, 1968).

Epiphyllae are dispersed by spores and gemmae; the former are presumably airborne but the latter are probably often carried by rain-splash. Algae and lichens are often the earliest colonists on leaves and in comparatively dry environments the succession may go no further. Under more favourable conditions, the liverworts follow and usually eliminate many of the smaller lichens and the algae. In very humid forests a whole leaf often becomes covered with a dense felt of bryophytes in which seedling orchids, Bromeliaceae and other macro-epiphytes may establish themselves, although they never grow to maturity. Ultimately the leaf falls off and the life of the little community comes to an end.

Winkler (1967) studied the colonization of leaves in montane rain forest in San Salvador and found that within three months young leaves had become colonized by various liverworts, but some of the epiphyllae died later during dry weather.

The abundance of epiphyllae on different types of leaf varies greatly. Leaves of some species are preferred to others for reasons which are not always apparent; this may depend on the structure of the cuticle or on exudates escaping through it. Epiphyllae do not avoid finely divided leaves such as those of the Mimosaceae, or hairy leaves, but are unable to form extensive colonies on them. Monge-Nájera & Blanco (unpublished, 1993)[5] found that on artificial (plastic) leaves of various shapes and sizes placed in rain-forest undergrowth in Costa Rica a cover of epiphyllous bryophytes developed within nine months. On plastic leaves with drip-tips the distribution of the epiphyllae was different from that on natural leaves but the relative cover was similar.

Epiphyllous bryophytes are firmly attached to the leaves on which they grow by rhizoids, but the extent to which they may be parasitic on them has been insufficiently studied. Berrie & Eze (1975) found that in the widespread epiphyllous hepatic *Radula flaccida* the rhizoids became attached to the leaf by an adhesive secretion; later they penetrated the cuticle and inserted themselves between the epidermal cells. Some of the latter died, apparently as a result of this. Experiments with radioactive tracers showed that water and phosphate can pass from the leaf cells into the epiphyll. With epiphyllous algae and lichens there is also evidence of parasitism. The alga *Cephaleuros virescens* is known to be partly parasitic and causes a virulent disease of tea and other plants. Other epiphyllous algae invade the tissues of the host phorophyte and are probably partial parasites, if only 'space parasites'. Fitting (1910) studied the relations of epiphyllous lichens to leaves and found that some species penetrated more or less deeply into the tissues, e.g. *Strigula*, in which the algal symbiont is *Cephaleuros virescens*; most penetrated the cuticle, but not the epidermis; and others did not penetrate the cuticle. Probably only the last group is entirely non-parasitic.

Whether or not the deleterious effects of epiphyllae on the leaves that carry them are significant, they are probably outweighed by the importance of their contribution to the nitrogen economy and nutrient conservation of the ecosystem as a whole. Ruinen (1956, 1974) and others show that leaf surfaces (the 'phyllosphere') have a large and varied population of micro-organisms which includes *Beijerinckia* and other nitrogen-fixers. These are found on leaves even in temperate regions, but as they are favoured by moisture and obtain energy from the photosynthates leached from leaves, they are particularly abundant in humid tropical forests. As well as bacteria, various epiphyllous Cyanophyta are able to fix atmospheric nitrogen. In rain forests nitrogen-fixing organisms seem to be particularly numerous on leaves colonized by algae, bryophytes and lichens, which probably provide them with a congenial environment. The observations of Edmisten (1970) in the El Verde forest (Puerto Rico) show that stemflow water is much richer in nitrogen compounds than rainwater. Experiments showed that an important source of these may have been Cyanophyta and other organisms on leaves, indicating that the contribution of the epiphyllae to the nitrogen budget of the forest may be significant.

Witkamp (1970) and others (also at El Verde) found that leaves covered with epiphyllae absorb considerable amounts of phosphorus and other elements as well as nitrogen; these are eventually recycled and made available to other plants. Jordan *et al.* (1980) believe that in this way epiphyllae play an important part in 'nutrient scavenging', especially in heath forests on very infertile soils such as podzols.

---

[5] Monge-Nájera, J. & Blanco, M.A. (1993). The influence of leaf characteristics on epiphyllic bryophyte cover: a test of hypotheses with artificial leaves. (Unpublished manuscript.)

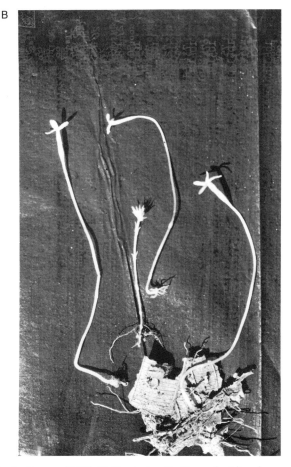

Fig. 6.21 Saprophytes. A *Voyria aphylla* on tree stump, Moraballi Creek, Guyana (1929). B *Voyria tenella* (three plants) and *Voyriella parviflora* on leaf litter, forest near Belém, Pará, Brazil (1969).

6.5 Saprophytes

In tropical, as in temperate forests, the vast majority of saprophytes are micro-organisms, mainly bacteria and fungi. In tropical rain forests saprophytic flowering plants are, however, often present. These are small and, easily overlooked, and though widely distributed are never, in the author's experience, very abundant. The abundance of large saprophytic fungi in rain forests is usually much underestimated, owing to the seasonal appearance of their fruit bodies (Corner 1935). Although the role of saprophytic flowering plants in the forest ecosystem is no doubt very small, they are of interest from several points of view. An account of the taxonomy and general biology of neotropical saprophytes has been given by Maas *et al.* (1986).

Only a very limited number of families are represented among saprophytes. In the monocotyledons there are Burmanniaceae (*Gymnosiphon*, *Thismia* and other genera), Petrosaviaceae (Malesia only), various Orchidaceae and Triuridaceae (*Sciaphila* etc.); among dicotyledons, Gentianaceae (*Voyria* etc.) and Polygalaceae (*Salomonia*). Several of these genera are pantropical in distribution. Many of these plants are of strange appearance and their morphology, anatomy and development have been much studied (Johow 1885, 1889, Ernst *et al.* 1910–14). In Guyana a small fern, *Schizaea fluminensis*, is often associated with 'saprophytic' flowering plants in deep shade and is possibly itself a partial saprophyte, at least in the gametophyte stage (Davis & Richards 1933–34, p. 372). Most of these are erect plants less than 20 cm high; the saprophytic orchid *Galeola* is unique in being a root-climber attaining a height of some 10 m or more. In colour rain-forest saprophytes are white (*Gymnosiphon*), bright yellow (*Voyria aphylla*) (Fig. 6.21), pink, blue or some shade of purple (*Sciaphila*). The majority of them are holosaprophytes nearly or entirely destitute of chlorophyll; some of the orchids may contain a little chlorophyll and be capable of a limited amount of photosynthesis.

The members of this synusia are found on the forest floor and occasionally on logs or stumps; a species found in the Amazon forest near Belém is said to occur only on termitaria (M. Pires). They prefer deep shade and places where dead leaves accumulate to a greater depth than usual, as in slight hollows. A favourite habitat for many species is in the corners between the buttresses of large trees. At Moraballi Creek, Guyana, saprophytes were especially abundant in the morabukea (*Mora gonggrijpii*) forest, the darkest of the five types of primary forest in the district. Their absence or scarcity in rain forests with a marked dry season, e.g. in southwestern Nigeria, suggests that they are not able to survive even a slight drying of the forest floor. The mode of occurrence of saprophytes where they are common, as in Guyana and Borneo, is very characteristic. Over large areas they are quite absent or represented only by occasional individuals of the commonest species; here and there, however, there are patches, often only a few square metres in extent, where many individuals of three or more species occur together.

The underground organs of these saprophytes contain mycorrhizal fungi on which they are dependent; in fact it would be more accurate to term them mycoparasites rather than saprophytes. Van der Pijl (1934) suggested that the patchy distribution of rain-forest saprophytes may be due to factors affecting their fungi. In some tropical saprophytes the root system is situated in the superficial humus layer, while in others it is in the underlying mineral soil, sometimes as deep as 15–20 cm (Johow 1889).

6.6 Parasites

As well as fungi and bacteria, there are two synusiae of parasitic higher plants in the tropical rain forest, the root parasites growing on the ground and the hemiparasites which are epiphytic on trees. No terrestrial hemiparasites comparable to the temperate *Striga* etc. seem to be known in tropical rain forests.

The epiphytic hemiparasites (mistletoes) of the rain forest all belong to two families, the Loranthaceae and Viscaceae. They occur throughout the tropical rain forest as well as in other tropical vegetation types. There are numerous genera and some hundreds of species. Many species can often be found in one area, and even on a single host tree there may be several species and many individuals.

Rain-forest mistletoes are more or less woody and vary in habit from erect shrubs to half-scramblers. Their leaves are entire, coriaceous and evergreen; they often closely mimic those of the host tree, so that when not flowering their presence can be easily overlooked. There is some evidence that this leaf mimicry protects them from herbivores (Ehringer *et al.* 1986). The flowers, especially those of the eastern tropical Loranthaceae, are often brilliantly coloured; a tree laden with flowering mistletoes is one of the most beautiful sights in the Malesian forests.

Most Loranthaceae and Viscaceae seem to have a wide range of hosts (Kuijt 1969). *Tapinanthus bangwensis*, a common West African species, occurs on hosts of at least twelve families and is often a pest on cocoa trees (Room 1973). Loranthaceae are sometimes hyperparasites, attacking members of their own family (Koernicke 1910) and 'hyper-hyperparasitism' (a parasite on a parasite on a parasite) has even been recorded. The seeds are dispersed by birds and have a sticky covering which holds them firmly where they are deposited. Large numbers of seedling Loranthaceae, like bromeliads (p. 140), are sometimes seen on telephone wires, but they do not survive for long.

Most probably all tropical Loranthaceae and Viscaceae are sun epiphytes; under forest conditions they have a well marked vertical distribution (Room 1973). In mature forest they occur on the twigs and branches of tall (A and B storey) trees and rarely at lower levels, except in gaps. In more open types of forest they may descend almost to ground level; thus in the well-illuminated heath forest of Borneo the author found

Fig. 6.22 *Thonningia sanguinea*, a parasite on tree roots; Omo Forest Reserve, Nigeria (1935).

Fig. 6.23 *Rafflesia gadutensis*, a parasite on roots of lianes, (Vitaceae), Ulu Gadut, Padang, Sumatera (type locality). (Photo. W. Meijer.)

In Ghana, Room (1973) found that *Tapinanthus bangwensis* was significantly commoner on unshaded than on shaded cocoa trees. He showed experimentally that germination, and the establishment and subsequent growth of seedlings, were more successful in full sunlight than in shade; when shaded the plants tended to become etiolated.

Mistletoes are serious pests of many tropical tree crops but their importance in natural forest ecosystems cannot be easily assessed. It is evident that apart from any damage they may do by depriving their hosts of water and nutrients, they compete with them for light. Room (1972, 1973) showed that on cocoa farms in Ghana *Tapinanthus bangwensis* is involved in a very complex web of interrelations in which capsids, mealy bugs, pathogenic fungi and viruses, as well as the cocoa tree itself, are components. On cocoa plantations Loranthaceae probably become much more abundant than in a natural community, but in native forests also they no doubt have similar complex interrelations.

The species of root parasite are not numerous and most of them are rare or local in occurrence. Apart from *Christisonia*, a genus of Orobanchaceae found in Malesia, mainly on bamboo roots, all the root parasites of lowland rain forests belong to two pantropical families, the Balanophoraceae and Rafflesiaceae. The former includes *Thonningia sanguinea* (Fig. 6.22) which is frequent in African rain forests, and the widely distributed tropical American *Helosis cayennensis*. In the Rafflesiaceae, besides the Malesian *Rafflesia*, famous for its gigantic flowers, there are several other genera that are rain-forest parasites.

Some of these parasites have a restricted range of hosts; the species of *Rafflesia* (Fig. 6.23), for example, occur chiefly on the roots of lianes belonging to the Vitaceae, but others, e.g. *Thonningia sanguinea*, are found on many unrelated host species. In any one district it is unusual to meet more than one species of parasite. The comparative scarcity of parasitic flowering plants in tropical rain forests is a surprising fact for which there is no obvious explanation.

Macrosolen beccarii on small trees and shrubs at 1–2 m from the ground, while in the more shady mixed dipterocarp forest no Loranthaceae occurred below about 20 m, except in clearings and by rivers (Richards 1936, p. 32). In West Africa Loranthaceae are common on isolated cocoa and cola trees at a height of a few metres.

Part II

The environment

Chapter 7

Climate

R.P.D. Walsh

THE CLIMATES OF tropical rain-forest regions are hot and wet, with little or no dry season. They are climates where sufficient water, heat and light are available for biological processes and production to take place at high rates throughout the year. This chapter examines in some detail not only rainfall and temperature, but the many other climatic factors that influence the character and dynamics of tropical rain forests. It stresses also the considerable diversity in climate found within the tropical rain forest zone. Average annual rainfalls range from 1700 mm to over 10 000 mm, with profound implications for nutrient cycling. Some areas experience a regular short dry season; others experience occasional but irregularly timed dry periods, whereas there are some very wet areas where soil moisture deficits have never been recorded. Climatic elements such as wind, humidity and cloudiness vary both between regions and with altitude up tropical mountains. Catastrophic events such as tropical cyclones and occasional long droughts affect some rain-forest areas, but not others. This climatic diversity has important ecological implications and is responsible for some of the differences in species composition, structure, productivity and dynamics found between rain forests.

7.1 Tropical rain-forest climates and their distribution

More precise definitions and subdivisions are considered later, but in general terms a tropical rain-forest climate can be defined as one with monthly mean temperatures of at least 18 °C throughout the year, an annual rainfall of at least 1700 mm (and usually above 2000 mm) and either no dry season or a short one of fewer than four consecutive months with less than 100 mm. The distribution of such climates over the Earth's surface (Fig. 7.1) is controlled primarily by the positions and intensities of the inter-tropical low pressure belt and the subtropical anticyclones during the year, the east–west dynamics of the tropical atmosphere and the distribution of land and sea. In some areas monsoonal wind circulations, the orientation of coasts and mountain ranges and the distribution of warm and cold ocean currents also play significant roles.

Rain-forest climates are chiefly found over much, but by no means all, of the areas astride the equator between 5°N and 5°S. These are affected either throughout or during most of the year by convection-enhancing weather disturbances of the inter-tropical low pressure area sandwiched between the subtropical high pressure belts of the northern and southern hemispheres. This low pressure zone extends well north of the equator especially over the continents in the northern summer, but only a few degrees south of the equator in the southern summer, the discrepancy reflecting the smaller land area in the southern hemisphere. In the winter months, areas poleward of 5°N and 5°S are dominated by the stable, descending air of the NE and SE trade winds on the equatorward sides of oceanic and continental subtropical anticyclones; the dry season becomes progressively longer and the wet season shorter with increasing distance from the equator. However, where coastlines and mountain ranges are aligned normal to the trades, onshore winds and orographic uplift can lead to considerable winter rainfall and result in ever-wet

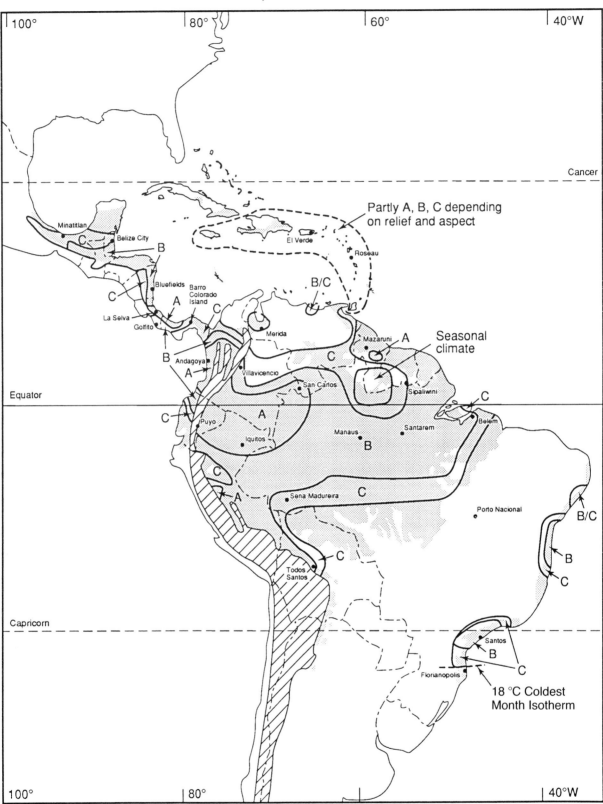

Fig. 7.1 World distribution of tropical rain forest and associated climatic types.

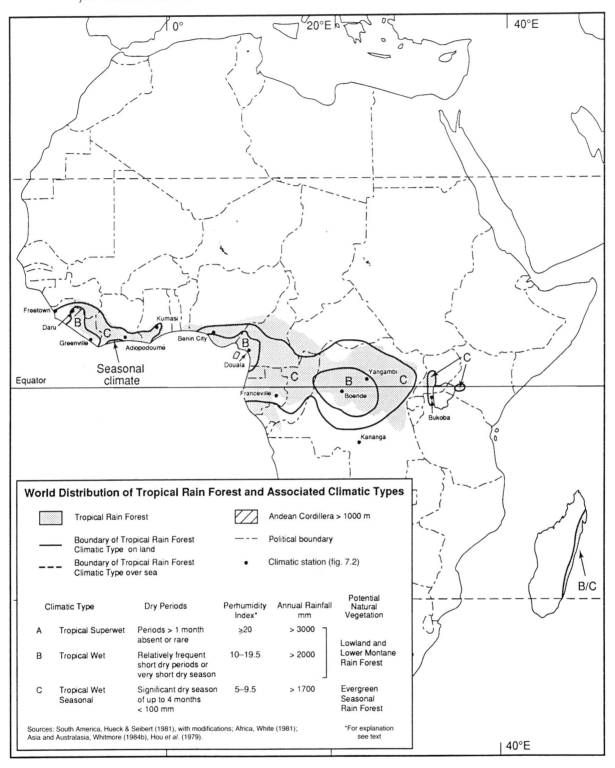

World Distribution of Tropical Rain Forest and Associated Climatic Types

Tropical Rain Forest

Andean Cordillera > 1000 m

Boundary of Tropical Rain Forest
Climatic Type on land

Political boundary

Boundary of Tropical Rain Forest
Climatic Type over sea

Climatic station (fig. 7.2)

Climatic Type		Dry Periods	Perhumidity Index*	Annual Rainfall mm	Potential Natural Vegetation
A	Tropical Superwet	Periods > 1 month absent or rare	≥20	> 3000	Lowland and Lower Montane Rain Forest
B	Tropical Wet	Relatively frequent short dry periods or very short dry season	10–19.5	> 2000	
C	Tropical Wet Seasonal	Significant dry season of up to 4 months < 100 mm	5–9.5	> 1700	Evergreen Seasonal Rain Forest

Sources: South America, Hueck & Seibert (1981), with modifications; Africa, White (1981); Asia and Australasia, Whitmore (1984b), Hou *et al.* (1979).

*For explanation
see text

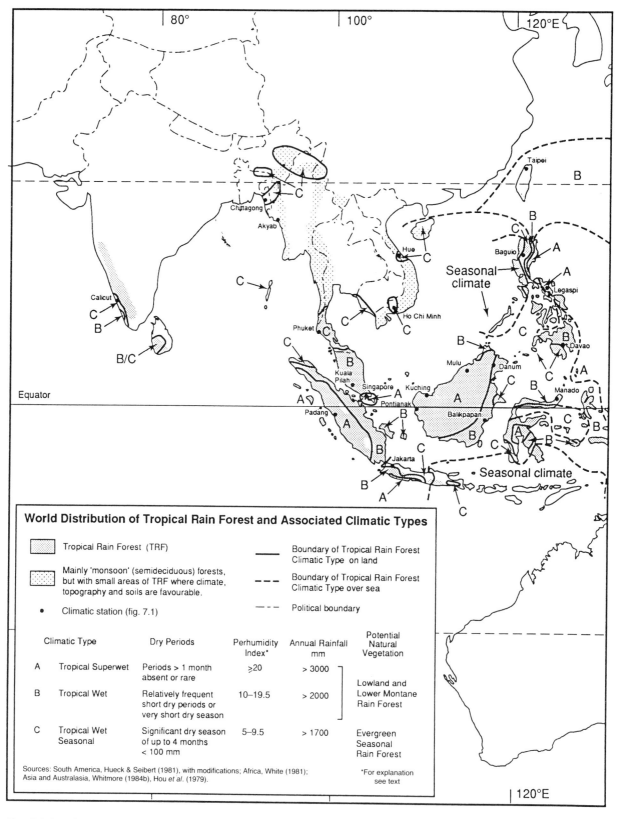

World Distribution of Tropical Rain Forest and Associated Climatic Types

Tropical Rain Forest (TRF)

Mainly 'monsoon' (semideciduous) forests, but with small areas of TRF where climate, topography and soils are favourable.

• Climatic station (fig. 7.1)

—— Boundary of Tropical Rain Forest Climatic Type on land

– – – Boundary of Tropical Rain Forest Climatic Type over sea

– · – Political boundary

Climatic Type		Dry Periods	Perhumidity Index*	Annual Rainfall mm	Potential Natural Vegetation
A	Tropical Superwet	Periods > 1 month absent or rare	≥20	> 3000	Lowland and Lower Montane Rain Forest
B	Tropical Wet	Relatively frequent short dry periods or very short dry season	10–19.5	> 2000	
C	Tropical Wet Seasonal	Significant dry season of up to 4 months < 100 mm	5–9.5	> 1700	Evergreen Seasonal Rain Forest

Sources: South America, Hueck & Seibert (1981), with modifications; Africa, White (1981); Asia and Australasia, Whitmore (1984b), Hou et al. (1979).

*For explanation see text

Fig. 7.1 *(cont.)*. For legend see p. 160.

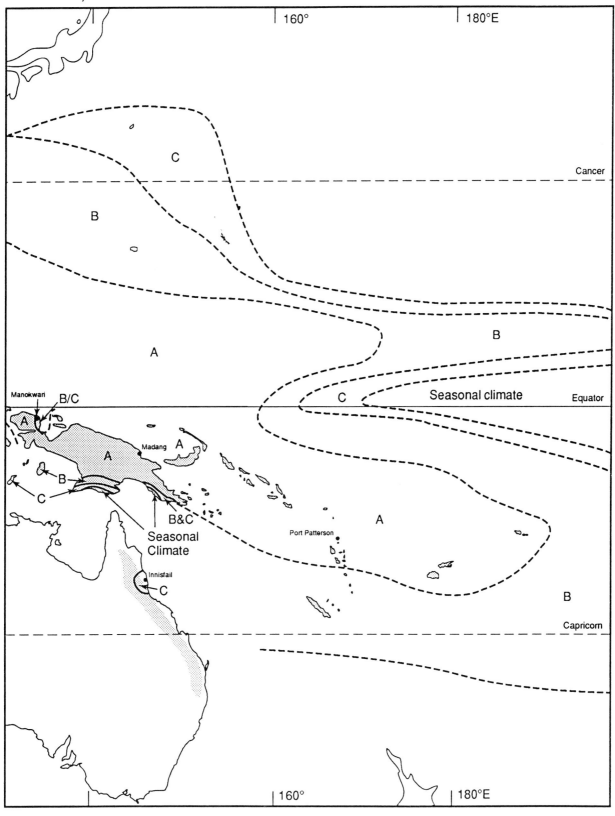

Fig. 7.1 (cont.).

rain-forest climates well north and south of the equator. Examples include parts of the Atlantic side of Central America, the northern parts of the Guianas, parts of eastern Brazil, part of the Queensland coast, eastern Madagascar and on the windward sides and in the mountain interiors of the Mascarenes, the West Indies and many Pacific islands.

Within the equatorial belt anomalously dry climates occur in western Ecuador and the central Pacific as a result of cold ocean currents and the east–west atmospheric circulation of the Walker Cell; in the Sunda Islands (affected by very dry air during the eastern monsoon from Australia during the southern winter); to the lee of mountain ranges aligned normal to monsoon winds, as in much of Java; and other areas where winds parallel coastlines and cause local low level divergence and air stability, as with the easterly trades along the north coast of Venezuela and the SE monsoon along the Port Moresby coast in southeastern New Guinea. The causes and controls of tropical weather and climate are more complex and varied than formerly thought and the reader is referred for more detail to recent texts by Nieuwolt (1977), Riehl (1979), Hastenrath (1985) and Jackson (1989).

There have been many attempts to correlate the boundaries of tropical rain forest and its subdivisions with climatic parameters. In the long-established climatic classification developed for ecological purposes by Köppen (1918, 1936), tropical rain forest is considered the characteristic vegetation of the ever-wet parts (Af) of his Tropical Rain (Class A) group of climates. In these the mean temperature (t) of the coldest month exceeds 18 °C and mean rainfall of the driest month exceeds ($2t+14$) mm (usually around 60–65 mm). The Af zone was bordered by either an Am climate (in which a dry season was considered to be partly offset by very high wet season rainfall) with evergreen seasonal forest and 'monsoon forest' (see Chapter 16) or an Aw climate (dry season and only moderate annual rainfall) with distinctly seasonal forest formations and savanna. In Zaïre, Bernard (1945) considered that this scheme worked reasonably well, but most botanists have found the Af/Am–Aw boundary, with its emphasis on the 60 mm mean rainfall of the single driest month, unrealistic and have considered the length and severity of the dry season more important. In addition, Beard (1945) in his classification of West Indian vegetation considered 100 mm (rather than 60 mm) to be a more accurate figure for monthly rain-forest transpiration, a suggestion that has been recently supported by transpiration measurements in Amazonia (Shuttleworth *et al.* 1984a) (see Chapter 8).

The empirical classifications developed by Thornthwaite (1948) and Holdridge (1947, Holdridge *et al.* 1971) are also of limited use within the humid tropics. Thornthwaite's widely used classification is based upon an empirical assessment of the water balance using potential evapotranspiration (PE) estimates and rainfall figures. Although it is capable of distinguishing broad climatic zones at a global level, its crude basis of calculating PE from temperature and day-length alone is unsatisfactory within the tropics, where humidity, wind and other factors are important. The complex formulae used tend to hide its lack of a theoretical basis and have no advantage over simpler systems in delimiting the boundaries of different tropical vegetation types.

Holdridge's system of 'life zones' subdivides the tropics into a series of ecological zones, but the divisions themselves and the method by which they have been derived are questionable. The system divides the earth into life zones using simple temperature and humidity criteria. The temperature divisions are based on Holdridge's 'biotemperature' concept, in the derivation of which any temperatures above 30 °C are reduced to 30 °C on the rather dubious assumption that biological processes become less productive above that temperature. His 'Tropical Zone' is delimited by an annual mean biotemperature of 24 °C, a rather high isotherm that would exclude some lowland tropical rain forests in the trade wind zones. Altitudinal 'premontane', 'lower montane', 'montane', 'subalpine', 'alpine' and 'nival' zones are recognized at successively lower biotemperatures (see Chapter 17). The humidity province divisions are based on the 'evapotranspiration ratio' (the ratio of potential evapotranspiration (PE) to mean annual precipitation). PE (mm) is calculated empirically as 58.93 × mean annual biotemperature and takes no account of wind or humidity. Successively lower evapotranspiration ratios are then used to divide the Tropical Zone into eight life zones of increasing wetness, of which three, Tropical Moist, Tropical Wet and Tropical Rain Forest, occupy the tropical rain-forest area. The Dry–Moist, Moist–Wet and Wet–Rain Forest boundaries are set at evapotranspiration ratios of 1.00, 0.50 and 0.25, respectively, values equivalent to annual rainfalls of 2000 mm, 4000 mm and 8000 mm at an annual biotemperature of 24 °C. A serious defect of the system is that it takes no account of rainfall seasonality and the length of dry season (if any), which are assumed to vary systematically with annual amount; such an assumption may have some local validity, but none generally within the tropics. The system has been applied to the forests of Costa Rica (Holdridge *et al.* 1971), Peru (Tosi 1960) and elsewhere, but the life zones and their boundaries appear to be highly arbitrary and the system has not proved useful in climate–ecological mapping in the tropics as a whole. As noted in later chapters of this book, Holdridge's conception of rain forest is much narrower than that used in this book and it excludes several areas regarded as tropical rain forest by Schimper and many others.

The theoretical approach to evaporation developed by Penman (1948) and refined by many later workers takes into account those factors (radiation, windspeed and humidity) ignored by the empirical approaches of Thornthwaite and Holdridge. However, the data requirements for their use in a water balance approach to climatic classification in the tropics are prohibitive. Furthermore, recent research in temperate forests (Shuttleworth & Calder 1979) and Amazonian rain forest (Shuttleworth et al. 1984a) has demonstrated the inadequacy of such equations in modelling total evapotranspiration and the need to use separate equations to assess the interception and transpiration components (see Chapter 8).

A classification of tropical climates using climate diagrams and the perhumidity index. A simple and effective way of depicting differences in tropical climate, and especially the average length and intensity of any dry season, is to use the Climate Diagram (Klimadiagramme) system of Gaussen (1955) developed for ecological purposes by Walter & Lieth (1960). In these diagrams, regimes of monthly rainfall and monthly temperature are displayed on a single graph, in which the temperature scale (from 0 to 50 °C) is twice as large as the rainfall scale (from 0 to 100 mm); to save space the rainfall scale above 100 mm is reduced by a factor of 10 (Figs. 7.1, 7.2). Three types of month are distinguished: 'wet' months, in which rainfall exceeds 100 mm (taken as a rough guide to potential evapotranspiration in the tropics); 'drought' months, in which rainfall falls below the plotted temperature graph (normally 40–60 mm rainfall, depending on temperature); and 'dry' or 'intermediate' months, in which there is a less serious deficit of rainfall below PE. This differentiation of intermediate and drought months had previously been employed by Mohr & Van Baren (1954). The diagrams can also be used to display other climatic statistics of the station. The climate diagram system is adopted in this book in a slightly modified form. Apart from monthly regimes of rainfall and temperature, the diagrams show station altitude; mean annual temperature and rainfall; the annual range of temperature; and, where available, the mean diurnal temperature range and the absolute highest and lowest temperatures recorded.

The 'perhumidity index' (PI) (Walsh 1992), which measures the degree of continuity of wetness (or perhumidity) of a tropical climate, attempts to summarize the dry and wet season characteristics of the climate diagrams into a single index value. This index incorporates some elements of the Q index of Schmidt & Ferguson (1951), who devised a classification of climates of Indonesia based on the ratio of dry (<60 mm) to wet (>100 mm) months, but gives greater weight to longer dry periods and different weights to wet months of greater than 100 mm and 200 mm rainfall. In calculating the index, each month is given a simple points score depending on its mean rainfall; the monthly scores are then summed to obtain the perhumidity index value. A very wet month (>200 mm) is given a score +2, a wet month (100–199 mm) +1, a dry month (50–99 mm) −1, and a drought month (<50 mm) −2. The first dry or drought month following a wet month scores −0.5 and −1.5, respectively, on the basis that soil water availability is still high. The scoring system thus allows for: (1) the lesser impact of two short dry seasons than one long one, via the −0.5 and −1.5 scores; (2) the greater impact of drought than dry months; and (3) greater (but not limitless) weight (+2) to be given to very high monthly means, which are more effective at maintaining high soil moisture levels and are more reliably above the 100 mm threshold than months with means between 100 and 199 mm. The perhumidity index is here applied to the average rainfall regime for data availability reasons, but, as with the Q index, it would be preferable to use monthly rainfall series covering individual years. The use of averages leads to PI values a few points higher than if monthly series were used and it would be misleading to construct maps with data based partly on averages and partly on monthly data. Perhumidity index values at tropical rain forest stations are shown on the climate diagrams. The index, which can vary from −24 (all monthly means less than 50 mm) to +24 (all monthly means over 200 mm), is very sensitive to differences in rainfall regime and correlates well with the boundaries between different tropical forest formations. Within the rain-forest zone, values range from around 5 at the margin to 24 in perennially very wet areas.

A broad classification of tropical climates in relation to vegetation based on the perhumidity index, annual rainfall and temperature is outlined in Table 7.1. Tropical lowland rain forest climates are divided into broad classes, (i) Superwet, (ii) Wet and (iii) Wet Seasonal, on the basis of perhumidity index values. The temperature criteria of a mean annual temperature of at least 20 °C and a coldest monthly mean of at least 18 °C apply to

A. Superwet Localities

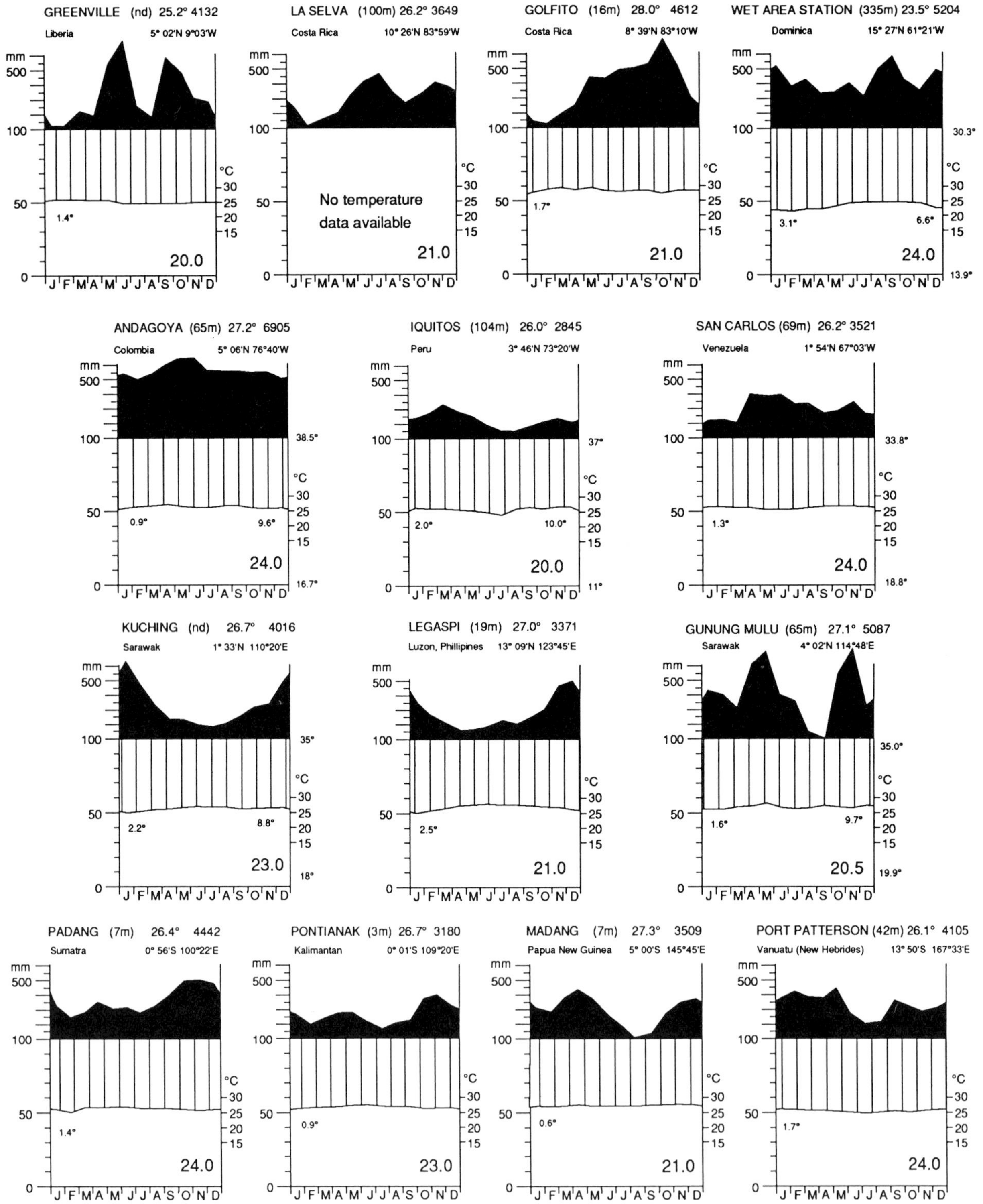

Fig. 7.2 Climate diagrams for tropical rain forest stations. Locations are shown in Figure 7.1. The diagrams mostly follow the conventions of Walter & Lieth (1960). The station name line also gives station altitude (m), mean annual temperature (°C) and mean annual rainfall (mm). The annual and mean daily ranges in temperature (°C) are shown towards the left and right, respectively, of the graph. The highest and lowest recorded temperatures are given (where available) above and below the temperature scale to the right of the graph. The perhumidity index value (see text) is given in the bottom right-hand corner of the graph.

B. Tropical Wet Localities

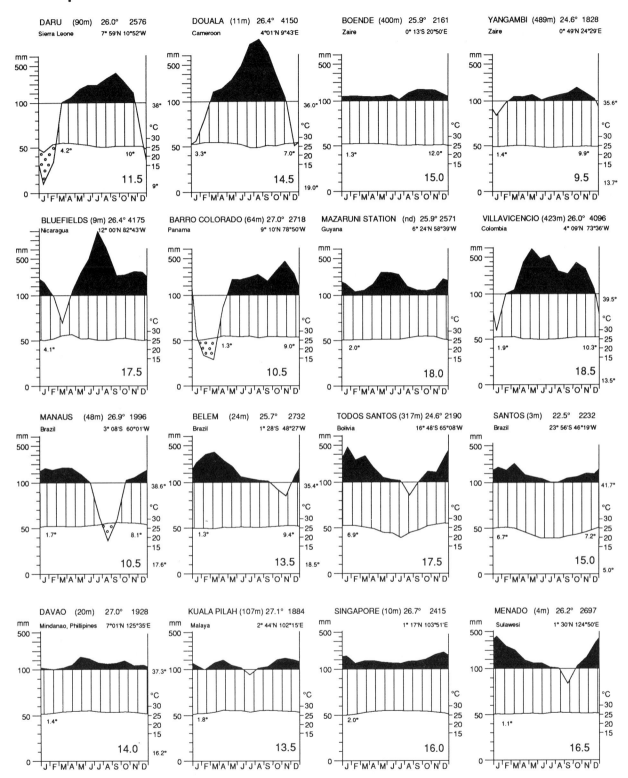

Fig 7.2 (*cont.*)

B. Tropical Wet Localities (continued)

C. Tropical Wet Seasonal (Rain forest marginal) Localities

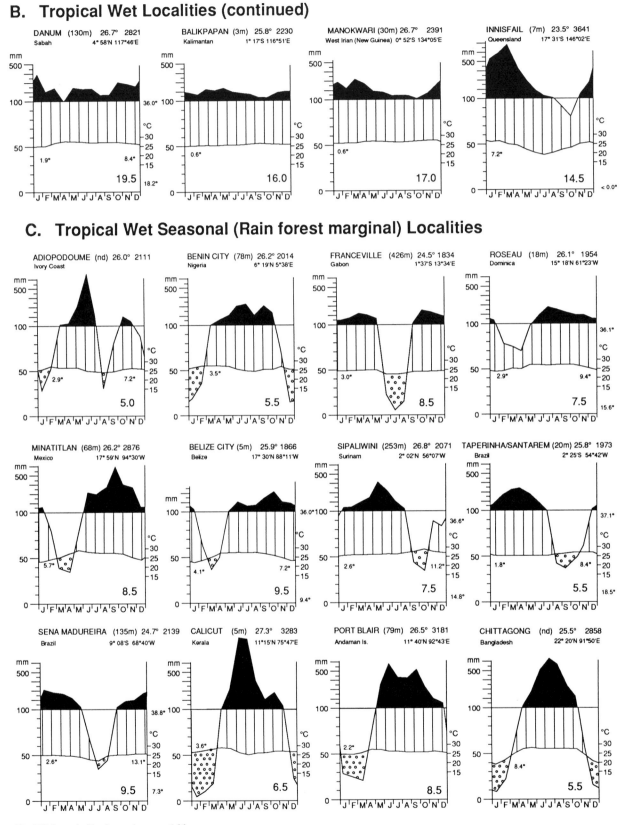

Fig 7.2 (*cont.*). For legend *see* p. 166.

C. Tropical Wet Seasonal (Rain forest marginal) Localities (continued)

D. Tropical Wet-Dry Localities

E. Montane & F. Subtropical Margin Localities

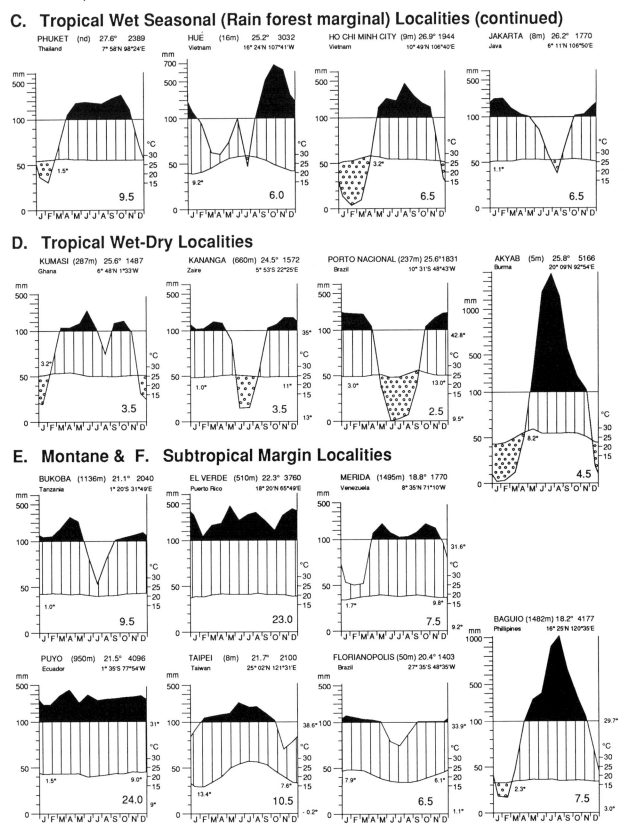

Fig 7.2 (*cont.*).

Table 7.1. A climatic classification for tropical rain-forest regions

AR=Mean annual rainfall; PI=Perhumidity Index (see text); DM=number of dry months (<100 mm) per year.

| Climatic class | Main distinguishing features | Rainfall | | | Temperature | | Solar radiation (langleys d^{-1}) | Relative humidity (%) | Characteristic vegetation |
		AR (mm)	PI	DM	Annual (°C)	Coldest month (°C)			
Tropical Montane Climates	Higher relative humidity and cloudiness and lower sunshine than TRF climates. Lower temperatures and often very high rainfall	≥1500	variable	<5	<24	≥18	<400	≥90	Upper montane and lower montane rain forests
Tropical Superwet	Very high annual rainfall. No dry season, though occasional short dry periods <2 months may occur	≥3000	≥20	0	20–28	≥18	350–500	70–90	Lowland moist rain-forest formations
Tropical Wet	Moderate–high annual rainfall with normally a very short dry season or frequent short dry periods	≥2000	10–19.5	0–3	20–28	≥18	350–500	70–90	Lowland moist rain-forest formations
Tropical Wet Seasonal	Long wet season but significant dry season	≥1700	5–9.5	3–5	20–28	≥18	350–500	70–90	Evergreen seasonal forest
Tropical Wet–Dry	(a) Marked dry season	≥1200	−4 to 4.5	4–6	20–29	≥18	>500	Decreasing →	Semi-evergreen seasonal forest
	(b) Long dry season and long wet season	≥700	−12 to −4.5	6–8	20–29	≥18	>500		Deciduous seasonal forest
Tropical Semi-arid	Short wet season and long intense dry season	<700	<−12	>8	≥20	≥18	>500		Thorn forest

all three classes. Tropical montane climates are treated as a separate, complex class in which temperature, humidity, sunshine and cloudiness are the criteria distinguishing them from lowland climates. The lowland climatic types (i)–(iii) grade into each other. The Tropical Superwet type (i) is characterized by a perhumidity index of at least +20, a very high annual rainfall and only rare and very short dry periods; the vegetation is usually dominated by species that are ill-equipped to withstand significant drought. The Tropical Wet class (ii) has a perhumidity index of 10–19.5 and either a much lower annual rainfall (but still exceeding 2000 mm) or a short dry season of 1–2 months. Dry periods, whether seasonal or more random, are more frequent and longer than in the Superwet zone, and the rain-forest vegetation tends to reflect this in its species composition. Class (iii), the Tropical Wet Seasonal, has a perhumidity index of around 5–9.5 and lies at the seasonal margin of the tropical rain-forest zone. It is characterized by a more or less regular, short to moderate dry season and moderate to high annual rainfall; the longer or more intense the dry season in this zone, the higher the annual rainfall that is needed to help offset it. This area is characterized by evergreen seasonal rain forest or allied formations (Chapter 16) and possesses many species able to withstand significant drought periods.

The distribution of these climatic types is tentatively shown in Fig. 7.1. Superwet climates, which in total occupy a relatively small area, are found in parts of Malesia, in the western or upper Amazon Basin, in western Colombia, in parts of Atlantic Central America and in the Guiana coastlands. Tropical Wet climates characterize the Malay Peninsula, eastern Borneo and the middle Amazon Basin. The Tropical Wet Seasonal climatic type is particularly extensive in Africa and the eastern Amazon Basin. The wettest parts of the Indian Monsoon area, despite their sometimes very high annual rainfalls, tend also to fall into this class. Thus Cherrapunji, despite receiving an annual average rainfall of 11 420 mm, has a perhumidity index of only +7.5 as a result of its long and intense dry season with four drought months.

It should be stressed that the boundaries are only approximate as no classification based on climatic averages is capable of delimiting vegetation boundaries with precision for several reasons. First, climatic factors other than rainfall and temperature, such as relative humidity and wind, and non-climatic factors (soil, lithology and relief) also influence soil moisture and the ecological impact of a dry period. Second, such a classification implies that vegetation is related to climatic means; it

may be, however, that extremes of drought, wind, cold or continuously saturated conditions are more responsible than means for vegetation characteristics and distribution. Although spurious correlations between vegetation and climatic averages are often evident at a regional scale, where means and extremes may be strongly correlated, such correlations often do not hold when applied to other regions. Third, as mentioned previously, the use of rainfall averages understates the true frequency of dry periods (Brünig 1969). Finally, the climatic averages used may be unrepresentative of the long term: not only are records at tropical stations often short and incomplete, but there is growing evidence that rainfall, drought and cyclone frequency have varied significantly even within the past century in some rain-forest regions (see section 7.9).

7.2 Temperature

Data on monthly and annual mean temperature, mean annual range, mean diurnal range and (where available) the absolute maximum and minimum temperatures during the period of record are given in the climatic diagrams of Fig. 7.2.

Rain-forest regions, though perennially hot, are less so than the drier tropics. Mean annual temperatures, which by definition exceed 20 °C, are generally between 24 and 28 °C in lowland forests within 10° latitude of the equator, but decline with increasing altitude and latitude. A key feature is the absence of a cool season; all monthly means exceed 18 °C. Annual ranges of temperature between hottest and coldest months are very small, particularly in maritime and equatorial localities. Only in continental areas poleward of 15° latitude, such as Bolivia (Todos Santos, annual range 6.9°), southeast Brazil (Santos 6.7°), south China (Taipei, 13.4°) and Queensland (Innisfail, 7.2°), do annual ranges exceed 5 °C as a result of reduced insolation of the 'winter' months and incursions of temperate continental air. At some maritime equatorial localities, annual ranges are less than 1 °C, as for example at Balikpapan (Borneo), Manokwari (New Guinea) and Andagoya (Pacific coast of Colombia). In equatorial areas (as at Manaus and Belém in Amazonia), drier months tend to be slightly hotter months, because of the increased frequency of sunny days with higher maximum temperatures. In areas poleward of 10°, however, the dry season is the cooler season, as it tends to occur at the winter solstice when reduced insolation leads to lower maximum and minimum temperatures, as at Roseau (Dominica) and Todos Santos (Bolivia).

Diurnal ranges of temperature exceed the annual range and are generally 6–10 °C, though rather higher in some continental localities (e.g. Sena Madureira, 13 °C). However, diurnal ranges are significantly lower than in the seasonal and arid tropics, where day maxima are higher and night minima lower, reflecting the clearer skies and lower dew points of such areas during the long dry season. Maximum temperatures, which tend to occur between 1200 and 1500 hours depending on diurnal patterns of cloudiness, rarely exceed 35 °C even on cloudless, sunny days, in part because of the cooling effect of intense evapotranspiration. Highest recorded temperatures at most rain-forest locations (see Fig. 7.2) are generally 35–40 °C. Minimum temperatures vary rather more between rain-forest regions. At many lowland equatorial locations, night minima (which tend to occur just before dawn) rarely fall below 20 °C and the lowest recorded temperatures tend to be over 15 °C, as at Manaus (17.6 °C), Taperinha/Santarem (18.5 °C) and San Carlos (18.8 °C) in Amazonia, Andagoya (16.7 °C) in western Colombia, Douala (19.0 °C) in Cameroon and Kuching (18 °C) in Sarawak. There are two reasons for these high minimum temperatures. First, the high absolute humidities mean that dew points are also very high; the release of latent heat with condensation once the dew point is reached (usually at 21–22 °C) effectively prevents temperatures falling much further. Second, there is the greenhouse effect of the clouds, high water vapour and carbon dioxide content of the wet tropics.

Areas affected by airflows from extratropical land masses that cool down in winter experience much lower minima. Of particular note are the *friagems* (cold waves) that affect the upper Amazon Basin, in which cold air from southern South America passes northwards between the Andes and the Mato Grosso (Ratisbona 1976). Because of their infrequency and short duration (normally 3–5 days), they affect mean temperatures very little, but result in much lower extremes than in the lower Amazon Basin. Minima recorded during such cold waves are 7.3 °C at Sena Madureira (9°S) and 11 °C at Iquitos (4°N) and over the seasonally dry Mato Grosso temperatures close to freezing have been recorded at Cuiaba (1.2 °C) and Corumba (0.8 °C). *Friagems*, which occur between May and August, vary in frequency, with up to five intense cold invasions in some years, but none in others. Particularly long ones occurred in May 1878 (15 days) and August 1882 (10 days) and on rare occasions, as in the 'Great *Friagem*' of 1933, their impact, albeit much reduced, reached the lower Amazon at Manaus and Santarem (Ratisbona 1976). In areas poleward of 15° latitude, extreme minima can be even lower and frost has been recorded at sea level at Innisfail (Queensland) (Tracey 1982) and Taipei (Taiwan) (Watts 1969).

The small day-to-day and seasonal variations in temperature typical of equatorial lowland rain-forest areas are demonstrated by the records at G. Mulu (4°N) in Sarawak in 1977–78 (Fig. 7.3) (Walsh 1982a). Night minima were almost always in the range 20.5–23.0 °C with a lowest recorded temperature of 19.9 °C during a January night. By comparison daily maxima varied rather more depending on weather during the day, and the morning hours in particular. Three types of day were identified: days with clear, uninterrupted sunshine, when maxima (at 1300 hours) were 32–34 °C; days with hazy sunshine or frequent cumulus cloud with maxima around 30 °C; and overcast rainy days, when the maximum only reached 26–28 °C. Most days fell in the second category. The maximum recorded was 35.0 °C during an exceptionally dry period in September 1977. The diurnal range at Mulu (9.7 °C) is somewhat higher than at the coast at Kuching (8.8 °C). Differences in

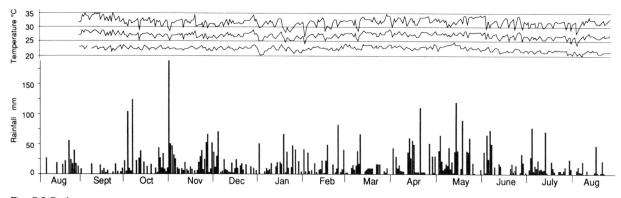

Fig. 7.3 Daily maximum, minimum and mean temperatures and daily rainfall at Long Pala, Gunung Mulu, Sarawak, during 1977–78. After Walsh (1982a).

mean diurnal range from month to month reflected differences in the relative frequency of the three types of days, with the highest diurnal ranges of 11.0 °C and 10.7 °C occurring in the driest, sunniest months of September 1977 and August 1978, respectively.

Temperature and altitude and the *Massenerhebung* effect. Temperature declines with altitude but the rate of fall (or lapse rate) varies regionally, vertically, seasonally and even from day to day with variations in cloudiness, rainfall and condensation levels as well as the humidity and temperature structure and direction of airflows. Generally, below the condensation level (marked by the cloud base) the rate of decline of temperature is high and either approximates to or exceeds the 'dry adiabatic' lapse rate (the rate at which unsaturated air cools as it rises) of 0.95 °C fall in temperature per 100 m increase in altitude. Because of the release of latent heat with condensation, lapse rates of saturated air above the cloud base tend to be much lower and incline to the 'saturated adiabatic' lapse rate (0.4 °C per 100 m at high temperatures, but increasing as temperature falls and the water vapour capacity of the air and rate of release of latent heat are progressively reduced). Changes in rainfall, cloudiness and sunshine up tropical mountains lead to considerable deviations from these rates.

Fig. 7.4 shows how mean annual temperature declines with altitude in Costa Rica, where mean temperature is around 2 °C higher on the Pacific side of the Cordillera than on the wetter Atlantic side, which is affected by onshore NE Trades throughout the year. Both slopes show a pattern of high lapse rates (1.2 °C per 100 m) at low altitude below the cloud base, very low lapse

rates within the cloud belt, and a moderate (0.6 °C per 100 m) lapse rate above the cloud belt. The threshold altitudes are lower on the Atlantic side, with its more humid air and hence lower cloud base levels. On the eastern slopes of the Andes in Ecuador, five lapse rate zones have been identified (A.M. Johnson 1976). Normal lapse rates of 0.65–0.68 °C per 100 m apply on the lower slopes up to 1000 m, between 1500 and 2000 m and again above 2500 m, but lapse rates are near zero or temperatures may actually rise between 1000–1500 m and 2000–2500 m. Both near-zero lapse rate zones are apparently associated with increases in sunshine, reduced cloudiness and reduced rainfall as one emerges above successive cloud belts.

In many parts of the tropics, lapse rates vary with season. In trade wind areas, lapse rates below the condensation level tend to be much higher in the dry season, when the contrast in cloudiness and rainfall (and hence difference in temperature) between dry coastal or lowland stations and wet mountain areas is greatest. Thus in New Caledonia, the lapse rate between Nouméa (10 m altitude) and Col d'Amieu (320 m) varies from 1.13 °C per 100 m in August (dry season) to 0.65 °C per 100 m in February at the height of the wet season. Similar seasonal contrasts occur in Central America (Portig 1976) and Dominica. In areas where the coastlands are humid and wet throughout the year, differences in cloudiness between lowland and mountain are less marked and lapse rates are low throughout the year. Thus, in western Sumatera, lapse rates between the very wet (4423 mm) coastal station of Padang and the rather less wet mountain station of Fort de Kock (920 m altitude, annual rainfall 2295 mm) average 0.58 °C per 100 m and range only between 0.55 °C per 100 m in in October–November and 0.62 °C per 100 m in January.

Large mountain masses, particularly plateaux, tend to be warmer at comparable altitudes than isolated mountains. Whereas isolated mountains merely protrude into the free atmosphere and temperatures fall according to the local atmospheric lapse rate, large upland surfaces present extensive areas for heating by solar radiation receipts that increase significantly with altitude, as transmission losses decline with a progressively thinner atmosphere. This effect of large mountain masses in raising temperatures above the values found at similar heights in the free atmosphere and on isolated peaks has been given the German term '*Massenerhebung*' (literally 'mass raise'). In the European Alps the snow line, the tree limit and the zones of vegetation are all higher on the large central massifs such as the Pennine and Engadine Alps than on the fringing '*Voralpen*'.

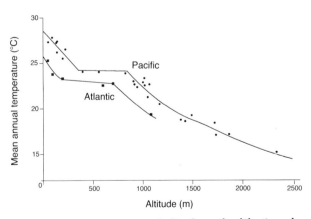

Fig. 7.4 Annual temperature and altitude on the Atlantic and Pacific sides of Costa Rica (based on the data of Holdridge *et al.* 1971 and Wernstedt 1972).

Table 7.2. *Seasonal changes of day-length, midday sun elevation and solar radiation at the top of the atmosphere at different latitudes*

Maximum solar radiation occurs at the equator at the equinoxes, in April and August at 10°N, and in June at 20°, 30° and 40°N.

Latitude	Day-length (h)	Midday sun elevation (degrees)	Solar radiation at top of atmosphere (langleys d^{-1})		
			Max	Min	Mean
Equator	12.1	66–90	898	791	853
10°N	11.5–12.7	57–90	899	734	842
20°N	10.9–13.3	47–90	935	608	806
30°N	10.2–14.1	37–84	970	465	no data
40°N	9.3–15.0	27–74	991	317	no data
60°N	5.9–18.9	7–54	979	49	no data

Plant geographers from Brockmann-Jerosch (1913) and Schröter (1926) onwards have applied the term *Massenerhebung* to this effect, although the limits are not simple functions of temperature and in the Alps seem to depend on length of lying snow and other factors only partly controlled by temperature (Ellenberg 1988).

In the humid tropics, as shown in Chapter 17, the altitudinal limits of the vegetation zones vary in a comparable way to those on temperate mountains and lowland tropical rain forest is replaced by montane rain forest at a higher elevation on large mountain ranges such as the Andes than on isolated mountains, particularly those near the coast, where the zones are considerably compressed. As in temperate regions this *Massenerhebung* effect is not directly related to the lapse rate. As Grubb & Whitmore (1966) showed in Ecuador and Grubb (1974) in New Guinea, the altitudinal distribution of the montane forest zones probably depends on the incidence and frequency of fog (ground-level cumulus and stratus cloud), which reduces illumination and depresses plant transpiration and photosynthesis rates (see pp. 224 and 226). Montane rain forests, especially upper montane forest on exposed ridges with shallow soils, though normally very humid, are subject to short periods of drought in sunny spells. Their short stature and often streamlined canopy, as well as their xeromorphic characteristics (see Chapter 17), have been attributed to these two influences, although soil nutrient deficiencies and other factors may also be involved. The altitude at which fog (cloud) forms depends on the humidity of the rising airflow over the mountain areas: the moister the airflow, the lower the altitude at which clouds form. This accounts for the compression of the vegetation zones on coastal mountains and foothills of large ranges, where air tends to be very humid, as compared with intra-montane valleys and in continental interiors, where absolute humidities are much lower (see Chapter 17).

In extensive upland massifs, notably in East Africa, temperatures are anomalously high. For example, Masindi (Uganda) (altitude 1146 m) has a mean annual temperature of 21.8 °C, over 2 °C higher than at the mountain (as opposed to plateau) station of Takengou (19.7 °C, 1205 m altitude) in Sumatera and only marginally lower than at Bandung (22.1 °C, 730 m altitude), over 400 m lower in altitude in Java. This *Massenerhebung* effect may be one reason why relatively tall rain-forest relics survive at comparatively high altitudes on the East African plateau.

7.3 Solar radiation, sunshine and cloudiness

In tropical latitudes, the high elevation of the sun at noon and the small seasonal variation in day length result in the solar radiation arriving at the top of the tropical atmosphere being high throughout the year (Table 7.2). The cloudiness and high water vapour content of the atmosphere above wet tropical areas, however, mean that solar radiation received by the forest canopy, though higher and less seasonal than in temperate climates, is considerably less than in drier tropical environments where average hours of sunshine are much higher (Table 7.3). At most locations mean values fall within the range 350–500 Gcal cm^{-2} d^{-1} (or langleys d^{-1}),[1] which represents 40–60% of the solar radiation

---

[1] 1 langley d^{-1} = 4.1868 J d^{-1}.

Table 7.3. *Solar radiation at tropical rain-forest and temperate locations (langleys d^{-1})*

Locality	Jan	Feb	Mar	Apr	May	Jun	Jul	Aug	Sep	Oct	Nov	Dec	Monthly average
Tropical lowlands													
Benin, Nigeria	340	360	390	400	420	380	300	300	310	360	380	370	360
Yangambi, Zaïre	411	447	454	446	441	391	345	363	409	412	422	373	410
Barro Colorado I., Panama	489	503	507	490	440	372	417	405	420	415	364	402	435
Manaus, Brazil	295	277	305	323	335	432	404	462	486	499	407	363	373
(% of radiation at top of atmosphere)	33	31	34	38	42	57	52	56	56	56	46	41	—
Tropical montane													
Puyo, Ecuador (950 m)	244	243	224	248	259	255	252	280	296	292	286	245	260
Mérida, Venezuela (1495 m)	509	537	562	565	501	487	521	544	542	492	492	492	520
Temperate lowland													
Kew, UK	66	60	150	330	383	471	443	390	304	137	67	49	238

Sources: Benin, Griffiths (1972); Yangambi, Bultot & Griffiths (1972); Barro Colorado, Dietrich *et al.* (1983); Manaus, Ratisbona (1976); Mérida, Snow (1976); Puyo, A. M. Johnson (1976).

reaching the upper atmosphere. Rainfall and cloudiness regimes and latitude-dependent seasonal variations in day-length and elevation of the sun are the main factors influencing seasonal regimes of solar radiation (Fig. 7.5). At the equator, peaks of solar radiation should occur around the equinoxes. This is evident at Yangambi (Zaïre), although radiation is noticeably lower at the rainier September equinox, but at Manaus, Benin and Barro Colorado, greater rainfall seasonality results in a pattern of low solar radiation in the wet season and high solar radiation in the dry season.

An important feature of rain-forest areas is the high proportion of diffuse radiation, a consequence of the cloudiness of such regions. At Yangambi, diffuse radiation accounts for 55% of the annual solar radiation, with monthly proportions ranging from 64% in wet August to 47% in the drier, sunnier month of February (Bultot & Griffiths 1972). In most rain-forest areas, skies tend to be partly cloudy even in any short dry season and the diffuse component remains high, whereas in the seasonal tropics diffuse radiation drops sharply in the dry season, for example to 20% at Lubumbashi (Zaïre) in July (Bultot & Griffiths 1972).

Not surprisingly, seasonal variations in the duration of bright sunshine are very similar to those of solar radiation (Fig. 7.6, Table 7.4). Most stations receive around 2000 h a^{-1} (or 5.5 h d^{-1}), considerably less than in the seasonal tropics, where 3000 hours per year is characteristic. However, some very wet sites and mountain sites in the cloud belt receive much less. Douala (Cameroon), which has an annual rainfall of 4115 mm, receives only 1274 h a^{-1}, with only 1.3 h d^{-1} in the peak

rainfall months of July and August, whereas Padang (Sumatera) with an even higher annual rainfall of 4442 mm, receives 2578 h. The main reason for the contrast is differing diurnal distributions of rainfall (and cloud amounts): Padang experiences late afternoon and night rainfall maxima and sunny daytime conditions (Braak 1931, p. 70), whereas Douala receives most of its rain during the day.

Changes in solar radiation and sunshine with altitude are very marked in the tropics (Tables 7.3 and 7.4, Fig. 7.5). Solar radiation at the Earth's surface in a cloudless atmosphere would tend to increase with altitude because of the smaller scattering, absorption and reflection losses associated with a progressively shorter path through the atmosphere, but on most tropical mountain ranges rainfall and cloudiness tend to increase with altitude; solar radiation and sunshine therefore tend to decline with altitude with a sharp drop as one enters the cloud belt. Above the cloud belt, however, solar radiation and sunshine rise sharply. The altitudes of the cloud base, zone of maximum rainfall and upper limit of the cloud belt vary regionally with airflow moisture and dynamic properties and the height, continuity and orientation of mountain masses. They also vary seasonally with changes in wind direction and the vertical moisture and temperature structure of the regional atmosphere and from day to day with the passage of tropical weather systems. Condensation levels marking the cloud base are at around 900 m on the eastern slopes of the Andes in Ecuador (A.M. Johnson 1976), around 1200 m on G. Mulu in Sarawak (Walsh 1982a) and around 700 m (but varying with weather and season) in the trade wind

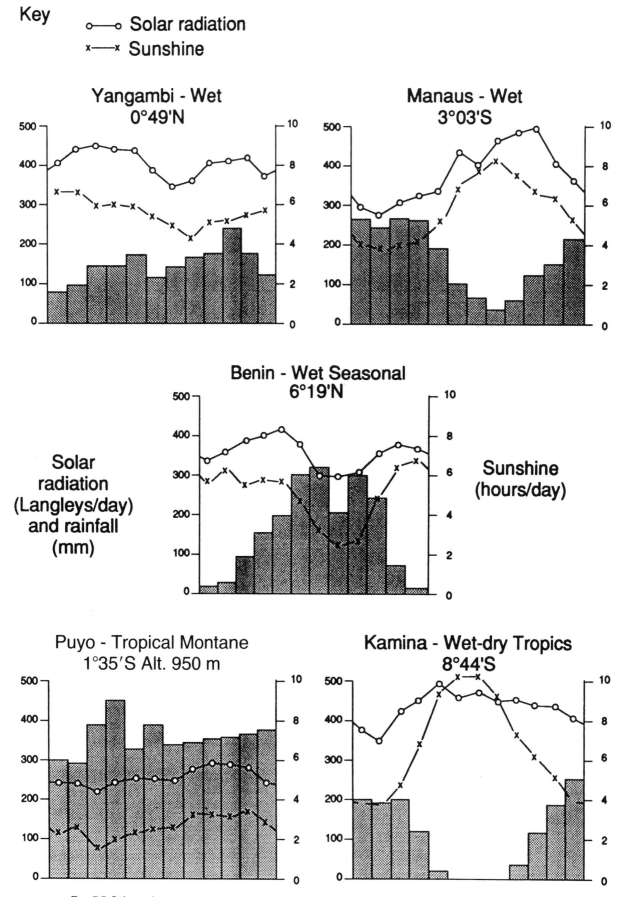

Fig. 7.5 Solar radiation and duration of bright sunshine in relation to rainfall regime at tropical stations.

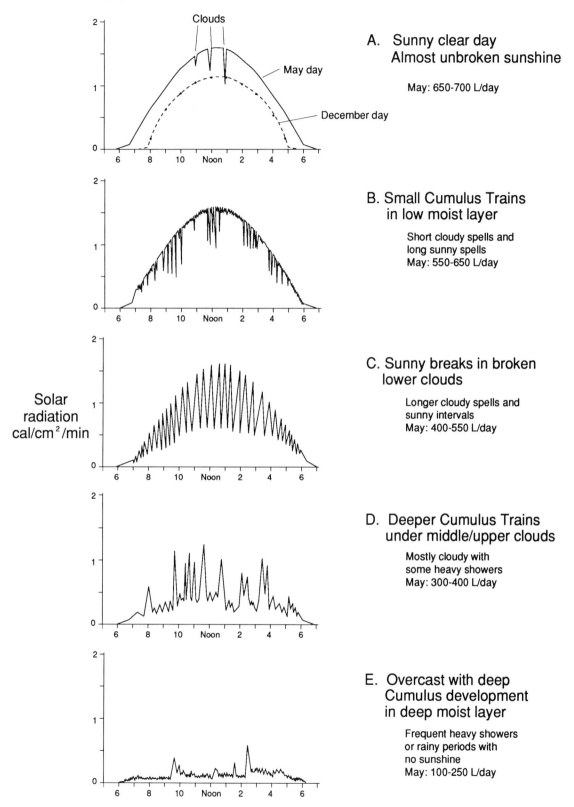

Fig. 7.6 Insolation patterns and totals on different types of day in May at El Verde (altitude 443 m), Puerto Rico (based on data and diagrams of Odum *et al.* 1970a).

Table 7.4. *Hours of bright sunshine at tropical rain-forest stations*

Station and territory	Jan	Feb	Mar	Apr	May	Jun	Jul	Aug	Sep	Oct	Nov	Dec	Year
(a) Lowland rain forest													
Douala, Cameroon	124	140	134	149	132	88	41	39	68	110	122	127	1274
Uaupés, Brazil	161	154	157	143	144	144	163	196	196	188	187	165	1995
Pontianak, Borneo	146	183	186	193	193	215	219	201	164	168	175	179	2229
Padang, Sumatera	208	223	223	226	248	230	230	226	208	201	186	190	2597
Kuala Lumpur, Malaya	190	200	156	197	179	187	214	180	176	167	184	153	2179
(b) Tropical rain forest (evergreen seasonal) and marginal													
Abidjan, Côte d'Ivoire	187	196	228	284	184	110	130	116	130	189	214	197	2086
Yangambi, Zaïre	207	189	186	182	184	164	156	135	155	157	165	176	2056
Fort de France, Martinique	237	227	257	245	242	211	224	256	219	223	220	226	2787
Manaus, Brazil	126	108	123	125	161	203	238	259	225	208	188	163	2125
Jakarta, Java	139	146	190	215	219	219	230	237	226	208	179	150	2326
(c) Seasonal and monsoon tropics													
Kupang, Timor	188	192	222	266	276	276	289	302	306	289	263	203	3072
Pasuruan, Java	179	190	208	266	277	299	299	299	307	277	252	190	2996
(d) Tropical montane													
Puyo, Ecuador (950 m)	74	75	50	61	74	79	80	101	98	98	105	90	985
Mérida, Venezuela (1495 m)						n/a							2475
Baguio, Philippines (1482 m)	148	150	145	151	123	107	75	70	101	115	119	124	1424
Pangerango, Java (3023 m)	80	51	88	99	161	172	219	186	172	131	110	73	1522

Sources: Puyo, Johnson (1976); Indonesia, Malaya and Philippines, Braak (1931); Abidjan, Douala and Yangambi, Bultot &
Griffiths (1972); Fort de France, Portig (1976); Manaus and Uaupés, Ratisbona (1976); Mérida, Snow (1976).

environment of Puerto Rico (Odum *et al.* 1970a) and
Dominica (West Indies). At Puyo in the cloud belt of
eastern Ecuador (1°S, 950 m altitude, annual rainfall
4096 mm), solar radiation and sunshine duration average
just 260 langleys d^{-1} and 2.7 h d^{-1}, respectively, with
little seasonal variation (Fig. 7.5), compared with 373
langleys d^{-1} and 5.8 h d^{-1} in the Amazonian lowlands
at Manaus. Rainfall and cloudiness start to decline (and
sunshine and solar radiation start to rise) above 1000 m
in the eastern Andes; no data are available for Ecuador,
but at Mérida (Venezuela) (1495 m altitude, 8°35'N),
solar radiation and sunshine duration average 520 langleys d^{-1} and 6.8 h d^{-1}, respectively. In Puerto Rico,
solar radiation is considerably lower (321 langleys d^{-1})
at the cloud-forested peak of El Yunque (altitude
1050 m) than at the lower montane rain forest site of
El Verde (382 langleys d^{-1}) below the daytime cloud
base (Odum *et al.* 1970a).

Monthly averages of solar radiation and sunshine
give no indication of day-to-day and diurnal variations.
Detailed continuous records of solar radiation were kept
at El Verde, Puerto Rico (18°N) above the canopy of
lower montane rain forest (Odum *et al.* 1970a). Days

were classified into five types on the basis of their
insolation pattern and associated weather and cloud situations (Fig. 7.6). Radiation totals on overcast, rainy
days were about 15% of the values recorded on clear,
sunny days. The daily insolation totals associated with
each type varied seasonally with the overhead sun, but
monthly insolation totals (and hence the radiation
regime) were largely the product of differences in the
relative frequency of each daily insolation type. The
seasonalities of peak instantaneous, peak daily, and peak
monthly solar radiation, sunshine and light at El Verde
are consequently different. Peak instantaneous radiation
will occur around noon at times of maximum sun elevation, at El Verde in May and July; maximum daily
solar radiation is recorded on a sunny day at the time
of year when day length is longest, at El Verde in June,
whereas the highest monthly totals at El Verde occur
during the somewhat drier and sunnier months (from
March to May). Such differences need to be considered
when possible links between peak radiation or light
and plant behaviour are being investigated. Day-to-day
differences in solar radiation of a similar order to those
at El Verde have also been reported at Yangambi and

other lowland equatorial stations in Zaïre (Bultot & Griffiths 1972).

7.4 Rainfall

Rain-producing weather systems. Old ideas of an equatorial convectional rain belt, produced by the convergence of the NE and SE trade winds, migrating northwards and southwards with the overhead sun have been found to be of only limited validity (for a review see Jackson 1989). Atmospheric conditions even in the wettest equatorial climates do not always lead to convectional rainfall. All tropical and equatorial areas experience considerable day-to-day variations in weather, with wet and drier spells evident even in very wet months, as demonstrated by the daily rainfall pattern for the Superwet station at G. Mulu in the year 1977–78, when the annual rainfall total was 5087 mm (Fig. 7.3). Surface heating and low-level convection occur daily in the tropics even in desert areas, but the cumulus clouds produced by such updraughts must be able to develop vertically to a sufficient height for cloud droplets to grow into raindrops; clouds generally need to reach a height of at least 3000 m for any rainfall to occur and 5000–6000 m for heavy rain to be produced. For this, a deep moist layer, low-level convergence of air to sustain updraughts and upper air conditions that allow high-level outflow (divergence) are usually necessary. These conditions are by no means usual even during wet seasons; rainy weather tends to be associated primarily with the passage of transient weather systems of a variety of types and spatial scales, although in some areas either orographic uplift or local wind systems play a significant role.

Satellite photography has demonstrated the varied structure and discontinuous nature both in time and space of the inter-tropical belt of rainy weather that is usually termed the inter-tropical convergence zone (ITCZ). It is rarely a continuous west–east cloudy zone, but usually consists of cloud clusters separated by large areas of clear sky; the feature tends to develop and decay *in situ* for periods of a few days repeatedly but irregularly during the year, with the position shifting jerkily with each redevelopment. Northward or southward movement is achieved by irregular net shifts in the axis of redevelopment, rather than by simple migration. Some of the cloud clusters are associated with westward-moving waves and closed low-pressure systems of the upper easterly winds of the tropical atmosphere and their passage tends to produce heavy convectional showers for a few days (Chang 1970, Reed

1970), whereas the areas between them are characterized by subsiding, stable air and clear, dry weather. Much of the rainfall of monsoonal areas is also produced by westward-moving lows in the upper easterlies. In West Africa, the cloud clusters take the form of 'line squalls' several hundreds of kilometres long, that develop and travel westward in the moist air during the rainy seasons and account for a substantial proportion of the rainfall (Barry & Chorley 1982, Omotosho 1985).

In the trade-wind zone, there is usually a temperature inversion at a comparatively low level in the atmosphere produced by subsiding air of the subtropical anticyclones; this normally prevents substantial vertical cumulus development and leads to spells of dry weather, except where orographic uplift occurs. Westward-moving waves in the upper easterlies, however, lead to low-level divergence (and dry cloudless weather) in front of the wave, but low-level convergence, high-level divergence and pronounced convectional ascent behind it, producing heavy showers over a two- to three-day period (Nieuwolt 1977). The passage of these 'easterly waves' is sometimes associated with the passage of waves in the ITCZ equatorward of them. A few easterly waves develop into tropical cyclones (p. 194), which affect some tropical areas poleward of 5° latitude and are characterized by violent winds and high rainfall totals and intensities, with often 150–600 mm over a 6–12 h period.

Orographic uplift can both trigger convection in an otherwise stable atmosphere and increase rainfall amounts. This is important not only in the trade wind zone, but also in equatorial and monsoonal areas where winds rise over coastal mountain ranges aligned normal to the prevailing wind. The zone of maximum rainfall up tropical mountains varies, but generally occurs at 1000–1400 m (Nieuwolt 1977, p. 107, Jackson 1989, p. 38), after which rainfall tends to decline, unlike up temperate zone mountains, where it increases up to much higher altitudes.

Annual rainfall totals and their variability. Details on rainfall at rain-forest stations are given in the climate diagrams in Fig. 7.2. Annual rainfall ranges from 1700 mm to over 10 000 mm. Areas with very high rainfall (>3000 mm) are comparatively small, the most notable being (1) the upper Amazon Basin, (2) the northern parts of the Guianas, (3) western Colombia, (4) much of New Guinea, Borneo and Sumatera, (5) parts of the Atlantic slopes and coastlands of southern Central America, (6) western Cameroon, southeast Nigeria and Equatorial Guinea in Africa and (7) mountainous interiors of numerous islands in the trade-wind zones. Over the greater part of the tropical rain forest, including much of the Amazon Basin, the entire Zaïre (Congo)· basin and most of the Malay Peninsula, annual rainfall is less than 2500 mm. Areas with more than 5000 mm are very small and are normally associated with the foothills and slopes of mountains in the tropics. Among the wettest localities are Lloro (13 473 mm in 1952–54) in the Chocó of western Colombia (Snow 1976), Cherrapunji (11 420 mm) in Assam, Ureka (10 450 mm) in Equatorial Guinea and Debundscha (10 299 mm) at the foot of Mt Cameroon (Bultot & Griffiths 1972), Mt Bellenden Ker summit (8679 mm) in Queensland (Herwitz 1987a) and Lakeside (8459 mm) in Dominica (Walsh 1980a).

These differences in annual rainfall are of both direct and indirect ecological significance. First, the higher the annual rainfall, the greater is the leaching potential, particularly as most rain-forest topsoils are permeable and most rainwater is able to pass readily vertically and laterally through the soil (see Chapter 8). To some extent the higher leaching potential of very high rainfall areas, particularly on nutrient-poor soils, may be offset by plant species that have root systems that are particularly efficient at 'filtering' nutrients from percolating water and which have leaves that decay and release nutrients more slowly. They can also be offset by higher inputs of nutrients as rainfall increases. This was demonstrated clearly by Herwitz (1987a) along an altitudinal transect on Mt Bellenden Ker in Queensland. Inputs of calcium, magnesium, sodium and potassium increased from 8.0, 11.4, 92 and 6.9 kg $ha^{-1}a^{-1}$ at the mountain base (80 m altitude) to 16.1, 18.7, 155 and 8.5 kg $ha^{-1}a^{-1}$, respectively, at 1561 m close to the summit, as annual rainfall increased from 4400 to 8200 mm. A further indirect consequence of high annual rainfalls is that solar radiation, sunshine and light receipts tend to decline with increased cloudiness (except in regions with pronounced night-time rainfall maxima, such as at Padang in Sumatera and in western Colombia); this can lead to reduced photosynthesis rates, transpiration and evaporation.

Also important are the year-to-year variability and reliability of the annual rainfall. In absolute terms year-to-year variability (as measured by the standard deviation) in rain-forest regions is higher than in the seasonal tropics and increases with annual rainfall, but the reverse is true when variability is expressed in relative terms as the coefficient of variation (standard deviation as a percentage of the mean annual rainfall) (Table 7.5). Coefficients of variation of less than 20% are typical of most rain-forest stations including most of the Amazon Basin (Ratisbona 1976) and the Zaïre

(Congo) basin (Riehl 1954) and are less than 15% at Superwet locations. However, at Menado in northeast Sulawesi, which is subject to occasional long droughts, variability is anomalously high in relation to its high annual rainfall. Of note also is the greater reliability of annual rainfall at wetter rain-forest locations. The minimum annual rainfall recorded over a forty-five-year period at Padang was 3464 mm, whereas at marginal rain-forest sites annual totals of 1100–1200 mm have occurred.

Recent research has pointed to the importance of El Niño–Southern Oscillation events (ENSO) in being responsible for anomalously wet or dry years in parts of the tropics (see section 7.6). The Southern Oscillation refers to year-to-year variations in surface pressure distribution between the eastern and western equatorial Pacific. These variations were found to be related to variations in sea surface temperatures and rainfall in the eastern equatorial Pacific, notably the periodic appearance of the warm El Niño current much further south than normal along the Peruvian coast. The reasons for these ENSO events are far from understood, but their consequences are quite well known. When an ENSO event occurs, the normal equatorial atmospheric circulation in the Pacific is disturbed. Normally, the cold seas of the eastern Pacific reduce convection and rainfall in the region and excessive rainfall occurs instead in the western Pacific, especially over Malesia, aided by a strong east–west air circulation (the Walker Cell). When a major ENSO event occurs, anomalously high rain occurs for a year or more over the eastern Pacific and in northeast Brazil, but rainfall over the western Pacific, especially eastern Indonesia, is much reduced and major droughts can occur. Recent major events occurred in 1972–73, 1976 and 1982–3 (Jackson 1989, p. 24), the last resulting in the devastating drought in Eastern Borneo, described on pp. 191–3 below. Links with the strength of the trades and rainfall patterns over the tropics in general have been widely investigated (see Jackson 1989).

Rainfall regimes. The climate diagrams show the rainfall regimes of representative stations within the wet tropics (Fig. 7.2). The generalization that equatorial regions exhibit two rainfall peaks associated with the passage of the overhead sun, whereas tropical regions poleward of 10° experience a single peak around the summer solstice, has some applicability, but there are many exceptions. Two wetter seasons roughly at the times of passage of the overhead sun do characterize much of the African rain forest, the upper Amazon Basin, Malaya and rain-forest areas of Sri Lanka, Indo-

Table 7.5. *Variability and reliability of annual rainfall at some tropical rain-forest stations*

Climate	Station	Period of record	Mean (mm)	Maximum (mm)	Minimum (mm)	Standard deviation (mm)	Coefficient of variation (%)
Superwet	Gleau Manioc, Dominica	1906–1916	6436	7685	5080	753	11.7
	Padang, Sumatera	1879–1923	4442	5220	3464	420	9.4
	Pontianak, Borneo	1879–1923	3203	4244	2521	463	14.4
Wet	Colon, Panama	1863–1925	3247	4659	2198	496	15.3
	Menado, Sulawesi	1879–1923	2635	4036	1590	555	21.1
	Colombo, Sri Lanka	1870–1920	2035	3548	1161	481	23.6
	Yangambi, Zaïre	1930–1959	1828	2629	1220	—	—
Wet Seasonal	Roseau, Dominica	1893–1973	1954	2734	1024	304	15.7
	Jakarta, Java	1864–1923	1832	2460	1166	333	18.2
	Bambesa, Zaïre	1930–1959	1781	2396	1442	—	—

Sources: Indonesia, Panama and Sri Lanka, Clayton (1927); Zaïre, Bultot & Griffiths (1972); Dominica, Author.

China and the Pacific side of Central America. Parts of Indonesia and East Malaysia also exhibit a double peak, although local relief in relation to the monsoon wind circulation tends to disrupt the pattern. However, some equatorial areas (e.g. parts of the eastern Amazon Basin, Manaus and Belém) experience a single maximum and some double peaks at equatorial stations are not associated with the overhead sun; thus the December–January peaks in rainfall in northern Guyana and northern Borneo result from the peak in strength of onshore trades and northern monsoon winds respectively. Also, some trade wind locations (e.g. Wet Area and El Verde in the Eastern Caribbean) experience a double rather than a single peak.

The length and timing of dry seasons also varies greatly within the rain-forest zone. A regular dry season of short to moderate intensity (and in parts of West Africa, the Guianas and Sri Lanka, two dry seasons) characterize the Wet Seasonal zone. In parts of West Africa, edaphically favourable soils and high humidity and cloudiness apparently enable rain forest to thrive where there are up to 5 successive dry months (<100 mm) on average each year, but more normally, a maximum dry season length of 3–4 months is the tolerance limit for evergreen seasonal rain forest. In the Wet zone (perhumidity index 10–19.5), dry periods are shorter and often less regular in timing and year-to-year occurrence. In the Superwet zone, dry periods of greater than a month are either infrequent or irregular or both. Dry season length as deduced from monthly rainfall averages is an unreliable guide, however. The question

of dry periods and droughts is considered separately later in the chapter.

Rainfall intensity and frequency. A high proportion of tropical rainfall occurs in large storms of high intensity (Jackson 1989, p. 100). Hudson (1971) considers that around 40% of tropical rain occurs at intensities greater than 25 mm h^{-1} compared with 5% in temperate localities. In Indonesia, Mohr *et al.* (1972, p. 24) found that 22% of the annual rainfall occurred at intensities of 60 mm h^{-1} compared with 1.5% in Bavaria. Dale (1959) found similar results for the Malay Peninsula. In New Guinea, Turvey (data cited in Jackson 1989) found that 34% of rain fell at intensities of 25 mm h^{-1} or greater and 47% in storms of at least 25 mm rainfall.

The size distribution and frequency of daily rainfalls vary considerably with rain-forest locality. Table 7.6 gives selected data for some contrasting stations in Borneo and Dominica (West Indies). The frequency of raindays is higher in relation to annual rainfall at the two trade-wind stations than at the two equatorial Borneo sites, reflecting the very showery nature of the former climate. The frequency of raindays at Roseau (Dominica) is anomalously high; at equatorial Wet Seasonal sites such as Balikpapan (eastern Borneo) and Manaus (Amazonia) values in the range 120–150 raindays per year are more typical. The frequency of large falls of rain is high at all sites compared with in temperate areas but increases sharply with annual rainfall. This has implications for rates of soil erosion and nutrient leaching from soils, as well as the frequency of high

Table 7.6. *Rainfall frequency and highest recorded daily falls at tropical rain-forest stations*

(a) Annual frequency of raindays and large daily falls

Station	Climatic type	Mean annual rainfall (mm)	Annual frequency of days		
			≥1 mm	≥25.4 mm	≥76.2 mm
Roseau, Dominica	Wet Seasonal	1922	182	18	1.4
Danum, Sabah	Wet	2821	189	33	1.7
Mulu, Sarawak	Superwet	5087	251	66	8.0
Wet Area, Dominica	Superwet	5432	298	70	7.0

(b) Highest recorded daily falls

Station	Years of record	Mean annual rainfall (mm)	Highest fall (mm)	Station situation
Malay Peninsula				
Penang	48	2736	241	W coast island
Cameron Highlands	26	2644	160	Interior
Kuala Lumpur	19	2441	145	Interior
Kuala Trengganu	15	2921	465	E coast
Borneo				
Kuching	23	4036	414	N coast
Marudi	43	2716	560	Close to N coast
Lio Matu	15	3456	136	Inland, NE Sarawak
Balikpapan	43	2228	88	Kalimantan, E coast
Pontianak	63	3175	74	Kalimantan, W coast
Amazonia				
Uaupés	10	2677	117	Interior
Manaus	25	1811	119	Interior
Santarem	22	1979	175	Coastal

(a) Periods of record: Roseau (1921–73); Danum 1985–90; Mulu 1977–8 (1 year); Wet Area 1971–5.

Sources: (a) Author; (b) Sarawak Drainage and Irrigation Department records; Meteorological Office (1958, 1967).

streamflows. At the two Superwet sites, 25 mm falls are thus around twice as frequent as at Danum Valley (Sabah) and three to four times as frequent as at the wet seasonal station of Roseau.

Highest recorded daily falls (Table 7.6) tend to be much higher in maritime areas, particularly where mountains parallel the coastline, than in continental localities. In times of onshore wind, the warm ocean provides an inexhaustible moisture source for sustained heavy rain in maritime localities, whereas convectional storms in continental areas soon run out of atmospheric moisture. Record daily falls inland in the Malay Peninsula (e.g. Cameron Highlands) are much lower than on the east and west coasts. Similarly, record falls in coastal Sarawak exposed to the NE monsoon are much higher than in the rest of Borneo and record falls at interior stations of Amazonia are also comparatively low. In trade wind areas, tropical cyclones may also be responsible for exceptionally large rainstorm totals.

Diurnal distribution. The diurnal distribution of rain can be of ecological significance through its influence on solar radiation, sunshine, light, temperature and interception losses. Other things being equal, areas with daytime rainfall maxima tend to have reduced solar radiation, sunshine and light and lower maximum temperatures than localities with night rainfall. In addition, saturation deficits and hence evaporation rates of rain intercepted by vegetation at night are lower than during the day and hence a greater proportion of night rain will reach the forest floor. Night rain is thus more effective in replenishing soil moisture.

Diurnal patterns of rainfall and the factors influencing them are rather more varied and complex (Table 7.7)

Table 7.7. *Mechanisms influencing diurnal patterns of rainfall in rain-forest areas*

Type of locality	Mechanism	Time of maximum rainfall
Continental areas	Daytime peak in land heating and convection	Afternoon
Oceanic islands/coastal areas	Steepened lapse rate and instability at night due to radiational cooling of cloud tops and heating of surface air by warm ocean	Night
Windward coasts	Convergence offshore of night land breeze and prevailing surface wind	Late night
Leeward coasts	Convergence onshore of daytime sea breeze and prevailing surface wind	Afternoon
Coasts with slack prevailing wind and backed by mountains	Sea breeze, orographic uplift and undercutting of warmer land air	Daytime
Equatorial lowland basins	Landward propagation of lines of instability initiated by regional sea breeze system due to friction over land	Daytime at coast, delay inland and night well inland
Large peninsulas and islands	Convergence inland of daytime sea breezes	Afternoon
Coasts adjacent to straits	Convergence of night land breezes in straits between two land masses and movement of showers across coasts by overlying prevailing wind	Night/dawn hours
Intermontane valleys and basins	Convergence of night katabatic winds in deep valleys and basins	Night
Sloping land	Nocturnal radiational cooling on slopes setting up strong horizontal temperature gradients and instability	Night

than suggested by early texts. The traditional simple classification into 'continental' and 'maritime' regimes, with afternoon and late night maxima respectively, is now considered of limited applicability (Gray & Jacobson 1977, Jackson 1989). Of greater significance are the roles played by local or regional scale land–sea and mountain–valley wind systems in relation to prevailing surface airflows. A few contrasting examples are now given.

One of the most striking examples of a marked night maximum is in the Superwet coastlands of the Colombian Chocó. At Quibdo (Snow 1976) and Andagoya (Knoch 1930) over 70–80% of the 6500–7000 mm annual rainfall falls during the night hours (Fig. 7.7). The reason is considered to be the convergence offshore of the easterly night land breeze with the prevailing southwesterly airstream to produce convectional clouds offshore that then blow eastward across the coast in the southwesterly airstream that overlies the land breeze. In contrast the days are relatively cloud-free and hence sunshine and solar radiation are very high for such high-rainfall locations.

In continental areas the simple pattern of afternoon rainfall as a result of maximum land heating and convec-

tion is not very common. In inland high relief areas, convergence of katabatic (downslope) winds often promotes considerable night rainfall in valleys, as noted in the deeper valleys around G. Mulu (Sarawak). Strong horizontal temperature gradients (and hence the conditions for instability) are also set up by nocturnal radiational cooling even on low-angle slopes (Lettau 1968) and have been cited by Snow (1976) as the likely cause of the night rainfall maximum at Sipaliwini, over 400 km from the coast in the central highlands of Surinam (Fig. 7.7). In the eastern Amazon basin, a slight night maximum at Manaus (500 km from the coast) is considered by Kousky (1980) to result from the continued propagation inland of sea-breeze-induced instability initiated at the coast during the previous day; he supports his theory by a sequence of infrared satellite pictures.

In coastal areas, whether the prevailing wind is offshore or onshore is often critical as regards the timing of convergence with a night land breeze (producing the late night – early morning rainfall maximum of windward coasts) or a daytime sea breeze (leading to an afternoon maximum). In localities where seasonal

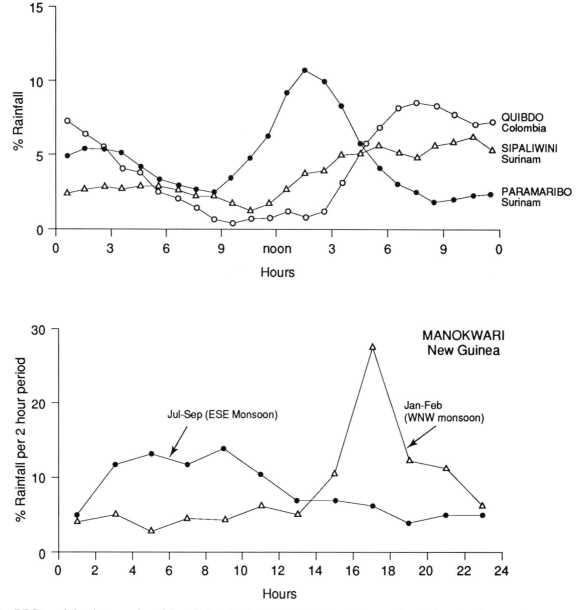

Fig. 7.7 Diurnal distributions of rainfall at Quibdo (Colombia), Sipaliwini and Paramaribo (Surinam) and seasonal variations at Manokwari (Western New Guinea). After Snow (1976) and Braak (1931).

reversals in wind occur, there is often a seasonal change in the diurnal pattern of rainfall. Thus Manokwari (Fig. 7.7), on the east coast of a peninsula in northwestern New Guinea, shows a late afternoon maximum during the offshore WNW monsoon months of January–February, but a late night–early morning maximum during the onshore ESE monsoon months of July–September. Afternoon solar radiation will be much reduced during the WNW monsoon than during the ESE monsoon and

this may have important implications for the seasonality of light and solar radiation receipts. Seasonal contrasts in diurnal rainfall regime associated with monsoonal wind changes have also been reported on the east and west coasts of the Malay Peninsula (Dale 1959, Ramage 1964), at Singapore (Nieuwolt 1968a) and in some of the Indonesian islands (Braak 1931). The famous 'Suma-tras' of the southwest Malay peninsula have been attri-buted in part to night-time convergence of land breezes from Malaya and Sumatera in the Straits of Malacca, producing a line of heavy convectional showers over the sea that then drift with the southwesterly monsoon winds over the Malayan coast in the early morning (Nieuwolt 1977, p. 76).

7.5 Humidity and evaporation

The humidity of the air is an important ecological factor in tropical forest areas because of its effect on potential evaporation rates and regimes, and hence also transpiration and photosynthesis rates of plants. The most useful and relevant measures of atmospheric humidity are the absolute humidity (often expressed as vapour pressure), the relative humidity and the saturation deficit. The saturation deficit is a measure of the evaporating power of the air.

Tropical rain-forest areas are characterized throughout the year by high absolute humidities (over 20 mmHg), high (60–80%) daytime and very high (95–100%) night-time relative humidities. Saturation deficits are therefore moderately high during the day, but zero or near it at night. Some typical seasonal and diurnal patterns in these interrelated variables at lowland tropical rain-forest stations and, by contrast, at tropical montane and seasonally dry locations are shown in Figs. 7.8 and 7.9. Seasonal patterns of daytime vapour pressure, relative humidity and saturation deficit vary with rainfall regime. At Eala (Zaïre), vapour pressure and relative humidity are high throughout the year and the saturation deficit at noon varies little around the 7.4 mm mean, reaching 9.2 mm in the somewhat drier period from January to March. At the Superwet lowland forest station of G. Mulu, vapour pressure and relative humidity are significantly higher than at Eala and saturation deficits are lower, particularly in the wetter months, where they fall below 5 mm. At the evergreen seasonal forest localities Manaus (Amazonia) and Bambesa (Zaïre), the impact of a three-month dry season on relative humidity and saturation deficit is clearly evident. At Manaus, although vapour pressure (absolute humidity) falls only marginally, the higher daytime temperatures of the sunnier dry season months result in much lower relative humidities and a rise in early afternoon saturation deficits to over 15 mm in September compared with 8–10 mm during the rainy months. At Bambesa, the seasonal contrast is even more striking, reflecting a sharper decline in absolute humidity (with vapour pressure falling below 15 mm) as well as increased temperatures in the dry season. Kananga (Zaïre), a seasonal tropical location, is characterized by

lower absolute and relative humidity and much higher saturation deficits during the marked dry season than occur in the rain-forest zone.

Diurnal variations (Fig. 7.9) in vapour pressure at rain-forest sites are small. Saturation deficits, however, which are zero or near-zero at night when temperatures fall close to or just below the dew point, become substantial as temperatures rise during the day. Deficits on fine sunny days are well above mean values; at Eala (Fig. 7.9A) such days occur 15–20% of the time (Bernard 1945).

Saturation deficits tend to decline sharply with increasing altitude for three reasons: the decrease in temperature and water-holding capacity of the air; an increase in rainfall and cloud cover; and most importantly, an increase in frequency and duration of cloud and mist as one moves upward above first the occasional and then the quasi-permanent condensation levels marked by the cloud base. This decrease with increasing altitude is evident in the Luquillo Mountain forests of eastern Puerto Rico (Fig. 7.9C). Midday saturation deficits in the lower montane rain-forest belt decline from 7.5 mm at 384 m altitude to 4.2 mm at 445 m altitude and in the elfin woodland at El Yunque peak (1043 m) remain below 1 mm throughout the day (Odum *et al.* 1970a). This decrease in saturation deficit with increasing altitude is an important factor in accounting for the reduction in stature and other differences in rain-forest structure that occur up tropical mountains (see Chapter 17).

It is important in considering water relations of tropical vegetation that saturation deficits are not taken in isolation, but in conjunction with wind speed and soil factors (notably depth, texture and horizonation) that influence soil drainage and the speed with which water deficits and plant water stress may develop. However, it is certainly true that the saturation deficit pattern of tropical rain-forest areas contrasts with those of both seasonal and montane forest areas in the tropics and is to a large extent responsible for the differences in forest character. Also within the rain-forest zone, smaller differences in the magnitude and seasonal pattern of saturation deficit (as demonstrated by the differences between high-rainfall Mulu, lower-rainfall Eala and somewhat seasonal Bambesa and Manaus in Fig. 7.8) may also have an influence on forest character and productivity.

Evapotranspiration in tropical rain-forest areas is considered in detail in Chapter 8. Here, data on the evaporation potential of the air at meteorological stations are briefly reviewed. There are many different measurement methods and empirical and theoretical formulae, recently

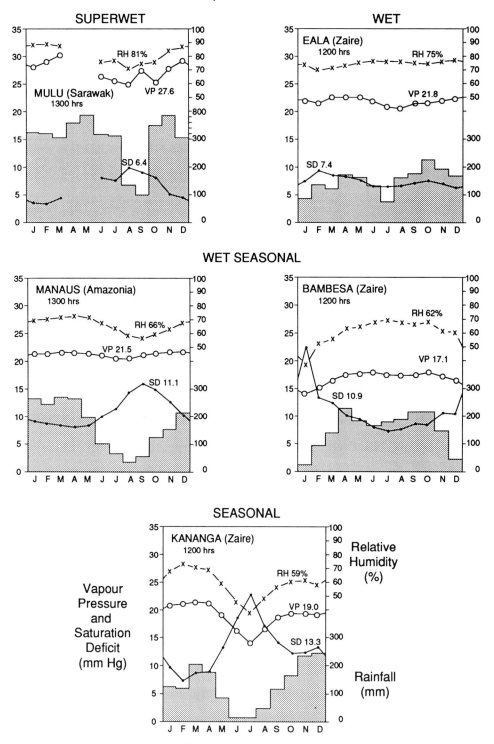

Fig. 7.8 Seasonal variations in saturation deficit (SD), vapour pressure (VP) and relative humidity (RH) around midday at rain-forest and seasonal-forest localities. Mean annual values of each variable are indicated on the diagram.

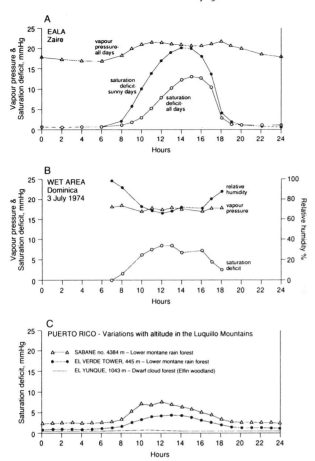

Fig. 7.9 Diurnal variation of saturation deficit and vapour pressure in rain-forest localities. A Eala, Zaïre (wet) (from Bernard 1945). B Wet Area, Dominica (superwet) (from Walsh 1980a). C Three altitudes in Luquillo Mountains, Puerto Rico (superwet) (from Odum et al. 1970a).

Class A evaporation pans gave results in close agreement with Penman open-water estimates (Blackie 1965) and in Nigeria British Meteorological Office evaporation tanks showed close agreement with evapotranspiration from a grass cover with non-limiting soil moisture (Stanhill 1963). Only data from evaporation pans or calculated using the Penman method and adjusted for rain-forest vegetation are discussed in this section.

Mean annual potential evaporation at most rain-forest locations lies in the range 1300–1900 mm (Table 7.8). Sengele (1981) calculated Penman evaporation for rain forest vegetation at Yangambi (Zaïre) by incorporating an adjustment for the lower albedo (reflectance) of rain forest. This has the effect of increasing Penman evaporation estimates considerably above the levels for the standard grass surface, as the albedo is typically 0.13 for rain forest, 0.23 for a cover of the grass *Paspalum notatum* and 0.18 and 0.23 for savanna in the wet and dry seasons, respectively (Bultot & Griffiths 1972). Evaporation at Yangambi for 1974–78 averaged 1478 mm per annum (or 4.0 mm d^{-1}) with little variation from month to month. At the wetter but rather more seasonal site of Barro Colorado Island (Panama), Penman evaporation, again adjusted for rain forest, is highest (4.5–4.9 mm d^{-1}) in the dry season months of January–April, when solar radiation, sunshine, saturation deficits and wind speeds are all significantly higher and relative humidity much lower than in the rest of the year; evaporation falls to values of 3.2–3.8 mm d^{-1} during the wet months (Dietrich et al. 1983).

The much lower evaporative demand of rain-forest areas compared with more seasonal climates in the tropics is well demonstrated by the pan evaporation data for stations in Nigeria and the West Indies. In Nigeria, pan evaporation increases from 1457 mm at Benin in the rain-forest zone to 3974 mm at Maiduguri in the savanna zone. The relatively low evaporation rates during the dry months result from the high humidity and cloudiness and low sunshine of southern Nigeria and are considered one of the reasons why tropical rain forest, albeit mostly of the evergreen seasonal type, is able to thrive in edaphically favourable areas in a region with two significant dry seasons and a comparatively low perhumidity index. Thus peak evaporation rates at Benin are only 4.8–4.9 mm d^{-1} in February and March compared with 12 mm d^{-1} at Jos and 18 mm d^{-1} at Maiduguri and annual evaporation differs little from pan measurements at much wetter stations at Wet Area (Dominica) and Georgetown (Guyana). The influence of wind and sunshine on evaporation is demonstrated by the contrast in pan evaporation at the Superwet rain-forest location of Wet Area, Dominica, and the long-

reviewed by Jackson (1989), but all have limitations and their results are often difficult to compare. The widely used Thornthwaite method, which calculates potential evaporation using empirical equations based only on temperature and day length, gives particularly poor results in the tropics (Blackie 1965, Riou 1984, Jackson 1989), where wind speed and saturation deficit are also important factors influencing evaporation. The physical equations developed by Penman (1963) and refined by later workers incorporate radiation, saturation deficit and windspeed terms and give rather more consistent and realistic results (Blackie 1965). The original Penman method attempted to assess open-water evaporation, but it is often modified to assess potential evaporation from a reference grass cover or other types of vegetation cover including rain forest (Sengele 1981). In East Africa, US

Table 7.8. *Evaporation at tropical rain-forest (TRF) and seasonal tropical stations*

Locality	Vegetation	Annual rainfall (mm)	Perhumidity Index	Evaporation (mm)												
				Jan	Feb	Mar	Apr	May	Jun	Jul	Aug	Sep	Oct	Nov	Dec	Year
1. Penman formulae calculations using field data																
Yangambi, Zaire	TRF	1828	9.5	121	127	130	135	133	120	121	127	117	124	105	118	1478
Barro Colorado, Panama	TRF	2718	10.5	139	133	152	142	129	100	119	112	114	115	96	111	1462
2. Evaporation pan measurements																
Benin, Nigeria	TRF	2014	5.5	117	135	152	142	137	119	99	91	102	122	127	114	1457
Makurdi, Nigeria	Seasonal	1377	0.5	185	221	300	246	218	224	157	132	163	178	178	183	2385
Jos, Nigeria	Seasonal	1414	−2.5	307	335	368	272	236	201	163	142	183	254	295	302	3058
Maiduguri, Nigeria	Seasonal	659	−13.5	320	396	559	526	500	295	216	165	185	251	279	282	3974
Georgetown, Guyana	TRF	2355	13.5	105	118	144	136	116	94	108	128	144	142	119	114	1451
Wet Area, Dominica	TRF	5204	24.0	106	82	99	106	131	134	126	134	113	108	95	98	1330
Husbands, Barbados	Seasonal	1393	0.5	148	146	189	206	210	188	178	189	138	152	137	143	2024

Sources: Yangambi, Bernard (1945); Barro Colorado Island, Dietrich *et al.* (1983); Georgetown, Snow (1976); Wet Area, Walsh (1980a); Husbands, Caribbean Meteorological Institute records.

cleared semi-evergreen seasonal forest site at Husbands, Barbados. Annual evaporation averages 1330 mm at Wet Area, where the mean wind speed is 2.9 knots, compared with 2024 mm at Husbands, where the mean wind speed at nearby Codrington is 11.2 knots. The differences also reflect the higher humidity, lower temperatures and greater cloudiness of the Dominica station.

Evaporation tends to decline with altitude with increases in cloudiness and relative humidity and reductions in solar radiation, sunshine, temperature and saturation deficits, though increased windspeeds can counter this tendency, especially on exposed ridges. On very high mountains, however, the increase in sunshine and decline in relative humidity that often occurs above the cloud belt may lead to an increase in evaporation compared with the cloudy zone.

It should be stressed that the above and all other 'potential evaporation' data give merely a rough measure of the drying power of the climate. Even in ever-wet rain-forest areas they do not give an accurate or reliable indication of either actual or potential evapotranspiration for reasons given in Chapter 8.

7.6 Droughts and dry periods

A general indication of the mean length and timing of the dry season (if any) at rain-forest locations is given in the climate diagrams (Fig. 7.2) by the pattern of dry (<100 mm) and drought (< about 50 mm) months. In lowland evergreen forest, there are normally two or fewer monthly means with less than 100 mm, but rain forests of the evergreen seasonal type extend into areas with three to four dry months and occasionally five dry months, as in parts of southern Nigeria where there are two dry seasons, but high humidity and comparatively low evaporation rates. However, monthly means give a very incomplete and often misleading picture of dry periods, not only because of the unequal and arbitrary calendar month divisions that are used (Brünig 1969), but also because averages tend to understate the true number of dry or drought months in each year. In addition, it may not be the mean dry season but the occasional extreme droughts that have the greatest ecological impact and influence on the rain forest.

To obtain a more comprehensive picture of dry periods, it is thus necessary to examine the rainfall records of a long series of years rather than averages. Twenty-year frequencies of dry periods of different lengths (as indicated by successive months with less than 100 mm rain) at twelve rain-forest stations are shown in Fig. 7.10. There are a few Superwet locations where a dry month has never or rarely been recorded. Thus only one dry month was recorded in fifteen years of record at Wet Area and nearby Gleau Manioc (central Dominica) and at Padang (western Sumatera) only twelve dry periods, each a single month in length, were recorded in forty-five years from 1879 to 1923. Arguably the impact of such short dry periods on plant water stress is, except on shallow or excessively-drained soil types, minimal. However, some Superwet stations are prone to an occasional longer dry period and Pontianak (west coast of Borneo) experiences around three dry periods of two months duration and one dry period of three months duration per twenty years. Some plant water stress may be experienced during these longer dry periods.

In more typical, less wet rain-forest locations with perhumidity index values of 14–19, but still less than two monthly rainfall means below 100 mm (and hence on average virtually no dry season), dry periods are of greater but contrasting frequency and degree of severity. At Manokwari (New Guinea) and Medan (Sumatera), dry periods of one to two months duration are common, but neither experienced a dry period exceeding four months in the forty-five years of record. In contrast, Menado (Sulawesi), though experiencing a similar mean annual frequency of dry months (2.8) to Medan (3.0) and Manokwari (2.7), is subject to fewer short dry periods but more frequent longer droughts, with five sequences of five to seven months duration occurring in forty-five years.

At the lowland evergreen – evergreen seasonal forest margin, where monthly averages indicate a three month dry season at the stations shown, there is again a contrast in the length and frequency of dry periods. At Colon (Panama), the dry season is remarkably consistent in length, mainly three to four months, occasionally less and only twice extending to five months in fifty-five years of record. At Calabar in southern Nigeria, there are two rainy seasons separated by a longer dry and a shorter, less rainy, but occasionally dry season. The main dry season varies from two to four months duration and only occasionally longer. The high frequency of single dry months reflects the times that the shorter drier season is actually dry. In contrast, Port Victoria (Seychelles) has a dry season of much more variable length and is prone both to frequent short dry spells and occasional long droughts, with five droughts of six months duration occurring in the eighty years of record.

The longer dry seasons and longer extreme droughts of the wet seasonal zone on the boundary between evergreen seasonal and monsoon or semi-evergreen seasonal forests are clear (Fig. 7.10), although the greater

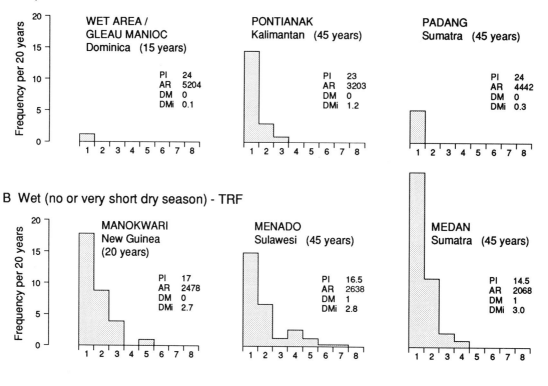

A Superwet - TRF

WET AREA /
GLEAU MANIOC
Dominica (15 years)

PI 24
AR 5204
DM 0
DMi 0.1

PONTIANAK
Kalimantan (45 years)

PI 23
AR 3203
DM 0
DMi 1.2

PADANG
Sumatra (45 years)

PI 24
AR 4442
DM 0
DMi 0.3

B Wet (no or very short dry season) - TRF

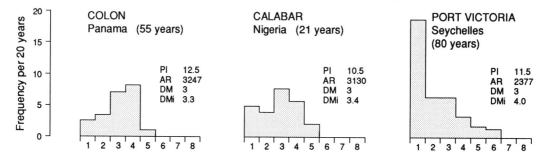

MANOKWARI
New Guinea
(20 years)

PI 17
AR 2478
DM 0
DMi 2.7

MENADO
Sulawesi (45 years)

PI 16.5
AR 2638
DM 1
DMi 2.8

MEDAN
Sumatra (45 years)

PI 14.5
AR 2068
DM 1
DMi 3.0

C Wet (with short dry season) - Evergreen Seasonal forest / TRF margin

COLON
Panama (55 years)

PI 12.5
AR 3247
DM 3
DMi 3.3

CALABAR
Nigeria (21 years)

PI 10.5
AR 3130
DM 3
DMi 3.4

PORT VICTORIA
Seychelles
(80 years)

PI 11.5
AR 2377
DM 3
DMi 4.0

D Wet seasonal - Evergreen Seasonal / Semi-Evergreen Seasonal and Monsoon forest boundary

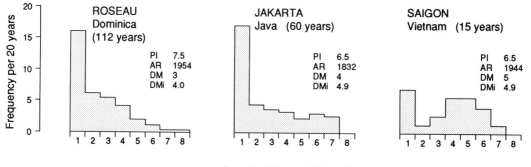

ROSEAU
Dominica
(112 years)

PI 7.5
AR 1954
DM 3
DMi 4.0

JAKARTA
Java (60 years)

PI 6.5
AR 1832
DM 4
DMi 4.9

SAIGON
Vietnam (15 years)

PI 6.5
AR 1944
DM 5
DMi 4.9

Length of dry period (months)

Fig. 7.10 Twenty-year frequencies of dry periods of different durations at tropical forest localities. PI, perhumidity index; AR, mean annual rainfall (mm); DM, dry months per year using rainfall averages; DMi, dry months per year using rainfall series (see text).

regularity of the dry season of the monsoon climate of Ho Chi Minh City (Saigon) (Vietnam) compared with northern Java (Jakarta) and the leeward side of Dominica (Roseau) is apparent from its contrasting histogram, with its four to six month modal peak and small number of single month dry periods.

It is clear from the above that the magnitude and frequency of dry periods varies greatly between different parts of the rain-forest zone. In addition, the extent to which the number of dry months per year indicated by monthly rainfall averages (DM) differs from the mean number of dry months calculated using individual year data (DMi) varies with the consistency in length and timing of dry periods. Where the dry periods are regular, there is little difference between DM and DMi (e.g. Ho Chi Minh City, Colon, Calabar), but where dry periods vary in timing (as at Medan and Manokwari) and/or length (as at Menado), then the difference between the index values can be 2 months or more.

In Superwet locations, although dry periods are shorter, less frequent and often aseasonal in nature, they may play important roles in rain-forest dynamics, as short dry spells are periods of unusually intense and uninterrupted solar radiation and light and high saturation deficit; they are thus also periods of enhanced transpiration and photosynthesis and may influence the timing of leaf flushes or flowering in some trees or of leaf fall if the dry periods are longer. Several studies have demonstrated the importance of dry spells in ostensibly ever-wet climates using daily rainfall analysis, so as to overcome the crudeness and arbitrary divisions of calendar month data. At Singapore, dry spells exceeding six days (with less than 0.25 mm on each day) occurred in every month except November during the decade 1951–60, with a longest dry spell of twenty days, although on average each month received more than 100 mm (Nieuwolt 1968b). Furthermore, particularly in the drier months, much of the total monthly rainfall often fell on just a few days with the rest of the month virtually dry. At Alor Star (northwest Malaya) (Jackson 1989), where rainfall in the driest month averages 50 mm and only two months average less than 100 mm, significant rainless spells are frequent. In twenty-three years, there were forty-six periods with at least fourteen consecutive rainless days and sixteen rainless periods of at least three weeks duration; the maximum rainless period was forty-nine days in January–February 1940. Even at the Superwet station of G. Mulu, Sarawak, during the year 1977–78 when annual rainfall totalled 5087 mm, four significant dry periods in August 1977 and August–September 1978 were noted (Walsh 1982a).

In Sarawak, Brünig (1969) demonstrated clearly how calendar month data understate the true frequency of dry months (<100 mm) by using 30-day running means. During the two-year period July 1963–June 1965, two periods exceeding 30 days with less than 100 mm were identified at Kuching and seven periods at Bako National Park when there were no calendar months with less than 100 mm. Brünig considered the possibility that such short dry periods, which never exceeded two months in duration, could lead to plant water stress in the Sarawak rain forest. Using pan evaporation and sunshine data, he estimated that evapotranspiration from dipterocarp forest during a 30 d period of predominantly fine weather (sunshine >7 h per day) would be 160–200 mm, a figure that compares with plant-available soil water capacities of 100–150 mm for rain-forest soils, but much lower capacities on sandy alluvial fan soils covered by kerangas (heath forest) (see Chapter 12) and on shallow mountain soils. He considered that if monthly rainfall fell below 60 mm, then plant water stress could well develop, especially on the latter soils, and that the streamlined canopy of the kerangas forest, which would tend to reduce air turbulence and hence transpiration, may represent an adaptation to the frequent drought conditions.

The impact of dry periods varies not only with their severity but with rain-forest type and its degree of adaptation to dry seasons and drought. What is a severe drought for a rain forest in a Superwet climate may be a normal dry season in the evergreen seasonal forest zone. In all rain forests, however, dry periods play an important role in the life of the forest (see Chapter 9).

The Borneo drought of 1982–83 and its ecological impact. Whereas 'normal' dry periods and their magnitude–frequency, which varies greatly between different rain-forest sites (Fig. 7.10), play an essential role in the life of a rain forest and probably help to account for its particular species composition, forest structure and dynamics, the role of exceptionally severe droughts is more difficult to evaluate. This is best considered through the example of the 1982–83 drought in eastern Borneo (Fig. 7.11).

The impact of an unusually long and severe drought in eastern Borneo in 1982–83 on rain-forest vegetation has been assessed in detail by Leighton & Wirawan (1986). The drought lasted from June 1982 to May 1983 and was linked to a particularly strong El Niño–Southern Oscillation (ENSO) event. At Balikpapan only 703 mm fell during the whole period compared with a long-term annual average of 2230 mm. In the Indonesian province of East Kalimantan about 8000 km² of undisturbed rain forest, 12 000 km² of selectively logged rain forest,

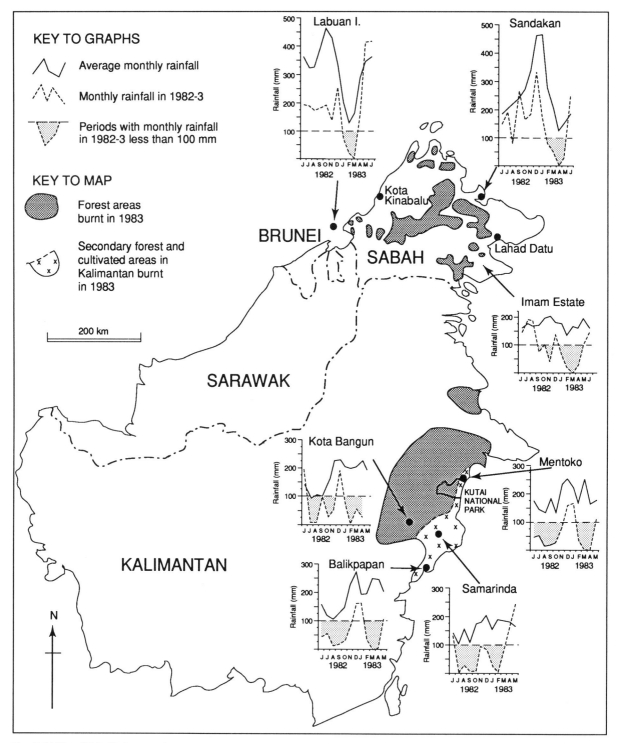

Fig. 7.11 The 1982–83 Borneo drought and areas affected (based in part on Beaman *et al*. 1985 and Leighton & Wirawan 1986).

5500 km² of secondary forest with shifting cultivation and 5500 km² of peat swamp forest were severely damaged by the drought and drought-linked fire. Severe drought and fires also occurred in Sabah, northern Sarawak, southern Sumatera, the Western Ghats of India and parts of southern Malaya.

Leighton made a detailed study of the nature of forest damage in the Kutai National Park, where plots sampled for phenology and forest structure in 1977–79 were resampled after the drought and fires. Some of the plots had been burnt but others only affected by drought. The normally evergreen canopy trees began shedding their leaves in February and March 1983, at which time daily maximum temperatures were 3 °C above average, and were leafless by the time of the fires in April 1983. Most canopy trees were killed by drought rather than fire, as there was no difference in percentage mortality between resampled burned and unburned rain-forest plots. Drought impact was greatest on steep slopes and ridges, where 37–71% of canopy trees (d.b.h. >50 cm) were killed, a consequence of the smaller soil water reserves of the shallow and often excessively well-drained soils of such sites. Drought damage was least (11% of canopy trees killed) in a valley bottom plot, which had both deeper and wetter soils. Small trees and lianes were largely unaffected by the drought, but were more easily killed than canopy trees by fire. Thus, areas affected by drought alone witnessed a decline in importance of larger trees, whereas areas affected by drought and fire saw an increase in their relative importance, because of their greater survival rate.

Eastern Borneo is prone to occasional damaging droughts, but the 1982–83 drought is thought to be the most severe at least since the very bad drought of 1877–78, and perhaps longer (Leighton & Wirawan 1986). It is currently unclear whether such an exceptional drought is to be viewed as an integral part of the current climate – vegetation system, in which case one would perhaps expect the pre-1983 rain forest in eastern Borneo to exhibit features attributable to previous severe droughts, or as an event heralding climatic change and instrumental in producing radical vegetational change. The example of the 1983 Borneo drought demonstrates how potent an agent the severe drought could be in promoting a change in forest type in any long-term change to a drier climate, although repeated droughts would be necessary to complete and maintain forest transformation. On the other hand, in a change to a moister climate it may be the long-term absence of a severe drought that would allow shade-tolerant or drought-susceptible rain-forest species to invade, survive and thrive.

7.7 Lightning and fire

Lightning, as noted in Chapter 3, can play two roles in rain-forest dynamics. First, a lightning strike of an individual tree (or small group of trees) produces a small canopy gap when the tree falls (either at the time or after decay), perhaps bringing down adjacent trees also. Second, lightning can be the instigator of fire in the rain forest, particularly in drier forests and after a dry period. The significance of this second role in humid forests is sometimes questioned because lightning is generally a wet-season phenomenon when the canopy is wet; it is generally accepted that humans are the main instigators of fires. However, forest on the limestone mountains of Gunung Api and Benarat in the Mulu mountains of Sarawak is occasionally damaged by fire, reportedly started by lightning (Whitmore 1984a, p. 175), and lightning fires have also been recorded elsewhere in Sarawak (Anderson 1966) and in the Solomon Islands in trees where dry epiphyte detritus had accumulated and despite rain (Whitmore 1984a, p. 82).

Brünig (1964, 1973a; see also Brünig & Huang 1989) used successive aerial photographs between 1947 and 1968 of an area of *Shorea albida* peat swamp forest close to the Karap river in Sarawak to plot and monitor gaps caused by lightning and windfall. In one 184 ha plot where the trees were higher and the canopy uneven, he identified forty-nine canopy gaps caused by lightning, covering 3% of the plot; thirty-three of the gaps had occurred between 1947 and 1968. In contrast, lightning-caused canopy gaps were of lesser importance in two nearby plots with more even canopies, where they covered just 0.2 and 1.0% of their plot areas. The average size of the gaps was higher in the taller forest (0.113–0.138 ha) than in the lower, even canopy forest (0.052–0.078 ha), but the size of individual gaps varied both sides of the 1000 m² (0.1 ha) critical gap size dividing small gaps favouring shade-bearers and large gaps favouring light-demanding species (Whitmore 1982). Overall, lightning was found to be of comparable importance to wind-throw in causing canopy gaps in Brünig's forest plots.

Lightning frequency probably varies greatly within and between rain-forest regions, but the data available are very subjective, being either days with thunder or days with thunderstorms, the recording of which depends much on the individual observer. At most rain-forest stations the annual number of thunder days ranges between 50 and 150. At Mulu in inland Sarawak there were 61 days with thunder in the year 1977–78 (Walsh 1982a) and at Labuan Island off the northern

Borneo coast there were on average 56 days with thunderstorms during the period 1924–27 (Braak 1931); it would appear, therefore, that there is no reason to regard Brünig's findings on lightning-caused canopy gaps as being exceptional.

The importance of fire in rain-forest areas is still being evaluated. Certainly it has increased in importance through time with intensification of human interference; Leighton & Wirawan (1986) have shown how the impact of the Eastern Borneo fires of 1983 was (a) greater in the disturbed (selectively logged and secondary) forest than in the primary forest, because of the denser undergrowth of the former, and (b) increased in the primary forest by the proximity of human-disturbed forest. Recent pollen and sediment studies of vegetation change in rain-forest areas during the Quaternary have demonstrated a much longer time-scale and intensity of burning history by people than formerly thought (see also Chapters 16 and 18). In his review of 'savannization' in Africa, Kadomura (1989) stressed the role of human-induced fire in exacerbating the impact of the aridifying trend of climate after 5000 yr BP. In the closed forest areas of southern Cameroon, radio-carbon dates of charcoal collected from soil profiles date back to nearly 3000 yr BP. Charcoal fragments in these soils increased in frequency after 1000 yr BP and have been interpreted as indicating more frequent burning of the forest with an intensification of shifting cultivation (Kadomura 1984).

7.8 Wind and tropical cyclones

Wind is ecologically important in several respects. Wind speed controls the rate at which humid air within and above the forest canopy is replaced by new, less humid air and hence strongly influences potential rates of evaporation of intercepted and transpired water and the speed with which soil water shortages will develop. It is one reason why in the trade-wind zone, where mean wind speeds can be 4–8 m s^{-1}, evaporation rates and rainfall limits between various forest types are higher than in equatorial rain-forest areas (notably the Zaïre basin and Amazonia) where mean wind speeds are usually less than 2 m s^{-1} (Table 7.9). An even rain-forest canopy is sometimes interpreted as a means of reducing the impact of wind on evapotranspiration losses in areas of high wind, periodic climatic drought or areas of shallow or excessively drained soil.

Wind direction and speed and relative humidity often change with day-to-day weather and with season. Desiccating conditions thus may be a seasonal or episodic phenomenon. In trade-wind areas, wind strengths are highest and relative humidities lowest in winter when the subtropical anticyclones are strong and displaced towards the equator. In some areas desiccating winds such as the Harmattan in West Africa and the 'east monsoon' of the Sunda Islands are associated with particularly low relative humidities and influence, together with local topography, the distribution of forest types within those areas (see Chapter 16).

Wind is also important in creating canopy gaps at various scales, by causing the fall of large trees by uprooting or snapping of the stems or by crown damage. Mean wind speed data are of no guide in this respect, as it is only extreme gusts or the sustained high winds of tropical cyclones that are capable of felling trees. All tropical areas are subject, though to varying extents, to squalls of strong wind both immediately prior to and during convectional rainstorms and thunderstorms. During these squalls, winds may reach sufficiently high velocities to cause tree-fall over areas varying from a single tree to many hectares in extent. Bultot & Griffiths (1972) provide some interesting and rare data on the frequency and magnitude of such gusts for Yangambi and Bambesa in the Zaïre rain forest (Table 7.10). Gusts of at least 17 m s^{-1} occur over seven times per year at both stations, largely in the wet season at Bambesa. The highest recorded gusts during the 3–11 years of record were 26.5 m s^{-1} (95.4 km h^{-1}) at Yangambi and 25.8 m s^{-1} (92.9 km h^{-1}) at Bambesa. In Sarawak, squalls are apparently more frequent inland than at the coast (Baillie, cited in Whitmore 1984a, p. 81). The size of wind-gaps varies greatly from 0.04 ha when a single tree is felled to over 80 ha in a *Shorea albida* peat swamp forest in Sarawak (Anderson 1964); any gap once formed presents an edge to future squalls and is often extended in the direction of the wind to form linear gaps (Whitmore 1984a, p. 81). In *Shorea albida* peat swamp forest close to the Karap river in Sarawak, Brünig (1973a), using aerial photographs for 1963 and 1968, found that gaps due to wind-throw covered about 1% of the total area (about the same as for lightning-caused gaps).

Although squalls play a significant role in canopy gap formation in all rain-forest areas, the impact of tropical cyclones on rain forests affected by them are much greater. Tropical cyclones, defined as closed-circulation, low-pressure systems with sustained surface wind speeds of at least 17 m s^{-1}, only tend to develop polewards of about 10° latitude and hence many rain-forest areas – including the entire South American and African rain forests and most of Malesia – are unaffected by them. Areas affected by cyclones include the West Indies and, though only occasionally, much of Central America;

Table 7.9. *Mean wind speeds (m s⁻¹) at tropical stations*

Station	Territory	Mean wind speed	Minimum and month	Maximum and month	Source
Yangambi	Zaïre	0.8	No data	No data	1
Monrovia	Liberia	2.2	1.7 (Dec–Apr)	3.6 (Jul/Aug)	1
Barro Colorado	Panama	1.2	0.8 (Sep/Oct)	1.8 (Mar)	2
Fort de France	Martinique	6.1	4.9 (Sep)	7.0 (Jun)	3
Wet Area	Dominica	1.5	1.1 (Sep)	2.0 (Jan)	4
Rochambeau	French Guiana	3.5	2.8 (Jun/Jul)	4.0 (Mar)	5
Uaupés	Brazil	1.0	0.9 (Jan–Jun)	1.2 (Oct)	6
Manaus	Brazil	1.6	1.4 (Apr)	1.8 (Aug/Sep)	6
Belém	Brazil	1.1	0.8 (Feb)	1.4 (Nov)	6
Porto Nacional	Brazil	0.7	0.6 (Various)	0.8 (Various)	6
Sena Madureira	Brazil	0.5	0.4 (Various)	0.6 (Jul/Aug)	6
Montane					
Puyo (0700 h)	Ecuador	0.1	0.1 (Various)	0.2 (Various)	7
(1300 h)	Ecuador	1.4	1.2 (Jun)	1.6 (Nov)	7
Mérida	Venezuela	1.8	1.5 (Nov)	2.4 (Mar)	5

Sources: 1, Bultot & Griffiths (1972); 2, Dietrich *et al.* (1983); 3, Portig (1976); 4, Walsh (1980a); 5, Snow (1976); 6, Ratisbona (1976); 7, Johnson (1976).

Table 7.10. *Gust frequency and extreme gusts at rain-forest stations in Zaïre*

Station	Record (years)	Mean speed (m s⁻¹)	Monthly gust frequency						Annual gust frequency (m s⁻¹)			Highest recorded gust (m s⁻¹)
			Wet season			Dry season						
			≥10	≥13	≥17	≥10	≥13	≥17	≥10	≥13	≥17	
Yangambi	11	0.8	6.0	2.4	0.6	No dry season			72.0	28.8	7.2	26.5
Bambesa	3	0.8	8.2	3.6	0.7	3.6	1.1	0.3	84.6	35.7	7.2	25.8

Source: Bultot & Griffiths (1972).

Madagascar, Mauritius, Réunion and the Comoros in the southwest Indian Ocean; the northern Philippines, northern Sabah, Taiwan and Indo-China in the Far East; and the Solomons, Vanuatu, Fiji and the tropical Queensland coast in the Southwest Pacific Ocean. The mean annual frequency for each major area is given in Table 7.11, but the frequency with which any given locality within these areas is struck is much less. Tropical cyclones are divided into tropical storms (17–32 m s⁻¹ or 39–73 m.p.h.) and hurricanes (33 m s⁻¹ or 74 m.p.h. and higher), but severe hurricanes can have sustained wind speeds exceeding 67 m s⁻¹ (150 m.p.h.). Cyclones tend to occur mostly between June and November in the northern hemisphere and December and May in the southern hemisphere.

As tropical cyclones are generally small in diameter compared with temperate depressions and their wind speed declines with distance from the storm centre, vegetation damage is localized. The damage zonation concept applied by Stoddart (1971) to the impact of hurricanes on coral islands and their vegetation can also be applied to forest damage in terrestrial environments, though impacts will vary with local topography (Walsh 1982b, Bellingham 1991) and vegetation type (Lugo *et al.* 1983). Montane forests, with their streamlined and lower canopies, appear to suffer less damage and recover quicker from hurricanes, as demonstrated in Jamaica following Hurricane Gilbert in September 1988 (Bellingham 1991).

Table 7.11. *Mean annual cyclone frequency in tropical regions*

Region	Annual frequency			Principal rain-forest areas affected
	Storms	Hurricanes	All cyclones	
North Atlantic/Caribbean	4.2	5.2	9.4	Caribbean islands and Atlantic coast of Central America
Eastern North Pacific	9.3	5.8	15.2	None
Western North Pacific	7.5	17.8	25.3	Philippines, Taiwan, southern China, extreme north of Borneo
Southwest Pacific	10.9	3.8	14.8	Queensland, Fiji, Solomons, Vanuatu
Southwest Indian Ocean	7.4	3.8	11.2	Mauritius, Réunion, east Madagascar
North Indian Ocean	3.5	2.2	5.7	Andaman Islands

Source: Crutcher & Quayle (1974).

Three zones of rain-forest damage were identifiable on Dominica after it was struck by the exceptionally severe Hurricane David in August 1979 (Fig. 7.12). A zone of catastrophic damage (absent in less severe cyclones), in which over 60% of canopy trees were felled by wind or landslides, occurred close to the path of the hurricane over the south of the island. A zone of major vegetational damage, with 20–60% of canopy trees uprooted or snapped and extensive defoliation, covered much of the centre of the island. Further north, damage to the rain forest was relatively small, although there were numerous smaller canopy gaps (Evans 1986). Ecological surveys carried out at intervals after the hurricane (Evans 1986) indicated that, in the areas of minor damage, species diversity increased through the invasion of canopy gaps by pioneer species. In the more damaged parts of the island, recovery was much slower. In 1982, flowering or fruiting was restricted to herbs and small shrubs, although many trees at first thought to be dead were found to have survived and had begun to produce new shoots. In 1982 gaps were being rapidly colonized by pioneer species, many of which were flowering and fruiting by 1984.

Few assessments of the longer-term impacts of cyclones have been made. In Puerto Rico, parts of the lower montane rain forest at El Verde suffered extensive windfall, stem breakage and defoliation when three hurricanes struck in quick succession in 1928, 1931 and 1932 (Odum 1970b, Crow 1980). This thinning process led to an influx of new species and a rapid increase in biomass and basal area until Hurricane Betsy in 1956 (Crow 1980). Odum (1970b) contrasted the streamlined canopy, small-girthed but many-stemmed nature of the lower montane forest at El Verde with the irregular, emergent canopy of equatorial forest with far fewer stems per hectare, attributing the contrast to cyclones selectively felling or snapping the crowns of large trees in the Caribbean forests. Webb (1958), however, describes 'hurricane forests' in Queensland as having broken and irregular canopies.

Whitmore (1974, 1984a, 1989) studied rain-forest plots on Kolombangara, Solomon Islands, prior to and following a cyclone that caused much more damage to the exposed north coast than sheltered west coast forests of the island. Differences in species composition and structure between the northern and western forests suggested that greater cyclone damage had occurred on the north coast also in the past. The large trees (>1.5 m girth) of the north coast forests were dominated by species requiring or favoured by canopy gaps, including pioneer species; in contrast, the west coast forests contained many species that do not require gaps to regenerate (shade-bearers). Like Crow (1980) in Puerto Rico, Whitmore (1989) found that the proportion of pioneers or light-demanders in the canopy declined with time since a major cyclone. Climbers and epiphytes were less frequent in the cyclone-prone northern forests.

Wyatt-Smith (1954) found that the impact of a rare storm (presumed to have been a cyclone) in Kelantan, which completely devastated hundreds of square kilometres of forest in November 1880, can be very long-term. By the early 1950s, the regrowth was still very different in species composition from normal dipterocarp forest, with the commonest species being the strongly light-demanding *Shorea parvifolia*. The long-term survival of the light-demanding forest may reflect the

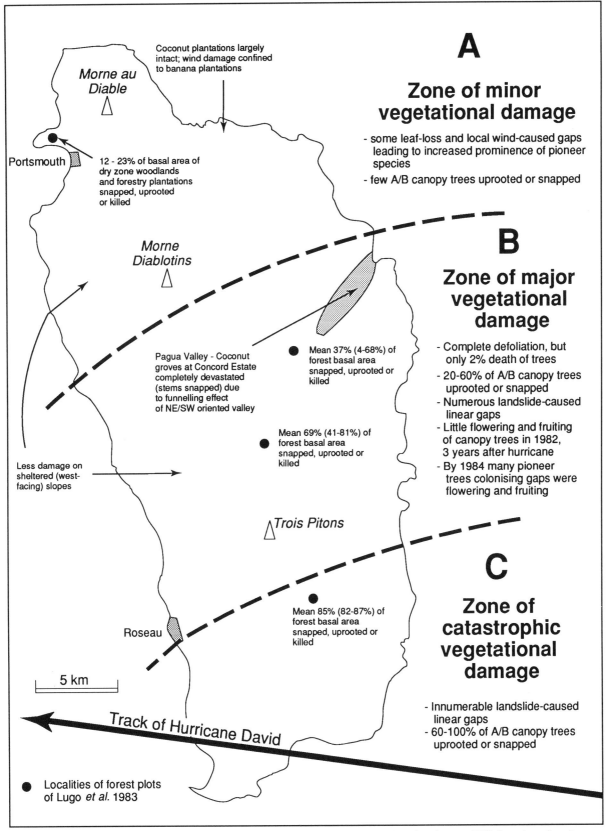

Fig. 7.12 Zonal nature of vegetational damage in Dominica caused by Hurricane David in August 1979 (based on data from Walsh 1982b, Lugo *et al.* 1983, Evans 1986, Reading 1986).

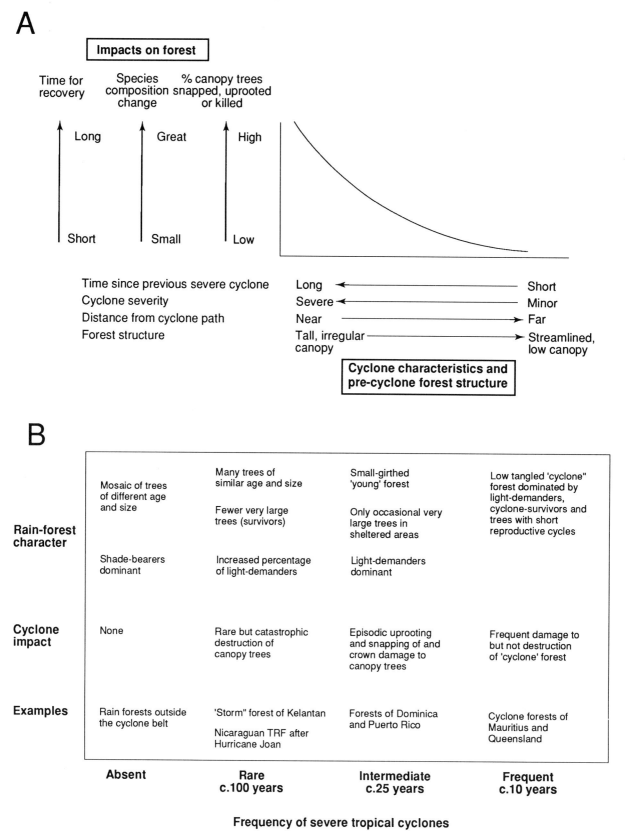

Fig. 7.13 A model of the (A) immediate and (B) long-term impact of tropical cyclones on rain forest in relation to cyclone frequency.

massive size of the gap created and the time required for shade-bearers to invade such a gap after forest regeneration. A more recent example of the catastrophic effect of a rare hurricane is that of Hurricane Joan, which in October 1988 badly damaged 500 km² of lowland rain forest near Bluefields, Nicaragua, with 80% of trees felled (Boucher 1990, Yih et al. 1991).

Fig. 7.13 attempts to summarize effects of cyclones on tropical rain forest (for fuller reviews see Brokaw & Walker (1991) and Tanner et al. (1991)) in relation to cyclone characteristics. The impact of an individual cyclone (Fig. 7.13A) and cyclones in general (Fig. 7.13B), it is suggested, will vary not only with cyclone severity, but also with their frequency and the time since the previous severe cyclone. The impact of a cyclone will be greatest where they are rare, as the forest structure and species composition are least capable of withstanding severe wind, and least where they are frequent and the vegetation becomes adapted to cyclones by being dominated by species that are less easily damaged, as suggested by Vaughan & Wiehe (1937, pp. 301–2) in Mauritius, and more quickly recover.

7.9 Climatic change

Long-term climatic change. Until comparatively recently it was believed that tropical rain-forest areas had experienced the same ever-wet tropical climate for millions of years and had escaped the large-scale climatic changes that affected higher latitudes during the Pleistocene (Douglas 1969). More recently, however, several independent lines of evidence have shown that the area covered by rain forest changed significantly in response to Quaternary climatic change. Much drier and colder conditions than at present occurred over most equatorial and tropical areas during the last glacial (especially towards its end from 18 000 to 12 500 yr BP) with the rain-forest belt contracting latitudinally into equatorial refugia and altitudinally down tropical mountains. On the other hand somewhat wetter conditions and more extensive forest than now occurred during the early Holocene from 10 000 to 5000 yr BP.

The biological evidence for climatic change has already been considered (Chapter 1), but additional information has been gained in geomorphological studies on dating and interpretation of changing lake levels (e.g. Street-Perrott & Roberts 1983, Street-Perrott et al. 1985), relict sand dune systems (Tricart 1985), fossil glacial and periglacial features in tropical highlands (Williams 1985) and river terrace and delta sediment characteristics (Thomas & Thorp 1980, Tricart 1985). Some of this is briefly alluded to here in relation to Africa and Malesia. For a more comprehensive review, the reader is referred to Thomas (1994).

Africa has the most complete and best-dated late Quaternary record in the tropics (for reviews see Williams (1985), Street-Perrott et al. (1985) and Kadomura (1989)). Much of the evidence, which mostly comes from places peripheral to the rain-forest areas, is geomorphological. Five climatic phases have been identified for Africa based on lake level evidence (Street-Perrott et al. 1985):

Climatic epoch (yr BP)		Tropical lakes	Equatorial lakes
A	pre-17 000	low	low
B	17 000–13 000	very low	very low
C	12 500–10 000	low, but some rising	rising/high
D	10 000–5000	high	high
E	5000–present	low	intermediate

There is strong evidence of widespread colder and more arid climates both during and particularly towards the end of the last Glacial (phases A & B). Both tropical and equatorial lakes were very low, indicating very dry conditions (Street-Perrott et al. 1985) and the firn-line was 600–1000 m lower, the treeline 1000 m lower and temperatures 4–8 °C colder than today in the East African mountains (Williams 1985). In contrast, after a transitional phase, the early Holocene climate from 10 000 to 5000 yr BP (phase D) was wetter and warmer than today and the rain forest in both East and West Africa was significantly more extensive than today. For example, Lake Bosumtwi in southern Ghana was high from 12 500 to 11 000 yr BP, very high from 10 000 to 8300 yr BP and high from 7500 to 5000 yr BP compared with today (Talbot & Delibrias 1980) and river flows in Sierra Leone, as indicated by their deposits, were also much higher than currently (Thomas & Thorp 1980). A somewhat drier phase commenced after 5000 yr BP. Because of the widespread appearance of humans during this period (for a review see Kadomura 1989), there is uncertainty as to the relative importance of the drier climate and anthropogenic forest clearance and use of fire in the decline in forest cover and retreat of the rain forest limit that has occurred since. In Rwanda, Roche & Van Grunderbeek (1985) report a contraction of the rain forest with a cooler, drier climate since 3000–2500 yr BP, but consider that recovery of the forest has since been hindered by people despite increased raininess. At Lake Bosumtwi, pollen evidence from near the present edge of the Ghana 'closed forest' (Chapter 1) shows a big increase in grass pollen after 4000 yr BP, ascribed both to a drier climate and to forest clearance (Talbot *et al.* 1984).

In the highlands of New Guinea, pollen analysis of sediments of Lake Inim, altitude 2550 m, indicated a lowering of the montane rain-forest limit by at least 1500 m during the last Glacial until 12 000 yr BP, whereas geomorphological evidence showed that the snowline fell by only 1000 m (Walker & Flenley 1979, Flenley 1979, 1985). They interpreted this as indicative of a drier as well as cooler climate. Their findings, when compared with a much smaller drop in sea surface temperature off New Guinea of around 2 °C, suggested a steeper lapse rate of around 0.8 °C per 100 m compared with 0.6 °C per 100 m at present, a suggestion that would also tally with a drier atmosphere. The montane rain forest at Lake Inim quickly re-established itself around 10 000 yr BP, although there was a small decline in temperature after about 5000 yr BP.

Recent climatic change. Significant changes in some ecologically important climatic variables, notably rainfall amount and seasonality, drought magnitude–frequency and tropical cyclone frequency, have occurred over much of the tropics, including some rain-forest areas, during recent historical time including the period of meteorological observations. Changes have been particularly noteworthy in some rain-forest and marginal rain-forest areas in the trade wind belt. In the West Indies, successive periods of low (mid-nineteenth century), very high (late nineteenth century), rather low (1899–1928), rather high (1929–58) and low rainfall (since 1959) can be identified, in which shifts in annual rainfall of over 20% have occurred (Walsh 1980a, 1985, Stoddart & Walsh 1992). These shifts have taken large areas across the climate–vegetational boundaries identified in the islands by Beard (1949) and Stehlé (1945), particularly in the drier parts. Even in the very wet interior of Dominica, rainfall at Shawford Estate fell from 4694 mm in the late nineteenth century to 4153 mm in the period 1899–1928. Records are particularly long at Roseau, on the leeward coast of Dominica, and Fig. 7.14 shows changes in annual rainfall, seasonality, annual number of dry months and dry period magnitude–frequency from 1865 to 1989. As Beard's suggested evergreen/semi-evergreen seasonal forest rainfall boundary lies at 2000 mm, the changes would result in Roseau having a semi-evergreen seasonal forest climate in the drier epochs, but an evergreen seasonal forest climate in the wetter periods. Of perhaps greater ecological significance are the changes in drought frequency, with Roseau experiencing an annual average of 5 dry months and longer extreme droughts in the recent very dry epoch compared with just 3.4 dry months and shorter droughts in the late nineteenth century. Similar increases in drought magnitude–frequency have occurred in Barbados.

Changes in equatorial rain-forest areas, where there are few long-term rainfall series, are less well documented, but appear to have been rather different in character than in tropical latitudes. Annual rainfall at most equatorial stations has fluctuated much less than at tropical localities, but there are exceptions. In the Seychelles, six alternating wet and dry phases have occurred at Port Victoria since records began in 1891 (Fig. 7.14) (Stoddart & Walsh 1992). Annual rainfall in the current dry phase is around 480 mm (20%) lower than in the wetter phases and has been accompanied by increases in dry month frequency and drought lengths. In parts of the Equatorial Lakes region of East Africa an increase in rainfall in the 1960s was indicated by a sudden marked rise in the level of Lake Victoria and the flow of the White Nile (Lamb 1966).

The history of droughts at Sandakan (Sabah) since 1879 (Fig. 7.15) is of considerable interest in view of the ecological consequences of the 1982–83 drought in eastern Borneo. Significant droughts of four months duration or longer occurred four times between 1885 and 1915 and five times in the period 1968–90, but in the intervening period of over 50 years there were none. The most severe drought (six consecutive months with less than 100 mm rainfall) occurred in 1903. Although the spate of droughts early this century was associated with lower annual rainfall, the recent series of droughts was not. In the absence of long-term forest plot data, the ecological consequences of the pattern of droughts at Sandakan can only be inferred. Given the large-scale death of canopy trees that a serious drought like the 1982–83 east Borneo drought can bring (Leighton & Wirawan 1986) and the severity of the 1903 and 1915 droughts at Sandakan, it would appear likely that the rain forests of drought-prone parts of east Borneo prior to the 1982–83 event already exhibited scars of previous droughts (perhaps in terms of relatively few old, very large trees and a large number of young, small girth canopy trees of similar age dating back to the previous severe drought). Whether the fifty-year period free of severe drought from 1916 to 1967 resulted in a gradual invasion or preferential growth of shade-tolerant species is not known, but such a period is not long enough to eliminate any impact of a previous severe drought.

Tropical cyclones were noted in a previous section to have major short- and long-term influences on forest structure and species composition in areas where they

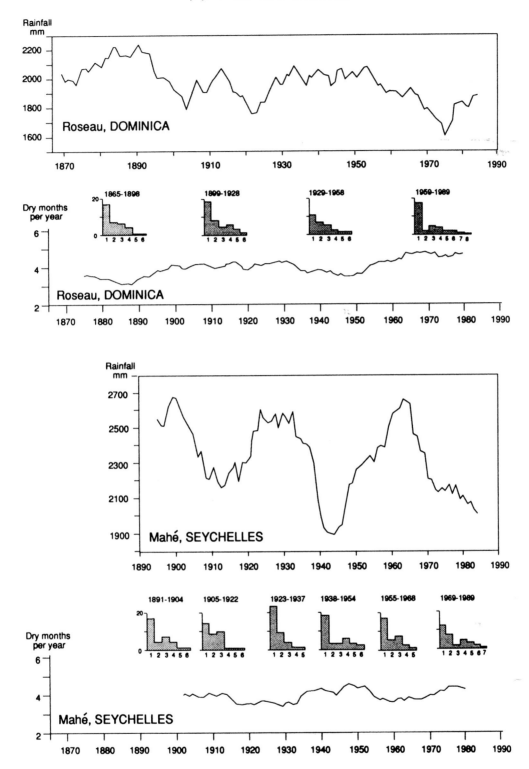

Fig. 7.14 Changes in annual rainfall, dry month (<100 mm rain) frequency and the length and frequency of droughts at Roseau (Dominica) 1865–1989 and Victoria (Seychelles) 1891–1989. The histograms show twenty-year frequencies of dry periods of different durations (as in Fig. 7.10). After Stoddart & Walsh (1992).

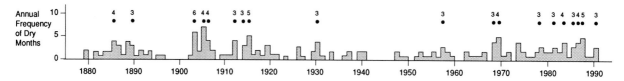

Fig. 7.15 Major droughts and annual frequency of dry months (<100 mm rain) at Sandakan, Sabah, 1879–1990. There are no data for the years 1897–1901 and 1941–46.

occur. The nature of this impact depends partly on the frequency and intensity of cyclone damage. Several recent studies have demonstrated changes in cyclone frequency and tracks both during this century and in the longer term. The cyclone belt over the southwest Pacific has been noted by Whitmore (1974) to have moved slightly closer to the equator since 1950, with the result that devastation is becoming more frequent in the Solomon archipelago between 7° and 11°N. On a longer time-scale, significant changes in the tracks and frequencies of cyclones within the Caribbean have occurred between 1500 and the present day (Walsh & Reading 1991). In the Lesser Antilles, cyclone frequencies in 1600–1770 and the mid-nineteenth century were remarkably low in comparison to the late eighteenth, early nineteenth, late nineteenth and mid-twentieth centuries and frequencies (especially of those of hurricane intensity) have fallen again since 1959 (Fig. 7.16). Broadly similar patterns apply to Hispaniola, Puerto Rico, Cuba and Jamaica. Shifts in cyclone frequency and tracks during the current century have also been noted in the South West Indian Ocean (Stoddart & Walsh 1979) and in the Pacific (Eyre & Gray 1990).

Little is known about the appropriate time-scales of climatic change relevant to rain-forest vegetation. It may be that some of the climatic epochs (drier, wetter, drought-prone, drought-free, cyclone-frequent, cyclone-rare) identified above were too short to bring about long-term changes in forest type or indeed that such short-term fluctuations should be regarded merely as an integral part of a long-term climate. Nevertheless, in a climate prone to occasional droughts (or cyclones), the time since the last severe drought (or cyclone) would clearly be a very important factor in accounting for forest character at any one time. Also, in longer term climatic changes, it may be the rare event (the extreme drought or cyclone) or its continued absence (as in a wet epoch or a cyclone-free period) that may bring about vegetational change.

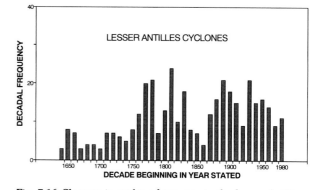

Fig. 7.16 Changes in cyclone frequency in the Lesser Antilles during the period 1640–1989. After Walsh & Reading (1991).

7.10 Deforestation and climatic change

With the world-wide increase in concern about global climatic change and the rate of destruction of tropical rain forests, attention has focused on the influence that rain forest has on regional and global climates and the impact that large-scale forest clearance might have on them. A further question concerns the possible impact that future human-induced global climatic change (whether or not partly caused by rain-forest clearance) might have on the remaining tropical rain forest. Our understanding of relations between rain forests and climate, however, remains poor and this has been one of the reasons for the highly variable and unreliable forecasts of tropical and global climatic change made by climatic modellers. Many of the uncertainties involved in attempting to assess and model the impact of rain-forest clearance have been pointed out by Dickinson & Virji (1987) and Henderson-Sellers (1987).

Rain-forest clearance might contribute to regional or global climatic change in five main ways: (i) by affecting the composition of the atmosphere; (ii) by affecting evaporation–rainfall patterns through its impact on the hydrological cycle; (iii) through changes in surface albedo; (iv) by affecting atmospheric turbulence by causing changes in the aerodynamic roughness of the vegetation canopy; and (v) by adding dust particles to the atmosphere as a result of biomass burning and increased wind-blown dust from drier and more exposed soil surfaces (Henderson-Sellers 1987). Although there is general agreement that rain forest disturbance affects all these variables to some degree, there is considerable debate about the scale and nature of their effect on regional and world climate.

The influence of tropical rain forest on various atmospheric trace gases has been discussed by Crutzen (1987) and Seiler & Conrad (1987). A major concern is the effect on carbon dioxide concentrations, as tropical rain forests are both a major user (in photosynthesis) and producer (in respiration and plant decay processes). Rain-forest removal, it is argued, will both increase carbon dioxide supply to the atmosphere via biomass burning and decay and reduce the rate at which it is removed from the atmosphere because of the lower transpiration rates of replacement vegetation. Fearnside (1990, p. 186) has estimated that, if the Amazonian rain forest were cleared over the next 50 years, the carbon dioxide released would total 50 billion tonnes (50 Gt), equivalent to 20% of that in the whole of the Earth's atmosphere (but a much smaller increase than that caused by fossil fuel burning). The tropics are also a major source of the trace gases methane (CH_4), nitrous oxide (N_2O), and carbon monoxide, with biomass burning a particularly important component (Seiler & Conrad, 1987). These, like carbon dioxide, are 'greenhouse effect' gases, although it is interesting that methane is in fact useful in countering ozone depletion in the stratosphere (Crutzen 1987). Burning of the rain forest and repeated burning of the secondary vegetation replacing it would increase levels of these gases further. Finally, with an estimated 60% of the total global biomass, the tropics is the main producer of hydroxyl (OH), which is important for removal of the above trace gases; destruction of rain forest would, it is argued, lead to a decline in hydroxyl production and accelerate the build-up of trace gases in the atmosphere towards higher equilibrium levels. The impact of rain-forest removal on trace gas levels will clearly partly depend upon the nature of the replacement vegetation (see Chapter 18). As it is estimated that 95% of forest clearance in Amazonia at present is for cattle ranching (Hecht 1983) and furthermore burning, soil erosion and leaching in many areas reduce soil nutrient levels severely, one concern is that the replacement vegetation will

be mainly grass and subject to frequent burning. The impact of rain-forest clearance would probably be much less if the long-term land use was selective logging, tree crops or shifting agriculture.

The argument that hydrological changes accompanying rain-forest clearance will induce major changes in the areal pattern and amounts of rainfall in the rain-forest region, in the tropics in general or globally, is very questionable. Salati (1987) has summarized the main points of this thesis. The first part of the argument, that forest removal would reduce evapotranspiration and increase streamflow, is probably to some extent valid (see Chapter 8), although much would depend upon the replacement land use and the extent of soil erosion. The few available data from the Amazon Basin suggest that currently about 50% of the annual precipitation of the basin is recycled water evapotranspired by rain forest rather than derived from the ocean and that about 50% of solar radiation received by the rain forest is used in evapotranspiration (and the other half in air heating and other processes). Salati therefore argues that less heat will be required to accomplish the reduced evapotranspiration and that higher temperatures, lower humidity, reduced cloudiness and a reduction of rainfall over the Amazon Basin will result. Changes in the nature and intensity of tropical atmospheric circulations might also result and bring about larger scale changes in climate and rainfall distribution in the tropics.

It is usually argued that the ground surface albedo (reflectivity) should increase with rain forest clearance from the low value (0.13) typical of rain forest to values of 0.18–0.23 typical of savanna vegetation. However, clouds are very effective reflectors of incoming radiation; changes in cloudiness, which are very difficult to predict or incorporate in climatic modelling, could more than offset the impact of ground albedo changes.

Recent attempts to model the impact of rain-forest clearance on climate, reviewed by Henderson-Sellers (1987), have produced very different results (Table 7.12), with predictions for the Amazon basin varying between a fall of 100–800 mm and an increase of 75 mm in annual rainfall and between a drop of 0.4 °C and a rise of 0.55 °C in surface temperature. Some of the models, moreover, fail even to simulate the current spatial pattern of rainfall with any accuracy. Models suffer from a dearth of good input data even on current climate, hydrology and land use, crude assumptions about the workings of the atmosphere–ocean–land surface interfaces and problems of trying to predict regional climate from essentially global models.

There is thus still great uncertainty about the effects of rain-forest clearance on tropical climates and rainfall.

Table 7.12. *Comparison of model simulations of climatic change resulting from tropical deforestation*

Model attributes	Throughout humid tropics		Amazonia only	
	Potter *et al.* (1975)	Wilson (1984)	Lettau *et al.* (1979)	Henderson-Sellers & Gornitz (1984)
Features				
Areal coverage	Global two-dimensional statistical dynamic model (tropical deforestation)	Global climate model (tropical land-type change)	Amazon Basin	Global climate model (deforestation in Amazonia)
Spatial resolution	10° latitude	2.5° latitude×3.75° longitude	5° longitude	8° latitude×10° longitude
Perturbation				
Albedo increase	0.18	0.08	0.03	0.06
Hydrology change	Dry case had increased runoff and evaporation	Soil moisture capacity reduced by 32, 20 and 12 cm for fine, medium and coarse textured soils	Evaporative flux from soil	Soil moisture capacity reduced from 200 to 30 cm and from 450 to 200 cm in upper and lower ground layers
Other changes	None	None	Infra-red emissivity changed	Roughness length decreased
Results				
Atmospheric circulation	Weakened Hadley cell	No significant change detected	Assumed regional trade winds unchanged	Walker circulation showed no significant (>1%) disturbance; no Hadley cell change
Annual rainfall	Decrease 230 mm in 5°N–5°S zone	Decrease 100–800 mm over Amazonia; decrease 200–600 mm over Zaïre	Increase 75 mm over Amazonia	Decrease 220 mm over Amazonia
Surface	Decrease 0.4 °C in 5°N–5°S zone	No systematic change	Increase 0.55 °C	No significant change

Source: Henderson-Sellers (1987).

There is also uncertainty about the scale and nature of climatic change that human activities outside the rain forest may be causing. There has been an increase in carbon dioxide concentration in the Earth's atmosphere, mainly due to the burning of fossil fuels, from around 275 ppm in the mid-nineteenth century to 340 ppm in the mid-1980s; levels may reach 500–600 ppm late in the next century (Dickinson & Virji 1987). With increases in other 'greenhouse' gases, tropical temperatures could rise by 1–5 °C in response (WMO/UNEP/ICSU 1986). Even if the question of rainfall changes is left aside, such warming and increases in carbon dioxide would themselves affect the rain forest. The higher temperatures would lead to higher evaporation demand and the rain forest in marginally wet areas might change to a more seasonal type of vegetation, because species better adapted to water stress could compete better in drought periods. Dickinson & Virji (1987) have questioned the suggestion that, because plant respiration is temperature-dependent, temperature increases of several degrees would cause many species to be severely stressed, die or be non-viable as a result of increased respiratory losses of carbohydrate. They point out that increased carbon dioxide concentrations may promote their survival by permitting increased rates of photosynthesis to balance the higher respiration and that rain forest thrived in the Cretaceous when temperatures were 5 °C warmer than today and carbon dioxide levels an estimated 4–10 times their current values.

Chapter 8

Microclimate and hydrology

R.P.D. Walsh

THE PREVIOUS CHAPTER was concerned with the general climate of regions occupied by tropical rain forest; the climate described was essentially the 'standard' one as recorded by instruments located in meteorological enclosures set in cleared land. In reality, however, rain-forest vegetation produces a complex three-dimensional mosaic of microclimates, with those beneath the forest canopy differing significantly from the 'standard' climate. Rain-forest microclimates vary vertically from canopy top to the forest floor and horizontally from point to point beneath the canopy. At a larger scale microclimates vary between canopy gaps of differing sizes, between building and mature forest phases, and between different rain-forest types. Each of the often contrasting microclimates has a role to play in accounting for the biology of individual plants and animals in the forest. These microclimates form the main theme of this chapter. The chapter also examines the hydrology of rain-forest areas, focusing on the interrelations between hydrological factors and forest structure and species composition, nutrient cycling and soil erosion.

8.1 General features of rain-forest microclimates

Despite some very detailed and valuable studies, our knowledge of rain-forest microclimates and their ecological implications remains very incomplete. Most studies have investigated vertical variations in climatic variables on observation towers sited in rain forests, but comparatively little attention has been given to spatial variations within and beneath forest canopies and to the micro-

climates of canopy gaps. The practical difficulties and cost of instrumenting more than a single site are important reasons for the scarcity of three-dimensional information. The number of detailed studies is still too small to permit an objective assessment of microclimatic differences between rain forests of different stature, canopy structure and density and different climatic environments. In addition, most studies have been carried out in somewhat seasonal locations and there are few data from truly aseasonal or very wet environments.

Some of the main features of microclimatic change from canopy top to forest floor are common to most rain forests. There is always a sharp decline in solar radiation and light received below the canopy with only a small fraction reaching the forest floor. Maxima, means and ranges of temperature also decrease but minimum temperatures are very similar at all levels of the forest. Wind speeds decline to near zero within the forest and daytime relative humidities are higher and saturation deficits (and hence evapotranspiration potential) much less at lower levels than in the upper canopy. Carbon dioxide concentrations are significantly greater close to the forest floor than higher up, but at all levels there are diurnal variations in response to daily patterns of plant transpiration, photosynthesis and respiration.

8.2 Solar radiation and light

Solar radiation and light vary both vertically and horizontally within rain forests, with important consequences for life-forms and their distribution within the

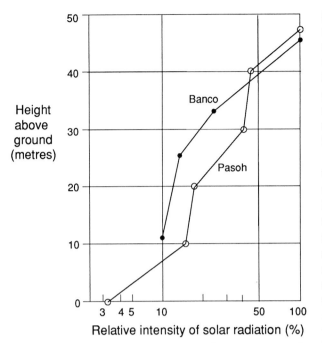

Fig. 8.1 Vertical profiles of solar radiation (as a percentage of that above the canopy) at rain-forest sites, Banco, Côte d'Ivoire, and Pasoh, Malaya. After Cachan & Duval (1963) and Yoda (1978).

gaps beneath emergent trees into the lower forest (Yoda 1978). Similarly at Banco, the average proportions of solar radiation reaching 11 metres were 13% from 0900 to 1000 and from 1600 to 1700 h, compared with 7% in the middle of the day (Cachan & Duval 1963, p. 94). Much of the solar radiation on sunny days and all on overcast days reaching the forest floor is diffuse (i.e. scattered or reflected) rather than direct and is in the form of moving sun-flecks.

At Reserve Ducke (Fig. 8.2) solar and net radiation were measured above and below the 35–40 m forest canopy (Shuttleworth *et al.* 1984b, Molion 1987). Below-canopy solar radiation peaked in absolute as well as relative terms in the morning because of low-angle sun penetration, but afternoon cloudiness prevented a second peak. Net radiation includes downward long-wave radiation either emitted or reflected from the canopy above as well as solar radiation penetrating the canopy, with the two components at Reserve Ducke of roughly equal magnitude but with contrasting diurnal regimes. Downward long-wave radiation beneath the canopy follows the above-canopy solar radiation, but with a lag

forest as well as for rain-forest dynamics. The forest canopy and its spatial pattern of gaps, building and mature phases are responsible for most of the variations, but slope angle, aspect and seasonal variations in the path and elevation of the sun also influence regimes of light and solar radiation received by the lower forest and on the forest floor.

Only a small fraction of the solar radiation arriving at the top of the canopy reaches the forest floor. Values of 0.4 and 3.2% were recorded in two short-term studies at Pasoh (Yoda 1978, Aoki *et al.* 1978) and 1.2% at Reserve Ducke, near Manaus (Molion 1987). Variations in mean solar radiation with height in the forest canopy are shown in Fig. 8.1 for rain forests at Banco in Côte d'Ivoire (Cachan 1963) and Pasoh (Aoki *et al.* 1978). At Pasoh, about 60% of incoming solar radiation was absorbed by the uppermost 5 metres of the forest canopy and 3.2% reached the forest floor. At Banco, a broadly similar pattern was recorded with 10% reaching 11 metres above ground level; no forest floor measurements were made. At Pasoh, although at most levels solar radiation tends to reach a maximum around noon, the proportion penetrating to lower levels tends to be rather higher early and late in the day, when the low angle of the sun permits sunlight to pass through 'horizontal'

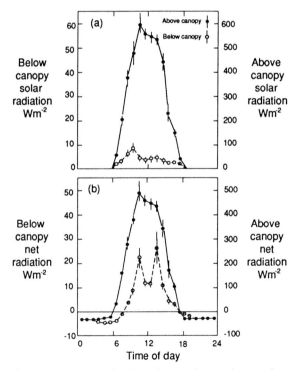

Fig. 8.2 Daily march of solar and net radiation above and beneath the forest canopy at Reserve Ducke, Amazonia. After Shuttleworth *et al.* (1984b).

The plotted points are means of six days of continuous observations. The vertical bars indicate the standard error for each plotted flux.

of about an hour, and therefore peaks in early afternoon on sunny days. The main peak in net radiation beneath the canopy is produced largely by this peak in downward long-wave radiation, but the first is partly the result of the morning peak in solar radiation.

Measurements in the lower montane rain forest at an altitude of 445 m at El Verde (Puerto Rico) (Odum *et al.* 1970a) showed that there the northerly aspect and relatively high latitude (18°N) result in marked seasonal differences in forest floor insolation. This varied from 1% of above-canopy values in February to 6% in early April; the latter period was therefore considered the most favourable for seed germination. Very different seasonal regimes and values of forest floor insolation would apply to steep south-facing slopes in the same area.

The visible light fraction of solar radiation reaching rain-forest floors is likewise small, but both receipts and regimes vary much with forest type and density and from point to point on the forest floor. There have been several attempts at quantifying these variations (Evans 1956, Evans *et al.* 1960) and a particularly detailed study was made in lowland forest at Pasoh (Yoda 1978). Here emergent trees (A storey) at the study plot reach a height of 50–60 m but do not form a continuous canopy and there is a closed canopy (B storey) at 25–35 m above the ground. The small tree (10–15 m) and shrub (2–6 m) layers are ill-defined, merging with each other, the overlying canopy layer and the ground cover. Illuminance was measured using two selenium photo-electric cells, one of which was fixed at 50 cm above a 55 m tall *Koompassia malaccensis* emergent, while the other was moved three-dimensionally within the forest stand. Values of relative illuminance (RI) were obtained by simultaneous readings of the two ammeters. Observations were made at twenty-eight heights at two-metre intervals from the ground surface to 54 metres at 0900 and 1500 h and continuously on selected days.

Diurnal patterns beneath the canopy were not only influenced by passing clouds, but also by the westerly aspect of the site and an opening in part of the main canopy; these factors combined to allow some late afternoon sunlight to reach the emergent tree base, so that afternoon light intensities were significantly higher than those of the morning. Mean relative illuminance during the study period declined in a stepped fashion through the canopy to 0.3% of above canopy values at the forest floor (Fig. 8.3). Yoda identified four microclimatic strata corresponding to the four log-linear segments of the log (RI) – height graph plot. The uppermost stratum (48–55 m) consists of the crowns of emergent trees, which (where they occur) intercept about 70% of incident

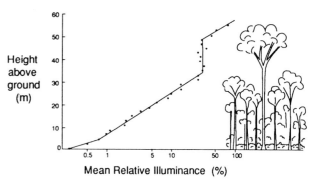

Fig. 8.3 Vertical profile of mean relative illuminance (expressed as a percentage of light received above the canopy) at a rain-forest site at Pasoh, Malaya. After Yoda (1978).

light. The second layer, where RI remains constant, is the leafless space between the emergent crowns and the upper surface of the main canopy at about 32 m height. Relative illuminance then falls at a log-linear rate (though a steadily decreasing rate arithmetically) within and beneath the closed canopy, before falling at a higher log-linear rate in the comparatively dense layer of palms, shrubs and tree saplings near to the ground.

Differences in horizontal variability in RI with height at Pasoh are summarized in Fig. 8.4, which is a three-dimensional version of the type of diagram developed by Evans (1966) for forest-floor light intensities. The curves on the diagram represent the isopleths where the cumulative frequency of observations reaches specified RI values (from 0.2% to 70%). Thus, about 50% of forest floor sites have RI values less than 0.2%, 40% of sites between 0.2 and 0.5%, 9% between 0.5 and 1%, and bright spots of over 1% RI occupy only about 1% of the ground.

Amounts and percentages of light reaching the forest floor vary between different rain-forest stands. In the 43 m high evergreen seasonal forest at Banco (Côte d'Ivoire), relative illuminance declined to 9.65% at 33 m height and 0.69% (twice as great as at Pasoh) at ground level (Cachan 1963). In Guyana, Richards relates differences in forest floor illuminance in five forest types to variations in forest density and structure (pp. 296–302). In closed lowland rain forest at Danum (Sabah), Brown (1990) found using hemisphere photographs (Fig. 8.9) taken close to ground level that around 6% of the sky was visible.

Although mean values of relative illuminance close to the forest floor are low, values in passing sun-flecks often exceed 5–10% (Carter 1934, Evans 1939, Cachan & Duval 1963) and in Guyana were measured as high as

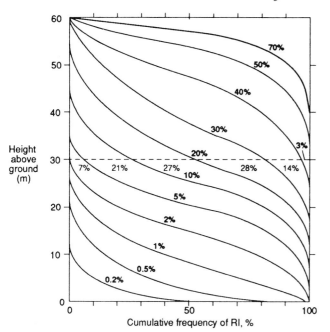

Fig. 8.4 Horizontal distribution of relative illuminance at different heights beneath a 60 m high rain-forest canopy at Pasoh, Malaya. After Yoda (1978).

The curved lines are of progressively higher relative illuminance. The figures along the dashed line indicate the percentage of points along the monitored horizontal transect at 30 m height above the ground within each relative illuminance class. See text (p. 208).

72% (Carter 1934). Sun-flecks are intense, but generally short-lived (except in canopy gaps). However, although Carter (1934) found that sun-flecks affected only 0.5–2.5% of the forest floor at any one time, they can account for a substantial proportion of the total light arriving at the ground. In a lowland rain forest in Singapore, where the mean relative illuminance at the forest floor was 2%, Evans (1966) found that *ca.* 50% of the light came from sun-flecks, which occurred from 2 h after sunrise to 2 h before sunset, compared with 6% through canopy holes and 44% as reflected or transmitted diffuse light. These figures, however, refer specifically to Singapore, which is on average 47% cloudy; higher sun-fleck percentages will occur in sunnier climates and diffuse light would be more important in areas with greater daytime cloudiness.

Sun-flecks are also important because of their spectral composition. Visible light covers a spectral range of 300–2400 nm, but only a small fraction of this within the red band from 400 to 700 nm can be used in photosynthesis. Much of the light arriving at the forest floor – and all of it when the sun is obscured by cloud –

is diffuse, having been reflected from leaf and bark surfaces or transmitted through leaves. Compared with sunlight, diffuse light contains higher proportions of blue, green and near infra-red light (Evans 1966), not used in photosynthesis because such wavelengths are outside the range of absorption of chlorophyll. Sun-flecks, which are similar in spectral composition to sunlight, account therefore for most of the photosynthetic light arriving at the forest floor.

At Danum (Sabah), Brown (1990) recorded how the ratio of red (655–665 nm) to far-red (725–735 nm) light, which is of particular importance to plant growth and development (Hart 1988), differed at ground level between closed canopy forest, an open site and in canopy gaps of different sizes. At the open site the ratio exceeded 0.8 all the time during sunny weather, whereas under closed canopy forest the ratio was 0.3–0.4 during the diffuse light conditions of most of the day and exceeded 0.6 only for an hour during passing sun-fleck conditions. In the small gap the ratio exceeded 0.8 for 20% of the time, but in the large gap for 87% of the time, reflecting the longer period in which direct sunlight can enter a large opening. In lowland tropical forest in Costa Rica, Chazdon & Fetcher (1984) measured photosynthetic photon flux density (PPFD) (i.e. the photosynthetically useful light fraction) and found that the understorey received on average only 1–2% of that recorded in a clearing. They also found that, although percentages received by the understorey were greatest on cloudy, overcast days and least on sunny days, absolute values of PPFD were highest on sunny days in the dry season.

Sunshine regimes, which vary considerably between rain-forest areas, are therefore very important both to the amount and percentage of light penetrating forest canopies that is available for photosynthesis. Localities with mainly nocturnal rainfall have a significant photosynthetic light advantage over forests with daytime rainfall and cloudiness. The light arriving at forest floors of montane cloud forests will be almost entirely diffuse on most days and hence particularly deficient in photosynthetic wavelengths.

8.3 Temperature

Maximum, mean and diurnal range of temperature are significantly lower beneath the rain-forest canopy than within or just above it or in gaps or clearings. This is so because the forest canopy, rather than the ground, is the main heating and radiation surface. In lowland rain forest at Mulu, Sarawak, temperatures were recorded in 1977–78 at heights of 1.3 m in Stevenson Screens beneath the forest canopy and in an adjacent clearing (Walsh 1982a). Mean maxima and diurnal ranges were on average 4.5 °C lower in the forest than in the clearing, with little variation in the difference during the year (Fig. 8.5). As in the clearing, maxima and diurnal ranges beneath the forest canopy were somewhat higher in the drier months of September 1977 and August 1978.

Several studies have examined vertical differences in the daily march of temperature at various heights on towers within rain forests. In rain forest 47 m high at Pasoh (Malaya), the diurnal range declined from 10 °C at the canopy top (and 6 °C at 6 m above the canopy) to 3 °C at the ground surface; temperature differences and hence gradients (Fig. 8.6) were large during the day, but small during the night with minima throughout the profile of between 20 and 21 °C, but lowest in mid-canopy (Aoki *et al.* 1978). At Banco (Côte d'Ivoire), diurnal ranges at all levels were over twice as great in December (dry season) than in June (wet season), but always with differences between the canopy and the lower forest (Table 8.1) (Cachan & Duval 1963). The reasons for the seasonal differences are (1) the lower frequency of sunny days with high maximum temperatures in the wet season and (2) the smaller proportion

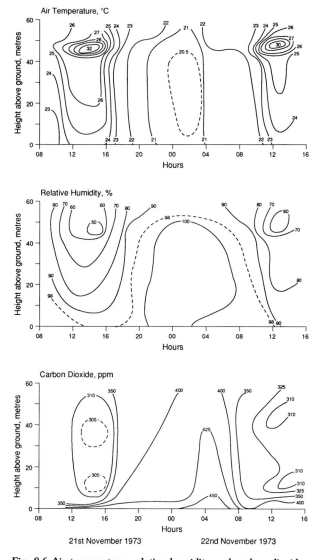

Fig. 8.6 Air temperature, relative humidity and carbon dioxide concentration at various heights in high forest, Pasoh, Malaya, 21–22 November 1973. After Aoki *et al.* (1978).

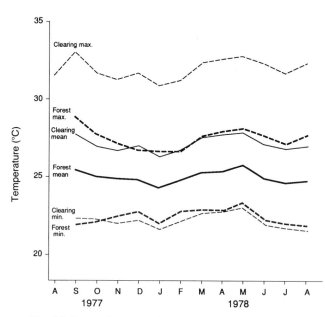

Fig. 8.5 Seasonal regime of mean maximum, mean minimum and mean temperature measured in Stevenson Screens at 1.3 m height in a clearing and beneath the alluvial rain forest at Long Pala, Gunung Mulu (Sarawak) in 1977–78. After Walsh (1982a).

of radiation used in evaporation (thereby cooling the canopy) in the dry season compared with the wet season.

During rainstorms the temperature profile of the forest may be significantly disrupted, as demonstrated at the Reserve Ducke (Brazil) by two days – one sunny, one with two rainstorms – in September 1983 (Shuttleworth *et al.* 1985, Molion 1987). In the first rainstorm at 0600 h on the rainy day, temperatures at all levels dropped sharply by around 3 °C (Fig. 8.7), probably because of cool downdraughts penetrating the canopy and perhaps some cooling by evaporation. After the second storm it is noticeable that temperatures in

Table 8.1. *Diurnal ranges in temperature (°C) at different heights in the forest canopy in the wet and dry seasons at Banco (Côte d'Ivoire)*

Height (m)	December (dry season)		June (wet season)	
	Mean	Range	Mean	Range
46	10.8	8.8–12.9	4.0	1.5–5.9
33	10.0	8.3–11.9	3.8	1.7–5.2
26	9.9	7.2–11.4	3.4	1.4–4.7
11	6.6	5.1–8.4	2.8	1.1–4.6
6	5.2	3.9–6.9	2.2	1.0–3.3
1	4.4	2.9–5.9	1.7	0.7–2.5

Source: Cachan & Duval (1963).

the upper canopy continued to decline after rain had stopped, whereas at lower levels recovery was swifter; this might suggest that evaporative cooling of intercepted water was important at upper levels, but ineffective at lower levels.

In montane forests, differences in temperature between canopy and ground and diurnal ranges beneath the forest canopy tend to be smaller than in lowland rain forests, reflecting the increased frequency of cloudy, humid conditions. Thus in the lower montane rain forest at El Verde (Puerto Rico), mean diurnal ranges were 4.5 °C at the tower top (27 m), 3 °C at 16 m and 3 °C close to the ground (Odum *et al.* 1970a) and in Borneo diurnal ranges below the forest canopy up G. Mulu were 3.2–4.4 °C at altitudes of 500–1800 m compared with 5 °C at the lowland site (Table 8.2) (Walsh 1982a). The higher range at 1800 metres compared with the two intermediate sites may be due to the low stature (5–8 m) of the upper montane forest at that site. However, although mean ranges are low in montane forests, on sunny days canopy maxima (and thus leaf temperatures and transpiration potentials) can be almost as high as in the lowland forest.

Little is known about horizontal variations in temperature at different heights in rain forests, but the variations in light and radiation amounts and regimes along traverses at Pasoh (pp. 208–9) suggest that they are considerable, although their ecological significance has not been assessed.

Soil temperatures under rain forest tend to be very close to mean annual temperature and vary even less both diurnally and seasonally than temperatures beneath the canopy. In an experiment at Okomu Forest Reserve in Nigeria, soil temperatures at 5 cm and 20 cm depth and air temperature at 1 m height were measured under

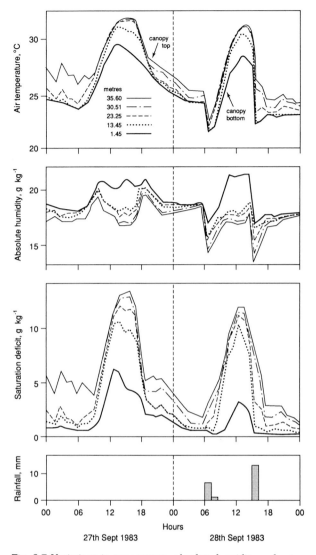

Fig. 8.7 Variations in temperature, absolute humidity and saturation deficit at five heights in rain forest at Reserve Ducke, near Manaus, Amazonia, 27–28 September 1983. After Shuttleworth *et al.* (1985).

a tall rain forest (over 40 m high) and in a forest clearing (Ghuman & Lal 1987). Mean temperatures under the forest cover (28.2–28.3 °C) were 1.7–3.1 °C lower than their counterparts in the clearing, but variations in the forest were significantly less in the soil than in the lower canopy and at 50 cm depth the soil temperature was constant. On a sunny day, the diurnal range was 5 °C at 1 cm in the soil compared with 10 °C in the lower canopy and 15 °C in the open.

Bernard (1945, p. 91) reported soil temperature results in the Yangambi area of Zaïre. Beirnaert (1941) showed that soil temperature at 1300–1400 h during a dry spell

Table 8.2. *Mean temperature under the forest canopy at four altitudes on Gunung Mulu (2386 m), Sarawak, during 1977–8*

Location	Altitude (m)	Mean maximum (°C)	Mean minimum (°C)	Mean (°C)	Daily range (°C)
Long Pala	65	27.4	22.4	24.9	5.1
Camp 2	500	23.9	20.7	22.3	3.2
Camp 3	1320	20.8	17.2	19.0	3.7
Camp 4	1800	18.6	14.1	16.3	4.4

Source: Walsh (1982a).

was 0.5 °C lower in primary than in secondary rain forest and that temperature varied little with depth under both types of cover. Louis (1939) found that diurnal ranges in soil temperature beneath rain forest were usually less than 2 °C at 4 cm depth and less than 1 °C at 10 cm depth; the most significant changes were caused by infiltrating water during rainstorms, when soil temperatures could drop by 1–2 °C in a few minutes. The high and near-constant soil temperature beneath rain forest permits litter decomposition and respiration by soil organisms to continue at high rates during the night and is an important reason for the high carbon dioxide concentrations of the soil and near ground zone (p. 214). Soil temperatures are close to mean annual air temperature and decline with altitude. They are therefore significantly lower in montane forests. In the cloud forests at Pico del Oeste (altitude 1004 m, annual rainfall 4530 mm) in Puerto Rico, soil temperature at 8 cm depth is identical to the annual temperature of 18.3 °C and varies very little (Howard 1970).

8.4 Relative humidity and saturation deficits

There are marked vertical differences in humidity and saturation deficit within rain forests. Diurnal variations in relative humidity within the lowland rain forest at Pasoh (Aoki *et al.* 1978) closely follow vertical differences in air temperature (Fig. 8.6). Diurnal variations at the canopy top are the most marked (and nearest to those measured in a standard meteorological enclosure), with relative humidity on sunny days dropping from 100% at night to less than 60% in the middle of the day, mainly as a result of the rise in temperature to 30–32 °C. Beneath the canopy, diurnal variations decline with decreasing height above the ground, with humidities remaining above 90% throughout the day close to the forest floor. All heights in the forest tend to exper-

ience saturated conditions at night, but above the canopy mixing with the atmosphere keeps humidity below 95%.

The consequence of this pattern, which can be regarded as typical of lowland rain forests, is a sharp contrast in daytime saturation deficits (and hence the evaporating power of the air) between the upper canopy and the lower forest, a contrast that has fundamental repercussions for the species composition, leaf character, transpiration rates and water relations of the latter. In the Pasoh example, saturation deficits at the canopy top reached 13–16 mmHg at midday on the two fine days shown in Fig. 8.6, but remained less than 3 mmHg close to the forest floor throughout the day.

Broadly similar results have been recorded at other rain-forest sites, but with some differences. In the rather seasonal Omo Forest Reserve (Nigeria) (Evans 1939, Richards 1952), maximum saturation deficits in the undergrowth are 3.5 times higher in the dry than in the wet season (Fig. 8.8). Similar seasonal contrasts were found at Banco (Côte d'Ivoire) (Cachan 1963). There is also often a disparity between localities in the gradient of changes within the forest and it is usually unclear whether these reflect differences in forest structure or merely the siting of the observation tower within forest stands of uneven canopy. Thus, at Pasoh, there are two sharp gradients in daytime relative humidity: (a) in the upper canopy above 40 m and (b) close to the ground below 10 m, with comparatively little change in relative humidity between those two heights (Fig. 8.6). In contrast, at the Reserve Ducke site near Manaus, Amazonia (Shuttleworth *et al.* 1985, Molion 1987), the main change was in the lower forest, with only a small gradient in the upper canopy between 23 and 35 m (Fig. 8.7). In a much lower rain forest in western Uganda, the main change occurred from 5 to 10 m, with little change from 10 m to the canopy top at 17 m (Haddow *et al.* 1947).

In montane forests, daytime relative humidities at all

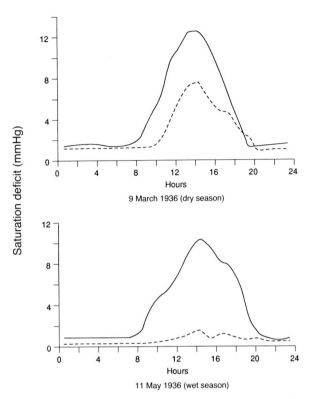

Fig. 8.8 Daily march of saturation deficit at two levels in old secondary rain forest, Omo Forest Reserve, Nigeria. After Evans (1939, fig.8). Continuous lines, 24 m above ground on tree; broken lines, 0.7 m above ground in undergrowth. Each point is a mean of four readings at 15 min intervals.

heights above the ground tend to be significantly greater than in lowland rain forests because of the frequent misty conditions and often higher rainfall; vertical differences within the canopy are only significant on sunny days. Montane forests need to survive not only long periods with zero or near-zero saturation deficit, but also occasional sunny spells with moderately high daytime saturation deficits. In the cloud forest on the summit of Pico del Oeste (Puerto Rico), where the mean relative humidity is 98.5%, spells of weather included a very wet, cloudy period of 16 d with only 6 h of clear conditions and relative humidity at 100% throughout twelve of the days, but also three weeks with cloud-free conditions averaging 15 h per day (Howard 1970) (see Chapter 17).

8.5 Wind and air movement

Rain-forest canopies react to wind, but also strongly influence air movement above and within the forest.

Mean wind speeds are invariably lower beneath the forest canopy than above it. At Pasoh (Malaya), wind speed in a short-term study in 1973 was found to decline from 2.7 m s^{-1} at the canopy top (47 m) to 0.5 m s^{-1} at 20 m height in the middle of the day and from 0.5 m s^{-1} at canopy top to 0.1 m s^{-1} at 20 m during the night (Aoki *et al.* 1978). At Reserve Ducke, near Manaus (Amazonia), mean wind speeds were over 2 m s^{-1} above the canopy compared with 0–1 m s^{-1} in the lower forest (Molion 1987). In the El Verde tower in lower montane rain forest, mean wind speeds declined from 1.1 m s^{-1} just above the canopy to 0.33 m s^{-1} at 16.5 m in the densest part of the canopy, before rising somewhat at lower levels where the leaf density was less (Odum *et al.* 1970a). During a period of twenty-three days in the dry season at Barro Colorado Island (Panama), when the trade winds are at their strongest, mean wind speeds in the undergrowth (2 m height) and in the middle of the forest canopy (23 m height) were 0.02 and 0.2 m s^{-1}, respectively, compared with 4.4 m s^{-1} in an unobstructed site outside the rain forest (Allee 1926, p. 279).

Vertical air movements within rain forests have sometimes been deduced from micrometeorological patterns of temperature and humidity, but have not been systematically recorded, mainly because the speeds involved are well below the recording capabilities of most anemometers and because of the problems of recording air movements in three dimensions. However, the sharp drop in temperature and humidity recorded at all levels in the Reserve Ducke (Amazonia) during two rainstorms on 28th September 1983 (Fig. 8.7) can be regarded as conclusive evidence of the significance of downdraughts of cool (and hence drier) air during rainstorms (Shuttleworth *et al.* 1985, Molion 1987). During nighttime cooling, the sinking of cool air from the canopy top and rise of warm, humid air from below has been held to be one reason for the lack of vertical variation in minimum temperatures within rain forests. In addition, higher wind speeds above the canopy, by producing continual mixing and replacement of canopy-top air with free-atmosphere air above, are considered an important reason why night-time temperatures and humidities just above the canopy do not fall and rise, respectively, more than they do.

8.6 Carbon dioxide

A knowledge of carbon dioxide concentrations and their dynamics above and within the rain-forest canopy is important because at times they become a limiting factor for photosynthesis and because three-dimensional assessments of carbon dioxide fluxes and changes can be used in the estimation of the productivity of the rain forest and its components. With rising carbon dioxide concentrations in the world atmosphere, an understanding of carbon dioxide use by plants is becoming of key importance in attempts to assess the likely impacts of climatic change on rain forests and of rain-forest clearance on climate.

Several processes combine to influence carbon dioxide concentrations at different levels in the rain forest. During the day photosynthesis by leaves, which increases with light intensity and temperature and is most active in the upper canopy, removes carbon dioxide from the air. Plant respiration, which also increases with temperature but which unlike photosynthesis continues at moderately high rates at night, produces carbon dioxide. Decomposition of plant debris and associated respiration of animals and micro-organisms are also major sources of carbon dioxide and are largely responsible for the comparatively high concentrations of carbon dioxide close to the ground and within the soil. Further important factors are the nature and three-dimensional pattern of air movements and mixing both within and above the forest. It is generally supposed that beneath the canopy mixing is poor because of very low wind speeds, but that at canopy level respiration- and photosynthesis-induced diurnal fluctuations in carbon dioxide are reduced by more efficient mixing with the air above.

Detailed data on diurnal variations in carbon dioxide at eight heights within primary rain forest 53 m tall at Pasoh (Malaya) were gathered by Aoki *et al.* (1978) for a 26 h period in November 1973 (Fig. 8.6) and these are summarized here.

(i) Above canopy (above 53 m). Carbon dioxide concentrations just above the canopy were at a minimum (310–330 ppm) throughout the middle of the day when photosynthesis is active, but replenishment from respiration and the atmosphere above prevented concentrations from falling lower. During the night, when photosynthesis had ceased but respiration continued at a moderately high rate, carbon dioxide concentrations rose progressively, reaching their maximum (420 ppm) about dawn before photosynthesis recommenced.

(ii) Intermediate levels (10 m, 40 m and 47 m). Diurnal patterns were similar to those above the canopy, but more pronounced, probably because of reduced mixing with fresh air from above. Carbon dioxide minima were around 305 ppm and maxima around 425 ppm. The morning fall in carbon dioxide at 40 m height and lower occurred later than above the canopy because of the lack of penetration of sunlight into the lower canopy at low angles of elevation.

(iii) Undergrowth (3 m). There was a very sharp gradient in carbon dioxide concentrations from the ground to 10 m during the afternoon, probably because the peaks in photosynthesis in the lower vegetation layers and of respiration in the soil coincided. At night, however, respiration alone occurred at all levels and there was little variation in carbon dioxide concentration with height.

(iv) Ground level. Carbon dioxide concentrations were higher here than at all other levels throughout, with values varying from 390 to 478 ppm. Highest values were recorded in the late afternoon and in the middle of the night. The afternoon peak may be the result of a peak in litter decomposition and soil animal activity at the warmest part of the day.

These patterns at Pasoh, which are relatively modest deviations from the atmospheric norm of the early 1970s, tend to confirm the early work of Stocker (1935) in Javanese primary rain forest, Evans (1939) in old and young secondary forest and Bünning (1947) in Sumateran rain forest and the conclusions of Richards (1952, p. 181) that carbon dioxide concentrations in the rain forest, even close to the ground surface, are not exceptionally high and are unlikely to compensate for low light intensities in encouraging plant growth beneath the canopy, as was at one time supposed (Schimper 1935, p. 147).

Broadly similar results to those at Pasoh were obtained by Odum *et al.* (1970a) in cooler and less tall (27 m) lower montane rain forest at El Verde (Puerto Rico). At the ground surface, respiration maintained carbon dioxide concentrations of 400–620 ppm compared with 320 ppm in the free air above the forest. Photosynthesis in the canopy steepened daytime gradients to a lesser degree than at Pasoh, probably because the trade winds were more efficient at replenishing carbon dioxide concentrations in the canopy. Odum *et al.* also noted that (1) on exposed ridges, where mixing was thorough and the forests of smaller stature, carbon dioxide concentrations showed little diurnal variation compared with the taller and more sheltered rain forests of the valleys and lower altitudes and (2) 'successional' (secondary) forest exhibited lower daytime carbon dioxide concentrations because of its higher net production and photosynthesis.

Stephens & Waggoner (1970) noted how rates of assimilation of carbon dioxide via photosynthesis varied greatly with the leaf-area index of a rain-forest stand and with tree species. In old secondary ever-wet rain forest at Bosque de Florencia (Costa Rica), they also showed how photosynthetic capability (per unit area of leaf) was greater for pioneer species than mature forest trees and least for species of the undergrowth. Medina *et al.* (1986), working in the upper Rio Negro basin near San Carlos (Venezuela) have linked differences in the isotopic composition of carbon in plant tissues between canopy and lower forest trees in 'caatinga' rain forest to contrasting origins of the carbon dioxide used in photosynthesis. They suggest that lower $\delta^{13}C$ values of the shade flora (−32.9 to −36.0‰ for small trees and saplings compared with −28.8 to −32.6‰ for leaves in the upper canopy) result primarily from assimilation of carbon dioxide depleted in $\delta^{13}C$ originating from soil respiration, whereas at canopy level the main sources of carbon dioxide are from leaf respiration and the free atmosphere. They also found higher soil and near-ground carbon dioxide concentrations in the flood-prone caatinga forest, which has ill-drained, poorly aerated soils, than in adjacent tierra firme forest on land above the flood level. Higher soil respiration rates and lower soil conductance rates of carbon dioxide in the caatinga forest soils were suggested as the main factors responsible for this contrast.

Micrometeorological information on carbon dioxide and other variables has been employed in Costa Rica using an aerodynamic method (Lemon *et al.* 1970) and at Pasoh using the energy balance (Bowen Ratio) method (Yabuki and Aoki 1978) to assess primary production rates. Assumptions involved in such calculations are usually rather sweeping, however, and the results, especially as they are usually, as at Pasoh, based on a few days' observations up a single meteorological tower, are at best only indicative unless combined with transpiration measurements around single leaves or plants and with litterfall, tree growth and forest density information.

8.7 Microclimates of canopy gaps and secondary forest

Canopy gaps of varying size and frequency within rain forests are produced by natural tree death, windthrow, lightning and fire, landslides, tropical cyclones and also because of shifting cultivation and timber exploitation by humans (Chapters 2, 3 and 18). Gaps are characterized by microclimates very different from those of the lower forest and the canopy top. In general, compared with the microclimate beneath a closed canopy, gap microclimates have the following main features:

(a) a greatly increased amount, altered regime and changed spectral composition of solar radiation (especially the photosynthetic light fraction) reaching the forest floor;

(b) increases in maxima, means and ranges of air and soil temperature, especially on sunny days;

(c) reduced daytime relative humidities and increased saturation deficits, again especially on sunny days;

(d) an increase in wind speed and turbulence.

Gap microclimates, however, vary greatly with size of gap, within individual gaps, and with age of gap as regrowth proceeds. In general, differences from beneath closed-canopy forest increase with size of gap and towards the centre of gaps and decrease as vegetation recovers. Although canopy gaps (and particularly gap size) are considered of fundamental significance in seed germination, seedling development and the nature of forest regeneration (see Chapter 18), there have been few systematic assessments of their microclimates. Among the most significant are those of Raich (1989) in Malaysia, Barton *et al.* (1989) and Fetcher *et al.* (1985) in Costa Rica and Brown (1990) in Sabah.

Of particular interest are the findings of Brown (1990) who, in a study of dipterocarp regeneration, recorded variations in microclimate both between and within ten canopy gaps of different size in lowland rain forest at Danum Valley (Sabah) (4°58'N). The gaps were artificially created by cutting down trees and non-dipterocarp saplings taller than 2 m. Gap size was measured as the 'percent area of the sky hemisphere that is unobscured'; this was assessed by taking a vertical, hemispherical, black and white photograph using a 180° field of view 'fish-eye' lens (Fig. 8.9). The area of open sky is expressed as a percentage of the total photograph area and the analysis was made fast and accurate by digitizing a video image. Microclimatic measurements were made over a period of a year at the centre of each gap and along north − south transects across it; measurements were also made in two control plots beneath closed-canopy forest.

Measurements over a year are summarized in Table 8.3. There are strong relationships between gap size and the microclimate recorded at the centre of a gap, but changes tend to be greatest at the small end of the gap size range, as the incremental increases in duration of intense sunshine become progressively less as gaps become larger. Mean photosynthetically active radiation, which is strongly correlated with duration of bright sunshine, averaged 19.2 mol m^{-2} d^{-1} in the largest gap, compared with 1.0–1.1 mol m^{-2} d^{-1} in the forest and around 35 mol m^{-2} d^{-1} in an open meteorological plot. Maximum air temperatures averaged 38.2 °C in the largest gap, compared with 28.4 °C in the undisturbed forest, while maximum soil temperatures at 10 mm and 50 mm depth were 8.2 °C and 4.4 °C, respectively above

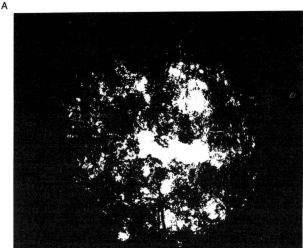

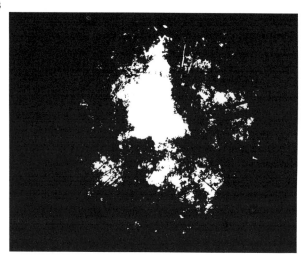

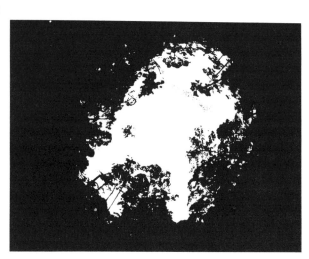

values in forest plot 2. Minimum temperatures were very similar to those in closed canopy forest and did not vary with gap size. Significant changes in the spectral composition or quality of light accompanied the great increase in solar radiation received in canopy gaps. The ratio of the irradiance of the red (655–665 nm) and far-red (725–735 nm) wavebands (see also section 8.2) was used to assess these differences. As gap size increased, the duration of peaks with high ratios (associated with direct and reflected sunlight) also increased, whereas tiny gaps were dominated by long periods of mostly transmitted light deficient in the 655–665 nm band.

Brown's studies also showed how microclimate varies within gaps. The duration of direct sunlight decreases away from the centre of gaps, but in a complex manner. At this low latitude, direct sunlight at low sun angles early and late in the day is able to reach and penetrate east – west sides of gaps more effectively than north – south edges of gaps. In addition, because afternoons are cloudier than mornings, the eastern sides of gaps receive considerably less direct solar radiation than the western sides. Because of penetration of both early morning and late afternoon low-angle light obliquely beneath the canopy at the gap sides, the microclimatic effects of gaps can extend considerable distances into the surrounding forest.

Gap microclimates and their vertical variation change rapidly as the vegetation regrows. The swift recolonization of gaps by lianes, grasses and young trees (Chapter 18) means that in most cases the bare soil phase is replaced after a few weeks by a set of microclimates in which the developing canopy is now the main radiation surface and the forest floor is progressively shielded from solar radiation and light. Thus Louis (1939, cited in Bernard 1945) found at Yangambi (Zaïre) that maximum soil temperatures at 4 cm depth in a secondary rain forest 1 year old were only 2 °C higher than in the mature rain forest, compared with 20 °C higher at a bare soil site. As the forest in the gap goes through the building phase, the microclimatic stratification characteristic of the mature phase will be progressively re-established.

Fig. 8.9 Hemispherical photographs of closed mixed dipterocarp forest and canopy gaps of different sizes at Danum, Sabah. (With kind permission of N.D. Brown.) A Closed canopy forest (control plot 2), % canopy gap = 3.8. B Gap 2 (small gap), % canopy gap = 8.1. C Gap 4 (large gap), % canopy gap = 29.7. For explanation of the technique, see text (pp. 215–17).

Table 8.3. *Microclimate of gaps and adjacent closed canopy forest over a period of a year at Danum Valley, Sabah*

Plot	Percentage canopy gap	Mean photosynthetically active radiation (mol m^{-2} d^{-1})	Air temperature (°C)		Mean soil temperature (°C)	
			Mean maximum	Mean minimum	10 mm depth	50 mm depth
Gap 4	29.7	19.2	38.2	21.6	33.4	29.0
Gap 6	17.8	13.5	34.6	21.3	27.4	27.4
Gap 1	17.4	14.4	35.3	21.5	31.5	27.0
Gap 5	12.6	9.9	35.4	21.4	31.3	25.4
Gap 7	10.6	10.6	34.9	21.3	—	—
Gap 9	8.4	3.1	33.6	21.5	29.4	26.0
Gap 2	8.1	4.9	33.5	21.5	30.0	25.6
Gap 8	7.3	6.4	34.2	21.3	28.9	25.9
Gap 3	6.4	3.1	31.1	21.3	29.4	25.8
Gap 10	6.1	2.3	30.3	21.2	29.5	25.6
Forest 1	2.4	1.0	—	—	27.0	25.2
Forest 2	3.8	1.1	28.4	21.2	25.2	24.4

Source: Brown (1990).

8.8 General hydrological features of rain-forest regions

The hydrological cycle of tropical rain-forest regions is characterized by high rainfall inputs and high evapotranspiration losses. River runoff increases with annual rainfall but usually accounts for a smaller percentage of rainfall than does evapotranspiration. Except in areas of impermeable lithology (such as shales and mudstones) and impermeable or shallow soils, overland flow on slopes is normally infrequent and localized because of the very high permeabilities of most rain-forest topsoils. Most water reaching the forest floor enters the soil, where it is either stored and used in transpiration or drains as throughflow, soil pipeflow or groundwater to stream channels. Streamflow responses even to heavy rainstorms tend to be small compared with other environments because of the dominance of subsurface flows. Because of the frequent rainfall and rarity of long dry periods, soil water shortages and serious plant water stresses tend to be infrequent except in areas with shallow or excessively drained soils.

8.9 Rainfall interception, throughfall and stemflow

Interception is that fraction of precipitation which falls on (i.e. is intercepted by) plant surfaces and is evaporated without reaching the ground surface. Throughfall is the fraction that reaches the ground surface by falling through the canopy either undisturbed or by dripping or splashing from vegetation surfaces. Stemflow is that part of rainfall that reaches the ground by flowing down the trunks of trees or the stems of other plants.

These three processes are of ecological significance for several reasons: (1) intercepted rainfall is normally regarded as a 'loss' as it fails to replenish soil moisture and hence is unavailable for uptake by roots in transpiration or for generation of streamflow; (2) intercepted water stored in organic accumulations, however, is used in transpiration of many epiphytes; (3) throughfall and stemflow play major roles in nutrient cycling by washing nutrients from plant surfaces to the ground and determining the spatial pattern with which nutrients arrive at the ground surface (see section 8.13); (4) drip-tips and stemflow also tend to concentrate water and energy at particular points on the ground and can lead to significant soil erosion (see section 8.13); (5) because intercepted water is more easily evaporated (mainly because of the large wetted surface area and its presence on the upper rather than lower surface of leaves) than transpired water, it is sometimes considered that in very rainy or misty environments, high interception may restrict transpiration.

Interception losses are assessed by measuring the amount of rainfall that penetrates the vegetation canopy and reaches the ground as throughfall or stemflow, and

Table 8.4. *Stemflow yields from different trees in the montane forest at 1000 m on Mt Bellenden Ker (Queensland)*

Species	Crown area (m²)	Basal area (m²)	Stemflow[a]		Funnelling ratio[b]
			litres	(%)	
Balanops australiana	27.2	0.061	53 384	25.2	112
Balanops australiana	24.4	0.120	6 239	3.3	7
Cardwellia sublimis	36.2	0.127	10 829	3.8	11
Ceratopetalum virchowii	26.2	0.049	38 045	18.6	100
Ceratopetalum virchowii	21.0	0.167	42 993	26.2	33
Ceratopetalum virchowii	38.5	0.147	23 019	7.7	20
Elaeocarpus foveolatus	22.7	0.182	70 319	39.7	50
Elaeocarpus sp.	45.8	0.159	11 332	3.2	9

[a] Percentage stemflow was calculated using the data of Herwitz as the percentage of rainfall falling on the crown of the tree flowing as stemflow.
[b] The funnelling ratio is the ratio of stemflow to rainfall falling on the trunk cross-sectional area.
Source: Herwitz (1986b).

then subtracting this from rainfall measured in a clearing or above the forest canopy. Throughfall is measured by siting water collectors (raingauges, troughs or tanks) beneath the forest canopy. Stemflow is separately assessed using either collars or open plastic bags coiled round the trunks of trees. Throughfall and stemflow (and hence also interception) are more difficult to measure in tropical rain forests because of the sampling problems associated with their complex species composition.

Throughfall. The rain-forest canopy not only reduces the amount of rainfall reaching the ground but also redistributes it. Throughfall varies across the forest floor, with local concentrations below drip points from canopy branches and leaves. In small, low-intensity rainfalls, throughfall tends to increase away from the trunks of trees. In large tropical rainstorms (>25 mm d^{-1}) work by Herwitz (1987b) in rain forests in northeast Queensland demonstrated, however, that throughfall is more variable close to tree trunks than beneath the periphery of tree crowns. Close to tree trunks lower throughfall occurred below branches acting as shelters, whereas the high throughfall catches were the result of flow down branches detaching itself, sometimes as concentrated waterspouts, before reaching the trunk. This concentration affects the patterns of nutrient supply and soil erosion (section 8.13) on the forest floor.

Stemflow. Studies in lowland rain forests in Ghana (Nye 1961), Tanzania (Jackson 1971), Venezuela

(Steinhardt 1979), Malaya (Manokaran 1979) and Amazonia (Franken *et al.* 1982) have all found that less than 1% of annual rainfall travels to the ground down the trunks of trees. However, stemflow tends to increase with storm size. In Jackson's study in the West Usambara Mountains (Tanzania), stemflow ranged from zero for rainstorms of less than 10 mm to 0.9 mm (about 2%) for storms of 40–50 mm. Similarly at Pasoh (Malaya), where stemflow was measured at eight trees in a 100 m² plot daily throughout 1973, stemflow varied with storm size from zero to 2.65% of rainfall and occurred on 133 out of 166 days with rain (Manokaran 1979). Stemflow varied between trees in the plot; of the eight trees monitored, the two largest trees, *Dipterocarpus cornutus* and *Shorea leprosula*, accounted for 19% and 10% respectively of total plot stemflow, whereas one of the smallest trees, *Knema malayana*, accounted for 53%.

Stemflow may be of greater significance in montane rain forests, where trees tend to branch lower down and often have upwardly inclined branches that are more capable of funnelling rainwater towards the trunks of trees. Herwitz (1986b) measured stemflow in a montane forest, 15–25 m high, classified by Webb (1959) as 'simple microphyll vine forest', at 1000 m altitude on the slopes of Mount Bellenden Ker in Queensland. Mean annual rainfall was estimated at 6500 mm. During nine months of monitoring in the wet seasons of 1981 and 1982, 13.5% of the total rainfall of 7800 mm was recorded as stemflow. Amounts were so great that surface flow was generated at the bases of the trees leading to significant local erosion. Stemflow amounts, however, varied greatly from tree to tree even for the same species (Table 8.4). Two *Balanops australiana* trees, although of similar crown area, generated stemflow of 53 384 and 6 239 litres and gave funnelling ratios (ratio of stemflow volume to the rainfall volume falling on the tree-trunk basal area) of 112 and 7, respectively. One consequence of the stemflow was the concentration of nutrient supply around the base of the trees (see section 8.13). Although the massive volumes of stemflow are partly the result of very high rainstorm amounts and intensities of the Mount Bellenden area, which on average experiences 15 daily rainfalls exceeding 100 mm per year and lies in the cyclone belt of Queensland, the main reason was considered to be the converging, steeply upwardly inclined branches of the trees.

Interception. Interception losses in rain-forest areas (Table 8.5), when expressed as a percentage of rainfall, average around 20%, but range widely from 7% at Reserve Ducke (Amazonia) (Lloyd & Marques 1988) to 37–42% at Pasoh (Malaya) (Low 1974). It is significant that in absolute terms interception losses tend to increase with annual rainfall and that percentage loss shows no evidence of decline in very wet rain forests. In addition, interception at sites of similar annual rainfall varies considerably (Fig. 8.10). For explanations of these points, the nature of the interception process and the factors influencing it need to be considered in more detail.

Jackson (1971, 1975) examined interception losses associated with individual rainstorms over a six month period in rain forest in the West Usambara Mountains (Tanzania). He showed (Table 8.6) that percentage interception loss declined sharply with storm size once the interception capacity (calculated at just over 2 mm) was exceeded. Thus the interception loss of a 1 mm shower was 80% (0.8 mm), compared with 5.5% (2.2 mm) for a 40 mm storm. Because evaporation rates of intercepted water during a storm tend to be low, the magnitude of interception per storm tends to be limited by the interception capacity of the vegetation.

The clear implication of Jackson's results is that the *frequency* of rainstorms (rather than rainfall amount, intensity or duration) and the interception capacity of the vegetation are the main factors controlling interception losses. For areas of similar vegetation and annual rainfall, therefore, interception losses should be higher in areas experiencing frequent short-lived showers separated by sunny intervals than in areas where most rain falls in a smaller number of prolonged storms of high intensity. This may partly explain the high interception

losses (Table 8.5) of the montane rain forests of Puerto Rico (Kline & Jordan 1968) and Mauritius (Vaughan & Wiehe 1947), both of which are characterized by showery trade-wind climates. The results from central Dominica (annual rainfall 5204 mm) show how high in absolute terms interception losses can be in a very wet locality where rainfall frequency averages 3.7–4.1 events per day, sunshine averages 5 h per day and there is a constant trade-wind breeze (Walsh 1987).

Attempts have been made to relate interception losses to measures of vegetation density. Crowther (1986), for example, compared mean interception loss with weighted Basal Area Index (a measure of the trunk occupance of ground per unit area of forest) for rain forests on limestone terrain at various sites in Malaya (Table 8.7). The two densest forests were characterized by the highest interception losses of 32.5 and 34.3%, but the overall correlation was not perfect. Herwitz (1985) has questioned both the attention given to interception by foliage rather than by stem surfaces and whether interception capacity is quickly reached in storms. Using a combination of field and laboratory measurements, he assessed the rainwater interception storages of individual mature canopy trees in tropical Queensland and showed that bark storage either equalled or exceeded leaf surface storage (Table 8.8). Interception capacities (2.2–8.3 mm rainfall depth), particularly for flaky-barked species, are above previously reported values and Herwitz points out that, because bark absorption is a relatively slow process, these storage capacities are filled relatively slowly and that interception continues to occur in prolonged storms for longer than Jackson (1971) envisaged.

8.10 Transpiration

By definition, transpiration is the evaporative process by which plants lose water, mainly through stomata on their leaves. The water involved originates in the soil and is taken up by the roots and then via the xylem to the branches, twigs and leaves. Transpiration losses are an inevitable consequence of carbon assimilation and photosynthesis, because the paths by which water is lost from plants are the same as those by which carbon dioxide is taken in. Although mineral uptake and distribution by the transpiration stream is also involved, this is not now regarded as a significant influence on transpiration rates.

Two of the principal objectives of transpiration studies in rain forests have been to assess:

(1) Transpiration rates and water requirements of individual rain-forest plants at different stages of their

Table 8.5. *Rainfall interception, throughfall and stemflow in tropical rain-forest areas*

Key: $, extrapolated estimate from short period study; *, throughfall figure includes measured stemflow <1%; RF, tropical rain forest; MRF, montane rain forest; SF, seasonal forest.

Locality	Vegetation	Length of study (months)	Rainfall (mm)	Throughfall (mm)	Interception (mm)	Interception (%)	Annual rainfall (mm)	Annual interception (mm)	Source
West Usambara Mts (Tanzania)	RF	6	839	702*	137	16	1338$	218$	1
Banco (Côte d'Ivoire)	RF	36	nd	nd	nd	12	1800	215	2
Yapo (Côte d'Ivoire)	RF	24	nd	nd	nd	23	1950	440	2
Kade (Ghana)	RF/SF	nd	nd	nd	nd	16	1649	264$	3
Mauritius	MRF	12	3094	2152	942	30	3094	942	4
El Verde (Puerto Rico)	MRF	7	1840	1370	470	26	3154	806$	5
Wet Area (Dominica)	RF	1	271	199	72	27	5204	1382$	6
Hacienda Paiva (Colombia)	RF	6	673	526	147	22	1600	349$	7
Venezuela	RF	12	1576	1260*	316	20	nd	nd	8
Manaus area (Amazonia)	RF	12	nd	nd*	nd	22	nd	nd	9
Ducke Reserve (Amazonia)	RF	24	4804	4376*	428	9	2391	273 (11.4%)	10
Pasoh (Malaya)	RF	12	2382	1863*	519	22	2382	519	11
Sungei Gombak (Malaya)	RF	6	nd	nd	nd	19	2500	476$	12
Janlappa (Java)	RF	18	4480	3484	996	20	2987	591	13
Western Java	RF	12	3007	2394	613	20	3007	613	14
Danum (Sabah)	RF	12	3103	2699*	569	17	3103	569	15

Sources: 1, Jackson (1971); 2, Huttel (1975); 3, Nye (1961); 4, Vaughan & Wiehe (1947); 5, Kline & Jordan (1968); 6, Walsh (1987); 7, Vis (1986); 8, Steinhardt (1979); 9, Franken *et al.* (1982); 10, Lloyd *et al.* (1988); 11, Manokaran (1979); 12, Kenworthy (1971); 13, Calder *et al.* (1986); 14, Gonggrijp (1941, in Bruijnzeel 1983): 15, Sinun *et al.* (1992).

Table 8.6. *Rainfall interception loss and storm size under tropical rain forest, West Usambara Mountains,*
Tanzania

Storm size (mm)	1.0	2.5	5.0	7.5	10.0	15.0	20.0	30.0	40.0
Interception (mm)	0.8	1.1	1.4	1.6	1.7	1.8	1.9	2.1	2.2
Interception (% of rain)	80	44	28	21	17	12	9.5	7.0	5.5

Source: Jackson (1971).

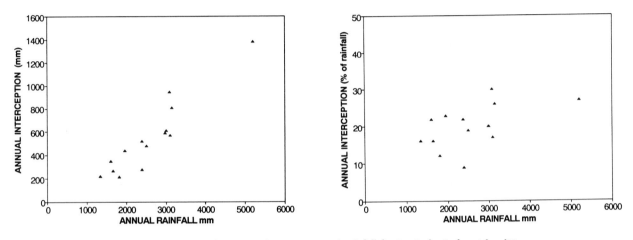

Fig. 8.10 Interception losses in relation to annual rainfall for tropical rain-forest localities.

life cycles and positions within the forest. Such data can be used to assess their 'transpiration' (and photosynthetic) competitiveness in different climatic and edaphic environments and perhaps help to explain variations in their relative abundance, both locally and regionally, between rain-forest communities.

(2) Transpiration rates and regimes of rain-forest stands as a whole and how they vary with climatic factors, soil factors and vegetation attributes such as forest structure, density and species composition.

Studies have been carried out at different spatial scales (ranging from the individual leaf or plant to an entire forest stand) using both field measurement techniques and laboratory experiments. The results of many laboratory or field experiments are often difficult to interpret because of the artificial or unrepresentative environmental conditions created by the experiments. In addition, results obtained at the leaf or plant scale are often difficult to extrapolate to the larger-scale context of the forest as a whole. Estimation of forest transpiration by

simple summation of measured transpiration rates of sample forest trees is unreliable because rates vary not only between species but also for the same species when in the understorey, in a gap or as a canopy tree. Furthermore, transpiration rates are expressed in different ways. Leaf and plant results are often given in grams (or litres) per unit leaf surface area or per plant, whereas at the forest scale transpiration rates are expressed per unit ground area. Rates cannot be converted from one system to the other unless leaf-area index and plant canopy overlap information are also available. Detailed consideration of transpiration at the individual plant scale is beyond the scope of this book; the interested reader is referred to recent plant physiology texts on water relations (e.g. Sutcliffe 1979, Kramer 1983, Medina *et al.* 1984). Here only the main influences on rain-forest transpiration rates and the results of some recent field studies are reviewed.

There are three main factors influencing transpiration rates of rain forests: (1) the drying power of the air

Table 8.7. *Rainfall interception loss and forest density on limestone terrain in the Malay Peninsula*

Locality	Topographic unit	Basal area index (%)	Interception loss (as % of rain)
Boundary Range, Perlis	hillslope	0.70	34.3
Kinta Valley, Perak	footslope	0.54	32.5
Anak Batu Takun, Selangor	footslope	0.38	26.9
Boundary Range, Perlis	footslope	0.25	18.9
Kinta Valley, Perak	hillslope	0.22	27.6
Anak Batu Takun, Selangor	footslope	0.11	24.6

Source: Crowther (1986).

Table 8.8. *Average storage capacities of intercepted rainfall of five canopy tree species of tropical rain forest in the Atherton Tableland (Queensland)*

F, Flaky-barked tree; S, smooth-barked tree.

Species	Sample size (n)	Mean crown area (m²)	Leaf surface[a] storage (l)	Bark storage (l)	Total[a] storage (l)	Interception storage[a] (mm)	
						Still-air	Turbulent
Aleurites moluccana (S)	10	62	55	81	136	2.2	1.6
Argyrodendron peralatum (F)	10	64	150	378	528	8.3	6.8
Castanospermum australe (S)	10	40	43	67	110	2.8	2.0
Dysoxylum pettigrewianum (F)	10	79	183	189	372	4.7	3.2
Toona australis (F)	11	38	38	115	153	4.0	3.4

[a] Leaf and interception storage capacities are for still-air conditions; in turbulent air conditions leaf storages (but not bark storages) are reduced and interception storages are correspondingly lower.

Source: Herwitz (1985).

(a function of radiative heat, saturation (or humidity) deficit and wind speed and turbulence); (2) vegetation attributes; and (3) soil moisture availability. Vegetation attributes, some of which may change with season or in dry periods, include variables of the individual leaves (leaf area, orientation, size, shape, surface characteristics, regime of stomatal opening, stomatal frequency and size), of the whole plant (rooting characteristics, plant leaf-area index and aggregated data for its leaves) and at the vegetation or plant community scale (e.g. canopy roughness, vegetation leaf-area index and aggregated data at the other two scales). Apart from leaf fall, the primary dynamic plant 'control' of transpiration is the stomatal (or internal) resistance, where the degree of opening of stomata is influenced by various factors (depending on species) such as light, humidity, temperature and water deficit.

In general, transpiration rates of rain forests are high because of the combination of high leaf area, soil moisture availability and high temperature throughout most or all of the year. Because the stomata of many plants open and shut in response to changing light conditions, transpiration tends to follow the same daily pattern as photosynthesis, with high rates during the day and very low rates at night. This pattern is reinforced by the diurnal pattern of relative humidity with saturated air conditions at night but substantial saturation deficits, particularly at canopy level, during the day (see Chapter 7). Both light intensity and the factors influencing the drying power of the air vary with position within the forest, with daytime values much higher at canopy level and within canopy gaps than in the forest undergrowth. Transpiration rates, therefore, vary greatly with the exact position of a plant within a forest. Although

Table 8.9. *Transpiration rates of trees in lower montane rain forest (Puerto Rico) using the tritium injection method*

Species	Profile position	Height (m)	d.b.h. (cm)	Basal area (cm²)	Mean daily transpiration		Mean water residence time (d)
					(l)	(l cm⁻² basal area)	
Dacryodes excelsa	Canopy	19.8	54.9	2367	372	0.157	11.0
Sloanea berteriana	Canopy	18.3	26.8	564	140	0.248	3.9
Dacryodes excelsa	Understorey	7.3	5.9	27	1.75	0.064	9.6

Source: Kline *et al.* (1970).

daily transpiration amounts vary with sunshine levels, windiness and humidity, the process is more continuous than rainfall interception. Formulae such as that of Penman (1948, 1963) and later refinements have been used extensively in transpiration estimation but recent field measurements of transpiration have demonstrated the need for their reassessment for use in rain-forest areas (q.v.).

The tritium tracer technique has been used to assess transpiration rates of both individual rain-forest trees and rain-forest plots (Kline *et al.* 1970, Jordan & Kline 1977). This involves (1) injecting tritiated water of known specific activity into a tree at points about a metre above ground level, (2) collecting leaf samples from the canopy at three-hourly intervals for a first week and then at progressively longer intervals up to 70 days, (3) analysis of the tritium content of these samples and (4) construction of a specific activity/time curve, which is then integrated to yield an estimate of flow or transpiration. In eastern Puerto Rico, transpiration rates of two canopy trees and one understorey tree were assessed in a somewhat low stature lower montane rain forest (Table 8.9). Transpiration rates not only increased with size of tree, but were also much higher (when expressed in per unit basal area terms) for the two canopy trees (0.16–0.25 l cm⁻² d⁻¹) than for the understorey tree (0.06 l cm⁻² d⁻¹). This latter finding tallies with transpiration experiments around individual canopy and understory leaves in the same area (Odum *et al.* 1970b) and with the lower wind speeds, more humid air and reduced solar radiation of the lower forest.

The same method was used to measure transpiration of twenty individuals of rain-forest trees of several species at a site 4 km east of San Carlos (Venezuela) (Jordan & Kline 1977). The site has a Superwet climate,

with an annual rainfall of 3500 mm and a Perhumidity Index of 24. In this experiment, the trees were injected on 11 September 1975 and twigs were sampled for eight days and then weekly for six weeks. The weather during the period was representatively rainy, averaging 6.9 mm d⁻¹ (or over 200 mm per month). Transpiration rates varied from 2.7 l d⁻¹ for small understorey trees to 1180 l d⁻¹ for a very large canopy tree. Although there is considerable scatter in the results, transpiration was found to increase linearly with size of tree as measured by cross-sectional trunk area at breast height and the values found for trees in the Puerto Rico study also fall within the same range (Fig. 8.11a). However, Jordan & Kline found a closer, and apparently linear, relationship between transpiration rate and sapwood area (Fig. 8.11b), where the latter was determined using an increment borer and recording the distance to heartwood from the bark of the tree. There was no evidence (though the number of trees sampled was small) of significant differences in transpiration rate between species.

Jordan & Kline (1977) measured the canopy area occupied by each tree and then estimated forest transpiration by averaging the transpiration rates per unit canopy area of individual trees; this yielded a rate of 2.76 mm d⁻¹ (or around 84 mm for a 30.4 day month). However, this method of calculation is arguably unsound as it failed to give greater weight to the larger trees with their greater crown areas. A recalculation weighting trees according to their crown areas yields the somewhat higher value of 3.18 mm d⁻¹ (or 96.7 mm per month). With both methods, however, there is a problem of overlapping tree crowns. A better approach, suggested by Jordan & Kline, would be to measure the sapwood area of all trees in an enumerated plot and to use the the sapwood area – transpiration rate correlation (Fig.

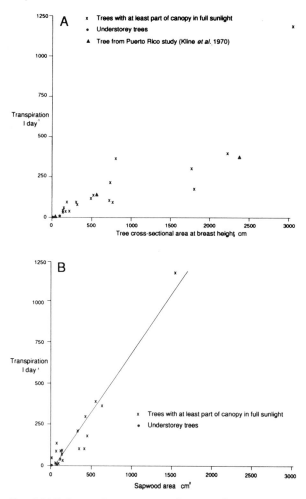

Fig. 8.11 Relation of transpiration of trees to the cross-sectional area of their trunk at breast height in rain forest at San Carlos, Amazonas, Venezuela. After Jordan & Kline (1977, Table 1 and Fig.2). Transpiration was measured by the tritium injection technique (p. 223). Sapwood area was measured from the inner limit of the sapwood-heartwood transition in trees where a transition existed. The regression equation for the transpiration (*T*)/sapwood area (*S*) relation is $T = 0.73S - 31$ ($r = 0.96$).

8.11b) to predict transpiration rates for all the trees of the plot. Forest transpiration would then be the sum of the transpiration rates of all trees (in litres per day) divided by the enumeration plot area.

A similar average transpiration rate of 3.45 mm d^{-1} (or 104.9 mm per month) was obtained for lowland rain forest near Manaus by calculating fluxes of water vapour above the canopy using data recorded by micrometeorological sensors located within and above the 35 m canopy on a 45 m tower (Shuttleworth *et al.* 1984a). It was found that measured daily evapotranspiration from the

well-watered and actively transpiring rain forest on eight fine days in September 1983 was consistently well below potential evapotranspiration (PE) values calculated using various formulae (Table 8.10). Actual evapotranspiration (which can be assumed to be transpiration, as there would be no intercepted water to evaporate on fine, sunny days) ranged from 1.75 to 4.22 mm d^{-1} compared with predicted PE values of 2.02 to 6.72 mm d^{-1}. The PE formulae were over-estimating transpiration, probably because they are calibrated to include evaporation of intercepted water, of which on fine days there is little or none to evaporate.

In the Janlappa Nature Reserve (Java), transpiration was measured using neutron soil moisture probes and diffusion porometry (Calder *et al.* 1986). Data for 1980–81 yielded a mean transpiration rate of 2.6 mm d^{-1} (or 79 mm per month). Annual transpiration was 949 mm (or 34% of the annual rainfall of 2755 mm). Transpiration can also be estimated, using a catchment water balance approach, by subtracting interception from evapotranspiration (Kenworthy 1971). Kenworthy's study in the Central Range of the Malay Peninsula and nine studies reviewed by Bruijnzeel (1989) yield annual transpiration estimates for lowland rain forest of 885–1300 mm (or 2.4–3.6 mm d^{-1}). As all the transpiration estimates reviewed above are within the range 73–108 mm per month, they give some support to the use of the 100 mm 'dry month' definition in the analysis of dry period sequences in rain-forest climates (pp. 165 and 189).

Although such values may be representative of rain forests on soils with good moisture retention in areas with no marked dry season, transpiration is lower in montane rain forests and rain forests subject to seasonal or frequent episodic water shortages for climatic or edaphic reasons. Thus Bruijnzeel (1989) in his review cites annual transpiration estimates of 600 mm for semideciduous forest, 560–830 mm for lower montane forests and 285–515 mm for montane cloud forests. Difficulties in estimating occult precipitation (condensation of cloud droplets on vegetation) affect the reliability of the cloud-forest studies, but the results of a recent study on G. Silam (Sabah), which used a more reliable soil water budget approach and yielded a transpiration rate of 0.85 mm d^{-1}) (or 310 mm a^{-1}) in the cloud-shrouded upper montane forest (870 m altitude), compared with 2.1 mm d^{-1} (766 mm a^{-1}) in the cloud-free lower montane forest (670 m altitude) (Bruijnzeel *et al.* 1993), suggest that they are probably of the right order.

These lower transpiration rates are the outcome of an interplay not only of climatic and edaphic factors but also of vegetational attributes (Fig. 8.12), which

Table 8.10. *Actual evaporation (mostly transpiration) and water equivalent of radiation for Amazonian rain forest near Manaus for eight fine days in September 1983 and comparisons with calculated evapotranspiration rates using various estimation formulae*

P, Penman (1948); TO, Thom–Oliver (Thom & Oliver 1977); PT, Priestley–Taylor (Priestley & Taylor 1972); E, Equilibrium (McNaughton & Jarvis 1983).

Date	Actual evaporation (mm)	Water equivalent of radiation (mm)	Calculated evaporation rates (mm d^{-1})			
			P	TO	PT	E
Sep 6	3.11	5.16	5.21	5.35	5.02	3.99
Sep 7	4.22	5.80	6.12	6.51	5.57	4.46
Sep 9	3.32	4.49	4.74	5.04	4.32	3.27
Sep 10	1.75	2.74	2.71	2.74	2.54	2.02
Sep 17	4.18	6.12	6.38	6.74	6.00	4.76
Sep 18	3.76	4.52	4.62	4.79	4.38	3.48
Sep 25	3.70	5.79	5.78	5.87	5.63	4.46
Sep 27	3.54	5.04	4.85	4.84	4.90	3.89
Mean	3.45	4.96	5.05	5.24	4.80	3.80

Source: Shuttleworth *et al.* (1984a).

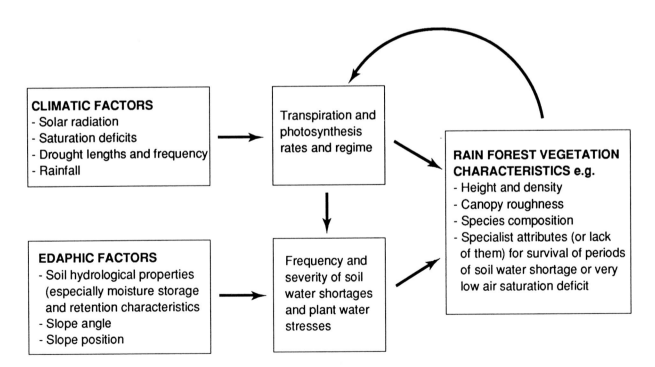

Fig. 8.12 Interrelations between climatic and edaphic factors, rain-forest character, transpiration and photosynthesis rates and plant water stresses.

Table 8.11. *Forest type and structure, aerodynamic roughness length, evapotranspiration, soil characteristics and drought susceptibility near Kuching, Sarawak*

Aerodynamic roughness length (z_0) was calculated using the equation $z_0 = \log(h-d) - 0.98$, where h=height of canopy and d= height of zero plane. Water availability is expressed as a depth in millimetres.

Forest type	Aerodynamic roughness length (cm)	Soil type	Water availability (mm)		Estimated ET (mm)		Minimum time to permanent water stress (d)
			Soil	Crop	Daily	Annual	
Mixed dipterocarp forest height 45 m (irregular canopy)	209	Red–yellow podzols (modal and clay kaolisols)	180	20	4.5–5.5	2000	33
Kerangas forest height 28 m (streamlined canopy)	31–66	Shallow humus podzol	60	5	2.5	900	16

Source: Brünig (1970, 1971).

may act either to reduce or enhance transpiration in less favourable climatic or edaphic conditions. In forests with seasonal water shortages, leaf-fall in the dry season may contribute to reduced transpiration, but, as shown in Chapter 9, deciduousness of canopy tree species is not simply related to liability to drought.

In montane rain forests, transpiration rates are usually low because of (1) very low saturation deficits during the frequent periods when the forest is shrouded in cloud, (2) streamlined canopies and (3) lower forest densities (as measured by bole area or leaf area). The streamlined canopies of many montane forests have sometimes been interpreted as the result of the mechanical action of wind, but an alternative interpretation is that they are a response to episodic water shortages, which can develop very quickly in spells of sunny weather because of the shallow soils and steep slopes of mountain areas. Montane forests may be adapted therefore to occasional water shortages as well as frequent saturated air conditions (Grubb & Whitmore 1966). They may exhibit attributes that increase transpiration and photosynthesis in wet, cloudy conditions, such as larger and more frequent stomata (Cintron 1970), or glossy leaves and drip-tips to drain leaves efficiently, so that transpiration and photosynthesis are less hindered by evaporation of intercepted water, as well as attributes that reduce transpiration and susceptibility to plant water stress in dry periods (e.g. streamlined canopies and thick waxy leaves). The question of interrelationships between vegetation attributes and water relations of upper montane forests, however, remains far from resolved and for a recent discussion the reader is referred to Bruijnzeel *et al.* (1993).

Brünig (1970, 1971) calculated using theoretical equations that forest types of greatly contrasting canopy roughness found in a Superwet climatic environment near Kuching (Sarawak) would result in greatly differing annual evapotranspiration (and transpiration) and he interpreted the differences in canopy structure as being responses to variations in soil moisture storages and depletion times of different soils during dry periods (Table 8.11). Mixed dipterocarp forest with a very uneven canopy and towering emergents and an estimated annual evapotranspiration of 2000 mm is associated with red–yellow podzols (modal and clay kaolisols[1]) with very high soil water availability (180 mm), whereas the poorest type of heath forest with a lower, smoother canopy and a predicted annual evapotranspiration of 900 mm is associated with shallow humus podzols with a much smaller soil water availability (60 mm) (see Chapter 12).

8.11 Evapotranspiration

Evaporation from all surfaces of the soil–plant complex is commonly termed 'evapotranspiration' (Jackson 1989, p. 146) and therefore comprises transpiration, interception and direct evaporation of water from the soil, surface pools and swamps. There are three approaches to estimating evapotranspiration: (1) summation of measurements of each component process; (2) theoretical or empirical formulae (including transformations of

---

[1] Nearest soil category of classification of Chapter 10 (see pp. 257–63).

evaporation pan data); and (3) catchment water balance. Because of the difficulties of obtaining accurate and spatially representative measurements, particularly of transpiration, estimating evapotranspiration by summing its components has rarely been achieved. Exceptions are the studies carried out at Janlappa Forest Reserve, Java (Calder *et al.* 1986) and Reserve Ducke (Lloyd *et al.* 1988, Shuttleworth *et al.* 1984a). At Janlappa, out of an annual rainfall of 2755 mm, evapotranspiration was estimated at 1500 mm, comprising 949 mm transpiration and 551 mm interception. Theoretical formulae for estimating PE, such as those of Penman (1948, 1963), have been widely used, but tend to produce inaccurate results partly because the equations, though appropriate for the transpiration component, fail to model interception. As shown in the previous section, such formulae in Amazonia tend to underestimate evapotranspiration in wet weather, when interception is important, and to overestimate evapotranspiration in dry weather, when transpiration alone is occurring (Shuttleworth *et al.* 1984a). If formulae are used, then the interception and transpiration components need to be predicted using separate equations and summed to obtain evapotranspiration. The success obtained, however, will only be as good as the data used.

An alternative method of assessing rain-forest evapotranspiration is the catchment water balance approach. For catchments with accurate long-term rainfall and streamflow records and where the underlying rocks are sufficiently impermeable and the catchment sufficiently large for groundwater outputs to be assumed negligible compared with stream discharge, simple water balance calculations of evapotranspiration can be made using the formula below:

$$ET = P - Q \pm {}^*S,$$

where ET = evapotranspiration (mm), P = catchment precipitation (mm), Q = runoff (or streamflow per unit area) (mm), $^*S$ = change in catchment storage. Although lags between rainfall and runoff make calculations on a monthly basis difficult, estimates of annual evapotranspiration are more reliable (Bruijnzeel 1983).

Catchment water balance data from tropical rain-forest regions are given in Table 8.12. Annual evapotranspiration is very high compared with other types of vegetation cover as a result of constantly high temperatures and rainfall, the lack of a marked dry season and the infrequency of significant or prolonged soil moisture deficits. However, although all catchments experience annual ET in excess of 1000 mm, amounts vary from 1079 mm for the Sungei Lui in Malaya to over 1700 mm for the upper Gombak in Malaya and the Melinau in

Sarawak. There appears to be no systematic variation in absolute losses with annual rainfall, but, when expressed as a percentage of annual rainfall, ET losses decline with increasing annual rainfall (Fig. 8.13). Absolute losses may vary in part because the climatic and vegetational factors controlling the interception and transpiration components of ET are rather different (see sections 8.9 and 8.10). Also, in Amazonian water balance studies, inaccuracies involved in flow and rainfall measurement and basin leakage (unmeasured groundwater flows) have been raised as possible sources of error (Rodda 1987, Bruijnzeel 1989).

The impact of rain-forest disturbance on evapotranspiration can be profound, but will vary considerably with the type of disturbance and the replacement land use. Logging, particularly if relatively light, may cause only a temporary and relatively small reduction in evapotranspiration, where this occurs chiefly through quicker runoff from the areas of compacted and eroded soil along skid tracks and log marshalling areas. Unless erosion and loss of nutrients are so severe and widespread as to restrict regrowth to scrub, then forest regrowth and soil recovery will soon lead to evapotranspiration losses little different from those of the original rain forest. On the other hand, replacement of forest by artificial grassland for cattle ranching (as in parts of Amazonia) would lead to a large and long-term reduction in evapotranspiration because of the lower interception and transpiration rates of grassland and the longer-term reduction in soil infiltration capacity and soil moisture and increase in overland flow. Where rain forest is replaced by agricultural crops, the impact will vary with the type of crop. In Malaysia, evapotranspiration (as estimated by rainfall minus streamflow) was only 995 mm in the Damansara catchment, which was covered by rubber plantations, compared with 1000–1200 mm in three adjacent rain-forest catchments (Low & Goh 1972).

8.12 Soil hydrology

A knowledge of soil hydrology in rain-forest environments is important in considering soil water availability for transpiration, with its implications for forest structure, density and species composition (section 8.11), nutrient cycling and availability (section 8.13) and soil erosion (section 8.13). Detailed studies of soil hydrology are few in comparison to the great variety of soil and terrain types in rain-forest areas.

The very high permeability of most rain-forest topsoils means that overland flow on slopes is rare and is

Table 8.12. *Water balance in tropical forested catchments*

TRF, tropical rain forest; MRF, montane rain forest; ESF, evergreen seasonal rain forest.

Catchment	Catchment area (km²)	Vegetation	Length of record (years)	Annual rainfall (mm)	Annual runoff (mm)	Annual evapotranspiration (mm)	(%)	Source
Mbeya (Kenya)	0.16	MRF	10	1924	541	1381	72	1
Kericho/Lagan (Kenya)	5.4	MRF	13	2131	788	1337	63	1
Guma (Sierra Leone)	8.7	ESF	7	5795	4649	1146	20	2
North Oropouche (Trinidad)	—	ESF	2	3116	1905	1211	39	3
Luto/Barro Colorado Island (Panama)	—	TRF	3	2425	969	1455	60	4
Taruma-Acu (Amazonia)	—	TRF	2	2089	541	1542	74	5
Barro Branco (Amazonia)	—	TRF	nd	2075	400	1675	81	6
Sungei Lui (Malay Peninsula)	68.9	TRF	1	2156	1077	1079	50	7
Ulu Langat (Malay Peninsula)	74.1	TRF	1	2483	1219	1263	51	7
Sungei Langat (Malay Peninsula)	76.4	TRF	1	2822	1152	1670	36	7
Upper Gombak (Malay Peninsula)	0.3	TRF	1	2540	762	1778	59	8
Sungei Gombak (Malay Peninsula)	41.5	TRF	1	2588	1118	1468	70	9
Melinau (Sarawak)	250.6	TRF	1	5200	3457	1743	33	10
South Creek (Queensland)	0.26	TRF	6	4037	2616	1421	35	11

Sources: 1, Edwards & Blackie (1981); 2, Ledger (1975); 3, Walsh (1980a); 4, Dietrich *et al.* (1983); 5, Salati (1987); 6, Leopoldo *et al.* (1982); 7, Low & Goh (1972); 8, Kenworthy (1971); 9, Douglas (1968); 10, Walsh (1982c); 11, Gilmour (1977).

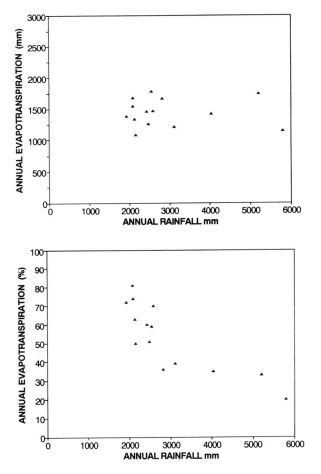

Fig. 8.13 Evapotranspiration losses in relation to annual rainfall for tropical river catchments. Partly based on Walsh (1987). Sources as in Table 8.12.

confined to small areas around channel heads and in valley bottoms where soil moisture converges and saturation occurs more readily. Thus, in Reserve Ducke near Manaus (Amazonia) (Nortcliff & Thornes 1981, 1988), the allophane latosolic (andosol)[1] and kandoid (modal kaolisol)[1] soil zones of Dominica (Walsh 1980c), the red-yellow podzolics (modal and clay kaolisols)[1] of the lower slopes of G. Mulu in Sarawak (Walsh 1982c) and ferrallitic soil (clay kaolisol)[1] areas on mica schists in French Guiana (Boulet et al. 1979), overland flow was absent from valley-side slopes and only recorded in valley bottoms and channel head hollows. Overland flow recorded by Leigh (1978a, b) at Pasoh was also at the base of slopes. No overland flow was observed on slopes even in intense downpours in rain forest in Puerto Rico (Lewis 1976) and in the more seasonal forest at Adiopodoumé (Côte d'Ivoire) (Dabin 1957). Runoff plot

[1] See pp. 257–63.

experiments in rain forest on Maracá Island (Amazonia) (Nortcliff et al. 1990) and at Danum (Sabah) (Sinun et al. 1992) also found that only 4–6% of rainfall travelled as overland flow. There are some exceptions, however. Widespread overland flow on slopes (produced by ponding of infiltrated rainwater above less permeable subsoils) was found to occur frequently at sites at Babinda (Queensland), where cyclone-related rainstorms, often exceeding 250 mm d^{-1}, were the cause (Bonell & Gilmour 1978), and in western Amazonia, where an impeding layer at an unusually shallow depth in the soil was responsible (Elsenbeer & Cassells 1990). Frequent overland flow, but resulting not from soil saturation but from an anomalously low (7–12 mm h^{-1}) infiltration capacity, has also been reported in the Tai rain forest of Côte d'Ivoire (Wierda et al. 1989).

In most rain-forest areas, therefore, water arriving at the ground surface generally enters the soil and either (1) replenishes soil moisture storage and is taken up and transpired by the vegetation, (2) percolates to the groundwater table, or (3) flows laterally as 'throughflow' or through soil tunnels as 'pipeflow' downslope towards stream channels. The relative importance of these hydrological pathways varies with the depth of the soil and its various horizons, changes in permeability with depth, and the water retention properties of the soil. Throughflow tends to be important in soils where a permeable layer overlies a horizon of much lower permeability, such as at the boundary between topsoil and subsoil (the A–B horizon boundary) and at the base of the soil. Substantial throughflow at these two levels has been recorded in studies in the allophane latosolics (andosols)[1] of Dominica (Walsh 1980b), the kaolisols of Babinda (Queensland) (Bonell & Gilmour 1978) and the red-yellow podzolics (modal and clay kaolisols)[1] of Mulu, Sarawak (Walsh 1982c), although the percentages of water which follow these routes varies. In montane rain-forest areas, throughflow can be of even greater importance. In the peaty gley (Tumau) soil at 1300 m altitude on G. Mulu (Sarawak), the contrast in permeability between the organic topsoil (2667 mm h^{-1}) and the underlying grey gleyed horizon (7.4 mm h^{-1}) and red-yellow subsoil (1.5 mm h^{-1}) leads to large amounts of rapidly moving throughflow in the organic layer during storms (Walsh 1982c); similar situations are reported on montane rain-forest slopes on G. Silam (Sabah) (Bruijnzeel et al. 1993) and in the very wet (>7000 mm) allophane podzolic zone of Dominica (Walsh 1980b, Rouse et al. 1986). Flow through soil pipes (or tunnels) has been reported in rain-forest localities in Sarawak (Baillie 1975, Walsh 1982c), Sabah (Sinun et al. 1992) and Dominica (Walsh & Howells 1988), where it accounted for at least 16–20% of baseflow, but its importance in generating runoff has never been properly assessed. A more detailed appraisal of slope hydrology in the humid tropics is given by Bonell with Balek (1993).

The availability of soil moisture to plants varies greatly between soils in the rain-forest zone with differences in soil depth, texture and organic matter content, as well as in relation to local drainage and topography. Generally, soil moisture reserves (and hence susceptibility to drought) will be less for shallow soils, excessively clayey or sandy soils, on steeply sloping terrain and in soils deficient in organic matter. Brünig (1970) presented data on available soil water and water stored in the vegetation on different soil types and rain-forest types in Sarawak (Table 8.11). Soil water availability varied from 180 mm for mixed dipterocarp forest on red-yellow podzols to 60–66 mm for shallow humus podzols and podzolic gleys and as high as 552 mm on deep humus podzols. Allowing similar rates of evapotranspiration, he calculated the minimum times to permanent water stress for each soil and forest type. Because transpiration rates tend to be much lower for forest types on the soils with low water availability (Table 8.11) and because transpiration rates tend to fall with decreasing soil moisture (Lockwood 1974, p. 202), actual times to water stress will be much longer. The significant point, however, is that the differences in water availability strongly influence forest structure and species composition.

Logging and clearance of rain forest radically change soil hydrology. Removal of the natural litter cover and ground vegetation allows raindrop impact to reduce infiltration capacities by destroying surface aggregate structure, loosening fine soil fractions and blocking soil pores. At Maracá Island, 16% of rainfall occurred as overland flow in completely cleared plots compared with 6% in undisturbed forest and partly cleared plots with the litter cover left intact (Nortcliff et al. 1990). Mechanical clearance methods have a more catastrophic impact because of the massive compaction of the soil that occurs. In a logged area of Sabah, tractor disturbance reduced infiltration capacities from 154 to 0.28 mm^{-1}, with such low rates persisting at least 5–9 years after logging (Malmer 1990). Subsoil permeability can also be affected, with reductions in porosity and bulk density recorded to a depth of 80 cm after bulldozer clearing in Surinam (Van der Weert 1974) and Dominica (Walsh 1980a). The generation of overland flow by soil saturation is rendered easier both by this subsoil compaction (Anderson & Spencer 1991) and by erosion of topsoil during logging.

8.13 Hydrological influences on nutrient movements and soil erosion in primary and disturbed rain forests

Hydrological processes play fundamental roles in the movement of nutrients into, within and out of rain-forest ecosystems. These are summarized in Fig. 8.14. Incoming rainfall not only contains nutrients, but becomes enriched as it washes minerals from vegetation surfaces and travels to the ground as throughfall and stemflow. Infiltrating and laterally moving water transports nutrients down the soil and slope profiles and transpiration facilitates the uptake and distribution of nutrients in plants. Nutrient losses occur through soil erosion, in solution in slope drainage waters into the streams, and, where forests have been burned, as gases and smoke particles into the atmosphere. This section considers some hydrological aspects of nutrient cycling and soil erosion in primary and disturbed forests. For more detailed examination of nutrient cycling and the impacts of forest clearance see also Chapter 10 and recent specialist texts and reviews (e.g. Jordan 1985, Vitousek & Sanford 1986, Proctor 1987, 1989, Brown & Lugo 1990, Anderson & Spencer 1991).

Nutrient fluxes in undisturbed rain-forest areas. Atmospheric inputs have been assessed by collecting and analysing rainfall samples. Comparisons of results between studies (Table 8.13) is difficult because of different methodology in rainfall sampling, notably as to whether bulk precipitation, i.e. including 'dry fallout' (dust etc.) into gauges between storms, or 'wet-only' rainfall is being collected. A major problem in montane environments is 'occult' deposition of cloud droplets and nutrients on plant surfaces in misty weather. As far as is known, all those results listed in Table 8.13 are of bulk precipitation. Atmospheric inputs vary greatly between sites. Inputs tend to increase with annual rainfall, as demonstrated in a local study on Mt Bellenden Ker in Queensland (Herwitz 1987a) and by the El Verde data, while sodium inputs (derived from sea salt) tend to be higher in maritime areas, but there are many anomalies. Potassium inputs at the sites in Malaya vary from 2.7 to 12.5 kg ha^{-1} a^{-1}. The high concentrations and inputs of calcium at his sites in Perlis, Perak and Selangor were ascribed by Crowther (1987) to limestone quarrying activities in those areas.

Details of the concentrations and fluxes of nutrients by throughfall and stemflow and of net canopy leaching (nutrient fluxes by throughfall and stemflow minus bulk deposition) for some rain-forest studies are given in

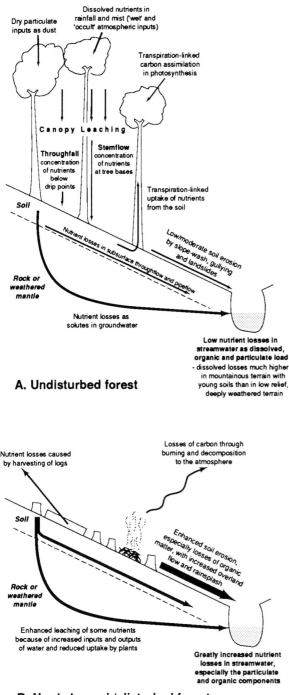

A. Undisturbed forest

B. Newly logged / disturbed forest

Fig. 8.14 Some of the roles of hydrological processes in nutrient movements in primary and logged rain forest.

Table 8.14. All studies show significant increases in concentrations in throughfall and stemflow (where measured) compared with rainfall, but the scale of canopy leaching varies greatly. Stemflow, though minor

in volume compared with throughfall and hence responsible for only a small proportion of leaching from the canopy (Table 8.14c), does concentrate nutrient supply at the bases of individual trees (McColl 1970, Clements & Colon 1975, Jordan 1978, Manokaran 1980). In the very wet montane rain-forest site on Mt Bellenden Ker (Herwitz 1986a), where stemflow volumes in heavy storms are unusually large, stemflow inputs of magnesium and potassium (expressed in per unit trunk basal area terms) in a 38 mm storm in 1982 were 5.55 g m^{-2} Mg^{2+} and 9.12 g m^{-2} K$^+$, rates that were 2–3 orders of magnitude higher than average throughfall per unit area inputs. The ecological impact of such spatial concentration of nutrient inputs (and less dramatic concentrations in lowland rain forests) remains unassessed for tropical forests, though in temperate forests their influence on soil chemistry (Gersper & Holowaychuk 1971), soil organisms and understorey herbs (Bollen et al. 1967, Crozier & Boerner 1984) has been demonstrated. Stemflow higher up the trunks of rain-forest trees may play a significant role in supplying nutrients to epiphytes and mosses.

The lack of overland flow in many rain forests noted earlier is very important to nutrient cycling, as it means that almost all the water reaching the forest floor infiltrates, thus giving maximum opportunity for the near-surface root mat to absorb nutrients washed down from the forest canopy or released from decaying litter on the forest floor. Furthermore there is now considerable evidence of direct uptake of K, Mg, Ca and P by roots from decaying litter (Herrera et al. 1978, Cuevas & Medina 1988, Medina & Cuevas 1989). Anderson & Spencer (1991) have pointed to growing evidence of the role of throughflow (lateral water flows through the soil) and river flooding in producing local differences in soil nutrient levels and rain-forest species composition. Thus, in the upper Rio Negro basin (Venezuela), Medina & Cuevas (1989) associated high base cation concentrations in highly leached sandy soils in valley bottoms with flushing of nutrients from the surrounding slopes. In Sarawak, Newbery & Proctor (1984) found Ca concentrations twelve times higher in soils subject to river flooding than on podzols above flood level with contrasting species composition in the two areas.

There are still few data on losses of nutrients from soils by leaching as indicated by dissolved material carried by streams. Coupled with the problems with measurement and contamination of atmospheric inputs, this severely limits the use of a rainfall input – streamwater output approach in assessing the degree of closure of rain-forest nutrient cycles. Anderson & Spencer (1991) stress the contrast in nutrient dynamics in rain-

Table 8.13. *Atmospheric input of nutrients as bulk precipitation at tropical rain-forest sites*

Locality	Annual rainfall (mm)	Concentrations (mg l^{-1})				Annual deposition (kg ha^{-1})				Source
		Ca	Mg	K	Na	Ca	Mg	K	Na	
El Verde (Puerto Rico)	3760	—	—	—	—	21.8	4.7	18.2	57	1
San Carlos (Venezuela)	3521					8.8	2.4	10.6	—	2
Central Amazonia	—	—	—	—	—	0.26	0.18	—	—	3
Pasoh (Malay Peninsula)	2382	0.18	0.03	0.29	0.98	4.2	0.7	6.4	23	4
Boundary Range (Malay Peninsula)	2089	1.02	0.09	0.13	0.26	21.3	1.9	2.7	5.4	5
Selangor (Malay Peninsula)	2441	1.48	0.14	0.15	0.21	36.1	3.4	3.7	5.1	5
Kinta Valley (Malay Peninsula)	2847	0.40	0.05	0.12	0.28	11.4	1.4	3.4	8.0	5
Mt Bellenden Ker (Queensland), summit	9140	0.20	0.22	0.09	1.84	16.1	18.7	8.5	155	6
Mt Bellenden Ker (Queensland), Wyvuri	4140	0.18	0.26	0.16	2.09	8.0	11.4	6.9	92	6

Sources: 1, Jordan (1970); 2, Jordan (1989); 3, Brinkmann (1983); 4, Manokaran (1980); 5, Crowther (1987); 6, Herwitz (1987a).

Table 8.14. *Canopy leaching in three tropical rain-forest localities*

(a) Lower montane rain forest at El Verde (Puerto Rico)

Process	Annual nutrient flux (kg ha^{-1})				
	Ca	Mg	K	Na	P
Rainfall (P)	21.8	4.7	18.2	57.2	94.1
Stemflow (S)	7.3	1.6	71.8	15.1	40.6
Throughfall (T)	23.9	6.8	80.1	58.2	100.9
Canopy leaching (S+T−P)	9.4	3.7	133.6	16.1	47.3

(b) Rain forest (six sites) on limestone terrain in Perak, Perlis and Selangor (Malay Peninsula). Stemflow was found to be insignificant

Process	Annual nutrient flux (kg ha^{-1})			
	Ca	Mg	K	Na
Rainfall (P)	11.4–36.1	1.4–3.4	2.7–3.7	5.1–8.0
Throughfall (T)	46.8–125.6	14.0–35.5	65.2–138.4	7.5–11.8
Canopy leaching (T−P)	25.5–98.9	12.6–32.1	62.5–135.0	−0.5–5.9

(c) Rain forest in central Amazonia

Process	Annual nutrient flux (kg ha^{-1})	
	Ca	Mg
Rainfall (P)	0.26	0.18
Stemflow (S)	5.79	2.05
Throughfall (T)	10.50	6.78
Canopy leaching (S+T−P)	16.03	8.65

Sources: (a) Jordan (1970); (b) Crowther (1987); (c) Brinkmann (1983).

forest systems on fertile or shallow, actively weathering soils compared with those on deep, thoroughly weathered regoliths. In rain forest on fertile soils, base cation (K, Mg and Ca) concentrations in soils are high, easily replenishable and accessible to vegetation and losses to streamwaters tend to be high. On thoroughly weathered soils in excess of 2–3 m depth, however, unweathered parent material is inaccessible to tree roots; base cations tend to be tightly locked into the vegetation – surface soil system and losses to streamwater are small, as in the eastern Amazon basin (Gibbs 1967). Rain forests on such 'closed cycle' soils are consequently more vulnerable to disturbance, as the nutrient pool, once lost through burning, soil erosion and leaching, is irreplaceable.

Soil erosion in undisturbed rain-forest areas. The principal soil erosion processes in rain-forest areas are rainsplash, overland flow erosion, subsurface tunnelling, soil creep and (in steeply sloping terrain) landslides. Rates of soil erosion by surface wash (which comprises detachment and downslope movement by rainsplash and overland flow) are given in Table 8.15. Rates under primary forest, which tend to increase with annual rainfall, are high compared with other forested zones of the world, mainly because of higher rainfall intensities and annual totals in the ever-wet tropics. Rates of splash erosion by raindrops are high, not only because of high rainfall intensities, but also because leaves and branches tend to concentrate throughfall at drip-points and the height of the canopy means that many large drops reach the ground at close to terminal velocities. Although the ground vegetation is usually sparse, the leaf litter and surface roots of the forest floor protect the soil to a large extent and keep rainsplash erosion rates well below those on bare ground.

Sinun et al. (1992) noted how soil animals provide areas of bare or loosened soil for detachment by splash or overland flow under rain forest at Danum Valley (Sabah). As many as 292 soil 'chimneys' (casts) produced by cicadas were counted in a 20 m^2 erosion plot on one rain-forest slope and differences in chimney frequency between plots were reflected in parallel contrasts in erosion rate. Localized overland flow produced by matted leaf litter (Ruxton 1967), by stemflow at the base of some rain-forest trees (Herwitz 1986b) and in easily saturated areas may also add to surface wash in some rain forests. Widespread overland flow on slopes, although not common (see section 8.12), was observed to contribute to erosion at Bukit Timah, Singapore (Chatterjee 1989) and Pasoh (Leigh 1978a), but was never recorded at a site in central Dominica with a much higher surface wash rate (7.6 mm a^{-1}), where high rainsplash due to the very high annual rainfall (5204 mm) was considered responsible (Walsh 1993).

The foregoing studies have all measured soil erosion under closed-canopy forest; erosion in canopy gaps has received little attention. Bursts of enhanced soil erosion certainly characterize fresh canopy gaps but are probably very short-lived, as the patches of bare ground are quickly covered by a profusion of vegetation. Where gaps have been caused by a landslide and much of the regolith has been stripped, however, vegetation may be much slower to recolonize and erosion rates may remain high for much longer, as on some landslide scars in Dominica following Hurricanes David and Frederic in 1979 (Reading 1986).

Effect of rain-forest disturbance on nutrient losses and erosion. The impact of rain-forest disturbance on nutrient losses and status varies with clearance or logging practices and the subsequent land use. Losses of nutrients may occur through timber harvesting, burning, leaching and soil erosion. Losses due to timber extraction itself tend to be minor. Thus, Poels (1987) found that with logging in Surinam, where about 8% of the timber was extracted, losses due to timber harvesting amounted to less than 3% of the pre-logging nutrient pools of all the elements examined (C, N, P, K, Ca, Mg). Although effects vary with the intensity of a fire, burning of biomass results in major losses of the more volatile elements of nitrogen, sulphur and carbon, but most other nutrients are deposited on the soil as ash. In 'slash and burn' clearance in wet rain forests in Costa Rica, Ewel et al. (1981) recorded losses of nitrogen, sulphur and carbon of 22%, 49% and 30%, respectively. Losses of other nutrients tend to be increased indirectly by fire, as the reduction in living vegetation cover increases the effectiveness of both leaching and soil erosion. Losses of nutrient stocks with clearance for agriculture will thus be much reduced if felled vegetation is allowed to decompose naturally, because of increased protection against erosion and more gradual release of nutrients to the soil and the replacement agricultural ecosystem.

Physical erosion of soil and particularly soil organic matter during logging or forest clearance for agriculture is potentially the most serious component of nutrient loss, although losses can be greatly reduced if techniques are used that retain as much leaf litter and soil organic matter on site as possible after tree removal. Effects have been measured at the slope scale using erosion plots and repeat survey stakes and at the catchment scale by monitoring changes in transport rates of suspended

Table 8.15. *Rates of soil erosion by surface wash (by rainsplash and overland flow) in rain-forest areas*

Key: TRF, tropical rain forest; ESF, evergreen seasonal forest; *, annual rate extrapolated from 3 months' data; +, mean rate calculated from Leigh's published data for individual sites.

Locality	Vegetation	Slope (degrees)	Technique	Rate of erosion by surface wash		Source
				(mm a^{-1})	(kg ha^{-1} a^{-1})	
Undisturbed rain forest						
Côte d'Ivoire	ESF	4–50	Stakes	1.5–3.5	–	Rougerie (1960)
Wet Area (Dominica)	TRF	17	Stakes	7.6	21 300	Walsh (1993)
Rupununi (Guyana)	TRF	10–15	Plots	0.28–0.32	7290–8550	Kesel (1977)
Maracá (Amazonia)	TRF	12–13	Plots	–	5120–9680*	Nortcliff et al. (1990)
Bukit Timah (Singapore)	TRF	20+	Stakes	4.0–4.7	–	Chatterjee (1989)
Pasoh (Malay Peninsula)	TRF	8–30	Pins	2.61+	–	Leigh (1978a)
Danum (Sabah)	TRF	7–40	Plots	–	155–253	Sinun et al. (1992)
Southeast Mindanao (Philippines)	TRF	14	Plots	–	46	Kellman (1969)
Disturbed rain forest						
Maracá: partly cleared, litter and undergrowth left intact		12–13	Plots	–	9480–20 680*	Nortcliff et al. (1990)
Maracá: completely bare ground		12–13	Plots	–	15 720–209 700*	Nortcliff et al. (1990)
Tawau (Sabah): clear-felled		32	Stakes	35.6	–	Liew (1974)
Southeast Mindanao (Philippines): logged-over forest		14	Plots	–	836	Kellman (1969)

Table 8.16. *Effects of logging on sediment yields*

(a) The impact of logging activities on the sediment yields of the Baru catchment, Sabah

Date	Logging activity	Ratio of monthly sediment yield of Baru to monthly sediment yield of undisturbed forest 'control' catchment
Pre-August 1988	Prior to logging	1 : 1
August–September 1988	Logging road construction	4 : 1
December 1988	Removal of trees 2 chains from road, including some high lead logging with some skid tracks	5 : 1
May–June 1989	Installation and then use of high lead machine for pulling logs. Main logging phase	18 : 1
August 1990	One year after logging had finished	3.6 : 1

(b) Suspended sediment yields for primary and logged rain-forest catchments in Malaysia

Catchment/locality	Vegetation	Catchment area (km^2)	Suspended sediment yield (t km^{-2} a^{-1})	Source
W8S5, Danum, Sabah	Tropical rain forest	1.1	312	Douglas *et al.* (1992)
Baru, Danum, Sabah	Logged	0.56	1600	
Sipitang, Sabah	Secondary rain forest	0.15	60	Malmer (1990)
	Logged	0.15	300	
Tekam, Pahang	Secondary rain forest	0.47	35	Malaysia Department of Irrigation (1986)
	Logged	0.47	660	
Bukit Berembun,	Logged (supervised)	0.13	22	Zulkifli *et al.* (1990)
Negri Sembilan	Logged (normal)	0.30	189	

Source: (a) Douglas *et al.* (1992).

sediment in rivers. Both have important limitations. Erosion plots and stakes can only measure erosion *after* rather than *during* logging or clearance as they would be destroyed by bulldozers and dragged logs; thus they may miss the main erosional phase. In addition, logged-over forest is very difficult to sample effectively, as it is a mosaic of patches of relatively undisturbed forest, patches where the canopy has been removed but the herbs, litter and saplings are left more or less intact, and a variety of areas of bare, compacted soil (logging tracks, roads, marshalling areas, etc.) (Anderson & Spencer 1991). Catchment monitoring, on the other hand, may underestimate the scale of erosion on slopes within a catchment because much of the eroded material may be stored at the base of slopes or choke headwater tributaries and thus fail to reach the catchment monitoring station.

At Maracá Island (Amazonia), hydrology and erosion at virgin forest, partly cleared and totally cleared plots of size 2 m × 5 m were monitored over a three-month period in 1987 at top, middle and base of slope positions (Ross *et al.* 1990). The terra firme forest of the area had an average canopy height of 20–25 m. In the partly cleared plots, all trees and undergrowth were cut to a height of 1.5 m, but vegetation and litter below this were left intact. In the totally cleared plots, all trees, undergrowth, ground layer and litter were removed. Whereas partly cleared forest resulted in only minor increases, at totally cleared forest sites overland flow and mineral soil erosion increased fivefold and sevenfold, respectively, above levels in virgin forest. Nutrient losses were found to occur mainly as eroded particulate organic matter rather than as mineral sediments or as solutes in runoff waters; amounts as high as 2.5 kg N, 0.45 kg P and 4 kg K per hectare were washed from totally cleared sites in a single month of the wet season, compared with 0.3 kg N, 0.04 kg P and 0.2 kg K at the virgin forest sites and 0.9 kg N, 0.08 kg P and 0.2 kg K at the partly cleared sites. The dramatic increases at the totally cleared plot were the result of greatly increased rainsplash detachment and removal by overland flow without the protection afforded to the forest floor by leaf litter and near-ground vegetation.

The soil erosion rate recorded by Fournier (1967) in secondary rain forest in Côte d'Ivoire (25 kg ha^{-1} a^{-1}) differs little from rates at primary rain-forest sites elsewhere in Côte d'Ivoire on similarly low-angle slopes (Table 8.15). In clear-felled rain forest near Tawau (Sabah), where annual rainfall is *ca.* 1800 mm, the six-month rate of 35.6 mm a^{-1} is an order of magnitude higher than at the primary rain-forest sites in Singapore and Pasoh, which experience similar annual rainfalls. In the paired plot experiment in southeast Mindanao (Philippines), the rate of surface wash in the logged-over forest was twenty times as high as at the primary rain-forest site (Kellman 1969). The impact of rain-forest clearance on soil erosion will vary with the method of clearance and the subsequent land use, land management practices and any conservation measures taken.

At the catchment scale, Douglas *et al.* (1992) from June 1988 monitored sediment and solute outputs of the Baru catchment in Sabah before, during and after selective commercial logging and compared results with an adjacent undisturbed forest catchment. They were able to relate changes in the ratio of monthly suspended sediment yield in the logged and undisturbed catchments to different phases of logging operation (Table 8.16a). The construction of a logging road across the head of the catchment and later logging within 37 m of the road led to increases in the ratio from 1:1 prior to logging to 4:1 and 5:1, respectively. During the main phase of logging in May and June 1989, the ratio reached 18:1. Some recovery was recorded during the first two years after logging ceased, with ratios falling to 3–4:1. Further recovery is likely to be slow as sediment accumulated in the channels and at the base of slopes on the narrow floodplain is progressively evacuated and as gullies linked to the main stream on abandoned logging tracks continue to erode headward. Changes in dissolved load transport included an increase in the ratio of output of Ca and Mg to Na and K following logging.

Some other sediment yield data for cleared and uncleared catchments are given in Table 8.16b for comparison. The scope for reducing erosion by adopting more careful logging practices is demonstrated by the sixfold difference between the sediment yields of two catchments, one logged normally, the other supervised, at Bukit Berembun (Negri Sembilan, Malaya) (Zulkifli *et al.* 1990). A more comprehensive review of sediment yields of tropical catchments of contrasting land use is given by Douglas & Spencer (1985).

Chapter 9

Phenology

IN TEMPERATE REGIONS the vegetation changes regularly with the seasons and its annual cycle closely follows the changes of temperature. In the humid tropics, on the other hand, there is no winter and seasonal changes in the vegetation are much less evident. The tropical rain forest has no obvious 'aspects' or 'resting periods'. As a whole the forest is always a rather sombre green, though at some times of year trees of some species or single individuals are conspicuous because they are bare of leaves or decked with brightly coloured young foliage. In flowering and fruiting, as in leaf-changing, individuals of the same species are often less synchronous in behaviour than temperate trees. There are always some trees in flower, but flowers are generally more abundant in some months than others: mass flowering of many individuals or species is seen occasionally. Towards its climatic limits, where the annual dry seasons are relatively severe, seasonal changes in the rain forest become more noticeable, though chiefly in the forest canopy rather than in the undergrowth, which has a more uniform microclimate.

Although the appearance of rain-forest communities as a whole varies little throughout the year, few of their component species are continuously active. In the majority of evergreen rain-forest trees, as will be seen later, periods of vegetative and reproductive activity alternate with quiescent periods. In most species flushing of young leaves, flowering and fruiting take place intermittently, at intervals which may be very irregular but in some species are remarkably regular. Certain species are so predictable in their behaviour as to provide fixed dates in the calendar. In Vanuata (New Hebrides) seasonal changes are so little evident that the inhabitants

never reckon their ages in years, but they plant their crops at times determined by the flowering or fruiting of certain trees, yams when *Erythrina indica* flowers, sweet potatoes when *Alphitonia zizyphoides* comes into fruit (Baker & Baker 1936) (see also Fig. 9.6). In Malaya the flowering of *Sandoricum koetjape* was formerly the signal for planting rice (E.J.H. Corner).

The fungi, as well as the higher plants, of the rain forest are seasonal in behaviour. Corner (1935, 1978; see also Hong *et al.* 1984) found that in Malaya and on Kinabalu (Sabah) the fruit bodies of the larger fungi appear at the end of dry spells. In normal years a succession of fungi develops over about three months, each species having a season of about a week.

For many rain-forest animals the availability at all times of year of the leaves, fruits and flowers on which they feed is of great importance. As noted in Chapter 5, the abundance in rain forests of flower-pollinating birds and bats and of herbivores requiring a continuous supply of young leaves depends on the forests' relatively non-seasonal character. However, this does not mean that particular plant foods are equally abundant at all times. In the rain forests of Borneo, which are less seasonal than most, M.R. & D.R. Leighton (1983) found that the preferred foods of frugivorous birds and mammals vary greatly in abundance from month to month and year to year. Some animals such as hornbills, feeding on large fleshy fruits, migrate when their food becomes scarce; others have fixed ranges and adapt their diet to what is available. *Ficus* species, which as a group fruit almost continuously, are important when other foods are in short supply.

There are many examples among rain-forest plants

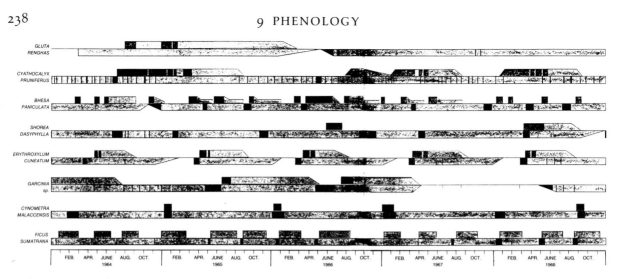

Fig. 9.1 Phenology of leafing, flowering and fruiting of trees in mixed dipterocarp forest observed from platform at 43 m above ground, Ulu Gombak, Selangor, Malaysia. After Medway (1972, Fig. 2).

of animals utilizing groups of species which flower and fruit in succession over almost the whole year. In the Arima valley in Trinidad, Snow (1966) found that the berries of twenty-two species of small trees of the genus *Miconia* had staggered flowering and fruiting seasons and together provided a year-round food supply for unspecialized frugivorous birds. A similar sequence of eighteen *Miconia* species exists in the 'pre-montane' rain forest in Colombia (Hilty 1980). In Costa Rica a succession of nectar-producing flowers provide food for hummingbirds, but in this case the plant species involved are more numerous and are not taxonomically related (Stiles 1978). Even in the relatively seasonal forest of Barro Colorado (Panama) young leaves and fruits of *Ficus* spp. and other Moraceae are available throughout the year (Milton 1991).

9.1 Leaf-change

9.1.1 Leaf-change in rain-forest communities

In tropical rain forests leaf-fall and the expansion of young foliage may take place at any time of year, but are seasonal in many species. Only in the most extreme non-seasonal climates is leaf-change generally continuous. For example, in the forests of the western Solomon Islands no regular periodicity in flushing or leaf-fall can be detected (Whitmore 1984a). In Borneo, where there are no regularly recurrent dry seasons, flushing does not in general seem to be markedly seasonal, though

leaf-fall may be so to some extent. In four types of primary forest at Gunung Mulu (Sarawak) leaf-fall in 1978 reached a maximum in April–June, perhaps because of exceptionally strong winds at that time (Proctor *et al.* 1983b).

In rain forests where dry periods are fairly regular both the flushing and leaf-fall tend to be seasonal, though somewhat variable from year to year. Thus data on leaf-change for some 2000 trees in forty sample plots in the chief forest types of Sri Lanka for the years 1941–52 were compiled by Koelmeyer (1959). Eleven of the fifty-eight species studied in the 'Tropical Wet Evergreen forest' (lowland rain forest) were obligately or facultatively deciduous and some were 'semideciduous', losing a quarter to half of their leaves during dry periods. In evergreen species times of leaf-fall and 'reflush' were often not sufficiently well defined to record. Except in the deciduous trees, there was much lack of synchrony between species and individuals. The monthly phenological indices showed that leaf-fall was mainly in the relatively dry period, November-February and all the deciduous species were leafless in January. Flushing occurred mainly in March–April. Comparison of phenological and climatic data showed that leaf-fall follows periods of low rainfall and/or low relative humidity and that flushing follows (or slightly precedes) high rainfall and humidity.

In mixed dipterocarp forest at Ulu Gombak (Malaya), where McClure (1966) and Medway (1972) (Figs. 9.1, 9.2) made observations over six years on 61 trees (45 species) seen from a platform 43 m above ground, they

Fig. 9.2 A and B. Forest seen from observation platform as in Fig. 9.1. (Photos. Lord Cranbrook.) (Opposite.)

found that on the average the expansion of young leaves reached a maximum in February–June, a period extending from the latter part of the main dry season to the beginning of the first rainy season. There was a second maximum of young leaf development about September–December, at the end of the second dry season and lasting into the wettest part of the year; this was more variable from year to year. Leaf-fall took place in some species at the same time as flushing, but in many others it was not noticeable and probably occurred through most of the year.

In the Douala-Edea Reserve (Cameroon), which has a very high annual rainfall (*ca.* 3000–4000 mm) but several relatively dry months (usually December–February), flushing takes place throughout the year, but reaches a peak early in the dry season and again, as in Ulu Gombak, towards its end; it continues at a high level into the early part of the wet season (J.S. Gartlan, unpublished).

The data of Frankie *et al.* (1974) (Fig. 9.3) for two localities in Costa Rica, La Selva, which has an annual rainfall of *ca.* 4000 mm with no month having an average of less than about 200 mm, but with two somewhat drier periods in the year, and Comelco Ranch with a mean rainfall of 1533 mm and a dry season of five consecutive months each with less than 100 mm, show very clearly the difference between leaf-changing behaviour in a slightly seasonal rain-forest climate and a deciduous forest climate with a severe dry season. At La Selva, although the forest as a whole is evergreen, 8% of the tree species are deciduous. Among the taller trees flushing reaches a major peak early in the long (first) dry season and a smaller one during the short dry season in September; leaf-fall is at a maximum early in the long dry season. Twenty-eight per cent of the species put out new leaves all through the year. In contrast, at Comelco, except in the riparian forest, the canopy becomes completely leafless at the beginning of the long dry season. Flushing does not reach its peak until the main rainy season (late April–June).

The leafing behaviour of the treelets and shrubs at the two sites was recorded over a three-year period by Opler *et al.* (1980a). In the 'Wet forest' at La Selva they were all evergreen and neither leaf production nor leaf-fall was seasonal; 40% of the species produced young leaves continuously. In the deciduous forest at Comelco about half the treelets and shrubs were deciduous and 78% produced new leaves synchronously at the beginning of the wet season.

At Barro Colorado Island (Panama), where the climate is more seasonal than at La Selva but considerably less so than at Comelco, the rhythm of leaf-change is much

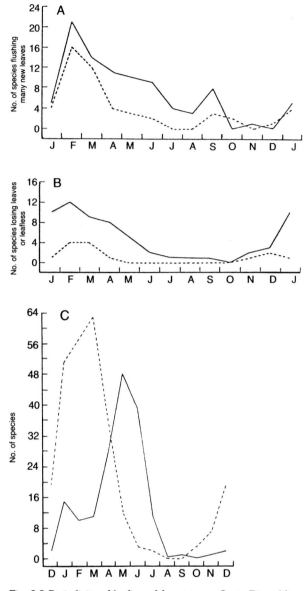

Fig. 9.3 Periodicity of leafing of forest trees, Costa Rica. After Frankie *et al.* (1974). **A** Leaf flushing; **B** leaf fall. Continuous lines, overstorey trees; broken lines, understorey trees. Wet (rain) forest, La Selva. **C** All species, dry (deciduous) forest, Comelco Ranch. Continuous lines, leaf flushing; broken lines, leaf fall.

like that at La Selva, but flushing starts rather later and does not reach its main annual peak until well into the rainy season (May–June). There is a secondary peak in September, associated with the second (shorter) dry season (Leigh & Windsor 1983). Data on leaf-fall and flushing in lower montane forest in Espiritu Santo (Brazil) are given by Jackson (1978) and on flushing in seasonal forest in Gabon by Hladik (1978).

9.1.2 Leaf-change in individual species

Leaf-changing in rain-forest communities as a whole is the resultant of widely different patterns of behaviour among the species and individual trees of which it is composed. In the upper strata of the forest evergreen species grow side by side with a small number of deciduous species which become leafless for varying lengths of time.

In the older literature from Scheffler (1901) and Schimper (1903) onwards there is much information on the behaviour of individual species mostly based on trees in botanical gardens. Particularly interesting are the data of Holttum (1931, 1938b, 1940, 1953) and Koriba (1958) from Singapore, where many species and individual trees have been kept under observation for many years.

The production of young leaves takes place continuously over most of the year in many rain-forest trees, but often they expand together in conspicuous flushes. These may occur at intervals which can be long or short, regular or irregular, as Holttum's data show. *Parkia roxburghii* produced young leaves regularly every twelve months with a standard deviation of only 0.24 in ten years; the interval between flushes was also about a year in *Cratoxylum polyanthum*, *Tamarindus indica* and several other species. *Terminalia catappa*, *Ficus variegata* and *Peltophorum ferrugineum* produced leaves approximately every six months; one tree of *F. variegata* had a mean period of 6.6 months for ten years, but other individuals of the same species had a period of almost exactly six months. A number of species were observed with cycles which were not annual or aliquot parts of a year; a tree of *Delonix regia* had regular periods of nine months and one of *Heritiera macrophylla* had three successive flushes at intervals of two years and eight months. At Bogor (Java), Smith (1923) found that the shrub *Breynia cernua* shed its old leaves and twigs, and expanded new leaves, every 5.5 months on the average, the plant passing through eleven complete cycles in five years.

Flushing is often not simultaneous on neighbouring trees of the same species and sometimes not even on different branches in the same crown. Lack of synchrony within the crown is often seen in the mango (*Mangifera indica*) in which the leaf-changing of the whole tree may extend over some weeks. In *Ceiba pentandra* it is common for the same tree to have some branches with mature foliage, some leafless and some with a flush of young leaves (Fig. 9.4). In Singapore the Asiatic cultivar behaves irregularly in this way, but the African – American variety is regularly and evenly deciduous (E.J.H. Corner).

Fig. 9.4 *Ceiba pentandra* showing lack of synchrony in leafing, Achimota, Ghana. December 1968.

As noted earlier, the life-span of the leaves of evergreen rain-forest trees probably varies in different species from about 18 months to several years; it is less than a year only in species that are regularly deciduous every year. Old leaves may be shed more or less continually, or during periods as short as a few days; they may fall fully green, e.g. in *Ficus glabella* (Volkens 1912), but more often they turn yellow, or sometimes, as in *Terminalia catappa*, bright red. The colours assumed by senescent leaves may be useful in identifying genera and species (Corner 1938, Whitmore 1972a).

In typically evergreen trees the old leaves do not drop until after the young ones have expanded, but in many species there is a period in which the total number of leaves is considerably reduced before flushing begins, though the crown never becomes completely bare. In deciduous species all the old leaves fall well before the breaking of the buds, so that the tree is leafless for a time. Some species of evergreen rain forests, e.g. *Bombax* spp. and *Cedrela toona*, are bare every year for several weeks or months, others, e.g. *Hymenaea courbaril* in Guyana, only for a few days. The relation between leaf-fall and flushing is related to weather conditions; in *Dyera costulata* in Singapore if it is wet, the new leaves expand shortly before the old ones drop, but in dry weather the old leaves fall before flushing and the tree is leafless for a while (Holttum 1953).

In many tropical trees leaf-changing habits vary with age. For example, seedlings and juveniles of *Mora excelsa* in Trinidad (and probably elsewhere) are evergreen, but trees that have reached the forest canopy are generally leafless for at least a week before the flush of pink young leaves appears (Bell 1971). The rubber tree, *Hevea brasiliensis*, is also more or less evergreen when young, but in older trees periods of defoliation ('wintering') often occur; in cultivation wintering is important because it is followed by a temporary reduction in the latex yield while the new leaves are expanding. In Malaya the tendency to wintering depends on the clone and usually occurs in the dry spells which are common early in the year. In *Hevea* there is also a gradual leaf-change, not leading to complete defoliation, some 5–6 months after wintering (Wycherley 1973).

Some tropical trees are more plastic in their leaf-changing habits than others. Beard (1944b) divided the deciduous trees of Trinidad into obligate and facultative. The former are characteristic of the deciduous forest and have a leafless period every year. In the latter, which are most abundant in the semievergreen seasonal forest (p. 394), leaf-fall varies from year to year with the severity of the dry season and in a wet year there is no marked leaf-fall. In the seasonally dry regions of Southeast Asia where they are native, teak (*Tectona grandis*) and *Bombax malabaricum* are both leafless for many weeks in the dry season, but when grown in the more uniformly wet climate of Singapore teak becomes almost evergreen, while the less plastic *B. malabaricum* continues to have a leafless period of about three months. Coster (1923) found that the buds on leafless branches of teak could be induced to break at any time by placing the branches in water, but in *B. malabaricum* bud-break could not be induced artificially until near the end of the normal dormant period.

It is interesting to note that the temperate deciduous trees beech (*Fagus sylvatica*) and oak (*Quercus robur*) when cultivated at *ca.* 1500 m at Cibodas (Java) (*ca.* 6°48'S) remain periodic in leaf production and growth, although the length and timing of the dormant period changes and branches of the same tree become out of phase with each other, much as in tropical trees such as *Ceiba* (p. 241) (Coster 1926b, Schimper 1935). *Quercus* spp., *Platanus hybrida*, etc., grown at sites in the mountains of Sri Lanka, were distinctly periodic in growth and leafing, the majority completing their normal cycle in the course of a year (Dingler 1911b).

It will be evident that in the tropics the terms 'evergreen' and 'deciduous' become difficult to define. A large proportion of rain-forest trees are obviously evergreen; they are never bare of leaves and the fall of the old leaves is either more or less continuous or is completed soon after the flushing of young leaves. Others are deciduous much as in broad-leaved temperate trees; they become bare every year and remain so for several weeks or months before the buds begin to break. But a considerable number of tree species do not fit exactly into either of these groups. In these trees, termed 'leaf-

exchanging' by Longman & Jeník (1987), leaf-fall is completed not more than about a week before the young leaves begin to expand and appears to provide a cue for bud-break.

9.1.3 Leaf-change and stem growth

Bud-break, leaf expansion, and the elongation and radial growth of stems, all result from meristematic activity. They are thus interdependent processes which can only be understood if considered together, as Borchert (1978), Tomlinson & Longman (1981), and others have pointed out. In some tropical trees, e.g. the mangrove *Rhizophora mangle*, stems and young leaves grow continuously (Gill & Tomlinson 1971b), but in others, probably a large majority, both stem growth and leaf expansion are intermittent and, though usually associated, are not necessarily simultaneous. For example, in most of the tree species at Ulu Gombak observed by Medway (1972; see p. 238), leaf expansion and stem elongation occurred at the same time, but in some, e.g. *Erythroxylon cuneatum*, the branches did not elongate until after leaf expansion had been completed and in three *Shorea* spp. the expansion of new leaves did not take place until several months after the elongation of the branches.

Cambial growth in trees in Java was investigated by Simon (1914) who found that in *Dillenia indica* and *Ficus variegata*, which are evergreen but flush intermittently, leaf expansion was preceded by a pause in cambial activity. In deciduous species, such as *Tetrameles nudiflora*, the cambium was inactive during the leafless period and secondary growth was not resumed until the new leaves had expanded. The more detailed and extensive work of Coster (1927–28), also in Java, gave similar results. From this and more recent research it can be concluded that in tropical trees the cambium is generally inactive when the buds are dormant and that bud-break is accompanied by renewed growth in the cambium and apical meristems. It is not always under dry conditions that the cambium becomes inactive: in Nigeria Njoku (1963) and Amobi (1973) found that in various tree species cambial growth slowed down or ceased during the rainy season, and the buds became dormant.

As mentioned in Chapter 4, very few tropical trees form annual growth-rings from which their age can be estimated. Coster (1927–28) found that in many species of non-seasonal climates no rings of any kind are formed; others form rings, but not at yearly intervals, and in some the cross-section of the stem shows 'growth zones' which are not complete rings. Even species that are deciduous every year often do not form distinct growth-rings. It is only a comparatively small number

Table 9.1. *Patterns of leafing in tropical trees*

Group	Growth	Foliage type	Life-span of leaves (months)	Examples
1	Continuous	Evergreen	3–15[a] (variable)	*Dillenia suffruticosa* *Trema orientalis*
2	Periodic	Evergreen	7–15[a]	*Celtis mildbraedii* *Mangifera indica*
3	Periodic	Leaf-exchanging	*ca.* 6 or 12	*Ficus variegata* *Terminalia catappa*
4	Periodic	Deciduous	4–11	*Bombax* spp. *Terminalia ivorensis*

[a] Longman & Jeník say '3–15 months' and '7–15 months' for Groups 1 and 2, respectively. These upper limits may be too low, especially for lower-storey species.

Source: From Longman & Jeník (1987, pp. 154–7), modified.

of species from seasonal tropical climates which produce rings comparable to those of temperate trees. Coster also found that the anatomy of the rings, when present, varied greatly within and between species. In some tropical trees rings are not formed until the tree has reached a certain age.

The formation of growth-rings is clearly associated with flushing and periodic extension growth in at least some species. Rubber (*Hevea brasiliensis*) in cultivation has up to four flushes in a year, each correlated with ring formation in the xylem (Hallé & Martin 1968). The 'wintering' of rubber is marked by rings in the stem (Wycherley 1963).

In Florida, where there is a considerable difference between summer and winter temperatures, annual rings are formed in native trees belonging to temperate genera such as *Acer* and *Fraxinus*, but not in tropical genera (with a few exceptions, e.g. *Swietenia mahagoni* (Tomlinson 1980).

Since leaf-change and stem growth are closely linked, a number of 'patterns of leafiness' can be recognized among tropical trees (Longman & Jeník 1987) (Table 9.1).

9.1.4 Climate and leaf-change

The proportion of evergreen, leaf-exchanging and deciduous trees in tropical forests is obviously related to climate, particularly to the length and severity of the dry seasons. Most tropical rain forests consist mainly of evergreen and leaf-exchanging species, but some deciduous species are usually present. The native tree

flora of the Malay Peninsula, where the climate ranges from ever-wet in the south to moderately seasonal in the north, includes about twenty-five species (*ca.* 0.3%) which have obligate annual leafless periods, the commonest being *Bombax valetonii*, *Intsia palembanica* and *Koompassia excelsa* (F.S. Ng), although the last named behaves differently in Sarawak (P.S. Ashton). Although lowland rain forests entirely lacking in deciduous trees seem to be rare, except on podzols, many tropical montane, and some temperate, rain forests probably consist only of evergreen species.

Towards the climatic limits of the tropical rain forest, as rainfall becomes increasingly seasonal, the proportion of deciduous trees becomes greater, first in the upper, and then in the lower strata, and semideciduous and deciduous forests eventually replace evergreen rain forests (Chapter 16). This gradient is well seen in Africa: in the forests of the Zaïre (Congo) basin the proportion of deciduous species increases from the centre towards the periphery (Lebrun 1936a). In Ghana there is a similar increase from the wettest to the driest of the four forest formations (Hall & Swaine 1976, 1981). Similar gradients are found in the American tropics (Chapter 16) and in Southeast Asia.

Leaf-changing habits reflect microclimate as well as macroclimate. In primary tropical rain forests deciduous and leaf-exchanging trees are found mainly in the emergent (A and B) storeys, which are those most exposed to variations in temperature and humidity. In the lower layers where the atmosphere is more constantly humid, most of the woody plants, including the juvenile trees, are evergreen. This is well shown by the differences in leaf-changing behaviour between the taller trees and the shrubs at La Selva (Costa Rica) (p. 240 above). In Malesian rain forests the C and D strata probably always consist only of evergreen species, but in the evergreen seasonal forests of West Africa, which have a well-marked dry season, the lower storeys, though mainly evergreen, include a few species that are at least facultatively deciduous, e.g. *Monodora* spp. and the treelet *Schumanniophyton problematicum* (Hutchinson & Dalziel 1954–72).

9.1.5 Edaphic factors and leaf-change

Although climate is undoubtedly the most important determinant of leaf-changing behaviour in tropical trees, edaphic factors also play a part. In semideciduous and deciduous forest areas the larger proportion of evergreen trees in the gallery forests along rivers makes them conspicuous in the dry season (Figs. 16.6 and 16.8). However, even in ever-wet climates deciduous species are more abundant on some soil types than others. In Borneo deciduous trees are found, though not in large numbers, in mixed dipterocarp forests on kaolisols, as well as on soils derived from limestone and basalt, but, surprisingly perhaps, are absent in heath forests on white sands (podzols) and in peat swamps (Chapter 14), which are entirely evergreen (P.S. Ashton). In the rain forests of Guyana and Amazonia there is a similar lack of regularly deciduous trees in the white-sand forests. The complete dominance of evergreens on podzolic sands seems to be due to their oligotrophic character[1] rather than to their physical properties. Similarly in Mato Grosso (Brazil), where there is a long dry season, the forests on the prevailing reddish sandy dystrophic soils are almost wholly evergreen, while on the more eutrophic 'terra preta' on adjoining sites the forests are completely leafless in the dry season (Ratter *et al.* 1973). Outside the tropics there is a similar situation in Florida where Monk (1965, 1966) found that in forests which are mixtures of evergreen and deciduous trees the proportion of evergreens is greater on nutrient-poor sands than on more fertile soils. Monk suggests that evergreen trees are better adapted than deciduous trees to oligotrophic conditions because nutrients are less easily leached from their leaves and are retained longer in the resulting litter.

In riverain forest in the tropics trees may become leafless during seasonal floods. For example, in *Irvingia smithii* and *Lannea welwitschii* on river banks in the Zaïre (Congo) basin the leaves fall and the buds become

dormant during the annual high water period (Lebrun 1968). In the *Oubanguia africana – Guibourtia demeusei* swamp forest association (p. 363), many species, including the two dominants, become leafless during the yearly floods. After the water level has fallen massive flushes of bright red young leaves are produced and there is an outburst of flowering (Evrard 1968).

In the várzeas and igapós of Amazonia (p. 356) some of the trees also lose their leaves during the annual flood. Takeuchi (1962) noted that in the igapó *Eugenia inundata* sheds its leaves as the water begins to rise, but the associated woody plants retain their foliage though submerged for several months. Worbes (1986) found that near Manaus most of the common species in both the upper and lower storeys of the várzea lost their leaves during the floods, but the trees and shrubs of the igapó, which had more sclerophyllous foliage, were more or less evergreen, although *Aldina latifolia* and a few others became leafless for a short time. P. de T. & R. Alvim (1978) found that the flushing and shoot growth of cocoa at Manaus ceased while the roots were submerged during the floods, presumably because of oxygen deficiency.

Seasonal waterlogging of the soil may be the explanation of the paradoxical behaviour of *Faidherbia albida*, a small tree of the Sahel region of Africa which sheds its leaves at the onset of the wet season but remains leafy through the dry season. Lebrun (1968) suggests that leaf fall is induced by temporarily anaerobic soil conditions and that the tree is able to retain its leaves in the dry season because the roots are deep enough to be able to supply them with water even during the long and severe annual drought.

9.1.6 Role of external and internal factors in leaf-change

The factors controlling the timing of leaf-change and periodic growth in tropical trees are not well understood and can be discussed only briefly here. Until recently experimental work has been chiefly on cultivated trees such as cocoa and coffee; experiments on a wider range of species, and under forest conditions, are needed.

There is much to suggest that the availability of water and water stress, which can sometimes be demonstrated by measurements of shrinkage in trunk circumference (Daubenmire 1972), is a major factor. Holttum (1953) observed that in many species in Singapore changes from wet weather to dry, or the reverse, seemed to trigger leaf-fall or flushing. In the Douala-Edea forest in Cameroon, where there is a short annual dry season, Gartlan (1983) also found that flushing seemed to be

---

[1] On the nutrient status of these soils, see Chapter 10.

Fig. 9.5 *Tabebuia* sp. flowering when leafless. Xavantina – Cachimbo Expedition Base Camp area, Mato Grosso, Brazil. August 1968.

a response to sudden decreases or increases in water availability.

In tropical deciduous trees it is clear that dormancy is generally induced by water stress and that bud-break depends on rehydration. In *Tabebuia ochracea* subsp. *neochrysantha* in Costa Rica, Reich & Borchert (1982) found that leaf-fall and flowering are closely related to the decline in soil moisture early in the main dry season; recovery from water stress is necessary for bud-break. Local and year-to-year variations in the time of leaf-fall could be explained by differences in the incidence and intensity of rainfall at different sites.

In a later paper (1984) in which they deal with twelve species in dry and moist sites in the seasonally very dry forest of Guanacaste, Costa Rica (cf. data for Comelco, p. 240), the same authors conclude that, although water stress and rehydration control the phenology of all these trees, the correlation with the availability of water in the environment is only indirect. Rehydration of water-stressed trees can often take place after heavy rain, but in some species it can occur during continuing drought. Water stress can sometimes be caused by shoot extension even during the rainy season. Reich & Borchert (1984) found that in the species they studied there was no evidence that the seasonal pattern of development was controlled by changes of temperature or photoperiod.

In evergreen and leaf-exchanging, as in deciduous, trees, water stress undoubtedly plays a major part in determining periodic growth, but other factors also seem to be important. In cocoa, a periodic evergreen rain-forest tree (*sensu* Longman & Jeník 1987), the availability of water seems to be the main factor influencing leaf-changing, although, there is experimental evidence of the role of endogenous factors. In cultivation near Bahia (Brazil), where there is no pronounced dry season, P. de T. & R. Alvim (1978) concluded from extensive observations and experiments that leaf-fall and flushing were dependent mainly on soil moisture and ultimately on rainfall. Under field conditions bud-break and flushing appeared to be induced by alternating spells of wet and dry weather. But it was found in a number of forest tree species in the same area, some of which were deciduous, that leaf-change was more closely correlated with the change of day-length at the September equinox than with water availability (except in an introduced species, *Artocarpus integer*, in which leaf-changing was continuous). The data of Coster (1923) for tree species in Java also show peaks of flushing at the equinoxes (Longman & Jeník 1987).

In the tropical zone, the variation in day-length is about three hours at the tropics and virtually nil at the equator, but though the seasonal differences are small, there is good evidence (*pace* Reich & Borchert 1984) that they affect the phenology of some tropical plants. Experiments by Njoku (1964), for example, showed that in the West African deciduous tree *Hildegardia barteri* bud dormancy is induced by days 11.5 h long and prevented by 12.5 h days. A review of the significance of day-length for tropical plants is given by Stubblebine *et al.* (1978) in a paper on photoperiod experiments on the widespread tropical American tree *Hymenaea courbaril*. In rain forests it is evergreen, but in more seasonal climates it may shed a large proportion of its foliage in the dry season and under some conditions it is truly deciduous. North of the equator, flushing normally occurs after the solstice when the days are lengthening. Experiments and observations showed that, although the behaviour of this tree could be modified by water stress, it was mainly determined by photoperiod. The response was of the 'high energy' type: light at low intensities was ineffective. It was also found that there were

genetically determined ecotypic differences in photo-
periodic response between populations that were correlated with their latitude of origin.

Whether endogenous factors, in addition to the
environment, play an important part in controlling the
phenology of tropical plants has been often debated.
Klebs (1926 and earlier papers) attempted to show that
even in the least seasonal tropical climates leaf-change,
periodic growth and flowering were determined entirely
by environmental changes. Some later work supports
this view, but there are various facts that are difficult
to reconcile with it, for instance the lack of synchrony
in many species between different parts of the same
tree and between different individuals in similar environments, and also that in seasonal tropical climates plants
are sometimes active in dry conditions when they might
be expected to be dormant or inactive when conditions
appear to be favourable.

Convincing evidence for the endogenous control of
flushing and intermittent growth in cocoa was provided
by the experiments of Greathouse et al. (1971), who
found that individual shoots pass through alternate
phases of growth and dormancy, even when kept at a
constant temperature and 12 h photoperiod. During
growth phases the stems elongate and new leaves
expand, while during dormancy the number of leaves
and leaf primordia remains constant. The mechanism
by which endogenous controls operate is not fully understood, although it is probable that flushing and shoot
elongation are regulated by a hormonal system. In temperate trees mature leaves produce inhibitors that control
bud-break and shoot growth; there is also evidence
that stimuli originating from the roots can affect shoot
growth. Similar controls probably exist in tropical trees.
Thus in the mango (Mangifera indica) Holdsworth
(1963) found that artificial defoliation (and sometimes
ringing the stem below its apex) led to an immediate
flush of young leaves. Longman (1969) found that
mature leaves inhibit shoot growth in the deciduous
tree Hildegardia barteri.

From similar experiments Huxley & van Eck (1974)
concluded that flushing and shoot growth in tropical
trees depend on leaf ageing. As the leaves mature, the
growth rate of the shoots decreases and when they
become senescent their inhibiting action ceases: the
meristems then resume activity and become able to
produce new leaf and stem tissues. The time taken for
leaves to become senescent varies with the species and
is also affected by environmental stresses such as water
deficits. Though the inhibiting effect of the leaves is
removed when they become senescent, leaf-fall may not

follow immediately, and bud-break may also be delayed
until environmental conditions are favourable. It is also
possible that bud-break sometimes does not occur until
a stimulus is received from the root system.

Further evidence supporting the latter hypothesis is
desirable, but it has the advantage that, in addition to
suggesting a plausible mechanism for maintaining the
photosynthetic efficiency of tree crowns by regulating
the replacement of old senescent leaves by younger
ones, it could account for many of the phenological
peculiarities of tropical trees, for instance the differences
in leaf-changing behaviour between different individuals
and branches of the same tree, and the differences in
behaviour of the same species in ever-wet and seasonal
climates. It also makes understandable the different patterns of growth and leaf-changing in the relatively
exposed upper strata of the forest and in the more
sheltered undergrowth.

The internal factors involved in leaf-changing and
periodic growth are certainly to a large extent genetically
determined. This is indicated by the fact that in some
families of trees, e.g. Bombacaceae and Sterculiaceae,
there are many deciduous species, while in others, e.g.
Lauraceae and Myristicaceae, almost all the species are
evergreen. It is well known that the 'wintering' habit
of Hevea and the periodicity of flushing in cocoa differs
between clones. The different behaviour of Ceiba varieties (p. 241) is another example. Many forest tree
species probably have phenologically different ecotypes
like Hymenaea courbaril (p. 241).

9.1.7 Adaptive value and evolution of deciduous habit

The varied and sometimes enigmatic leaf-changing habits
of trees and shrubs in rain forests and some more
seasonal tropical ecosystems prompts questions about
the adaptive value and evolutionary origin of various
'patterns of leafiness', in particular of the deciduous
habit. It seems obvious that in climates with cold winters
and tropical climates with long and severe dry seasons
a dormant leafless period is adaptive by reducing transpiration under conditions in which water lost may be
hard to replace; yet various facts cast doubt on the
critical importance of leaf-shedding as a strategy for
drought avoidance in tropical trees.

For example, in the West African evergreen seasonal
and semideciduous forests many trees produce flushes
of young leaves at the driest time of year. These leaves
are soft and apparently vulnerable to desiccation. For
example, the brilliant red young foliage of Lophira alata
and many other species is conspicuous during the har-

mattan season in January when very low relative humidities are often experienced. In seasonal forests and savanna woodlands in many parts of the tropics, flushing reaches a peak late in the dry season, well before appreciable amounts of rain have fallen. In Sri Lanka Dingler (1911a), by pruning them late in the previous wet season, induced certain normally deciduous trees to remain leafy during the dry season, but he observed no damage to their leaves.

These facts suggest that in the humid tropics the advantage of leaflessness during dry periods may be relatively small. It is also possible that for some species the deciduous habit may have advantages unconnected with water stress. Janzen (1970) observed that the shrub *Jacquinia pungens* of deciduous and semideciduous forests in Central America is leafless in the rainy season and leafy during the dry months (cf. *Faidherbia albida*, p. 244). He suggests that the advantage of this behaviour is that the plant can escape competition for light with taller vegetation.

In a climate which is usually favourable for photosynthesis and growth at all times of year the rain-forest flora has been able to evolve a variety of leaf-changing strategies, probably in response to selection pressures of several kinds. The possible adaptive value of different strategies must be assessed in relation to the fitness of the plant as a whole (see Gould & Lewontin 1979) and in deciduous trees must take account of the energetic costs of a leafless period.

The deciduous habit has probably evolved many times among dicotyledonous trees, as well as more than once among gymnosperms. In the modern flora it is particularly conspicuous in north temperate broad-leaved forests and in the seasonal tropics, but it did not necessarily arise where it is now most prevalent. The presence of deciduous trees in humid rain forests has sometimes been regarded as anomalous. Von Ihering (1923) suggested that in South America deciduous trees invaded the rain forest from adjacent seasonally dry regions. This may have been so, but it is also possible that they evolved in the rain-forest areas where they are now found during a dry climatic period when the evergreen forest was reduced to refugia, or at least less extensive than in recent times.

There is some palaeobotanical evidence suggesting that the ancestors of modern broad-leaved deciduous trees arose in the early Cretaceous in climates marginal to the tropical zone (Axelrod 1966). From there they migrated polewards and also into the mountains of the tropics. However, it should be remembered that most temperate deciduous trees are frost-tolerant and most tropical

deciduous trees are probably not. This may indicate that the deciduous habit arose independently in the temperate and tropical zones and that temperate deciduous trees are not directly derived from deciduous tropical taxa.

9.2 Flowering and fruiting

In tropical rain forests trees can be seen in flower or fruit at any time of year, but the reproduction, like the vegetative activities, of species and individuals is generally not continuous. 'Ever-flowering' species are few, the great majority flowering intermittently, one or more times a year or at longer intervals. In some rain-forest trees many individuals flower simultaneously over large areas, but in many species there is a wide range in the time of flowering of individuals and populations. Rain forests as communities have times of maximum and minimum flowering and fruiting. The peaks and troughs, though not always clearly defined, recur at about the same time every year and can be correlated with seasonal variations in climatic conditions. Even in the New Hebrides (Fig. 9.6) and the ever-wet 'premontane' forest of the Chocó (Colombia), where no month has an average rainfall of less than 400 mm, some periodicity in flowering and fruiting can be observed (Hilty 1980). In regions where there are well-marked annual dry seasons the peaks of flowering, as might be expected, are much more pronounced and the behaviour of individual trees is more synchronous than in ever-wet climates.

9.2.1 Flowering and fruiting of rain-forest communities

Koelmeyer's records (1959–60) for a large number of emergent and 'main canopy' trees on sample plots of 'Tropical Evergreen forest' in Sri Lanka for 1941–52 show little regularity in flowering and fruiting periods, but owing to considerable year-to-year variations the ten-year averages conceal the fact that in any one year flowering is usually seasonal. Thus in one locality there was a peak of flowering in September in 1941, but in the following year there were peaks in April, August and December. Maximum flowering most often occurs in March–April, after what are usually the two driest months; the minimum is generally in the wet months July–August. Fruiting reaches a maximum about 2–4 months after flowering. There was some variation in flowering and fruiting behaviour between different localities; it was most regular in those with the most marked annual dry periods. The observations of McClure (1966) and Medway (1972) from mixed dipterocarp

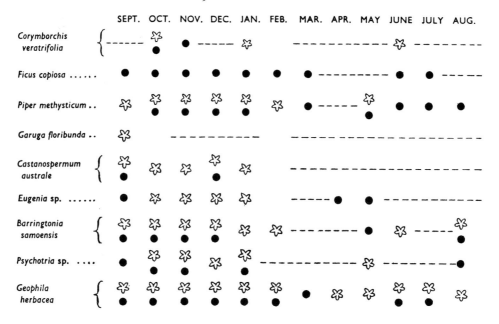

Fig. 9.6 Flowering and fruiting of nine common plants in a non-seasonal climate, Vanuata, New Hebrides. From Baker J.R. & Baker I. (1936, p. 508). The selected plants include trees and herbs. For four species no observations were made in February. Rosettes, flowering; black dot, fruiting.

forest at Ulu Gombak (Malaya) (p. 238 above) record flowering and fruiting of sixty-one trees (forty-five species) from 1960 to 1969. The trees, all of which belonged to the 'canopy' (A and B storeys) were very varied in behaviour. In the group as a whole, flowering reached a peak in February–June, a period including the early dry season and the first (short) rainy season. From July onwards flowering declined, reaching a minimum in December–January, the wettest part of the year. Ripe fruits were most plentiful in September–November. Flowering and fruiting, as well as rainfall, varied considerably from year to year: in 1960–69 flowering and fruiting were most abundant in 1963 and 1968, the years in which the early dry season was most pronounced.

The peak of flowering of emergent trees at Ulu Gombak during the early months of the year is probably typical of lowland rain forests in Malaya generally, although, according to Whitmore (1984a), there are indications of a second peak later in the year which is not shown in these observations. In the rain forest at Bukit Timah, Singapore, where the climate is almost completely non-seasonal, there are super-annual 'bursts of flowering' but no regular recurrent annual flowering seasons (Corlett 1990). The special features of flowering behaviour in the Dipterocarpaceae, of which twelve species are included in the Ulu Gombak group of trees, are discussed later. Little information is available on the flowering of lower-storey trees in Malaya, except for a

few observations by Yap (1980) and Appanah (1982).

At La Selva, Costa Rica, Frankie et al. (1974) and Opler et al. (1980a) made observations on the flowering and fruiting of a large number of trees and shrubs over a two-year period (Fig. 9.7). Here the climate is perhaps somewhat more seasonal than at Ulu Gombak, though the annual rainfall is higher. The 'overstorey' (A and B storey) trees have two, not clearly separated, peaks of flowering. The main one is at the beginning of the first rainy season (May–June) and there is a well-marked minimum at the beginning of the second rainy season (November). The 'understorey' (C storey) trees were less clearly seasonal in flowering than the larger trees, showing three apparent peaks, two of them out of phase with the 'overstorey'. Periodicity among the treelets and other shrubs was even less marked; they had no consistent peaks in flowering during the years of observation, but showed a tendency to fruit during the second half of each year.

Although the La Selva forest as a whole was to some extent seasonal in flowering and fruiting, the maxima and minima were much less well defined than in the 'Dry' (deciduous) forest (Tropical Dry forest of Holdridge) at Comelco Ranch in western Costa Rica (Guanacaste Province) where Frankie et al. (1974) also made observations. Here the annual rainfall is about 1533 mm and there is a long dry season of five months, each of which has an average of less than 100 mm. All

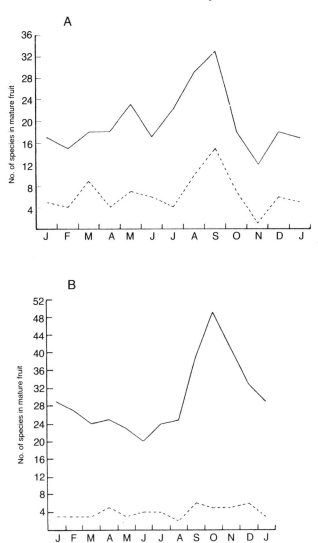

Fig. 9.7 Periodicity of flowering and fruiting of overstorey (A) and understorey (B) species in wet forest at La Selva, Costa Rica. From Frankie *et al.* (1974). Continuous line, all species; broken line, seasonally flowering species.

secondary forest. The climate (Fig. 7.2), though more seasonal than that of La Selva, is less so than that of Comelco Ranch; there is only one dry season and the average rainfall in the dry months is considerably more than at Comelco. Croat (1969, 1975) compiled data on flowering and fruiting during 1967–74 for the whole flora (which includes many herbs and non-forest species of swamps and open ground) and found that 171 species flowered and fruited only in the wet season and 132 only in the dry, but there were many other types of behaviour. Some 295 species flowered and fruited all through the year; many of these were understorey trees, early seral species and herbs.

Foster (1983) investigated flowering and fruit-fall over two years in eighty-three 1 ha plots of 'relatively mature' and 'younger' forest in the highest part of Barro Colorado Island. In the plots as a whole there was a peak of flowering in the first half of the rainy season (April–June) and two peaks of fruit-fall, a major one in March–June and a minor one in the wet months September and October, but there were interesting differences between the behaviour of certain groups of species: those fruiting between July and February mostly flowered just after the beginning of the rainy season and those fruiting between March and June flowered over a longer period, with a peak in the latter part of the dry season (about March), many of the second group taking nearly a year to mature their fruits. As at La Selva, different synusiae also differed in behaviour; thus 'canopy trees' had fruiting peaks in April–June and September–October, while 'understorey' trees, treelets and other shrubs had peaks in June and November–December. There was a striking difference between animal-dispersed and wind-dispersed species: the former had two peaks of fruit-fall, during the rainy season, but the latter had one clear peak in the transition from the dry to the rainy season (which, it may be noted, is also the period when the greatest number of trees are leafless). The differing phenological patterns seemed to be adapted to several factors, including availability of dispersal agents and of pollinators.

For African rain forests precise information on flowering seasons is scarce. The fragmentary information for Yangambi (Zaïre) (Capon 1947) suggests that in the least seasonal African areas tree flowering is spread through the year and irregular. In the Douala-Edea Reserve (Cameroon), which has an annual rainfall of 3000–4000 mm with three 'dry months', J.S. Gartlan's data (unpublished) for 1975–78 show that the percentage of individual trees in flower is highest in the late dry and early rainy season (February–May) and lowest in August, a very wet month. Ripe fruits are said to be

strata of the forest, including the treelets and shrubs, were strongly seasonal in behaviour, the peak periods for flowering being the long dry season (December–April) and at the onset of the main rainy season (May–June). Relatively few species were in flower during the last three months of the year. Maximum fruit production was in the latter part of the long dry season.

Much phenological information is available for the semideciduous forest of Barro Colorado Island (Panama), which consists partly of very old and partly of young

seldom produced, but the percentage of trees with immature fruits is highest in the wet months May–July, and
lowest in the late rainy and early dry season (October–
January). Year-to-year differences in reproductive
behaviour were considerable and there were also differences between observations at a low-lying and a slightly
more elevated site in the forest.

In most of the West African evergreen seasonal forest
the dry season is longer and more severe than in the
Douala-Edea area. Kennedy's data for southern Nigeria
(Njoku 1963) show that the maximum number of tree
species flower in the early dry months (November–
January), but some flowering occurs all through the
year. Taylor's (1960) data for Ghana forest trees (which
include species of semi-evergreen as well as evergreen
seasonal forest) also suggest a peak of flowering early
in the dry season. In the moderately seasonal Budongo
forest (Uganda) Eggeling (1947) found that the main
flowering season of the large trees was at the beginning
of the rainy season. Many species ripened their fruit
in the dry weather about nine months after flowering.

The flowering and fruiting, like the leaf-changing,
behaviour of rain-forest communities seems to be affected directly or indirectly by soil conditions as well as
by climate. Striking instances of this are the swamp
forests of Amazonia and Zaïre, which are annually
flooded for long periods. In the Amazonian igapó and
some of the Zaïre swamp forests mass flowering, as
well as the flushing of young leaves mentioned on pp.
238ff., occurs when the annual floods subside. In some
swamp forests, such as those of South Johore (Malaya),
which are liable to flooding for a short time only,
flowering follows a more or less similar pattern to that
of terra firme forests in the same area (Corner 1978).
It may, however, be true that riparian and swamp forests
and other communities of wet habitats tend to be less
seasonal in flowering than those of well-drained land
(see the data of Frankie et al. (1974) for riparian forest
in Costa Rica and those of Croat (1975) for aquatic
plants at Barro Colorado Island).

Little has been recorded about the phenology of
flowering in tropical montane forests. Nevling (1971)
found that in the 'elfin forest' at ca. 1000 m in Puerto
Rico (which is always very wet owing to cloud
condensation), although there is some monthly variation
in rainfall, flowering is somewhat seasonal. It reached
a maximum in July when forty-five out of fifty-five
species were in flower; eighteen species flowered all the
year round and most of the remainder for more than half
the year. In the trees of the Blue Mountains (Jamaica)
flowering occurs all through the year, but with a peak
in the late dry and early wet seasons (Tanner 1982).

9.2.2 Flowering of individual species

Most tropical trees flower intermittently, but some tree
species in Malaya are, according to Corner (1988) 'ever-
flowering', beginning to flower at an early age and
continuing until they die. Most of these are early seral
(pioneer) species. One of them, *Adinandra dumosa*,
begins to flower when 2–3 years old and flowers without
interruption for many years; similarly, *Dillenia suffrut-
icosa* starts flowering at about eighteen months and
continues for perhaps forty or fifty years.

These and other species are ever-flowering in the
sense that flowering individuals can be found at any
time of year. It is less certain that the same tree flowers
continuously, as Corner implies. *Dillenia triquetra* was
the only one of the 125 tree species in Sri Lanka studied
by Koelmeyer (1959–60) which could be classed as ever-
flowering, but even in this, individuals that were continuously flowering in one year flowered in one or two
definite periods in other years. In populations of the
mangrove *Rhizophora mangle* in Florida flowers are
always present, but no individual flowers continuously.
Both the expansion of flower buds and fruiting occur
mainly in summer (Gill & Tomlinson 1971b).

No large primary forest trees are known to be ever-
flowering. Among the shrubs of the 'Wet forest' in
Costa Rica, Opler et al. (1980b) found that continuous
flowering was uncommon and occurred only among a
few pioneer (seral) species. It is probable that in primary
rain forests in general some of the pioneer species
colonizing gaps are the only truly ever-flowering species.

Among intermittently flowering rain-forest trees there
is much inter-specific variation in the frequency, length
and regularity of the flowering periods. Records of
flowering over several successive years are available for
relatively few species and mainly for specimens in botanical gardens or arboreta, which may differ in behaviour
from trees in natural habitats. Ashton (1982) says that
two dipterocarp species flower almost annually in the
Kepong arboretum (Malaya), but in the forest only at
intervals of several years.

According to Corner (1988) some intermittently
flowering 'wayside' trees in Malaya flower once a year,
e.g. *Koompassia excelsa*, *Parkia speciosa*, others twice,
e.g. *Albizia falcataria*, and some several times, e.g.
Syzygium grandis, *Rhodamnia cinerea*. Holttum (1931,
1935, 1940, 1953) kept records over many years of the
flowering of various native and introduced species at
Singapore, showing that some flowered at regular intervals of about twelve months, but several such as *Delonix
regia* and *Lagerstroemia speciosa* flowered regularly
every 7–10 months. There were also species in which

the intervals were very irregular. In some of these flowering might occur quite rarely. An extreme example was *Homalium grandiflorum*, which Ridley (1922) saw flowering only once in twenty-six years and Holttum (1940) once in twelve years.

In species in which flowering is intermittent, lack of synchrony among individuals may considerably extend the flowering period of a whole population. Individual trees of *Dipteryx panamensis* at La Selva (Costa Rica) flower for about six weeks, but not simultaneously, so the flowering of the whole population lasts about twice as long (Perry & Starrett 1980). Ng (1980b) observed two trees of *Peltophorum pterocarpum* in the arboretum at Kepong (Malaya) for seven years and found that although flowering always followed leaf flushing, the phenological cycles of the two individuals were different, apparently determined by intrinsic genetic factors rather than climatic conditions.

Year-to-year variations in the timing and abundance of flowering in rain-forest trees are no doubt very common, but have seldom been accurately recorded, except in species with conspicuous mass flowering such as dipterocarps. Piñero & Sarukhán (1982) found that in the Mexican understorey palm *Astrocaryum mexicanum*, the population as a whole flowered and set seed every year, but individual plants did not reproduce annually or on any regular pattern. Populations of *Mora excelsa* in Trinidad seem to behave similarly (Bell 1971).

9.2.3 Mass flowering

The most striking examples of synchronous intermittent flowering in rain-forest trees are mass flowerings in which most of the individuals of one species flower at the same time, sometimes over hundreds of square kilometres of forest. This gregarious flowering contrasts remarkably with the more sporadic flowering characteristic of many rain-forest plants. It is well known in Dipterocarpaceae but occurs in trees of other families, also in bamboos, some Malayan orchids and in lianes, e.g. in the African *Calycolobus heudelotii* (Rees 1964b). The quantity of flowers produced can be prodigious and flowering may be followed by equally spectacular crops of seeds and seedlings. At the height of flowering of the dipterocarps in Sepilok Reserve (Sabah) in 1955, 'the ground appeared to be carpeted in snow and the scent of the flowers pervaded the jungle' (Wood 1956). Gérard (1960) estimated that in a mass flowering of *Gilbertiodendron dewevrei* in Zaïre 11 042 325 flowers and 10 721 seeds fell on a single hectare, mostly within a few metres of the parent trees.

Each mass flowering may last a few weeks or be very short-lived. Spruce (1908, vol. 1, p. 257) says of the Amazonian Myrtaceae: 'On a given day all the myrtles of a certain species, scattered throughout the forest, will be clad with snowy fragrant flowers; on the following day nothing of flowers appears save withered remnants. Hence it comes that if the botanist neglect to gather his myrtles on the very day they burst into flower, he cannot expect to number them among his "laurels" '.

In tropical trees that flower gregariously the flower buds are initiated a considerable time before anthesis and become dormant. Thus in coffee (*Coffea* spp.) cultivated in Malaya all the bushes of one species flower at the same time in the same district, different species usually flowering on different days (Corner 1988). The flower buds, after reaching a certain size, remain dormant until stimulated to open. In the wild *C. rupestris* in Nigeria during the dry season flowering normally takes place 3 ± 1 days after an adequate fall of rain, and can be induced by watering the base of the plant (Rees 1964a). In the cultivated *C. arabica*, flower bud dormancy can also be broken by release from water stress.

Some of the best-investigated examples of gregarious flowering among rain-forest plants are *Dendrobium crumenatum* (Pigeon orchid) (Fig. 9.8) and certain other Malesian epiphytic orchids (Coster 1926a, Wycherley 1973). The flowering periods of the various species are not coincident, but overlap; in any one species the flowers last one day or less. Flowering often occurs after

Fig. 9.8 Mass flowering of pigeon orchids, *Dendrobium crumenatum*, Malaya. (Photo. Ivan Polunin.)

a thunderstorm following dry weather. The interval between the shower and flowering may be 8, 9, 10 or 11 days, depending on the species; in some it may vary by ±1–2 days. As in *Coffea* the buds remain dormant for some time, but experiments show that the releasing stimulus is a sudden fall of temperature (usually 5° or more after a thunderstorm), not the direct effect of rain. In some Malayan trees anthesis also seems to be a response to temperature (Corner 1988).

In tropical trees gregarious flowering may be more or less annual, as in *Gilbertiodendron dewevrei* (Gérard 1960), or at irregular intervals of some years, as in most Malesian Dipterocarpaceae. In the latter the years of heavy fruiting, following gregarious flowering, are comparable to the 'mast years' of European and North American species of *Fagus* and other genera.

9.2.4 Flowering of dipterocarps

Much has been written about the flowering behaviour of the Malesian Dipterocarpaceae, in which, at intervals of some years, many species of various genera flower massively at about the same time over large areas (see reviews by Ashton 1982, 1989 and Whitmore 1984a). The biological significance of 'mast fruiting' in dipterocarps has been discussed by Janzen (1974a), who regards it as being, like monocarpy (p. 253), a strategy for avoiding seed predators. From a silvicultural point of view it is important because the seeds remain viable for a very short time and a good supply is available only after mass flowering.

Most Malesian rain-forest dipterocarps, unlike those of seasonal forests in other parts of Southeast Asia, which flower annually, generally flower only in certain years. In 'heavy' flowerings, which occur at intervals of about 9–11 years on average, flowers are abundant over very large areas. The exceptionally heavy flowering of 1955, for example, took place through the whole of Sabah, except the extreme southeast, and was also observed in Sarawak (Wood 1956). 'Light' flowerings occur between heavy flowerings. In a heavy flowering many species of several genera flower together, but not all the mature individuals of any species flower (Burgess 1972). In Borneo in 1955 over a hundred species flowered. In light flowerings fewer species are involved. Occasionally individual trees may flower in a year when there is no general flowering. It has been noticed that intermittently flowering trees of other families, e.g. *Koompassia excelsa*, often flower at the same times as the dipterocarps.

In the Malay Peninsula most, but not all, dipterocarps generally flower between March and May, i.e. after the driest part of the year (Ashton *et al.* 1988, Burgess 1972, Ng 1984). During a flowering the beginning of anthesis and the length of the flowering period vary from species to species. At Andulau (Brunei) Ashton (1982) found that the species of *Shorea* sect. *Richetioides* flowered in succession with hardly any overlap and at Pasoh (Malaya) Chan (1980) found a similar sequence in *Shorea* spp. of sect. *Mutica*: each species flowered for *ca.* 2–3.5 weeks with so little overlap that when the flowering of any one species was at its height no related species was in flower. In dipterocarps individual flowers open in the evening and the corolla falls the following morning. Synchrony in flowering between trees of the same species in the same locality is fairly exact.

There are differences in phenology between the dipterocarps of different habitats. Sarawak peat-swamp species, e.g. *Shorea albida*, do not flower in the relatively dry part of the year like those of better drained sites (Anderson 1961a). Wood (1956) found that in the great 1955 Sabah flowering the dipterocarps at higher elevations began to flower about two weeks later than those in the lowlands.

The fruits of rain-forest dipterocarps take about three months to ripen. Chan (1980) found that the fruits of the later-flowering species developed more quickly than those of species flowering earlier, so that although the species flowered in sequence they all ripened and shed their fruit at about the same time (August at Pasoh). Losses of flowers and fruits from predators and rain damage are very heavy. 'Mast fruiting', followed by large crops of seedlings, generally results only from 'heavy' flowerings, little good seed being produced at other times.

The physiology of flowering in dipterocarps has not been studied experimentally and is not well understood. It is evident that climatic factors must be involved, but the connection between flowering and weather conditions is not a simple one. Poore (1968) and others suggested that dipterocarp flowering depends on water stress following spells of dry weather, but both in Borneo (Wood 1956) and in Malaya (Burgess 1972) it has been found that gregarious flowering is not consistently preceded by periods of low rainfall. There is, however, some evidence that suggests a connection between flowering and high levels of insolation. It may be noted also that dipterocarps growing in rain forest do not flower until they are tall enough for at least part of the crown to be exposed to direct sunlight (Ashton 1982), and under forest conditions emergent species probably do not reach flowering age for many years.

Wycherley (1973) found that, although dipterocarp flowering in Malaya and Sarawak could not be precisely related to low rainfall in the previous year, there was a highly significant correlation with annual mean temperature and wide diurnal temperature range, both of which are related to of high insolation. In the Kepong arboretum (Malaya) dipterocarp flowerings take place consistently in March or April (Ng 1977a). The two previous months are normally a period during which the daily hours of sunshine rise following the very wet months November and December. Ng found evidence from the records that gregarious flowering took place when this increase in insolation was unusually large.

From the climatic data, and the fact that Malesian dipterocarps flower only at long intervals, Wycherley concluded that a build-up of assimilates to a threshold level may be necessary for forming inflorescence initials. Anthesis itself could be triggered by a different factor, which might be a period of water stress following heavy rain; small changes in weather conditions such as these are not easy to detect in meteorological records as usually presented.

9.2.5 Monocarpy

A few rain-forest trees are monocarpic[2]: they flower only once, usually very profusely, and die after the fruit has ripened. An example is *Spathelia excelsa* (Rutaceae) of central Amazonia, which is a slender unbranched tree, 15–20 m high (Holttum's model, see p. 74), with large compound leaves; it superficially somewhat resembles the palm *Euterpe*. After some years an enormous panicle of flowers and fruits is produced; the apical meristem then becomes inactive and the tree dies. This remarkable plant is widespread in 'rather open terra firme forests' (Rodrigues 1962).

At least two species of *Tachigali* (Leguminosae; Caesalpinioideae) are also monocarpic. *T. versicolor*, which occurs in evergreen and semideciduous forests in Central America, is an emergent tree reaching over 40 m. Foster (1977) studied 430 individuals on Barro Colorado Island and found that after flowering the leaves are shed and the tree usually dies within a year. Flowering does not occur every year; in 1967–69 no trees flowered, but in 1970 eighteen flowered and died. In 1974 three trees flowered and fruited only on a few branches and these trees did not die. The biological significance of monocarpy in such species is hard to understand: Foster suggests that its value may be to provide gaps in the canopy favourable for the survival of juvenile trees. *T. myrmecophila*, a species found in

rain forest near Belém (Lower Amazon) is also a very large monocarpic tree (M. Pires).

Monocarpic plants, though rare in lowland rain forest, are more common in other tropical environments, especially in Southeast Asia. Examples are the Asiatic talipot palms (*Corypha*) of seasonal climates. The species of *Strobilanthes* (Acanthaceae), tall woody herbs mainly of montane forests in seasonally dry tropical climates in the Indo-Malayan region, flower gregariously and die after fruiting, leaving the ground carpeted with seedlings (Fig. 9.9). Flowering takes place at intervals of 6–12 years, depending on the species and the locality (Petch 1924, further references in Whitmore 1984a); on Gunung Gede (Java) *S. cernua* flowered regularly every nine years from 1902 to 1956 (van Steenis 1972). Some species of the related African genus *Mimulopsis* are also monocarpic and flower gregariously.

Many bamboos of seasonal tropical and temperate regions grow vegetatively for long periods (up to about 120 years in some species); they then flower, fruit and die (references in Janzen 1976a). In some Asiatic climbing palms the flowering shoots die after fruiting and are replaced by new shoots from the base, but others are truly monocarpic (Uhl & Dransfield 1987). The flowering of most bamboos in the rain forests of Malaya is annual and not gregarious (Holttum 1953), but a few species, e.g. *Schizostachyum* spp., are monocarpic and die after fruiting (Whitmore 1984a). Little is known about the flowering of African and tropical American bamboos, but some at least of the latter die after flowering (Janzen 1983, pp. 330–1).

9.2.6 Flowering and vegetative growth

In tropical trees a relation between flowering and vegetative activity is most obvious in deciduous species, many of which flower when bare of leaves, e.g. *Bombax* and *Tabebuia* spp. (Fig. 9.5) or while the young leaves are expanding.

In Neotropical Lecythidaceae, which are mostly trees of lowland rain forest, there are three types of behaviour (Prance & Mori 1979). In one small group of *Lecythis* species, leaf-fall is followed by a flush of young leaves and then by flowering. In another group, of which *Couratari* spp. are examples, all the leaves fall in a short time and flowering occurs while the tree is bare; afterwards there is a flush of leaves. In the third type of behaviour, found in many members of the family, flowering is unconnected with leafing and leaf production is more or less continuous.

In *Ceiba pentandra*, in which some parts of the crown retain their leaves while the rest is leafless (p. 241), it

---

[2] See Hallé *et al.* (1978) and Simmonds (1980) on the use of the term monocarpic.

Fig. 9.9 *Strobilanthes* sp., a monocarpic shrub that has recently flowered at *ca.* 2100 m, between Pattipola and Ambarwela, Sri Lanka, March 1956. Dead stems and young seedlings cover the ground.

can often be seen that flowers and fruits are borne only on leafless branches. The curious Central American treelet *Jacaratia dolichaula* loses its leaves during flowering, in the lower part of the crown, while the upper part retains them (Bullock & Bawa 1981).

In leaf-exchanging and evergreen trees the connection between flowering and leaf-change is more variable. Many leaf-exchanging trees of the Nigerian evergreen seasonal forest, such as *Lophira alata* and *Terminalia* spp., flower when the young leaves are expanding. In some evergreen trees in Malesia, e.g. Dipterocarpaceae and oaks (*Quercus* and *Lithocarpus*), flowering does not occur at the same time as flushing, although in others, e.g. *Saraca* and *Myristica* spp., flowers and young leaves are produced together.

Close study of tropical trees shows that the relations between their flowering and vegetative activity are complex, depending on their shoot architecture and the behaviour of the apical meristems. This has been emphasized by Borchert (1983) whose views may be briefly summarized here.

In some tropical trees, e.g. *Cordia* and *Tabebuia* spp., the shoots are hapaxanthic, i.e. their growth is determinate, usually ending in the production of a terminal inflorescence, but growth is continued from lateral meristems (except in monocarpic species such as *Spathelia*). However, the majority of tropical trees are pleionanthic: in these the apical meristem continues to function indefinitely, inflorescences arising from lateral meristems.

More important for the phenology of tropical trees is the distinction between what Borchert calls sylleptic and proleptic flowering (extending the terms originally applied to branching systems by Hallé & Oldeman 1970). In the former the meristems are continuously active from flower initiation to the actual opening (anthesis) of the flowers, but in proleptic flowering there is a dormant period between flower initiation and anthesis. In species in which flowering is proleptic the factors controlling flower initiation and anthesis are different. Little is known about the factors controlling initiation in tropical trees. Borchert believes that it probably depends on internal correlative factors, such as carbohydrate levels in the meristems and the balance between growth regulators. External factors affect flower initiation only indirectly (and in Borchert's opinion those doing so do not include day-length). On the other hand, as shown earlier, there is much evidence that changes in water stress, and in some cases in temperature, control anthesis in many tropical trees. Leaf-fall, by temporarily reducing water stress, can induce anthesis.

Borchert's conclusions are based mainly on a small number of tree species, mostly deciduous, in seasonal forests in Costa Rica, and need to be tested on a much larger and more diverse sample of species before it will be possible to generalize about their ecological significance.

9.2.7 Fruiting

Peaks and troughs in the flowering of rain-forest trees are usually followed, as might be expected, by similar peaks in fruiting. In Medway's observations (1972) in Malaya (Fig. 9.1) the percentage of trees flowering

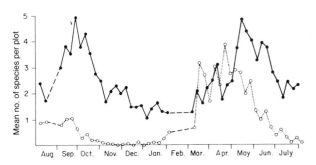

Fig. 9.10 Seasonality of animal- and wind-dispersed species in forest, Barro Colorado Island, Panama. From Foster in Leigh *et al.* (1983). Filled circles, animal-dispersed; open circles, wind-dispersed.

during 1963–69 was on the average greatest in June–July, and the maximum of fruiting was about eight weeks later. Similarly, at La Selva, Costa Rica (Fig. 9.7) the main flowering peak of the 'overstorey' trees in May is followed by a large peak of fruiting in September, ripening and seed-fall taking place some months later; the fruiting peak of the 'understorey' species is a little later than that of the 'overstorey' trees (October) (Frankie *et al.* 1974). However, for several reasons there is no exact correspondence between flowering and fruiting seasons.

In some individual trees or species large crops of flowers are not necessarily followed by large crops of fruit. In many cases a large proportion of the flowers and immature fruits may be destroyed by heavy rain, fungi or predators. There is also much variation between species in the interval between pollination and seed ripening, and in the course of fruit development. Information on how long the fruits of tropical trees take to mature is scanty: for many species it seems to be about 2–3 months and even some of the heaviest fruits do

not take longer. *Durio zibethinus* (average fruit mass *ca.* 2–3 kg) is said to take three months to ripen, mango 2–5 months and cocoa 4–5 months (Purseglove 1968). The extremely large seeds of *Mora megistosperma* (Fig. 5.4) are shed *ca.* 12 weeks after flowering (Janzen 1983) and the slightly smaller ones of *M. excelsa* some fourteen weeks after (Bell 1971). Some fruits and seeds require much longer than three months to mature; the seeds of *Carapa guianensis* ripen about eight months after flowering and the development of some tropical fruits is said to take more than a year. Most light wind-dispersed fruits and small berries probably develop more quickly, but some mature as slowly as large heavy fruits; the very small wind-dispersed fruits of *Cordia alliodora* are shed 2–3 months after flowering (Janzen 1983). Some trees which flower continuously or at relatively short intervals bear two or more generations of fruits simultaneously.

The development of a fruit from a pollinated flower is not necessarily continuous. In *Hymenaea courbaril* the seed pods expand to full size by 1–2 months from flowering, but remain on the tree for several months before the ripe seeds drop in the following dry season, while those of *Albizia saman*, which in Costa Rica flowers late in the dry season, become dormant while still very small: they do not reach full size until late in the following rainy season and drop in the next dry season (Janzen 1983).

Although there appears to be little correlation between size of the fruits and the time they take to ripen, Foster (1983) found that in a normal year at Barro Colorado Island there is a striking difference in the seasonal rhythm of fruitfall between wind-dispersed fruits and those dispersed by other means; the former have a single well-marked peak of abundance at the end of the dry season, but in animal-dispersed fruits distribution is bimodal (Fig. 9.10).

Chapter 10

Soils of the humid tropics

I.C. Baillie

10.1 General soil features and soil-forming processes

SINCE THE FIRST edition (Richards 1952) it has become increasingly appreciated that the soils of the humid tropics are very heterogeneous (see Nicholaides 1978, van Wambeke 1992, Ahn 1993). Their diversity notwithstanding, many of the soils have a number of characteristics in common. These give the region its distinct pedological identity, and arise from its characteristic combination of environmental features.

As was seen in Chapter 7, the macroclimates of the humid tropics are characterized by high, even temperatures and abundant rainfall. This combination of warmth and moisture means that climatic constraints do not significantly retard the rates of a wide range of soil processes for all or much of the year. In the absence of non-climatic constraints the leaching of solutes, the weathering of primary rock minerals, the evolution and weathering of secondary clay minerals, and many biological processes such as the comminution and decomposition of organic materials, are able to proceed rapidly and without interruption. The region has a high proportion of old soils because there are large areas of tectonically and geomorphologically stable land surfaces, and only small areas in high mountains were directly affected by glaciations or periglacial activity during the Quaternary.

The result is a high proportion of old and intensively processed soils that are visually striking; they appear deep and fairly uniform, with little organic darkening of the topsoil and with bright reddish or yellowish colours throughout the subsoil. Weathering is far advanced and few of the original rock minerals remain in the upper horizons. The secondary clay minerals are also well weathered, and they include substantial quantities of free oxides and hydroxides of iron, aluminium and manganese, collectively referred to as the sesquioxides. The ferric sesquioxides give the soils their distinctive bright colours. The kandites (kaolinites) are the main group of aluminosilicate clay minerals. They have low permanent electrical charges on their surfaces, so that they are chemically and physically rather inactive compared with other clay minerals. Surface litter layers are thin or absent, as fresh organic inputs are rapidly comminuted by animals or decomposed directly by microbes. However, the organic matter content of the underlying horizons is not as low as the colours suggest. The soils have been subject to prolonged and intensive leaching and contain relatively small amounts of mineral nutrients, either in the soil solution or as exchangeable ions on the soil's clay–organic exchange complex. The soils are acid, and aluminium is often the dominant exchangeable cation. They have open porous structures, and aeration is generally good; water infiltrates rapidly at the surface and percolates down the profile fairly freely.

This is a conventional portrayal of the typical (zonal) soil of the humid tropics. It masks much variation that is important in the functioning of landscapes and the ecology of rain forests: it also excludes many important soils that are not deep, reddish, highly weathered, nutrient-depleted, acid, or well drained. The main soils are described in more detail below (section 10.3), after a brief discussion of soil classification and nomenclature (section 10.2).

10.2 Soil classification and nomenclature

Unfortunately there is no generally agreed system of soil classification. The present confusion is further exacerbated by rather user-hostile terminologies. This makes soil identification and characterization difficult and ambiguous, especially for non-pedologists. As discussed elsewhere, ecologists have long recognized soils as an important feature of the rain-forest environment, but have lacked a generally accepted and comprehensible system for classifying them. There are two major international systems of soil classification and nomenclature in wide use at present; the legend of the *Soil Map of the World* (FAO–UNESCO 1974, 1988) and *Soil Taxonomy* (Soil Survey Staff 1975, Soil Management Support Services 1990). There is also the French system and its derivatives (Duchaufour 1982), which are widely used in the tropics, especially in Africa, and important national systems, such as those of Indonesia (Whitten *et al.* 1984) and Brazil (Brown 1987), which classify wide ranges of humid tropical soils.

The subdivision of the deep and reddish mature soils of the humid tropics has been a considerable problem for soil taxonomists. Early attempts emphasized single characteristics such as consistency or colour (Thorp 1936). Soviet pedologists used subdivisions based on the elemental composition of the clays, especially the sesquioxides (Gerasimov 1973). The modern international systems originally emphasized the vertical distribution of clay within the profile and the exchangeable base status (FAO–UNESCO 1974, Soil Survey Staff 1975). The recent revisions of these systems continue with these criteria but now give more weight to the nature of the clay fraction, especially its cation exchange capacity (FAO-UNESCO 1988, van Wambeke 1989, Soil Management Support Services 1990).

In order to emphasize differences that appear to be ecologically important, this chapter uses its own loose grouping of the main soils of the region. This is not intended as a formal classification system and the groups are not rigidly defined. They are described in terms of their central concepts rather than their outer limits. The level of detail is intermediate between the very broad climatic zonal approach and the intricacy of the international systems. The scheme is similar in many ways to that used by Burnham in Whitmore (1984a). The names of the groups are either traditional or descriptive. The groups and their approximate equivalents in the international systems are listed in Table 10.1. The profile characteristics of some of the more important soils are depicted in Fig. 10.1. The lithology of soil parent mater-

ials figures more prominently than in most classification systems, as many ecologically important differences appear to be associated with the geological substrate, even in deep and highly weathered soils (Gerasimov 1973, Lu 1989).

Many of the soil parent materials in the region are allochthonous and were emplaced by creep, wash, mass movement and other transportation processes, during which they were considerably churned and mixed. Many soils have therefore developed in materials of heterogeneous lithology. The mixing, combined with the intense and deep weathering, can blur parent materials and make their identification difficult, but it is usually possible to distinguish broad lithological classes.

10.3 Main soil groups of the humid tropics

10.3.1 Kaolisols

This group includes most of the intensively weathered and leached zonal humid tropical soils as briefly described above (section 10.1). It corresponds more or less with the 'tropical red earths' of the first edition of this book and with broadly defined groups such as lateritic soils (Robinson 1932, Li Jin *et al.* 1988), kaolisols (Young 1976), and the red soils of Indonesia (Buurman 1978) and southern China (Thorp 1936). More recent names emphasize particular characteristics of these soils, such as the importance of iron and aluminium in the elemental composition, e.g. ferrallitic soil (Duchafour 1982), and ferralsol (FAO–UNESCO 1974); or of sesquioxides (oxisol) (Soil Survey Staff 1975) and kanditic aluminosilicates (kandisol) (Bleeker 1983) in the clay minerals. As a general name, 'kandisol' has much to commend it, but recent revisions of the American Soil Taxonomy use the prefix 'kandi' at a fairly low taxonomic level and with a restricted meaning (Soil Management Support Services 1990). The older name 'kaolisol' (Sys 1960) is therefore preferred, to avoid confusion with others in current use.

Deep mature kaolisols predominate in stable landscapes in the interiors of tectonic plates, such as much of humid Africa, India, and South America. They are less important in the more unstable landscapes of tectonic plate margins, such as Central America, the Caribbean, Melanesia, and Malesia. The modal kaolisols are derived from moderately quartziferous parent materials. They are described below in some detail and the other kaolisols are treated more briefly, emphasizing their differences from the modal group.

Table 10.1. *Soils of the humid tropics*

	Approximate equivalents in international systems of soil classification	
	Soil Taxonomy (Soil Survey Staff 1975, Soil Management Support Services 1990)	Soil Map of the World (FAO-UNESCO 1974, 1988)
Kaolisols		
Modal kaolisols	Udult, Perox, Udox	Ferralsol, Lixisol, Acrisol, Alisol, Ferralic Arenosol
Clay kaolisols	Udult	Acrisol (also Alisol, Lixisol, few Nitosol)
Basic oxidic clays	Perox, Udox (Eutric groups)	Nitosol (also Ferralsol)
Ultrabasic oxidic clays	Perox, Udox (Eutric groups)	Nitosol (also Ferralsol)
Limestone oxidic clays	Eutric Perox, Udox (also Udalf)	Nitosol, Ferralsol (also Luvisol)
Alluvial kaolisols	Udult (also Udalf, Tropofluvent)	Acrisol (also Luvisol, Fluvisol)
Ferricrete kaolisols	Plinthic groups in Ultisols, Oxisols, Udalfs	Plinthic units in Acrisols, Ferralsols, Alisols
Non-kaolisol mature terra firme soils		
Podzols	Tropohumod (also Tropaquod)	Carbic (also Gleyic) Podzol
White sands	Tropopsamment	Albic Arenosol
Acid planosols	Aquult	Planosol (mostly Dystric)
Immature terra firme soils		
Andosols	Andisol (formerly Andept)	Andosol
Recent colluvial soils	Tropept, Troporthent	Cambisol, Leptosol, Dystric Regosol
Recent alluvial soils	Fluvent	Fluvisol
Skeletal soils	Tropept	Cambisol, Leptosol
Shallow calcareous clays	Rendoll, Eutropept	Rendzic and Mollic Leptosol
Poorly drained soils		
Freshwater gleys	Tropaquent, Tropaquept	Gleysol
Ferruginous semi-gleys	Aquic and Plinthic suborders and groups of Oxisols and Ultisols	Plinthosol, Gleyic Acrisol
Saline gleys	Tropaquent, Tropaquept	Gleysol (including Thionic)
Acid sulphate soils	Sulfaquent, Sulfaquept	Thionic Gleysol and Fluvisol
Peats	Histosol (Tropic groups)	Histosol (all units except Gelic)
Montane soils		
(not subdivided)	Histosol, Inceptisol (and others, especially Humic suborders and groups)	Histosol, Cambisol, Leptosol, Gleysol (especially Humic units and subunits)

Modal kaolisols

These soils are widespread on a range of sedentary and colluvial parent materials of intermediate and moderately acid mineralogical composition, such as granites, gneisses, acidic granodiorites, feldspathic sandstones and sandy shales. They also develop in old and highly weathered alluvia and volcanic ash with similar mineralogies. The full development of these soils takes a long time and therefore requires a reasonably stable site.

Table 10.2 summarizes the morphological characteristics of a moderately deep example of these soils. They are generally deep, with two metres or more of true soil (solum) above the weathered parent material. This has been greatly softened but retains traces of its original rock structure. However, it is usually chemically weathered to considerable depth, often many metres. The solum appears rather uniform. There is little surface litter and the darkening of the topsoil by organic matter is weak in most profiles but is more pronounced in coarser textured soils. The predominantly red–yellow subsoils tend to become redder with depth, but only gradually. Some sola are completely unmottled but other profiles have weak or moderate pale yellow and light grey mottles in the subsoils (e.g. Nyalau family profile in Table 10.2). These may be pockets of intermittently

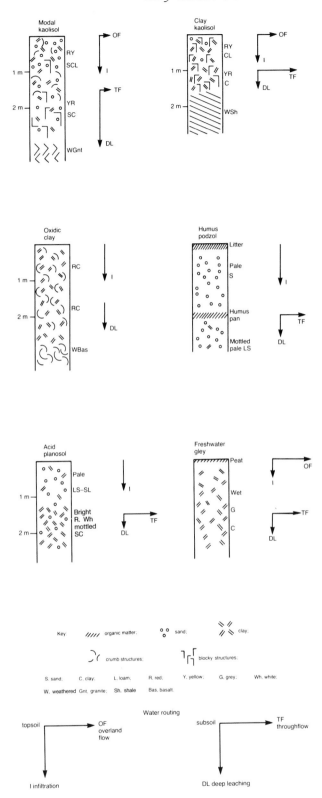

Fig. 10.1 Morphology and hydrology of major soil groups of the humid tropics.

poor aeration or incomplete weathering. The colours of the soft weathering rock can be variegated, ranging from those of the fresh rock through purple, red and brown to yellow and white. Comparisons between areas of similar lithology suggest that the iron sesquioxides are more hydrated, giving yellower matrix colours, or more likely to be patchily reduced, giving more pronounced subsoil mottles, in areas of high rainfall. However this effect is often masked by more striking differences due to variations in parent materials.

The main chemical and physical characteristics of the example are summarized in Table 10.3. The parent materials contain substantial quantities of quartz, which is relatively unweatherable and remains to form a significant sand fraction. All of the other common primary minerals weather to secondary clay minerals. The topsoils are usually of coarse or medium texture, with sandy loams most common (Lal 1986). The topsoils generally have moderately developed crumb or fine subangular structures. There is generally a gradual but clear increase in clay content with depth, and subsoils are generally of medium texture, sometimes as fine as clay loam. The main structures in the subsoils are moderate or weak subangular blocks of medium size. These are sometimes compound structures, which crumble to finer and more rounded crumbs or granules. The blocky structures may have moderately thick and almost continuous clayskins, but these are weak or absent in most profiles. The consistence is usually quite friable in the topsoil and becomes firmer with depth, but the subsoils are rarely so compact as to mechanically hinder root penetration.

Modal kaolisols are fairly porous and permeable throughout. There does not appear to be any significant impedance to the free vertical percolation of water, even in medium textured and more compact subsoils. About 40% of the throughfall reaching the surface of one of these soils under forest in Central Amazonia penetrated to a depth of two metres (Forti & Moreira-Nordemann 1991), the remainder presumably being diverted by root uptake and transpiration. Some of the water flows through the macropores, which drain rapidly and act as the aeration network except briefly after heavy rain (Greenland 1977). This water has only limited opportunity for equilibration with the soil solids and has lower solute concentrations than the water in micropores, which remains longer in the soil and is the main source of water and nutrients for roots and microbes (Nortcliff & Thornes 1989). However, some soils appear to drain by complete displacement 'piston flow' (Bruijnzeel 1990). Whichever route is followed, there appears to be little lateral diversion as subsoil through-

Table 10.2. *Profile of modal kaolisol*

Nyalau family, Bakam Road Silvicultural Reserve, northern Sarawak (4°N, 113°E). Steep upper slope under recently logged dipterocarp forest, over fine-grained sandstone. Annual rainfall *ca.* 3000 mm.

0–8 cm	Light yellowish brown sandy loam with angular quartz grit, fine subangular blocky structure, friable and porous, abundant roots
8–30 cm	Brownish yellow sandy clay loam, medium subangular blocky, friable and porous, many roots
30–68 cm	Yellow and reddish yellow sandy loam, medium subangular blocky, slightly firm, few roots
68–120 cm	Reddish yellow with faint yellow and light grey mottles, sandy clay loam, medium subangular blocky with clayskins, firm, few roots
120–160+ cm	Mixed yellow, reddish yellow and light grey sandy clay loam, massive, firm, no visible pores, no roots

Source:–From Baillie (1971).

Table 10.3. *Chemical and physical characteristics of Nyalau modal kaolisol[a]*

	Depth (cm)		
	0–8	30–68	120–160
pH (in water)	4.2	4.6	4.5
Organic carbon (%)	2.4	0.1	0.1
Total nitrogen (%)	0.14	0.04	0.05
C:N ratio	17	3	2
Exchangeable Ca (milli-equivalents per 100 g fine earth)	0.2	0.1	0.1
Exchangeable Mg	0.4	0.1	0.1
Exchangeable K	0.1	Tr	Tr
Exchangeable Na	Tr	Tr	Tr
Cation exchange capacity[b] (meq per 100 g)	9.1	7.3	8.7
Base saturation (%)	9	4	3
Total P (ppm)	80	60	70
Total Ca (ppm)	60	160	60
Total Mg (ppm)	650	730	1260
Total K (ppm)	1700	1930	3210
Coarse[c] sand (%)	2	3	2
Fine[c] sand (%)	63	64	54
Silt (%)	17	15	16
Clay (%)	18	18	28
Texture[d]	SL	SL	SCL

[a] See Table 10.2 for morphology.
[b] CEC determined at pH 7. May exceed field values owing to pH-dependent deprotonation. Base saturations may be depressed accordingly.
[c] Coarse sand, 0.2–2.0 mm; fine sand, 0.05–0.2 mm.
[d] SL, sandy loam; SCL, sandy clay loam.
Source: From Baillie (1971).

flow (Nortcliff & Thornes 1978), and streams off these soils are mostly fed by groundwater flow. The free drainage is partly due to the pronounced and stable aggregation of the clay particles so that physically they behave rather like silt or fine sand. Bonding by the sesquioxides contributes to this aggregation (Ahn 1979), but organic matter is probably equally or more important (Soong 1980, Lal 1986). The mainly vertical drainage contributes to the deep leaching and weathering of these soils.

Quartz is the only important primary mineral remaining in quantity, mostly as coarse particles in the sand fraction, but also with some finer particles in the clay. The secondary minerals are well weathered, with few 2 : 1 lattice aluminosilicates. As the group name indicates, the clay minerals are dominated by 1 : 1 kandites (especially kaolinites). There may also be some 2 : 1 : 1 hydroxy-interlayered minerals, especially in the upper horizons.

Because of the advanced weathering, much of the original iron is present as free sesqioxides, giving the the soils their bright colours. However, the limited quantities of ferromagnesian and other iron-bearing minerals in the moderately acid parent materials means that contents of free iron sesquioxides are not very high, rarely exceeding 10% of the fine earth. In some soils there are hard, dark, round ferrimanganiferous concretions, resembling shotgun pellets (Ojanuga & Lee 1973). Many of these are inactive remnants of former ferricrete sheets, but they may not be completely inert. Some are currently undergoing isomorphic transformations, with ferric iron, manganese and aluminium substituting in each other's sesquioxides and in aluminosilicate structures. Little of the original aluminium is present as free gibbsite, much of it remaining in the octahedral layers of the 1 : 1 kandites, deposited as interlayers in the 2 : 1 : 1 minerals, or in aluminized goethite (Tardy & Nahon 1985). There is also labile aluminium, either in the soil solution or adsorbed as hydrated polycations on the cation exchange complex. Many of the free sesquioxides are poorly crystallized and occur as very fine particles or as coatings on, and bridges between, particles of other minerals.

The kanditic and sesquioxide clay minerals carry few permanent negative charges; hence the collective name of 'low activity clays' (Buol 1985). They and the organic matter have some capacity to adsorb and desorb protons according the pH of the soil solution. As the pH decreases some of the negatively charged sites on their surfaces are blanked off by these protons, and the cation exchange capacities are reduced accordingly. In very acid soils enough protons may be adsorbed to give the surfaces a net positive charge. However, most soils retain a low net negative charge and some cation exchange capacity, although a few positively charged sites may remain and give localized spots capable of anion adsorption (Sollins 1989).

The intensive leaching makes these soils quite acid, with a pH range of 3.5–5.5 (as measured in water). It also leads to considerable depletion of the basic cations, so that the cation exchange complex is of low base saturation, despite its limited capacity. There is a tendency for leaching, acidity and base deficiency to be increased with rainfall but, as with colour, the effect is often masked by variation in the parent materials. Aluminium dissolves from gibbsite and other aluminium sesquioxides in the acid conditions prevailing, and becomes the dominant cation in the soil solution and on the exchange complex. By releasing hydroxyl ions, as well as aluminium, into the soil solution the solution of gibbsite may act as a partial buffer for the soil pH, which rarely falls below 3.5. The pH and exchangeable base status are often highest close to the soil surface, owing to the biological recycling of the basic cations. There may also be a slight increase at depth, close to the underlying parent material, due to the release of fresh cations by weathering. Phosphate concentrations are low, especially the available forms. Because of their polyvalency, phosphate ions are preferentially and strongly adsorbed onto anion exchange sites. This and co-precipitation with labile forms of iron and aluminium remove much of the phosphate in these soils from biological circulation.

Most litter overlying modal kaolisols disappears within a few weeks or months (Madge 1969), but some persists for up to a year (Anderson & Swift 1983, Anderson & Spencer 1991). Much of it is grazed or comminuted by the soil and litter fauna prior to microbial decomposition. The animal groups involved vary between areas. Termites are generally the most important, but worms, ants, springtails, myriapods and mites are all significant in some forests (Leakey & Proctor 1987). Some of the termites and their symbionts, especially fungi nurtured in their nests, are able to utilize cellulose. In some West Malaysian forests termites directly consumed over 10% of the litterfall (Matsumoto & Abe 1979). In Sarawak and most other areas the proportion is lower but termite frass is highly susceptible to microbial colonization and decay, so that the effect of these animals on the disappearance of litter is still considerable (Collins 1983). Despite its generally rapid initial processing, organic matter does not get completely respired and some humus persists. The organic carbon and total nitrogen contents of the topsoils are often moderately low, but higher than the absence of dark colours suggests (Sanchez 1976). The light colours are attributed mainly to the masking effect of the brightly coloured ferric sesquioxides. Subsoil contents of organic matter are low and little appears to be eluviated, either vertically or laterally. Streams draining these soils are clear or turbid with suspended mineral material, but lack the weak tea ('blackwater') colours characteristic of streams draining podzols and white sands (see Chapter 14).

The coarser textures in the topsoils indicate that some of their original clay has either been weathered to destruction or been washed out. The general increase

Table 10.4. *Profile of clay kaolisol*

Piedra series, Topco National Land, southern Belize (16°N, 89°W); crest of low wide interfluve; semideciduous broad-leaved forest with high proportion of cohune palm (*Orbignya cohune*), over mudstone. Annual rainfall *ca.* 3800 mm.

0–21 cm	Dark yellowish brown silty clay loam, coarse crumb structure, porous, moist and friable, earthworms, many roots
21–38 cm	Yellowish red clay with many red and yellow mottles, fine blocky with weak clayskins, moist and firm, porous, many roots
38–70 cm	Pale brown clay with light grey and reddish mottles, medium blocky with strong clayskins, wet and sticky grading to moist and friable at depth, few pores, few roots
70–95+ cm	Light grey soft weathering mudstone with brown and yellow mottles and black manganese stains, blocky–platy with moderate clayskins, dry and crumbly, few roots

Source: From Baillie & Wright (1988).

in clay content with depth and the existence of patchy clayskins suggest that at least some has been eluviated vertically and redeposited in the subsoil. There are suggestions that in some areas this process was more active in the past and that the clayskins are palaeoclimatic relics (Dijkerman & Miedma 1988).

Other kaolisols

Clay kaolisols

These are very extensive soils on stable sites over a wide range of micaceous metamorphic and non-calcareous, fine-grained sedimentary rocks. Shale is the commonest parent rock but they also occur on mudstone, phyllite, slate, schist and gneiss, and on derived fine-grained alluvial and colluvial deposits. These soils resemble modal kaoldisols in having little organic darkening in the topsoils and predominantly yellow and red subsoils that become redder with depth. Clay content increases with depth, and low pH and exchangeable base status. However there are important differences, which stem from the parent materials and the aluminosilicate clay minerals generated by their weathering.

Table 10.4 summarizes the morphology of a clay kaolisol. The main differences between these soils and the modal kaolisols are: subsoil mottling is more common, although reddish or reddish-yellow matrix colours still predominate; textures are finer, mostly clay loam or clay; the clay minerals include more 2:1:1 interlayered and some 2:1 aluminosilicates, especially potassium-bearing illites, as well as the dominant kandites; subsoils have blockier structures and firmer consistence; clayskins are more pronounced, suggesting that clay translocation contributes to the finer textures of the subsoils; subsoil permeability tends to be lower and some of the mottling is due to locally and intermittently

impeded drainage; and sola tend be shallower, with weathered parent material often at less than two metres. Some of the subsoil mottles may be patches of incomplete weathering.

Table 10.5 summarizes the analytical data of the Belizean example, and shows that the pH, exchangeable base status, and available phosphate contents of these soils are still low by global standards but are slightly higher than in the modal kaolisols. However, the ranges overlap considerably and aluminium is still the dominant exchangeable cation in the subsoil. Total contents of potassium, magnesium and, to a lesser extent, phosphorus are considerably higher, mostly held in non-exchangeable forms by the non-kandite aluminosilicates, such as interlayer potassium in the 2:1 and 2:1:1 minerals (Yew 1979).

The clay kaolisols differ hydrologically from the modal group. The pronounced blocky structures tend to emphasize the bifurcation of percolating water into rapid flow between peds and much slower movement in the fine pores within them (Sollins 1989). The combination of fine texture and the partial constriction of the macropores by illuvial clay renders the subsoils only moderately permeable. Infiltrating water tends to stagnate slightly in the middle horizons or is diverted laterally as subsoil throughflow. The wet horizon at 38–70 cm in the Piedra series profile in Table 10.4 is thought to be the main throughflow transmission zone. Substantial quantities of throughflow were measured at similar or shallower depths in other clay kaolisols in southern Belize (Baillie *et al.* 1991). The throughflow can cause subterranean erosion by enlarging macropores and forming pipes (Walsh & Howells 1988). The low subsoil permeability of clay kaolisols over basic schists in northeastern Queensland causes the surface horizons to

Table 10.5. *Chemical and physical characteristics of Piedra clay kaolisol*[a]

	Depth (cm)		
	0–21	21–38	38–70
pH (in water)	4.2	3.8	3.7
Organic carbon (%)	3.6	1.2	0.4
Total nitrogen (%)	0.36	0.16	0.06
C:N ratio	10	7	7
Exchangeable Ca (milli-equivalents per 100 g fine earth)	5.2	3.9	5.8
Exchangeable Mg	2.5	0.8	1.5
Exchangeable K	0.2	0.2	0.3
Exchangeable Na	0.1	0.3	0.0
Exchangeable Al	0.9	13.7	21.0
Cation exchange capacity[b] (meq per 100 g)	15.3	24.8	34.8
Base saturation (%)	52	21	22
Available P (ppm)	6	3	4
Total P (ppm)	230	140	80
Total Mg (ppm)	3100	6250	9100
Total K (ppm)	1560	3450	6600
Coarse[c] sand (%)	4	1	2
Fine[c] sand (%)	30	11	5
Silt (%)	34	18	22
Clay (%)	31	70	71
Texture[d]	ZCL	C	C

[a] See Table 10.4 for morphology.
[b] CEC determined at pH 7. May exceed field values owing to pH-dependent deprotonation. Base saturations may be depressed accordingly.
[c] Coarse sand, 0.2–2.0 mm; fine sand, 0.05–0.02 mm.
[d] C, clay; ZCL, silty clay loam.

Source; From Baillie & Wright (1988).

become rapidly saturated and generates considerable overland flow during prolonged heavy rainfall in the monsoon (Bonnell *et al.* 1983). Similar mechanisms operate in what appear to be clay kaolisols in Western Amazonia (Elsenbeer & Cassel 1991). These lateral diversions reduce the quantity of water that reaches and leaches the lower subsoil and saprolite, and may contribute to the shallower sola and less advanced weathering of these soils. As streams off these soils are partly fed by shallow throughflow, sometimes also with overland flow, they tend to have rapid response hydrographs.

Basic oxidic clays
These are not extensive soils in rain forests but are widely scattered throughout the humid tropics. They are among the most fertile soils of the region and are important for agriculture. They are developed on stable sites from basalt and other basic rocks, and old basic volcanic, colluvial and alluvial deposits. They can also occur on schists and, less commonly, gneisses with high contents of ferromagnesian minerals. The profile of an

example of these soils is summarized in Table 10.6. The distinctive differences from the modal kaolisols are: the intense, warm and usually unmottled red and brown colours; the lack of horizonation; the uniformly high clay contents; and the crumb-like aggregation and friable consistence of the subsoils. The main clay minerals are kandites and sesquioxides, especially those of iron. The contents of iron sesquioxides are much higher than in the modal or clay kaolisols, accounting for up to half of the clay fraction in some horizons (Curi & Franzmeier 1987). They occur as large and well-crystallized discrete particles, as well as forming coatings and bridgings on other minerals (Juo 1981). Gibbsite is absent or of minor importance. Some of these soils have subordinate contents of expansible 2:1 aluminosilicates. These may give the soils quite a dark colour, ranging from dark red to very dark brown.

The well-aggregated crumb structures in the subsoils are attributed to the plentiful sesquioxides. They make these soils very porous and freely drained, considering their fine textures. Drainage is mainly vertical with

Table 10.6. *Profiles of oxidic clays*

(a) *Basic oxidic clay*

Tarat series, Kuching, Sarawak (1°N, 110°E); hillslope under old rubber, over basalt. Annual rainfall *ca.* 4000 mm.

0–8 cm	Dark brown loam, crumb structure, porous, moist and friable, many earthworms, many roots
8–42 cm	Red clay loam-clay, crumb, moist and friable, many roots
42–80 cm	Red clay, crumb, moist and very friable
80–150+ cm	Red clay, crumb, moist and friable, fragments of weathering basalt increase with depth

(b) *Limestone oxidic clay*

Putput red clays, Putput, New Britain (5°S, 152°E). Flat raised shoreline, altitude 80 m; under cacao plantation with *Leucaena glauca* shade, over coral. Annual rainfall *ca.* 2700 mm.

0–13 cm	Dark brown clay, fine angular blocky, slightly friable, earthworms, common roots
13–36 cm	Yellowish red clay; weak angular blocky, with clayskins; firm; few roots
36–70 cm	Red clay; coarse blocky; wet, plastic and sticky; no roots
70–110+ cm	Reddish yellow clay with red, yellow and increasing light grey mottles; massive; wet, sticky and plastic; rare roots

Sources: (a) Andriesse (1972); (b) Bookers (1981).

little lateral diversion as throughflow or overland flow. In profiles with significant contents of 2:1 expansible minerals there is a tendency to compound structures, with the crumbs aggregated into blocks. These soils are highly leached and acid, and contents of exchangeable bases and available phosphate are modest by global standards but higher than in the modal and clay kaolisols. The total contents of the main nutrients, usually but not alway including calcium, are considerably higher. Organic matter contents are higher than the lack of topsoil darkening indicates. Decomposition appears to be rapid and animal activity is high. The C:N ratios are generally less than 15 and most of the organic nitrogen is readily mineralized. However, in some soils part of the humus appears to be complexed by the plentiful sesquioxides and stabilized against further decomposition, thus retarding the mineralization of its nutrients.

Ultrabasic oxidic clays

These soils are found in small areas over serpentinite, peridotite and other ultrabasic (ultramafic) rocks. Although not extensive they are worth separating from the soils on basic rocks because of some ecologically significant differences. Morphologically they are somewhat similar to the basic oxidic clays, with weak horizon definition, fine textures, fairly high porosities and free or moderate drainage. Some profiles have bright red and unmottled subsoils, but most are rather more brownish and mottled (Fox & Tan 1971). Ferruginous sesquioxides are very abundant and may account for up to 70% of the fine earth (Schwertman & Latham 1986, Curi & Franzmeier 1987). The weathering of the abundant ferromagnesian minerals releases magnesium in large quantities and it may be antagonistic to the uptake of other cations (Alexander *et al.* 1985). These soils also contain significant quantities of a range of heavy metals, some at concentrations high enough to be toxic to non-adapted plants. The commonest are nickel, chromium, cobalt and copper, but others occur on specific lithologies (Proctor *et al.* 1988).

Limestone oxidic clays

These are extensive in some parts of the humid tropics. Much limestone topography in the region is rugged karst, in which there is a preponderance of very steep slopes with very shallow and stony dark clays (Section 10.3.3). However, there are also deeper soils on gentler and more stable slopes. Many of these are oxidic clays, with some resemblances to those on basic rocks, as can be seen in Tables 10.6 and 10.7. Features in common include uniformly high clay contents, intense reddish colours, high contents of ferruginous sesquioxides, high porosity and free drainage. Differences include more pronounced blocky structures and firmer consistency in the subsoils. There is a thin horizon of sticky and massive olive-coloured clay between the blocky red clay subsoil and the underlying limestone in many of these

Table 10.7. *Chemical and physical characteristics of oxidic clays[a]*

	Tarat series Depth (cm)			Putput red clay Depth (cm)		
	0–8	42–80	110–160	0–13	13–36	36–70
pH (in water)	4.7	4.5	4.8	6.8	6.5	6.3
Organic carbon (%)	6.1	0.6	0.02	4.6	1.0	0.7
Total nitrogen (%)	0.50	0.06	0.02	0.60	0.13	0.08
C:N ratio	12	8	1	8	8	9
Exchangeable Ca (milli-equivalents per 100 g fine earth)	1.3	0.7	0.5	29.1	17.9	16.4
Exchangeable Mg	0.6	0.2	0.2	4.2	2.1	2.0
Exchangeable K	0.5	0.1	Tr	1.7	0.8	0.7
Exchangeable Na	0.1	0.1	Tr	0.2	0.2	0.4
Exchangeable Al	3.3	Tr	Tr	0	0	0
Cation exchange capacity[b] (meq per 100 g)	23.6	8.5	7.8	38.0	22.9	22.8
Base saturation (%)	11	12	10	93	92	86
Available P (ppm)	—	—	—	5	1	0
Total P (ppm)	340	260	110	—	—	—
Total Ca (ppm)	1270	1800	1210	—	—	—
Total Mg (ppm)	2040	1800	2100	—	—	—
Total K (ppm)	580	420	<80	—	—	—
Coarse[c] sand (%)	11	7	12	2	1	0
Fine[c] sand (%)	7	6	16	6	2	0
Silt (%)	23	29	31	25	13	4
Clay (%)	59	58	41	67	84	96
Texture[d]	C	C	C	C	C	C

[a] See Table 10.6 for morphologies.
[b] CEC determined at pH 7. May exceed field values owing to pH-dependent deprotonation. Base saturations may be depressed accordingly.
[c] Coarse sand, 0.25–2.0 mm; fine sand, 0.05–0.25 mm.
[d] C, clay.

Sources: Tarat series, Andriesse (1972); Putput clay, Bookers (1981).

soils in Belize (King *et al.* 1986). The depth of these soils is variable, and the boundary with the underlying limestone is often abrupt and pocketed.

The main clay minerals are kandites and sesquioxides. Gibbsite is more common than on basic parent materials, and some of the world's most important workable sources of bauxite have formed from soils like these. The high contents of sesquioxides stabilize soil structures and contribute to the free drainage (Scholten & Andriesse 1986), which may be enhanced by high porosity in the underlying limestone. The chemical properties depend considerably on the depth to limestone. Where it is shallow enough to be tapped by root systems, biological recycling via root uptake and litterfall ensures a continual supply of calcium to the upper horizons. This tends to offset leaching losses, so that the soil pH and exchangeable base status remain high, with calcium as the dominant exchangeable cation (as in the Putput

example in Table 10.7), replaced or joined by magnesium where the limestone is dolomitic. In deeper soils, where the limestone is below rooting depth, little calcium is recycled. Such soils are subject to predominantly vertical leaching, and become acid and base-deficient (pp. 334–5, Chapter 12). Phosphate contents can be very variable, ranging from moderately low to exceptionally high (Schroo 1963).

Alluvial kaolisols

These soils occur in well-drained sites on the older deposits of emergent coastal plains, river floodplains and terraces. They resemble sedentary and colluvial kaolisols except that they tend to have higher contents of silt, which is often micaceous. The micas give high total contents of potassium and magnesium, and the silt may make the subsoils firmer and more compact than in other kaolisols. The deeper subsoils may be mottled and

Table 10.8. *Profiles of podzol, white sand and acid planosol*

(a) Podzol

Miri series: Kuala Nyabau, northern Sarawak (3°N, 113°E). Raised beach under kerangas (heath forest), over old marine alluvium. Annual rainfall *ca.* 3800 mm.

0–8 cm	Dark reddish brown mor litter
8–12 cm	Dark reddish grey sand, single grain structure, moist and loose, few roots
12–25 cm	Light brownish grey sand, single grain, wet–moist and loose, no roots
25–52 cm	White and pale brown sand, single grain, wet–moist and loose, no roots
52–75+ cm	Black sand, massive, wet and extremely hard, no roots
	The humic pan of this profile was too hard and wet to dig further. In a nearby exposure the pan was more than 1.5 m thick and highly indurated.

(b) White sand

Tika family, Long Tegoa, north Sarawak, Malaysia (5°N, 115°E), flat terrace under primary forest, over old river alluvium. Annual rainfall *ca.* 3000 mm.

0–41 cm	Light yellowish brown fine sand, moist and very friable, few roots
41–102+ cm	White fine sand, distinctly mottled towards base, moist and very friable

(c) Acid planosol

Puletan suite, Stann Creek district, southern Belize (17°N, 88°W). Coastal plain under pine savanna, over Pleistocene alluvium. Annual rainfall *ca.* 2500 mm.

0–10 cm	Very dark greyish brown loamy sand, moderate blocky, moist and friable, common roots
10–27 cm	Very pale brown coarse sand, very weak blocky, moist and friable, few roots
27–48 cm	Pale brown and light grey loamy sand, weak blocky, moist and friable, few roots
48–86 cm	Contrasting patches of pale yellow and red sandy clay loam, weak coarse blocky with weak clayskins, moist and firm, rare roots
86–165+ cm	Strongly contrasting reticulate pattern of white and red ('corned beef') sandy clay, strong blocky with strong clayskins, moist and extremely firm, no roots

Sources: (a) Baillie (1970); (b) Eilers & Looi (1982); (c) King *et al.* (1988).

imperfectly drained because of the proximity of the watertable.

Ferricrete kaolisols

These are well-drained soils in which ferrimanganese concretions have at some stage been cemented together to form a continuous slag-like hardpan. This material is known as ferricrete, ironpan, cuirasse and by various other local names. The earlier term 'laterite' came to be used for such a wide range of materials that it is best avoided. The active formation of the concretions and ferricretes is associated with relatively shallow fluctuating water tables in low-lying and seasonally wet sites (see section 10.3.4). However, there are now many ferricretes in sites that were originally low-lying but have since been been topographically elevated and subsequently dissected. The ferricretes rendered the sites relatively resistant to erosion (hence the name 'cuirasse'), so that they end up as the high ground in the new topography. Inverted landscapes with ferricrete-capped plateaux are widespread in some tectonically stable plate interiors, especially in Africa (MacFarlane 1976) and parts of South Asia. The interfluve terra firme soils at San Carlos in southern Venezuela appear to be ferricrete kaolisols (Medina & Cuevas 1989), but shallow and surface ferricretes do not appear to be generally widespread in Amazonia (Sombroek 1966).

These soils are worth distinguishing from other kaolisols because of the effects of shallow ferricretes on the vegetation. They act as a physical barrier to root growth when intact, the concentrated free sesquioxides tend to immobilize phosphorus, and the fine earth is diluted by the gravel of fragmented ferricrete, making the soils rather droughty.

Table 10.9. *Chemical and physical characteristics of podzol, white sand and planosol*

	Podzol, Miri family[a] Depth (cm)			White sand, Tika family[a] Depth (cm)		Planosol, Puletan suite[a] Depth (cm)		
	0–8	25–52	52–75	0–13	76–102	0–10	27–48	48–86
pH (in water)	3.2	5.7	3.8	4.4	5.1	5.2	5.0	5.0
Organic carbon (%)	40.0	0.1	3.5	1.2	0.1	1.3	0.1	0.1
Total nitrogen (%)	0.54	0.01	0.04	0.1	Tr	0.09	0.01	0.02
C:N ratio	74	10	88	17	7	15	10	6
Exchangeable Ca (milli-equivalents per 100 g fine earth)	1.4	Tr	Tr	0.2	0.2	0.0	0.0	0.0
Exchangeable Mg	10.4	Tr	Tr	3.5	0.2	0.1	0.0	0.3
Exchangeable K	0.4	Tr	Tr	0.1	Tr	0.0	0.0	0.0
Exchangeable Na	0.1	Tr	Tr	0.1	0.1	0.0	0.0	0.0
Exchangeable Al	—	—	—	—	—	1.0	0.2	0.5
Cation exchange capacity[b] (meq per 100 g)	85.7	0.6	14.8	4.6	3.8	3.1	0.7	2.2
Base saturation (%)	14	?<25	1	86	14	3	0	14
Available P (ppm)	—	—	—	8	3	2	1	1
Total P (ppm)	180	Tr	60	52	53	70	50	70
Total Ca (ppm)	1110	60	60	57	Tr	Tr	Tr	50
Total Mg (ppm)	1360	Tr	Tr	25	710	300	300	750
Total K (ppm)	Tr	Tr	Tr	250	1610	1500	2150	4550
Coarse[c] sand (%)	—	19	26	—	—	29	39	37
Fine[c] sand (%)	—	75	63	—	—	41	35	18
Silt (%)	—	1	1	—	—	23	17	13
Clay (%)	—	5	10	—	—	7	89	32
Texture[d]	P	S	LS	—	—	S	S	SCL

[a] See Table 10.8 for sources and morphologies.
[b] CEC determined at pH 7. May exceed field values owing to pH-dependent deprotonation. Base saturations may be depressed accordingly.
[c] Coarse sand, 0.2–2.0 mm; fine sand, 0.05–0.2 mm.
[d] P, peat; S, sand; LS, loamy sand; SCL, sandy clay loam.
Sources: (a) Baillie (1970); (b) Eilers & Looi (1982); (c) King *et al.* (1988).

10.3.2 Non-kaolisol mature terra firme soils

Podzols

These soils develop from very sandy and quartzose parent materials on the gently sloping tops of old alluvial terraces, and on plateaux and gentle dip slopes underlain by siliceous sandstones and quartzites. They also occur on old volcanic deposits of very siliceous composition.

Table 10.8 summarizes the morphology of a moderately shallow podzol in Sarawak. Some profiles are very deep, with individual horizons several metres thick, giving rise to the name 'giant podzols'. Characteristic features of podzols include very sandy textures, the accumulation of mor surface litter, a pale leached horizon, and an illuvial humic horizon in the subsoil. The mor may be wet, verging towards peatiness in the more developed profiles, in which the humic pan may be over a metre thick, and indurated to a rock-like hardness. Unlike many temperate podzols, they are not usually underlain by much of an illuvial iron horizon, because of the very low iron contents of the parent materials. The deeper subsoil is usually considerably drier than the upper horizons.

The analytical data from the Sarawak example are summarized in Table 10.9. The quartziferous parent materials yield low contents of clay, free sesquioxides and total nutrients. Because of the very low contents of aluminium sesquioxides, the possible pH-buffering mechanism of gibbsite dissolution is weak or absent. These soils can therefore be very acid, with pH values as low as 3.0 and with hydrogen rather than aluminium as the main exchangeable and solution cation. The total and available contents of most nutrients may be very low but the contents of exchangeable calcium and available phosphate in the more organic horizons, both mor topsoils and humus pans, are similar to those in kaolisols (Medina & Cuevas 1989, Proctor *et al.* 1983a). The capacity to hold water reserves available to roots and microbes is low because of the virtual absence of clay.

The mobile organic material contains many secondary metabolites, especially polyphenolics. It is mostly derived from the surface litter layers, but some may be leached directly from the canopy. Most is deposited in the humic pan but some drains through to give streams the 'blackwater' colours that are characteristic of podzol catchments. Immature podzols are initially very permeable, and drainage is rapid and completely vertical. At this stage of development these are very droughty soils. However, as the humus pan develops, it becomes an increasing barrier to the downward movement of water. Eventually the drainage becomes impeded and the upper horizons of mature profiles can be saturated for long periods, particularly on flat alluvial terraces. Some studies attribute the deposition of the humic pan to initially high groundwater tables (Schwartz 1988), making poor drainage a cause rather than an effect of podzol formation. Podzols are widespread in South America and Southeast Asia but are much less common in tropical Africa.

White sands

These soils have upper profiles similar to those of giant podzols, with thick mor surface litter and deep, bleached, sandy subsoils, but they lack the underlying humus pan, as can be seen in the summarized profile of an example from Sarawak in Table 10.8. However, distinction of these soils can be difficult because of the great thickness of the bleached eluvial horizons in giant podzols and the possibility of missing a deep humic pan in routine field observations. Like the podzols, these soils have low contents of nutrients (see Table 10.9). They also tend to be excessively drained and liable to drought.

Acid planosols

These soils are locally important in humid savanna areas. They are found mainly on stable sites over coarse- or medium-grained Pleistocene alluvia but also occur on acid crystalline rocks, such as granite (Wright *et al.* 1959). They occur in wet seasonal climates, with mean annual rainfalls of up to 4000 mm and a dry season of at least three months. The profile of an example from Belize is summarized in Table 10.8. The dominant morphological feature of these soils is the sharp discontinuity, normally at about 30–60 cm, from pale, loose and sandy upper horizons to brightly mottled, compact and medium textured subsoils. The analytical data from the Belizean profile are summarized in Table 10.9. These soils are only moderately acid by humid tropical standards, but some of them have exceptionally low contents of exchangeable bases. The complete absence of any measurable exchangeable bases in the second sample in Table 10.9 is not unique, and has been encountered in these soils elsewhere in Belize (King *et al.* 1986). The total contents of the main nutrients are also very low. An exception is total potassium in the medium-textured subsoil, due to the persistence of old alluvial muscovite. The fourfold increase in clay content with depth in the profile in Table 10.9 is not exceptional and wider ratios have been encountered. The subsoils are very compact and impermeable, with little penetration by roots or water. In the wet season the upper horizons are flooded for long periods but in the dry season the sandy topsoils are droughty, with few reserves of available moisture.

These soils occur in Central America and the Caribbean (Wright *et al.* 1959, Alexander 1973) and there are extensive areas of somewhat similar soils in the llanos of Venezuela and Colombia (Sarmiento 1983). The Grey White Podzolic soils of Sarawak have pale-coloured upper horizons and considerable increases in clay content with depth, but they do not have such abrupt textural discontinuities nor such brightly mottled and extremely compact subsoils (Andriesse 1972).

10.3.3 Immature terra firme soils

The mature soils described so far have developed in parent materials that have been stable for considerable periods. However, there are also extensive areas of recent parent materials or older materials that have been recently truncated by erosion, giving rise to a range of immature soils.

Andosols

These are immature soils on recent volcanic deposits. Because of their association with active vulcanism, they tend to occur in steep mountainous terrain, but some also occur on gentler slopes in lowland areas. They have variable morphologies, depending on the mineralogy and age of the tephra. Some deposits are so siliceous that podzols rather than andosols are formed. The profile of an andosol on an ash deposit of intermediate composition in lowland New Britain is summarized in Table 10.10. Distinctive features include the field texture and consistency of the subsoil. The soil hand textured as very sandy at first but with continued kneading it felt increasingly silty. The subsoil is very friable and porous, owing to very loose packing, as is confirmed by the low

Table 10.10. *Profile of andosol*

Deep pumice soil, Putput, New Britain (5°S, 152°E); crest of broad ridge under logged evergreen rain forest, over pumice of about 1500 years age, over clastic sediments of volcanic origin. Annual rainfall *ca.* 2500 mm.

0–18 cm	Very dark greyish brown silty loam, medium crumb, moist and slightly firm, common roots
18–31 cm	Yellowish brown silty loam, weak subangular blocky, moist and friable, common roots
31–105 cm	Olive yellow fine sandy loam, weak subangular blocky breaking to 'single grain', moist and very friable, rare roots
105–150+ cm	Brown sandy clay with reddish brown mottles, medium subangular blocky, moist and slightly firm, rare roots
	0–105 cm is developed in the young pumice, below that is developed in the older volcanic sediments

Source: From Bookers (1981).

mass : volume ratios. The soil has nearly neutral pH and high exchangeable base status. Total contents of the cations are also high, especially potassium. Although the available phosphate in this example is low, quite high total phosphates are found in some andosols.

The distinctive features of the andosols are mainly due to their clay mineralogy. The original ash contains volcanic glass and a variety of poorly ordered silicates (Parfitt 1985). These weather to form allophanes and imogolites, which are secondary aluminosilicates of variable composition and microcrystalline structure. They form a range of microspherical, -fibrous and -cylindrical particles (Wada & Kakuto 1985), the outer layers of which are composed of the octahedral aluminium components. These largely determine the physicochemical behaviour of these soils. They have the ability to form complex bonds with some organic matter and partly protect it from microbial decomposition, so that it persists longer than in most kaolisols (Boudot *et al.* 1988). Organic matter contents are therefore often relatively high throughout the solum (Edwards & Grubb 1977) and the topsoils are often quite dark. The mineralization of nutrients, especially the quite substantial nitrogen contents, is inhibited in some of these soils, although very high rates of nitrogen mineralization have been recorded in a lowland andosol under forest at La Selva in Costa Rica (Vitousek & Denslow 1986). There is little permanent negative charge on these minerals, so that their cation exchange capacities are low in acid conditions. However, the variable charges are high, giving high CEC (cation exchange capacity) values at higher pH, as shown in Table 10.11. The external octahedral aluminium layers also have considerable ability to adsorb phosphate, some of which is firmly bonded and unavailable to biota. They also cause the rather weak inter-particle bonding, which contributes to the

low packing densities and allows some andosols to adsorb large quantities of water and form slightly thixotropic masses that are prone to landslips.

As these minerals age, they weather and crystallize to kandites. Andosols are young soils but vestigial andic characteristics may persist for some time. Quite old kaolisols on non-acidic volcanic parent materials are often still relatively fertile, particularly in terms of the cationic nutrients. The andosols are important in areas close to tectonic plate margins, such as Indonesia, Central America, and the Lesser Antilles.

Recent colluvial soils

These are immature soils developed in recent hillwash deposits on lower slopes. These materials have generally moved short distances from their source, and have been only slightly abraded in transit. They may accumulate gradually or as discontinuous pulses (Moeyersons 1988). Colluviation often involves mixing and lithological heterogeneity, and profile development is complicated by the periodic deposition of fresh material on the surface. The profiles of young colluvial soils tend to be deep but poorly horizonated (Tie *et al.* 1979) and often contain scattered and randomly oriented fragments of partly weathered rock, sometimes of mixed lithology. The textural profile is largely determined by the vagaries of deposition. These soils are often slightly more fertile than the corresponding sedentary soils, especially in terms of the total contents of lithogenic nutrients.

Recent alluvial soils

These are freely drained immature soils on recently deposited alluvia. Many alluvia are laid down in low-lying and poorly drained sites, and most of their soils qualify as gleys (see section 10.3.4). The non-gley soils considered here are only moderately weathered and have

Table 10.11. *Chemical and physical characteristics of Putput andosol[a]*

	Depth (cm)		
	0–18	18–31	31–105
pH (in water)	6.6	6.8	6.9
Organic carbon (%)	4.0	0.5	0.1
Total nitrogen (%)	0.45	0.07	0.01
C:N ratio	10	7	8
Exchangeable Ca (milli-equivalents per 100 g fine earth)	25.9	8.7	4.0
Exchangeable Mg	6.1	3.3	0.9
Exchangeable K	3.9	5.5	2.1
Exchangeable Na	0.4	0.8	0.6
Cation exchange capacity[b] (meq per 100 g)	38.4	18.3	7.6
Base saturation (%)	95	83	100
Available P (ppm)	5	1	0
Coarse[c] sand (%)	4	6	21
Fine[c] sand (%)	11	23	43
Silt (%)	72	63	33
Clay (%)	14	8	3
Texture[d]	ZL	ZL	SL
Disturbed mass per volume (g cm^{-3})	0.78	0.67	0.86

[a] See Table 10.10 for morphology.
[b] CEC determined at pH 7.
[c] Coarse sand, 0.2–2.0 mm; fine sand, 0.05–0.2 mm.
[d] ZL, silt loam, SL, sandy loam.

Source: From Bookers (1981).

low contents of sesquioxides. They generally lack the bright colours of plentiful free ferric sesquioxides, so that browns and yellows predominate. Most of them have some mottling due to slightly impeded drainage in the lower subsoil. Layers of glinting silvery flakes of muscovite mica, which is resistant to abrasion during transport, are a striking feature in some profiles. Textures are largely determined by the depositional stratification of the alluvium, and have been little modified by clay translocation or podzolization. Silt contents are often high, giving a tendency to topsoil capping and subsoil compaction. Because of their youth, these soils are usually only moderately leached and acid (Edelman & Van der Voorde 1963). Their chemical fertility is further enhanced if they are still being replenished with primary minerals in fresh deposits or nutrients dissolved in floodwater.

Skeletal soils

These soils are very extensive in steep and rugged terrain, and are formed where erosion is rapid and surface material is removed before weathering and pedogenesis are far advanced. Profile development is continuously or spasmodically set back, and has to restart in fresh

material. Many of these soils are shallower, stonier and less developed versions of the mature profiles that would develop on parent materials of similar lithology. They usually have fragments of unweathered or partly weathered rock in the solum, and more or less continuous hard rock at shallow depths. There are still primary minerals in the solum and contents of free iron sesquioxides are relatively low, so that reddish colours tend to be less intense. The limited duration of leaching means that the pH and the exchangeable bases are usually higher than in older soils. Leaching-driven translocation processes, such as clay illuviation or podzolization, are less advanced and the resultant features are less pronounced or absent. There are shallower, stonier, less horizonated, and less acid and base-depleted skeletal versions of most kaolisols, planosols and podzols.

In kaolisolic skeletal soils on steep slopes in western Amazonia the pedological immaturity includes poorly developed soil structures and pore systems in the surface layers. The resultant low infiltration rates lead to high frequencies and volumes of overland flow (Elsenbeer & Cassel 1991). However, most skeletal soils are as permeable as, or more permeable than, their mature counterparts, aided by the high stone contents.

Shallow calcareous clays

These soils over limestone are not merely shallow and rudimentary versions of their red oxidic clay mature counterparts. Many of them have distinctively very dark colours (black, dark brown, or very dark grey) which arise from of the interaction of organic matter, calcium and smectoid 2:1 aluminosilicate clay minerals. These minerals are uncommon in the intensively leached soils of the humid tropics, as they require soil solutions with high concentrations of basic cations. They are able to survive close to weathering limestone because of its abundant and continually recharging supply of calcium. The ability of the smectoid minerals to shrink and swell according to their moisture content gives these soils very pronounced blocky structures with shiny pressure faces. With prolonged leaching and increasing solum thickness the cations are eventually depleted, and the smectites are destabilized and weather to kandites and free sesquioxides. The soils then undergo considerable morphological changes and assume the redder colours, less blocky structures and lower base status of the mature limestone oxidic clays (see 10.3.1).

10.3.4 Poorly drained soils

In non-terra firme soils, watertables are high for all or part of the year and impede soil drainage. The main effects are restricted leaching, poor aeration, the reduction of free iron from its trivalent ferric to divalent ferrous state, and a reduction in the decomposition rate of organic matter, sometimes leading to peat formation (see below).

Freshwater gleys

These are mineral soils formed in sites with almost permanent waterlogging and reducing conditions. Virtually all of the free iron is in the divalent ferrous state and the characteristic colours of the soil matrix are grey, white and pale yellow, sometimes with distinctly bluish or greenish tinges. In some horizons there are networks of rust-brown venose mottles. These are formed by localized oxidation of free iron in the vicinity of roots and they tend to outlast the roots responsible for their formation. These soils are mostly formed in young alluvial parent materials in floodplains and swamps, and often inherit textural layering from the depositional stratification. As leaching is very limited, there is little horizon development generated by the vertical translocation of clay or other components. The soils are often only slightly acid and base-deficient. This is partly due to the restricted leaching, but periodic enrichment by nutrients dissolved in fresh floodwaters also contributes. These soils often accumulate dark peaty topsoils, owing to the slow decomposition of litter.

Ferruginous semi-gleys

These soils occur in sites with fluctuating watertables in which the upper parts of the profile are waterlogged only intermittently. The definitive feature of these soils is the marked segregation of iron sesquioxides in the zone of alternating wet and dry conditions. When these horizons are waterlogged, the iron is reduced to its soluble ferrous state, a microbial process requiring a source of energy. As the soils dry, aerobic conditions are re-established and the iron in solution is oxidized to the less soluble ferric state. Ferric sesquioxides precipitate as coatings around extant sesquioxide particles rather than forming fresh nuclei. Repeated redox cycles lead to the particles becoming centres of iron concentration at the expense of the surrounding soil. Eventually the matrix between the concentrations becomes so depleted of iron that it stays pallid even when dry and oxidized. This gives the soil a mottled appearance, with bright reddish blotches set in a pale, often creamy, matrix. With time the contrast between the mottles and the matrix becomes very intense, eventually giving the soil a distinctive 'corned beef' appearance.

In sites where the watertable fluctuates within the same range for long periods and where there is a sufficient supply of iron, the concentrations can grow very dense and form concretionary gravels. These may eventually coalesce and the soil changes fairly rapidly from a soft, plastic mass with discontinuous concretions to a continuous rock-like mass of ironstone, with lacunae of soil material. The plastic precursor is now usually known as plinthite (Soil Survey Staff 1975) and the indurated material as ferricrete, petroferric, ironpan, ironstone or cuirasse (see p. 266 above). The induration process is accelerated by exposure to the atmosphere. Partly because of the more intense dehydration, which causes more oxidation and precipitation, and partly because of increasingly frequent cycles of wetting and drying, plinthite appears to have been much more extensive during the Pleistocene than now. Its uplift, dissection, and drainage gave rise to extensive plateau ferricretes and derived kaolisols in parts of Africa and South Asia (see section 10.3.1).

Saline gleys and acid sulphate soils

These soils develop in marine muds, particularly in mangrove swamps. Some are similar to the freshwater gleys except that the soil solution is saline or brackish. They tend to be fine-textured, but sandy variants also occur (Scott 1985). Many contain ferrous and other sulphides, formed by the microbial reduction of the marine sulphates after burial. At high concentrations these are toxic to a wide range of plants (Koch & Mendelssohn 1989). Acid sulphate soils are formed when sulphidic saline gleys dry out and become more aerobic. This may occur naturally due to local tectonic uplift, a drop in sea-level, or a change in the course of a current or channel. Increasingly, however, it is due to artificial drainage in coastal land reclamation schemes. However initiated, the change causes the sulphides to be oxidized to sulphates. In the process considerable acidity is generated and the soil pH may fall to well below 3. This makes them the most acid mineral soils in the humid tropics, and renders them sterile to all but highly adapted plants and microbes (Rorison 1973). If there are aluminiferous minerals present, aluminium in the soil solution can rise to high concentrations.

Peat

Shallow peat is formed on the surface of many mineral gley soils. In some sites this tendency is intensified to give rise to deep peats. These develop where vigorous and bulky vegetation grows in shallow water, such as swamps or infilling lakes. The undecomposed litter accumulates and can eventually reach depths of several metres. As the peat gets deeper it takes over from the mineral substrate as the main rooting medium and the conditions for root growth become less favourable. Nutrient supply becomes particularly meagre as it depends almost wholly on the releases from the very slow decomposition process. The vegetation becomes quite stunted but adapted species continue to grow and produce yet more litter. Peat therefore accumulates further, reaching depths of over ten metres in some pericoastal swamps in Borneo (Tie 1990). The vegetation on the extensive peat swamps of Malesia is forest, and the peat is mostly woody, consisting of a tangle of hard fragments in a semi-liquid matrix of more decomposed material (Anderson 1964) (see pp. 365–71). These peats mostly have low ash contents, and are very acid and oligotrophic. In other areas there are less acid and oligotrophic peats, derived from herbaceous swamp vegetation (e.g. King et al. 1986).

10.3.5 Montane soils

On tropical mountains temperatures decrease and rainfall generally increases with altitude, giving moisture balances with large excesses for soil leaching and waterlogging. The low temperatures retard some inorganic reactions, especially the chemical weathering of minerals. The cool conditions, in combination with the excessive wetness, also inhibit the microbial decomposition of organic matter. Many montane soils are formed in unstable sites on steep slopes and are subject to rapid erosion. They tend to be skeletal with shallow, stony and rudimentary profiles. Compared with skeletal soils at lower altitudes, they have darker and more organic topsoils, and greyer, less weathered and more gleyed subsoils.

There are some areas of deeper and more mature soils, which are very variable, according to altitude and parent material. On the lower slopes they are recognizably related to the lowland soils on similar parent materials, but tend to be slightly more gleyed, less weathered, more leached, and have higher contents of organic matter and darker colours. The dark colours are particularly noticeable in lower montane andosols and oxidic clays over basic rocks (Vitousek et al. 1983).

Further up the mountains the mature soils tend to lose their tropical characteristics, and many of them resemble wet and leached temperate soils. They are neither intensively weathered nor very red, and the clays are not dominated by sesquioxides and kandites. The hydromorphic inhibition of organic decomposition is accentuated by the low nutrient concentrations in the sclerophyllous litter (Grubb 1977b). Peaty gleys and blanket peats occur on argillaceous and other impermeable rocks (Tie et al. 1979). On coarser-grained and more permeable materials, such as sandstones and granites, leaching is more intense and saturation less prolonged, so that the surface organic layers are dry mor rather than wet peat (Tanner 1977). The mineral profiles beneath are similar to temperate humus–iron and peaty podzols. In contrast to the podzols of the lowlands, they have subsoil accumulations of illuvial iron as well as humus (Whitmore & Burnham 1969).

10.4 Edaphic characteristics of rain-forest soils

Many of the soils of the humid tropics impose few physical limitations on plant growth but are of low nutrient fertility. From an agricultural point of view it has been estimated that about 90% of the soils of the Amazon basin are deficient in nitrogen and/or phosphorus, almost 80% have aluminium toxicity, but only 24% are poorly drained or liable to flooding, the commonest physical deficiencies (Nicholaides et al. 1984). Like many other generalizations about soils of the region, this one applies best to the modal kaolisols. Other soils have their own combinations of edaphic characteristics and constraints, as summarized in Table 10.12. Some important soils present considerable

Table 10.12. *Edaphic characteristics*

Soil group	Edaphic characteristics and constraints
Modal kaolisols	Good depth, drainage and aeration. Moderate–good available moisture. Limited supplies of most nutrients, especially P, Ca. Moderate–strong Al-acidity
Clay kaolisols	Good–moderate depth and available moisture, moderate drainage and aeration. Moderate supplies of K and Mg, but limited P and Ca. Moderate–strong Al-acidity
Basic oxidic clays	Good depth, drainage and aeration and available moisture. Moderate supplies of all nutrients, but strong P fixation
Ultrabasic oxidic clays	Variable depth and available moisture, good drainage and aeration. Mg antagonism to Ca and K, excess heavy metals, strong P fixation
Limestone oxidic clays	Variable good–moderate depth and available moisture, good drainage and aeration. Good supplies of Ca and Mg, limited available K, variable P with strong tendency to fixation
Alluvial kaolisols	Good depth and available moisture, moderate drainage and aeration. Some tendency to cap and erode if bared. Moderate–poor supplies of most nutrients but K generally moderate–high. Moderate Al-acidity
Ferricrete kaolisols	Limited depth and available moisture, good–moderate drainage and aeration. Low–moderate supplies of nutrients, strong P fixation. Strong–moderate Al-acidity
Podzols	Limited depth, available moisture, drainage and aeration. Limited supplies of all nutrients, especially N. Strong H-acidity
White sands	Limited available moisture and all nutrients, including N. Good aeration and rooting depth. Moderate H-acidity
Acid planosols	Limited depth, limited available moisture in dry season, drainage, aeration in wet season. Very limited supplies of all nutrients. Strong Al-acidity
Andosols	Good depth, available moisture, drainage and aeration. Some instability. Moderate supplies of Ca, Mg, K and P, but moderate–very strong P fixation. Some N immobilization
Recent colluvial soils	Good–moderate depth, available moisture, drainage and aeration. Some instability. Limited–moderate supplies of nutrients
Recent alluvial soils	Moderate depth, available moisture, drainage and aeration. Tendency to cap and erode if bared. Moderate–good supplies of nutrients, especially K
Skeletal soils	Limited depth and available moisture, good drainage and aeration. Limited stability. Variable supplies of nutrients (generally better than deeper equivalents)
Shallow calcareous clays	Limited depth and available moisture, good–moderate drainage and aeration. Limited–moderate stability. Good supplies of Ca, Mg, moderate P, limited K
Freshwater gleys	Depth limited by very poor drainage and aeration. Abundant moisture. Variable nutrients
Ferruginous semi-gleys	Intermittently poor drainage and aeration. Ample moisture. Variable nutrients with some P fixation
Saline gleys	Very poor drainage and aeration. Intermittent salinity. Variable–moderate nutrients. Variable sulphide toxicity
Acid sulphate soils	Variable–moderate depth, available moisture, drainage, aeration and nutrients. Very strong Al- and H-acidity
Peats	Very poor drainage and aeration. Abundant moisture. Very limited nutrients. Strong organic acid and H-acidity
Montane soils	Moderate–limited depth, available moisture, drainage and aeration. Variable stability. Variable nutrients with tendency to N immobilization. Variable Al- or H-acidity

physical limitations (Lal 1986) and some are chemically very fertile (Ahmad et al. 1968). It is not always possible to distinguish the effects of several interacting constraints and they are best considered holistically. However, the main limitations are examined separately before considering rain-forest adaptations to them.

10.4.1 Nutrient deficiencies

Although the higher plants need broadly the same suite of nutrients, the demands of species-rich natural communities such as rain forests are variable and complex, as there are interspecific differences in the quantities, proportions, spatial placement, and timing of nutrient requirements (Jordan 1977). The nutrients are best treated indivdually, as each is essential and non-substitutable, and has its own dynamics in the ecosytem. In particular it is worth separating the lithogenic nutrients, such as the basic cations, for which the weathering of rock minerals in soil parent materials is the main primary source, from those for which atmospheric sources are important, such as nitrogen and sulphur. Although phosphorus is basically lithogenic, it is so important and its reactions in the rain-forest environment so complex, that it is best considered separately.

Lithogenic nutrients may be deficient in soils because of poor initial stocks, as in soils derived from very siliceous parent materials such as podzols, white sands and coarse-textured modal kaolisols. Other soils may have been originally better endowed but have been intensively weathered and leached for so long that they have lost many of their initial stock of lithogenic nutrients. This may account for the low total contents of these nutrients in many kaolisols, especially on old land surfaces such as the planaltos of Amazonia (Sombroek 1966) and the Brazilian cerrado (Goedert 1983). Some of the deficiencies in planosols can also be attributed to terminal weathering and leaching.

Nutrients may be present in some soils in sufficient total quantities but in forms that are not accessible for plant uptake. The distinctions between available and unavailable forms are not sharp, and some apparently unavailable nutrients come into biotic circulation over long time-spans. Slow-growing forest trees may have access to nutrients that are effectively unavailable to annual plants (Mullette et al. 1974).

The removal of nutrients from the soil solution and their immobilization in forms that are only slowly available, if at all, is widespread in many soils of the humid tropics. The commonest example is the fixation of phosphorus in insoluble forms with iron and aluminium. This may occur when ionic phosphate co-precipitates with ionic aluminium or iron from the soil solution as virtually insoluble variscite (Al) or strengite (Fe) salts. Phosphate may also be sorbed from the soil solution onto positively charged sites on the exchange complex, from which it is only very slowly released. These processes are pronounced in soils with much labile aluminium or high contents of sesquioxides, such as most kaolisols and acid sulphate soils. The adsorption of phosphate onto the outer aluminium layers of the allophanes and associated minerals in andosols is basically similar.

Very slow decomposition of organic matter can also act to immobilize nutrients. This is most extreme in the oligotrophic woody peats, where nutrients reside in organic matter for years before mineralization (Tie 1990). Many of the scarce nutrients in the podzols are similarly locked up in the mor litter and humus pan, and nutrient deficiencies in many montane soils are partly due to immobilization in mor or peat (Grubb 1977b). In all of these soils there appears to be an element of positive feedback, in that the low availability of nutrients due to sequestration in soil organic matter induces the vegetation to produce litter that decomposes slowly, thus reinforcing the mechanism for the nutrients' immobilization. Soil reaction and drainage are also important in inducing this kind of immobilization. Litter tends to decompose more slowly in acid soils, particularly if the pH falls below 4. Most free-living soil saprophytes are aerobic, and decomposition rates are also low in the anaerobic conditions prevailing in wet and poorly drained soils.

Another form of nutrient sequestration is specific to potassium. Because of its ionic dimensions this nutrient is immobilized between the layers of some $2:1$ and $2:1:1$ aluminosilicates, especially illites. In the humid tropics these minerals and this process are of greatest importance in the clay kaolisols and some alluvial soils. The slow release of this potassium means that trees may be adequately supplied on clay kaolisols with exchangeable contents that would normally indicate deficiency (Ahmad & Jackson 1965).

Inhibition by antagonistic ions can restrict nutrient uptake in soils that appear to have adequate and available supplies. The commonest example in the humid tropics is the inhibition of the uptake of the basic cationic nutrients, especially calcium, by labile aluminium. This is important in many kaolisols and especially in some acid sulphate soils. The dominance of magnesium in some ultrabasic oxidic clays can also be antagonistic to the uptake of calcium and, to a lesser extent, potassium. There may also be effective molybdenum deficiencies in these soils due to antagonism by other heavy metals (Alexander et al. 1985). Crops may experience potassium stresses due to calcium or magnesium antagonism in soils over limestone (Baseden & Southern 1959).

The effect of soils on the distribution of various types of rain forest is discussed elsewhere in this book. It is clear that no single nutrient determines the distribution of all tree species and associations. However a number of detailed studies (Austin et al. 1972, Gartlan et al. 1986) and general surveys (Vitousek 1984) have indicated that phosphorus appears to be of prime importance on a wide range of kaolisols, reflecting the role of free sesquioxides and labile aluminium in the immobilization of this nutrient. Other lithogenic nutrients, such as calcium or magnesium, can also be critical in specific types of kaolisol (Baillie et al. 1987, Johnston 1992). The availability of nitrogen becomes a major constraint where recalcitrant litter and soil organic matter are the main immobilizing nutrient sinks, and is important in podzols, peats, some gleys and probably in some planosols.

10.4.2 Acidity and toxicities

Owing to its solubility at low pH, aluminium is the dominant cation in the soil solution of many soils in the humid tropics. This facilitates the immobilization of phosphorus and is antagonistic to the uptake of the basic cationic nutrients. It is also moderately toxic in its own right to a wide range of plants. These effects are most acute in the modal kaolisols, some acid sulphate soils and some limestone oxidic clays. In soils without significant sources of aluminium in their parent materials, such as podzols, peats and some acid sulphate soils, acid toxicity is mostly due to the direct activity of hydrogen ions (Rorison 1973) but there may also be specific effects by some organic acids. Other kinds of toxicity in humid tropical soils include sulphides in some saline gleys, and heavy metals in some ultrabasic oxidic clays.

10.4.3 Rooting depth and moisture availability

Deep rooting may be mechanically constrained by obstructions such as hard rock, ferricrete, cemented humic pans, or compact claypans (Dabral et al. 1984). Poor drainage and aeration may also preclude deep rooting except by specially adapted systems, such as aerenchyma. Even in soils which are well drained and free of mechanical impediments, the lower horizons may be inhospitable to sensitive roots because of acidity or extreme nutrient deficiencies (Jordan 1982). The amount of moisture held in the soil in forms available for plant uptake is mainly a function of rooting depth and soil texture. Moisture stress is most likely to occur in soils

in which roots are largely confined to shallow sandy surface layers, as in many podzols and planosols. Deep-rooted soils of medium and fine texture can generally retain substantial reserves of available moisture, although some highly aggregated fine-textured soils, especially oxidic clays, have lower reserves than expected from their high clay contents (Goedert 1983). Even in well-watered soils, canopy trees may suffer moisture stress for a few hours on dry sunny days, owing to the inability of diffusion through the soil, uptake by the roots, or translocation through the vascular system to keep pace with the transpiration demand of the canopy (Chiariello et al. 1987). It has been estimated that dry spells lasting weeks rather than days are likely to exhaust the available reserves of almost all soils, even in ever-wet climates like that of Sarawak (Baillie 1976). Cessation of water uptake by the trees is likely to exacerbate any nutrient deficiencies that exist, as mass flow is one of the mechanisms by which soil nutrients reach root surfaces, and nutrient translocation from roots to crown is mainly in the transpiration stream.

10.4.4 Aeration and drainage

Poor drainage and aeration may be due to low-lying topographic position and high watertables, as in gleys and other swamp soils. In some other soils the upper horizons may become waterlogged because of the development of impermeable pedogenic layers in the subsoils, such as humic pans in podzols and clay pans in planosols. Similar but less intense surface gleying can occur in clay kaolisols. As well as the direct effects of inadequate oxygen supplies for root respiration, poor drainage can exacerbate nutrient deficiencies, by retarding litter decomposition and causing nutrient immobilization. Restriction of rooting depth by poor aeration can also lead to moisture deficiency in dry spells, when watertables fall, the topsoils dry out, and shallow root systems are left stranded. This appears to be a common occurrence in planosols during the dry season.

10.4.5 Mechanical support and site stability

This is a plant requirement that tends to be overlooked, but physical stability is important for plants of great size and long life-spans, such as rain-forest trees. Many site disturbances in rain forests are the inevitable results of the life and death cycles of trees. Some have climatic origins, especially in hurricane-prone areas in latitudes greater than 5 degrees; others can be attributed to combinations of soil and topographic conditions. Any soils on very steep slopes may be prone to disturbance by mass movements (Guariguata 1990). These may be triggered by earthquakes in tectonically active areas (Garwood *et al.* 1979) or by prolonged saturation after very heavy rainfall (Walsh 1982b). The soils most affected are andosols, skeletal and montane soils. Slower but more continuous surface wash erosion is also active on steep forested slopes (Richardson 1982), but is not as catastrophic for forest stability as rapid mass movements.

Gleys and young alluvial soils are formed in active or recent alluvium, and are liable to erosion by stream scouring or to burial by fresh deposits. Similarly colluvial soils can be disrupted by the same hillwash processes that emplaced the original parent material. Some alluvial soils can also be destabilized by erosion if exposed to direct rainfall by the removal of vegetation or litter. Because of their high silt contents, surface structures are fairly readily disrupted by raindrop impact and tend to form impermeable caps. These reduce infiltration and increase the potential for erosive overland flow. The young alluvial soils are more susceptible than the older alluvial kaolisols in which some of the inherited micaceous silt has been weathered to clay.

10.5 Soil–forest interactions

There are a number of features in the structure and functioning of rain forests that appear well suited to their soil environments. As the edaphic characteristics of the soils vary considerably, so do the forest features. However the most widespread soils are the kaolisols, which combine low nutrient fertility with few physical constraints (section 10.4). The most striking features, operating at community and individual tree levels, relate to this combination, especially the nutrient deficiencies.

10.5.1 Nutrient scarcity and nutrient cycling

The main flows and stores of nutrients in rain forests are depicted in Fig. 10.2. Not all the flows and stores shown operate in all soils and forests or for all nutrients. Mineral weathering and adsorption on to the exchange complex are unimportant for nitrogen and sulphur, and atmospheric additions are usually a minor input for the cationic nutrients. In general the cycle is similar to that of temperate forests but there are some distinct tropical features. One is the size of the biomass of mature rain forests and the relatively high proportion of the total stock of nutrients that is held in aerial tissues. This may be a safeguard against losses from soils by leaching and erosion, although a high proportion of most nutri-

ents are concentrated in leaves and reproductive tissues (Golley *et al.* 1980), where they are most vulnerable to loss from the biomass by abscission or leaching.

Another distinctive feature of some rain forests is the virtually 'leak-proof' efficiency of the forest floor and soil compartments of their nutrient cycles. Nutrients leave the aerial biomass as litter or as leachates in throughfall (Jordan *et al.* 1980) and stemflow (Herwitz 1986a) (see Chapter 8). These losses are reduced by the withdrawal of nutrients back from senescing tissues prior to abcission, although this is not effective for all nutrients (Vitousek 1982) and usually appears to be negligible in the case of potassium. The reverse flow may also occur, when senescent leaves become loaded with metabolically inert or toxic elements such as aluminium or heavy metals, and are then jettisoned on abcission (Haridasan 1982, Proctor *et al.* 1989). Most of the nutrients leaving the biomass reach the forest floor rapidly, although some shoot litter gets caught up in forks between branches to form an 'aerial soil' (see pp. 139–40, Chapter 6). The nutrient content of the litter is mineralized by microbes, mostly free-living saprophytes, which attack the litter directly or after it has passed through an animal food chain. Although fluctuations in moisture influence short-term variations in decomposition activity, the overall rate at low altitudes is largely determined by the characteristics of the litter, the ambient pH, and aeration. Eventually the litter nutrients are mineralized to inorganic ions in the soil solution. In some temperate and subtropical forests there can be substantial withdrawals of nutrients from the soil solution back into the microbial biomass, reversing the general trend of mineralization (O'Connell 1988). This form of nutrient immobilization appears to be less important in tropical forests, but has been noted in *Shorea robusta* litter in lower montane forest in central Himalaya (Upadhyay & Singh 1989).

Some of the mineralized nutrients in the soil solution are adsorbed on to the clay and organic exchange complexes, but these are of limited capacity in soils with predominantly kanditic clays. Many of the nutrients are rapidly readsorbed back into the tree biomass by dense networks of fine roots, which tend to be concentrated near the soil surface. Uptake by the roots of many of the trees is enhanced by mycorrhizas, which are symbioses with fungi growing on the surface and within the cortex of the host root. Surface-growing ectomycorrhizas are found on the roots of many dipterocarps (Smits 1983), Caesalpinioideae (see Chapter 12), and some other families. However the more embedded vesicular–arbuscular (VA) habit appears to predominate in most rain forests (Alexander 1989). The hyphae of both

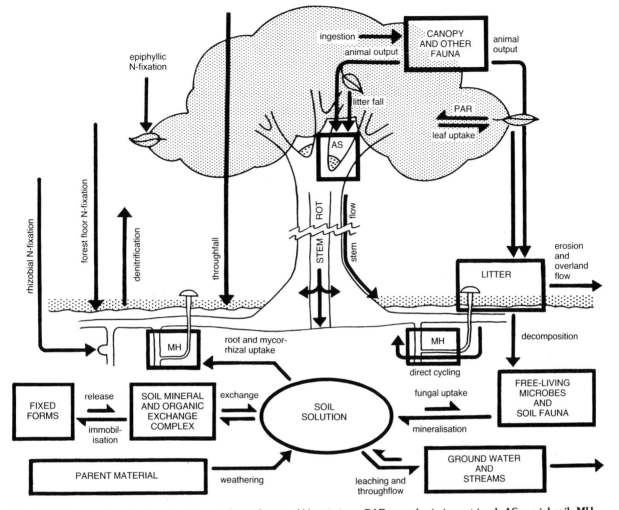

Fig. 10.2 Diagram of nutrient cycles in tropical rain forests. Abbreviations: PAR, pre-abscission retrieval; AS, aerial soil; MH, mycorrhizal hyphae; N, nitrogen.

types extend out into the surrounding soil and probably take over much of the nutrient scavenging and uptake role from the host's root hairs, especially for phosphorus. The hyphae ramify through great volumes of soil and may be able to adsorb phosphorus from very dilute soil solutions and other sources that are inaccessible to non-mycorrhizal roots (Bowen 1980). Ecto-, but not VA, mycorrhizas may also enhance nitrogen uptake (Högberg 1989). The nutrients are transferred at the fungus–plant interface, giving the host the benefit of an enhanced nutrient uptake system. In return the fungi receive energy in the form of photosynthates. Many rain-forest tree species appear to be facultative rather than obligate mycotrophs, and levels of mycorrhizal infection tend to decrease in soils of higher phosphate status, in which the plant's own roots can meet its

requirements (Janos 1983). Mycorrhizas may also contribute to the parsimonious cycling within a tree by transferring nutrients from senescent into more vigorous roots. As mycelial systems may be attached to more than one individual, there is also the possibility of mycorrhizal transfer of nutrients between neighbouring trees (Alexander 1989).

Fig. 10.2 shows an alternative and more direct nutrient pathway from litter to roots, going through the mycorrhizas. It has been suggested that some mycorrhizal fungi are facultative saprophytes, with extra-cellular enzymes that enable them to decompose some components of litter. The hyphae grow into the litter and participate in its decomposition, and the nutrients mobilized are adsorbed and passed on to the host roots (Herrera *et al.* 1978). This kind of direct cycling enables nutrients to bypass the soil solution, where they are vulnerable to leaching. Mycorrhizal mycelia are often heavily concentrated around decomposing organic matter, but this may be caused by attraction to nutrients mineralized by free-living saprophytes. It has also been suggested that ectomycorrhizal mycelia may be able to utilize protein nitrogen directly. However, the importance of direct cycling in forest conditions is still unclear (Alexander 1989).

There may be small but significant additions to the nutrient stock of some forest ecosystems from the atmosphere. This may be a passive process, relying on inputs of rainfall solutes and deposits of aerosols and dust. However, the ecosystem is also actively involved in the fixation and conversion of atmospheric nitrogen to organic forms. This is mostly achieved by symbioses between rhizobial bacteria and leguminous roots. Legumes are common in the tree flora of many rain forests, but many of the important Caesalpinioideae do not nodulate with *Rhizobia* (Högberg 1989). Adequate phosphate nutrition of savanna legumes appears to be important for their fixation of nitrogen, which seems to benefit from mycorrhizal as well as rhizobial infection (Högberg 1989). This may also apply to rain forests on kaolisols in which phosphate is scarce. Nitrogen is also fixed by free-living organisms, including microbial components of the epiphytic, especially epiphyllic, flora (Goosem & Lamb 1986), and blue-green algae growing on soil surfaces (see Chapter 6), but these are probably minor contributors to the overall nitrogen cycles of most rain forests.

Fixation of atmospheric nitrogen increases the ecosystem's stock, but making it available depends on the mineralization of organic matter, the first nitrogenous product of which is ionic ammonium. Some rain-forest ecosystems appear to inhibit nitrification, by which the ammonium is oxidized to nitrate ions. Some of the uptake of nitrogen by rain-forest trees may be as ammonium rather than nitrate ions. Other possible advantages of this to the ecosytem include the avoidance of acidification and the ability to adsorb nitrogen as cationic ammonium onto the exchange complex. Anionic nitrate can only be adsorbed by the soil's limited anion exchange capacity, so that it mostly remains vulnerable to leaching in the soil solution. Nitrates are also vulnerable to denitrification by which they are microbially or physico-chemically reduced to the gaseous element and oxides and lost from the forest floor to the atmosphere (Robertson 1989). There may also be minor losses of nitrogen by the volatilisation of gaseous ammonia from the ammonium ions in the soil solution.

Some rain-forest ecosytems have closed and virtually leak-proof cycles in which the forest subsists almost entirely on the stock of nutrients circulating through the soil, biomass and litter. Most nutrients do not stay in the soil solution long enough to be leached, there are few erosion losses, and streams draining these forests contain few solutes (Bruijnzeel 1990). Inputs from the weathering of soil parent materials are negligible (Burnham 1989), and additions from the atmosphere are the only significant source of nutrient replenishment. This kind of cycle is often associated with high levels of nutrient storage in the aerial biomass, small, hard, phenolic and long-lived sclerophyllous leaves, the accumulation of recalcitrant litter on the forest floor, and dense shallow root systems with much mycorrhizal infection. There was a tendency to assume that such closed cycles operate in all rain forests. In fact this extreme parsimony with nutrients appears to be a response to their scarcity, so that closed cycles are best developed on soils with severe nutrient constraints, such as white sands, podzols, planosols, peats, and deep, coarse-textured modal and ferricrete kaolisols (Medina & Cuevas 1989, Kellman 1989, Tie 1990).

More open nutrient cycles operate in forests on less nutrient-deficient soils (Jordan & Herrera 1981, Golley 1986). Streams draining from such forests may carry significant solute loads (Bruijnzeel 1990), which represent losses to the ecosystem's stock of nutrients. These are greater than the rates of atmospheric replenishment and are made good by inputs from weathering. The forest characteristics associated with this kind of cycle are: lower levels of nutrient storage in the biomass (Whitmore 1984a); larger, softer, less phenolic and more ephemeral leaves; little accumulation of litter; and deeper root systems, possibly with less dependence on mycorrhizae. Partly open cycles are common in forests on andosols and many immature soils (Lambert *et al.* 1984, Bruijnzeel 1990). Many mature oxidic clays also appear to sustain partly open cycles, despite conventional soil analyses that indicate low contents of nutrients in available forms. The total contents of many lithogenic nutrients are moderate or higher in these soils, and the forests appear to be able to utilize some of these reserves. Clay kaolisols are transitional, with apparently closed cycles operating in the dipterocarp forest on these soils at Gunung Mulu in Sarawak (Proctor *et al.* 1983), but partly open cycles on similar but slightly more fertile soils in southern Belize (Baillie 1989).

There are a number of features of rain forests which can be interpreted as beneficial to nutrient conservation and therefore adaptive to infertile soils. However, these features also have other functions or causes, and their roles in nutrient cycling may be incidental.

Butt and stem rots

In some rain forests the stems of many of the larger trees are riddled with fungal infections, although the crowns are still intact and flourishing. The rots involve modification, mechanical weakening and eventual disappearance of the heartwood, and may leave the stem as little more than a hollow cylinder of sapwood. Infections usually start at ground level as butt rots and ramify upwards. They may be beneficial to efficient nutrient cycling in the tree while it is living. The rot is, in effect, the initiation of the decomposition of stem tissue and the mineralization of its nutrient content while the tree is still alive. This spreads the decomposition process in time, thus damping surges in the supply of nutrients to the forest floor and reducing the risk of temporarily swamping the microbial community and exchange complex. The stem rot also takes place in sheltered conditions, where the danger of losing nutrients by leaching is low. The effectiveness of these infections in nutrient cycling is improved by the growth of internal adventitious roots into the rotting tissue (Fisher 1976). Initially the rots interfere little with the metabolism of the tree, as they mainly infect the inert heartwood, and encroach into the sapwood only in the terminal stages (Bakshi & Singh 1970). There is some loss of strength and an increased risk of the stem snapping in high winds, but a hollow tube of slightly elastic tissues is mechanically quite robust. High levels of decay are sometimes associated with low soil fertility, possibly owing to the greater susceptibility of nutrient-stressed trees to infection.

The hollows in rotten stems are often inhabited by complex communities of animals and microbes (Dickinson & Tanner 1978). Foraging by the tenant ants and termites tends to concentrate extraneous litter from the surrounding forest within the stem. Its decomposition there makes its nutrient content exclusively available to the host tree (Janzen 1976b). In myrmecophilous trees, which have hollow stems or other organs that can be colonized by ants, the debris harvested by the ants contributes to the nutrient stock.

Root distribution

The concentration of fine roots in the surface soil and litter is often associated with closed nutrient cycles, and is seen as a response to infertile subsoils and dependence on biological recycling. However, there are disadvantages in such distributions. As well as nutrients, roots are also responsible for the uptake of water and the tree's mechanical support, neither of which benefit from excessive concentration near the surface. Most rain forests experience periods of drought, either irregularly or

seasonally. At such times it is advantageous to be able to exploit the soil's available moisture reserves from as deep as possible (Landsberg 1984). Similarly, rooting deeply into subsoil that is stabilized by overburden pressure gives a more secure anchorage than shallow roots, even if these are robust and extensive. Forests on less infertile soils and with partly open nutrient cycles have depth distributions of roots similar to those of temperate forests. Many of the trees have central tap-roots when young but these atrophy and disappear as the trees mature. Shallow lateral roots grow to substantial girths and radiate widely, so that the final diameter of the root system often exceeds that of the crown. The absence of a tap-root and the spread of the laterals give an impression of root concentration close to the soil surface. However, many laterals develop substantial vertical sinkers at some distance from the stem (Longman & Jeník 1987). Sinkers from dipterocarp laterals in Sarawak are up to 20 cm in girth and some penetrate between the bedding planes of the underlying shales to depths of several metres (Baillie & Mamit 1983).

10.5.2 Adaptation to adverse soil moisture conditions

Nutrient deficiency is not the only edaphic constraint, and some rain forests grow on soils that provide either excessive or insufficient moisture. The main adaptations to waterlogging are in the root system. In freshwater swamp trees there are specialized aerial roots called pneumorhizae (breathing roots) (see pp. 89–90, Chapters 4 and 15). These have a system of internal air spaces and numerous lenticels on the surface which allow gas exchange between the atmosphere and the parts of the root system that are submerged in water or anaerobic mud.

Forests on soils that are very sandy or in which rooting is shallow are liable to moisture stress during dry spells. Some forests on such soils exhibit a number of xeromorphic features (see Chapter 12). Such features are most pronounced on podzols and white sands, peats, peaty upper montane soils, and planosols. These are all acid soils, and most of them lack significant contents of aluminium sesquioxides and their possible pH-buffering capacity. They also have limited total stocks of nutrients and/or tardy mineralization of litter. The chemical constraints on these forests may be exacerbated by intermittent moisture supplies, either edaphically or climatically determined. These characteristics seem to be associated with a combination of H-acidity, nutrient constraints, and moisture stress, although some of the soils, such as the peats, appear well watered. There are extensive forests in which these characteristics are absent or only weakly developed on shallow skeletal soils. These have limited available water capacities but are only moderately acid and deficient in nutrients.

10.5.3 Soil effects of undisturbed forests

One of the main effects of undisturbed forest on soils is the return of nutrients from the subsoil to the surface. This counters the downwards movement by the leaching and accounts for the higher pH and exchangeable base status of surface horizons in most forests. On the other hand the evolution of carbon dioxide by the respiration of roots and microbes reduces the pH of the soil solution, increases its leaching potential, mobilizes aluminium, and blanks off cation exchange sites with hydrogen ions (Sollins 1989). The nitrification of ammonium also generates hydrogen ions and appears to significantly increase acidity, but this is countered by other decomposition reactions involving nitrogen which raise soil pH (Robertson 1989). Differences between individual trees in nutrient uptake and litterfall contribute to the short-range spatial variability of soils under undisturbed forest (Kang & Moormann 1977, Baillie & Ahmad 1984), as do the harvesting activities of nest-building insects (Salick *et al.* 1983).

10.6 Soils and forest disturbance

10.6.1 Natural disturbances

Rain forests have their most dramatic effects on soils when they are disturbed. The commonest form of natural disturbance is the death of trees and the formation of gaps. This affects soils in four main ways: the microclimate at the soil surface is altered, the subsoil becomes moister, there is a substantial and sudden input of litter, and the soil may be physically disturbed. The removal of the canopy exposes the soil surface to direct sunlight and unintercepted rainfall. Surface soil temperatures fluctuate more widely and rise to higher maxima, occasionally up to 45 °C. The increased evaporation leads to considerable drying of litter and surface layers (see Chapter 8). In contrast subsoils tend to become moister owing to the loss of the main transpiration route back to the atmosphere (Denslow 1987).

Many soil biological activities are inhibited by the higher temperatures and frequent desiccation, and some mycorrhizae may die. However, there may also be a 'Birch effect' after rainfall, in which the rehydration of dry litter causes a surge of microbial activity that is more intense than if the litter had remained moist throughout, so that there may be little or no net decrease in mineralization and nitrification (Birch & Friend 1956). The high temperatures may also cause some physico-chemical oxidation of soil organic matter.

The surface is exposed to the direct impact of rainfall, although the loss of the canopy may be partly offset by the increased litter on the soil surface. In topsoils with reasonably robust structures, infiltration capacities are not reduced enough to greatly increase overland flow. However, some alluvial soils with high contents of silt and fine sand may form surface caps, lose infiltration capacity and become vulnerable to increased overland flow and erosion.

The litterfall associated with tree death differs in type as well as amount from that under undisturbed forest. There is a higher proportion of coarse woody components that decompose only slowly. The trunks and larger branches of some species may remain on the forest floor for years after their fall. The litter is unevenly distributed on the surface, with piles of woody and leafy litter close to patches that are almost bare. The patchy distribution of the litterfall causes the organic material in the soil to be unevenly distributed. The patchiness of litterfall is one of the ways in which forest gaps contribute to short-range heterogeneity in soils.

The fall of large branches from trees that die and disintegrate while snagged or still standing may gouge and compact the soil surface on impact, but the overall physical disruption is slight. Soil disruption is mainly caused by trees that topple over when they die. If the butt of the stem rotates it may break the shallow lateral roots at some distance from the stem. These tear up an irregular plate or lens of soil, which may be several metres in diameter and over a metre thick at the centre. Much of the excavated soil slumps back into the hole but some is washed away and lost from the site. The process mixes up the original soil horizons, and disrupts the soil structure and pore system. In shallow soils fresh minerals from the underlying parent material are brought up and incorporated into the new solum. Despite the input of root and butt litter, these raw soils may be initially less fertile than before disturbance, owing to the burial or loss of the topsoil. In the long term, however, treefall may enhance the fertility of the site as the nutrients released by the weathering of the excavated primary minerals may contribute small but significant recurrent additions to the mineral cycle. Treefall can thus be seen as a means by which the forests regenerate their soils and top up their nutrient resources. Clearly this can only happen if there are weatherable minerals within rooting and excavation depth.

This kind of soil rejuvenation is most likely to be effective on skeletal soils, and the shallower of the andosols and clay kaolisols. These occur mainly on steep slopes, which have a tendency to induce asymmetric crowns. The resulting imbalance increases the likelihood of dead trees toppling over rather than decaying and disintegrating while standing. Shallow root systems due to closed nutrient nutrient cycles, mechanical barriers or poor subsoil drainage also predispose trees to topple when they die (Mueller & Cline 1959) so that uprooting is common on podzols, peats, planosols and some clay kaolisols. The proportion of trees disintegrating while still upright tends to be higher on gently sloping sites with deeply rooted soils such as oxidic clays, deep andosols, medium-textured modal kaolisols, and some clay kaolisols (Brokaw 1983). The mixing of soils by treefall tends to increase their spatial variability.

Despite the frequency of treefalls in tropical rain forests, there are few reported cases of the pronounced hummock and hollow terrain that characterize some temperate wind-prone forests (Denny & Goodlett 1968). Their absence is attributed to the vigour of soil creep, surface wash and faunal activity (Herwitz 1981b).

The intensity and relative importance of the microclimatic, litter input, and physical disruption effects on the soil, vary with the way the tree dies and the size of the gap. Gaps are small and mild for trees that disintegrate while standing or snagged, and more intense for trees that snap off but are not uprooted. The greatest impact on soils – literally as well as environmentally – occurs when trees are uprooted. The gaps created tend to large, up to 0.1 ha (Richards & Williamson 1975), although they are often quite narrow and elongated.

The effects vary considerably in different parts of the gap (Bazzaz 1984). The centre of the gap, along the line of the main stem, is the least shaded and experiences the most drastic microclimatic changes. Most of the litter there is woody and decays slowly, so that topsoil organic matter contents may show a net decrease. The physical disruption is slight, often little more than some compaction and minor gouging. The microclimatic effects are less pronounced at gap edges. At the crown end there is a large and sudden input of mainly fine litter. This tends to be non-woody and easily decomposed, and has relatively high concentrations of nutrients. The organic matter content and nutrient status of the soil is therefore likely to be enhanced. At the root end the effect of the stump and root litter tends to be obliterated by the physical disruption of the soil and the incorporation of raw material from the subsoil (Vitousek & Denslow 1986). In reality treefall gaps are usually chaotic, and the zonation is rarely clear. In particular the centre of the gap is often strewn with medium and fine litter from trees and lianes that are collaterally damaged by the main fall.

The effects of forest gaps on soils are modified if they are accompanied by mass movement of the regolith or by fire. Natural fires are infrequent and small in most evergreen forests, although some stunted communities on limestone are vulnerable. Fires are ecologically important in some deciduous forests and moist savannas in climates with marked dry seasons. The effects on soils are most pronounced in the surface horizons, but there may be indirect effects on subsoils. Faunal populations are reduced in size and vigour, and decreased physical mixing by animals may lead to an increase in leaching, clay translocation and textural segregation of the soil profile. Microbial populations are also likely to decline in the short term but appear to recover quickly. The high temperatures can affect the mineral constituents of the topsoil, and the surfaces of burnt ground often have characteristic rounded, baked crumb structures. The impact on the organic matter is more pronounced and contents decline substantially. The nutrients of the surface litter, and some of those in the soil organic matter and the aerial biomass, are mineralized by burning. The non-volatile nutrients, such as the basic cations and most of the phosphorus, remain in soluble forms as ash, and there are increases in topsoil pH and exchangeable bases. These nutrients are vulnerable to leaching and erosion unless they are rapidly adsorbed by the regrowth vegetation. They may be removed in solution by overland flow, in a suspension of ash, or by downward leaching. Volatile nutrients such as nitrogen, sulphur and some phosphorus, are lost to the atmosphere as gaseous oxides. The risk of nutrient loss from the site is higher than in unburnt gaps.

Landslips and other mass movements of the regolith create substantial gaps in the forest. The regolith is often destabilized by prolonged saturation following very heavy rainfall (Walsh 1982b), but the actual slip may be triggered by a physical jolt such as an earthquake (p. 460, Chapter 18). However initiated, the resultant gap is often of considerable size: over a hectare in some cases. Most of the original soil profiles are either truncated or buried. At the top of the slip the remaining soil cover is often thin, possibly with a veneer of the original subsoil material, or it may be bare rock or saprolite. There is little input of litter, and the surface is fully exposed to the sun and direct rainfall. At the base of the slip there is an accumulation of debris but it tends to be raw and poorly structured, and the original topsoil is churned and buried. As in the uprooting of single trees, landslips are a way of bringing fresh nutrients into the ecosystem. However, this is achieved at considerable cost: the loss of vigorous trees, microclimatic deterioration, and the loss of topsoil and litter. In addition, the freshly weathered nutrients are liable to erosion because slips mainly occur on steep slopes. Landslips are of local occurrences, but they tend to recur in vulnerable areas, and can be important in the dynamics of these landscapes and the environment of their forests.

During the regrowth after natural disturbance, soil conditions return quite rapidly to those under high forest. The surface microclimate soon reverts to one of low insolation, steady temperatures, and high humidities, and the subsoil becomes drier as the new root systems exploit its moisture reserves. Microbial and faunal populations recover and resume their forest rhythms. The high density and low canopy of the early regrowth provide good protection against rain splash and other forms of erosion. The nutrients mineralized during the disturbance are mostly taken up by the regrowth. Some may be leached to the subsoil during the lag between disturbance and regrowth but they are mostly retrieved later by deep roots (Harcombe 1980). Soil pH and exchangeable base status gradually revert to their pre-disturbance ranges. During the early stages of succession litterfall tends to be light, but soil organic matter contents are likely to be maintained by the continuing decomposition of the more recalcitrant litter from the original disturbance. Litterfall and soil organic matter contents increase as succession proceeds (Weaver *et al.* 1987).

Soil recovery depends on the rapid regrowth of forest. If forest is not re-established, the nutrients are not sequestered and can be lost by erosion and leaching. This occurs to some extent on the sites of recurrent landslips, which may remain under low fern vegetation for years and even decades after disturbance (Guariguata 1990). The inflammability of this and other types of secondary vegetation further increases the risk of nutrient loss.

10.6.2 Soils and anthropogenic disturbances

Very large areas of rain forest are currently being disturbed by a range of human activities, to the great concern of environmentalists and conservationists. Some artificial disturbances, though drastic, are temporary and the site may eventually return to forest. In others the disturbance is permanent and the forest is 'converted'. The effects of these disturbances on soils include elements of microclimatic alteration, physical disruption, and changes to litter and nutrient dynamics similar to those that characterize natural gaps, but they are more severe, extensive and prolonged. Their relative contributions and duration vary according to the use to which the disturbed land is put, the method of forest clearance, and especially the extent to which the site has been affected by burning either during or after disturbance.

Logging

This is a major cause of forest disturbance in all rainforest formations. Logging practices vary widely but in most rain forests they involve the selective extraction of only a proportion of the larger trees. This leaves some of the canopy in place, so that the soil surface is only partly exposed to direct sunlight and rainfall. The benefits to soil conservation of retaining a partial canopy tend to be outweighed by the damage to soils and surviving vegetation by heavy machinery. The passage of tractors, skidders and other heavy plant scrapes off litter and compacts topsoils, reducing their porosity, aeration and infiltration capacity. The skidding of logs smears the surface layers and closes pores, further reducing infiltration capacity and increasing the likelihood of overland flow and erosion. Physical damage and erosion hazards are exacerbated at sites where traffic is concentrated, such as major skidding trails and log landings, and is extremely intense where earth moving has occurred, such as cuttings on slopes and especially at embankments across streams. A high proportion of the increased erosion takes place on these relatively inextensive but very disturbed and vulnerable sites.

As most logging in rain forests harvests only a few stems and leaves the crowns, butts and roots as litter, the proportion of the site's stock of nutrients removed is limited. However, the erosion associated with logging can be a considerable additional drain on the site's nutrient stocks. Further losses, especially of the volatile nutrients, can occur if fires follow logging. Fire is not normally an intentional part of logging practice, but the large quantities of dry litter greatly increase the risk of accidental conflagration, especially in forests close to slash-and-burn agriculture. The extensive forest fires in Borneo during the drought of 1982–83 were probably exacerbated by large quantities of dry logging litter on the forest floor (see pp. 191–3, Chapter 7).

Large areas are supposed to revert to forest after logging. With varying degrees of silvicultural treatment, they are usually left to regenerate naturally and grow another crop of timber. Soil recovery during the regeneration seems to be similar to that after a natural disturbance. Topsoil organic matter contents and porosities increase, and some of the leached nutrients are recycled from the subsoil. Rates of recovery vary greatly, largely according to the intensity of disturbance. Badly damaged

Table 10.13. *Changes in topsoil (0–25 cm) nutrients during shifting agriculture*

Semongok, Sarawak (2°N, 111°E), clay kaolisol over shale. Annual rainfall *ca.* 3800 mm.

| | (t ha^{-1}) | | (kg ha^{-1}) | | | | |
| | | | Organic P | Available | | | |
	Organic C	Total N		S	Ca	Mg	K
Before burn	68	5.5	360	90	1170	220	260
After burn	87	6.8	420	170	1710	330	590
After harvest	77	8.0	380	120	1520	310	370
After 1 year fallow	78	6.4	410	120	1520	270	380
After 3 years fallow	76	5.8	360	150	1700	270	380

Source: From Andriesse (1987).

sites such as roads and landings appear to recuperate very slowly. Some have compact, bare and erodible soil surfaces for years after logging and traffic have ceased (Hamilton 1985).

Shifting agriculture

This is a very widespread form of land use in which forest is allowed to re-establish itself after temporary disturbance. The term covers a wide range of farming practices but with common basic principles. The cycle of forest clearing, litter burning or mulching, cropping, and forest fallow is described in Chapter 18. As shifting agriculture is practised almost entirely by small farmers, no heavy machinery is used and the physical disruption of the soil is slight, restricted to the impact of falling timber and unintercepted rainfall. Runoff and erosion may be intensified after burning, but for only a few weeks at most, as the crops and weeds quickly establish sufficient cover for protection. Similarly, there is little risk of serious soil erosion after harvest, as the other species of crops in the polyculture, agricultural weeds, or the already sprouting woody fallow species soon make good any gaps and provide a protective cover.

Shifting agriculture is usually regarded as an adaptation to the low availability of nutrients in kaolisols. The conventional model has been that the nutrient stock of the substantial aerial biomass is suddenly and massively mineralized by the burn. Most of the volatile nutrients – nitrogen, sulphur and a little of the phosphorus – are lost from the site as gaseous oxides, but the ash contains most of the cationic nutrients and the bulk of the phosphorus in available forms, such as soluble salts. The pH, exchangeable bases and available phosphate concentrations in the topsoil are raised sub-stantially. This flush of nutrients is short-lived, as they are rapidly depleted by leaching, erosion and uptake by crops and weeds. Soil fertility declines during the cropping phase to levels below those under the original forest. The decision to cease cropping is usually attributed to declining crop yields owing to falling soil fertility. Under the fallow the site's nutrient stock is replenished, mainly in the accreting biomass but also in the topsoils. Recovery tends to be most rapid in the early stages but then tails off; it may take decades for topsoil nutrients to regain their pre-clearing levels.

There have been numerous studies of various aspects of soil conditions during the shifting agricultural cycle. It is difficult to compare their results because of differences in climate, soil, agricultural practice and research methods. There is also a tendency for some of the subtler changes to be masked by the inherent spatial and temporal variability of the soil characteristics being monitored (Andriesse 1977). Andriesse (1987) overcame some of these problems with strictly uniform methods and intense sampling in a comparative study of sites in Sri Lanka and Sarawak. The Sri Lankan site was in a seasonally dry climate and the previous vegetation was probably an old secondary semideciduous woodland, not rain forest. Although its soils contain few exchangeable bases, this is mainly due to their low clay content rather than to very intense leaching, as the pH is above 6 and the base saturation above 40%.

Table 10.13 summarizes some of the results from the ever-wet rain-forest site at Semongok in West Sarawak. They show that all of the nutrients monitored peaked after the burn, with some decline during the subsequent cropping. However, the decrease was not generally dras-tic, except in potassium. Even this nutrient was present at higher contents after harvest than before the burn. This persistence of elevated nutrient contents right through the cropping phase may have been due to incomplete mineralization by the burn. It appears that nutrients continue to be slowly released from large fragments of partly burnt timber by normal decomposition processes during cropping and into the fallow. This is more likely in wet and non-seasonal climates where the drying of litter is a recurrent problem for farmers, and a good spell of dry weather during the pre-burn period may be the most important determinant of yields in the subsequent crops. Andriesse (1987) found that maximum surface temperatures during the burn at Semongok were below 250 °C whereas temperatures under unpiled litter at the seasonally dry Sri Lankan site rose to over 300 °C.

The persistence of high concentrations of nutrients after harvest suggests that their depletion is not necessarily the chief reason for the decision to stop cropping. Increasing problems with agricultural weeds, especially the appearance of *Imperata* and other grasses, appear to be the main cause in many cases. It has also been suggested that farmers recognize the importance of viable stumps for the rapid regrowth of a vigorous and effective woody fallow. If cropping is too prolonged and the stumps are repeatedly slashed back during weeding, they eventually sprout less profusely and the fallow has to be recruited more from the soil seed bank. This may give a slower initial regrowth and a less desirable floristic composition.

Table 10.13 shows that the buildup of nutrients in topsoils during the early years of the fallow is patchy. Available S, Ca and K show slight increases but total N, organic P and available Mg decline slightly. Similarly erratic results were obtained at the Sri Lankan site and accord with findings at Ile-Ife in Nigeria (Table 10.14) and elsewhere. At other sites in West Sarawak, Andriesse (1977) traced soil nutrient contents for much longer into the fallow phase. He found no significant increases, even after twenty years, possibly because of high topographic and stochastic variability within and between sites under fallows of similar age. He concluded that a site's nutrient recovery on these soils, also clay kaolisols, was mostly by storage in the accreting biomass rather than in the soil. He also noted that the vigour of the regrowth, and therefore of the nutrient recovery, was greatly influenced by the long-term history of the site. Lands in the fallow of their first cycle after the clearing from high forest had more profuse regrowth than those which had been through several cycles.

On sites which have been through several cycles with over-extended cropping and curtailed fallows the capacity

Table 10.14. *Concentrations of water-soluble nutrients in topsoil (0–15 cm) leachates during forest fallow*

Ile-Ife, Nigeria (7°N, 5°E), (modal) kaolisol over acid Basement Complex. Annual rainfall *ca.* 1400 mm. Concentrations in milligrams per litre.

	Undisturbed forest	Age of fallow (years)			
		1	2	3	6
N	3.12	0.38	0.38	0.41	1.03
P	0.36	2.71	2.05	1.73	0.41
Ca	1.61	3.01	2.59	2.46	2.06
Mg	0.93	2.32	2.15	1.93	1.25
K	1.51	0.96	0.85	0.68	0.56

Source: From Adedeji (1984).

of the woody vegetation to regenerate may be so weakened that the succession is deflected towards persistent herbaceous communities such as *Imperata* grassland or low fern bush. While intact this kind of vegetation may provide adequate protection against erosion and may operate a fairly efficient nutrient cycle, although with small stocks in the aerial biomass (Nakano & Syahbuddin 1989). However, these communities are prone to desiccation and fires, during and after which the site is vulnerable to erosion and nutrient loss. Recurrent fires also further weaken the potential regeneration by any remaining forest propagules. Although this kind of deflected succession need not start from impaired soil fertility, it may soon induce it (Trenbath 1989). Land covered with these fire climax communities is often perceived as degraded, but demographic and socio-economic pressures necessitate that it often continues to be cultivated.

The sustained viability of shifting agricultural sytems depends on a balance between short periods of cropping, preferably in small plots surrounded by primary or successional forest, and sufficient periods of woody fallow. The actual length of fallow required varies greatly. Fallows of twenty or more years, allowing high forest to return before starting the next cropping cycle, are now but a fond memory and rare ideal in many areas. Pressures on land resources dictate shorter fallows, often progressing only to medium-growth secondary communities.

In some areas this appears to be sufficient. On skeletal dark clays and oxidic clays over limestone and clay kaolisols over slightly calcareous shales and sandstones in southern Belize, modern Amerindian farmers successfully operate shifting agricultural cycles of 2–3 years cropping and 4–8 years fallow (King *et al.* 1986). The pre-Columbian Maya in the same and adjacent areas were able to sustain large populations and elaborate civilizations for many centuries with a combination of intense shifting agriculture, permanent orchards and wetland agriculture (Turner 1986). Shifting agriculturalists in the same area recently attempted to expand on to nearby modal kaolisols and related skeletal soils derived from granite. Despite the promising appearance of the undisturbed forest, the first crop was disappointing and the second hardly worth harvesting. The recovery of the fallow, as indicated by its floristic composition, was too unfavourable to prompt another cycle even after more than a decade (King *et al.* 1989).

In general the more fertile soils such as many oxidic clays, andosols, alluvial soils and some clay kaolisols can be farmed with short fallows. At the other end of the scale, the Iban name 'kerangas' for heath forest on podzols in Sarawak denotes land that cannot be used for even a single crop of hill rice.

Pastures

In the Neotropics large areas of forest have been cleared for pastures, often because of government land and fiscal policies rather than expectations of high productivity (see Chapter 18). Felling of the forest has sometimes been followed by bulldozer windrowing to ensure large, hot and consuming fires. This may cause considerable compaction of topsoils, resulting in deteriorating porosity, aeration, and infiltration. This damage is often exacerbated by soil poaching by the grazing animals once the pasture of introduced non-native grasses, often *Panicum maximum*, is established. The topsoil pH and exchangeable bases are temporarily higher than under the original forest in the early years, owing to the mineralization of biomass nutrients during the burn. However, this benefit is dissipated after a few years and the resulting soils tend to be chemically as well as physically infertile (Fearnside 1980), although the overall nutrient losses are not always severe (Buschbacher 1987).

Pastures appear to be viable in the long term only on the more physically robust and nutrient fertile soils, such as some oxidic clays, andosols, and dark calcareous clays over limestone. Many alluvial soils are also satisfactory but those with high silt contents are susceptible to structural degradation by poaching. Pastures on the less fertile modal and clay kaolisols appear to be productive for a few years. Then the nutrient flush from the burn fades and the palatability, digestibility and nutritional value of the forage decreases. Many such pastures are abandoned and revert to woody vegetation. However, if soil structure was badly damaged by mechanized land clearing or other operations, or if the pasture has been maintained for many years, the normal succession to forest may be deflected and scrubby savanna or sedge-rich herbaceous vegetation is established. This is vulnerable to recurrent fires, further retarding reforestation (Uhl *et al.* 1988a). On the least fertile soils, such as podzols, white sands and the more dystrophic planosols, the pastures tend to be unproductive from the start (Buschbacher 1987) and are likely to be quickly abandoned.

Permanent cultivation

Rain forests are being cleared for permanent as well as shifting agriculture. The productivity of a site is considerably influenced by the type of soil and by the kind of clearing techniques used. A number of studies and much empirical experience have shown that the use of heavy plant and the attendant topsoil compaction depress crop yields for some years after clearing (Alegre *et al.* 1986). Avoidance of bare soils is also important, and cover crops, often legumes, are an integral part of successful crop establishment in many areas. Once established most crops require some fertilization in order to supplement nutrient concentrations and replace losses by leaching, erosion and harvest offtake. In rain-forest climates perennial crops are generally more successful than annuals. They obviate the need for frequent disturbance of soil structure by cultivations. They also provide a permanent canopy to protect the soil from direct exposure to the atmosphere, so that they simulate physical conditions under natural forest to some extent. Productive plantations of tree crops for industrial uses (rubber and oil palm), beverages (tea, coffee, and cacao), fruits (cashew, citrus, and coconuts), fast-growing timber (pines, eucalypts, *Gmelina arborea*, etc.) and many minor products have been established on a wide range of kaolisols and related immature soils. Although not trees, sugar cane, bananas and, to a much lesser extent, pineapples are similar in their physical effects on soils, especially if many ratoons are taken and the trash is used as a mulch.

There have been many failures in establishing continuous annual arable cropping on a wide range of kaolisols. However, it appears that this form of agriculture can be viable on some of these soils provided that a number of conservation practices are incorporated into the farming sytem. These include mixed cropping, minimal tillage, mulching, alley cropping, and contour planting (Juo & Kang 1989). Even with these, annual arable agriculture is problematic and is vulnerable to serious soil erosion in exceptionally heavy or intense rainfalls. The conservation practices can also introduce agronomic complications with pests, weeds, diseases and mechanization. This type of agriculture is more likely to be viable smallholdings on oxidic clays, dark calcareous clays over limestone, andosols, and non-capping alluvial soils. It is more hazardous on most modal and clay kaolisols and is likely to fail on poorer soils such as podzols and planosols.

Part III

Floristic composition of climax communities

Die uebergrosse Mannichfaltigkeit der bluethenreichen Waldflora verbietet die Frage woraus diese Wälder bestehen.
The excessive diversity of the flora and its richness in flowering plants forbid one to ask 'What is the composition of these forests?'
Humboldt (1808)

Chapter 11

Composition of primary rain forests: I

11.1 Species diversity

11.1.1 Number of species

The most important single characteristic of tropical rain forests is their astonishing wealth of plant and animal species. In samples of one or two hectares as well as in vast forested regions such as Amazonia or the island of Borneo, the number of different kinds of plants is enormous: far larger than in any temperate forests. Species of animals, especially insects, are even more numerous, but there are few (if any) reliable estimates of the total number of animal species in even a single hectare of rain forest. It is certain that it generally much exceeds the number of plant species in the same area.

It is the very large number of species of trees, large and small, which is particularly significant ecologically (Tables 11.1 and 11.2). In primary lowland forest the number of species ≥10 cm d.b.h (*ca.* 30 cm girth) is often about 60–150 per hectare; in very rich areas, such as western South America and parts of Malesia, it can exceed 200 or even 300 per hectare. As the size of the sample is increased the number of tree species also rises and species – area curves (Chapter 12 and Figs. 11.1 and 11.2) usually show little tendency to 'flatten' even at areas of 4–5 ha. For this reason plots of 1 ha or less, though often used, are inadequate as sampling units (see Appendix 2). Most rain-forest enumerations include at least a few unidentified species.

The number of tree species per unit area is always greater in the smaller than in the larger diameter-classes, though less relative to the number of individuals. This is because there are more trees in the lower strata and because the latter include many young individuals of species that may reach the canopy when mature as well as species that will not do so. For example, in a 1.5 ha plot of mixed forest in Guyana (Table 11.3) in the 10–41 cm d.b.h. class there were 86 species and 554 individuals (ratio 1:6.4), but in the ≥41 cm class only 32 species and 90 individuals (ratio 1:2.1) (see also Tables 11.1 and 11.2).

Klinge *et al.* (1975) counted and identified as far as possible all dicotyledonous woody plants and palms, including juveniles, on a 0.2 ha plot of primary forest near Manaus (Amazonia) and found that the A and B storeys (mean heights 23.7–35.4 and 16.7–25.9 m, respectively[1]) contained 85.6% of the biomass, but only about 10% of the species; the remaining 90% were in the C, D and E storeys. The total number of vascular species on the plot, including lianes, epiphytes and parasites, was estimated to be about 500. This figure, if not an overestimate, is probably exceptional, judging from data for similar forests in Guyana and elsewhere.

Only a few counts have been made of the entire flora of rain-forest plots. A count by Whitmore *et al.* (1986) of species in all synusiae in 100 m^2 of lowland forest in the building phase at Las Horquetas (Costa Rica) gave a total of 233 species of vascular plants and 32 species of bryophytes (Table 11.2 A). At Neung (Ghana), which like other West African forests is relatively poor floristically (pp. 292–3), Hall & Swaine (1981) found 350 species on 500 m^2 of wet evergreen forest.

In the wetter types of Neotropical rain forest the

---

[1] Strata as defined by Klinge (1973).

[289]

Table 11.1. *Number of tree species in sample plots of primary lowland tropical rain forest*

Data for trees 10 cm (4 in) d.b.h. (ca. 30 cm g.b.h.) except where otherwise indicated.

Locality	Forest type	Size of plot (ha)	Number of individuals		Number of species		Source of data
			On plot	Per hectare	On plot	Per hectare	
Tropical America[a]							
Cuyabeno Reserve, Ecuador	Mixed	1.0	697	697	313	313	Korning & Balslev (1994)
Yanamono, Peru	Mixed	1.0	580	580	283	283	Gentry (1988b)
Manaus–Itacoatiara road, Brazil	Mixed	1.0	350[b]	350	179[b]	179	Prance et al. (1976)
Mocambo, Brazil	Mixed	2.0	897	449	153	144	Cain et al. (1956)
Moraballi Creek, Guyana	Mixed	1.5	644	432	91	—	Davis & Richards (1933–4)
Moraballi Creek, Guyana	Wallaba forest	1.5	919	617	74	—	Davis & Richards (1933–4)
Moraballi Creek, Guyana	Mora forest	1.5	462	310	60	—	Davis & Richards (1933–4)
Belém, Brazil	Igapó (swamp)	1.0	430	430	60	60	Black et al. (1950)
La Selva, Costa Rica	Mixed	2.0	845	423	118	—	Hartshorn (1983)
Africa							
Okomu, Nigeria	Mixed	1.5	582	390	90	—	Richards (1939)
Southern Bakundu F.R., Cameroon	Mixed	1.5	551	367	109	—	Richards (1963a)
Korup F.R., Cameroon	Mixed	0.64	—	471[c]	75[c]	—	Gartlan et al. (1986)
Douala–Edea F.R., Cameroon	Mixed	0.64	—	376[d]	39[d]	—	Newbery et al. (1986)
Akilla, Nigeria	Swamp	1.5	536	360	38	—	Richards (1939)
Asia and Australia							
Sungei Menyala, W. Malaysia	Mixed dipterocarp	1.6	791	488	210	ca. 155	Wyatt-Smith (1966)
Pasoh, W. Malaysia	Mixed dipterocarp	4.0	2280	570	328+	—	Wong & Whitmore (1970)
Wanariset, E. Kalimantan	Mixed dipterocarp	1.6	866	541	239	173	Kartawinata et al. (1981)
G. Mulu, Sarawak	Mixed dipterocarp	1.0	778	778	ca. 223	ca. 223	Proctor et al. (1983a, b)
G. Mulu, Sarawak	Heath forest	1.0	708	708	ca. 105	—	Proctor et al. (1983a, b)
Toraut, Sulawesi	Mixed	1.0	408	408	109	—	Whitmore & Sidiyasa (1986)
Hydrographic Range, Papua New Guinea (Plot 1)	Mixed	0.8	528	636	122	—	Paijmans (1970)
Davies Creek, Queensland	Mixed with *Agathis*	1.7	1372	807	121	—	Williams et al. (1969b)

[a] Further counts are given by Campbell *et al.* (1986).
[b] Trees ≥15 cm d.b.h.
[c] Mean of 135 plots (40 668 trees, 411 species).
[d] Mean of 104 plots (24 997 trees, 230 species).

Table 11.2. *Number of vascular plant species in synusiae of sample plots in lowland tropical rain forest*

A. 100 m², Las Horquetas, Rio Puerto Viejo, Costa Rica (Whitmore *et al.* 1986).
B. 1000 m², Rio Palenque, Ecuador (Gentry & Dodson 1987).
Note: some species occur in more than one synusia.

A. Las Horquetas

	Individuals	Species
(a) Plants ≤1 m high		
Herbs, Cyclanthaceae, palms	218	30
Juvenile trees, shrubs and woody climbers	1132	102
Subtotal	1350	132
(b) Plants 1–3 m high		
Palms, shrubs and woody climbers	10	8
Trees	134	28
Subtotal	144	36
(c) Plants ≥3 m high		
Palms	3	2
Trees (≥17 m)	35	16
Subtotal	38	18
(d) Climbers	301	44
(e) Epiphytes	339	61
Total vascular plants	2171	233

B. Rio Palenque

	Individuals	Species
(a) Trees		
(i) ≤2.5 cm d.b.h. (juveniles)	559	87
(ii) ≤10 cm, ≥2.5 cm d.b.h.	217	86
(iii) ≥10 cm d.b.h.	52	32
Total trees	653	114
(b) Shrubs	531	39
(c) Herbs (including small palms)	1220	50
(d) Climbers (excluding hemi-epiphytes)	117	36
(e) Epiphytes	127	63
Total vascular plants	7210	365

number of species of herbs, lianes, epiphytes and other 'non-trees' may exceed the number of tree species. At Rio Palenque (Ecuador) Gentry & Dodson (1987) found that the 'florula' of 'wet' lowland rain forest (Premontane Wet forest in Holdridge's terminology; see pp. 164–5) included 1033 species of vascular plants of which only 154 (15%) were trees. The composition of a 1000 m² sample is shown in Table 11.2B.

For whole countries or large regions the size of the rain-forest flora is in most cases known only approximately. In Peninsular Malaysia and Singapore (area

132 000 km²), which 100 years ago were mostly forest-covered and are well known botanically, there are about 3000 tree species of 'timber size' (≥ 90 cm girth = *ca.* 30 cm d.b.h.) (Whitmore 1972a), but many more of smaller dimensions.

In a recent survey of a 50 ha plot in undisturbed primary forest in Pasoh Forest Reserve (Negeri Sembilan), 802 species of dicotyledonous trees and shrubs over 1 cm d.b.h. were found (Kochummen *et al.* 1990, 1992). The plot was fairly typical of mixed forest in south-central Malaya and it is remarkable that it contained 25% of the flora of the Peninsula in these categories (or 50%, if species such as mangroves, which are unlikely to occur in mixed lowland rain forest, are excluded from the total).

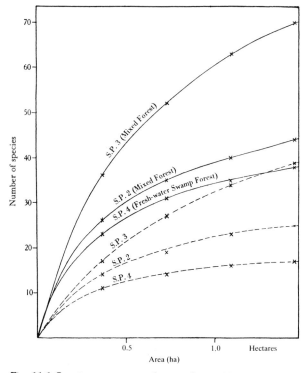

Fig. 11.1 Species–area curves for rain forest, Nigeria. From Richards (1939, Fig. 7). Continuous lines, trees ≥10 cm d.b.h.; broken lines, trees ≥30 cm d.b.h. S.P. 2–4 refer to sample plots 400ft × 400ft (ca. 1.5 ha). The points shown on the curves are for contiguous subplots.

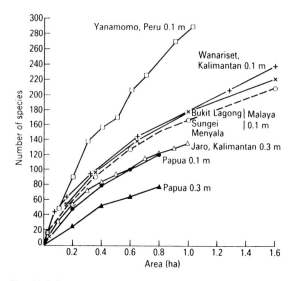

Fig. 11.2 Species–area curves for lowland tropical rain forest for trees ≥0.1 m or ≥0.3 m d.b.h. as indicated. Whitmore (1990), from data of Whitmore (1984a) and Gentry (1988a).

In Sarawak and Brunei (137 347 km^2) there are 2540–3040 tree species (Brünig 1973b). Jacobs (1974) estimates the number of species of vascular plants in the whole of Malesia as 25 000–30 000 species. A considerable proportion, perhaps more than a third, of these are forest trees.

In tropical America, which includes large areas of savanna and montane vegetation, there are believed to be at least three times as many plant species as in Malesia (Gentry 1992). The Amazon rain forest is so vast (ca. 650 × 10^6 km^2), and so inadequately explored botanically, that it would be difficult to make even an informed guess on how many plant species it contains. There are considerable regional differences and endless local variations in the forest flora (Pires & Prance 1977) and the number of tree species is probably considerably more than in Malesia. As many as 313 tree species ≥10 cm d.b.h. have been recorded on one hectare in Ecuador (Korning & Balslev 1994) and a still larger number from a relic of the Atlantic rain forest in Bahia.

Many South American rain-forest species seem to be restricted to quite small areas: in the western Chocó (Colombia) about a quarter of the species are probably endemic. The rain forests of Central America are much less extensive than those of South America, and for historical reasons are floristically less rich (Gentry 1982b).

African rain forests stand apart from those of the American and eastern tropics in their comparatively poor flora (Richards 1973a), although the difference may have been somewhat exaggerated. Brenan (1978) estimated the number of vascular plant species in the whole of tropical Africa (a large part of which is savanna) at 30 000. West Africa (to as far east as the former British Cameroons) has fewer than 7500 species (Hutchinson & Dalziel 1954–72), of which about 806 are rain-forest trees ≥10 cm d.b.h.; (F.N. Hepper) but its rain forests are not very extensive and have for some centuries been fragmented (and probably impoverished) by agriculture. The forests of central Africa (Cameroon, Zaïre etc.) are less well investigated and richer in species (Evrard 1968, Lebrun 1960b). Although African rain forests as a whole are certainly relatively poor, the number of species in small sample plots may be as great as in other parts of the tropics (Table 11.1). Gentry (1992) says that 0.1 ha plots in Gabon contained as many or more species than those in similar climates in South America. The richness of small sample areas seems to reflect local site characteristics rather than the species diversity of the region as a whole.

It is relevant to note here that there are other important floristic differences between the African rain forest and that of other continents. Many of its tree species range very widely, some throughout its area, e.g. *Lophira alata, Piptadeniastrum africanum*; localized endemics occur in some places, e.g. in the 'Western Block' forests of West Africa, but are relatively few. Another difference is the poor representation of several families that are characteristically tropical (though not confined to rain forests), e.g. Lauraceae, orchids and palms. The small number of genera and species of palms is particularly remarkable. In mainland Africa there are only about sixty species, although in Madagascar there are over 130, most of which are endemic (J. Dransfield). In both tropical America[2] and Malesia there are over 1000 species of palms.

Axelrod (1952, 1972; see also Raven & Axelrod 1974) attributed the comparative poverty of tropical Africa to the uplifting of its land surface in the Tertiary, which caused a desiccation of the climate, especially during the pliocene. Another possible cause may be that, for topographical reasons, the refuges available for the African rain-forest flora in the arid periods of the Pleistocene were fewer and smaller than those in other parts of the tropics (See Fig. 1.10).

How far present, as well as past, climatic conditions have contributed to the relative poverty of African forests is uncertain. Nearly all the rain forests have a distinctly seasonal climate, seldom with less than three consecutive months with less than 100 mm rainfall (Chapter 7). In all continents the species diversity of tropical forests decreases with increasing seasonality and increases with the total annual precipitation. Tropical deciduous and semideciduous forests are thus generally less rich in species than rain forests. From data for a large number of 0.1 ha plots (Gentry 1982a, 1992) showed that in lowland tropical America 'Dry forests' usually have about 50 plant (angiosperm) species, 'Moist forests' *ca.* 100–150 and 'Wet forests' over 200. The species – precipitation curve reaches an asymptote of *ca.* 250 species per 0.1 ha at 4000 mm (Fig. 11.3). A similar correlation between species richness and rainfall is shown by Hall & Swaine's (1981) ordination of data from 0.1 ha plots in a range of forest types in Ghana. In Southeast Asia also, species richness is related to rainfall and is less in monsoon forests than in rain forests, but quantitative data are not available.

In a given climate species richness varies with soil characteristics and, as might be expected, is lower in single-dominant than mixed rain forests (Chapter 12).

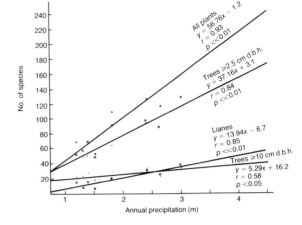

Fig. 11.3 Species richness (vascular plants) in tropical American lowland rain forest in relation to annual precipitation (samples 1000 m²). From Gentry (1982a, Fig.1).

Although the diversity of seasonal forests is less than that of rain forests, it is interesting that some non-forest communities of seasonally dry tropical climates rival the latter in species richness. In the cerrados (savanna woodlands, pp. 397ff) of Brazil, for example, woody species may be very numerous. In the Distrito Federal (Brasília) 230–250 species of vascular plants per 0.1 ha and 300–350 per ha have been found (Eiten 1984) and on 1 ha at Botucatu (S. Paulo) Silberbauer-Gottsberger & Eiten (1987) recorded fifty-four species of trees and 'thick-stemmed shrubs' (defined as ≥0.3 m tall and ≥10 cm at base).

Tropical rain forests have many more species of trees, as well as of other vascular plant than even the richest extra-tropical forests. The mixed mesophytic (broad-leaved deciduous) forests of the southern Appalachians, the floristically richest in North America, have only about thirty (or rather more) 'canopy tree' species, of which about twelve are constantly present (Braun 1964). In the Great Smoky Mountains National Park, the richest area, the maximum number of woody plant species ≥10 cm d.b.h. is about 20 in a 0.1 ha plot, and in a 1.0 ha plot *ca.* 25–30. The maximum total number of vascular species is about 100 on 0.1 ha and 125–150 on 1.0 ha (P.S. White). The mesophytic forests of the upper Chang Jiang (Yangtze) region of China (Wang 1961) probably have more tree species per unit area than those of America, but fewer than even relatively species-poor tropical rain forests. Further data for species-rich temperate forests are given by Gentry (1988b, 1992).

Some heathlands of temperate Mediterranean climates, especially in the Southern Hemisphere, are also

---

[2] Kahn & de Castro (1985) found thirty-two species of palms in one hectare in Amazonia.

very rich in species, but less so than tropical rain forests. In South African fynbos 52–128 species per hectare (average 75) have been recorded, and in Chilean mattoral 108 and 109 species on 0.1 ha samples (Naveh & Whittaker 1980). Cowling *et al.* (1992) have made a detailed comparison of species richness and diversity in fynbos and tropical forests. In the very species-rich heaths in southern Australia there are not as many species per 0.1 ha as in Australian tropical rain forests (Parsons & Cameron 1974; Rice & Westoby 1983).

11.1.2 Species richness and species diversity

The terms species richness and species diversity are often treated as interchangeable, but the latter should be regarded as involving the relative importance as well as the number of species in a community or stand (Greig-Smith 1983). There is no general agreement on how diversity should be measured. According to MacArthur (1965) counts of species are the simplest measure of diversity and are adequate for comparing some 'patterns of diversity'. Data on species richness have the disadvantage that they are not independent of the size of the area sampled and do not take account of the biomass of the organisms involved. The latter shortcoming has particularly to be borne in mind in the case of rain-forest communities, in which it is characteristic for a large proportion of the species to be represented by very small numbers of individuals per unit area (Chapter 12). Indices of diversity involving the relative density of individuals or biomass have not been widely used in studying the floristic composition of rain forests and will not be discussed here. It should be noted that *any* measure of plant or animal diversity will show that the diversity of tropical rain forests is very great compared to most other terrestrial ecosystems.

11.1.3 Theories of species richness and diversity

Why there are so many plant and animal species in tropical rain forests has long been debated, especially by zoologists. A satisfactory theory must explain how their great species richness is maintained, as well as how it originated. According to some theories climax rain forests are closed systems that have reached an equilibrium; others assume that they are open systems and that the number of species can continue to increase. The many hypotheses put forward have been discussed by MacArthur (1965, 1972), Baker (1970), Ricklefs (1973), Leigh (1983) and others (see also Wilson & Peter 1988, Gentry 1988b). These hypotheses are not all mutually exclusive; several factors may be respon-

sible for the species richness of tropical rain forests. Only some general comments can be made here.

One factor often invoked is time. This hypothesis assumes that tropical rain forests have developed over a longer period, and are more stable, than other ecosystems. The supposedly primitive trees and other organisms found in rain forests (p. 19) support this view and fossil evidence shows that in some parts of the tropics forests floristically, and perhaps also physiognomically, similar to modern rain forests, existed in the Cretaceous and early Tertiary periods some 50–80 million years ago (pp. 13–15). The time theory has been criticized by Pianka (1966). It cannot be readily tested and it does not by itself explain how the high species richness of tropical forests is maintained.

The continuity of tropical forests through geological time is only relative. Some of the land surfaces on which rain forests are now found are very ancient, e.g. tropical Africa, but others, such as a large part of Amazonia, are of more recent origin. Much of the present-day rain forest may not have developed *in situ* from Tertiary forests. Even if this were so, the once common belief that most of the humid tropics escaped drastic climatic changes during the Pleistocene and Holocene can no longer be upheld, as it is now clear that the alternating humid and arid phases of those periods probably caused forest areas to fragment and rejoin (Chapter 1). It has been suggested that these changes might have encouraged speciation, but from wide experience in tropical America Gentry (1992) considered that the importance of Pleistocene refuges for speciation has been exaggerated. He believed that most of the botanical evidence for the existence and location of refuges can be explained by present-day climatic and soil conditions.

Whether or not the age and chequered history of tropical rain forests are the chief causes of their great richness in species, there can be little doubt that another, and probably more important, factor is their present climate. Tropical ecosystems illustrate the general rule that in all classes of organisms, with a few exceptions, species numbers increase from the poles towards the equator (MacArthur 1965, Bourlière 1983). These latitudinal gradients are well documented for birds, mammals and other animal groups and are also demonstrable for plants and plant communities. Kira *et al.* (1962) and Kira (1991) have pointed out that on the western borders of the Pacific, the only part of the world where there are uninterrupted gradients of humid forest climates from the Subarctic to the Tropical zones, the forests increase in species diversity from high to low latitudes (see also Pianka 1966 and Chapter 17).

Within the tropics, species diversity of plants is

strongly correlated with annual precipitation (and probably even more closely with non-seasonality of rainfall (Fig. 11.3) (Gentry 1988b). Although moisture is no doubt an important determinant of high species diversity, the fact that lowland rain forests are in general richer in species than montane forests, which have a similar rainfall regime and also lack a cold winter (Chapter 17), suggests that the high mean temperatures at low elevations may also be a contributing factor.

There is no evidence that rain-forest climates increase mutation rates, but there are several ways in which they might increase rates of speciation. For example, owing to the lack of cold or very dry seasons resources such as flowers, fruits and young leaves are available, though not equally abundant (see Chapter 9) at all times of year, making possible a denser 'packing' of niches among animals using them (the 'finer adaptation' of Dobzhansky 1950) than in other terrestrial environments. Partly because of this, plant–animal non-trophic (mutualistic) interactions can be more varied and complex than in temperate climates. In rain forests long series of species have evolved with staggered flowering and fruiting periods, e.g. in genera such as *Miconia* (p. 238). There is of course a limit to the possible subdivision of plant and animal niches. In South American forests Gentry (1992) found that in any one locality there are about twenty species of Bignoniaceae, each with different pollination mechanisms and flowering seasons; in another locality there is a similar number of niches, but the species may be different.

The climate no doubt also affects survival and extinction rates of plant and animal species in various ways. Wallace (1878, p. 66) considered that in equatorial climates the severest part of the struggle for existence is between species, and not, as in extra-tropical regions, against cold or drought. In modern terms, selection in the former is density-dependent, involving mainly characters such as fecundity and ability to interact and compete with other organisms, while outside the tropics it is mainly density – independent and related to ability to survive climatic hazards (Dobzhansky 1950).

It has sometimes been asserted that in the 'permissive' climate of tropical rain forests competition between species is less rigorous than in temperate climates or even non-existent (van Steenis 1969). This belief seems to rest on some misunderstanding: in a tropical forest it is impossible not to be aware of the severity of competition between plants for light or between animals for food. It is, however, conceivable that the selection pressures may in some way give relict and newly evolved species greater chances of survival in rain forests than in other ecosystems.

An aspect of species diversity in primary rain forests which has often been discussed is the frequent occurrence of numerous apparently very similar sympatric species of the same genus (see Richards 1969). Examples are species of *Shorea* (and other Dipterocarpaceae) and *Eugenia* in Malesia (Ashton 1988) and of *Eschweilera*, *Licania* and *Protium* in tropical America. In African rain forests such species groups are less common and contain fewer species, but they exist in *Drypetes*, *Rinorea* and other genera. The presence of these apparently sympatric species groups seems at variance with Gause's hypothesis that in equilibrium communities more than one species cannot occupy the same niche, but this may merely reflect our ignorance of the autecology of the species concerned. Rogstad (1990) found that species of *Polyalthia* (Annonaceae), which often grow together in Malesian rain forests, differed in characteristics such as requirements for seed germination, height at maturity, and responses to drought and flooding.

Fedorov (1966) argued that the great species diversity of rain forests is due not to selection but to non-adaptive variation; this arises, he suggests, from genetic drift due to isolation in very thinly scattered populations. Ashton (1969) disputed this conclusion and maintained that among dipterocarps morphologically very similar species occupy slightly different physical and biotic environments. He found no evidence that speciation processes in rain forests are basically different from those in other ecosystems. Similar conclusions were reached by Gentry (1992) about large Neotropical rainforest genera such as *Passiflora* and *Psychotria* in which close investigation shows that sympatric species differ in floral biology, phenology and 'micro-habitats'.

Grubb (1977a) suggests that the key to the apparent paradox in applying Gause's hypothesis to species-rich rain forests lies mainly in what he terms the 'regeneration niche'. In earlier chapters it was seen that the conditions needed for producing viable seeds, germination, and all the later stages in the replacement of old by young individuals, vary enormously among rain-forest trees, even between closely related species. These requirements are adequately known for only a very few rain-forest species. When more species have been investigated, and more is known about the dynamics of rain forests generally, it may be easier to understand how so many superficially similar tree species can exist together in rain-forest communities.

It has been claimed (Fittkau 1973, Fränzle 1977) that great species richness in tropical forests is correlated with low concentrations of soil nutrients, but the evidence is not conclusive. Hall & Swaine (1976) found that the species richness of the seasonally dry forests of Ghana was inversely related to total exchangeable bases and Faber-Langendoen & Gentry (1991) found that the nutrient content of the soils in the extremely species-rich rain forests of Bajo Calima in the Chocó region of Colombia was low. Huston (1980) showed that in Costa Rican forests there was a correlation between high species richness and low concentrations of certain soil nutrients, but in the author's own experience in Guyana and Sarawak (see pp. 296–307 below) forest types on soils known (or presumed) to be nutrient-poor had fewer tree species per unit area than those on more fertile soils. In the G. Mulu National Park (Sarawak) Proctor et al. (1983a) found no clear relationship between species richness and nutrient concentrations: the nutrient-rich limestone forest was relatively poor in species, while both the nutrient-rich forest on alluvial gleys and the nutrient-poor mixed dipterocarp forest were very rich in species.

11.2 Phytosociological analysis

Because of the very large number of tree species and the difficulty of identifying them in the field (see Appendix 1) the analysis of rain forests into phytosociological units poses formidable problems. To early botanical explorers such as Humboldt, lowland rain forests appeared to be vast mixtures of species which varied unpredictably in composition with little obvious relation to soil or other environmental factors. In some habitats such as swamps, distinctive assemblages of species were recognized. Spruce's (1908) account of the caatinga forests of the Rio Negro (p. 322) is one of the earliest descriptions of an edaphically determined type of terra firme rain forest. Distinct associations comparable to those of temperate regions did not seem to be recognizable.

Later on, as knowledge of forest floras improved and policies for forest exploitation and management began to be developed, empirical classifications of 'forest types' became a necessity. A pioneering attempt to apply current Anglo-American synecological terminology to tropical forests was Chipp's (1927) study of the Ghana forest, in which he recognized one climax association and four preclimax associations. Chipp's concept of the association was later criticized by Aubréville (1950–51) who believed that no true associations were distinguishable in the West African rain forest. At Moraballi Creek (Guyana) Davis & Richards (1933–34) recognized one mixed association and four associations with single dominants (consociations) (see below). Marshall (1934) recognized three lowland rain-forest associations in Trinidad:

all of these, including the single-dominant *Mora excelsa* association, were, however, regarded by Beard (1946b) as faciations of one association.

After World War II the phytosociological treatment of rain-forest communities became for some years a matter of considerable controversy. Mangenot (1950, 1955) and others applied the methods and terminology of the Braun-Blanquet (Zürich-Montpellier) school to the 'forêt dense' of Côte d'Ivoire and concluded that (in addition to freshwater swamp forest and semideciduous forest) it consisted of two associations, the Diospyro-Mapanietum and the Turraeantho-Heisterietum, the former occurring on the more clayey and the latter on the more sandy soils. Both were rain-forest communities of mixed composition. According to Emberger et al. (1950), these two associations differ in the same structural and analytic characteristics as temperate forest associations and are fully comparable with them. The Braun-Blanquet system was later applied to the forest communities of Zaïre by Lebrun & Gilbert (1954). Because of their floristic complexity there are great difficulties in applying Braun-Blanquet methods to tropical forest communities and these methods have been used only to a limited extent.

In the past twenty years ordination and other mathematical methods such as principal component analysis have been widely used in studying the floristic composition of rain forests (Appendix 2). These techniques, combined with more critical sampling methods, have led to a much improved understanding of the variations in the floristic composition of rain forests and their relation to environmental and other factors. They have also provided a sound framework for classifications of rain-forest communities. For practical purposes the need for classification still remains.

Before discussing the composition of rain forests generally (Chapter 12), the lowland forest communities of areas in Guyana and Borneo which the author has studied at first hand will be described.

11.3 Primary rain-forest communities in Guyana

11.3.1 Moraballi Creek

A detailed description of the rain forest of Moraballi Creek, based on the work of the Oxford University Expedition to British Guyana in 1929, was given by Davis & Richards (1933–34). The following is a revised version of the shorter account in Edn 1 of this book. Ogden's (1966) account of the Twenty-four Mile Reserve in the 'Bartica Triangle', where the climate, topography, soils and vegetation are very similar, updates and amplifies the Moraballi Creek results and is discussed on pp. 302–4. The Moraballi Creek area is described here as it was in 1929. It was later declared a nature reserve; when the author revisited Guyana in 1979, he was told that it had not greatly changed.

Moraballi Creek is a tributary, some 24 km long, which joins the Essequibo river from the east *ca.* 80 km from its mouth. The area studied was within a radius of a few kilometres from the base camp of the Oxford Expedition (6°11′N, 58°32′W). The terrain is low and undulating, but most of it is not flat or swampy. From the creek the ground slopes up to a plateau *ca.* 100 m above sea-level from which steep-sided ridges extend towards the valley.

Throughout the area the underlying rocks are granite and gneiss, but the soils are remarkably varied. In the flood-plain of the creek and its chief tributaries the soil is a fine-textured whitish alluvial silt liable to short periods of flooding at any time of year; even in dry weather the water-level is close to the surface. Above the flood-plain the soils are mostly kaolisols (latosols) (Chapter 10), varying in texture from very heavy, sticky, bright red clays to more friable and porous loams; these lighter soils tend to occur mainly at a higher level and farther from the streams than the clays. The other soils are the white and brown sands. The former (Tiwiwid Series of Stark *et al.* 1959) are pure white, coarse-grained, very porous, quartz sands (tropical podzols and white sands *sensu* Baillie, pp. 267–8), which cap the plateau and higher ridges often to a considerable depth. The brown sands (sandy kaolisols, probably the Kasarama Series of Stark *et al.* 1959) are also coarse-grained and porous, but are stained with iron oxides; at Moraballi Creek they are found chiefly as narrow bands on the slopes of the steep-sided ridges and occupy only a small area. In ascending from the creek to the watersheds, the catena of soils is usually but not invariably, in the order: silt, red clay, red loam, brown sand, white sand. The similar soils of the 'Bartica Triangle' have been fully described by Stark *et al.* (1959) and Ogden (1966).

The climate is typically equatorial and, as far as is known, similar throughout the area. The annual rainfall is over 2500 mm; there are two maxima in the year and no month has an average of less than 100 mm, although droughts occur occasionally. Climatic data for Mazaruni Station (27 km NNW) are shown in Fig. 7.2. Figures for rainfall at Bartica (24 km NW) are given by Ogden (1966).

Rain forest, much of it primary, covers the whole creek basin and until comparatively recently extended uninterruptedly westwards to the Orinoco and east, through Surinam and French Guiana, to Amazonia. About 50 km north of Moraballi Creek the forest gives way to the intensively cultivated and thickly populated alluvial coastal belt of Guyana.

Much of the Moraballi Creek forest appeared (in 1929) to be primary and in an almost entirely natural state, although there were a few easily recognizable patches of secondary forest ('mainap'). Logging on a small scale had been going on for about 80 years, but it had mostly consisted of extracting individual trees of greenheart (*Ocotea rodiei*), the most valuable timber species. There had also been some shifting cultivation by native people, but probably of limited extent.

Five main primary forest communities could be distinguished. Some have minor variants (facies), but on the whole they were well characterized and fairly constant in composition. The boundaries between adjoining communities were sometimes very abrupt, but more often gradual. In the order they were met with from the creek to the plateau they were the mora association (dominant *Mora excelsa*) the morabukea association (dominant *Mora gonggrijpii*), the mixed forest association (no single dominant), the greenheart association (dominant *Ocotea rodiei*) and the wallaba association (dominant *Eperua falcata*).

Communities similar to all of these occur elsewhere in Guyana (Fanshawe 1952) and in Surinam (Schulz 1960, Maas 1971). Some are also widespread in other parts of South America, but others are more local. Mora forest has a wide range in northern Guyana and is found inland to the Pakaraima mountains where it occurs as gallery forest in the savannas (Myers 1936). Outside Guyana it extends east to western Surinam and west to the Orinoco, reappearing in a modified form in Trinidad (Beard 1946a, Bell 1969, 1971). The morabukea association occurs in various parts of Guyana and very locally in western Surinam. The mixed forest of Moraballi Creek is one variant of the mixed association which is the most widespread type of rain forest in the Guianas and through much of Amazonia. Greenheart as a species occurs only in Guyana and sparingly in adjoining parts of Venezuela and Surinam; it probably occurs as a single dominant only near the centre of its range. As will be seen in Chapter 12, forests resembling wallaba forest in structure and composition are widespread in the Guianas on white sands and some other non-kaolisols; they are sometimes dominated by *Eperua* spp., sometimes by other trees. Similar forest types, known as caatinga or campina forests, are also found in parts of Amazonia. The term heath forest, originally used for rain forests

Table 11.3. *Forest communities at Moraballi Creek, Guyana*

	Forest type				
	Mora	Morabukea	Mixed	Greenheart	Wallaba
Habitat					
Texture of soil	Fine silt	Heavy silt	Light loam	Sand	Light sand
Index of texture (Hardy) (lower sample)	40	44	18	5	0
Illumination at breast height (%)	1.33	0.61	0.67	0.81	1.43
Floristic composition					
Dominant or most abundant species	*Mora excelsa*	*Mora gonggrijpii*	*Eschweilera sagotiana, Pentaclethra macroloba*	*Ocotea rodiei*	*Eperua falcata*
As percentage of trees ≥ 4 in (10 cm) d.b.h.	23.4	26.6	6.1, 11.3	9.4	21.1
As percentage of trees ≥ 16 in (41 cm) d.b.h.	67.2	60.7	15.6, 6.7	43.4	67.0
Approx. no. of tree species ≥ 4 in (10 cm) d.b.h.	60	71	91	95	74
Approx. no. of tree species ≥ 16 in (41 cm) d.b.h.	11	21	32	33	16
Other features					
No. of trees ≥ 4 in (10 cm) d.b.h. (per hectare)	310	309	432	519	617
No. of trees ≥ 16 in (41 cm) d.b.h. (per hectare)	45	60	60	87	67

on white sands in Malesia (see below, pp. 305–7 and Chapter 12), can be applied to wallaba and other comparable forest types in South America.

The quantitative data in the following descriptions are based on single sample plots (400 ft × 400 ft, *ca.* 1.5 ha) which were selected as typical of each community.

Mora forest. This occupies the lowest ground, forming broad strips on the silt of the flood-plain along the creek and its larger tributaries. It is also found, but less commonly, above flood level on steep rocky slopes with an impermeable and very shallow clay soil. The feature common to the two habitats seems to be shallowness of soil, in one case due to the high watertable, in the other to erosion which prevents a deep soil from accumulating. Mora forest has much in common with the Central American riverain forests in which *Prioria copaifera* (cativo) is a single dominant (Holdridge *et al.* 1971).

The structure of the mora forest was not examined in detail, but it is probably similar to that of the Trinidadian mora forest described by Beard (see Figs. 2.8B, p. 39 and 12.8, p. 328), except that the upper storeys are less regular and more open; the undergrowth is consequently lighter (average illumination at breast height 1.32%[3]). The A storey is over 30 m high and is composed chiefly of the dominant species; below this there are mixed B and C storeys, a shrub storey (D) and a herb layer which is indistinctly separable into strata of tall and dwarf herbs (Fig. 1.4). The luxuriance of the tall herbs, such as *Carludovica* sp. and various Scitamineae, is characteristic. *Mora excelsa* forms over half the total stand in the larger diameter-classes and a considerable percentage in the smaller classes, as the following figures show.

Diameter-class	Percentage of *Mora excelsa*
4–7 in (10–19 cm)	13.6
8–11 in (20–29 cm)	13.5
12–15 in (30–40 cm)	35.4
16–23 in (41–60 cm)	54.6
≥24 in (61 cm)	91.3
All trees ≥4 in (10 cm)	23.4
All trees ≥16 in (41 cm)	67.2

Mora excelsa is in fact the most abundant tree species in every diameter-class except the 8–11 in; young moras are also abundant and on the plot there were 2.7 juveniles under 3 ft (91 cm) high and 2.0 juveniles 3–6 ft

(91–183 cm) high per square yard (0.84 m²) on the average in the drier parts, although in the wetter parts there are fewer. The commonest trees after mora are *Pterocarpus officinalis* and *Pentaclethra macroloba*, both B storey species. Each formed 10.2% of trees ≥4 in (10 cm) d.b.h. Altogether there were about sixty species of trees reaching 4 in (10 cm) diameter on the plot. The greater number of these, including mora itself and *Pentaclethra*, are found in most of the other four primary forest communities. *Pterocarpus* and a few other species seemed to be confined to the mora association.

Morabukea forest. This community, dominated by another species of *Mora* (Fig. 11.4), has a habitat less well defined than that of *M. excelsa*. Characteristically it occurs on the slopes of the lower flat-topped hills, sometimes extending over their summits; it avoids the flood-plains and prefers heavy red clay soils, although in some situations, such as on low-lying ground at the foot of slopes, it grows on loamy or almost sandy soil.

A distinctive feature of this type of forest is the density of the A and B storeys. Because of this illumination in the undergrowth is very poor (average 0.61%); this is the darkest of the five types of primary forest. Characteristic features are the dense thickets of juveniles (1.0–1.5 m high) of the dominant species, the comparatively thick litter of dead leaves, the abundance of small saprophytes and the scarcity of all other ground herbs.

Mora gonggrijpii is the most abundant species in almost every diameter-class, but in the larger classes its dominance is slightly less complete than that of *M. excelsa* in the Mora association. The percentages in the various classes are as follows.

Fig. 11.4 Morabukea (*Mora gonggrijpii*) forest, Moraballi Creek, Guyana (1929). The undergrowth consists largely of juveniles of the dominant tree species.

---

[3] The light measurements in the sample plots were made by moving an actinometer over a constant distance, so as to give a rough integration of the light in the shade and sun-flecks. The figures are expressed as a percentage of full daylight.

Diameter-class	Percentage of *Mora gonggrijpii*
4–7 in (10–19 cm)	17.6
8–11 in (20–29 cm)	16.5
12–15 in (30–40 cm)	19.6
16–23 in (41–60 cm)	61.3
⩾24 in (61 cm)	59.3
All trees ⩾4 in (10 cm)	25.9
All trees ⩾16 in (41 cm)	60.7

After the dominant the most abundant species are *Eschweilera sagotiana* (7.4% of trees 4 in and over) and *Catostemma commune* (5.0%). Both these belong to the A storey. No other species reaches 5%. The total number of tree species on the plot is seventy-one (⩾4 in). Few, if any, of these are peculiar to morabukea forest; the list of species as a whole closely resembles that for the mixed forest, the main differences being the smaller total number of species and the high percentage of *Mora gonggrijpii*.

Mixed forest. The mixed forest association (*Eschweilera sagotiana* faciation of the *Eschweilera–Licania* association of Fanshawe 1952), which has no single dominant, is characteristic (at Moraballi Creek) of rather sticky loamy soils, intermediate between the heavy clays on which the morabukea association is found and the brown sands of the greenheart forest. It covers much of the lower hilly land of the district and occurs on hilltops when they are not capped with bleached sand.

The structure has already been described in detail (pp. 30ff., Fig. 2.3). A noteworthy feature is that the number of trees per unit area is greater than in either of the two preceding communities (*Mora* association 310 (⩾4 in) per hectare, morabukea association 309, mixed association 432). The illumination in the undergrowth is poor (0.67%), but somewhat greater than in the morabukea forest; because of this, probably, herbaceous undergrowth is more abundant, though less so than in the greenheart and wallaba forests.

The most characteristic feature of the floristic composition is the large number of tree species; the total on the plot was ninety-one, the largest in any of the communities except the greenheart forest. Among trees ⩾16 in (41 cm) d.b.h., the most abundant species is *Eschweilera sagotiana*, which forms 15.6% of the stand, but in the 4 in (10 cm) class, it forms only 6.1%, while a smaller (B storey) tree, *Pentaclethra macroloba*, forms 11.3%. *Licania alba* forms 11.1% of trees ⩾16 in d.b.h. and 9.5% of those ⩾4 in: four other species each form over 5% of trees ⩾4 in d.b.h. Most of the

species found in the other forest types also occurred in the mixed plot with the exception of some mora and wallaba forest species and a few seen only in morabukea and greenheart forest.

Greenheart forest. As has been said, this community is limited at Moraballi Creek to a single type of soil, a light reddish-brown sand, occurring chiefly on the sides of ridges, the tops of which are covered with white sand bearing wallaba forest. Where greenheart and wallaba forest adjoin, the boundary may be very sharp.

The structure of greenheart forest is probably very like that of the mixed forest, although trees of large dimensions are somewhat more frequent. The number of trees ⩾4 in d.b.h. per hectare is 519, i.e. considerably higher than in the mixed plot, but much lower than in the mora and wallaba plots. The herb layer is well developed and has a rather distinctive composition.

Ocotea rodiei is much the most abundant species in the larger diameter-classes, but in the smaller classes it forms only a small proportion of the stand, as the figures show.

Diameter-class	Percentage of *Ocotea rodiei*
4–7 in (10–19 cm)	0.5
8–11 in (20–29 cm)	3.3
12–15 in (30–40 cm)	1.1
16–23 in (41–60 cm)	38.4
24 in (61 cm)	60.0
All trees ⩾4 in (10 cm)	9.4
All trees ⩾16 in (41 cm)	43.4

The most abundant subordinate species are *Pentaclethra macroloba* (9.1% of trees ⩾4 in), *Eschweilera sagotiana* (7.6%) and *Licania alba* (5.6%). All three of these are also among the most abundant species of the mixed forest; greenheart forest in fact resembles the mixed association so closely in composition that it could be regarded as a special facies or 'lociation' of it rather than a distinct association (see above). The total number of tree species on the plot is 95, the highest number of all the sample plots, though perhaps not significantly greater than in the mixed forest. An extensive investigation might show that a few of these are characteristic of the greenheart forest.

Wallaba forest. The last of the five forest types is the association of *Eperua falcata*, which, like the two moras, belongs to the Leguminosae. In structure, floristic composition and habitat, wallaba forest is very distinctive (Figs. 2.9, 11.5, 11.6) and differs more from the other

Fig. 11.5 Wallaba forest (*Eperua falcata* association), Moraballi Creek, Guyana (1929). Two trees of *Eperua* sp. in centre. To the left of the nearer *Eperua* a plant of the bromeliad *Tillandsia* sp. which has fallen from a tree is growing on the ground; another *Tillandsia* is growing epiphytically at *ca.* 1 m on the small tree to the left.

Fig. 11.6 Wallaba (*Eperua*) by road, Kwapau, R. Mazaruni, Guyana (1979). Note the uniform height of the A storey trees.

four communities than they differ from one another. It is strictly limited to white sand soils and stops sharply at the boundary of this soil type. At Moraballi Creek, wallaba forest was the only type of primary forest occurring on white sand, but elsewhere in the Guianas other communities occur on similar soils.

The structure of wallaba forest, in which a B storey is almost absent, was described in Chapter 2 (Fig. 2.9). A remarkable feature is the extremely large number of trees

per unit area (617 per hectare ≥4 in d.b.h.), nearly twice as many as in the mora forest. The shrub layer is very dense. Herbaceous plants are few in species and individuals, but highly characteristic, a proportion of the species being probably confined to this type of forest. The poor development of the herb layer is not due to lack of light; illumination in the undergrowth is greater than in any of the other forest types (1.43%). Other characteristic features of wallaba forest are the scarcity of buttressed trees (Fig. 11.5) and of large lianes. Some species of the latter are probably characteristic of this association.

Diameter-class	Percentage of *Eperua falcata*
4–7 in (10–19 cm)	10.0
8–11 in (20–29 cm)	17.3
12–15 in (30–40 cm)	32.7
16–23 in (41–60 cm)	66.7
24 in (61 cm) and over	75.0
All trees ≥4 in (10 cm) and over	21.1
All trees ≥16 in (41 cm)	67.0

Eperua falcata is the most abundant species in every diameter-class, except the smallest, and in the larger classes it is overwhelmingly dominant. It also regenerates abundantly; six random counts on the sample plot gave an average of 1.8 juveniles under 2 m high per square metre.

In the wallaba forest there is a tendency to 'gregariousness', several species besides the dominant being represented by a comparatively large number of individuals. Two of them, *Catostemma fragrans* and *Licania buxifolia*, form 15.1 and 12.2%, respectively, of trees ≥4 in d.b.h.: both are medium-sized (B storey) trees and form less than 5% of trees ≥16 in d.b.h. Only 18.8% of the species on the plot were represented by only a single individual; this is a smaller proportion than in the other forest types (Table 11.3). There were in all seventy-four tree species ≥4 in d.b.h. on the plot, many fewer than in the mixed and greenheart forest, but somewhat more than in the mora and morabukea forest. About three-quarters of the species, among them *Eperua falcata* itself, are also found (though in smaller numbers) in the other forest types. The remainder, including *Catostemma fragrans* and *Licania buxifolia*, are apparently confined to wallaba forest, at least in the Moraballi Creek district. The floristic composition of the *Eperua* association is thus very distinctive. Not only are certain species nearly or quite confined to it, but those species that occur in other types are found here in very different relative proportions. It is also of interest that there are shrubs, herbs, and probably lianes, peculiar to wallaba forest.

Comparison of communities. The chief characteristics of the five primary forest communities at Moraballi Creek are summarized in Table 11.3. From a comparison of their floristic composition a number of facts emerge. A large number of species are common to all five communities (see Davis & Richards 1933–34, Tables II–VI), but their relative proportions are different in each. Some species are characteristic of, or perhaps even confined to, each of the five communities, but since only one plot of each was listed, it would be misleading to attempt to enumerate them here. The differences in composition between the communities are not equally great; the morabukea, mixed and greenheart forests (the 'central' types) are much more similar to each other than they are to the mora and wallaba forests.

The climate of all five types is alike and all are primary communities which have probably been little affected by logging or other human activities. It is evident that the composition of each type is determined by the characteristics of the soil, or by a combination of soil and topography. The correspondence between the boundaries of the plant communities and the soil boundaries is often strikingly exact. The degree of similarity or difference between the soils is correlated very closely with the degree of difference in floristic composition; the morabukea, mixed and greenheart soils have much in common, but the mora and wallaba soils are very different both from each other and from those of the 'central' forest types.

It is also striking that the five types form a regular series, or catena, so that when arranged in the order in which they normally occur – mora, morabukea, mixed, greenheart, wallaba – their characteristics vary in a regular way. Some increase or decrease more or less steadily through the series; for example, the number of trees per unit area increases steadily from a minimum in mora to a maximum in wallaba. Other characteristics reach their maximum or minimum towards the middle of the series and fall or rise towards the two ends. Thus the percentage of the most abundant tree species is greatest at the two ends of the series; it falls regularly towards the middle and reaches a minimum in the mixed type with no single dominant. The number of species per plot, on the other hand, is greatest in the middle and falls towards the two ends. The degree of dominance of single species and the total number of species are related and vary more or less inversely. Both characteristics seem to depend on the nature of the soil. The mora and wallaba soils, it seems reasonable to assume,

are relatively unfavourable to plant growth, the one because it is shallow, subject to flooding and poorly aerated, the other because like other strongly leached podzolic soils it is probably deficient in nutrients; it is also excessively porous and may be liable to dry out in drought periods (pp. 267–8). The mixed forest soil, on the other hand, is probably the optimum soil of the catena; that this is so is suggested by the fact that the native people, when choosing a site for shifting cultivation, prefer sites in mixed forest. Attempts at permanent agriculture in wallaba forest in Guyana have completely failed.

It thus appears that the most 'mixed' type of forest is found on the most favourable soil and that single-species dominance depends on unfavourable soil characteristics of one kind or another. It may be supposed that on optimum soils a great variety of species, with differing degrees of tolerance for unfavourable soil characteristics, flourish together, while on less favourable soils competition excludes the more exigent species and gives a selective advantage to those that are relatively tolerant. Single-species dominance is further discussed in the next chapter, where it is shown that in primary rain forests generally single-species dominance seems to be usually, but perhaps not invariably, associated with unfavourable soil characteristics.

11.3.2 Twenty-four Mile Reserve

This reserve, which Ogden (1966) studied in 1963, is a square mile (2.6 km²) of almost undisturbed primary forest in the triangle between the Essequibo and Mazaruni rivers, about 39 km south of Bartica. It is *ca.* 30 km west of Moraballi Creek as the crow flies. Ogden used much more thorough sampling procedures than was possible in 1929 and analysed his data by ordination methods.

Twenty-five 'stands' of 1 ha were located on a grid pattern, and not, like the sample plots at Moraballi Creek, selected subjectively. In each stand there were 25 sampling points at which, following the 'point-centered quarter' method of Cottam & Curtis (1956), the three nearest trees ≥2 in (5 cm) d.b.h. in each quarter were measured and identified (See Appendix 2). The data were analysed by ordination of basal area and density. In addition, counts of 'seedlings' and saplings were made in each stand and 10 samples of soil (to a depth of 10 cm below the surface root-mat) were collected for analysis. Ogden also carried out an ordination including the sample plot data of Davis & Richards (1933–34) from Moraballi Creek and compared the results with his own.

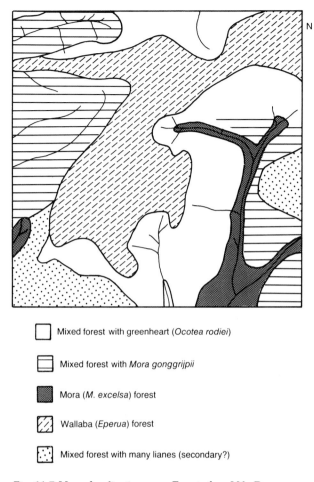

Mixed forest with greenheart (*Ocotea rodiei*)

Mixed forest with *Mora gonggrijpii*

Mora (*M. excelsa*) forest

Wallaba (*Eperua*) forest

Mixed forest with many lianes (secondary?)

Fig. 11.7 Map of ordination area, Twenty-four Mile Reserve, Bartica, Guyana. After Ogden (1966, Fig. 8).

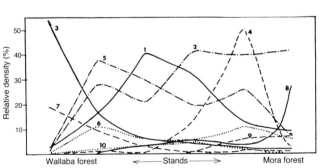

Fig. 11.8 Density of ten 'most important' species along the continuum, Twenty-four Mile Reserve, Bartica, Guyana. After Ogden (1966, Fig. 71). The stands were grouped in fours and the mean density in each group plotted. 1 *Ocotea rodiei*, 2 *Eschweilera sagotiana*, 3 *Eperua falcata*, 4 *Mora gonggrijpii*, 5 *Licania heteromorpha*, 6 *Licania alba*, 7 *Aspidosperma excelsum*, 8 *Pentaclethra macroloba*, 9 *Mora excelsa*, 10 *Swartzia leiocalycina*.

In the Twenty-four Mile Reserve, mora, morabukea, mixed, greenheart and wallaba forest types could be recognized. Their composition was very similar to, but not quite identical with, that of the corresponding types at Moraballi Creek. For example, in the mixed forest the most abundant large tree species were *Eschweilera sagotiana* and *Licania* spp., but *Pentaclethra macroloba*, which at Moraballi Creek was one of the two most abundant mixed forest species, was less important here and found mainly in wetter sites on hydromorphic soils. Although there were other differences, Ogden (1966, p. 25) concludes that 'data from all available sources confirm Davis & Richards' classification into five major "forest types".' The ordination methods, as might be expected, throw much new light on this catena of plant communities.

In the ordination (Fig. 11.9) the wallaba forest is separated from the other forest types by a well-marked discontinuity, confirming Davis & Richards' earlier conclusion that it differs more from the other communities than the latter do from each other. The mora forest in the Twenty-four Mile Reserve is small in extent and limited to alluvial soil by creeks; it is also separated in the ordination from the 'central' types, morabukea, mixed and greenheart. These three form a continuum, with no clear discontinuities. Although stands of each occurred resembling the corresponding sample plots at Moraballi Creek, there were large transitional areas described as 'mixed with morabukea' and 'mixed with

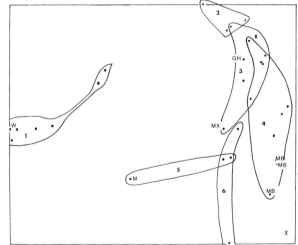

Fig. 11.9 Ordination of forest types, Twenty-four Mile Reserve, Bartica, Guyana. From Ogden (1966, Fig. 16). See text, pp. 302–4. The numbers refer to groups of stands in: M, Mora (*Mora excelsa*) forest; MB, morabukea (*M. gonggrijpii*) forest; MX, mixed forest; GH, greenheart (*Ocotea rodiei*) forest; W, wallaba (*Eperua*) forest.

greenheart'. A map of the distribution of the five forest types is shown in Fig. 11.7 and the density of the ten 'most important' tree species through the catena in Fig. 11.8.

Ogden's ordinations showed that, as at Moraballi Creek, the composition of the forest is closely related to soil characteristics. His soil data cannot be adequately summarized here, but it may be noted that they support the view that in tropical rain forests soil moisture is generally the most important edaphic factor. Ogden (1966, p. 79) considers that a basic division can be made into communities of wet, mesic and dry habitats, occupied in the Twenty-four Mile Reserve by mora forest, the 'central types' and wallaba forest, respectively. Schulz (1960) came to a similar conclusion in the Mapane region of Surinam, although there the forests of the extreme wet and dry habitats are not usually single-dominant communities. A similar classification probably applies to rain forests in many parts of the tropics.

However, Ogden suggests that chemical soil characteristics, especially nutrient status, are probably also important. Unfortunately he examined only the superficial horizons (to about 10 cm depth) of the soils and his data are insufficient to show significant correlations between chemical factors and the species composition of the forest, such as have been found in Cameroon and other rain-forest areas. In his samples, however, he found a strong positive correlation between the clay fraction of the soil and the concentration of nutrient elements, especially phosphorus, probably due to the adsorption of cations on the clay minerals. From this it may be inferred that the soils of the 'central' forest types and the mora forest have a higher base status and base exchange capacity than the white sands of the wallaba forest, which have only a very small clay fraction, 89–90% of the soil consisting of sand.

Ogden's soil data, incomplete though they are, support the conclusion of Davis & Richards (p. 302 above) that the mora and wallaba soils are the less favourable, and those of the 'central' types the more favourable, soils of the catena.

11.4 Primary rain-forest communities in Borneo

The two types of primary lowland forest studied by the author in 1932 (Richards 1936) near Gunung Dulit (3–4°N, 114°E) and at Marudi (4°15′N, 114°19′E) in the Baram basin in northern Sarawak are instructive to compare with the forest communities at Moraballi Creek

considered above. The climate here (p. 427) is similar in essentials to that of Moraballi Creek, but the flora, as might be expected, is entirely different, almost the only species of flowering plants common to the two areas being a few weeds.

The two forest types, mixed dipterocarp forest and heath (kerangas) forest, were studied on the foothills and lower slopes of Gunung Dulit, a Lower Miocene sandstone escarpment rising to some 1100–1400 m to the south of the Tinjar river, a tributary of the Baram. Heath forest was examined at ca. 750 m in the Ulu Koyan on the dip-slope of the escarpment and in low-lying country at Marudi on the Baram, ca. 80 km to the north.

Until quite recently most of the lowlands of Borneo were forest-covered, except near the rivers where the natural vegetation had been cleared for cultivation. In the G. Dulit district in 1932 shifting cultivation of rice had already destroyed much of the primary forest on more or less level ground and there were large areas of secondary forest, but undisturbed forest could still be found on the steeper slopes. Four types of primary forest were recognized: (1) mixed dipterocarp forest, (2) heath forest, (3) lower montane forest, (4) upper montane forest. Types (3) and (4) were found only above about 450 m; these types are dealt with in Chapter 17. Mixed dipterocarp and heath forest are both types of lowland rain forest and occur under similar climatic conditions, but on different soils.

Mixed dipterocarp forest was found on sticky red and yellow loams (kaolisols) and was the most widespread type of lowland forest, while heath forest was confined to sandy soils. At Marudi it occurred on pure white sand of considerable depth, a tropical podzol very similar in its characteristics to the wallaba sands of Guyana. The heath forest in the Ulu Koyan was on a slightly sticky brownish-grey sand overlying clay. At Marudi and on G. Dulit where there was a sharp boundary between clayey or loamy soil and sand there was an abrupt transition from mixed dipterocarp forest to heath forest (Fig. 11.10).

The mixed dipterocarp and heath forests were strikingly different in physiognomy, structure and floristic composition, but because the flora was very rich and (in 1932) comparatively little known, their composition could be compared only in general terms. The forest types described here are in general similar to those of the Gunung Mulu National Park, some 200 km to the northwest, of which there is an account by Proctor et al. (1983a; see also p. 321).

Mixed dipterocarp forest. The structure and physio-

Fig. 11.10 Aerial photograph of mixed dipterocarp forest and heath forest, Sabal Forest Reserve, Sarawak (*ca.* 1981). (Photo. E.F. Brünig.) The mixed dipterocarp forest occupies the kaolisols on the scarp slope (right foreground); the heath forest with more even canopy and smaller crowns, the dip slope (top left).

gnomy of the G. Dulit mixed dipterocarp forest was described in Chapter 2. The outstanding features of its floristic composition were the very large number of tree species and the small number of mature individuals of any one of them: it was thus a rather extreme example of a mixed rain forest association.

On the sample plot (400 ft × 400 ft, *ca.* 1.5 ha) the number of species ≥8 in (20 cm) d.b.h.[4] was estimated to be more than 98 and there were at least thirty-two species ≥16 in (41 cm) d.b.h. The larger diameter-classes are therefore about as species-rich as in the Guyana mixed forest, but the smaller classes are much richer. Out of 261 trees ≥8 in d.b.h., not more than 12 (4.5%) belonged to any one species. The most abun-

[4] It was impracticable in the time available to include trees of 4–8 in (10–20 cm) d.b.h. in the enumeration as at Moraballi Creek.

dant tree species on the plot was in the B storey, all the individuals being under 6 in d.b.h. In the class ≥16 in the two most abundant species (both unidentified) were each represented by six individuals, out of a total of sixty-two (9.7%), but of the whole stand ≥8 in d.b.h. they formed as little as 3.4 and 2.7%, respectively. The Dulit mixed forest thus shows even less single-species dominance than the mixed forest in Guyana, but it was noteworthy that, although no one species formed more than a very small proportion of the whole stand, one family, the Dipterocarpaceae, formed at least 17% of trees ≥8 in d.b.h. and at least 44% of those ≥16 in. There is thus a 'family dominance' of Dipterocarpaceae, as in most primary lowland forests in Malesia. Family dominance seems to be a rather common feature of tropical rain forests, both with and without single domi-nant species (see pp. 309ff.).

Heath forest. Heidewald (heath forest) was the term used by H. Winkler (1914) for a characteristic type of 'subxerophilous primary forest' on sandy soils near Buntok (1°45'S, 114°47'E) and elsewhere in southeastern Borneo. He gave it this somewhat incongruous name because, like the heaths of north Germany, it was associ-ated with infertile sandy soils and many of its trees and shrubs had small sclerophyllous foliage with a xero-morphic appearance. The forests on bleached sandy soils at Marudi and Ulu Koyan agreed with Winkler's descrip-tion in many features of their physiognomy and floristic composition, so the same name was adopted for them (Richards 1936). Since then the term heath forest has been applied to similar forest communities in various parts of the Old and New World tropics (Chapter 12). The Iban (Dyak) word kerangas, meaning 'land on which rice cannot be grown', is often used for Malesian heath forests.

There were striking differences between the Marudi and Ulu Koyan heath forests and the mixed dipterocarp forest. The undergrowth (D and E storeys) was much denser and the illumination near the ground stronger; the stratification of the taller trees also appeared to be different (Chapter 2). The trunks were much less mark-edly buttressed (p. 320) and lianes, including climbing palms, were relatively scarce. Ground herbs, both as species and as individuals, were very few.

The sample plots at Marudi and Ulu Koyan (both 400 × 400 ft, *ca.* 1.5 ha) were somewhat different in composition, as might be expected from the difference of altitude and locality, but agreed in important respects (Table 11.4). In both there were fewer species than in the Dulit mixed plot: it was estimated that on the Marudi plot there were fifty-six tree species ≥8 in

Table 11.4. *Comparison of forest communities in Guyana and Borneo*

	Guyana		Borneo		
	Mixed	Wallaba	Mixed	Marudi Heath	Koyan Heath
Soil (lower samples)					
Texture	Light loam	Light sand	Loam	Sand	Sand
Index of texture (Hardy)	18	0	18–27	0	5–13
Illumination (breast height)	0.67%	1.43%	Moderate	Good	Good
Structure and physiognomy					
No. of trees ≥ 8 in (20 cm) d.b.h. per hectare	232	325	184	246	232
No. of trees ≥ 24 in (61 cm) d.b.h. per hectare	6	2	29	7	29
Woody undergrowth	Thin	Dense	Thin	Dense	Dense
Herbaceous undergrowth	Few species, many individuals	Very few species, few individuals	Many species and individuals	Few species	Few species
Lianes	Large and numerous	Few and scarce	Large and numerous	Few and scarce	Few and scarce
Buttressing of trees	Many strongly buttressed trees	Few strongly buttressed trees	Many strongly buttressed trees	Few strongly buttressed trees	Few strongly buttressed trees
Floristic composition					
Approx. no. of tree species ≥ 8 in (20 cm) d.b.h.	55	49	98	56	55
Approx. no. of tree species ≥ 16 in (41 cm) d.b.h.	32	16	32	12	18
Percentage of most abundant species among trees ≥ 8 in (20 cm) d.b.h.	*Pentaclethra macroloba* — 13	*Eperua falcata* — 32	Medang lit — 5	Tekam — 12	*Agathis borneensis* — 15
Percentage of most abundant species among trees ≥ 16 in (41 cm) d.b.h.	*Eschweilera sagotiana* — 16	*Eperua falcata* — 67	Meranti daging and Marakah batu each — 10	Tekam — 36	*Agathis borneensis* — 35

d.b.h., and on the Koyan plot fifty-five, as against about ninety-eight on the mixed plot. In both heath forest plots there was a tendency to single-species dominance, the most abundant species forming a considerably larger proportion of the stand than any one species in the mixed forest, though not such a large proportion as the dominants of the Guyana associations. In the Marudi plot the most abundant tree species was an unidentified dipterocarp[5] which formed 12.0% of trees ≥8 in (20 cm) d.b.h. and 35.9% of those ≥16 in (41 cm): in the Ulu Koyan plot the conifer *Agathis borneensis*, which was also plentiful at Marudi, was the most abundant species (15.2 and 35.2%, respectively). As in the Guyana single-dominant plots, the tendency to gregariousness was shared by more than one species; at Marudi the second most abundant species was a dipterocarp[6] forming 11.4% of the ≥8 in diameter-class, but with only one individual in the ≥16 in (41 cm) class. On the Ulu Koyan plot a species of Anacardiaceae formed 12.4% of the ≥8 in class and 30.8% of the ≥16 in class. In the Marudi plot several tree species, including the two most abundant, were strikingly patchy ('clumped') in distribution. About 43% of the tree species occurred in both sample plots.

At Marudi, family dominance of the Dipterocarpaceae was well marked (40% of trees ≥8 in d.b.h., 51.3% of those ≥16 in d.b.h.), but at Ulu Koyan this family was poorly represented and estimated to form only 5.8% of trees ≥8 in d.b.h. and 2.2% of those ≥16 in d.b.h. This was no doubt an effect of the higher altitude; Dipterocarpaceae are mainly a lowland family (p. 311).

Enough species were identified on the two heath forest plots to indicate that the flora was very different from that of the mixed forest. It probably included many shrubs, lianes and herbs, as well as trees confined to heath forest. A conspicuous example is the pitcher plants (*Nepenthes*) which are common in moist places in heath forest, but absent in the Dulit mixed forest. *Agathis* and other trees common at Marudi are also common in the heath forest in southern Borneo (Winkler 1914, Diels & Hackenberg 1926).

[5] Tekam, probably *Shorea materialis* (P.S. Ashton).
[6] Mang or Chengal, probably *Hopea pentanervia* (P.S. Ashton).

A comparison of the heath and mixed forests at Gunung Dulit shows that in Borneo, as in Guyana, the forest on bleached sandy podzols is poorer in species than the corresponding type of forest on kaolisols. In both countries the community on sand showed a tendency to single-species dominance, less pronounced in the heath forest perhaps because of the very great floristic richness of Borneo as a whole. In Borneo, as in Guyana, the sandy soils can be regarded as less favourable to plant growth than the kaolisols, the name kerangas (see above) indicates their reputation for infertility. They also have a tendency to dry out in drought periods. As will be seen later (pp. 326–8) the relative importance of nutrient deficiencies and drought in determining the special characteristics of heath forest is still debatable. In any case, it may be assumed that the heath-forest soils, like the wallaba white sands, act selectively, favouring species tolerant of low nutrient concentrations (and perhaps also of occasional water stress) and excluding those intolerant of such conditions.

It is very significant that the heath forest resembles the wallaba forest in many characteristics besides the relatively low number of species and the tendency to single-species dominance, as can be seen in Table 11.4. The two communities are so similar in general aspect that in the heath forest it seemed to the author as if every individual plant of the wallaba forest had been replaced by one of similar life-form but different systematic affinity. Not only do the two communities resemble each other closely, but each differs from the mixed rain forest of its own region in the same characteristics. Since no species, and very few genera, are common to both, the resemblances must be due entirely to soil and climate. The similarity between the vegetation of distant regions with similar climates is well known and is illustrated by the likeness in physiognomy and structure between Mediterranean sclerophyll vegetation in different continents or between mixed rain-forest communities in tropical America, Africa and Southeast Asia; but the similarity between the Bornean heath forest and the wallaba forest of Guyana is a different phenomenon, a parallel edaphic differentiation of plant communities in similar climates.

Chapter 12

Composition of primary rain forests: II

12.1 Mixed rain forests

Communities in which no single species forms more than a small proportion of the whole stand are the most widespread type of primary rain forest in all its main geographical sectors. Mixed forests are extremely variable in composition on both a local and regional scale, making the variants difficult to define and to classify. On terra firme sites it is only on certain soils, such as podzols, white sands and soils overlying limestone or ultrabasic rocks, that mixed forests are found which can be regarded as distinct associations. On podzols and white sands both mixed and single-dominant forests are found, but even the mixed communities show some tendency to single-species dominance (see pp. 300ff. above). On sites liable to frequent flooding, or with permanently impeded drainage, communities are found which usually have more or less marked dominance of one species; these grade into freshwater swamp forests (Chapter 14).

In mixed rain forests on terra firme there are very large numbers of tree species per unit area and most species occur at very low densities. Species richness was considered in the previous chapter. The number of species of trees ⩾10 cm d.b.h. per hectare may be over 200 but is generally in the range 100 to 150 (Table 11.1), but is very variable, depending on the proportion of the sample in the gap and mature phases, as well as on soil and other environmental factors.

The large proportion of tree species in mixed forests which are represented by very small numbers of individuals is shown in Table 12.1. In plots of *ca.* 1–2 ha there are not more than about ten trees ⩾10 cm d.b.h.

of the majority of the species; of many there is only one. Thus in the author's mixed plots in Guyana and West Africa there were not more than five individuals of at least half the species. In very species-rich forests in Malesia the proportion of species occurring at very low densities is even greater; in a 0.61 ha plot at Rengam (Malay), for example, 56% of species were represented by a single individual (Cousens 1951). The occurrence of species at such low densities implies that the presence or absence of a given species in a small sample is mainly a matter of chance: the large number of 'rare' species has important consequences both for sampling methods (Appendix 2) and for conservation.

Though most species are present at very low densities and one species seldom forms more than 10–15% of all trees ⩾ 10 cm d.b.h., a few relatively abundant species often form a considerable proportion of the whole stand, both as individuals and as basal area. In the mixed plot at Moraballi Creek, three species together form 27.3% of the stand and in the Southern Bakundu (Cameroon) plot (Table 12.2) the five species represented by twenty or more individuals formed 36.8% of trees ⩾10 cm d.b.h. In a 10.5 ha plot at Mocambo (Amazonia), partly on terra firme and partly on swampy ground, which had 255 species ⩾10 cm d.b.h., the nineteen most numerous species contributed over 50% of the stand in both the drier and the wetter parts of the plot (Pires 1984).

In the mature phase of mixed forests, in contrast to single-dominant forests, the most abundant species in the ⩾10 cm (or 10–41 cm) d.b.h. class is usually different from that most abundant in the ⩾41 cm class; for example, in the Moraballi mixed forest the B storey species *Pentaclethra macroloba* was the most abundant

Table 12.1. *Composition of mixed tropical rain forest*

Sample plots of *ca.* 1.5 ha.

	Locality			
	Asia Mt. Dulit, Sarawak[a]	S. America Moraballi Creek Guyana[b]	Africa Okomu Forest Nigeria[c]	Africa S. Bakundu Forest Reserve, Cameroon[d]
No. of trees per hectare:				
≥4 in (10 cm)	—	432	390	367
≥8 in (20 cm)	184	232	223	—
≥16 in (41 cm)	44	60	47	109
No. of species:				
≥4 in (10 cm)	—	91	70	109
≥8 in (20 cm)	98	55	51	—
≥16 in (41 cm)	32	32	31	42
Ratio species:individuals:				
Spp. ≥4 in (10 cm)	—	7.1	6.4	5.1
Spp. ≥8 in (20 cm)	2.7	6.0	5.6	—
Spp. ≥16 in (41 cm)	1.9	2.8	2.3	3.9
Percentage of most abundant species:				
	—	*Pentaclethra macroloba*	*Strombosia retivenia*	*Diospyros xanthochlamys*
≥4 in (10 cm)	—	11	30	—
	'Medang lit'	*Pentaclethra macroloba*	*Strombosia retivenia*	—
≥8 in (20 cm)	5	13	35	*Cynometra hankei*
	'Marakah batu', 'Meranti daging', each	*Eschweilera sagotiana*	*Pausinystalia* sp. (or spp.)	—
≥16 in (41 cm)	10 10	16	14 (or less)	—

Sources: [a] Richards (1936), [b] Davis & Richards (1933–4), [c] Richards (1939), [d] Richards (1963a).

tree ≥10 cm d.b.h., but in the ≥41 cm class *Eschweilera sagotiana* formed the largest percentage of individuals. This is because the ≥41 cm class consists mainly of tall A and B storey species which are often poorly represented in the lower storeys. In the ≥10 cm class the majority of the individuals belong to B and C storey species. These, although more numerous in individuals than the A storey species, contribute less to the basal area of the stand.

12.1.1 Geographical distribution of taxa

The seemingly endless variations in the composition of mixed rain forests reflect the geographical and ecological ranges of their component families, genera and species. The floristic differences between the American, African, and Indo-Malayan and Australasian rain-forest floras probably depend more on continental movements, cli-

matic changes and other past events than on differences in their present environments. On smaller scales, on the other hand, the composition of the flora seems to be correlated to a large extent, though perhaps not entirely, with existing climatic and edaphic factors.

Families and family dominance. In the tree floras of the main rain-forest formations the families present, and their relative importance, differ considerably, although many are well represented throughout the tropics, e.g. Annonaceae, Euphorbiaceae, Moraceae, Myrtaceae and Rubiaceae. Comparatively few families are entirely restricted to one continent, but some very large families, although present in more than one continent, are not equally abundant in each. For example, the Leguminosae, especially Caesalpinioideae, are more important in the rain forests of tropical America and Africa than in those of the eastern tropics and the

Table 12.2. *Species representation in mixed and single-dominant rain forests*

	1	1–5	6–10	11–20	21–30	31–40	41–50	51–60	61–70	71–80	81–90	91–100	Over 100	Total individuals	Total species
Mixed forests:															
1. Moraballi Creek, Guyana	21	68	79	4	4	1	2	–	1	1	–	–	–	644	91
2. Southern Bakundu Forest Reserve, Cameroon	44	61	95	6	1	1	1	–	1	–	–	–	–	551	109
3. Okomu Forest Reserve, Nigeria	23	48	58	7	2	1	–	–	1	–	–	–	1	–	–
4. Gunung Dulit, Borneo[a]	41	85	95	1	–	–	–	–	–	–	–	–	–	261	98
Single-dominant forests (or associations with tendency to single-species dominance):															
5. Mora (*Mora excelsa*) forest, Moraballi Creek, Guyana	22	41	47	9	1	2	–	–	–	–	–	–	1	462	60
6. Morabukea (*Mora gonggrijpii*) forest, same loc.	25	49	61	3	3	1	–	–	–	–	–	–	1	460	71
7. Greenheart (*Ocotea rodiei*) forest, same loc.	31	63	78	4	–	3	2	2	–	2	–	–	–	773	95
8. Wallaba (*Eperua falcata*) forest, same loc.	14	45	55	11	4	–	1	–	–	–	–	–	3	536	38
9. Freshwater swamp forest, Omo Forest Reserve, Nigeria	11	19	24	6	2	3	–	–	1	–	–	1	–	–	
10. Heath forest, Marudi, Borneo[a]	13	42	49	3	1	2	1	–	–	–	–	–	–	350	56

[a] For trees ≥8 in (20 cm) diameter (other data for trees ≥4 in (10 cm)).

Sources: From Davis & Richards (1933–4), Richards (1936, 1939, 1963a).

Lecythidaceae are also much more numerous in the New World than the Old. On the other hand the Dipterocarpaceae, which are very numerous and important rainforest trees in Southeast Asia, are represented in Africa only by the Monotoideae, which are small savanna trees, and in the Neotropics by the problematical *Pakaraimaea* (Maguire *et al.* 1977) and perhaps by an as yet undescribed species in northwestern Amazonia (Gentry 1993). In Australia they are absent, though plentiful in New Guinea.

The family dominance of Dipterocarpaceae throughout Malesia, referred to on p. 305, is discussed in detail by Whitmore (1984a). It is characteristic of almost all rain forests, including heath and swamp forests, in Malaya, Borneo and Sumatera; it is also a feature of many

Fig. 12.1 Tropical rain forest on steep ridge, Malay Peninsula. From Foxworthy (1927, Pl. 1). 'Hill dipterocarp forest' at 350–550 m, with society of *Shorea curtisii* (light-coloured crowns) in centre.

seasonal forests from India to the Philippines. Most, but not all, dipterocarps are tall trees and their dominance is particularly marked in the A storeys; in the smaller size-classes, and sometimes in the stand as a whole, dipterocarps may be outnumbered by other families. In extensive samples of lowland forests, dipterocarps often form over 50% of trees ≥2.4 m g.b.h. in Malaya and over 75% in Borneo (Whitmore 1984a, p. 220). They commonly form a much larger proportion of the species and individuals of trees ≥10 cm d.b.h. and of the basal area than any other family. On 1 ha of mixed forest in the Gunung Mulu National Park (Sarawak), for example, they were represented by many more trees than any other family and formed 43.2% of the basal area (Proctor *et al.* 1983a). In contrast to this at Wanariset (Kalimantan), although the dipterocarps formed the largest fraction of the basal area, Euphorbiaceae were more numerous in species and individuals (Kartawinata *et al.* 1981). According to Ashton (1982) dipterocarps reach their maximum in both species and individuals on deep well drained yellow/red soils (kaolisols) of intermediate fertility where nutrient levels are limiting, but not severely so.

It is interesting that though Dipterocarpaceae show strong family dominance, they rarely become single-species dominants, except in some types of heath forest and peat-swamp forest.

Dipterocarp forests are the most striking example of family dominance, but they are not the only one. It would be possible to speak of family dominance of Leguminosae in parts of tropical America and Africa. In African rain forests the Leguminosae subfamily Caesalpinioideae are the most numerous group of large trees, with 115 species in West Africa alone (F.N. Hepper). In Africa they play an important part in mixed forests; in all African terra firme single-dominant rain forests, some of which are very extensive (pp. 331ff.), the dominants are Caesalpinioideae. The possible ecological significance of the gregarious habit of the Caesalpinioideae and Dipterocarpaceae is discussed later (pp. 337–8).

Species distribution. Only in a few taxonomically well studied genera, especially those such as *Hevea* (Fig. 12.2) and *Theobroma* which are economically important, is the distribution of individual rain-forest species fairly completely known. A few have been mapped: Prance (1977) and Prance & Mori (1979) have published maps of the distribution of some Lecythidaceae, Chrysobalanaceae and members of several other families in South America and Hopkins (1986) of Neotropical *Parkia* species. The distribution of over 600 forest species in Ghana has been mapped by Hall & Swaine (1981).

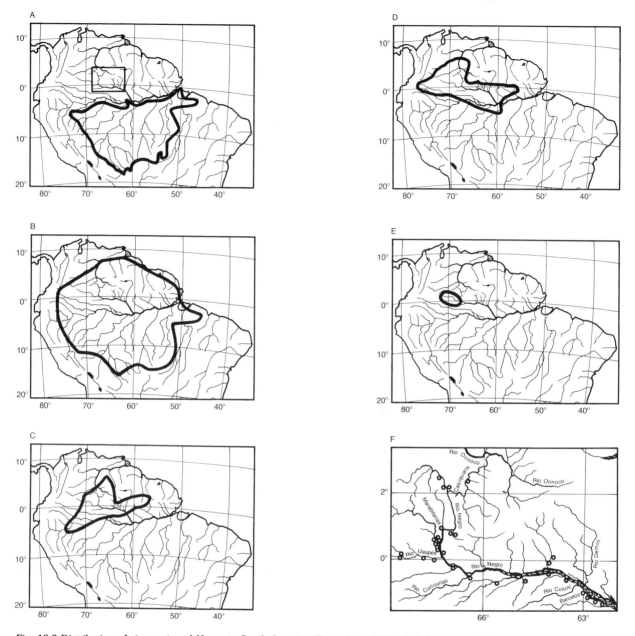

Fig. 12.2 Distribution of six species of *Hevea* in South America. From data of R.E. Schultes, unpublished. A *H. brasiliensis*, B *H. guianensis* (including var. *lutea*), C *H. nitida*, D *H. benthamiana*, E *H. rigidifolia*, F *H. microphylla*.

A few coastal and secondary forest trees are pantropical in distribution, but no primary rain-forest species occurs in all the continental areas and very few in more than one. *Symphonia globulifera* is found in both the South American and African rain forests, and the tropical American *Carapa guianensis* is doubtfully distinct from the African *C. procera*. *Ceiba pentandra*, although undoubtedly introduced in Asia, is native in tropical America and probably in Africa (Baker 1965). Much

more numerous are species which extend throughout the rain forest of one continent, e.g. *Hymenaea courbaril* (tropical America) and *Pometia pinnata* (Malesia to Solomon Islands). In Africa, as noted earlier (pp. 292–3), many tree species are wide-ranging.

The boundaries of the less widely distributed rainforest species often appear to be climatically determined. Whitmore (1984a) notes the dramatic decrease in the number of dipterocarp species in Malaya north of the

Isthmus of Kra where the rainfall pattern becomes much more seasonal. Some rain-forest trees do not extend even into the mildly seasonal climate of the northern Malaysian states of Kedah and Kelantan. Similarly, in Africa and tropical America many species disappear or become rarer towards the climatic boundary of the rain forest.

However, the distribution of some species does not appear to be related to any obvious climatic or geographical boundaries. *Koompassia excelsa*, a common and conspicuous Malesian tree, is inexplicably absent from the Malay Peninsula south of a line from Kuala Lumpur to Kuantan, although in Sumatera it occurs further south. In Ghana, Hall & Swaine (1981) found that the distribution of *Diospyros chevalieri* was almost congruent with that of the wet evergreen forest, but many of the other species of the genus had patchy distributions with no apparent environmental explanation. Aubréville (1938) noted similar problematic species distributions in the forest of Côte d'Ivoire.

Endemic taxa confined to a small region or single locality may be evidence of past changes in climate, land level or land connections or they may be species which for some reason have not been able to spread from where they originated. In the humid tropics they tend to be aggregated in certain areas, e.g. the Guiana Highlands and parts of Amazonia, where Prance (1977) recognizes seven 'major phytogeographic regions', each of which has many endemic species (and genera). It is perhaps suggestive that the distribution of the three species of the primitive palm genus *Chelyocarpus* in Amazonia corresponds with three of Haffner's postulated Pleistocene refuge areas (Chapter 1) (Uhl & Dransfield 1987). Davis (1941) examined the distribution of forty common Guyana, rain-forest trees and found that many of them were surprisingly local, indicating, in his opinion, that they evolved on islands in the sea which covered the present peneplain in the Tertiary.

Parts of the Malesian rain forest also have numerous endemics. Ng & Low (1982) have published a list of the 654 endemic trees of the Malay Peninsula. Three tree species are known only from the small island of Penang (Whitmore 1984a). Many dipterocarps of the non-seasonal (but not of the seasonal) parts of Southeast Asia are endemic to one island or an area within it: in Borneo 155 of the 267 species (57%) are endemic (Ashton 1982). It is interesting that endemic dipterocarps are absent in the large area of eastern Borneo affected by the disastrous drought and fires in 1982–83 (p. 191), perhaps indicating that similar disasters might have occurred in this area in the past (P.S. Ashton).

12.1.2 Variations in composition of mixed forests

Large-scale variations. The variability in the composition of mixed rain forest is well shown by the largest continuous area of lowland rain forest in the world, the 'hylaea' of the Amazon and Orinoco. Unfortunately this vast forest, now rapidly decreasing in extent owing to clearances, is still very inadequately known floristically, especially in the interfluval regions. Quantitative data on forest composition are available for only a few localities. The area is not uniform climatically, the central and much of the southern parts having a lower and more seasonal rainfall than the eastern and western parts (Chapter 7; see also Pires & Prance 1977). Much of it is less than 100 m above sea-level, but in the west and on the fringes of the Guiana Highlands the ground rises and the lowland forest extends upwards to its climatic limit. The soils are varied and there are large areas of non-kaolisols, mainly white sands and podzols, especially in the basin of the Rio Negro; on these caatingas (heath forests) and other edaphic communities are found (pp. 322ff. below), but from the accounts of Ducke & Black (1953) and Pires (1984) most of the primary vegetation on terra firme, except on the white sands, is mixed forest.

The structure and physiognomy of this mixed forest is probably very similar throughout Amazonia, although in the southern part there are more than 10^5 km^2 of 'mata de cipo' (liane forest) characterized by a great abundance of lianes. This may not be primary (see p. 133). All the mixed forest is extremely variable floristically and its composition is said to be different in the basins of all the major tributaries. According to Ducke & Black (1953), the composition of the forest changes more with longitude than with latitude, there being a greater difference between the forest at Belém and *ca.* 750 km westwards at Santarem than between Belém and Cayenne, *ca.* 900 km to the north. Ducke & Black recognized seven main phytogeographical sectors in Amazonia. These have been adopted, with minor alterations of boundaries, by Prance (1977).

The factors responsible for the great regional variability in the composition of the primary terra firme forest on mainly kaolisol soils have been little investigated. They are certainly partly climatic and related to gradients in the annual total and seasonal distribution of rainfall (Chapter 16). The importance of non-environmental factors, such as the slow migration rate of species, especially heavy-seeded trees, is harder to assess. It has long been known that there are considerable floristic differences north and south of the Amazon and Solimões rivers: some species reach their limit on the south bank and some on the north. This line does not correspond to a climatic boundary and as these rivers are in most places more than 10 km wide and bordered by a broad zone of annually flooded várzea, etc., it is plausible to suggest that the main river and its larger tributaries may act as barriers to dispersal. It is also possible that the composition of the mixed forest is related to the location of Pleistocene refuges (Prance 1977, see also pp. 13–18), although Gentry (1992) and others have disputed this, believing that it is unnecessary to invoke historical factors of this kind. On a small scale, as the example given below shows, the heterogeneity of the Amazon forest reflects small differences of land level, topography and soil.

The mixed rain forests of Southeast Asia and Africa also vary regionally in composition, although not to the same extent as in Amazonia. In the Malay Peninsula, for example, there is a general change in the composition of the lowland forest from south to north which reflects a gradient of seasonality in rainfall. In Borneo there are also regional differences in forest composition. These and other variations in Malesian mixed forests are dealt with in detail by Whitmore (1984a) and need not be further discussed here.

The regional differences in African rain forests also appear to be in part related to the seasonality of the rainfall; these will be discussed later (pp. 409ff. and Chapter 16).

Small-scale variations. Within areas a few hectares or less in extent, which can be assumed to be climatically uniform, the composition of primary mixed forest often varies in a complex and bewildering fashion. As at Moraballi Creek, direct observation and sample plot data suggest that the composition of mixed forests is related to environmental factors such as topography, drainage and soil characteristics. In low-lying sites, there are usually marked differences between the composition of the forest on well-drained and swampy land and in more hilly country the forest on ridge tops often differs in composition from that on the slopes and valley bottoms. But superimposed on variations which appear to be determined by the environment there are patterns of mature, gap and building phases due to the regeneration cycle as well as to the dispersion patterns of individual species (pp. 25–6). It is therefore very difficult to show unambiguously whether observed variations in community composition or species distribution are due to environmental or other causes, or to 'chance', without recourse to appropriate methods of sampling and multivariate analysis (see Appendix 2).

The Rio Guamá Research Area. A remarkable record of species distributions and forest composition on a small scale is provided by the Rio Guamá Research Area near Belém (Amazonia), where several permanent plots have been protected and kept under observation for more than twenty years (Pires & Prance 1977, Pires 1984). On these all trees and large lianes ≥30 cm g.b.h. (ca. 10 cm d.b.h.) have been identified, measured and mapped (Fig. 12.3). The largest plot, consisting of the Mocambo Reserve and parts of the Catú Reserve, is a rectangular area 450 m × 250 m (10.5 ha), of which the central part (5.7 ha) is mixed forest on terra firme. The rest (4.8 ha) is at a slightly lower level and is swamp forest which is mostly flooded daily to a varying depth when the water of the Catú stream is backed up by the tide. The whole plot is subdivided into 10 m × 10 m quadrats and could provide data for analysis of variations in community composition. The maps show very clearly the response of a wide range of species to differences in ground level of a few metres or less, and to the incidence of recurrent flooding.

Maps of the eight most abundant species have been published (Pires & Prance 1977); four of these, illustrating some common types of distribution, are reproduced in Fig. 12.3 A–D. *Eschweilera odora* (A) is an example of a species fairly evenly distributed over the terra firme, where it is the most abundant species (287 trees); it also extends into the less deeply flooded marginal parts of the swamp forest (70 trees). *Goupia glabra* (B) is another species which is found mainly on the terra firme (69 trees), but it is found mainly towards its edges and extends a short distance into the swamp forest. For this species moist but not deeply flooded soil appears to be optimal. *Theobroma subincanum* (C) is about equally frequent on the terra firme (44 trees) and in the swamp forest (49 trees). The fourth example, *Virola surinamensis* (D), is a swamp species widespread in tropical America; in this plot it is almost confined to the swamp forest (205 trees), but two trees were found on the terra firme.

In the whole plot there were 282 species; of these 224 species occur in the mixed forest on terra firme and 180 in the swamp forest. One hundred and twenty-two species (43%) are found in both types of forest.

There appear to be few other maps of species distributions in large permanent rain-forest plots comparable to those for the Rio Guamá area, but similar maps are being made for a 50 ha plot in the semideciduous forest at Barro Colorado Island (Panama) (S.P. Hubbell, unpublished).

Dispersion patterns. In all rain forests, as noted in

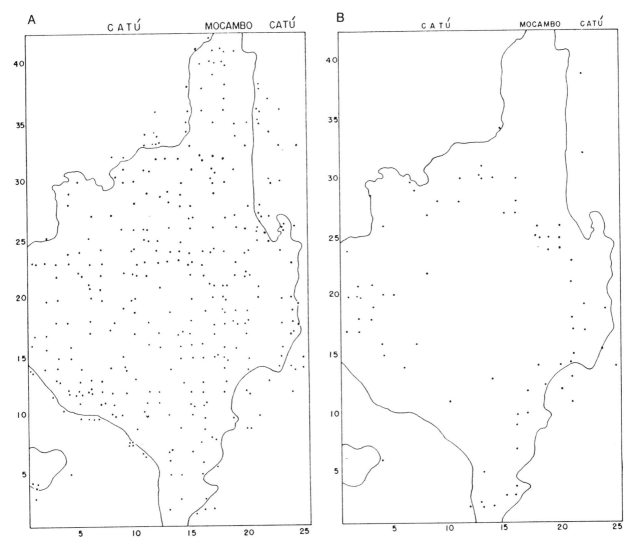

Fig. 12.3 Distribution of tree species in Rio Guamá Research Area, Brazil. From Pires & Prance (1977, Figs. 10, 12, 15 and 16). The central area (Mocambo) is mainly terra firme and the marginal areas (Catú) are swamp forest. A *Eschweilera odora*, B *Goupia glabra*, C *Theobroma subincanum*, D *Virola surinamensis*.

Chapter 2, the trees are unevenly, but usually not randomly, spaced. This applies to individuals of one species as well as to the tree population as a whole and seems to be due to various causes, some intrinsic to the trees themselves and some environmental. Poore (1968), Ashton (1969), Herwitz (1981a), D.W. Thomas (unpublished)[1] and others have shown by pattern analysis (Appendix 2) that in mixed forest, although the dispersion pattern of some species does not depart significantly from random, many are clumped (contagious)[2]

[1] Thomas, D.W. Sub-optimal establishment and the maintenance of species diversity in an African rain forest (Unpublished manuscript.)

[2] 'Clumped' or 'contagious' seems preferable to 'over-dispersed', which is sometimes used in the opposite sense to what is meant here. Similarly, 'regular' or 'even' is better than 'under-dispersed'. See Greig-Smith (1983, p. 60).

to various degrees. Regular (under-dispersed) patterns are rare: they are sometimes found in peat swamp forests (pp. 365ff.), but probably never in mature terra firme rain forest. Clumps of one species vary from groups of three or four mature individuals to stands occupying a hectare or more. The latter differ from single-dominant associations only in scale. Clumping is conspicuous in many large dipterocarps in Southeast Asia, especially in those such as *Shorea curtisii* (Fig. 12.1) which have characteristically coloured foliage. West African Caesalpinioideae (pp. 331ff.), and trees of many other families, also have a strong tendency to grow in clumps.

Clumping of individuals of the same species may be due to opportunity or 'chance', as when numerous

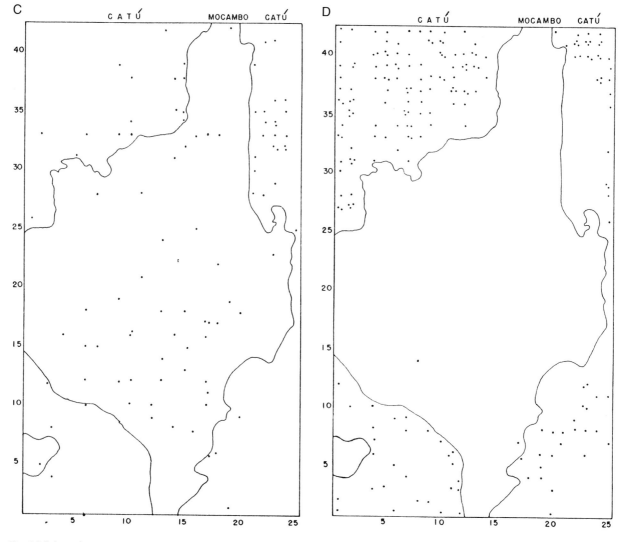

Fig. 12.3 (*cont.*)

saplings are able to grow up where a large tree has died or in a larger gap due to windfall, and it is often clearly related to the dispersal mechanism of the species. Poore (1968) showed that at Jengka (Malaya) trees with small wind-dispersed fruits had a random (or nearly random) distribution pattern, but those with large heavy fruits were clumped. Ashton's (1969) comparison of the dispersion patterns of groups of tree species in the Sarawak rain forest with different methods of dispersal was referred to earlier (see p. 111 and Fig. 5.6). In Dipterocarpaceae, in which the fruits, though winged, are relatively inefficiently dispersed (p. 114), the dispersion pattern showed pronounced clumping.

In the 'dry' (deciduous) forest of Costa Rica, Hubbell (1979) also found that clumping was related to the seed dispersal mechanism. Rare species were more clumped than common ones. Armesto *et al.* (1986) compared the

dispersion patterns of trees in tropical and temperate climates in various parts of the world and concluded that clumping was characteristic of forests in which the formation of canopy gaps was the chief source of disturbance and random patterns of those subject to frequent large-scale disturbance.

The effects of different dispersal methods interact with heterogeneities of soil and microclimate, and with competition, to produce the enormous variety of species distributions found among rain-forest trees. A single example will suffice to illustrate this. Pemadasa & Gunatileke (1981), in a study of the distribution of six tree species in Kottawa forest (Sri Lanka), found that two showed no evidence of pattern, i.e. they were randomly distributed, but the rest had contagious (clumped) distribution patterns on various scales. In *Dipterocarpus zeylanicus* the size of pattern was about half the study area

(4.1 ha) and the distribution was related to a gradient in soil potassium. In *D. hispidus*, the largest tree in the area, the pattern was much smaller (*ca.* 50–100 m) and was also related to soil nutrients, in this case nitrogen and phosphorus. In the two other trees, *Artocarpus nobilis* and *Campnosperma zeylanicum*, the pattern was also small in scale and appeared to be determined by competition with *D. hispidus*, with which these species could not co-exist.

Ordination studies. In recent years principal component analysis and other ordination methods have proved valuable in studying the composition of mixed rain forests. Some of the results are dealt with briefly here: the techniques are discussed in Appendix 2.

Ashton's (1964) pioneering ordination work on mixed dipterocarp forests at Andulau and Kuala Belalong in Brunei suggested that their composition was strongly correlated with topography and soil characteristics; in particular there were marked differences between the composition of stands on ridge tops, slopes and valley bottoms. The data were later re-examined by Austin & Greig-Smith (1968) and Austin *et al.* (1972), using more efficient analytic methods. Ashton's conclusion that variations in species composition (at the scale of the analysis) were more closely related to gradients in soil characteristics than to opportunities for species establishment or 'chance', was in general confirmed; even within plots as small as 0.08 ha meaningful relations were found between forest composition and soil.

Later work on lowland mixed dipterocarp forests in Borneo has confirmed and extended these conclusions. Baillie & Ashton (1983) compared the distribution of some of the most frequent tree species in relation to soil characteristics in 291 samples from Sarawak. They found that their occurrence was more closely correlated with the reserve cationic soil nutrients rather than with the exchangeable nutrients and was also related to soil parent material and depth. In central Sarawak Baillie *et al.* (1987) examined the forests on modal and clay kaolisols (pp. 258–63) and obtained the best correlation of species distributions with the magnesium content of the soil. More recently Ashton & Hall (1992) have reported on the relation to soil and other site characteristics of forest stature, stratification, growth increments and and other features in groups of permanent sample plots at thirteen sites in northwestern Borneo. Since the plots had been monitored for twenty years, the results (which are too complex to summarize briefly) give considerable insight into the dynamic changes in the mature phase of mixed lowland rain forests.

Analyses of lowland mixed dipterocarp forest by ordination techniques have also been made of forest communities in Peninsular Malaya at Jengka (Pahang) and Pasoh (Negeri Sembilan). Both areas have a slightly more seasonal climate and are more uniform in soil and topography than Ashton's Brunei forests. The former, unfortunately, has since been felled, but the latter, which was intensively used for research for the 1964–74 International Biological Programme, is still preserved.

In the Jengka Reserve Poore (1968) selected for study 1 km² of undisturbed primary forest chosen for its apparent uniformity. The area was one of gently sloping hills intersected by a network of swampy stream valleys; the difference in height between the valleys and the hill tops was never more than 30 m. The swamps occupied 1.07 ha (4.6%) of the sampled area. The soils were yellow clay kaolisols derived from Triassic shales.

Within the study area 23.04 ha were used for intensive work. This area contained 2773 'canopy trees' (equivalent to the A and B storeys), belonging to 375 species; 8% of the species and 27.8% of the individuals were dipterocarps.

The distribution of the species was sampled systematically by transects and plots and on a subsample 620 m × 420 m (2.6 ha) all trees ≥91 cm g.b.h. were mapped. The techniques used in analysing the data were association analysis and covariance analysis of pattern (Appendix 2). For reasons of space, the results can be dealt with here only very briefly. The analysis of the data indicated that some species had a marked preference for swampy ground, e.g. *Myristica elliptica*; these were the most distinct of the twenty-four species groups resulting from the inverse analysis. In the area as a whole there was a difference between the ecology of the 'rarer' and the 'commoner' species (the latter being the twenty most abundant species). Many of the former, in addition to the swamp group, seemed to have distinct habitat preferences. The commoner species, on the other hand, appeared to be more tolerant of small differences of soil and microclimate and showed no evidence of habitat preferences. Covariance analysis showed no consistent associations between species. Their distribution was distinctly patterned, but the pattern did not appear to be controlled environmentally; it seemed to be related to the filling of gaps and the availability of seedlings and juveniles. Poore thus regards the community as consisting of a matrix of common, tolerant, species whose distribution depends mainly on chance, within which there are groups of more exacting species with more limited habitat preferences.

Further evidence on the relative importance of

environmental and 'chance' factors in determining the composition of little disturbed lowland dipterocarp forest was provided by the work at Pasoh. This reserve is *ca.* 650 ha in extent and is mainly on gently undulating land *ca.* 300 m above sea-level, though it contains a range of hills rising to *ca.* 570 m. The soils, derived from shales and granite, are base-poor kaolisols varying in texture from clay to sandy loams, differing in drainage characteristics and in the presence or absence of a concretionary layer; they include most of the soil types found in Malayan lowland dipterocarp forests. In a later paper (Ho *et al.* 1987) the dynamics of the Jengka forest are discussed and it is suggested that part of it (that on Batu Aram series soils) may be in a state of non-equilibrium due to recovery from past disturbance.

Wong & Whitmore (1970) laid out ten 200 m × 20 m plots within an area of *ca.* 560 ha, lying astride the granite–shale boundary, chosen for its uniform relief and variety of soils. The plots furthest from each other were 2.7 km apart. In the plots there were 2280 trees ⩾12 in (*ca.* 30 cm) g.b.h., belonging to 328 species. The ordination was by principal components analysis using Orloci's coefficient for presence and absence of the 190 species found in two to eight plots; those found in one, nine or ten plots were excluded as being too rare or too common to provide evidence of soil preferences.

In spite of the variety of soils, Wong & Whitmore found no evidence of groups of species having clear-cut correlations with soil characteristics; many species seemed to have overlapping, if not identical, soil preferences. There also seemed to be no species with narrow ecological ranges such as Poore (1968) had found at Jengka. No clear relation was found between forest composition and soil, parent materials or drainage conditions, the clearest correlation being between floristic composition and distance apart of plots, the nearest being the most similar and the furthest apart the most unlike.

Later work at Pasoh by Ashton (1976) modified these conclusions considerably and showed that there was in fact a consistent pattern of floristic variation, although it was overlaid by a small-scale pattern due to 'chance' factors such as the reproductive behaviour of individual species and the incidence of gaps. Ashton criticized the sampling methods of Wong & Whitmore, considering that the size and total area of their plots was too small and the plot shape insufficiently compact to reveal correlations between small variations in forest composition and differences in soil and relief. Ashton used for association analysis larger plots, subdivided into units comparable to Wong & Whitmore's and up to three

times the size. In part of one plot the position of every tree ⩾10 cm d.b.h. was mapped and plotless samples analysed by Crawford & Wishart's (1967) group analysis technique to investigate smaller-scale variation.

By these procedures Ashton was able to show that there was a consistent pattern of species groups correlated with environmental factors. The chief differences in forest composition were between the hilly areas, the old alluvium in the valleys and the transitional areas. Within this environmentally determined pattern he found variations on a smaller scale due to the dispersion patterns of the individual species. One area which appeared to be anomalous in Ashton's classification proved to be one that was constantly disturbed by wild pigs and other large animals.

Ordinations of mixed rain forests other than dipterocarp forests are still rather few. At La Selva (Costa Rica) Lieberman *et al.* (1985a) used an ordination method to study forest composition in relation to small differences of level ('altitude'). The forest at La Selva has been described by Hartshorn (1983) who classifies it as 'Tropical Wet forest'. The higher ground is intersected by streams and seasonally flooded valleys. The soils are acid, nutrient-poor kaolisols, mostly clayey in texture; in the lower-lying areas the soil profile has a hardened ferruginous layer and the water table is permanently high. The forest has a fairly continuous A storey of very mixed composition. In the B storey the leguminous tree *Pentaclethra macroloba*, which formed 13.6% of all trees ⩾10 cm d.b.h. in the samples, is by far the most abundant tree species. A notable feature of the La Selva forest is the large number of species and individuals of palms: *Welfia georgii*, *Socratea durissima* and *Iriartea gigantea* were respectively the second, third and fourth most abundant species in the samples.

Lieberman *et al.* (1985a) counted and identified all trees, including tree ferns and lianes ⩾10 cm d.b.h., on three permanent sample plots less than 1.5 km apart, totalling 12.4 ha and extended over an altitudinal range of 32–71 m above sea-level. Each included a variety of terrain: Plot 1 was plateau with some swamp, Plot 2 swamp and rolling hills and Plot 3 steep hills and plateau. Each plot was divided into 20 m × 20 m subplots, which were treated as samples for ordination using the DECORANA FORTRAN programme of Hill & Gauch (1980). The total number of live stems in the samples was 5530 and of species 269.

The ordination showed that the species composition of the samples varied continuously with altitude, although the most striking differences (as at Jengka, p. 317 above) were those between the seasonally flooded sites and the higher ground. In the former there was an abrupt reduction in the number of species and stem density (Fig. 12.4). Some species, e.g. *Pterocarpus officinalis*, were confined to swampy sites; others, e.g. *Astrocaryum alatum*, *Carapa guianensis*, were much more abundant on swampy than on upland sites. The disproportionate abundance of species tolerant of water-logged soils and the absence of many upland species gives the swamp forest a distinctive composition (see Chapter 14).

At higher levels the composition of the forest changed even with quite small height differences, but there was also considerable variation at any one level which did not seem to be correlated with environmental factors. The ordinations showed no evidence of gradients related to succession or to treefall gaps or other temporary openings in the canopy. It was concluded that edaphic factors affecting drainage and soil aeration controlled the altitudinal gradient and that variations at any given altitude depended on chance. There seemed to be a large

pool of species able to grow at a particular site; which of them actually succeeded in occupying it was assumed to depend on such factors as variations in canopy cover, transient soil disturbances and accidents of seed dispersal.

The work of Gartlan *et al.* (1986) on Korup F.R. (4°25'–5°54'N, 8°42'–9°9'E) and of Newbery *et al.* (1986) on Douala-Edea F.R. (3°13'–3°49'N, 9°33'–9°54'E) shows the influence of soil nutrients as well as drainage and other topographical factors on forest composition. Both reserves are large, relatively undisturbed, rain-forest areas in western Cameroon; they are mainly mixed forests of the 'Forêt littorale' type of Letouzey (1968).

In both reserves the mean annual rainfall is high (Korup 5460 mm, at Douala-Edea *ca.* 3000–4000 mm), with a dry season of two to three months. The southern part of Korup F.R. is flat and *ca.* 50 m above sea-level; towards the centre and the north the reserve becomes hilly and rocky, reaching 1079 m at one point. The

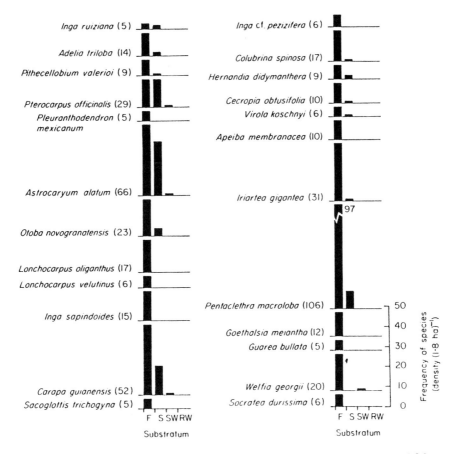

Fig. 12.4 Species frequency in relation to increasingly poor drainage on four types of substratum in a 1.8 ha sample, La Selva, Costa Rica. From Lieberman *et al.* (1985a, Fig. 5). Species listed in order of descending axis 1 ordination scores. Number of individuals given in brackets. Species with fewer than five individuals omitted. Substratum categories: F, firm sediments; S, soft sediments; SW, standing water; RW, running water.

soils, primarily kaolisols derived from granite and gneiss, have sand, silt and clay contents of 60–91, 0–24 and 4–20%, respectively. The Douala-Edea Reserve is near the Atlantic coast; all of it is low-lying, but not flat; parts become flooded when the level of the adjoining rivers is high. The soils are derived from ancient sand dunes and marine sediments which are still accreting; they are predominantly sandy in texture.

Korup is unusually rich in species; 411 tree species ⩾30 cm g.b.h. were found in 135 plots 80 m × 80 m (86.4 ha). Douala-Edea, perhaps in part because of its less varied topography is less rich, having 230 species in 104 plots 80 m × 80 m. There were considerable floristic differences between the two reserves, although many species were found in both. The most abundant species ⩾30 cm g.b.h. in Korup was *Oubanguia alata*, a B storey tree not more than *ca.* 20 m high; in Douala-Edea it was *Coula edulis*, also a medium-sized tree. In both forests Caesalpinioideae played an important role mainly as tall emergents: in the Korup plots there were 43 species forming 17.5% of the basal area; at Korup there were 27 species, forming 16.3% of basal area. Letouzey (1968) regards the abundance of Caesalpinioideae as characteristic of the Cameroon 'Forêt littorale'.

At both reserves the plots were sited on transect lines and subdivided into four subplots. In each of these all living trees and lianes ⩾ 30 cm g.b.h. were counted and identified, to species when possible. The altitude and slope of each plot was estimated and composite soil samples to a depth of 10 cm were collected.

At Korup detrended correspondence analysis of all plots showed a major indirect floristic gradient correlated with increasing altitude and slope and with soil phosphorus and potassium. When the topographic effect was removed by separate re-ordinations of four groups of plots at low, middle (two) and high altitude/slope the main floristic gradients were strongly correlated with concentration of available phosphorus. This also appeared in the second axis of the ordination of the high altitude/slope group, but was absent in the lower flat plots where the main correlation was with the organic content of the soil.

Direct gradient analysis with respect to available phosphorus using all plots confirmed the indirect analyses. Low available soil phosphorus (<5 p.p.m.) was strongly associated with the presence of Caesalpinioideae, especially the tribes Amherstiae and Detarieae, e.g. species of *Didelotia*, *Tetraberlinia* and *Microberlinia*. The possible significance of this is discussed on p. 338.

Classification of the Douala-Edea plots on the basis of six soil variables provided three large and distinct groups, swamp plots, non-swamp plots with high available phosphorus and non-swamp plots low in phosphorus. Ordination studies in the 'closed forest' of Ghana and Nigeria are discussed in Chapter 18.

12.2 Heath forests

Heath forest, as mentioned earlier, was a name originally applied to the kerangas forests of Borneo: now it is widely used as a general term for the characteristic types of rain forest found on podzols and white sands and sometimes on other coarse-grained nutrient-deficient soils. Such communities are widespread in South America as well as in the eastern tropics. In Africa primary heath forests are absent, although here and there along the west coast there are secondary communities suggesting that vegetation similar to heath forests may have formerly existed there (Richards 1961).

Heath forests are readily recognizable by their structure and physiognomy. As Brünig (1961, p. 28) remarks, even a botanically inexperienced wanderer is at once aware of a change of vegetation when passing from mixed forest into heath forest. Some of the distinctive features of heath forests, such as the density of the undergrowth and the scarcity of strongly buttressed trees and large lianes, were mentioned earlier (p. 305) and their characteristic stratification was described in Chapter 2 (Fig. 2.11). Heath forests are easily recognizable from the air (Fig. 11.10) because, owing to the scarcity of large emergent crowns, the canopy is smoother and more even than in most types of rain forest; it also differs in colour and is greyish green owing to the more sclerophyllous foliage. Heath forest is completely evergreen, deciduous and leaf-exchanging trees are generally completely absent.

As was seen in Chapter 11, heath forests are usually single-dominant communities or have a tendency to single-dominance. The number of tree species per unit area is generally smaller than in mixed forests; where heath and mixed forests adjoin, the difference in floristic composition is striking. Many species appear to be confined, or nearly so, to heath forests and others are more abundant there than in other kinds of rain forest. Some genera are represented in both heath and mixed forests, but by different species, e.g. *Hevea* in South America and several Malesian genera of Dipterocarpaceae.

At family level, as might be expected, the composition of heath forests in general resembles that of other rain forests in the same geographical region. Thus families characteristic of the Neotropics, such as the Chrysobalanaceae, Malpighiaceae and Vochysiaceae, are well represented in South American heath forests and characteristically Palaeotropical families are rare or absent. Leguminosae (especially Caesalpinioideae) play a much larger part in South American than in Malesian heath forests; in the former most of the single dominants and many of the subordinate species belong to this family. The Myrtaceae, though well represented in South American heath forests, are much less abundant and diverse than in those of Malesia. It is interesting that the Ericaceae, which are mainly a montane family in the tropics, occur in lowland heath forests both in Amazonia (Anderson 1981) and in Malesia.

In Malesian heath forests the Dipterocarpaceae are usually as prominent as in other rain forests in the eastern tropics. On Gunung Mulu (Sarawak), for example, dipterocarps were represented by 92 individuals (42.9% of basal area) on heath forest plots and by 114 individuals (43.2% of basal area) on mixed forest plots (Proctor et al. 1983a). In some Malesian heath forests, however, dipterocarps are present only in small numbers, e.g. at Ulu Koyan (pp. 305ff.) or absent. An important feature of Malesian heath forests is that conifers (Agathis, Dacrydium, etc.) are often abundant and sometimes dominant. In mixed rain forests in Malesia they are relatively rare (apart from Araucaria in New Guinea) and in South American heath forests they are quite absent.

The Guyana wallaba forest and the Sarawak kerangas described in Chapter 11 both represent only one type of heath forest in their respective regions; there are many others differing from these in composition, and in many cases in structure and physiognomy. In both the American and the Asiatic tropics there are parallel catenas of communities grading from tall heath forest to low xeromorphic scrub. Some of the latter communities may be edaphic climaxes, but others are probably secondary and result from burning of primary forest. The secondary vegetation arising after the clearing and burning of heath forests is quite different floristically from that derived from mixed forests on kaolisols. In the recovery of heath forest from burning and other disturbances, 'resprouts' (from roots and stumps) appear to play a larger part than seedlings (Chapter 18).

South American heath forests. In the Guianas and Amazonia heath forests cover large areas, but they seem to be absent from Central America and the West Indian islands.

On sandy soils in both the peneplain and highlands of the Guianas there are extensive heath forests ('savanna forests' of some Dutch authors). The most widespread and best known type is the consociation dominated by

Eperua falcata (soft wallaba). This community varies considerably in composition, but the wallaba forest at Moraballi Creek and in the 'Bartica Triangle' discussed in the previous chapter is probably fairly typical. There are also other types of heath forest with different dominant species. In most of these E. falcata occurs as a subordinate species. It is not entirely restricted to heath forests or to podzols and white sand soils and is found in small numbers in other types of rain forest.

Fanshawe (1952) includes the wallaba forests of Guyana as the 'Eperua–Eperua association' in his dry evergreen forest formation-series. In this Eperua falcata and E. grandiflora are both constantly present, but the former is generally much the more abundant. The associates vary considerably, though some, such as Aspidosperma excelsum and Catostemma fragrans, have a high degree of constancy. Fanshawe recognizes a number of variants, some found over large areas ('faciations') and some more local ('lociations'). For example, in northwestern Guyana there is a faciation in which Manilkara bidentata is the chief subdominant. In some faciations and lociations another species replaces Eperua falcata as single dominant. One very characteristic faciation which is widespread in Guyana and also occurs in Surinam (Maas 1971) is dominated by Dimorphandra conjugata (dakama), a somewhat taller tree than E. falcata, which grows to ca. 38 m. It forms almost pure stands or may be mixed with D. hohenkerkii (which also sometimes occurs as a single dominant), Swartzia bannia, Catostemma fragrans and other trees. D. conjugata reproduces freely, usually from root suckers and epicormic shoots on fallen trunks rather than from seed. A striking feature of dakama forest is the layer of dead leaves and raw humus 40–50 cm deep which accumulates as mounds round the tree bases. In dry weather this dead material becomes inflammable and fires are frequent in this type of forest. Deep accumulations of litter and humus, and a tendency for the trees to reproduce vegetatively, are also characteristic of some Amazonian heath forests (Anderson 1981).

The most remarkable faciations of the Eperua association are perhaps those dominated by the clump wallabas, Dicymbe corymbosa and D. altsoni, which occur as co-dominants or form separate communities on podzols, white sands and sandy kaolisols over large areas in the interior of Guyana (Figs. 3.4, 4.1). Both species are self-coppicing, the primary trunk becoming gradually surrounded and eventually replaced by a 'clump' of younger stems developing from near its base (p. 72). The broad crowns of the clumps form a canopy at ca. 20–30 m, in which there are scattered emergents of other taller species.

In lowland *Eperua* forests local dominants include *Dimorphandra davisii, Pouteria* sp. and *Cassia apoucouita* in addition to those mentioned. Fanshawe (1952) also records several other species as dominants of faciations in the *Eperua falcata – E. grandiflora* association and other 'Dry Evergreen' communities on sands and sandstone in the Pakaraima upland region.

On the white sands of the Guianas scrub vegetation (muri bush) and open savannas occur under similar climatic conditions to wallaba and other kinds of tall rain forest. The white-sand savannas and transitional communities between forest and savanna are dealt with in Chapter 16. Muri, which is sometimes found as islands within closed forest and sometimes on the margin of savannas, consists mainly of evergreen sclerophyllous shrubs about 2–3 m high. The most characteristic, but not necessarily dominant, shrub is *Humiria balsamifera* var. *floribunda* (muri), a species which also occurs in forest as an erect tree. Muri bush typically consists of circular patches of various shrubs separated by sandy spaces thinly covered with grasses, lichens and other small plants. At the centre of each patch one or two trees of *Clusia nemorosa* are usually found and there is an outer belt of semi-prostrate *Humiri* bushes. Muri bush is sometimes regarded as resulting from fires but there is some evidence that under very unfavourable edaphic conditions it may also occur as an edaphic climax (Cooper 1979).

In Amazonia, as in the Guianas, the vegetation on white sands (including podzols) varies from tall closed forest to open woodland and scrub similar to 'muri bush'. Forest on white sand is often called 'caatinga',[3] an Amerindian word meaning 'white forest'; this probably refers either to the colour of the soil or to the relatively well illuminated undergrowth. In recent literature the tall closed forest is termed 'high caatinga' and the less tall and more open types of vegetation 'low caatinga'. Klinge & Medina (1979) say that tall caatinga is found in the upper Rio Negro basin where the average rainfall of the driest month is at least 100 mm (Af climate), and the lower types further south, where the climate is more seasonal (Am climate). It is, however, questionable whether the determining factors are mainly climatic as both high and low caatingas occur at San Carlos, which

has an ever-wet climate, as well as at Manaus, where rainfall is somewhat seasonal. It is, perhaps, more probable that the differences are due to differences in ground level, soil and, in some cases, burning and other disturbances.

The caatinga forests of Amazonia were first described by Spruce (1908), who encountered them near São Gabriel (now Uaupés) and elsewhere on the Rio Negro. Ducke & Black (1953) added much further information about the caatinga flora and its regional variations. Caatinga vegetation (in the broad sense) is widely distributed in northern Amazonia from the foot of the Andes to Marajo island in the east. Projeto RADAMBRASIL (p. 10) has shown that it covers 'hundreds of thousands of square kilometres' in the Rio Negro and Rio Branco basins (Anderson 1981). Elsewhere it occurs in scattered patches, of which only a few are south of the Amazon and Solimões.

Most caatingas are on podzols or white sands (spodosols, tropaquods; Chapter 10) similar to the wallaba soils of the Guianas, but in some places they are found on hydromorphic soils (Anderson 1981). They mainly occur on well-drained terra firme, but sometimes also on low ground which is moister and may be liable to flooding.

Detailed studies of Amazon caatingas have been made in only a few places. Since 1975 a large amount of research, much of it concerned with biomass and nutrient cycles (pp. 276ff.), has been done on the caatingas at San Carlos de Rio Negro (1°54'N, 67°06'W) (Jordan 1985; see also references in Klinge & Herrera 1983). An ordination study of the floristic data has been undertaken (Brünig *et al.* 1979), but the full results are not available. The surroundings of San Carlos consist of rolling hills rising to not more than 50 m above mean river level, capped with lateritic material; in the intervening valleys, which are filled with sandy sediments, there are 'domes' 4–12 m high (Fig. 12.5). In a 10 ha study area Klinge (1978) and Medina & Cuevas (1989) recognize a catena of primary soil–vegetation units: (1) mixed forest on concretional oxisols (ferricrete kaolisols) on the hills, (2) forest at lower levels on greyish-white sandy loams over a yellow clay horizon (modal clay kaolisol); here the chief species were *Eperua purpurea* (yévaro) and near streams *Monopteryx uacu*; (3) forest on podzolized white sands ('tropaquods') in which *Micrandra spruceana* (cunuri) was dominant and *Eperua leucantha* (yaguacana) subdominant; (4) forest on similar soils at the lowest levels dominated by *Eperua leucantha*; and (5) low woodland or sclerophyllous scrub (bana), also on white sand, of *Aspidosperma album, Clusia* spp., the palm *Mauritia carana* and other small trees. Klinge

[3] More precisely, 'Amazonian caatinga' or 'pseudocaatinga' (Aubréville 1961), because (especially in the older literature) 'caatinga' is also used for the quite different, seasonally dry, thorn woodlands of northeastern Brazil (pp. 397–9). In Amazonia types of white-sand vegetation are also called 'campina', 'campina rana' and 'carrascal'; in Venezuela low scrub on white sand is called 'bana'. This very confused terminology has been reviewed by Anderson (1981), who recommends that all types of Amazonian white-sand vegetation dominated by trees or shrubs should be referred to as caatinga.

(1978) and Klinge & Medina (1979) regard (3) and (4), both of which are closed forest communities with trees up to 25.4 m high, as tall Amazon caatinga and (5) as low Amazon caatinga. Type (2) seems to be intermediate in composition between caatinga and mixed forest, although the boundary between it and the latter is sharp. Ecotones were found between (2), (3), (4) and (5), which form a catena on the 'domes' in the sandy valleys. The bana (5) resembles the Guyana 'muri bush' (p. 322) and the Malesian padang (p. 326) (Bongers *et al.* 1985).

The distribution of the forest types in the San Carlos catena, as in the Moraballi Creek catena, is closely related to relief and soil conditions. Herrera (1979) and Jordan (1982) have discussed the cycling and availability of soil nutrients and Medina & Cuevas (1989) nutrient accumulation and release. Franco & Dezzeo (1991) have described the soil types and the soil – water relations. The soils of the mixed and yévaro forests (1 and 2) at upper levels are relatively well drained; in the yévaro forest the watertable is at least 40 cm below the surface after rain, falling to below 100 cm in dry periods. The sandy podzols of the tall caatinga forests (3 and 4) are constantly wet; even in dry weather the watertable does not fall more than *ca.* 20 cm. The bana at the lowest levels is frequently flooded, but in dry periods the watertable falls to below 100 cm.

Although the San Carlos caatingas resemble the Guyana wallaba forest in many characteristics, such as their sclerophyllous foliage and thick layer of mor humus, it is remarkable that they are found at the bottom of the catena and not (like the wallaba forest) at the top.

Two samples of 'high campina' (high caatinga) near Uaupés, some 200 km south of San Carlos, were described by Takeuchi (1961a). In one the 'main' species in the A–B storeys in a strip 5 × 40 m were *Eperua leucantha, E. rubiginosa, E. purpurea, Aldina discolor, Pouteria* sp. and *Tachigali* sp., the first being much the most abundant (11.5 individuals per 100 m²); 73% of the trees were Leguminosae. The C storey was scanty and the herb layer consisted mainly of an aroid (*Anthurium* sp.) and a fern (*Adiantum* sp.).

Rodrigues (1961a) gives a more detailed description of a high caatinga on low-lying white sand at the Ilha das Flores, not far from Takeuchi's sites. Here the tallest trees were 22–27 m high and the C storey was dense (Fig. 12.6). As in wallaba forest, large lianes and strongly buttressed trunks were few. The composition appears to be rather similar to that of the yaguacana forest at San Carlos. *Eperua leucantha* was dominant and the subdominants were *E. purpurea* and *Micrandra sprucei* (*M. crassipes* auct.). Highly constant subordinate species included *Catostemma sclerophyllum, Hevea viridis* and *Macrolobium unijugum*. The number of species per unit area was less than in an adjoining area of moist mixed forest on a different soil.

Rodrigues (1961a) also described a low caatinga with a woodland rather than forest structure at Taracua, *ca.* 150 km west of Uaupés. In this the upper storey was 12–13 m high and formed mainly by *Aldina discolor*, with some *Lissocarpa benthami*. The *Aldina* trees had remarkably broad crowns (16–19 m wide) and somewhat contorted branches. Beneath this open canopy there was a very dense layer of shrubs and small trees up to *ca.* 7–8 m high. Among the numerous species in this a small palm, *Bactris cuspidata*, and *Sphaeradenia amazonica* (Cyclanthaceae) were conspicuous, and the cycad *Zamia ulei* also occurred, but less frequently.

Another example of a low caatinga, *ca.* 29 km NNE

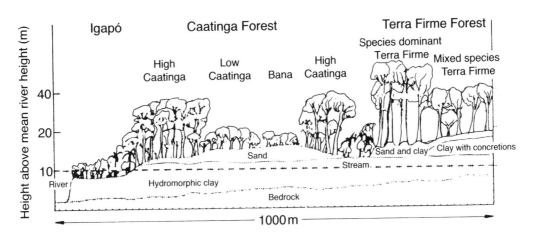

Fig. 12.5 Schematic diagram of catena of forest types, upper Rio Negro, Amazonia. After Jordan (1985).

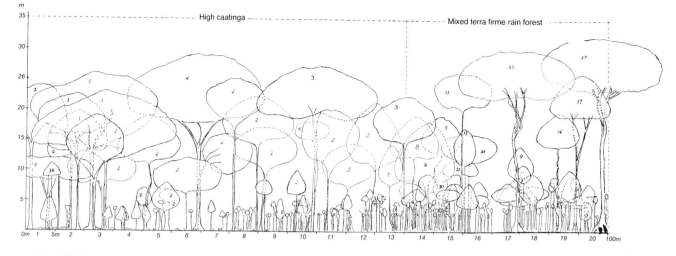

Fig. 12.6 Profile (100 m × 5 m) of transition from high caatinga to mixed terra firme rain forest, Ilha dos Flores, mouth of Rio Uaupés, Rio Negro, Brazil. After Rodrigues (1961a). Key: 1 *Micrandra sprucei*, 2 *Eperua leucantha*, 3 *Eperua purpurea*, 4 *Peltogyne caatingae*, 5 *Quassia* sp. cf. *amara*, 6 *Helicostylis* sp. cf. *asperifolia*, 7 *Macrolobium unijugum*, 8 *Kutchubaea* sp., 9 *Swartzia* sp., 10 *Trattinickia* sp., 11 *Hevea guianensis*, 12 *Ryania* sp. cf. *pyrifera*, 13 *Bombax* sp. cf. *spruceanum*, 14 *Aptandra spruceana*, 15 *Aldina heterophylla*, 16 *Iryanthera obovata*, 17 *Monopteryx uacu*, 18 *Oxytheca* sp. cf. *robusta*.

of Manaus, was described by Takeuchi (1961b); the structure and physiognomy was similar to the previous example but the species composition was somewhat different.

In spite of considerable structural differences between high and low caatingas, their floristic composition seems to be rather similar. In the Rio Negro area many species occur in both, e.g. *Aldina discolor*, *Eperua purpurea* and *Lissocarpa benthami*. All caatinga communities are probably characterized by the dominance of one or a few species. Caatinga sample plots also seem usually to be less species-rich than forest plots on kaolisols (oxisols) in the same region, although some accounts of caatinga vegetation (e.g. Ducke & Black 1953) emphasize its species richness. Anderson (1981), in his review of the still fragmentary information on the flora of the Amazonian caatingas, points out that it has much regional diversity. In some regions, such as the upper Rio Negro, the caatingas have long been known to be rich in endemic species. Anderson suggests that the species richness of the Amazonian caatingas as a whole, as distinct from small samples, may be partly due to its discontinuous distribution, which has favoured speciation.

The caatingas have a highly characteristic flora; in a caatinga in central Amazonia Anderson *et al.* (1975) found that 54.5% of the vascular species were confined to this type of vegetation. Species adapted to caatinga conditions have evolved in many families and genera, e.g. Bignoniaceae (Gentry 1979a), *Hevea*, *Parkia* (Hopkins 1986). Some characteristic caatinga species are

widely distributed and can be regarded as indicators of caatinga vegetation, e.g. *Gaylussacia amazonica*, *Glycoxylon inophyllum* and *Lissocarpa benthami*.

The flora of the Amazon caatingas is evidently somewhat similar to that of the wallaba and other heath forests of the Guyana white sands, but is probably richer in species; a detailed comparison has not been made. A striking similarity is the important role of the genus *Eperua* in both. Many other genera are represented in both floras by the same or related species. About 20% of caatinga species also occur in igapó (Anderson 1981), which is also an oligotrophic habitat (p. 355): this parallels the floristic similarity between kerangas and peat-swamp forests in Malesia (pp. 365–71).

Heath forests in the eastern tropics. The heath (kerangas) forests of Malesia, especially those of Borneo, have been much more thoroughly studied than those of tropical America. There is a general account by Whitmore (1984a). Brünig (1974) has made detailed studies of the kerangas forests of Sarawak and Brunei. Kartawinata (1978, 1980) and Riswan (1982) have described kerangas in Kalimantan.

The distribution of heath forests in the eastern tropics is wide, but very discontinuous (map in Whitmore 1984b). Scattered areas are found from Hainan and Thailand south to New Guinea, but heath forests are absent in Java, the Philippines and other parts of Malesia where there is a strong annual dry season. The greatest extent and diversity of heath forests in Malesia is in

Borneo where they cover thousands of square kilometres. In Peninsular Malaysia they are found only as narrow strips near the east coast (Beveridge 1953, Tomin 1991), which are now mostly replaced by secondary grasslands (Whitmore 1984a). In New Guinea small areas of heath forest have been reported (Paijmans 1976), but these do not seem to be extensive.

Within this large fragmented area heath forests occur mainly on podzols and white sands derived from siliceous materials, such as Quaternary raised beaches near the coast and further inland. In Borneo, but probably not elsewhere, extensive heath forests are also found on the dip-slopes of Tertiary sandstone escarpments (cuestas), as on Gunung Dulit (Chapter 11), where they extend up to the lower limit of lower montane forest (ca. 1000–2000 m). Although characteristically occurring on podzolized white sands, in Sarawak heath forests sometimes occur on other soils, including clays, but probably never on the red–yellow kaolisols typical for mixed dipterocarp forest (Brünig 1974). The sites are usually well drained, but some types of heath forest (kerapah; see below) are found on soils that are temporarily or even permanently waterlogged.

Malesian heath forests differ from mixed dipterocarp forests as much in floristic composition as in physiognomy. Out of an estimated 849 tree species recorded for heath forests (including kerapah) of Sarawak and Brunei only 220 (25.9%) also occur in mixed dipterocarp forest (Brünig 1973b). It is interesting that 146 heath-forest species (17.2%) also occur in peat-swamp forest (Chapter 14); most of these, e.g. *Shorea albida*, are found only in kerapah and the moister types of heath forest (Brünig 1973b). Some heath-forest species occur also in montane forests, e.g. *Eugenia* and *Vaccinium* spp. Some of these are found over a wide range of altitudes; others have a disjunct range, occurring in lowland heath forest and upper montane forest, but not in lower montane forest (see discussion in Whitmore 1984a, p. 252).

Dipterocarpaceae are usually the most numerous family in species and individuals, but dipterocarps the most abundant in heath forests are generally not species common in mixed forests; some, such as *Shorea materialis*, are confined to heath forest. The second most abundant family is often the Myrtaceae, represented by species of *Eugenia*, *Tristaniopsis* and other genera, the last named being conspicuous by their smooth peeling bark. The abundance of small-leaved Myrtaceae is partly responsible for the 'Australian' appearance of kerangas, but in fact a considerable proportion of heath-forest plants belong to genera in which the centre of distribution is in Australasia (Richards 1943). As in Ama-

zonian caatingas, many Malesian heath-forest species are localized endemics.

A notable feature of Malesian heath forests is the frequency of trees very different in physiognomy from the broad-leaved dicotyledonous trees which dominate most rain forests, in particular *Casuarina* and conifers. Among the latter, *Agathis* and *Dacrydium* spp. are locally dominant in some places. Small palms are common in the moister types of heath forest. The characteristic, rather sparse, ground flora of Bornean heath forests, which often includes *Nepenthes* spp., was mentioned in Chapter 11.

Kerangas forests usually, but not invariably, show some tendency to single-species dominance, as in the Sarawak sample plots described on pp. 305–7. Sometimes there are one or a few strongly dominant species, though always, as in most rain-forest communities, the majority of the species are present at very low densities.

Brünig (1974) proposed a classification of the kerangas forests of Sarawak and Brunei, based on fifty-five 0.4 ha plots selected to represent a wide range of localities, sites and soils. It uses a combination of topographical and floristic characteristics and includes forest types in which the 'leading dominant species' include *Agathis borneensis*, *Casuarina nobilis*, *Dacrydium* spp., *Dryobalanops* (three species), *Shorea materialis*, *Tristaniopsis* spp. and *Whiteodendron moultonianum*. In plots transitional to other forest types there may be other 'leading' dominants, e.g. *Shorea rugosa* in transitions to mixed dipterocarp forest, *Shorea albida* or *Combretocarpus rotundatus* in transitions to peat swamps. In some of Brünig's types no one species forms more than a small proportion of the stand.

On small sample plots of kerangas the number of tree species ≥10 cm d.b.h. is generally smaller than in mixed dipterocarp forest, but, as Brünig points out, in both formations there are wide variations in structure and number of trees per unit area; because of this, general comparisons of species richness between them are difficult to make.

To investigate the relations between the composition of kerangas communities and site and soil characteristics, Brünig used the ordination technique of Bray & Curtis (1957), treating the 55 plots as stands. The groupings of species in the ordination agreed well with the groups in the classification. One of the most important factors influencing the composition and structure of kerangas communities is drainage. Where the ground is liable to flooding, or the soil, because of an underlying impervious hardpan or other reasons, tends to become waterlogged, a variant of kerangas known as kerapah is found. In this the typical kerangas species are mixed with swamp-forest trees such as *Gonostylus* spp. and *Ploiarium alternifolium*. On shallow humus podzols, podzolic gleys and peats, small-leaved trees such as *Casuarina nobilis* and *Dacrydium* spp. are often abundant. Gradients in species richness, and in the rate at which the number of species increases with area, were closely related to features of the landform–soil–vegetation complex.

In the varied landscapes of Sarawak and Brunei, vegetation catenas are common in which the smaller differences in geology and soil are reflected in changes from one variant of kerangas to another and the larger differences in gradations from kerangas to mixed Dipterocarp forest or swamp forest.

On very shallow soils overlying sandstone, tall kerangas may be locally replaced by low, more or less open, woodland or by 'padang', a type of scrub or heath-like vegetation comparable to the muri bush and bana of South America (p. 322 above). Padang occurs in Sumatera (Hardon 1937, Whitten *et al.* 1987), on Jemaja island (Malaya) (Henderson 1931), Bangka, and elsewhere in Malesia. In Sarawak it is found here and there, in some places as small islands enclosed in tall forest. There are extensive padangs on the coast near the mouth of the Sarawak river in Bako National Park. This consists mainly of a sandstone plateau about 10 km across and *ca.* 50–100 m above sea-level, with low hills rising from it. The vegetation (Fig. 12.7), which was described and mapped in detail by Brünig (1961, 1965), is a mosaic of low open woodland and padang, with patches of tall kerangas and mixed dipterocarp forest on the deeper and less infertile soils. In some places the boundaries between the various types of vegetation are gradual, in others quite abrupt. The six types of kerangas, in which some of the commonest species are

Dipterocarpus borneensis, *Shorea* spp., *Casuarina nobilis*, *Cotylelobium burckii*, *Stemonurus umbellatus*, *Eugenia tetraptera* var. *pseudotetraptera* and *Calophyllum* spp., vary from closed forest 30–35 m tall to pole forests 5–12 m tall and open woodland of bushy *Dacrydium beccarii* 3–6 m high.

The padangs at Bako (Fig. 12.7) are extremely variable in structure and species composition; the vegetation ranges in height from 12 m to less than 1 m. Among the more abundant woody species are *Baeckea frutescens*, species of *Calophyllum* and *Cratoxylum*, *Dacrydium beccarii*, *Rhodamnia cinerea*, *Rhodomyrtus tomentosus* and dwarf individuals of many species also found in the kerangas forests. Among herbaceous plants various Cyperaceae are abundant; locally the orchid *Bromheadia finlaysoniana* and species of *Burmannia* and *Utricularia* are found. The presence in the padang of the insectivorous *Drosera spathulata* and of five species of *Nepenthes*, as well as the abundance of myrmecophilous epiphytes are indications of the extreme nutrient deficiency of the soil.

Most padangs are subject to occasional anthropogenic fires, and it has often been assumed, as with the South American muri bush, that their existence depends on frequent burning. Brünig (1965, pp. 304–5) believes, however, that at Bako there are both anthropogenic fire padangs and primary edaphically determined padangs. The latter show no evidence of burning: there is no charcoal in the soil and the plants show no coppicing or other indications of fire damage.

Physiological ecology of heath forests. The podzolized sandy soils on which heath forests are usually found differ from the red–yellow kaolisols commonly associated with most other types of rain forests in many characteristics, notably in their coarse texture and low water-holding capacity, their low content of sesquioxides and clay minerals, and their generally low levels of available nutrients, especially nitrogen and phosphorus. Their pH is low and they lack the aluminium buffering capacity of kaolisols (Chapter 10). All these soil properties affect the growth of vegetation, but there are different opinions on their relative importance in determining the very characteristic physiognomy and floristic composition of heath forests. According to one view the poor water-retaining capacity of the soil and recurrent periods of water stress are the most significant factors. An alternative hypothesis, supported by Givnish (1984) among others, is that the acidity and nutrient deficiency of the soil are more important than its physical properties. The two hypotheses are not altogether incompatible. Unfortunately there is little relevant experimental

Fig. 12.7 Padang vegetation, Bako National Park, Sarawak. (1983). (Photo. T.C. Whitmore.)

evidence and both views rely partly on *a priori* arguments based on the supposed adaptive value of the xeromorphic features of heath-forest plants and the structure of the community.

The drought hypothesis is strongly supported by Brünig (1974) in his work on the Sarawak kerangas. He argues that various features of heath forests tend to decrease water loss in dry weather by reducing evapotranspiration either directly or indirectly by their effect on the heat load of the foliage. The aerodynamic properties of the relatively smooth canopy would also tend to lower the rates of evaporation. The high reflectance of the canopy, due to its light colour and shiny leaf surfaces, would also lower the heat load. The relatively small size of the leaves and their tendency to be held obliquely also contribute to leaf temperatures, relative to the ambient air, being lower than in mixed forests with larger and more horizontal leaves; this would also decrease transpiration rates. All these features would be expected to diminish the effects of drought periods on the vegetation. Brünig's hypothesis applies equally to the South American heath forests, which are similar to kerangas in structure and physiognomy and are found in similar environments. As mentioned earlier, heath forests occur only in ever-wet climates without a severe annual dry season, but even where, as in Sarawak, the average rainfall of every month is over 100 mm, droughts of irregular incidence and varying intensity are quite common (p. 191). During these dry periods light sandy soils would probably lose a larger fraction of their available water than clays and loams, which are water-retentive, but actual measurements of the moisture regime in heath-forest soils are few. Baillie (1972, 1976) estimated that in shallow sandy soils near the Sarawak coast periods of water stress were fairly frequent, but inland on deeper soils they occurred only at intervals of some years. Though many heath forests seem to experience drought periods, some types, such as the Bornean kerapah, live under permanently wet conditions. The soil of the tall caatinga (yaguacana) forest at San Carlos, as noted above, is water-saturated for longer periods than that of the mixed or other types of forest in the catena.

Although ability to survive droughts may well be important in heath forests, there are several reasons why it may be doubted whether water stress is the chief factor determining their physiognomic and floristic characteristics. In the first place there is no physiognomic or floristic similarity between heath forests and the semideciduous and deciduous forests of seasonally dry climates on non-podzolic soils. In the latter most of the trees have thin mesophyllous leaves and the majority are deciduous or leaf-exchanging, shedding their leaves in the dry season (Chapter 4). Even in mixed rain forests on kaolisols some deciduous and leaf-exchanging species are present. Heath-forest trees, as mentioned earlier, are entirely evergreen and tend to be markedly sclerophyllous. Leaf-fall in heath forests does not reach a peak in dry weather; in the Gunung Mulu (Sarawak) heath forests the maximum leaf-fall, as in the mixed dipterocarp forest, was at the wettest time of year (Proctor *et al.* 1983b). Species of trees characteristic of seasonal forests are absent in heath forests: for example, Sterculiaceae, *Garuga floribunda* and the widespread *Tetrameles nudiflora* do not occur in kerangas, nor trees such as *Brosimum alicastrum* and *Guazuma ulmifolia* in South American caatingas.

Such experimental work on the water relations of heath-forest trees as is available does not support the view that their leaves are more drought-resistant than those of other rain-forest species. Ferri (1960), in a preliminary study of the stomatal behaviour, water deficits and transpiration of twenty-seven caatinga species on white sand at Taracua (Amazonia) (p. 323, above), concluded that for them water shortages could not be a limiting factor. He suggested that the xeromorphic facies of the vegetation depended on mineral deficiencies. A study of the sclerophyllous foliage of the Rio Negro caatingas by Medina *et al.* (1990) came to similar conclusions. A comparison by Peace & MacDonald (1981) of transpiration rates and sublethal water deficits in detached shoots of seven kerangas tree species from Bako National Park (Sarawak) and four mesomorphic species from mixed dipterocarp forest in Peninsular Malaysia showed no significant differences between the two groups.

There can be little doubt that the ability to survive short periods of water stress is a necessary part of the equipment of heath-forest plants, yet it is plausible to suppose that low concentrations of some nutrients rather than the physical features of the environment are responsible for the xeromorphic facies of the vegetation.

The unsuitability of heath-forest sites for agriculture and the relatively small animal populations found in heath forests (Janzen 1974a, Whitten *et al.* 1987) are clear evidence of nutrient poverty, but analyses of heath-forest soils and vegetation seem to indicate that it may be the availability of nutrients, especially nitrogen, rather than the size of the total stock which are significant (Jordan 1985). Heath-forest soils are certainly in general more acid, have a lower base-exchange capacity and are poorer in available nitrogen and phosphorus than most rain-forest soils (Chapter 10), but the amounts of nitrogen and other nutrient elements per unit volume are sometimes unexpectedly large. In Brunei, Richards (1965) found no significant differences in the percentage of nitrogen, phosphorus and bases between soil samples from heath and mixed dipterocarp forest. In heath forest at Gunung Mulu the amount of nitrogen (in tonnes per hectare) was in fact somewhat higher than in the mixed forest (Proctor *et al.* 1983a). It is, however, likely that a large part of this nitrogen is unavailable to the plants, perhaps because of the low rate of mineralization of organic matter. In the San Carlos caatinga forest, Herrera & Jordan (1981; see also Jordan 1985) found that the nitrogen in the soil solution was mainly in the -NH$_4$ form and nitrification was inhibited, probably by the high acidity and tannin content of the soil. Medina *et al.* (1990) also showed that the decay of organic material is slower in caatinga than in terra firme (mixed) forest soils, especially in the bana (low caatinga). Medina & Cuevas (1989) say that the soils of the tall caatinga at San Carlos accumulate and circulate relatively large amounts of phosphorus, but root growth and litter decomposition appear to be limited by nitrogen.

In the South American caatingas a thick mat of rootlets is found at the surface of the ground and roots form a high proportion (over 60%) of the phytomass; these and other features have been regarded as nutrient-conserving mechanisms (Herrera *et al.* 1978, Jordan 1985). Similar features are found in the Guyana wallaba forest and in Malesian kerangas. Ectotrophic mycorrhizas are very common in the root systems of heath-forest trees and the hyphae have been shown to provide a direct pathway from litter to roots (p. 338 below). Some heath-forest species, e.g. some South American Leguminosae (p. 338) and the Malesian *Casuarina nobilis*, have nitrogen-fixing root nodules.

The xeromorphy of heath-forest vegetation can be compared with that in other ecosystems that are oligotrophic but not normally water-stressed, such as tropical montane forests (Chapter 17) and acid peat bogs (mires) in temperate regions. Attempts to explain the 'bog xeromorphy' of the latter in terms of 'physiological drought' were abandoned long ago and it is now widely accepted that it is related to deficiency of nitrogen (and perhaps phosphorus). There is some experimental evidence for this view (Lötschert 1969), although the way in which xeromorphic leaf structure and nutrient deficiency are connected is still unclear. The replacement of mixed forest by more sclerophyllous heath forests on sandy podzols in the tropics is analogous to the replacement of broad-leaved deciduous trees by evergreens on poor sandy soils in subtropical Florida (Chapter 9, p. 244).

Janzen (1974b) suggested that, owing to nutrient shortages, the leaves of plants in the Sarawak heath forests and padangs were long-lived, slowly replaced and dependent for their survival on being well protected from herbivores by phenols and other secondary compounds. Later work (Proctor *et al.* 1983b, Anderson 1981, Anderson *et al.* 1983) has not supported this hypothesis.

12.3 Single-dominant forests in the Neotropics and Malesia

In tropical America there are numerous single-dominant rain-forest communities. Most of these have been mentioned earlier and are types of heath forest, transitional between heath and mixed forest or, like the Guyana mora forest, are usually found in flood-plains and sites with impeded drainage. However, in Guyana there are also a few inadequately known primary single-dominant types which do not seem to belong to any of these categories. On the lower Corentyne river and in the Mahaicony district rain forests dominated by *Aspidosperma excelsum* occur locally on the lower hill slopes (T.A.W. Davis). Near the interior savannas Myers (1936) found (in addition to mixed forest and associations of *Mora excelsa* and *Dicymbe corymbosa*) forest types dominated respectively by *Peltogyne* sp. and an unidentified tree with the native name asheroa.

In the island of Tobago, Beard (1944b) found a forest type he called 'xerophytic rain forest' on shallow soils overlying igneous rocks on exposed sites above 800 ft (240 m); in this nearly half the trees ≥4 ft (120 cm) g.b.h. were *Manilkara bidentata*.

The climax lowland rain forest of the Lesser Antilles, of which little now remains, is a *Dacryodes–Sloanea* community in which the chief dominants are *Dacryodes*

Fig. 12.8 Primary mora (*Mora excelsa*) forest, Matura, Trinidad. (Photo. T.I.W. Bell.) Compare with Fig. 1.4.

Fig. 12.9 *Gilbertiodendron dewevrei* forest bordering road, Zaïre. From Louis & Fouarge (1949, Pl. 3).

excelsa and species of *Sloanea* (Beard 1949). The former is the most abundant species throughout the community, but the dominance of *D. excelsa* is low in Dominica, where the community is richer in species: with increasing distance from the centre, the dominance increases and the number of associated species decreases.

In Malesian rain forests, primary single-dominant communities are very uncommon and cover only small areas. The gregarious tendency and family dominance of the Dipterocarpaceae have already been discussed (p. 311). Some species tend to grow in clumps or 'groves', but extensive terra firme communities dominated by one species are rare. However, one rain-forest dipterocarp, *Dryobalanops sumatrana* (Borneo camphor, kapur), a valuable timber tree of Malaya, Sumatera and northwestern Borneo, is a single dominant in what can be regarded as a distinct association. It also occurs as solitary individuals in mixed communities, but in localized areas it may form 60–90% of the timber volume. A well-known example, long since modified by silvicultural treatment, is a stand at Kanching (Selangor). On the east coast of Malaya *Dryobalanops* stands once covered several thousands of square kilometres; most of these have now been felled. Whitmore (1984a) says that some *Dryobalanops* stands consisted of apparently even-aged individuals, others of 'family groups' of trees of various sizes. Most stands had sharp boundaries, some consisting of small trees, suggesting expansion, and others mainly of very large individuals, suggesting a shrinking population.

Van Zon (1915) described *Dryobalanops* forests in Bengkalis (Sumatera). The largest patch was not more than 20 ha in extent and seven others of less than 5 ha were seen; all were situated in the midst of normal mixed dipterocarp forest. On a 0.25 ha plot out of thirty-one trees ≥20 cm d.b.h., ten were *Dryobalanops*.

Whitmore (1984a) attributes the dominance of *D. sumatrana* to 'sheer reproductive pressure'; its seedlings and saplings are very abundant and unusually shade-tolerant. Except that it usually occurs on sedimentary rocks, there is no obvious environmental factor to account for the patchy distribution of this association.

A similar, but less striking, gregarious habit is shown by another dipterocarp, *Shorea curtisii*, which has a geographical range like that of *D. sumatrana*. In hilly country in Malaya, particularly on ridge tops, it is dominant over considerable areas. These communities, which are conspicuous because of the greyish crowns of *S. curtisii* (Fig. 12.1), are probably best considered as societies rather than associations. The sites where they are found are liable to drought and, owing to leaching, are poor in nutrients. The ecology of *S. curtisii* has been discussed by Burgess (1975) and Whitmore (1984a).

One of the most remarkable primary single-dominant rain-forest communities in Malesia is that dominated by *Eusideroxylon zwageri*, the Borneo ironwood (belian), a tree valued for its very hard durable wood. It grows to 50 m high and 2.20 m d.b.h. and is common in Borneo and southern Sumatera, though absent in the Malay Peninsula. It often occurs as scattered individuals in mixed forests, but locally it may become strongly dominant (Gresser 1919, Koopman & Verhoef 1938). In some stands it may be almost the only large tree. In Jambi (Sumatera) it formed 96% of trees ≥15 cm d.b.h. on a 0.5 ha plot and 26% of trees ≥10 cm d.b.h. on another of 0.2 ha (Whitten *et al.* 1987). In Kalimantan *E. zwageri* may form up to 30% of the timber volume and 33% of the trees ≥20 cm d.b.h. per hectare (van der Laan 1926). It regenerates abundantly and is well represented in all diameter-classes. Gresser (1919) says that the patches of *Eusideroxylon* forest in Jambi (Sumatera) were mostly less than 100 ha in extent, though he found one of 373 ha and another of 1068 ha. Koopman & Verhoef (1938) reported 'complexes' of up to 10 000 ha in Jambi and Palembang. Most of these have now been felled and what is said to be the last remaining stand of *Eusideroxylon*-dominated forest in Jambi was described by Franken & Roos (1981).

The peculiar structure of ironwood forests was mentioned in Chapter 2: occasional tall trees of *Koompassia*, *Intsia* or *Shorea* stand out far above the *Eusideroxylon* crowns. In Sumatera the undergrowth consists mainly of young *Eusideroxylon* and small palms (*Licuala* and *Livistona* spp.).

Eusideroxylon often grows on clayey hillsides, but prefers sandy or loamy soils. Witkamp (1925) showed that in Borneo it could be used as a geological indicator. In the Kutei district (Kalimantan) it is found only on sands and sandstone and is absent on clays, shale and other rocks. The outcrops of various rock formations form parallel bands and the *Eusideroxylon* communities also occur in bands, corresponding to the sand and sandstone outcrops; their boundaries are sharp, except where the rocks merge gradually. In the 0.5 ha plot in Jambi of Whitten *et al.* (1987) the soil was a moderately acid, sandy silt-loam, low in phosphorus, sodium and magnesium, very low in total exchangeable bases, but fairly high in potassium; in the 0.2 ha plot the soil was a clay-loam with clay below 20 cm.

In Sarawak *E. zwageri* is a common associate of illipe nut trees (*Shorea macrophylla* etc.) in 'empran', a distinctive variant of mixed dipterocarp forest on fertile friable soils (old alluvial kaolisols) on raised levels by rivers (Whitmore 1984a).

12.4 Single-dominant forests in tropical Africa

It is a remarkable fact, the significance of which will be discussed later (pp. 337–8), that all the single-dominant species in climax African rain forests, like most of those in South American rain forests, belong to the Caesalpinioideae. Species of this subfamily, mostly large trees, are also common constituents of African mixed rain forests.

Gilbertiodendron dewevrei **association.** This community is remarkable for its very strong single-species dominance, its peculiar physiognomy and its great extent: it probably covers a larger area than any other type of single-dominant rain forest. As a species *G. dewevrei* extends only slightly more than 6° north and south of the equator, ranging from southwestern Nigeria, Cameroon and Gabon to the eastern frontier of Zaïre and south into northern Angola (maps in Lebrun 1936a, Letouzey 1968). It nearly always grows in groups rather than as single individuals, but it is only in the Congo basin (mainly in the peripheral areas) and in southern Cameroon that it dominates extensive communities (Gérard 1960, Evrard 1968, White 1981, 1983). Both on the ground (Fig. 12.9) and from the air *Gilbertiodendron* forests are conspicuous features of the landscape because of their dark-coloured, even canopy, which is very unlike the paler green, more irregular surface of other African closed forests.

G. dewevrei grows to 30–40 m, rarely to 45 m in height and to 0.5–1.5 m in diameter (Louis & Fouarge 1949, Gérard 1960); Letouzey (1968) records one tree of 3 m d.b.h. The crowns of mature individuals are densely leafy, 10–15 m wide and unusually deep, branching from low down, so that half to two thirds of the total height is occupied by the crown. The leaves are evergreen, but part of the crown sometimes becomes defoliated during droughts.

The structure of the community was briefly described in Chapter 2 (pp. 40–1). The continuous A storey consists mainly of crowns of the dominant, above which occasional individuals of *Tieghemella heckelii*, *Oxystigma oxyphyllum* and other tall trees stand out. In the Uele–Itimbiri district (Zaïre) Lebrun (1936a) found that *G. dewevrei* formed 94% of trees ≥20 cm d.b.h. in a sample hectare. In 14 ha in the same region Gérard (1960) found that the number of individuals of the dominant in the 'strates dominantes et sous-dominantes' (≥200 and 100–200 cm g.b.h., respectively) ranged from 127 to 130 (66.8–89.2%). A common associate is

another evergreen tree, *Julbernardia seretii*, which occurs in varying numbers and also forms pure associations on its own. *G. dewevrei* is very shade-tolerant; it is well represented in all size-classes and seedlings and juveniles are generally abundant (Gérard 1960). The few other tree species in the A storey are light-demanders that also occur in mixed semideciduous forest and are able to establish themselves in gaps.

The sparse middle storey consists almost entirely of young trees of the dominant (and sometimes also of *Julbernardia seretii*) and there is a moderately dense C storey of small trees such as *Isolona thonneri* and, on deep soils, *Scaphopetalum thonneri*. The scanty herbaceous flora includes *Geophila* and *Palisota* spp. and the grass *Leptaspis cochleata*. Scitamineae are present, but not usually plentiful. Both lianes and epiphytes are scarce. As far as is known no species of trees or other plants found in the association are confined to it; most are widely distributed in the evergreen and semideciduous forests of the region.

As would be expected in a community with very strong single-species dominance, the total number of species is much smaller than in mixed forests. This is illustrated by the following data from Louis' (1947a) profile diagrams for forests near Yangambi, Zaïre (data are numbers of species).

	Gilbertiodendron dewevrei forest, Ngula	Semideciduous (*Scorodophloeus*) forest, Isalowe
Tree strata (100 × 10 m plot)	28	71
Herbaceous stratum (0.5 × 10 m plot)	38	78

Near Yangambi Louis & Fouarge (1949) found 204 species in a hectare of *G. dewevrei* forest and 303 in one of mixed semideciduous *Scorodophloeus zenkeri* forest. The species–area curves of Gérard (1960) indicate that this community is also less species-rich than the *Brachystegia laurentii* association (below).

G. dewevrei does not occur above 1000 m and reaches its optimum in climates with only a short dry season. Extensive associations are found only where there are not more than three successive months with less than 100 mm rainfall. Typically they occur on undulating terrain where the land slopes down to a river; they do not extend on to the plateaux. Towards the edge of its range *G. dewevrei* is increasingly restricted to the neighbourhood of streams, but it avoids sites which are permanently swampy or liable to frequent flooding.

In favourable climates the *G. dewevrei* association is found on various types of soil. In the Kisangani-Yangambi region of Zaïre it occurs on pure white or yellowish sands which are unsuitable for agriculture (Louis & Fouarge 1949) and in the Central Congo Basin on leached colluvial sands (Evrard 1968), but in the Ubangi and Uele regions it is found on heavy soils.

Hart *et al.* (1989) studied the stands of *G. dewevrei* in the Ituri forest (eastern Zaïre) and their relation to the surrounding mixed forest. They found no consistent differences in soil conditions between the two forest types. The floristic composition of the two was also similar, except in the presence or absence of *G. dewevrei*. They suggest that the forests in which the latter is dominant may be more mature than the mixed forest because it has escaped serious disturbance in the past. Conway & Alexander (1992) also found that in the Ituri forest differences among soil samples within the *G. dewevrei* forest were greater than the differences between these and samples from the adjacent mixed forest. It should be noted that *G. dewevrei* is ectomycorrhizal and has very numerous mycorrhizal rootlets in the A1 soil horizon (Fassi 1963).

In the Uele region *Gilbertiodendron dewevrei* occurs only on deep soils and *Julbernardia seretii* or *Cynometra alexandri* replace it on more shallow soils. This difference in the soil preferences of these species is related to the structure of their root systems: *G. dewevrei* has a massive tap-root which reaches a depth of *ca.* 2 m and bears whorls of stout laterals with descending 'pivot roots'. It has no buttresses. The two other species have more superficial root systems with strong laterals near the surface and large buttresses.

The very strong dominance of *G. dewevrei* probably depends partly on its prolific reproduction and its capacity to cast and tolerate deep shade. Flowers and fruit are produced in enormous quantities, though not every year (p. 251). The seeds germinate immediately after falling, and, as noted earlier, regeneration of all ages is abundant. The seeds weigh *ca.* 25 g and have little dispersal capacity. In a stand in the Uele, Gérard (1960) could find no seeds on the ground more than 20 m outside the boundary, so migration must be extremely slow. However, at Yangambi Louis (1947a) observed a stand of *G. dewevrei* which was invading adjoining semideciduous forest.

The scattered distribution of this association might indicate that the stands are relics left by shifting cultivators because of their relatively infertile soil, but White (1983) thinks that this should not be assumed: he suggests that, because of its slow rate of spread, the species may have been unable to achieve its maximum potential

range during climatically favourable phases during the Pleistocene.

In some places in Cameroon Letouzey (1968) saw stands of *G. dewevrei* with many old and dead trees which were being replaced by herbaceous vegetation. He suggested that this might be due to a lowering of the water table caused by deepening of the neighbouring river beds.

Brachystegia laurentii association. This tree, like *Gilbertiodendron dewevrei*, is a member of the Caesalpinioideae; it has a scattered distribution, mainly in the Congo basin, to as far south as the Lusambo district (*ca.* 4°49'S). *B. zenkeri*, which differs little from *B. laurentii* and is probably conspecific with it, is widely distributed in the Cameroon rain forest where it forms 'pure populations' (Aubréville 1970). In Zaïre *B. laurentii* occurs sometimes as single individuals or in small groups, but in the central part of its range it becomes the dominant of an association. The stands occur as islands in semideciduous forest and usually occupy not more than a few hectares; their total area is much less than that of the *Gilbertiodendron* association. Most of the information given here is taken from a detailed account of the *Brachystegia* forest near Yangambi by Germain & Evrard (1956) and from Evrard (1968).

The structure of this community (Chapter 2, pp. 40–1, Fig. 12.10) differs only slightly from that of the *Gilbertiodendron* forest. *B. laurentii* grows to a height of 45–50 m; the trunk is unbuttressed and may be unbranched up to 20 m; it often exceeds 1.5 m in diameter. The crown, like that of *G. dewevrei*, is massive and 30–40 m wide. The forest canopy, formed mainly by the dominant species, has an average cover value of 65% and casts a less heavy shade than that of *G. dewevrei*: this is because it is less continuous and because the leaves of *B. laurentii* tend to be more crowded at the tips of the branches and have smaller leaflets (*ca.* 7–10 cm × 3–5 cm, compared with 9–30 cm × 5–18 cm in *G. dewevrei*). Measurements showed that the average illumination in the undergrowth, though greater than in the *G. dewevrei* forest, was less than in neighbouring semideciduous forest. The foliage is usually evergreen, although occasionally in the dry season the leaves are shed and the tree remains bare for a short time.

B. laurentii is shade-tolerant and is well represented in all size-classes. Seedlings and juveniles are abundant: on 11 samples near Yangambi the number of plants ≥50 cm high varied from 3 to 800. Some individual trees fruit annually, others only once in two or three years. The seeds have little dispersal capacity (average mass *ca.* 3.7 g) and lose their viability in a week or two when exposed to the air.

The A storey consists mainly of enormous trees of the dominant, with occasional individuals of other species. In eleven samples of varying sizes (0.25–0.10 ha) near Yangambi, Germain & Evrard (1956) found that the number of trees ≥30 cm g.b.h. (9.5 cm d.b.h.) per hectare was 519, of which 240 (46.2%) were *B. laurentii*. For trees ≥130 cm g.b.h. (41 cm d.b.h.) the corresponding figures were 55 and 40 (72.7%). The dominant represented 66% of the basal area (trees ≥20 cm g.b.h.). The B storey, which is somewhat denser than that of the *Gilbertiodendron* forest, as well as immature trees of *Brachystegia*, includes more or less shade-tolerant species such as *Annonidium mannii* and *Polyalthia suaveolens. Scorodophloeus zenkeri* and other semideciduous forest trees appear in gaps, but often fail to reach maturity.

Among the shrubs and tall herbs may be mentioned *Cola griseiflora, Haumania liebrechtsiana* and *Scaphopetalum thonneri.* Several species of shrubs, e.g. *Pavetta*

tetramera, seem to be restricted to, or at least characteristic of, this association[4].

Lianes, though numerous in species, have a low cover value (9%); many of them do not reach maturity and they rarely reach the A storey.

Germain & Evrard (1956) calculate from their data that the average number of species (all synusiae) per hectare in *Brachystegia* forest is 193. This would indicate that its species richness is about the same as that of the *Gilbertiodendron* association and less than that of the mixed (probably secondary) semideciduous forest in the same region; it is greater than that of mixed rain forest in Côte d'Ivoire and elsewhere in West Africa. The species–area curve for the samples rises steeply to an area of *ca.* 0.2 ha and then becomes almost level: Germain & Evrard suggest that the minimum area is about 0.01 ha. They found that up to 0.8 ha their curve agreed closely with values calculated from C.B. Williams' (1947) index of diversity.

[4] It is interesting that the fungus *Russula badia* also appears to be highly characteristic of it.

Fig. 12.10 Profile diagram of *Brachystegia laurentii* association, Zaïre. After Germain & Evrard (1956, Fig. 2). Trees ≥8 m high on strip 80 m × 10 m, trees 4–8 m on strip 80 m × 5 m. Crowns of *B. laurentii* stippled.

In the Yangambi region *Brachystegia* forest, unlike the *Gilbertiodendron* association, is usually found on the margin of low plateaux (a few hundreds of metres above river-level), rather than on slopes and valley bottoms. The soils are rather light loams, usually somewhat pale in colour (yellowish-red on the Munsell scale); they have very low total exchangeable bases (0.6–1.6 meq per 100 g in the upper and 0.9–1.9 in the lower horizons) and a pH of *ca.* 4.3–5.1 and are of 'mediocre' agricultural value. The organic matter content, C : N ratio and water-retaining capacity are somewhat greater than those of soils in adjacent semideciduous forests.

Germain & Evrard consider that the discontinuous distribution of the *Brachystegia* community is due to deforestation; the existing stands may have survived because they occupy relatively infertile sites unsuitable for cultivation. However, like that of *Gilbertiodendron*, the spread of *B. laurentii* may have been restricted by a very slow rate of dispersal.

Cynometra alexandri association. Forests dominated by *Cynometra alexandri* (ironwood) are found in northeastern Zaïre and western Uganda up to 1400 m. They occur mainly above *ca.* 700–800 m and in some parts of this region they are very extensive; in the Beni-Irumu district they 'constitute 50–70% of the forest on dry land' below 1200 m (Lebrun 1936a). The species also occurs as a co-dominant with *Julbernardia seretii* and *Staudtia stipitata* and in small numbers in mixed semideciduous forests (White 1981, 1983).

Eggeling (1947) has given a detailed account of the *C. alexandri* association in the Budongo forest (Uganda). It experiences a dry season of three successive months with an average rainfall of less than 100 mm and, though Eggeling regards it as typical lowland rain forest, Lebrun & Gilbert (1954) include *C. alexandri* among the characteristic species of the semideciduous forests (Piptadenio-Celtiditalia) of Zaïre. Langdale-Brown *et al.* (1964) classify the '*Cynometra–Celtis* forest of Uganda as Moist Semideciduous'.

At Budongo *C. alexandri* forms 20–35% of trees ⩾4 in (20 cm) d.b.h., and as much as 75–90% of those ⩾20 cm. In an area in Zaïre it formed *ca.* 50% of trees ⩾50 cm d.b.h. (Lebrun 1936a). The number of other tree species in the *C. alexandri* forest at Budongo is small; on one 400 ft × 400 ft (1.5 ha) sample there were only eleven species and on another 25, compared with fifty-eight and fifty-three, respectively, on two similar mixed forest plots. The A storey consisted entirely of very large *Cynometra* trees emerging above the B storey in which *Strychnos* sp. and three *Celtis* spp. were the most abundant trees. *Lasiodiscus mild-*

braedii, or sometimes *Alchornea laxiflora* are common in the C storey. The forest has a characteristic appearance; the scanty undergrowth makes visibility possible for long distances and lianes are very scarce. The herb layer is not well developed and in the dry season the ground herbs become half-wilted. The soil is a kaolisol, which down to the upper limit of the B2 horizon varies in texture from a heavy to a very sandy loam, but is more clayey below this depth. In the Bugoma forest the finest *Cynometra* stands are on valley bottoms with impeded drainage.

C. alexandri also occurs at Budongo as a constituent of a mixed forest community, where it reaches larger dimensions than where it is a single-dominant. Eggeling (1947) found good evidence that at Budongo, where the forest as a whole was extending at the expense of the adjacent grassland, the mixed forest was a seral community and the *C. alexandri* association the climax. The latter is slowly invading and replacing the former. According to Langdale-Brown *et al.* (1964) this invasion of mixed forest by *Cynometra* occurs on the poorer and not on the more fertile soils.

12.5 Forests on limestone

Outcrops of limestone of various geological ages are widespread in the seasonal tropics, but occur only locally in rain-forest regions. In Malesia, and northwards to Thailand and southern China, steep limestone hills are a prominent feature of the landscape in many places (map in Whitmore 1984b). Their total area is not large; in Malaya it may be rather over 260 km² (Scrivenor in Burkill 1966) but, as limestone is a scarce resource in the humid tropics, many of these hills are endangered by quarrying. Most of the outcrops are in the lowlands, but in a few places, as on Gunung Api (Sarawak), they occur at higher elevations. In tropical America limestone occurs in the Caribbean region mainly in seasonal climates where the climax vegetation is semideciduous forest or thorn woodland (but see Kelly *et al.* 1988). In the Guianas and Amazonia limestone outcrops seem to be absent. In the African rain-forest region, apart from a very small area of karst near Calabar (Nigeria) (Petters 1981), the vegetation of which has not been described, limestone outcrops seem to be entirely lacking.

In wet tropical climates limestone outcrops often occur as tower karst, forming precipitous hills 100 m or more high (Fig. 12.11). These are found singly or in small groups and are riddled with caves, dolines and fissures. The limestone may contain over 95% calcium carbonate, with very little silica, and the associated soils are variable in composition and usually shallow. On some limestone hills in Sarawak the soil is partly derived from igneous intrusions or residual material from other rocks (Anderson 1965).

The flora of tropical limestones in some areas is extremely rich. Chin (1977) lists 1216 species of angiosperms from limestone hills in Malaya, of which 261 (21.4%) are endemic to the Peninsula; 335 species are 'characteristic' of limestone and 254 (20.8%) confined to it. Anderson (1965) estimated that there are over 600 species on the limestone hills of Sarawak. Many limestone species are not trees; herbs (Acanthaceae, Gesneriaceae etc.) are very numerous.

As Anderson points out, the Sarawak limestone hills provide a great variety of plant habitats, ranging from exposed rock at all inclinations to scree slopes, fissures, shaded gulleys and dolines. Trees grow where there is soil or crevices in which they can root, but closed forest

Fig. 12.11 Forest on karst limestone, Bukit Takun, Selangor, Malay Peninsula (1974).

develops only on gentler slopes and on colluvial material at the foot of cliffs. On the summits of the hills a thick layer of acid humus (mor) (pH 5.0) accumulates, on which heath-forest trees such as *Casuarina nobilis* and *Tristaniopsis obovata* often occur, as well as presumably calcifuge plants such as *Rhododendron* and *Vaccinium* spp. During dry weather the vegetation on the exposed summits of limestone hills may become dry and inflammable and damage due to anthropogenic fires or lightning is frequent.

Little is known about the composition of limestone forests. Proctor et al. (1983a) describe a 160 × 60 m plot on a 20–30° slope on limestone at ca. 300 m in the Gunung Mulu National Park (Sarawak). About 9% of the ground surface was bare rock: the soil filled the interstices between boulders to an average depth of 11 cm. The mean pH was 6.1 higher than in their sample plots on other substrata: cation exchange capacity and the concentrations of plant nutrients were also high.

The forest structure was somewhat similar to that of heath forest, except that the mean height and density of trees $\geq$10 cm d.b.h. were less (18.8 m and 644 per hectare, respectively). Trees $\geq$100 cm d.b.h. were absent and the density of lianes 1–10 cm d.b.h. was about ten times greater than in the mixed and heath-forest plots.

No firm conclusions about floristic composition can be drawn from a single plot, but it may be noted that the Fagaceae and two other families represented in all the other Mulu plots were absent; one family, the Combretaceae, was found only in the limestone plot; and the Dipterocarpaceae, as in most of the Mulu plots, were the most numerous family in individual trees and the largest in basal area. The limestone forest thus appears to be floristically different from both the mixed dipterocarp forest and the heath forest. It is surprising that the species – area curves show that this limestone forest plot was considerably poorer in species than the sample plots on other substrates: the number of species $\geq$10 cm d.b.h. was about 75, about one third of the number in the mixed forest plot (see also Newbery & Proctor 1984).

12.6 Forests on ultrabasic rocks

In the tropics, as in temperate regions, the vegetation on serpentine, peridotite and other ultrabasic (ultramafic) rocks is often very characteristic (Fig. 12.12). The soils derived from these rocks vary, but usually have a high sesquioxide: silica ratio and are poor in plant nutrients generally, with other features unfavourable to most plants, such as high Mg:Ca ratios and relatively high concentrations of chromium, cobalt and nickel (p. 264; see also Proctor & Woodell 1975). In rain-forest climates ultrabasic rocks are found locally in Malaya, Borneo, New Guinea, Palawan and the Solomon Islands, but the total area is small. They seem to be absent in the tropical American and African rain-forest regions.

In Peninsular Malaysia the rain forest on ultrabasic rocks at low altitudes is only slightly different from that on other rocks. At Raub in Pahang (W. Malaysia) the forest looks similar in structure and physiognomy to other rain forest in the area (Whitmore 1984a, p. 176). Fox & Tan (1971) give a short account of the soils and vegetation at 518–762 m on an ultrabasic hill northeast of Ranau (Sabah). The B storey of the forest was 30–34 m high, with emergent trees up to 55 m. The composition was mixed and not particularly species-poor (81 tree species ≥30 cm g.b.h. on six 0.24 ha plots). As in hill forests elsewhere in Malesia, Fagaceae (*Castaniopsis, Lithocarpus* and *Quercus* spp.) were frequent. There were several dipterocarps; one of them,

Fig. 12.12 Aerial photograph of forest, Sabah (1971). (Photo. Jabatan Tanah dan Ukor, Kota Kinabalu.) Forest of trees with small crowns on ultrabasic rocks in central area; forest of larger-crowned trees on other types of rocks in surrounding area.

Dipterocarpus ochraceus, is known only from ultrabasic rocks and basalt (Ashton 1982). According to Meijer (1965) the dipterocarps on ultrabasic rocks near Labuk (Sabah) are very different from those in rain forest on other rocks in the area, but all the species he cites are known to occur on other rocks (usually on sandy soils) elsewhere in Borneo.

Proctor *et al.* (1988) have described the rain forest of Gunung Silam, a coastal mountain of ultrabasic rocks in Sabah. Although the surface soil horizons contained large amounts of nickel and had high magnesium/calcium quotients, the forest below about 600 m was tall, species-rich, mixed in composition and in general little different from mixed dipterocarp forest on other substrata. Only two tree species, *Borneodendron aenigmaticum* and *Buchanania arborescens*, seemed to be confined to ultrabasic rocks. Leaf analyses of a large number of species (Proctor *et al.* 1989) showed that they appeared to be able to tolerate low foliar nutrient concentrations and to concentrate calcium relative to magnesium. *Shorea tenuiramulosa* had up to 1.0 mg g^{-1} of nickel in dried leaf samples.

At *ca.* 2400 m on Kinabalu (a few km distant from Ranau) the differences between the flora and physiognomy of the vegetation of the ultrabasic and other rocks are very striking, although few species are confined to the former. Ultrabasic montane forests in Sabah are recognizable in aerial photographs by the small crowns of the trees (Fox & Tan 1971). On ultrabasic rocks in the Solomon Islands there is a very distinct forest type dominated by *Casuarina papuana* and *Dillenia crenata*, which can be easily mapped from the air. In the Solomons, however, only four species are confined to ultrabasics (Whitmore 1984a, pp. 175–6).

12.7 Relation of mixed and single-dominant communities

In this and the previous chapter, examples have been given of primary single-dominant rain-forest communities which exist along with mixed communities in both the New World and Old World tropics. Some single-dominant communities, such as the wallaba forest of the Guianas and the *Gilbertiodendron* forest of equatorial Africa, cover hundreds of square kilometres; others, for example the *Dryobalanops* forest of Malesia and some African rain-forest communities dominated by Caesalpinioideae, occupy quite small areas and amount to little more than societies of strongly gregarious species. The degree of dominance varies; no sharp distinction can be drawn between single-dominant and mixed for-

ests. The more extensive single-dominant communities have a very distinctive floristic composition, some species being confined to them and absent from mixed forests. Single-dominant communities are poorer in species than mixed ones, when areas of one or a few hectares are compared.

Information on the ecology of single-dominant forests is still very incomplete. They occur chiefly in two kinds of habitat: (1) sites liable to flooding or with impeded drainage; and (2) terra firme on white sands and certain other non-kaolisols. To the first group belong the Guiana mora forest, the African *Mitragyna stipulosa* association and other types of swamp forest discussed in Chapter 14. Examples of the second group are some types of heath forest and the African *Brachystegia laurentii* and *Gilbertiodendron dewevrei* associations. Connell & Lowman (1989) have discussed several possible mechanisms for explaining the occurrence of 'low diversity' (single-dominant) rain forests.

The evidence, though not complete, supports the view put forward earlier, that single dominance is associated with non-kaolisols and non-optimal conditions. The single-dominant forests of swampy sites are found on soils that are poorly aerated owing to waterlogging; in the terra firme group the soils are usually poor in nitrogen and perhaps in other nutrient elements, although a distinction must be made between total and available nutrients. Unfavourable soil conditions seem to act selectively, excluding many potential dominants and giving an advantage to other species, some of which may become single dominants. In mixed forests conditions seem to be almost equally favourable for a large number of species. Although single-dominant forests usually occur under non-optimal conditions, the converse is not necessarily true; on some relatively infertile soils, such as those derived from ultrabasic rocks, the forest is mixed and without single-species dominance.

Single-dominant species generally also occur as scattered individuals in mixed communities. Why these species are able to become dominant under certain soil conditions must be largely a matter for speculation, but they have certain biological traits in common which may be significant, and belong to a very small number of families. A large proportion are Leguminosae of the subfamily Caesalpinioideae and most of these are members of the tribes Amherstieae or Cynometreae. It is remarkable that species belonging to the leguminous subfamilies Mimosoideae and Papilionoideae, though well represented in mixed rain forests, do not become single dominants.

Most (and possibly all) single-dominant species share certain characteristics which are probably connected with a gregarious or strongly clumped dispersal pattern. Among these are prolific reproduction, heavy seeds without an efficient dispersal mechanism, and shade-tolerant seedlings and juveniles. Also, though data are scarce, it may be surmised that the leaves and seeds of at least the leguminous single-dominants have strong chemical defences against herbivores. Probably more important than any of these features are traits related to growth in soils poor in nitrogen, phosphorus and other nutrient elements.

Malloch *et al.* (1980) have drawn attention to the possible ecological importance for forest trees of different types of mycorrhizal and other symbiotic associations. They point out that most of the trees in species-rich forests have endomycorrhizas, but those of species-poor forests, such as temperate coniferous and broad-leaved forests, usually have ectomycorrhizas. Among leguminous trees in the tropics, the Caesalpinioideae, to which so many rain-forest single dominants belong, generally have ectomycorrhizas but lack bacterial nodules; the other leguminous subfamilies usually have endomycorrhizas and nodules, which thus seem to be alternative means of supplying nitrogen to the hosts that possess them (see Chapter 10). Malloch *et al.* note also that the fungi that act as mycobionts in endomycorrhizas are widely distributed Phycomycetes of which only about thirty morpho-species are known. On the other hand some 5000 fungal species are involved in the formation of ectomycorrhizas; they are all Basidiomycetes and some are specialized for associations with one (or a few) host species and are local in distribution. One tree species may form ectomycorrhizas with several fungal species.

Information on mycorrhizal associations is available for relatively few tropical forest tree species (see Redhead and other contributors in Mikola 1980, Jordan 1985). However, with some exceptions, notably the Dipterocarpaceae and Fagaceae, which are ectomycorrhizal, the trees of mixed lowland and lower montane rain forests seem to have mainly endomycorrhizas. It is significant that most (perhaps all) single-dominant species have ectomycorrhizas, as do many trees in the South American caatingas. In Amazonia, according to Singer (1984) the trees of the campinas (caatingas) and the igapós (pp. 355ff.) mostly have ectomycorrhizas, while in the terra firme (mixed) forests on latosols only endomycorrhizae are present. Basidiomycete fungi, which seem to be the most important litter decomposers in Amazonian forests, are much less abundant in the caatinga soils than in those of the terra firme forests; this probably accounts for the tendency of raw humus (mor) to accumulate in the former (Singer & Silva Araujo 1979).

In temperate forests ectomycorrhizas are well known to facilitate the transfer of nitrogen, phosphorus and other nutrients from the soil to tree roots and to have other beneficial effects on tree growth, perhaps by the synthesis of auxins and other growth regulators. The relation between mycorrhizal fungi and roots is probably mutualistic as the former seem to lack the capacity of other fungi for the enzymatic breakdown of polysaccharides which trees possess. The relation between tropical trees and ectomycorrhizal fungi is probably similar (Singer 1984). Smits (1983) found that in Borneo seedlings of a dipterocarp, *Shorea* sp. cf. *obtusa*, grown in soil inoculated with fragments of mycorrhiza from the parent tree, grew 40 cm taller in eight months than controls in uninoculated soil. It has been suggested that the clumped distribution of many dipterocarps is due to the slow spread of the mycobiont in the soil. This may also be true of Caesalpinioideae and single-dominant ectomycorrhizal rain-forest trees generally.

Herrera *et al.* (1978) showed by means of a radioactive tracer that there is a direct transfer of phosphorus by means of mycorrhizal hyphae from litter to living roots of *Eperua leucantha* and *Micrandra spruceana* in the San Carlos caatinga (pp. 322–3 above). Singer (1984) says that such direct transfer does not occur in terra firme (mixed) forest soils.

Part IV

Primary successions

Chapter 13

Primary xeroseres and the recolonization of Krakatau

The successions (seres) leading to the establishment of climax tropical rain forest are classified, somewhat arbitrarily perhaps, into primary successions (priseres) starting on soil not previously occupied by plants, for instance new soil formed from volcanic materials or arising by the silting up of lakes or seas, and secondary successions (subseres) starting where previously existing vegetation has been destroyed or damaged by felling, burning, flooding, etc. The subseres, which in the rain forest are mainly due to human interference, will be dealt with in Chapter 18; in this and the two following chapters the rather fragmentary information on the priseres will be considered.

13.1 The recolonization of Krakatau

The most spectacular example in the tropics of a primary succession of which there are records over a long period is the recolonization of the Krakatau Islands (Indonesia) after the catastrophic volcanic eruption of 1883. Many of the facts are fairly well known but a short sketch of the development of the plant communities will provide an appropriate introduction to successions leading to the establishment of tropical rain forest. A more detailed account, with full references, of the recolonization of the islands up to 1984 has been given by Whittaker *et al.* (1989).

History of the vegetation 1883–1932. The Krakatau Islands are a small volcanic group between Java and Sumatera and about 40 km from both. In 1883 the main island (Pulau Rakata or Rakata Besar) was about 9 km

long and 5 km broad, rising to a peak 822 m above sea-level. There were also two smaller islands, Sertung (Verlaten Island) and Rakata Kecil (Lang Island). All three were part of the caldera of a huge prehistoric volcano some 2000 m high which collapsed probably about 60 000 years ago. There were some eruptions in the seventeenth century, but Krakatau remained dormant from then until May 1883 when increasingly violent activity began. This reached a climax on August 26 and 27 when the famous eruption occurred, the sound of which was heard as far away as Australia and

Fig. 13.1 Anak Krakatau in 1989. (Photo. R.J. Whittaker.) In foreground, volcanic debris sparsely colonized by vegetation. To left, pioneer stand of *Casuarina equisetifolia*. This photograph gives some idea of what the vegetation of Rakata Island may have been like a few decades after the eruption of 1883.

Sri Lanka. The main island split into two, more than
half sinking below the sea, but leaving the highest peak
(since reduced to about 735 m by erosion). The remaining land surface of Pulau Rakata was covered with ash
and pumice to an average depth of *ca.* 60–80 m and a

A

B

Fig. 13.2 Rakata Island in 1989. (Photos R.J. Whittaker.) A
The forest-covered slopes and cloud-capped summit (780 m),
in foreground the eroding cliffs with hanging gullies. B A
closer view of the secondary forest and rugged surface.

new marginal belt 4.6 km in area was added on the
southern side.

Since 1883 there has been no major volcanic activity
on Rakata, but in 1927 a new volcano, Anak Krakatau
(Fig. 13.1), rose from the sea bed a few kilometres to
the north: this has erupted several times and is still
active.

It is now generally accepted that the great eruption
of 1883 destroyed all life on Rakata, Sertung and Rakata
Kecil, although Backer (1929) argued that some plants
may have survived as seeds, spores or underground
parts. On Sebesi, an island 19 km distant from Krakatau,
the vegetation was severely damaged but not completely
destroyed. Only the successions on Rakata, which seem
not to have been interrupted by major disturbances, are
considered in detail here. The eruptions of Anak Krakatau from 1927 onwards have done considerable damage
to the vegetation of the smaller islands and on them
the primary successions have been complicated by subseres on the damaged areas. For this and other reasons
their present vegetation is somewhat different from that
of Rakata.

Before 1883 the Krakatau Islands were covered with
luxuriant vegetation. Very little is known about its
nature or composition, but there is every reason to
suppose that most of it was forest similar to that now
existing in the neighbouring parts of Sumatera and Java.
Since at the present time the islands have an annual
rainfall of over 2000 mm with a dry season of about
three months, this forest was probably of the evergreen
seasonal type (pp. 36–8).

On Rakata there has been no further major volcanic
activity since 1883. Rain soon cut gullies into the thick
covering of volcanic debris, but for a while the island
remained without any vegetation. The first visitors after
the eruption (October 1883 and May 1884) saw no signs
of living plants.

By September 1884 'a few sporadic blades of grass'
had appeared (Verbeek, quoted by Whittaker *et al.* 1989)
and when the botanist Treub arrived in June 1886 there
was already a considerable amount of vegetation. In his
account of what he found Treub (1888) emphasizes the
sharp distinction which already existed between the flora
of the beach and that of the interior of the island. On
the beach there were nine species of flowering plant:
Calophyllum inophyllum,[1] *Cerbera manghas, Erythrina*
sp., *Hernandia nymphaeifolia, Ipomoea pes-caprae, Scaevola taccada,* the grass *Pennisetum macrostachyum* and
two unidentified Cyperaceae. In addition he found seeds

---

[1] These and following identifications in section 13.1 are as emended by
Whittaker *et al.* (1989, Appendix I).

or fruits of several other species. Except for the *Pennisetum*, all these are species common on recently emerged coral islands, as Treub noted; they are widely distributed pantropical or Indo-Malayan sea-shore species. Inland the most striking feature of the vegetation was the abundance of ferns, of which there were eleven species including *Acrostichum aureum*, *Pteridium aquilinum*, *Pityrogramma calomelanos*, *Pteris* spp. and other ferns. The following flowering plants were also found: *Blumea tenella*, two species of *Conyza*, *Scaevola taccada*, *Messerschmidia argentea*, *Wedelia biflora* and the grasses *Neyraudia madagascariensis* var. *zollingeri* and *Pennisetum macrostachyum*. These flowering plants grew scattered and as individuals were far less numerous than the ferns. Some of both the flowering plants and the ferns were species usually found near the sea. Docters van Leeuwen (1936) suggests that their occurrence inland may have been partly due to the absence of competition, but ecologically the conditions must have been similar to those on a sandy shore. Besides these vascular plants Treub found two unidentified mosses and noted that the surface of the ash and pumice was everywhere covered with a thin crust of blue-green algae (Cyanophyta); six species were collected, belonging to the genera *Anabaena*, *Hypleothrix*, *Lyngbya*, *Symploca* and *Tolypothrix*. The gelatinous and hygroscopic layer formed by these algae was probably important in providing a favourable substratum for the germination of seeds and spores. At this date, the inland vegetation thus consisted of a two-layered community, the upper layer formed chiefly of ferns, the lower of Cyanophyta. The total number of species of vascular plants recorded was twenty-six. Owing to the roughness of the ground Treub was unable to ascend far into the interior of the island.

The vegetation of Krakatau was not again examined until 1897, 14 years after the eruption. The party which then visited the island again included Treub and the results are described in a brief paper by Penzig (1902). Since 1886 the vegetation had made great progress and closed plant communities had begun to appear.

On the beaches there was a well-developed 'Pes-caprae formation' (associes dominated by *Ipomoea pes-caprae*), a characteristic community of sandy tropical shore (pp. 383–4). In this certain species belonging to the 'Barringtonia formation' (littoral woodland community; see Chapter 15) were found, including *Terminalia catappa* and *Barringtonia asiatica*. *Calophyllum inophyllum*, a characteristic tree of this community, was not recorded on this occasion, although Treub had found young specimens on the beach during his previous visit in 1886. *Casuarina equisetifolia*, another very characteristic tree of littoral woodland, had already established itself; on

Sertung at this date it was beginning to form a closed stand.

The interior of the island now looked very different from what Treub had seen 10½ years before. Instead of vegetation consisting chiefly of ferns, there was now a dense growth of grasses, in places taller than a person. The species forming this savanna or 'grass-steppe' were, on the lower parts of the hills, chiefly *Saccharum spontaneum*, *Neyraudia madagascariensis* var. *zollingeri* and *Pennisetum macrostachyum* and higher up *Imperata cylindrica* (the well-known alang-alang grass) and *Pogonatherum paniceum*. Associated with the grasses in the lower region, there were various dicotyledons, including several climbers. The dicotyledons included several beach plants, for instance *Scaevola taccada* and *Vigna marina*. Higher in the hills isolated shrubs, various dicotyledonous herbs and ferns were intermixed with the grasses. On steep tufa walls the vegetation remained much as it was before, consisting of ferns and a crust of algae. In addition to algae and some sixty-four species of vascular plant, the 1897 expedition found two mosses, one liverwort and one agaric. As in 1886, it was found impossible to penetrate very far into the interior of the island.

The next thorough examination of the vegetation was made in 1906, when Krakatau was visited by a party of botanists. The results of this expedition are given by Ernst (1908). Many further changes had by then taken place in both the shore and the inland vegetation.

The plant communities on the shore now covered a wider belt, probably because the beach itself had widened by accretion. Instead of consisting only of the 'Pes-caprae formation', the vegetation of the beach consisted of two distinct zones. The outer one was the *Pes-caprae* associes, better developed and richer in species than in 1897. Behind this was the 'Barringtonia formation' forming a discontinuous but well-developed belt of woodland, not everywhere closed, but similar in essentials to the corresponding community on the coast of Java. Much of the closed portion of this woodland apparently consisted of a more or less pure stand of *Casuarina equisetifolia*; elsewhere the chief trees were *Barringtonia asiatica*, *Calophyllum inophyllum*, *Hibiscus tiliaceus* and *Terminalia catappa*. As well as trees, the woodland included climbers, shrubs, grasses and other herbs. An interesting feature was the occurrence of coconut palms, but they may have been planted.

Ernst's description of the interior of the island is much less satisfactory than that of the beach, but it is clear that up to a considerable altitude the vegetation was still a savanna consisting mainly of the same species of grasses as those noted by Penzig. The associated plants included Cyperaceae, climbers such as *Vigna*, *Canavalia* and *Cassytha*, shrubs of *Messerschmidia* and *Scaevola* and various ferns.

Beyond the strand-forest the whole gently sloping surface of the southeast side of the island was covered with the steppe-like vegetation which we have described and this extended in a dense mass into the wild ravines and on to the steep sides far up on the cone. The uniformity of the jungle of fresh and decaying stems of grasses and reeds is only occasionally broken by the occurrence of a fallen tree or shrub.

(Ernst 1908, p. 31.)

A deep ravine about half-way up the slopes of the peak was particularly striking because of its luxuriant growth of trees and shrubs, but it proved impossible to reach it and, like most of the previous expeditions, Ernst's party could not approach at all near to the peak. Further details about the inland vegetation at this period were given by Backer (1909, 1929), who had been a member of Ernst's expedition and visited Krakatau again two years later. On the southeast of the island, behind the beach vegetation, he found here and there narrow strips of mixed woodland consisting of *Ficus fistulosa*, *F. fulva*, *Macaranga tanarius*, *Melochia umbellata* and *Pipturus argenteus*, growing from 5 to 15 m high. All these trees are species characteristic of the secondary rain forest of Malesia generally (Chapter 18). Above this mixed forest stretched the grass savanna dominated by *Saccharum spontaneum* where isolated specimens of all five of the mixed-forest trees were seen.

Backer succeeded in exploring the wooded ravine which Ernst had failed to reach and ascended to a height of about 400 m. The forest in ravines was best developed between 300 and 400 m above sea-level; it was formed by a very few species of trees with herbs, etc. beneath. Among the ten species not previously recorded from the island was the shrub *Cyrtandra sulcata*, which, as we shall see, later became extremely abundant. According to a member of the party that accompanied Backer, the vegetation above 400 m consisted of trees and ferns. It is thus likely that the fern community, which in 1886 had covered most of the island, had persisted at the higher altitudes longer than at the lower, where it had been ousted by the grass savanna.

From 1908 to 1919 no botanical explorations of Krakatau were made and the vegetation of the island suffered slight interference. In 1916 a German obtained a concession to work building stone and settled in the island where he remained for several years. He made a garden and was probably responsible for introducing a few new plants; otherwise his occupation seems to have left little trace. In 1919 a fire was started which burnt much dry grass, but according to Docters van Leeuwen (1936) the trees and shrubs were merely singed and the damage was negligible. It does not appear that the natural course of the plant successions on Rakata was ever seriously affected by man in the period 1883–1932 (see Whittaker *et al.* 1989).

In 1919 a party including Docters van Leeuwen made a more thorough investigation of the vegetation than any that had been made previously. All regions of the island were examined, including the upper slopes, and the summit of the peak itself was reached. A full account of conditions in 1919 and of subsequent changes up to 1932 was published (Docters van Leeuwen 1936).

The littoral plant communities were essentially the same in 1919 as they had been in 1906, but in the meanwhile there had been much erosion of the beach and both the *Pes-caprae* and the *Barringtonia* formations were reduced in area; the *Pes-caprae* formation formed only a very narrow strip in 1919 and the *Barringtonia* formation had also been washed away in many places. The composition of both these communities was fairly typical by 1919, though even by 1932 they did not include all the species which are found in similar vegetation in Java and Sumatera. Docters van Leeuwen regarded the *Pes-caprae* and *Barringtonia* communities as permanent, stable and not liable to change into anything else (see below). The *Casuarina* association was, however, clearly being replaced in many places by the mixed *Macaranga–Ficus* woodland which had first been noted by Backer in 1908. The tall old *Casuarina* trees were being overgrown by lianes; some had already fallen down and others were obviously dying. Beneath them, probably owing to the shade, no young casuarinas were coming up; instead there were young trees of *Macaranga*, *Ficus* spp., *Pipturus argenteus*, *Premna obtusifolia*, etc. Since 1919 the replacement of the *Casuarina* stands by other species has continued, although in some places new *Casuarina* woods have developed. According to Docters van Leeuwen (1936, p. 254), the *Casuarina* community was sometimes succeeded, not by the *Macaranga–Ficus* association, but by the *Barringtonia* community.

Until 1919 the vegetation of the lower slopes behind the beach had looked much as it had done since 1898, but by 1932 it was completely transformed. The strip of mixed *Macaranga–Ficus* woodland first noted in 1908 had developed and extended. It was now so shady that there were few ground herbs, though young trees were abundant. The greater part of the middle region in 1919 was, however, still a grass savanna dominated by

Saccharum spontaneum, associated with *Imperata* and various dicotyledonous herbs, but everywhere among the grass there were scattered trees growing singly or in groups. These trees were *Macaranga, Ficus*, etc. and foreshadowed the development of mixed woodland from the savanna. In some places the shade under the trees was sufficient to suppress the grasses and allow shade species, such as the ground orchid *Nervilia aragoana*, to grow. This tendency towards the development of woodland was much more marked in the ravines where a luxuriant growth of trees and shrubs replaced the grasses and there was even some accumulation of humus. At low altitudes in the ravines *Macaranga* was the commonest tree; higher up the composition was more mixed. On the steep tufa walls, on the other hand, the succession lagged behind; where the surface was sufficiently stable the grass *Pogonatherum* was dominant.

Between 1919 and 1932 the groups of trees steadily increased in size and number until nearly the whole savanna was converted into a mixed *Macaranga–Ficus* wood. By 1928 the whole southeastern side of the island was wooded up to about 400 m above sea-level. This woodland had spread both upwards from below and outwards from the ravines. During the transition from savanna to woodland the fern *Nephrolepis biserrata* was sometimes locally dominant for a time. The *Saccharum* community still survived on ridges here and there, but by 1931 it had almost gone, although whenever the soil was laid bare by a landslip it reappeared as a stage in the subsere. The mixed woodland seems to be similar in many respects to certain types of the secondary forest (Chapter 18) which appears everywhere in Malesia after the primary rain forest has been cleared. Though lianes were very abundant, the mixed wood was not impenetrable. In many places the chief plant in the undergrowth was *Nephrolepis*, either alone or mixed with *Selaginella plana*. Here and there *Costus speciosus* was seen. This mixed wood was richer in species than the grass savanna, but still far less rich than primary rain forest.

One of the most interesting results of Docters van Leeuwen's 1919 expedition was the exploration of the highest part of the island. Almost all of this was found to be covered with a dense growth of the shrub *Cyrtandra sulcata*, with very few other species intermixed. In the ravines above 400 m this plant became increasingly common and at still higher altitudes it was dominant everywhere except on the ridges, where *Saccharum*, ferns and a few other herbs were found. The branches of the *Cyrtandra* were loaded with epiphytic ferns and bryophytes, giving the community the aspect of a

'mossy' montane forest (Chapter 17). After 1919 the *Cyrtandra* consocies has changed as strikingly as the grass community at lower altitudes. The tree *Neonauclea calycina*, which in 1919 occurred scattered through the *Cyrtandra* scrub, especially in ravines, became more and more common. By 1929 it had become dominant in the *Cyrtandra* zone up to 750 m and old trees had reached a height of 25–30 m. In 1932 *Neonauclea* had spread still more and it would clearly not be long before most of the summit region consisted of a *Neonauclea* forest with *Cyrtandra* as an undershrub.

History of the vegetation 1932–84. After 1932 no detailed studies of the vegetation of Krakatau were made for nearly twenty years. In 1951 and 1952 short visits to Rakata and the smaller islands were made by van Borssum Waalkes (1960).

Since 1932 important changes had taken place: most of Rakata was now forest-covered. Forest dominated by *Neonauclea calycina*, a tree growing to *ca.* 15 m high and common in other parts of Malesia, occupied most of the surface up to *ca.* 750 m and had spread downwards, replacing much of the 1932 *Macaranga–Ficus* woodland. Much of it had a closed canopy and thin undergrowth. Van Borssum Waalkes distinguished four altitudinal zones: (1) 0–50 m, (2) 50–200 m, (3) 200–500 m and (4) 500–700 m. In (1) *Neonauclea* was mixed with *Terminalia catappa* and a few trees of *Macaranga tanarius* and *Ficus* spp., but in (2) it was almost the only tree species and its trunks were bigger. In zone (3) *Maranthes corymbosa* and *Ficus* spp. were present and there was a luxuriant undergrowth of ferns and *Selaginella*. Zone (4) was characterized by thick coverings of bryo-

Fig. 13.3 Interior of secondary rain forest dominated by *Neonauclea calycina*, southeast Rakata, 1989. (Photo. R.J. Whittaker.)

phytes on the trunks and branches: *Asplenium nidus* was an abundant and conspicuous epiphyte. *Neonauclea* trunks were larger here than in the lower zones. From 750 m to the summit the vegetation consisted mainly of shrubs with occasional tussocks of *Saccharum spontaneum*. *Schefflera polybotrya*, a shrub which had not previously been recorded for the island, was now commoner than *Cyrtandra* in this community. Bryophytes were very abundant both on the ground and as epiphytes.

The differences between the zones were attributed by van Borssum Waalkes to the increase of cloudiness and humidity with altitude, but the differences in the size of the *Neonauclea* trunks presumably depended partly on their age.

From 1979 onwards several expeditions have made investigations of the vegetation of the Krakatau islands which have been more thorough than those of earlier years and have used quantitative methods (Tagawa *et al.* 1985, Whittaker *et al.* 1989). New studies of the fauna have also been undertaken (Thornton *et al.* 1990).

In 1979–84 Hull University parties on Rakata, in addition to making a general survey of the vegetation, set up nineteen sample plots on various parts of the island on which all trees ⩾30 cm g.b.h. were counted. The data were analysed by the TWINSPAN classification (Hill 1979) and the DECORANA ordination method (Hill & Gauch 1980).

Although some thirty years had passed since van Borssum Waalkes' visits in 1951–52, the general aspect of the plant cover of Rakata had not greatly changed (Fig. 13.2). The trees had grown taller, some plant communities had extended their boundaries and others had contracted; there were also minor changes due to coastal erosion and accretion.

The sample plot results can be summarized only briefly here. The quantitative analysis clearly separated the three coastal communities (*Pes-caprae* associes, *Terminalia catappa* and *Casuarina* woodland) from the inland forests. The inland forest plots fell into four groups characterized especially by differences in the relative abundance of *Neonauclea calycina*, *Ficus pubinervis* and *F. tinctoria* subsp. *gibbosa*: (i) *Neonauclea* forest, (ii) *Neonauclea–Ficus* forest, (iii) *Ficus* forest, and (iv) 'cloud' or 'mossy' forest (Fig. 13.3). In (i), found from 40 to over 550 m, *Neonauclea* formed 52–89% of the basal area; the associated species varied with altitude, as noted by van Borssum Waalkes (1960). Group (iii) comprised three plots at various altitudes in which *Ficus pubinervis* had high scores in the TWINSPAN classification and *Neonauclea* low ones; some of the *Ficus* trees were of very large girth. The two highest plots

(iv), at 670 and 730 m respectively, were characterized by luxuriant coverings of bryophytes on trunks and branches. The shrubs *Schefflera polybotrya* and *Leucosyke capitellata* formed a low canopy, a few taller trees of various species rising above it. *Neonauclea* was almost absent. The summit vegetation was much as van Borssum Waalkes had described, a mixture of *Saccharum spontaneum*, *Schefflera*, *Cyrtandra* etc.

The plant successions. From the account above it will be seen that although the records for the first fifty years after the 1883 eruption are very incomplete, the course of the successions can be reconstructed fairly well (Fig. 13.4A). The successions on Sertung and Rakata Kecil have been described by Tagawa *et al.* (1985) and Whittaker *et al.* (1989) but are less well documented. On all the islands the vegetation has not yet reached a climax and is still changing, though more slowly than in the earlier years after the great eruption.

On Rakata an altitudinal zonation into coastal, lower, upper and summit zones is now evident, although the boundaries between them are not sharp. The succession in the lower zone is a good example of a primary xerosere in a rain-forest climate. Except in its early stages it is similar to the secondary successions on clearings and abandoned cultivated land discussed in Chapter 18. The primary succession in the lower zone of Rakata shows the successive dominance of cryptogams – chiefly algae and ferns – and later of grasses and other herbaceous plants and finally of trees. This is typical of primary successions in climates where the natural climax is forest. On solid rocks the first colonists are usually algae, lichens and bryophytes. On Rakata bryophytes were seen as early as 1886, but neither they nor lichens seem to have been of much importance on the loose and porous ash and pumice which covered the Krakatau Islands. The temporary dominance of ferns in the early stages of the primary succession on Rakata is seen in some secondary successions; here it may have been due partly to the early arrival of wind-blown spores, and partly to the tolerance of some ferns for nutrient-poor substrata. It is interesting that it was not a feature of the early colonization of Anak Krakatau (Whittaker *et al.* 1989).

Closed forest began to establish itself on Rakata after about fifty years and at about the same time on Sertung and Rakata Kecil. It seems to have developed unevenly, no doubt mainly because of the steeply dissected topography of the islands combined with accidents of plant immigration and seedling establishment. The first closed stands of trees seem to have been mainly in ravines.

The forest which by the 1950s covered much of the

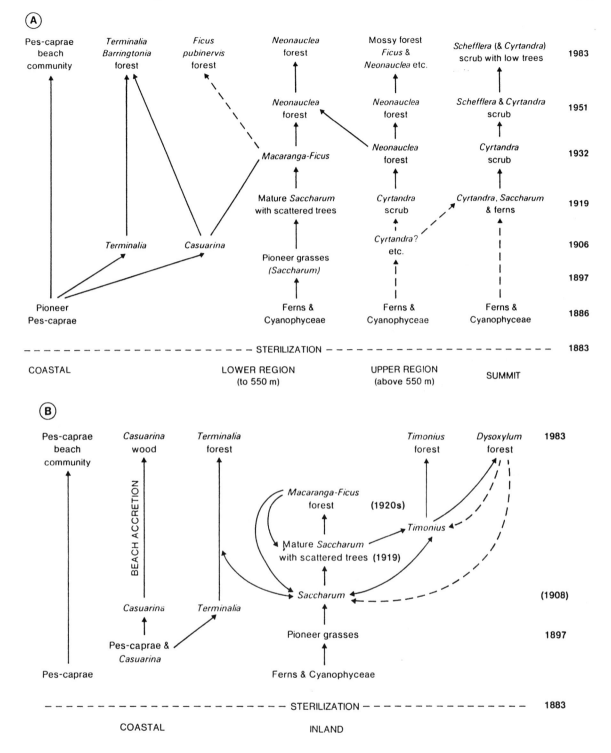

Fig. 13.4 Successions on Krakatau islands since eruption of 1883. After Whittaker *et al.* (1989), modified. A Rakata, B Sertung and Rakata Kecil. On Rakata the primary succession has been interrupted only by minor disturbances (landslips etc). On the other islands the development of vegetation has been interrupted since 1931 by damage due to the eruptions of Anak Krakatau.

lower zone in Rakata and large parts of Sertung and Rakata Kecil, though it now resembles young secondary rain forest, is much poorer in species than such forest elsewhere in Malesia. Its structure is very irregular and the tallest trees are not more than about 20 m high. In 1982 on the richest of the plots (0.25 ha) of Whittaker *et al.* (1989) there were only eighteen tree species ⩾30 cm g.b.h. and on the other plots even fewer. The dominant trees were *Neonauclea calycina* and *Ficus* spp. on Rakata and *Dysoxylum gaudichaudianum* and *Timonius compressicaulis* on the other islands. All of these are fast-growing, light-demanding species with the characteristics of early seral (pioneer) trees (see Chapter 18). *Neonauclea* has small wind-dispersed seeds and is common by rivers in Malaya and Borneo (T.C. Whitmore); the other species are animal-dispersed and frequent in Malesia in second-growth on abandoned farmlands, etc., as are *Macaranga tanarius* and other woody species of the inland forests. As in other young secondary forests, climbing plants are abundant.

Up to now no Dipterocarpaceae or other trees typical of primary rain forests in Malesia have not yet appeared in the islands.[2] Many other plants common in rain forest in Java, Sumatera, etc., including, surprisingly, the bird-dispersed mistletoes (Loranthaceae), are also absent.

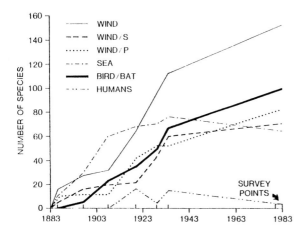

Fig. 13.5 Means of dispersal of the Krakatau flora (all islands except Anak Krakatau). From Whittaker *et al.* (1989, Fig. 21). The totals are for collations, rather than the dates indicated. The wind-dispersed component is divided into seed plants (S) and pteridophytes (P).

---

[2] The report of one primary rain-forest tree, *Dysoxylum caulostachyum*, in Krakatau seems to be due to a misidentification (Whittaker *et al.* 1989, p. 83). *Planchonella duclitan*, which occurs on Rakata and Rakata Kecil, might be regarded as a primary forest tree.

Species immigration and dispersal. Whittaker *et al.* (1989) give complete lists of the vascular plants recorded on the islands up to 1983 and a critical discussion of species numbers and species losses.

On Rakata the total (cumulative) number of vascular species recorded from 1883 to 1983 is 308 (232 seed plants and 76 pteridophytes) and it is still increasing (Fig. 13.5). On all the islands, excluding the still active Anak Krakatau, the cumulative total is about 455–488 species. Of the 252 species recorded on Rakata from 1883 to 1934, only 54% were re-found in 1979–83. Extinction of species is not easy to prove on such a difficult and broken terrain, but it is likely that some species have failed to maintain themselves because of unsuitable habitat conditions, e.g. various species found as juveniles on the sea-shore. The change over much of Rakata from open grassland to forest may have led to the loss of some light-demanding species, but for some plants, such as epiphytic ferns, it provided a more favourable environment.

But even if the number of species on Rakata in 1979–83 was somewhat greater than the number actually recorded, it is evident that for a tropical island with an area of 17 km² and reaching an altitude of over 700 m, the number of vascular plants is small. It is likely that this must be because many species have been excluded by the sea barrier rather than because suitable habitats are not yet available.

The probable method of transport of each species can only be inferred from whether its seeds, fruits or spores appear to be adapted to wind, sea or animal (bird or bat) dispersal. The cumulative totals of seed plants in each of these categories up to 1983 in the Krakatau islands are: wind 99 (27.7%), sea 103 (28.9%), animal 123 (34.5%). (Whittaker *et al.* 1989). In addition, 32 species have probably been human introductions, but only three of these were re-found in 1979–83. All pteridophytes are presumed to have been wind-dispersed. The relative proportions of species transported by different means has changed slightly since colonization began (Fig. 13.5). Animal-dispersed species, which were few at first, have become increasingly numerous relative to wind-dispersed species in later years.

The lack of trees and other plants characteristic of primary rain forests is presumably due to their means of dispersal. As was seen in Chapter 5, primary forest trees tend to have heavy seeds or fruits ill-adapted to long-distance dispersal by birds, bats or wind. Even the Dipterocarpaceae, which have winged fruits, do not seem to have efficient long-distance dispersal (p. 111).

The nearest sources for the colonization of the Krakatau islands are West Java and southern Sumatera. Since in recent years, a large part of the rain forests in these areas have been destroyed, it is doubtful whether many primary forest species which under natural conditions might eventually have reached the islands will now do so unless artificially introduced.

Changes in the environment. The slow rate of arrival of species has no doubt greatly influenced the successions on Rakata and the other islands, but the part played by soil development and changing microclimates must not be underestimated. Immediately after the great eruption the islands were not a favourable habitat for plant life. The volcanic materials that covered the old surface were loose, shifting and extremely porous, consisting of vitric ash and pumice blocks of all sizes. The unattractiveness of this substratum was increased by the complete lack of shelter from sun and wind. Chemically the ash was less unfavourable because it contained most of the ordinary plant nutrients, though not all in an immediately available soluble form, and it was also lacking in organic matter (Ernst 1908).

By 1928 the ash had begun to develop into a soil. Leaching had removed much sodium and potassium and the silica:alumina ratio was lower than in the parent material, showing that the lateritic weathering characteristic of volcanic rocks in humid tropical climates had already made progress (van Baren 1931).

Already by 1886 erosion had produced four different types of substratum from the original more or less uniform covering of ash: pumice sand, which still covered the greater part of the land surface; bare unweathered lava; sea sand mixed with pumice on the shore; and the soil of the ravines, which was partly old unweathered volcanic soil and partly eroded pumice sand. By the action of the climate, micro-organisms and the newly arrived plant life, the original volcanic ash had gradually become a more favourable medium for plant growth. Nitrogen compounds were added to it, first by rain alone, later also by nitrogen-fixing organisms. In 1906 Ernst (1908) found that bacteria were almost as numerous in soil samples as in soils from Java; several different types of organism, including nitrogen-fixers, were present. The abundance of Cyanophyta in the early stages of colonization was probably significant because many of them have nitrogen-fixing ability; they have been noted elsewhere as early colonists of volcanic material, e.g. on the Soufrière of Guadeloupe (Fritz-Sheridan 1987).

Newsome and others (references in Whittaker *et al.* 1989) have reported on the stage of soil development in the islands reached in the 1980s. Most of the soils of Rakata were still very permeable; the pore space was over 69% and the bulk density was low. The clay fraction of the soils was still small, but their water-holding capacity was then much greater than that of the original volcanic ash. The average solum depth was about 20–30 cm, but in places as much as 35–40 cm. Soil development was most advanced in the more stable gullies, where incorporation of organic material had given rise to distinct A horizons.

After a fairly continuous plant cover had developed the environment on the islands must have become much more favourable for seedling establishment. The shading of the soil surface by vegetation would have altered the microclimate near ground level, decreasing the range of temperature and raising the humidity. The roots made the soil less mobile and dead plant remains added humus and modified both the physical and the chemical properties of the soil.

A further important step in the succession was the development of closed forest. This changed the microclimate still further. It also provided habitats for epiphytes and for ferns and other plants intolerant of exposure to direct sunlight. The forest also provided suitable conditions for many species of bird and indirectly accelerated the immigration of new plant species.

Thus, as in all primary successions, the plants have modified the environment so as to make it more favourable to themselves. Less exigent species have given way to more exigent, with which they cannot compete under the conditions they have themselves helped to create. The general trend has been from a xerophytic to a more mesophytic environment, in other words, from more to less extreme conditions, but, owing to the isolation of the islands, further development to mature rain forest, similar to that in large land areas in Malesia, has not taken place.

13.2 Other xeroseres

Several other examples have been described of primary successions in rain-forest areas devasted by eruptions of volcanic ash and pumice. These are interesting to compare with the recolonization of Krakatau as in all of them there was intact forest, not far from the devastated area, which could have been a source of propagules.

Seral communities in the blast areas of three volcanoes in Papua New Guinea are described by Taylor (1957). The climate was ever-wet with an annual rainfall of over 2800 mm and the climax vegetation of the region was lowland and montane rain forest. Mt Lamington (1600 m) had last erupted in 1951, Waiowa (370 m) in

1944 and Mt Victory (1800 m) at a much earlier date, believed to be about 1870. In each case most of the former vegetation of a large area was destroyed but some plants probably survived.

There are many similarities between the seral communities in Papua in 1953 and those on Krakatau. The invading species in both were mostly plants common in second growth on abandoned cultivated land. In Papua New Guinea trees such as *Octomeles sumatrana*, *Tetrameles nudiflora* and *Trema*, which are common on the devastated areas, are also found in the regrowth on the sites of native 'gardens'. Many of the seral species are the same as those on Krakatau or are different species of the same genus.

In Papua, as on Krakatau, a grass community dominated by *Saccharum spontaneum* and *Imperata cylindrica* had become established on some substrata, for instance on the crater walls of Waiowa volcano, but in more favourable sites trees were usually the first colonists and closed forest soon began to appear. On the Mt Victory blast area a three-storeyed forest with trees over 50 m high had developed on mature soils. The chief species in the A storey were *Albizia falcataria* and *Octomeles sumatrana*, but in the lower storeys these were rare; this fact, together with the presence of many young individuals of species normally found in undamaged rain forest, probably indicated that this community will later develop into a typical mixed climax rain forest. But even after eighty years (estimated) the tree flora was much poorer in species than an equal area of primary forest. Thus even where there is no sea barrier to dispersal the re-immigration of primary forest is a slow process.

The deposits on the devastated areas consisted of volcanic ash of varying depth, mud flows (lahars) and much coarser material such as gravel and rock fragments. The different kinds of substratum seemed largely to determine the nature of the pioneer communities and the speed of succession in the earlier stages, but later on climatic factors became more important. On Mt Victory, where the successions had gone on longest and where the devastation is over a considerable range of altitude, the communities in the upper zones were probably stages in successions leading to montane rain-forest communities.

Other examples of successions where the previous vegetation was destroyed by volcanic ash are on volcanoes of the West Indian islands. Beard (1945b) briefly described the successions on the Soufrière of St Vincent (alt. 1246 m) where much of the vegetation was destroyed by an eruption in 1902. On the lower parts

of the mountain secondary forest had developed of trees such as *Cecropia* spp. and *Ochroma lagopus*, which are common pioneers on abandoned farmland on the island (Chapter 18). At higher elevations 'moss and lichen tundra' and other types of montane vegetation were found. Successions on the Soufrière of Guadeloupe have been described by Sastre *et al.* (1983).

On lava flows which solidify into hard rock plant colonization is very slow. In western Samoa there are lava flows of various ages, some of known date. Information about the seral communities and some idea of the time-scale of the successions is given by Vaupel (1910). He says that a lava flow on the north side of the island of Savaii formed during an eruption in 1905 was entirely bare of vegetation in 1906–07; other lava flows in the same island, about 10–150 years old, were already covered in favourable places with trees, rooted in the detritus in cracks of the lava, on soil accumulated on its surface or sometimes perhaps penetrating through it to the soil below. Older residents could remember that fifty years earlier the only plants on these tree-covered areas had been ferns and small bushes. A more recent account of the flora of the Samoa lava flows has been given by Uhe (1974).

Keay (1959a) has described the vegetation on the 1922 lava flow of Cameroon Mountain in West Africa as it was in 1951. Forest had not established itself after twenty-nine years even though the mean annual rainfall is *ca.* 9500 mm with only a short dry season. A few individuals of trees common on abandoned farmland, such as *Alstonia boonei*, *Milicia excelsa* and *Musanga cecropioides*, were present but most of the vegetation consisted of low bushes; the commonest species were *Hymenodictyon biafranum* (mostly about 2 m high but sometimes to 5 m) and *Harungana madagascariensis*. As in the succession on Krakatau, ferns and *Ficus* spp. were abundant, presumably dispersed by birds.

On rock outcrops in the West African closed forest and Guinea savanna zones (pp. 405ff) successions can be seen in which the arborescent sedge *Afrotrilepis pilosa* plays a major part. Such successions were first described by Chevalier (1909) in the Fouta Djallon mountains (Guinea), and are also found on granite inselbergs in Sierra Leone, Nigeria and Cameroon. Plants of *Afrotrilepis* become established beneath exfoliations of the rock surface and in cracks where water collects, but as long as they remain alive the succession makes little progress. If they are killed by drought or grass fires in the dry season seeds of various other plants germinate in the resulting debris. A community of herbaceous and woody plants then grows up and patches of scrub are formed which can perhaps eventually develop into forest. Some details of these successions are given by Richards (1957) and Hambler (1964). Another arborescent member of the Cyperaceae, *Microdracoides squamosus*, which physiognomically resembles a *Vellozia*, seems to play a similar part on rock outcrops in Cameroon.

Chapter 14

Hydroseres and freshwater swamp forests

In lakes and the slower reaches of rivers the land often tends to encroach on the water. In this process of *Verlandung*, the formation of new land from water, vegetation plays an important part. Aquatic plants offer resistance to water movement, slowing them down and so increasing the rate of deposition of suspended solids. Later on amphibious plants help to consolidate the sediments and, by adding to them their dead remains, change their texture and increase their bulk. In peat swamps, the deposits consist more of plant remains than of mineral materials. The rising ground level is accompanied by a hydrosere or plant succession which starts with various types of aquatic vegetation. In forest climates this usually ends with some type of freshwater swamp forest. Once the ground level has reached the height of the highest water level, accretion ceases. The watertable then remains at or near the soil surface during at least part of the year. In the tropics such land is capable of bearing high forest similar in many respects to the climax forest of well-drained sites, but different in floristic composition because most of the species are adapted to flooding or conditions in which the soil is permanently waterlogged. Such forests must be regarded as edaphic climaxes and there is little evidence that without some allogenic cause such as a general rise in the land level or a change in a river course the succession can ever reach the climatic climax of the region.

In the tropics examples of hydroseres, often on a grand scale, are plentiful, for example on the floodplains of the upper Amazon headwaters in Peru. By many rivers and in the low-lying depressions which occupy large areas in some tropical countries, all stages from open water to high forest can often be traced. The islands and meanders of large rivers such as the Amazon and the Zaïre (Congo) also show hydroseral stages, but here the vegetation is often modified or destroyed by human activities. In some places freshwater swamp forests, the climax of these successions, still cover very large areas, but they are being increasingly exploited for timber and wood pulp or converted to agricultural uses. In Southeast Asia, where wet rice cultivation has been practised for centuries, the deltas of large rivers such as the Mekong must once have had extensive swamp forests but they have been intensively cultivated for so long that little trace of the original vegetation now remains. In recent years swamp forests have been successfully converted to rice cultivation in Amazonia (Jari River Project) and in Africa.

The complete story of a hydrosere has been worked out in only a few places in the tropical rain-forest area but the course of the succession can often be inferred from the zonation of the plant communities or by piecing together fragmentary observations. Many of the species taking part in hydroseres range widely through the tropical zone. Some species and many of the genera of water plants are even found in both the temperate and tropical zones. Several important genera of tropical hydroseres are represented by closely related species in both the Palaeo- and Neotropics, e.g. *Carapa, Raphia, Symphonia*. Hydroseres therefore tend to be similar in regions far apart, not only in their general course, but in the structure and physiognomy of their successive phases. The earlier stages are to a considerable extent unaffected by climate and are similar in ever-wet and seasonal tropical climates.

14.1 Hydroseres and swamp forests in tropical America

14.1.1 Central America, Guyana, etc.

A very simple hydrosere was described by Kenoyer (1929) from Gatun Lake in the Panama Canal. Five stages are distinguished.

(1) Floating aquatic association. *Salvinia auriculata, Ludwigia repens, Utricularia mixta, Pistia stratiotes, Eichhornia azurea.*

(2) Water-lily association. *Nymphaea ampla*, together with the preceding species.

(3) Emergent aquatic association. Most abundant species: *Typha angustifolia* and the fern *Acrostichum danaeifolium*, associated with *Crinum erubescens, Hibiscus sororius* and *Sagittaria lancifolia.*

(4) Reed swamp association. *Cyperus giganteus, Scirpus cubensis* and other Cyperaceae, together with large grasses such as *Phragmites communis* and *Gynerium sagittatum*, the dicotyledonous herb *Ludwigia octovalvis* and ferns.

(5) 'Marsh scrub' association. The shrub *Dalbergia ecastophylla* and the tall aroid *Montrichardia arborescens.*

The 'marsh scrub' community would probably in time give place to swamp forest, but such a community had not yet developed.

This succession is very like some temperate hydroseres and a similar sequence of stages can be seen in many places in Central America, the Guianas and the West Indian islands, for instance associated with the now destroyed 'mesophytic forest' climax of Puerto Rico (Gleason & Cook 1926). Beard (1946b, 1955a), in his work on the vegetation of Trinidad and other parts of tropical America, later made a useful distinction between 'swamp forest' which is permanently waterlogged and 'marsh forest' which is flooded only seasonally.

A similar distinction is made by Lindeman & Moolenaar (1959) in their account of the coastal swamp vegetation of Surinam. Here the true 'swamp forests' are flooded for most of the year and even when the water level is lowest the ground is wet. The trees grow to about 20 m and there are two tree strata as well as a herb layer consisting mainly of Scitamineae. Palms of several genera (*Bactris, Euterpe, Mauritia, Maximiliana,* etc.) play a large part in the community along with dicotyledonous trees. An interesting feature is that in some places 'pegass' (peaty material) accumulates to a depth of up to 1.5 m, as in some of the swamp forests of Africa and Southeast Asia described later (pp. 360ff.).

Three types of swamp forest are recognized: (1) forest dominated by *Triplaris surinamensis*, which is poorer in species than the other two; (2) *Symphonia globulifera – Virola surinamensis* swamp forest, a more mixed type from which *Triplaris* is absent; and (3) a type with still more species and little tendency to single-species dominance. The composition of the 'swamp forest' seems to depend mainly on the length of time during which the ground is flooded with almost oxygen-free water: the longer the period of flooding, the poorer the forest is in species.

The 'marsh forests' in Surinam are also two-storeyed but are somewhat taller than the swamp forest. The herb layer is denser. The ground surface is conspicuously 'hog-wallowed' (the reason for this is not clear) and the trees grow mainly on the hummocks. There are various types of marsh forest, all considerably richer in species than any of the swamp forest types. It is interesting that three of the types of marsh forest, the *Carapa* (usually *C. procera*) type, the *Hura crepitans* type and the *Mora excelsa* type, show marked single-species dominance: the last named is similar to the mora forest at Moraballi Creek, described in Chapter 11.

Although the successional relations of these various permanently and seasonally flooded forest communities are not well understood, Lindeman & Moolenaar (1959) believe that open swamp communities of herbaceous plants such as *Typha, Montrichardia* and Cyperaceae develop first into what they call 'swamp wood' in which *Triplaris*, palms and other trees form a low single-storey forest 10–15 m high: this community probably later develops into swamp forest and eventually into marsh forest as the ground level rises. There are perhaps several seres converging to marsh forest, which can probably be regarded as an edaphic climax.

Similar successions take place in the swampy coastal region of Guyana and the Orinoco delta: some information is given by Davis (1929) and Myers (1935). On the slower-flowing parts of the Guyana rivers the early stages in the colonization of alluvium are similar to those of the Amazon estuary (p. 354ff.). *Montrichardia arborescens*, a robust aroid growing to a height of 2–3 m, is the first pioneer; it is followed by thickets of the shrubby *Machaerium lunatum* and other woody Leguminosae. As in parts of Amazonia, floating masses of grasses (here chiefly *Panicum elephantipes*) anchor themselves to the *Montrichardia* and the bushes. The later stages of the succession were not worked out, but Davis (1929) described various types of swamp forest in the North West District of Guyana which are probably late stages in hydroseres.

Fanshawe (1952) regards the fringing vegetation along

the Guyana rivers as forming four not very well defined 'longitudinal zones' from the estuary upstream and from one to three 'cross-sectional zones' at any particular point along the river. The latter, beginning at the margin, are 'aquatic swamp', 'arborescent swamp' and 'swamp forest'. The former, which are based on the tallest of the 'cross-sectional zones' are: (1) mangrove forest (Rhizophoretum and Avicennietum) in the estuaries; (2) *Pachira aquatica – Pterocarpus officinalis* woodland, usually associated with 'arborescent swamp' of *Machaerium lunatum*, *Montrichardia*, etc.; (3) swamp forest dominated by *Macrolobium bifolium* (on some rivers replaced by forest of *Inga* and *Ficus* spp. or other kinds of vegetation); (4) swamp forest of *Macrolobium acacifolium* or various other species. On the lower reaches of the rivers the fringing forest is usually disturbed or replaced by various secondary communities and in rocky situations, by rapids, etc., a different kind of fringing vegetation is found.

The 'longitudinal zonation' in the fringing vegetation is comparable to that found along rivers in Africa, Malaysia and elsewhere and probably depends on factors such as decreasing salinity, narrowing of the river channel, increasing speed of current and accompanying changes in the texture of the sediments. It may also be affected by chemical characteristics of the water.

Oldeman (1972) has analysed the structure of the forest edge along the rivers of French Guiana in terms of the Hallé–Oldeman models of tree architecture (p. 74). He suggests that the characteristic features of the river margin – the trunks leaning over the water, the asymmetric and often densely stratified branching and the plagiotropic growth even in species which in other habitats are orthotropic – are responses to gradients in 'plant crowding' (i.e. competition) and soil cohesion as well as to unilateral lighting.

In Central America swamps cover extensive areas, particularly in the lowlands bordering the Atlantic coast. Holdridge *et al.* (1971) have described examples of two interesting types of swamp forest from the lower reaches of the Rio Colorado in Costa Rica: that dominated by the large leguminous tree *Prioria copaifera* (cativo) and palm swamp. The former is found on terraces of alluvium and never extends far from river banks; it comes within Beard's definition of marsh forest (see p. 353 above) as it is only occasionally flooded, but in dry periods the watertable falls to 1 m below the soil surface. The A storey consists mainly of *Prioria* trees 40–50 m high, forming a closed even canopy. These make up over 95% of the basal area and only six species reaching 10 cm d.b.h. were found on 0.5 ha. Seen from within, the cativo forest is dark and has the aspect of 'cathedral

forest'. Whether it is a climax community is not clear: *Prioria* does not appear to reproduce freely in its own shade. *Prioria* forest is also found in Panama and has been briefly described by Golley *et al.* (1975).

The palm swamps consisted of dense closed stands of small palms with scattered emergent wide-crowned trees of which the most abundant were *Pentaclethra macroloba* and *Carapa guianensis*. In sites where the water level did not change appreciably, true swamp forest in Beard's sense, the dominant palm was *Astrocaryum alatum* but where the water was more variable in level and probably better aerated there were pure stands of *Raphia taedigera*, very similar to those formed by related *Raphia* species in West Africa.

14.1.2 Swamp forests and successions in Amazonia

Estuarine region. In the *furos*, the channels near the mouth of the Amazon, the water is more or less fresh, except close to the sea, but rises and falls with the tide by an amount varying from about 2 m at Belém to about 1 m at the confluence of the Xingú, some 350 km from the coast. The *pocoroca* (bore) which accompanies the incoming tide temporarily causes even greater fluctuations of water level. New islands of alluvium are continually being built up and consolidated by the growth of vegetation. The first plants to colonize the islands are replaced by others which can only establish themselves in their shelter. As the succession proceeds the island grows in area and the earlier colonists are pushed outwards to form a marginal belt. Round the larger islands there is thus a zonation representing successive stages in a hydrosere. A similar zonation can be observed on the convex banks of meanders in the channels. These successions were excellently described by Huber (1902) and further details are given by Bouillenne (1930).

As in Guyana, the earliest pioneers on the mud are *Montrichardia arborescens* and the shrub *Machaerium lunatum*. These form almost pure communities, one seeming to exclude the other; often the two alternate every 10 m or so. The seeds of both species float and are abundant in the river drift. Germination is exceptionally rapid and the seedlings can establish themselves between successive tides: in this respect these plants resemble mangroves. In the shelter provided by the stands of *Montrichardia* and *Machaerium*, vegetation of a quite different kind develops, a floating sward consisting of masses of the grass *Hymenachne amplexicaulis*, together with *Eichhornia spp*. Higher up the Amazon such grass swards are much more extensive than they are in the *furos*. This community is similar to the 'sudd' of the Nile and other African rivers (p. 360).

The second stage in the succession in the *furos* is the invasion of the *Montrichardia* and *Machaerium* communities by the mangrove *Rhizophora racemosa*. This tree is found as far upstream as the influence of the tides and is not, like many mangroves, confined to salt and brackish water (see p. 379). In places *Rhizophora* is replaced by other mangroves, *Avicennia* and *Laguncularia*; these are also said to grow here sometimes in water which is never salt or even brackish. The mangroves soon shade out *Montrichardia* and *Machaerium* and beneath them there is only a very scanty undergrowth.

After a time, seedlings of the tall palms *Euterpe oleracea* and *Mauritia flexuosa* and of the dicotyledonous trees *Cecropia palmata* and *C. paraensis* establish themselves beneath the mangroves. The genus *Cecropia* is one which is also characteristic of the secondary succession on terra firme throughout the tropical American rain forest (Chapter 18). The palms and cecropias soon overtop and suppress the mangroves; the next stage in the succession may then be the establishment of várzea (periodically flooded) forest. This includes tall trees and is similar in general aspect to climax rain forest (see below). Under some conditions it seems that the *Rhizophora* community is followed by one of the palm *Raphia taedigera* and not immediately by várzea. Behind the *Rhizophora* or *Raphia* belt there may be one dominated by *Mauritia flexuosa* growing as a pure community or intermixed with other palms such as *Euterpe oleracea*, *Manicaria saccifera* and *Maximiliana regia*, or a belt of Cecropietum dominated by the two species mentioned above. The trees *Virola surinamensis* and *Carapa guianensis* are characteristic of the palm and *Cecropia* belts. The reasons for these variations in the succession are uncertain, but whatever the course of events in the intermediate stages, várzea forest eventually develops.

The Amazonian várzea forests vary much in different places. Their average height is about 15–30 m. Very characteristic are their irregular profile and the huge emergent crowns of *Ceiba pentandra*. Other common species in this forest are *Carapa guianensis*, the rubber trees *Hevea brasiliensis* and *H. guyanensis*, *Symphonia globulifera*, *Virola surinamensis* and many Leguminosae. Palms are frequent and where the forest reaches the water's edge there is often a fringe of *Pachira aquatica* similar to that along the Guyana rivers.

Further information on the várzea forest of the furos region are given by Cain *et al.* (1956), who made a detailed analysis of the floristic composition of the forest

on Arapari Island near Belém as well as of the *Montrichardia* zone surrounding it. Aubréville (1961) and Pires (1973) have also published data on the composition of várzea forest in this area.

Most várzea forests are flooded only during periods of high water level and belong to Beard's category of 'marsh forest'. They are estimated to cover 55 000 km^2 in the whole of Amazonia (Pires 1973). The variations in their composition are due in part to differences in the period of flooding, but they probably also depend on tree felling and other kinds of disturbance. Because it is relatively accessible the várzea is much used as a source of timber, wild rubber and other products and as a site for habitations, particularly by (now extinct) aboriginal tribes (Meggers 1987). Frequent flooding makes the soil relatively fertile, and clearings are made in the várzea in which crops are grown; in recent years the várzea of the lower Amazon has been extensively cleared for planting jute. In structure and floristic composition the várzeas of the Amazon delta are similar to the mora forest of Guyana (p. 299) and the various types of marsh forest in Surinam, but in Amazonia *Mora excelsa* does not occur: another species of *Mora*, *M. paraensis*, is found but it has not been reported as forming single-dominant communities. Várzea forests, like mora forests, usually have fewer tree species per unit area than terra firme forests above flood level.

Besides várzea forests, there are areas of igapó[1] in the delta region, though they are probably less extensive than in the central part of the Amazon basin. Igapó is true freshwater swamp forest in Beard's sense; typically the water is acid and oligotrophic and its level remains high even when the river is not in flood. Organic matter tends to accumulate in the soil, forming peat, and the flora is quite different from that of the várzeas.

Upper and lower Amazon. Above the estuary the Amazon and its tributaries are fringed by large areas that are seasonally or permanently flooded. Conditions differ in various ways from those in the estuary. Tidal influence, although felt as far up as Obidos (*ca.* 1000 km from the river mouth), is unimportant, but there are very large seasonal differences of water level dependent more on rainfall in the distant headwaters than on conditions in the area through which the river flows. During the high water period (April to August) the maximum water level at Manaus reaches about 12 m above the lowest level (November): on the middle Solimões (upper Amazon) the difference may be as much

---

[1] The terms várzea and igapó are not always used as defined here. See Prance (1979) and Daly & Prance (1988, p. 409).

as 20 m. The flooding is so deep, and lasts so long, that it has profound effects on the vegetation. For the trees of the várzeas and igapós the high water season is a period of relative dormancy during which cambial growth ceases or slows down. Many várzea trees lose their leaves at this time (Worbes 1986).

The lower Amazon between the Xingú mouth and the Rio Negro is about 5 km wide and has numerous islands, which grow in length downstream by accretion: a typical section of this part of the river and its flood-plain is shown in Fig. 14.1. Bordering the main channel there are strips of várzea; the whole flood-plain may be up to 100 km wide. The várzeas are built mainly of relatively nutrient-rich sediments transported from the Andes and elsewhere, the highest part along the river bank forming *barrancos* (levées). These are not very stable, parts being liable to crumble away at the height of the floods, together with the trees growing on them. Further from the river banks, between the várzeas and the terra firme, the ground is very low-lying, much of it consisting of lakes and land which is swampy even when the river is low. Some of the latter consists of seasonally flooded grasslands (*campos de várzea*) which provide grazing for cattle when the ground is dry enough. Elsewhere there are lakes and permanently swampy forests (igapó) which receive little or no sediment during floods. Along the main stream above Manaus (R. Solimões) and the major tributaries the river channels are narrower and better defined.

One of the most striking features of central Amazonia (as of the Guianas) is the difference in the water and flood-plain forests between the 'whitewater', 'clearwater' and 'blackwater' rivers. The Amazon, Solimões and many of their larger tributaries are whitewater rivers; their water is more or less colourless but cloudy because of fine suspended material. Clearwater rivers such as the Tapajós are also colourless, but the water is quite transparent. The water of the blackwater rivers, the largest of which is the Rio Negro, resembles coffee; it is black by reflected light but orange-brown and transparent by transmitted light. There are important chemical differences between the three kinds of river water (Fig. 14.2). White water is circumneutral (pH 6.5–6.9) and is relatively rich in calcium, magnesium and other ions. Black water is strongly acid (pH *ca.* 4.0) and has a very low ion content and conductivity. The composition of clear water is variable but usually more or less intermediate between that of white and black water. Furch (1984) has given a general account of the chemistry of Amazonian waters (Fig. 14.6).

The differences between the three types of river water depend on the geology and soils of the catchments. White waters come mainly from the Andes and their foothills or other areas where the rocks are geologically young and easily eroded, giving rise to kaolisols (latosolic soils). Clear waters come from the ancient, very hard rocks of the Guiana Shield and the Brazilian Planalto. The blackwater rivers and streams rise within the Amazon basin itself from white sands and similar deposits with a low clay content (Richards 1941, Klinge 1967). The dark colour is due to fulvic and humic acids derived from mor humus and other dead plant material in caatinga (heath) forests and igapós.

Several types of seral community associated with 'whitewater' rivers have been described. (i) Various kinds of floating aquatic vegetation, including a com-

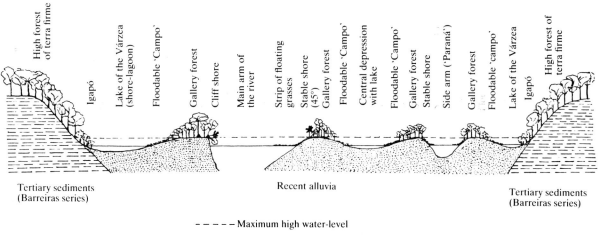

Tertiary sediments (Barreiras series)   Recent alluvia   Tertiary sediments (Barreiras series)

– – – – – Maximum high water-level

——— Minimum low water-level

Fig. 14.1 Schematic transect of lower Amazon valley. Based on Sioli (1964, Fig. 2). The heights are exaggerated.

Fig. 14.2 The confluence of the Rio Solimões (right) and the Rio Negro (left) near Manaus, Brazil. (Photo. G.T. Prance.) The 'blackwater' of the R. Negro drains a terrain predominantly of caatinga forest on podzolic soils. The 'whitewater' of the R. Solimões comes mainly from vegetation on lateritic soils (kaolisols).

munity dominated by the giant water-lily, *Victoria amazonica*, and 'floating meadows' formed by *Paspalum repens*, *Echinochloa polystachya* and other grasses. The ecology of the latter has been studied in some detail by Junk (1970) who has shown that they may be colonized by herbaceous plants such as *Eichhornia crassipes*, Cyperaceae and *Montrichardia* and even by trees such as *Cecropia* spp. which are able to maintain themselves on the 'meadows' for several years or even decades. (ii) The seasonally flooded *campos de várzea* mentioned above which are dominated by the same grasses as the 'floating meadows'. (iii) A community of the shrubs *Salix humboldtiana* and *Alchornea castaneifolia* fringing the river banks. (iv) A *Cecropia* community. At the edge of the *campos de várzea* the forest begins abruptly and this *Cecropia* zone, consisting of *C. palmata*, *C. paraensis* and other species growing in groups rather than as a continuous belt, forms its outer margin. These groups of trees last for one generation only and are then replaced by other species of the várzea forest.

Similar successional changes were briefly described by Ule (1908) on sandbanks in the Rio Juruá, a tributary of the upper Amazon, and by Kalliola *et al.* (1991a) and Campbell *et al.* (1992) on land being built up on the convex side of meanders of the Amazon and Rio Ucayali.

In central Amazonia two successional stages of várzea

forest can be recognized: young várzea in which the chief trees are quick-growing species such as *Calycophyllum spruceanum*, *Hura crepitans*, *Inga* sp. and *Triplaris surinamensis*, and mature várzea forest which is like that of the delta region, except that palms are relatively rare, the commonest species being *Astrocaryum jauari*. Huge emergent trees of *Ceiba pentandra* and *Maquira coriacea* are sometimes a conspicuous feature (Grenand & Grenand 1993). Like those of the delta region, the várzea forests higher up the Amazon and its tributaries have long been subject to disturbance and are probably rarely seen in a truly natural condition. The gallery forests (Chapter 16) which extend along rivers and permanent streams far into the seasonally dry cerrados (savannas) to the south have many species in common with the várzea forests of the Amazon (Ratter *et al.* 1973).

Campbell *et al.* (1986) compared a 0.5 ha sample of undisturbed várzea forest on the lower Rio Xingú with a 3.0 ha sample of evergreen seasonal forest on terra firme near by. The trees in the várzea forest did not exceed 25 m in height and the emergents in the terra firme forest were considerably taller. In the várzea plot there were forty tree species ⩾10 cm d.b.h. Although a direct comparison of species-richness was not made, species–area curves showed that it was less rich than the terra firme forest. In the várzea plot Leguminosae formed the majority of the species and more than half the basal area, but two non-leguminous trees, *Mollia lepidota* and *Leonia glycycarpa*, were the species with the highest importance values.

Fringing vegetation of the blackwater rivers. On the Rio Negro and other blackwater rivers the marginal vegetation is strikingly different in physiognomy and composition from that along the Solimões and other whitewater rivers. Plants common by the latter such as *Alchornea castaneifolia*, *Ceiba* and *Salix humboldtiana* are rare or absent and there is little grass. Igapós, which are flooded to a depth of many metres for a large part of the year, take the place of the várzeas of the whitewater rivers. These are composed mainly of trees and shrubs with a few lianes: herbaceous plants other than epiphytes are almost absent. Takeuchi (1962) and Keel & Prance (1979) have described igapó vegetation near Manaus; Worbes (1986) made a comparison between igapó forest at Tarumã mirim on the R. Negro and a whitewater várzea on an island in the R. Solimões.

Igapó forest consists mainly of trees (some of which grow to 25–28 m) and shrubs; there are a few lianes, but no herbs except epiphytes. Most of the trees are evergreen and have somewhat sclerophyllous leaves, in contrast to the várzea in which the foliage is mesophyllous and some species are leafless in the dry season. According to Junk (1989) the floristic composition of the igapós in central Amazonia depends chiefly on the depth and length of the annual flooding: he distinguishes three zones with average submersion periods of 270, 230 and 130 days. In the lowest zone *Eugenia inundata* and the shrub *Symmeria paniculata* are common. In the two higher zones the species are more numerous and include *Cecropia* spp. and *Hevea brasiliensis*. Worbes' samples of igapó and várzea were flooded to a depth of *ca.* 7 m for 40–60% of the year. The igapó, though much less species-rich than the terra firme forest, was about as rich in species as the várzea and had forty-three tree species ($\geqslant$5 cm d.b.h.) on 2100 m². *Aldina latifolia* was dominant and formed 47% of the basal area. In

Fig. 14.3 Freshwater swamp forest in Brazilian Amazonia. A várzea forest, R. Solimões. (Photo. G.T. Prance.) B Interior of seasonal várzea forest, at low water, R. Zinho, Pará. From Prance (1979, Fig. 2).

the igapó studied by Keel & Prance (1979) the shrub *Myrciaria dubia* was dominant (importance value 75.7) in the most deeply flooded area up to 10 m from the shore. The depth of flooding was shown by the growth of epiphytic sponges on the stem.

During the flood period the oxygen concentration in the water of the igapós falls very low. Worbes (1986) found that at Tarumã mirim it was 0.6 mg l^{-1} at a depth of 1.5 m and was zero round the tree roots. At this season the trees become dormant and cambial growth ceases; for this reason the stems show well-marked annual growth-rings. Flowering and fruiting are related to the flooding regime rather than to the local climate (Chapter 9).

The igapó vegetation of the Rio Negro has many similarities to the flooded forests of central Africa and to the peat swamps of Malesia (see below). In all these swamp forests the vegetation is adapted to nutrient-poor conditions and the roots are tolerant of long periods of submergence in oxygen-deficient water.

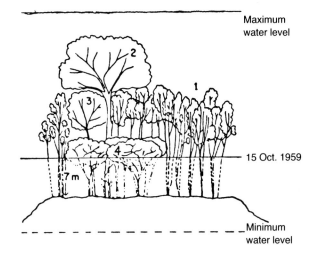

Fig. 14.5 Profile diagram of igapó on Rio Negro near Manaus. After Takeuchi (1962). Key: 1 *Eugenia inundata*, 2 *Campsiandra latifolia*, 3 *Symmeria paniculata*, 4 *Coccoloba* sp.

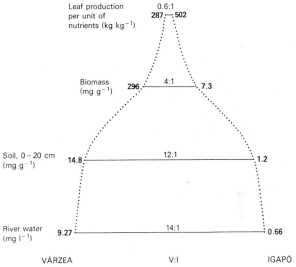

Fig. 14.6 Mineral elements, water, soil and biomass compartments of várzea and igapó swamp forest ecosystems in Amazonia. From Furch & Klinge (1989, p. 202). Mean values, drawn to horizontal scale, are shown in the centre.

Fig. 14.4 Freshwater swamp forest in Brazilian Amazonia. A Flooded igapó, Rio Negro near Manaus. (Photo. G.T. Prance.) B Igapó at low water, Rio Negro. (Photo. G.T. Prance.)

14.2 Hydroseres and swamp forests in tropical Africa

Much of the west coast of Africa from Senegal to Cameroon is fringed with shallow freshwater lagoons into which the rivers discharge. Bordering these were formerly large areas of permanently or seasonally flooded forests, now largely replaced by rice cultivation and other forms of land use. Hydroseral stages can also often be seen to great advantage, but the successions have not been studied in detail. Further inland and elsewhere in many parts of tropical Africa similar swamp vegetation is found by slow-flowing rivers and in low-lying basins, but it is only in the basin of the Zaïre (Congo) that hydroseral communities and swamp forests are developed on a scale and in a diversity at all comparable to those of Amazonia.

14.2.1 West Africa

In southwestern Nigeria the succession in freshwater lagoons and waterways begins with submerged and free-floating aquatic communities. The former consist of such plants as *Ceratophyllum demersum* and *Utricularia* spp., the latter often of a well-developed mat of grasses, *Pistia*, etc., similar to the sudd of the Nile. The sudd drifts about in large masses and may become an obstacle

A

B

Fig. 14.7 Freshwater swamp forest, Akilla, Nigeria (1935). In the opening in the foreground the dominant plant is *Cyrtosperma senegalense*. The broad-leaved tree is *Mitragyna ciliata* and the palm is *Raphia* sp.

Fig. 14.8 Freshwater swamp forest, Omo Forest Reserve, Nigeria (1935). The stilt-rooted tree in A is *Uapaca staudtii* and the dangling moss in B is *Floribundaria* sp.

to navigation. When the water becomes shallow enough, a rooted floating-leaf community appears, usually dominated by *Nymphaea lotus*. These purely aquatic communities are succeeded by herbaceous swamp consisting of grasses, *Cyperus papyrus* or stands of *Pandanus candelabrum*. Sometimes this may be followed by a stage dominated by shrubs such as *Alchornea cordifolia*, but generally the next stage seems to be palm swamp dominated by one (or more?) species of *Raphia*. Finally tall forest consisting chiefly of dicotyledonous trees may develop.

This swamp forest has a general resemblance to primary rain forest on sites with normal drainage, but is usually more open and irregular in structure. The number of species per unit area is smaller and though many of the tree species are also found in other habitats a considerable proportion are characteristic of waterlogged soils. Among the latter are some species with pneumorhizae, stilt-roots and other special features. An example of such forest was described by Richards (1939) from the Omo (formerly Shasha) Forest Reserve in Nigeria (Figs. 14.7 and 14.8) on land flooded in the wet season and waterlogged to within a few centimetres of the soil surface during the rest of the year. Tall trees were abundant but not evenly distributed, so that patches of deep shade with little undergrowth were mingled with gaps filled only with thickets of bushes and lianes. The dominant plant in these thickets was *Leea guineensis*, a straggling shrub supported by prop roots. The commonest tree was *Mitragyna* sp. (probably *M. ciliata*) which formed as much as 36% of trees ≥16 in (41 cm) d.b.h. and over and 13% of those ≥4 in (10 cm) d.b.h. This species and the closely related *M. stipulosa* are common in swamps and on river margins through much of tropical Africa. On islands and flooded river banks on the Bas Kasai and round Lake Leopold II in Zaïre *Mitragyna* sp. forms pure stands 'like veritable plantations' (Vermoesen 1931, p. 181, quoting E. & M. Laurent). *M. ciliata* is mainly found in rain-forest areas and *M. stipulosa* in more seasonal climates: Schnell (1952a, p. 95) regards the two as ecotypes.

Among the subordinate species in the Omo swamp forest which are not confined to waterlogged soils are some such as *Bridelia micrantha* which on drier ground are commoner in young secondary than in mature forest. The occurrence of such secondary forest species in swamp forests is probably accounted for by the relatively good illumination.

The succession from open water to tall forest can be seen in a telescoped form as a zonation on the banks of the larger Nigerian rivers. On the Osse, for instance,

the successive zones are: (i) submerged aquatics and sudd of *Pistia* etc.; (ii) Pandanetum; (iii) Raphietum; (iv) freshwater swamp forest.

Communities similar to those mentioned above have been described from other parts of West Africa, for instance by Guillaumet (1967) from the Côte d'Ivoire and by Taylor (1960) from Ghana.

Schnell (1952a) in a phytosociological study of West African vegetation divides the communities of waterlogged habitats of the forest and Guinea savanna zones into swamp forests (*forêts marécageuses*), fringing forests (*forêts ripicoles*) and seral communities (*brousses arbustives*) dominated by the shrub *Alchornea cordifolia*. In the freshwater swamp forests of the rain-forest region, Schnell recognized four associations belonging to the order Mitragyno-Raphietalia; in these the presence of *Mitragyna ciliata* and of several *Raphia* species is characteristic.

14.2.2 The Zaïre (Congo) Basin

Swamp forests and herbaceous swamp communities cover very large areas in the basin of the Zaïre and its many tributaries, especially in the low-lying central part (the 'Cuvette Centrale' of Belgian writers), a vast and until recently little known region crossed by innumerable rivers and streams. According to Evrard (1968), some 150 000 km² of this area consist of periodically or permanently waterlogged hydromorphic soils. The depth of the annual flood is not more than *ca.* 4–5 m (Germain 1965). Though not so large, the Central Basin of Zaïre is the African counterpart of Amazonia.

The water of the Congo and its large tributary the Ubangi is turbid, colourless and circumneutral, but most of the streams in the Central Basin are clear, acid and oligotrophic. Some are 'whitewaters' (pH 5.0–5.2, conductivity *ca.* 10 μmho cm^{-1}), but the majority are 'blackwaters' (pH *ca.* 3.6, conductivity *ca.* 130–140 μmho cm^{-1}) similar to those of Amazonia (Evrard 1968, p. 19); see also Germain (1965). The great variety of plant communities in this region probably depends partly on these differences in the characteristics of the water, though there is not such a clear-cut distinction between blackwater and whitewater rivers as on the Amazon. More important factors seem to be differences in the flooding regime and in the sediments, which vary from sands and silts of various ages and textures to mucks and peaty soils rich in organic matter.

Lebrun & Gilbert (1954) published a classification of the forest types of the hydromorphic soils. Five main types were recognized (excluding the mangrove forests of salt or brackish tidal swamps): (1) pioneer communities of shrubs and small trees continually or almost continually flooded and actively promoting soil accretion (*forêts ripicoles colonisatrices*); (2) riverain forests with long periods of submergence and short periods of emergence, accretion moderately active (*forêts riveraines*); (3) periodically flooded forests with long periods of emergence and little accretion (*forêts périodiquement inondées*); (4) swamp forests which never dry out, whether periodically flooded or not (*forêts marécageuses*); and (5) alluvial valley forests liable to flooding but normally more or less well drained. The herbaceous and forest communities of flooded and waterlogged soils were later dealt with more fully by Germain (1965) and Evrard (1968). Many references are given in the two latter papers as well as numerous *relevés* of the plant communities described.

A well-documented account of a hydrosere in this region is that of Louis (1947c) on the development of vegetation on the islands in the Zaïre near Yangambi (Fig. 14.9). In this part of the river there are many sandbanks from a few metres to several kilometres long which are more or less deeply submerged during the season of high water. When the water is low the islands become colonized by plants and all stages of succession from the establishment of the first pioneers to the development of dense forest can be seen (Fig. 14.9). The sandbanks are of various ages and are usually spindle-shaped, growing downstream as a long tapering tail: all

seral stages may be found on the same island. The soil, which consists of sand mixed with finer materials, is very fertile and the larger islands are sometimes cleared for growing crops. At river level a thick fringe of the shrubby *Alchornea cordifolia* (which plays a similar part on the Zaïre to *A. castaneifolia* on the Amazon) hides these clearings from view.

There are four main seral stages: (1) a pioneer association mainly of grasses and Cyperaceae with long rhizomes such as *Leptochloa coerulescens* and *Cyperus maculatus* together with mainly ruderal dicotyledonous herbs, e.g. *Ageratum conyzoides*, *Glinus* and *Ludwigia* spp.; (2) a sward of rhizomatous grasses (this includes *Echinochloa* spp. and others but *Vossia cuspidata* generally becomes dominant sooner or later); (3) a shrub community in which *Dichaetanthera corymbosa*, *Sesbania aegyptiaca* and other species are present at first but are soon suppressed by an impenetrable thicket of *Alchornea* interwoven by climbing plants and anastomosing branches; (4) seasonally flooded forest. The pioneer trees are *Bridelia micrantha* and *Ficus mucuso*, but later a stratified forest community in which *Chrysobalanus atacorensis* forms the upper storey establishes itself. Subsequently more species-rich forest develops in which trees such as *Mimusops andongensis*, *Myrianthus arboreus* and *Trichilia heudelotii* are the supporting pillars for a dense blanket of lianes about 10 m above ground level. The undergrowth now consists mostly of

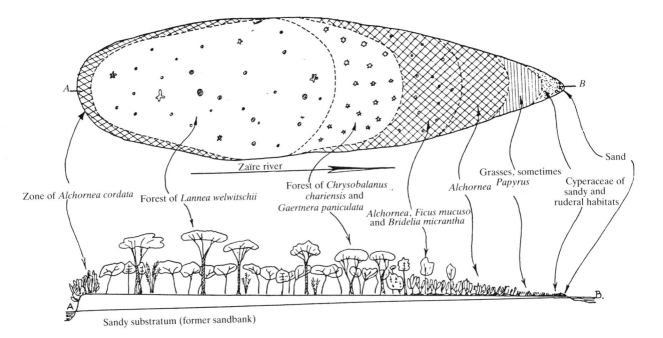

Fig. 14.9 Succession on island in Zaïre river, near Yangambi. Schematic. After Louis (1947c, Fig. 2).

a scanty growth of small woody plants with the herb *Pseuderanthemum ludovicianum* in the less shady places.

This is a good example of a succession in which each stage reacts upon the habitat in such a way as to produce physical conditions more or less unfavourable to its permanence but favourable to invaders of the next stage. The effect of the plants slows the speed of the current and increases the rate of sedimentation so that the surface level of the island rises. The grasses of stages (1) and (2) are particularly effective in this process and quickly cover and consolidate the sediments. During the succession there is a change from a very variable environment with long periods of submergence alternating with periods of emergence in which the surface temperature of the ground sometimes reaches 50 °C, to more constant conditions in which periods of submergence are shorter and the forest creates its own internal microclimate. On the older islands there is often erosion at the upstream end while at the lower end accretion provides new habitats for colonization.

In other situations different kinds of hydroseres are found depending on the type of water, variations in water level and other factors.

Evrard (1968) divides the woody plant communities of the hydromorphic soils in the Central Basin into seral stages and mature forests. The seral communities include: (1) *Alchornea cordifolia* thickets (similar to those on the islands described above); (2) a *Macaranga lancifolia – Harungana robynsii* association (with two subassociations (both characteristic species are small, fast-growing, stilt-rooted trees resembling those found in the early stages of secondary successions on well-drained sites); (3) an association of the palm *Raphia laurentii*; (4) a *Raphia sese* association; and (5) a *Uapaca heudelotii – Parinari congensis* association. The last named is often found on the sloping banks of the Zaïre and its larger tributaries; it is widely distributed, extending to the coastal region of West Africa (though with replacement of some species by others). The stilt roots of *U. heudelotii* and the exposed interlacing lateral roots of *P. congensis* give this community a characteristic appearance.

The main types of mature forest on the hydromorphic soils are the seasonally flooded *Oubanguia africana – Guibourtia demeusii* association and the permanently swampy *Entandrophragma palustre – Coelocaryon botryoides* alliance. The former association, which is comparable to the várzea forests of the Amazon, is found chiefly by the Zaïre and other large rivers on the banks of alluvium between the main channel and the wide, permanently wet depressions which lie behind, especially where the river course is straight rather than meandering. The soil is sandy to silty in texture and not particularly rich in organic material: flooding typically lasts for three to four months. The *Entandrophragma–Coelocaryon* forest, on the other hand, is comparable to igapó and occupies the wet depressions, often covering very large areas. The soil never dries out and contains 50–70% of organic material, which may be either well decomposed, forming muck, or little decomposed forming true peat (Evrard 1968, p. 44). Both associations seem to be considerably poorer in species than the forest on well-drained sites in the same region (Evrard 1968, p. 72).

The *Oubanguia–Guibourtia* association is distinctly stratified with an A storey at 35–40 m and a B storey at 20–25 m, the two together forming a cover of about 60–90%; below these is a close-packed layer of trees 8–15 m high and a scanty undergrowth. In places where the flood water flows rapidly *Oubanguia africana* and *Scytopetalum pierreanum* are the most abundant tree species, but where the current is less swift the chief dominant is *Guibourtia demeusii*. On islands in the Zaïre there is a facies in which the tall *Ceiba pentandra* is conspicuous, as in the várzeas of the Amazon. Physiognomically the association is characterized by the flanged rather than distinctly buttressed bases of many of the trees: stilt-roots and pneumorhizae are rare. It is interesting that as in the Rio Negro igapó (pp. 358–9) the phenology of the trees is dependent on the water level rather than the weather. High water is a period of relative rest when a large number of leaves are shed (Chapter 9).

The *Entandrophragma palustre – Coelocaryon botryoides* alliance is divided by Evrard into two associations, both found on permanently waterlogged organic substrata; one is characterized by *Rothmannia megalostigma* and the other by *Lasiodiscus mannii*. The former is typical of wide depressions where flood water is impounded by natural levées, having had most of its sediments already decanted so that little but organic matter is left. Under such conditions the soil is peaty and the ground surface more less even. The *Lasiodiscus* association prefers sites where streams drain into wide low-lying valleys, carrying sand in suspension during floods. The soil is a mixture of sand and organic matter and develops a hummocky surface which is stabilized by the tree roots. Both communities are less tall than the *Oubanguia–Guibourtia* forest, the taller trees reaching only 25–35 m. The root systems are shallow and many species have expanded bases with conspicuous flutings or indistinct buttresses. Pneumorhizae are common.

The evidence for successional relationships between these communities is inevitably fragmentary. Evrard (1968) suggests that in general there seem to be three principal seres, one leading to the *Oubanguia–Guibourtia* association, the others to one or other of the associations in the *Entandrophragma – Parinari* alliance. In all these associations accretion is very slow or has ceased altogether, so they can be regarded as permanent edaphic climaxes. At any particular place the shifting of river courses and the building up or erosion of sandbanks may lead to the succession being accelerated, retarded or taken back to an earlier stage. On mounds where the ground level is slightly higher than average, trees characteristic of the terra firme forest, such as *Strombosiopsis tetrandra*, can be found; this suggests that if some allogenic factor led to a general raising of the relative soil–water level the edaphic climax would be replaced by the climatic climax (here evergreen seasonal forest).

14.3 Hydroseres and swamp forests in the eastern tropics

In the flood-plains and deltas of slow-flowing rivers in the eastern tropics there are vast areas which are permanently seasonally waterlogged. Much of this swamp land can be made agriculturally productive and on the mainland of Southeast Asia (as well as in Java) most of it has been used for centuries for *padi* (irrigated rice) cultivation. In Borneo, Sumatera and New Guinea, swamp forests and seral swamp vegetation still cover extensive areas. Relics of similar communities can be found here and there over most of the eastern tropics.

The swamp communities of Malesia were briefly reviewed by van Steenis (1957) and Champion & Seth (1968a) include various kinds of swamp forest in their account of the forest types of India. Information on swamp forests in Malaya can be found in papers by Symington (1943), Wyatt-Smith (1959), Corner (1978) and Furtado & Mori (1982). A unique feature of Borneo, Sumatera and some of the neighbouring countries is the occurrence of extensive moor forests (peat swamps) which develop under very oligotrophic conditions and are the tropical equivalents of the raised bogs (*Hochmoore*) of the temperate zones. These forests have been relatively well studied (pp. 366–70), but detailed information on other eastern tropical swamp vegetation is scanty and can be discussed only briefly.

As in tropical America and Africa, eastern swamp forests have fewer tree species per unit area than terra firme sites. Many of these species are confined to swampy habitats. In an intensive survey of the Jengka Forest Reserve in Pahang (Malaya) (pp.317–18), Poore (1968) found that the flora of the narrow swampy valleys was quite distinct from that of the other rain forest. It included 73 tree species ⩾3 ft (91 cm) g.b.h. as against 357 in the non-swampy areas. The latter were of course much more extensive but the index of diversity was also only about 70 in the swamps compared with about 114.2 elsewhere. In the eastern tropics, as in other tropical regions, swamp forests show a strong tendency to single-species dominance, species of *Campnosperma*, for example, forming a large percentage of the total stand in swamp forests in parts of Borneo and New Guinea. Single dominance is most marked in moor forests where nearly pure communities of *Shorea albida*, *Koompassia malaccensis* and other species are found.

New Guinea. By far the largest swamp forests in Malesia now remaining in a more or less natural condition are in New Guinea, especially in West Irian and the deltas and lower courses of large rivers such as the Fly and Sepik in the eastern part of the island. Various parts of the CSIRO *Land Research* series of publications give a general account of the extent and composition of the Papua New Guinea swamp forests. Further information is given by Paijmans (1976), van Steenis (1954) and Taylor (1959). In many places a sago palm *Metroxylon sagu* (and perhaps other species) is abundant or locally dominant; such swamp communities occupy 500 square miles (800 km²) in the Ramu–Sepik delta alone and provide the staple food for a large population. Types of swamp forest without sago palms are little utilized and are usually so impenetrable that they are still in a completely natural condition. Van Steenis says that several of these types of swamp forest are characterized by single-dominant species (*Mitragyna speciosa*, *Barringtonia spicata*, *Campnosperma macrophylla* and others).

According to Robbins (1961), deltas and alluvial plains which are regularly flooded have a range of plant communities from forests of trees 30 m high on relatively well-drained alluvium, to less tall seasonally or permanently flooded forests of palms and dicotyledonous trees (Fig. 14.10). There are enormous areas of 'back swamps' which never dry out; these are dominated by Cyperaceae and grasses (*Phragmites*, *Saccharum*). The tall alluvial forests are different in species composition but similar in structure to the climax forest above flood level. They are comparable to the várzea forests of the Amazon. The 'back swamps' are perhaps similar to the *campos de várzea*. White (1975) has shown that such communities are very unstable, as rivers in the New Guinea

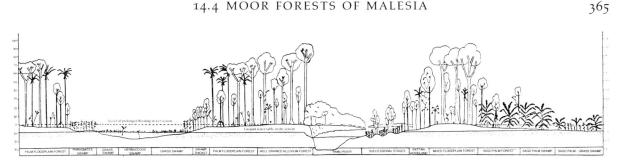

Fig. 14.10 Transect across Ramu river, Papua New Guinea. After Robbins (1961).

lowlands frequently change their course; the trees in established forests may then be killed by flooding while new alluvium becomes available for colonization.

The frequency, depth and duration of flooding seem to be the major factors controlling the occurrence of different types of swamp vegetation in New Guinea, but the successions have not been investigated. The longer the period of high water level, the larger is the part played in the vegetation by palms (including *Metroxylon*). In permanently swampy forests *Campnosperma* sp., or some other single species, is often dominant. It is interesting to note that, as in other parts of the tropics, tree species that are characteristic of secondary forest on dry land are often common in swamp forests.

Taylor (1959) attempted to classify the lowland swamp vegetation of north-eastern Papua using Beard's (1955a) classification of tropical American vegetation as a basis. He distinguished a number of 'sequences' (catenas) of swamp communities. In addition to salt water, brackish water and estuarine sequences, he recognized four found under freshwater conditions: (1) permanent swamp, (2) fluctuating swamp, (3) semi-seasonal swamp and (4) seasonal swamp. The first includes types of forest found on deep peat as well as types found on non-peaty or mineral soils.

On sand banks in rivers in New Guinea pure stands of *Casuarina* sp., *Eucalyptus deglupta* and *Octomeles sumatrana* are found (Robbins 1961, Chujo 1982). The last two also occur commonly in other seral communities.

14.4 Moor forests (peat swamps) of Malesia

The most remarkable of all tropical swamp forests are the moor forests of the coastal region of western Borneo and northern Sumatera, (the 'Tropical evergreen peat forests' of Ellenberg & Mueller-Dombois, 1967). These cover lens-shaped masses of peat, the lower surface of

which is below the level of the surroundings and the convex upper surface some metres higher at the centre than at the margin (Fig. 14.11). Smaller, less well developed moor forests are found on the east and west coasts of the Malay Peninsula; similar swamps may exist in western New Guinea (Whitmore 1984a) and in the Nam Can Peninsula (Vietnam). As swamp forests in which peat is formed occur in Amazonia and elsewhere in the tropics, it is surprising that there seem to be no exact counterparts of the eastern tropical moor forests anywhere in Africa or South America.

Moor forests were first described in detail by Polak (1933a,b), who studied the stratigraphy and hydrology of several moor forests in Sumatera and West Borneo (now Kalimantan) and showed that they were true ombrogenous mires, comparable to the raised bogs (*Hochmoore*) of Eurasia (see Tie 1990) except that their

Fig. 14.11 Aerial view of peat swamp, Sarawak. (Photo. E.F. Brünig.)

plant covering consisted of broad-leaved evergreen trees, sometimes over 60 m high. Polak described the living vegetation of these peat swamps only very briefly, but further details were given by Sewandano (1937). Symington (1943) later recorded the existence of peat swamps in the Malay Peninsula.

The moor forests studied by Polak were more or less circular and several kilometres across, with concentrically arranged zones of vegetation. At the margins the trees were as much as 30 m high, becoming gradually less tall towards the centre where in one moor there was dwarf forest ('Krüppelholz') of Tristaniopsis, interspersed with pools of water. The zones appeared to represent stages in a succession, the details of which were not fully worked out. The peat stratigraphy showed that woody plants were dominant from an early stage in the development of the moor.

More recently knowledge of these moor forests has been much increased by the careful work of Anderson in Sarawak and Brunei (1961b, 1963, 1964; see also Tie 1990 and long summary by Whitmore (1984a)[2]. In Sarawak moor forests occupy about 1 500 000 ha (about 12% of the land area) and the largest moor is 1070 km^2 in extent. They are found mainly in the delta of the Rejang and meanders of the Baram and other large rivers. The centre of a moor may be as much as 9 m higher than the margin, but the difference in height is not usually so great. Carefully levelled transects (Fig. 14.13) showed that the slope was steepest near the edges and there is generally a large, more or less flat central 'bog plain'; the surface is thus shaped more like an inverted saucer than a dome. The maximum depth of peat may be more than 15 m. As in a temperate raised bog, the convexity of the surface is due to peat accumulating faster during the earlier than during the later stages of moor development.

The peat consists of fragments of wood, roots and leaves in various stages of decomposition, suspended in a chocolate-brown matrix of the consistency of thin porridge. The percentage of organic material is always very high and in peat from the Baram moor forests is nearly 100%. The peat is everywhere very acid and in some moors the pH falls from a little over 4.0 near the edges to slightly over 3.0 in the central areas. Owing to the convexity of the surface, drainage is centrifugal and only the marginal zone (the 'rand') is affected by flood water bringing in small additions of nutrients. The central part of a moor thus depends entirely on rainfall

for its input of nutrients. A decrease in the available nutrients from the margins towards the centre is therefore to be expected and in some moors the trees near the centre are stunted (Fig. 14.14C) and show other symptoms of extreme nutrient deficiency. Analyses of nitrogen, potassium, phosphorus and other elements in peat samples from the various zones (Anderson 1961b, Tables 5 & 6, Richards 1965) do not, however, show a regular gradient and are not always consistent, though it is clear that the content of phosphorus (and probably of nitrogen) in the central zone is generally much lower than in the peripheral mixed swamp forest zone. It is possible that the analyses do not show the amounts of nutrients actually available to the vegetation; further investigations are needed. In all zones more nutrients are found in the top 15 cm of the peat, which contains most of the active tree roots, than deeper down.

Though the substratum is semi-liquid, tall trees are able to grow on it, their interlacing roots usually forming a firm network, thinly covered with dead leaves, on which it is possible to walk dry-shod. All the dominants are evergreen trees and are often surrounded at the base by thickets of loop-like or peg-like pneumorhizae up to 30 cm high. The undergrowth, which is seldom very thick, consists of small palms, Pandanus spp., Nepenthes spp. and Cyperaceae as well as young trees and other small woody dicotyledons.

Anderson recognizes the following six zones ('phasic communities') of vegetation from the margin towards the centre of the peat moors. Their boundaries are not always well defined and often not all the communities are found on any one moor.

(1) Mixed swamp forest. This is a marginal zone within reach of river floods. It is a mixed association, richer in species than any of the other zones: important species include Gonystylus bancanus, Copaifera palustris, Dactylocladus stenostachys, Dyera lowii, Shorea spp. (not S. albida) and in the B storey Neoscortechinia kingii. The canopy is uneven with trees up to 40–45 m high; the general aspect of the forest more or less resembles mixed dipterocarp forest.

(2) Alan forest (Fig. 14.14A). In this community the dipterocarp Shorea albida (alan) is the single dominant (though less strongly dominant than in the next zone). Gonystylus bancanus and the small tree Stemonurus umbellatus are the commonest associates. The older trees of S. albida are hollow and many appear moribund; young regeneration is almost absent.

(3) Alan bunga forest. Here S. albida is the only species in the A storey, forming a characteristically even canopy (Fig. 14.14B). It grows extremely tall; stands 55–60 m high are not uncommon. There is almost no

[2] In the following pages a few observations made by the author in Brunei and Sarawak in 1959 have been incorporated with data from Anderson's work.

A

B

C

D

Fig. 14.12 Peat swamp (moor forest) communities in Borneo.
A *Shorea albida* association (alan forest) bordering oil field,
Brunei. B *Shorea albida* association, road to oil well, Seria,
Brunei. (Photo. P.S. Ashton.) C High pole forest (*Litsea
palustris, Shorea albida*, etc.), Badas Forest Reserve, Brunei.
(Photo. E.F. Brünig.) D Open stunted forest (padang
keruntum), near Marudi, Sarawak. (Photo. E.F. Brünig.)

B storey but there is a moderately dense C storey of various species.

(4) High pole forest. This is a community of closely spaced, slender-stemmed trees up to 35–40 m high (Fig. 14.12C). The most abundant species are *Shorea albida* and *Litsea palustris*; *Parastemon urophyllum* is a characteristic associate.

(5) Low pole forest. This is a narrow zone, similar to the last, but the trees are less tall (to about 25 m high) and even more closely spaced. The most abundant tree species are *Palaquium* spp., *Parastemon spicatum* and *Tristaniopsis* spp.

(6) Padang keruntum (or padang paya) (Figs. 14.12D, 14.14C). Very open stunted forest mainly of *Combreto-carpus rotundatus* and *Dactylocladus stenostachys* beneath which there is a scanty growth of Cyperaceae, *Nepenthes* spp. and the small *Pandanus ridleyi* with occasional tufts of the moss *Sphagnum junghuhnianum*. Here and there are small pools of open water. The contorted branches of the trees and their habit of frequently coppicing from the base give this curious community a superficial resemblance to some types of savanna woodland.

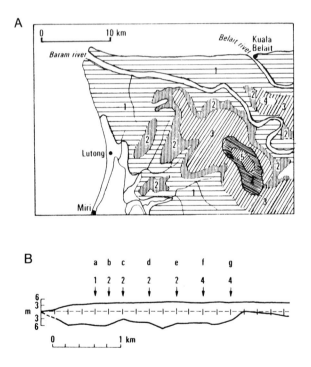

Fig. 14.13 Peat swamps (moor forests) in Sarawak. A Map of swamp forests in Baram estuary showing more or less concentric zones of 'phasic communities' 1–5 (see pp. 366–8). Based on Anderson (1961b). B Section along a transect at Rantau Panjang, Rejang delta, showing domed peat surface. The letters refer to sample plots and the numbers to the 'phasic communities'. Based on Anderson (1964, Fig. 2).

Many of the characteristics of the six zonal communities (summarized in Table 14.1) vary in a regular fashion. For instance, the number of trees per unit area rises to a maximum of 240–280 per 0.5 acre (0.2 ha) in the low pole forest (Zone 5) and falls to a minimum of 60 in Zone 6. The average height, diameter and basal area of the trees are greatest in Zone 3 and least in Zone 6. The number of species per unit area is maximal in the mixed swamp forest and falls to a minimum of six in the central zone.

An interesting feature is the increasing abundance of the insectivorous *Nepenthes* spp. from the outer towards the inner zones: this is probably mainly a result of increasing light intensity near ground level but may also be connected with increasing nutrient deficiency, as may also be the abundance of *Dischidia* spp. and other myrmecophytes in Zone 6.

Anderson (1963) records a total of 242 tree species (⩾30 cm girth) from the Sarawak and Brunei peat swamps. A few may prove to be confined to this habitat but most are also found in heath forest (pp. 305ff.) and other types of vegetation on nutrient-poor soils.

There is evidence from peat profiles to support the view that the zonal communities are stages in a hydrosere and it seems that the peat swamps developed from coastal mangrove forests or at their inland margins. Anderson (1964) suggests that toxic concentrations of sulphides and sodium which are often found in clay underlying mangrove swamps promoted peat accumulation by inhibiting the decomposition processes. The profiles showed the Baram peat swamps to be older than those of the Rejang delta.

The only peat profile from Sarawak in which the pollen content has so far been examined in detail is from Lubok Pasir in the Baram valley (Anderson 1964; see also Muller 1972). At the base (13 m), which is above present sea-level, there is an abundance of pollen of the mangroves *Rhizophora* and *Bruguiera* and of the palm *Nypa* (pp. 373ff.), followed at a higher level by an association of *Campnosperma coriaceum* with the palms *Cyrtostachys renda* and *Salacca*; this can be compared with the community dominated by the palm *Oncosperma tigillarium* which often occurs at the landward margin of present-day mangrove forests. Above this layer *Shorea albida*, *Combretocarpus* and other moor forest species are found, the peaks of pollen abundance following more or less the same order as their zonal distribution in the surface vegetation.

Radio-carbon dating shows that the peat at the base of the profile is 4270 ± 70 years old. From this and other data the rate of peat accumulation has been calculated as 47.7 cm per 100 years at 13 m depth and 22.2 cm at 5 m (age 2255 ± 60 years).

Table 14.1. *General features of zonal communites in moor (peat swamp) forests of Brunei and Sarawak*

	Zone 1	Zone 2	Zone 3	Zone 4	Zone 5	Zone 6
	Mixed swamp forest	Alan forest	Alan bunga forest	High pole forest	Low pole forest	Padang keruntum
	Gonystylus–Dactylocladus–Neoscortechinia association	*Shorea albida–Gonystylus–Stemonurus* association	*Shorea albida* association	*Shorea albida–Litsea palustris–Parastemon* association	*Tristania–Parastemon–Palaquium* association	*Combretocarpus–Dactylocladus* association
Trees						
Species per 0.5 acre (0.2 ha)						
Range	35–55	30–55	10–20	19–26	31–18	3
Average	46	42	14	22	—	—
Individuals per 0.5 acre (0.2 ha)						
Range	120–160	120–155	65–125	135–185	240–280	60
Average	135	131	102	153	—	—
Height of canopy (A storey)						
ft	120–150	120–170	160–190	90–120	40–70	40–50
m	36.6–45.7	36.6–51.8	48.8–57.9	27.4–36.6	12.2–21.3	12.2–15.2
Basal area per 0.5 acre (0.2 ha) (average)						
ft²	73	98	108	92	50–70	18
m²	6.78	9.10	10.03	8.55	4.65–6.50	1.67
Undergrowth	Abundant	Fairly abundant	Variable	Abundant	Sparse	Abundant
Lianes (large)	Abundant	Frequent	Very rare	Very rare	Absent	Absent
Epiphytes	Abundant	Frequent	Sun epiphytes rare: shade epiphytes occasional	Rare	Rare	Sun epiphytes abundant
Foliage		*Mesomorphic*			*Xeromorphic*	

Source: After Anderson (1961b), modified.

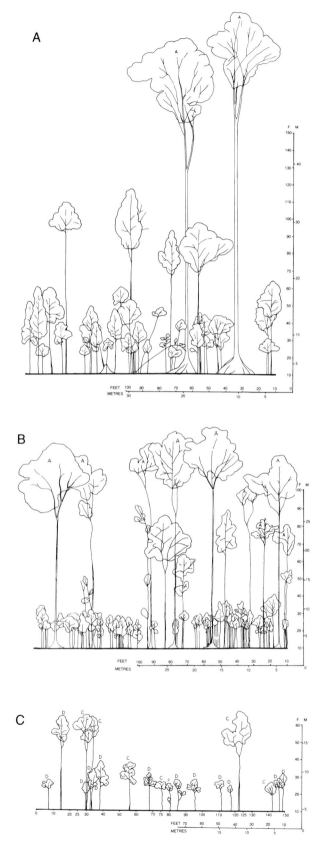

Anderson believes that the development of the peat swamps had four phases. (1) Colonization of deltas, shallow bays etc. by mangroves occurred, followed by a succession to 'intermediate communities'. (2) As the coast receded seawards, 'perimeter swamps' (perhaps similar to existing mixed swamp forests) replaced the mangroves; at the same time the rivers backed up, building levees of alluvium which were higher than the base of the original swamps and allowed peat to form in saucer-shaped depressions behind them. (3) Peat accumulation slowed down, giving the typical bog plain occupied mainly by stands of *Shorea albida*. (4) The bog plain extended and the rate of peat formation continued to diminish; simultaneously there was a rapid seaward retreat of the mangrove. The course of development may be checked by erosion: in 1961 it undermined part of the Baram levée 'causing half a mile of peat swamp to collapse' (Anderson 1964, p. 14).

A general account of the moor forests of the Malay Peninsula has been given by Wyatt-Smith (1959). They are less extensive than those of Borneo but peat 0.5 m or more deep occupies an area of some 3000 km². *Shorea albida*, which is so abundant in the peat swamps of Borneo, is absent. The clay underlying the moor forests on the west coast of the peninsula contains marine diatoms and remains of higher plants, indicating that they developed from mangrove forests. This may not be true of those on the east coast, most of which have developed in deep freshwater lagoons impounded between a series of raised beaches.

From this brief account it can be seen that there are many resemblances between tropical moor forests and temperate ombrogenous bogs, as Polak originally suggested. The peat of tropical moor forests is mainly composed of wood fragments and differs in structure and texture from most temperate peats. The successions involved are different. There are also resemblances between tropical moor forests and the 'cypress domes' in the *Taxodium* swamps of the southeastern United States. In the latter the water is oligotrophic and peat formation takes place. In the large Okefenokee swamp (Florida and Georgia) the peat is shallow near the edge and over 6 m thick near the centre (Cypert 1972).

Fig. 14.14 Profile diagrams of 'phasic' peat swamp communities. A Alan forest (phasic community 2) near Badas, Sarawak. Trees marked A are *Shorea albida*. There are about 14 other species in the diagram. B Alan bunga zone (phasic community 3), near Badas, Sarawak. Trees marked A are *Shorea albida*. About 12 other species in the diagram. C Padang keruntum (phasic community 6), near Marudi, Sarawak (cf. Fig. 14.12 D). C *Combretocarpus rotundatus*, D *Dactylocladus stenostachys*, P *Parastemon versteeghii*.

Why moor forests like those of Borneo and the neighbouring countries have not developed elsewhere in the humid tropics is not clear. It may be that in addition to a coastal plain with high rainfall and rivers with oligotrophic water other factors are required. If Anderson's suggestion of the importance of sulphide toxicity for initiating peat formation is correct, then nearness to an extending coastline with extensive mangrove forests may be another factor, though probably not an essential one.

Tropical moor forests are of great biological interest as examples of ecosystems existing under extreme oligotrophic conditions in which their survival must depend on a very efficient recycling of nutrient elements. The *Shorea albida* associations of the Borneo peat swamps, especially Zone 3, are among the purest single-dominant tropical rain-forest communities known (apart from short-lived seral stands of *Cecropia, Musanga*, etc.) and it is interesting that, like artificial monocultures, they are prone to damage by wind, lightning and insect pests (Anderson 1961a, Brünig 1964, Brünig & Huang 1989).

These remarkable forests have become of increasing economic importance. Though they occupy land which probably has very little agricultural potential, they are a valuable source of ramin (*Gonystylus bancanus*) and other useful timbers, which has been exploited on a large scale in recent years.

Chapter 15

Mangroves and other coastal vegetation

On tropical coasts the starting point of succession is a marine deposit such as mud, sand or coral fragments on which land plants can establish themselves. As in other primary successions there is a progression from conditions under which only a few, in this case salt-tolerant, plants (halophytes) can grow to less extreme conditions suitable for a much greater range of species. In estuaries and on coasts sheltered from wave action the first colonists are mangroves, trees which can establish themselves between the tidemarks and can survive long periods of immersion in salt or brackish water. On more exposed sandy shores colonization by herbaceous plants begins at or above the high tide mark: woody plants appear later and in time a distinctive community of evergreen trees, littoral woodland, may develop. Where the shore is not too steeply sloping, whether it is sheltered or exposed, the plant communities tend to form zones more or less parallel to the shore. If accretion is active and the coastline is advancing, the zones may represent actual stages in a succession, the early stages being the zones nearest and the later those further inland. However, zones of vegetation are also found on coasts that are stable or receding; the zones then represent a potential rather than an actual succession.

In the zones closest to the sea, factors such as wave, wind and tidal action, the instability of the substratum, and salinity have great effects on plant life. Further inland these factors become less important and the vegetation becomes more similar to the climatic climax of the region. Because they are so much influenced by salinity and other factors not dependent on climate, coastal plant communities tend to be very similar throughout the tropical zone and even beyond its limits.

Thus mangrove vegetation, although it becomes reduced in stature and poorer in species near its climatic limits, is found as far north as Bermuda (32 °18'N), the Red Sea (to ca. 29 °N) and southern Japan (to ca. 31 °N); southwards it extends to northern New Zealand and to 38 °45'S in Australia (Tomlinson 1986). It also ranges from ever-wet rain-forest climates to arid desert conditions. Similarly the 'Pes-caprae formation' of sandy beaches (see below) is found from the equator to far outside the tropics and in both eastern and western hemispheres.

In his classic memoir on the Indo-Malayan strand flora Schimper (1891) showed that two of the communities he described, the Mangrove and the Nypa 'formations', could begin to develop as far down as the low tide mark, while the Pes-caprae and the Barringtonia formations could become established only above the level of normal high tides. Schimper later described (1903) counterparts of all four Indo-Malayan formations in the western hemisphere. It is still convenient to follow Schimper's divisions, though in intermediate habitats mixed communities are found. Some types of tropical coastal vegetation are not included in this classification, e.g. the salt marshes colonized by *Batis*, *Suaeda* and other succulent halophytes; these and the vegetation of rocky shores and coral reefs as well as communities of algae and sea grasses (*Halophila*, etc.) will not be dealt with here.

15.1 Mangrove vegetation (mangal)

The word 'mangrove' is often used both for an ecological group of plants found in certain kinds of coastal habitat, chiefly in the tropics, and for communities formed by

Fig. 15.1 Pioneer mangrove, Leiden island, Djakarta (Batavia), Djawa (Java). (Photo. W.M. Docters van Leeuwen.) Showing a mature tree and seedlings of *Rhizophora stylosa*.

Fig. 15.2 Mature stand of *Rhizophora mucronata* at low tide, Aroe Bay, East Sumatera. The trees are *ca.* 40 m high. (Photo. Schreuder.)

them. In the second sense the word 'mangal' is now being generally adopted. The biological interest and economic importance of mangroves are so great that a vast literature has grown up: the UNESCO bibliography on mangroves (Rollet 1981) includes over 5600 entries and mangal is now one of the most studied of all tropical ecosystems. In this chapter only a few aspects of mangrove ecology which are particularly relevant to the main themes of this book can be discussed. Fuller information is available in the works of van Steenis (1958b), MacNae (1968), Chapman (1976), Lugo & Snedaker (1974), Teas (1983) and Tomlinson (1986).

15.1.1 Characteristics of mangroves

Mangroves, with the exception of the palm *Nypa fruticans* and the large fern *Acrostichum aureum*, are dicotyledonous trees and shrubs. According to Tomlinson (1986), 'true mangroves' world-wide include about fifty-four species, of which thirty-four are important components of mangal. They belong to sixteen families and are one of the most remarkable examples among plants of epharmonic convergence between unrelated taxa living in similar environments. Associated with 'true mangroves' are about sixty other flowering plants, some of them herbs, with similar but less strongly marked characteristics, the 'semi-mangroves' of Tansley & Fritsch (1905). These are found mainly in the 'back-mangal', on the landward side of the true mangal.

Mangroves are an ancient group. There is evidence from macrofossils and pollen that several existing genera of mangroves had evolved by the early Tertiary (p. 13 above; see also references in Ellison 1991). *Nypa* was present in the eastern tropics in the Maastrichtian (late Cretaceous) (Muller 1964, 1981) as well as in Europe in the Lower Eocene (p. 14).

The morphological and anatomical features of mangroves, especially of their root systems, have often been described (Schimper 1891, 1935, Chapman 1976, Tomlinson 1986). Most mangroves have aerial roots (pneumorhizae, pneumatophores) (see Chapter 4), which are above water at least at low tides. These are like those found in many tropical freshwater swamp forest trees (p. 361, above) and take various forms, e.g. stilt-roots or flying buttresses, (*Rhizophora*), fluted buttresses (*Pelliciera*), knee roots (*Bruguiera*), or rows of short erect branches from horizontal lateral roots (*Avicennia*, *Sonneratia*) (Figs. 15.1–15.3). In the bark of all these types of aerial roots there are numerous lenticels: through these, gases can be exchanged between the outside atmosphere and the system of gas spaces which extends throughout the plant, including the absorbing

Fig. 15.3 A Pneumorhizae of *Avicennia germinans* on shore near Lagos, Nigeria. (Photo. D.E. Coombe.) B Root system of *Avicennia marina* on eroding shore, Pantai Acheh, Penang, Malaysia, showing origin of pneumorhizae from lateral roots.

roots buried in a more or less anaerobic environment below the soil surface.

In addition to allowing atmospheric oxygen to reach the active roots, the pneumorhizae of mangroves have another important function: to enable the root system to keep pace with accretion (Troll 1930, Troll & Dragendorf 1932). In many mangals the soil level is continually rising, at a locality in Sumatera where Troll worked, accretion was estimated at not less than 15–35 mm per year. In *Bruguiera* and *Sonneratia* the absorptive rootlets are borne mainly on the underground parts of the pneumorhizae (Fig. 15.4). As the soil level rises these grow upwards and produce new rootlets at successively higher levels, thus keeping the active part of the root system at a constant depth below the surface. Comparable mechanisms for adjustment to accretion are found in other mangroves.

One of the most important physiological characteristics of mangroves, in addition to their ability to grow on poorly aerated unconsolidated mud and sand, is their tolerance of salt or brackish water (although where the salt concentration is very high they become dwarfed). They seem to be facultative rather than obligate halophytes, because as well as being able to grow on soils with high salt concentrations some mangroves are occasionally found in freshwater lakes. They often

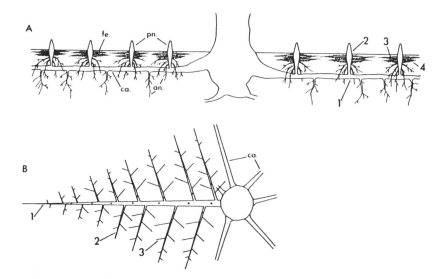

Fig. 15.4 Root system of *Sonneratia*. After Troll & Dragendorf (1931). A Profile, showing cable (lateral) roots (ca), anchoring roots (an) and feeding roots (fe). B Plan of lateral roots, showing branches of first (1), second (2) and third (3) order. The order does not determine the type of root.

ascend large rivers for some distance: on the Fly River (New Guinea) Brass found *Bruguiera* and *Sonneratia* 320 km upstream (van Steenis 1958b). They can be cultivated in normal non-saline soils. Their mechanism for regulating their internal concentrations of Na^+ and Cl^- ions has been much studied but is not yet completely understood (see Scholander 1968 and the discussion in Tomlinson 1986). Salt tolerance seems to vary between different mangrove species. *Avicennia* and some other mangroves have salt-excreting glands on the leaves, but most mangroves do not.

The reproductive biology of mangroves has several characteristic features. *Bruguiera* and *Rhizophora* are viviparous, the embryo developing into a large seedling (in *Rhizophora mucronata* over 70 cm long) before it is shed. Other mangroves (e.g. *Avicennia*, *Aegiceras*, *Nypa*) are 'cryptoviviparous': in these the developing embryo emerges from the seed-coat, but not from the fruit, before the fruit is shed. Various explanations have been suggested for the frequency of vivipary and cryptovivipary among mangroves, but their adaptive significance is still not entirely clear; it seems likely that they are connected with the problems of seedling establishment in an unstable tidal environment (see discussion in Tomlinson 1986). The propagules of all mangroves, including the large seedlings of the viviparous species, are dispersed by water and are subject to the hazards of tides and currents.

Mangroves colonize newly deposited mud and sand and might be expected to have a life strategy similar to that of pioneer trees. However, as Tomlinson (1986, Table 1.2) has shown they do not fit easily into the conventional *r* and *K*-selected categories. For example, like other pioneers they begin to reproduce early in life and produce propagules nearly continuously and usually in large numbers but they are hard-wooded and probably relatively long-lived like the dominants of mature rain forests.

15.1.2 Mangrove communities

Mangrove ecosystems are complexes of communities at various stages of development. Until recently they covered enormous areas in most parts of the tropics and subtropics, but in the past 100–200 years much mangal has been converted to rice cultivation and other forms of land use. The largest areas still remaining in a natural state are probably those in New Guinea. Mangal varies in stature from low scrub to forest over 30 m high. It is at once recognizable from the air or at ground level by its monotonous shiny dark green foliage and by its conspicuous aerial roots. In structure, as in floristic composition, it is very different from terra firme rain forest, though as might be expected it is less unlike some types of freshwater swamp forest.

Mangal is characteristically single-storeyed; in mature stands the undergrowth consists entirely of seedling and sapling mangroves. As mentioned above, nearly all mangroves are woody dicotyledons. In Malesia *Nypa fruticans*, the only palm that is a 'true mangrove', forms extensive pure stands, particularly in estuaries. Other palms, including rattans, may be found in the back-mangal. The tall herb *Acanthus ilicifolius* is an important semi-mangrove in the eastern tropics. Epiphytes, and sometimes lianes, occur in mangal and back-mangal.

Stands of pioneer and mature mangal may consist of only a few, sometimes only one, species. Mangrove vegetation, as mentioned on p. 372, extends far outside the tropics but reaches its greatest luxuriance and species richness where the tropical rain forest is the climatic climax, especially in western Malesia.

Floristically, two mangal formations can be recognized, the Western (Atlantic and Pacific coasts of Central and South America and West Africa) and the Eastern (Indian and western Pacific Oceans). The Eastern formation is much richer floristically (fifteen genera and over forty species of 'true mangroves') than the Western, which has only six genera and eight species. The genera *Avicennia* and *Rhizophora* are represented in both formations, but by different species. Most mangrove species are widely distributed, but the remarkable mangrove *Pelliciera rhizophorae* (Fig. 15.6) is confined to the Pacific coast from Costa Rica to western Colombia and a few small relict populations on the Caribbean coast (Jiménez 1984). The only mangrove occurring in both Western and Eastern formations is *Rhizophora mangle sensu lato*, a western species which mixes with eastern species in Fiji, New Caledonia and Tonga (van Steenis 1962b).[1] The reason why eastern and western mangroves are unable to cross the gap separating them near the Cape of Good Hope is probably because their dispersal is wholly dependent on marine currents and their seedlings are intolerant of sea surface temperatures below 24 °C (or 27 °C in some regions; see Tomlinson 1986, p. 57). Established trees can withstand much lower temperatures but cannot survive severe frosts.

Mangal is sometimes referred to as 'tidal forest'. Mangroves can grow from below the level of the lowest tides to above the level of the highest, but they can also grow in seas where there are no tides and, as noted

---

[1] It has been suggested that in Tonga *Rhizophora mangle* had been accidentally introduced in ballast, but fossil pollen of this species has been found in peat pre-dating the arrival of man by some 2500 years (Ellison 1991).

earlier, they penetrate long distances up large tropical rivers. Typically, mangal is found on sheltered muddy shores where the land is extending seaward by accretion. Mangroves do not initiate accretion, but where it is occurring their root systems help to stabilize the sediments and also add organic material to them.

Mangals are unusual ecosystems because of their links with both sea and land. They receive nutrients from sea water and from rivers and land drainage and have a high productivity. They consists mainly of evergreen trees which produce and shed leaves throughout the year (Tomlinson 1986). The annual production of leaf litter in western mangals is estimated at about 8 t ha^{-1} a^{-1} (Snedaker 1978). Much of this litter does not remain in situ, but is removed by ebb tides. In the lower zones, which are frequently flooded, the litter, after being carried away, is broken down into small particles by fungi and other organisms, providing food for a complex web of organisms ranging from small invertebrates to fish. In the upper zones, which are flooded less often, the dead leaves usually decay in situ and form soluble organic compounds, which are also eventually flushed away (Snedaker 1978), although in some situations, such as land-locked lagoon, the litter accumulates to form peat.

In contrast to terra firme rain forests, mangals thus have 'open' nutrient cycles. Because of their large contribution to the fertility of tidal waters, mangroves are of great importance for inshore and estuarine fisheries for prawns, shrimps and sea fish.

Data on the biomass and productivity of mangals are given by Lugo & Snedaker (1974) and Putz & Chan (1986). On protected plots of mature mangal on an island near Port Weld (west coast of Malaya) the latter authors monitored the growth of *Rhizophora apiculata* from 1920 to 1981, and of all trees from 1950 to 1981. The net primary productivity was 17.7 t ha^{-1} a^{-1} over the observation period and the biomass (above ground) ranged from 270 to 460 t ha^{-1}. Both figures are higher than those reported for other mangals, but comparisons must be made with caution because of differences in the maturity of the stands and in the techniques used.

15.1.3 Zonation and succession of mangrove communities

A striking feature of large mangal stands is the occurrence of the dominant species in zones more or less parallel to the shoreline. At their seaward margin there is usually a pioneer zone of young trees, often species of *Avicennia*, *Rhizophora* or *Sonneratia*, which are partly submerged by medium high tides and are colonizing bare mud or sand (Fig. 15.1). Inland from this there is

a transition to tall mangrove (forest) with different species arranged zonally (Fig. 15.1). On the landward side of this is the 'back-mangal' in which semi-mangroves such as *Conocarpus erectus* and *Exocoecaria agallocha* are mixed with true mangroves. Still further inland, out of the reach of normal tides, there is, in undisturbed areas, freshwater swamp forest or other types of non-marine vegetation. The zonation appears to be determined mainly by the land level and by the frequency, depth and duration of tidal flooding, although gradients of salinity, differences in soil texture and other factors may be involved.

It is often assumed that the zones represent stages in an actual or potential succession, but detailed studies show that many factors are involved and that there is no exact equivalence between the zonation and succession in time. The subject cannot be fully discussed here, but some insight into the problem can be obtained by focusing attention on a few areas which have been intensively studied.

Florida and tropical America. The mangrove vegetation of southern Florida, which lies just outside the tropics and close to its temperature limit, has been much studied. It formerly covered a large area, especially on the west coast, but much of it has now been destroyed by urban and other development. The complex pattern of mangal communities due to different frequencies of tidal flooding, variation in the substratum and the salinity of the ground water has been further complicated by anthropogenic fires and several kinds of natural disturbance, as Craighead (1971) has vividly described, especially recurrent hurricanes, in which large areas of mangal have been killed or severely damaged by the deposition of large masses of impervious mud and by salinity changes, as well as by wind. Alligators, which were formerly abundant, were also responsible for considerable disturbance. These changes have been superimposed on a rise of sea-level over a very long period, estimated at *ca.* 0.6 mm per year (Egler 1952).

In some parts of the Caribbean region four fairly clearly recognizable mangal zones are found, dominated, from the shore inland, by *Rhizophora mangle*, *Avicennia germinans*, *Laguncularia racemosa* and *Conocarpus erectus*. In the south Florida Everglades, according to Egler (1952), it would be more accurate to speak of a 'dragged out gradation from open sea water to freshwater swamp' than of a zonation.

Davis (1940), in a very detailed account of the south Florida mangrove vegetation, distinguished the following communities: (1) pioneer *Rhizophora* 'families', (2) a mature *Rhizophora* consocies, (3) an *Avicennia* salt-

Fig. 15.5 Mangroves, road to Flamingo, Everglades, Florida (1965). A Stand of *Rhizophora mangle*. B Small bushes of *Rhizophora* among *Cladium jamaicense*.

marsh associes, (4) a *Conocarpus* transition associes, and (5) mature mangrove forest. The ground level rises from (1) to (5). In the pioneer zone the vegetation is almost continually under water, but in the other communities the frequency of submergence diminishes; in (3), (4) and (5) tidal flooding is quite infrequent. The frequency of flooding is linked with differences in salinity (Table 15.1); in the pioneer *Rhizophora* 'families' the salinity varies very little from that of the surrounding sea water, but in the other communities it rises to high values between tides and falls very low after heavy rain.

Davis regarded the five main mangal communities as seral stages. The main line of succession, in his view, was from pioneer *Rhizophora* through the mature *Rhizophora* consocies to *Avicennia* salt-marsh and the *Conocarpus* associes to the subtropical 'hammock forest', the climatic climax of the region. The 'mature mangrove forest' he considered to be a stable 'subclimax' community which, in the absence of disturbance, was stable

and not part of the normal succession from open sea water hammock forest.

Davis' opinion that the Florida mangal communities were successionally related in time was supported by very little objective evidence and in the light of later work by Egler (1952), Craighead (1971), Lugo (1980) and others it is clear that the real relations of the communities must be much more complex.

In some places *Rhizophora* can certainly be observed colonizing unconsolidated sediments and the pioneer families gradually develop into mature *Rhizophora* stands, but this happens only on prograding shores, where for reasons unconnected with mangrove colonization accretion is taking place. The vegetational changes are therefore not an autogenic succession in Tansley's sense. The well-established fact, mentioned above, that in south Florida sea-level is slowly rising cannot be reconciled with a long-term successional trend such as that illustrated by Davis (1940). Successional changes in the Florida mangal are also much complicated by frequent disturbance by hurricanes; many mangrove stands are secondary and have replaced older stands which have been damaged or destroyed.

Egler (1952), in his study of the saline Everglades, argued convincingly that the observed changes in the vegetation are due to allogenic (external) factors. He pointed out that the level of fresh water in the Everglades has been falling in recent times owing to drainage and water abstraction, which has caused the upland vegetation to move seawards while the rise in sea-level is tending to push it inland, probably at a much slower rate. Egler regarded the mangal as forming a series of zones, each potentially stable, but actually in continual change owing to the effects of external factors.

On parts of the Atlantic and Pacific coasts of Central and South America where there is a high and well-distributed rainfall, large areas of mangal are found, for example, at the mouth of the Orinoco, Amazon and other large rivers and on the Pacific coast of Colombia, but large areas are being reclaimed for agriculture. Where undisturbed, the mature mangroves grow to 30 m or more high. In addition to *Rhizophora mangle*, *R. racemosa* and the putative hybrid, *R.* × *harrisonii*, which are absent in Florida, play an important part in these areas.

In tropical America *Rhizophora mangle* is generally the first colonist in sheltered situations, though in some places, especially where there is less shelter and the sediments are sandy, the pioneer is *Avicennia germinans*. Lindeman (1953) says that on accrescent shores in Surinam *Avicennia*, and sometimes also *Laguncularia*, are usually the pioneers, but where silting is very rapid the grass *Spartina brasiliensis* is the first colonist and is later replaced by *Avicennia* and *Laguncularia*. In the rain-forest regions generally the pioneers may be later replaced by complexes of mangrove species which locally on the Pacific coast may include *Pelliciera*. Well-defined belts of single species are not always evident.

At the landward margin of the mangal where the vegetation has remained relatively undisturbed, transitions can often be seen to swamp forest in which the ground water may be still tidal but is brackish or fresh. Thus in western Surinam tidal freshwater swamp forest dominated by *Mora excelsa* is found to the rear of the mangal (Lindeman 1953). In Guyana there is a transition from mangal to freshwater palm swamps. West (1956) described how in the very extensive mangal belt on the Pacific coast of Colombia between 2° and 3°N the semi-mangrove community dominated by *Conocarpus*

Fig. 15.6 Mangal of *Pelliciera rhizophorae* at low tide, Utria Sound, Utria National Park, Chocó, Colombia. (Photo. Alan Watson.) This species is found only on part of the Pacific coast of tropical America and in a few isolated stations on the Caribbean coast. It is one of the few mangroves with a restricted geographical distribution.

and *Acrostichum aureum* gradually gives way to freshwater swamp forest in which *Mora megistosperma* is dominant, associated with the palm *Euterpe cuatrecasana*. Similar transitions from mangal to pure stands of *Mora megistosperma* and *Pterocarpus officinalis* can be seen on the Osa Peninsula in Costa Rica. In none of these transitions is there evidence that an actual succession is taking place.

In pollen profiles from near Georgetown on the Guyana coast, van der Hammen (1974) found that in the early Quaternary (*ca.* 45 000 yr BP) mangal was replaced by freshwater swamp vegetation. Later (*ca.* 8600 yr BP) mangal returned. These changes were due to eustatic falls and rises in sea-level.

Mangal in West Africa. On the Atlantic coast of Africa the Western mangal formation is found on suitable sites from Senegal to Angola, but much of this coast is exposed to strong wave action and well-developed mangal is found only in the lagoons that fringe some stretches of coast and in the estuaries and deltas of large rivers such as the Cross River, Niger and Sanaga. In the Niger delta it still covers large areas. The West African mangrove forests have been exploited for firewood, pit props, etc. and, as in other parts of the tropics, much mangal has been drained and used for rice cultivation and other purposes. Schnell (1952a) and Lawson (1986) have given useful accounts of West African mangrove vegetation.

The mangrove species in West Africa are the same as those in tropical America, except that *Pelliciera* is absent. The palm *Nypa fruticans*, near relatives of which occurred in West Africa in the Tertiary, is not now native there but has been introduced in the Cross River estuary and elsewhere, where it is becoming naturalized. The zonation is more or less similar to that in America, but *Rhizophora racemosa*, not *R. mangle*, is usually the pioneer in coastal mangal. In West Africa the latter is found mainly at the inner margin of the *Rhizophora* zone where it grows to a height of only *ca.* 5 m (Keay 1953). The higher zones are dominated by *Avicennia, Laguncularia* and *Conocarpus* (Gledhill 1963). In closed lagoons *Rhizophora* spp. are generally absent.

In the Niger delta the mangal is extending seawards and *Rhizophora racemosa* is colonizing the mud flats. It becomes a tall tree, up to 40 m high, with a girth of up to 3 m. Seen from the air a striking feature of the mangal is that large trees are usually found only on the edge of the channels where silt is being actively deposited and that further away there is only a shrubby tangle which may be no more than 3 m high. Keay (1953, p. 12) attributed this to poor conditions of growth, but it is possible, as Chapman (1976) suggests, that this is at least partly due to exploitation and that the tangles are secondary communities. The details of the succession have not been worked out, but the *Rhizophora* communities are probably often replaced by *Avicennia germinans* and other mangroves when the ground level becomes high enough and tidal flooding becomes infrequent.

At the landward margin of the mangal in West Africa the vegetation is in most places much disturbed but under natural conditions transitions can be seen in some places to a brackish or freshwater swamp community dominated by *Pandanus candelabrum* (*sensu lato*) in which palms (*Phoenix reclinata*, *Raphia* spp. and climbing species (probably of several genera) are common. Elsewhere the mangal passes landwards into seasonally flooded grasslands or salt-marsh (Kunkel 1965b). Though these transitions may in some places represent an actual successional change, there is at present no objective evidence that this is so.

The vegetation fringing the estuaries of large West African rivers shows a change from mangal to brackish and freshwater communities, similar to that seen in the Guianas and elsewhere in South America, as the salinity of the water decreases. Gledhill (1963) refers to the 'succession' (longitudinal zonation) from the coast upwards in Sierra Leone. First *Avicennia*, then *Rhizophora*, followed by *Machaerium lunatum*, *Raphia* with *Pandanus* and finally *Raphia* alone, dominate the vegetation on the river banks. In some places in Africa, as in other parts of the tropics (see above), *Rhizophora* grows in quite fresh water, e.g. near the mouth of the Zaïre (Pynaert 1933).

Mangal in Malesia and Australia. The Eastern mangal formation is found on suitable coasts from the Red Sea and East Africa to the Western Pacific and southwards to Australia and New Zealand. Because the Eastern formation is much richer in mangrove species than the Western, its community structure is generally more complex. Where mangal is extensive, a considerable number of zones or communities can be recognized and the successional relationships are correspondingly complicated. It is on the Malay Peninsula and the neighbouring islands that the Eastern mangal formation reaches its optimum development. Here mangrove trees grow to a height of 40 m or more and the number of mangrove species is greatest.

On the east coast of the Peninsula, which is exposed to the northeast monsoon, mangal is restricted mainly to estuaries and lagoons, but on the more sheltered west coast until quite recently it formed an almost continuous

belt, in places nearly 20 km wide. Much of this has now been replaced by coconut plantations, etc., but some large areas of mangal still remain. The Malayan mangrove forests have for many years been carefully managed for timber and firewood production and it is doubtful how much of them is now in a natural condition. Watson's account (1928), which is summarized here, is probably still the best description of mangal in the eastern tropics, although it now needs some modifications.

Watson lists some seventeen 'principal' and twenty-three 'subsidiary' mangrove species. All these generally grow in fairly well-marked zones. A 'typical' distribution is illustrated in Fig. 15.7, but other distribution patterns are sometimes found. The following main types of mangrove communities can be recognized, but not all of them are necessarily present even in extensive mangals.

(1) The *Avicennia – Sonneratia caseolaris*[2] type. In Malaya the pioneers are not species of *Rhizophora*, but *Avicennia alba* and *A. marina*, or sometimes, on deep mud rich in organic matter, *Sonneratia caseolaris*. These pioneer stands establish themselves on shoals or sandbanks out at sea which are exposed at neap tides, or along the seaward edge of existing forests. *Avicennia marina* grows on a comparatively firm clayey substratum which is easy to walk on, *A. alba* and *Sonneratia* on softer and blacker mud. On clay *Avicennia marina* is normally succeeded by *Bruguiera cylindrica*, but where *Sonneratia* is the pioneer, *Rhizophora mucronata* usually follows on.

(2) The *Bruguiera cylindrica* type. This occurs at a higher level than the preceding type and along the west coast of the Peninsula forms a nearly continuous pure belt behind the *Avicennia* forest, interrupted only by small stretches where *Avicennia* forest merges directly into *Rhizophora* forest. The soil is a firm stiff clay above the reach of most tides and is flooded only for a day or two before and after spring tides. This type is found chiefly on the sea-face and is usually absent both on shoals and in forests by rivers.

(3) The *Rhizophora* type. The dominants here are *Rhizophora apiculata* and *R. mucronata*. This type occurs on land flooded by ordinary high tides and therefore at a somewhat lower level than that occupied by the *Bruguiera cylindrica* type. There are many small streams which assist in the dispersal of the seedlings. In the Malay Peninsula *Rhizophora* forest covers a larger area than any other type of mangrove forest. Typically

---

[2] Watson's names are corrected here. According to the reviewer of MacNae (1968) in *Flora Malesiana Bulletin*, No. 24, 1969, p. 1855, 'Watson muddled with the names of the species [of *Sonneratia*] (his *alba* = *ovata* and his *griffithii* = *caseolaris*)'.

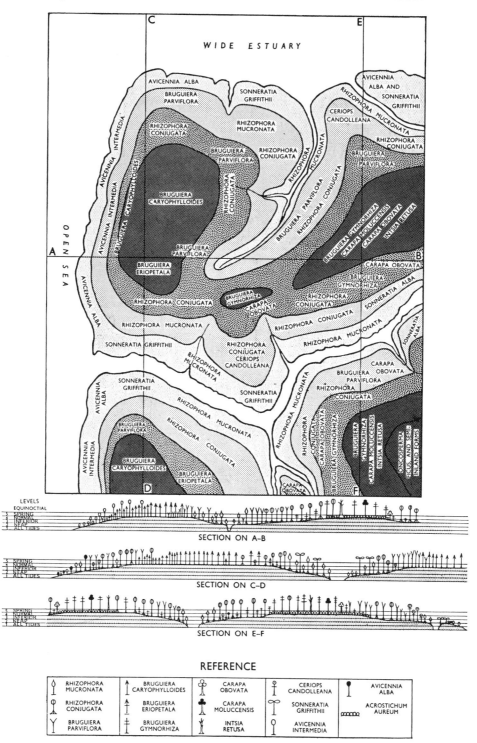

Fig. 15.7 Typical distribution of the more important species in the mangal of the Malay Peninsula (schematic). After Watson (1928). See footnote [2] on p. 379. Corrections to nomenclature of original diagram are as follows. *Avicennia intermedia*, now *A. marina*; *Bruguiera caryophylloides*, now *B. cylindrica*; *B. eriopetala*, now *B. hainesii*; *Carapa obovata*, now *Xylocarpus granatum*; *C. moluccensis*, now *C. mekongensis*; *Intsia retusa*, now *I. bijuga*; *Ceriops candolleana*, now *C. tagal*.

Table 15.1. *Inundation classes of Malaysian mangroves*

Inundation class Flooded by	Height above mean sea-level (Admiralty datum)		Days per month of tidal flooding	Typical species
	ft	m		
(1) All high tides	0–8	0–2.5	56–62	*Rhizophora mucronata* (sterile) or none
(2) Medium high tides	8–11	2.5–3.4	45–59	*Avicennia* and *Sonneratia* spp., *Rhizophora mucronata* (by streams)
(3) Normal high tides	11–13	3.4–4.0	20–45	*Rhizophora* spp., also *Bruguiera parviflora* and *Ceriops tagal*
(4) Spring high tides	13–15	4.0–4.6	2–20	*Bruguiera cylindrica* and other spp.
(5) Equinoctial and other abnormal high tides	15+	4.6+	0–2	*Bruguiera gymnorrhiza*, *Acrostichum aureum* and on landward side *Oncosperma filamentosa* (palm)

Source: Data of Watson (1928) for Kelang (Port Swettenham).

it grows on dark-coloured soil rich in humus and with an admixture of fine sand; it will not thrive on stiff clay, but is tolerant of quite dry soils and will maintain itself as far inland as any of the mangrove communities except the *Bruguiera gymnorrhiza* type. *Rhizophora apiculata* probably covers larger areas than *R. mucronata*, which prefers the wetter places such as the banks of streams and creeks. Associated species include *Bruguiera parviflora* and *Xylocarpus granatum*. The older *Rhizophora* stands are usually invaded by the fern *Acrostichum aureum*; this tends to establish itself on the mounds made by burrowing prawns, out of reach of any but the highest tides.

(4) The *Bruguiera parviflora* type. This type may grow as a pure stand in the wetter areas. Some believe that it precedes type (3), others that it follows it. Perhaps it may be regarded as an 'opportunist' community occupying either a higher or a lower position than the longer-lived and more shade-tolerant rhizophoras, which are liable to invade it where conditions favour the dispersal of their seedlings. It certainly often grows on areas formerly occupied by *Rhizophora*, but it may have colonized such areas after clear-felling.

(5) The *Bruguiera gymnorrhiza* type. This type is undoubtedly the last stage in the development of the mangrove forests and heralds the beginning of the transition to the inland rain forest. The dominant species is the largest and probably the longest-lived of the Rhizophoraceae; in the earlier stages of the succession its occurrence is markedly sporadic. It is very tolerant of shade and can establish itself in *Rhizophora* communities where the shade is too intense for the *Rhizophora* itself to regenerate. Its own seedlings, however, do not do

well under the shade of the parent tree. The undergrowth consists mainly of *Acrostichum aureum*. The ground level has by now been raised by deposits of sediments and humus, and by the activities of prawns, till it is permanently above the reach of even the highest tides. The transition to the inland forest is marked by the invasion of such species as *Xylocarpus moluccensis*, *Intsia bijuga*, *Ficus retusa*, *Daemonorops leptopus*, *Pandanus* spp. and many others, but the clearing of the land for agriculture often puts an end to the succession.

These communities can be subdivided. Chapman (1976) considers that as the *Avicennia* spp. and *Sonneratia caseolaris* often grow in pure stands and have different habitat preferences, the former usually growing on the firmer sediments of the sea-face and the latter on softer muds at river mouths, it is better to regard (1) as two communities, either of which can act as pioneers. Chapman also prefers to consider *Rhizophora apiculata* and *R. mucronata* as dominants of distinct communities.

Each mangrove community (and each species) seems to be adapted to a different tidal regime and consequently each tends to grow at different levels. Watson divided the Malayan mangroves into five 'inundation classes', based on tidal observations made near Port Klang (Port Swettenham) (Table 15.1).

The position of *Bruguiera cylindrica* in the zonation seems anomalous. It is found immediately behind pioneer stands of *Avicennia* but grows at a higher level than the *Rhizophora* spp., which are believed (and can sometimes be proved) to follow it in succession. The explanation of this is not entirely clear: Watson suggests that the structure of the root system of *B. cylindrica* leads to a great accumulation of organic debris and hence to a raising of the soil level. A system of streams subsequently develops draining *away* from the shore and these, by erosion, carry away the soil and lead to a fall in its level. When this subsidence has taken place a soil is left that is rich in humus and more suitable for the establishment of *Rhizophora* than before the colonization by *Bruguiera*.

The natural successions in Malayan mangal are by no means simple and they have been affected to an extent which is difficult to assess by both artificial and natural disturbances. The former include irregular exploitation for timber and firewood as well as large-scale silvicultural operations; natural disturbances include the effects of occasional severe flooding with fresh water (Watson 1937). There are many local variations in the succession but the main lines seem to be:

Avicennia–Sonneratia community (Type 1) → *Bruguiera cylindrica* community (Type 2) → *Rhizophora* community (Type 3) → *Bruguiera gymnorrhiza* community (Type 5)

Not all mangal areas show the whole developmental sequence.

According to Watson, the succession may lead eventually to the replacement of mangal by some type of non-halophytic rain forest. Transitions can indeed be observed in some places on open coasts from *Bruguiera gymnorrhiza* mangal to inland forest, but in estuaries and sheltered bays there are transitions from mangrove communities to pure stands of *Nypa* (Fig. 15.8) or to brackish swamp forests in which *Sonneratia* spp., *Heritiera littoralis* and the spiny palm *Oncosperma tigillaria* are common. These transitions reflect a gradient in environmental conditions: there is no convincing evidence that a change from mangal to a non-halophytic type of vegetation is actually proceeding.

On the other hand there is good stratigraphical evidence in both the Malay Peninsula and Borneo of replacement of mangal by peat swamps or other kinds of freshwater swamp forest. For instance, at Telok

Fig. 15.8 *Nypa fruticans* in Malayan mangal. A By estuary, Kelang (Port Swettenham) (1932). B With rain forest in background, Pantai Acheh, Penang (1974).

(Selangor) Webber (1954) found mangrove fragments and marine diatoms in clay below freshwater swamp forest 15 km inland from the present coastline; in Sarawak peat swamps overlie remains of mangrove communities (see p. 368 above). These changes seem to be due to long-term physiographic changes on a prograding coast, probably accompanied by a falling sea-level. Van Steenis (1958b) says that in some parts of Malesia the land is sinking and mangal has invaded and replaced rain forests. Where the coast is stable, mangrove forests have an all-aged structure and can maintain themselves permanently without developing into any other type of vegetation.

However, in the undisturbed mature mangal near Port Weld referred to on p. 376, Putz & Chan (1986) found much evidence of successional changes. During thirty-one years, many old trees of *Rhizophora apiculata* had died, but seedlings were abundant. The gaps were at first filled by *Derris uliginosa* (liane) and *Acanthus* spp (herbs). These were soon overtopped by young trees. *Rhizophora* declined in relative abundance and the more shade-tolerant *Bruguiera gymnorrhiza* and *B. parviflora* increased.

The Malayan mangal has been described at some length because its ecology, particularly its zonation, has been much studied. More extensive areas of mangal are found in other parts of the Indo-Malayan region. In India itself the largest area of mangrove vegetation is in the Sundabans (Ganges–Brahmaputra delta), but much of it has been destroyed or degraded by over-exploitation. In Sumatera and Borneo species-rich mangal, similar to that in Malaya, still occupies considerable areas.

In New Guinea mangal is estimated to cover over a million hectares; this is now probably much the largest area of unmodified mangal in the world. In the Purari delta alone there are believed to be some 136 000 ha of 'pure mangrove forest', as well as very large areas of mixed communities and back-mangal (Johnstone 1978). There are about thirty-seven species of mangrove. *Avicennia* and *Sonneratia* spp. are usually the pioneers. Seven principal zones of mangal can be recognized, but the zonation varies from one part of the coast to another and is often unclear (Percival & Womersley 1975).

A general survey of the mangal of Australasia is given by Chapman (1976) and Saenger *et al.* (1977). Mangroves are found where conditions are suitable in the islands of the western Pacific, in Australia and in northern New Zealand. They are absent on the arid south coast of South and Western Australia and reach their finest development in Queensland, especially near Cairns and Inisfail where there is a rain-forest climate

(Chapter 7). The number of mangrove species decreases with latitude from about thirty-six in tropical Queensland to the single species *Avicennia marina* var. *resinifera* at the climatic limit of mangroves in Victoria and New Zealand.

15.2 Vegetation of sandy shores

The eastern and western variants of Schimper's 'Pes-caprae formation' are the tropical equivalents of the 'strand vegetation' of temperate climates in which plants such as *Agropyron* spp and *Cakile maritima* are found; indeed, on the Caribbean coasts, and probably elsewhere, these two types of vegetation merge into each other. At least one sandy shore plant, *Salsola kali*, occurs in both tropical and temperate regions. The vegetation of sandy shores in the tropics has often been described: Asprey & Robbins (1953), van Steenis (1961b), Corner (1978), Lawson (1986), Lebrun (1969) and Whitmore (1984a) may be mentioned.

Pacific and Indian oceans. The typical 'Pes-caprae formation' of the eastern tropics is a zone of mainly low-growing herbaceous plants among which *Ipomoea pes-caprae* is often conspicuous but not always present. Other common species include dicotyledonous plants such as other *Ipomoea* spp., *Alternanthera maritima*, *Canavalia* spp. and *Sesuvium portulacastrum*, also grasses such as *Paspalum vaginatum*, *Spinifex* spp. and *Sporobolus virginicus*, and Cyperaceae. Many of the species have long runners like those of *I. pes-caprae*, which quickly cover the surface of the sand. Most of these shore plants are probably facultative halophytes and seem to be unharmed by occasional submergence

Fig. 15.9 *Ipomoea pes-caprae* on sandy shore near Limón, Costa Rica (1960).

in sea water during storms: like many halophytes, some of them are somewhat succulent. The seeds or fruits of most of them float in sea water and are dispersed by this means; they often germinate among the drift, which provides them with a fertile soil and a store of moisture.

The 'Pes-caprae formation' may be made up of several associations and on broad gently sloping beaches it may be zoned. In regions with a strong dry season, such as in Vietnam and on the south coast of Sri Lanka, sand may accumulate around the taller plants and dunes may be formed.

On most sandy tropical coasts, shrubs and trees increase in frequency with distance from the sea. There may be a gradual transition to woody vegetation, but in some places a belt of littoral woodland rises as a solid wall behind the low-growing vegetation nearer the foreshore. In the eastern tropics there is locally a zone of woodland, usually not more than 25–50 m wide, in which *Barringtonia asiatica* is a characteristic species, the 'Barringtonia formation' of Schimper. In Malaya this association is rarely seen in an undamaged condition, as the trees are often cut for firewood or cleared for planting coconuts, but in New Guinea and other western Pacific islands fine stands still remain. *B. asiatica* is a broad-crowned evergreen tree, which can grow to a height of 20 m or more. Some of its most common associates are *Calophyllum inophyllum, Hibiscus tiliaceus, Morinda citrifolia, Terminalia catappa* and *Thespesia populnea*. These all have broad evergreen leaves and resemble typical rain-forest trees except that their trunks are often gnarled or twisted and tend to lean seawards. Most of them rarely or never occur inland, except where planted; some at least are not tolerant of immersion in sea water. Burkill (1928) noted in Pahang (Malaya) that

Fig. 15.10 Dipterocarp forest on sea cliffs, Bako, Sarawak (1965).

after being flooded by an exceptional high tide the leaves of many trees in the *Barringtonia* zone turned yellow or dropped off.

The abundance of lianes and the density of the undergrowth sometimes makes mature *Barringtonia* woodland difficult to penetrate. Epiphytic ferns and orchids are common on the branches, especially in areas of high rainfall.

In some places a transition from the *Barringtonia* association to inland rain forest can be observed, but *Barringtonia* woodland also grows on sand bars behind which are swamps or lagoons. In the Micronesian islands forests of *Barringtonia racemosa*, *Hibiscus tiliaceus* and other littoral woodland species extend inland into swampy areas (Hosokawa 1957).

Casuarina equisetifolia, a tree in habit and morphology very unlike the members of the *Barringtonia* association, is very characteristic of sandy sea shores in the Indian and Pacific oceans and is also commonly planted throughout the tropics. It often grows in pure stands, especially on rapidly accreting shores and sand spits at river mouths. Because it is intolerant of shade and its seedlings have difficulty in establishing themselves on the litter under the parent tree (van Steenis 1961b) it is nomadic, behaving as a seral species rather than as a component of mature littoral woodlands.

Atlantic ocean. On the sandy shores of the Atlantic in tropical Africa and America two main zones of vegetation are found, the lower consisting chiefly of herbaceous plants and an upper of 'evergreen shrubs' or 'strand woodland' further from the sea. The former is is similar to the 'Pes-caprae formation' of the eastern tropics and the latter represents the '*Barringtonia* formation', although *Barringtonia* spp. are absent.

Lawson (1986) has given a general account of the vegetation of sandy shores in West Africa. He follows Boughey (1957) in dividing the lower zone into a 'pioneer zone' close to the shore and a 'main strand zone' at a higher level. The former has an unstable, highly saline substratum and the flora consists of a relatively small number of species which are adapted to extreme conditions. Some are stoloniferous (e.g. *Ipomoea pes-caprae*, *Canavalia rosea*) or rhizomatous, (e.g. *Cyperus maritimus*, *Remirea maritima*, *Sporobolus virginicus*). Some species are succulent (*Alternanthera maritima*, *Sesuvium portulacastrum*). In the 'main strand zone' the microclimate and soil conditions are less extreme (Jeník & Lawson 1968) and the flora includes many grasses and exotic weeds as well as some species of the 'pioneer zone'.

Further inland is the 'evergreen shrub zone', consisting of shrubs *ca.* 1–3 m high, severely pruned by wind and salt spray. The chief species are *Chrysobalanus orbicularis*, *Eugenia coronata*, *Sophora occidentalis* and the small palm *Phoenix reclinata* (Boughey 1957, Schnell 1977). Nearly everywhere on the West African coast this zone has been much disturbed by shifting cultivation of cassava and other crops and the existing 'evergreen shrub zone' is probably the much modified relics of former littoral woodland.

Lebrun (1969) has given a detailed account of the vegetation of sandy shores in Zaïre. Here, owing to the effects of the cold Benguela current, the rainfall is low (mean *ca.* 730 mm), but the beach flora is very similar to that of West Africa. The chief pioneers on the drift line are *Alternanthera maritima* and *Sporobolus maritima*, the *Ipomoea pes-caprae* community developing at a somewhat higher level; further from the shore line *Remirea maritima* and *Canavalia rosea* are dominant. Beyond this is a belt of littoral woodland consisting of two communities: on the exposed margin a wind-pruned thicket of *Chrysobalanus orbicularis* and *Dalbergia ecastophylla* and behind it a taller, relatively open woodland in which *Manilkara obovata* and the palm *Hyphaene guineensis* are characteristic trees. In the former, over half the species are confined to the coast and are probably salt-tolerant; in the latter, which can grow to a height of 10–15 m, only about 32% of the species are. Where the woodland has been destroyed the grass *Eragrostis prolifera* becomes dominant.

The vegetation of sandy shores in the American tropics has been described by Beard (1946a), Asprey & Robbins (1953) and many others. The zonation and floristic composition of the communities differ only slightly from those of shore vegetation on the west coast of Africa. Behind the herbaceous zones, where *Canavalia*, *Remirea* and the ubiquitous *Ipomoea pes-caprae* are characteristic, there is often a belt of scrub or low woodland. Towards the sea the shrub or small tree *Coccoloba uvifera* generally forms a belt which varies in width with the degree of wind exposure. *Chrysobalanus icaco* and *Hippomane mancinella* are also common. In Trinidad Beard (1946a) distinguishes two woody communities of sandy shores, an outer *Coccoloba–Hippomane* association forming a thicket and a taller association of *Manilkara bidentata* with the palm *Roystonea oleracea* found further inland. The latter is comparable to Lebrun's *Manilkara obovata–Hyphaene* association in Zaïre.

In Surinam (Lindeman 1953) *Coccoloba uvifera* and many of the species usually associated with it are absent and the zonation does not follow the pattern described above, probably cecause of some edaphic or climatic difference. *Hibiscus tiliaceus* is often dominant on sand bars above the level of high spring tides. This tree also occurs on the 'restingas' (consolidated sand ridges behind the shore) on the Brazilian coast. On these, communities dominated by *Alternanthera maritima, Iresine portulacoides*, etc. are found on the seaward side and stands of the dwarf palm *Allagoptera arenaria* with various small dicotyledonous trees further inland.

Part V

Tropical rain forest under limiting conditions

Chapter 16

Rain forest, deciduous forest and savanna

16.1 Classification and definitions

SCHIMPER distinguished four lowland tropical 'climatic formations' dominated by trees, which he defined as follows:

Rain forest is evergreen, hygrophilous in character, at least 30 m. high, but usually taller, rich in thick-stemmed lianes and in woody as well as herbaceous epiphytes.

Monsoon forest is more or less leafless during the dry season, especially towards its termination, is tropophilous in character, usually less lofty than the rain forest, rich in woody lianes, rich in herbaceous, but poor in woody epiphytes.

Savanna forest is more or less leafless during the dry season, rarely evergreen, is xerophilous in character, usually, often much, less than 20 m. high, park-like, very poor in underwood, lianes and epiphytes, rich in terrestrial herbs, especially grasses.

Thorn forest, as regards foliage and average height, resembles savanna forest, but is more xerophilous, is very rich in underwood and in slender-stemmed lianes, poor in terrestrial herbs, especially in grasses, and usually has no epiphytes. Thorn-plants are always plentiful.

In addition to these four lowland formations with a structure more or less that of forest or woodland, Schimper recognizes two others not dominated by trees: tropical grassland and tropical desert. The former occurs either as 'savanna' (grassland with evenly or unevenly scattered trees) or as 'steppe' (grassland without trees).

These six formations occur as analogous, but floristically different, communities in the tropics of America, Africa, Asia and Australia; Schimper regarded each as well-defined and readily recognizable by its physiognomy and structure.

With at least 180 cm. of rainfall, the high forest alone predominates. In regard to rainfalls of 150–180 cm. no data are available. With 90–150 cm. of rainfall there is a struggle between xerophilous woodland and grassland. Xerophilous woodland gains the victory when greater heat and more prolonged rainless periods prevail during the vegetative season; grassland succeeds when a milder temperature, a more even distribution of rainfall during the vegetative season, and windy, dry or frosty seasons prevail.

(Schimper 1898 (German), 1903, pp. 260ff.)

The six formations are each determined, in Schimper's view, by a distinct type of climate; within the tropical zone we can thus distinguish a rain-forest climate, a monsoon-forest climate, a grassland climate, etc.

Schimper recognized that edaphic factors could locally overcome or modify the influence of climate. For example, belts of evergreen forest (gallery or fringing forest) often penetrate far into regions of deciduous forest or savanna along rivers and streams (Figs. 16.5, 16.8, and 16.9). These are dependent mainly on permanent moisture in the soil, which compensates for the seasonal dryness of the climate.

Schimper's conceptions had the merit of simplicity and enabled much progress to be made in the description of tropical vegetation. From the ecological point of view their most serious defect was that one of the six climatic formations was not in fact climatic. Tropical grasslands, in which Schimper included both savannas with scattered trees and bushes and treeless 'steppes', occur under a wide range of climatic conditions and can be found in close proximity to luxuriant rain forest as well as to 'drier' (semideciduous and deciduous) types of forest. As Beard (1946b, 1953) pointed out, there is no such thing as a tropical grassland climate. It is, however, true that savanna grasses tolerate a lower annual rainfall than forest trees and are adapted to different water regimes (see Walter 1971, pp. 241–51). Most tropical grasslands and 'open savannas' are biotic climaxes maintained (though not necessarily originated) by recurrent fires. Some are probably edaphic climaxes due to soil conditions unfavourable to trees; others are hydroseral stages. Probably no lowland tropical grassland is a true climatic climax. It has long been evident that even in the most thinly populated parts of the tropics much of the vegetation has been greatly affected by burning, shifting cultivation and other human activities. In the present century these anthropogenic effects have vastly increased in extent and intensity.

Since Schimper's time there have been many attempts to improve the classification of tropical plant communities. The names of the plant formations as well as their classification have been much discussed but there is no generally agreed terminology. Even the term 'tropical rain forest' has been used in various senses. In this book, as in its predecessor of 1952, it is used fairly broadly, to include the evergreen seasonal forests of moderately seasonal tropical regions (Beard 1955a) as well as the 'true rain forest' of ever-wet climates. Some authors, notably Holdridge, use it in a more restrictive sense. In Holdridge's 'life zone system' (Holdridge *et al.* 1971) the tropical forest life-zones are defined by bioclimates (Chapter 7) rather than by the vegetation itself. Lowland tropical rain forest in his sense is found only in the relatively small regions such as Borneo and the Chocó of Colombia with a superwet climate. This excludes almost all the African rain forest (as generally understood) and a large part of the forests in tropical America and Asia which Schimper included in his definition. On the other hand, 'tropical rain forest' has sometimes been used in an extremely wide sense, especially in Australia (see p. 449), so as to include the strongly seasonal (or 'rain-green') forests classified by Schimper as monsoon forest. Some authorities believe that the name 'tropical rain forest' should be replaced by some other term, e.g. tropical ombrophilous forest (Ellenberg & Mueller-Dombois 1967), tropical evergreen forest (Champion & Seth 1968a) or complex mesophyll vine forest (Webb 1959). It must be admitted, however, that difficult as it may be to reach agreement on its use, the term tropical rain forest is so firmly established in scientific and popular usage that it is unlikely to disappear.

Savanna is another term which has been used in many senses and it is even harder to define than tropical rain forest (for varying views see Germain (1965), C.C.T.A. (1956), Walter (1971), White (1983), Bourlière & Hadley (1983) and Eiten (1986)). 'Savanna', like 'heath', 'moor' and 'prairie', is a vernacular word to which it is difficult to attach a precise scientific meaning. It is probably of Carib origin and in the West Indies it is used for almost any uncultivated non-forest vegetation, even for parks and grass-plots in towns, as well as for communities of tall grasses, etc., in swamps ('water savannas') which have little in common with the savannas we are concerned with here. Although some authorities, e.g. Lanjouw (1936) and Beard (1953), have tried to restrict the term to the Neotropics, its use has long since been extended to Africa and tropical Asia.

'Savanna' will be used here in a physiognomic sense for tropical vegetation in which trees and bushes in various patterns and densities are combined with a lower layer dominated by grasses. In some savannas the trees may be 10 m or more high and grow close enough together to form a not very dense canopy; more often they are scattered and less tall. There are all gradations from savannas to treeless grasslands and only an arbitrary distinction between them is possible. Sometimes the taller trees are in groups forming a park-like landscape; in 'orchard savannas' small crooked trees are spaced as in a derelict apple-orchard. Savanna trees (which include palms) are very different in physiognomy and biology from rain-forest trees (Chapters 4 and 5) and are generally more or less fire-tolerant (Fig. 16.2). Savannas, as will be seen later, have little in common floristically with rain forests. They also differ from them in their life-form spectrum, which has a smaller proportion of phanerophytes, a much larger proportion of hemi-cryptophytes and geophytes and also includes many therophytes (Fig. 16.1).

It should be recognized that entities as complex and variable as tropical plant formations can be defined only arbitrarily because transitions between them are so common. Their structure and composition should be described quantitatively as far as possible, but even the terms used are seldom precise. For example, as was seen in Chapter 9, 'evergreen' and 'deciduous' usually have to be used subjectively, as many tropical trees vary in their leaf-changing habits and others do not fall clearly into either category.

In this chapter the relations of tropical rain forest to 'drier' types of forest, and to savanna, in various parts of the humid tropics will be considered. It will be shown that in the tension belt near the climatic limits of the rain forest the landscape is often complex mosaics of

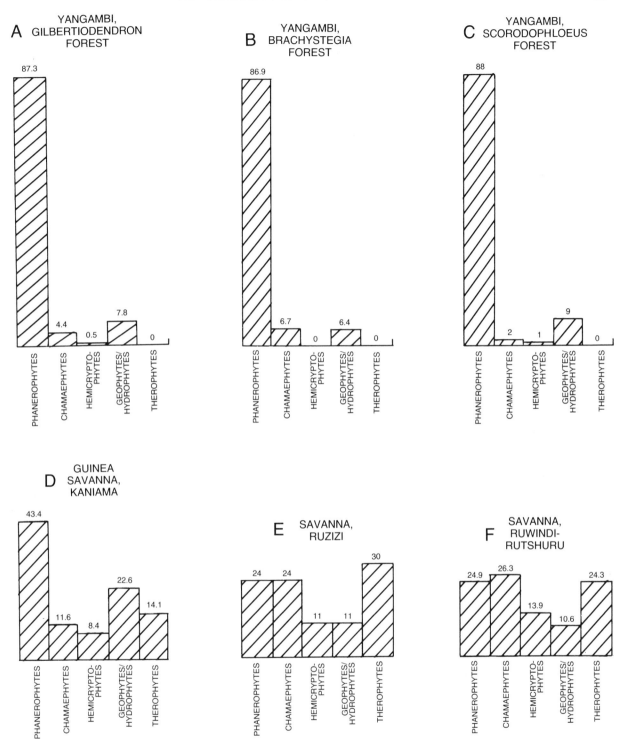

Fig. 16.1 Raunkiaer biological spectra for the flora of rain-forest and savanna regions in Zaïre. Data of Germain (1957). A Yangambi region (1°N), *Gilbertiodendron dewevrei* rain forest. B Yangambi region, *Brachystegia laurentii* rain forest. C Yangambi region, *Scorodophloeus zenkeri* (semideciduous) forest. D Kaniama region (7°30′S), Guinea savannas. E Ruzizi plain (3°S), Sudano-Zambezian savannas. F Rwindi–Rutshuru plain (1°S), Sudano-Zambezian savannas. Each column represents the number of species of vascular plants as a percentage of the total flora (A–C 2184 species, D 823 species, E 861 species, F 539 species).

A

B

C

Fig. 16.2 A Guinea savanna, Upper Ogun Reserve, Nigeria
(December 1977), before annual grass fire. Note habit of tree
and understorey of dry grass. B Isolated A storey tree in
cleared rain forest, Oban, Cross River State, Nigeria (1978).
C Trunk of large tree in Guinea savanna, Mole National Park,
northern Ghana (December 1967), showing thick bark and
effects of fir^

different plant communities, rather than a gradual or abrupt transition from one to another.

16.2 Rain forest, deciduous forest and savanna in tropical America

16.2.1 Forest types of Trinidad

The island of Trinidad, separated from the mainland of South America by straits 11 km wide, is particularly favourable for studying rain forest at its climatic limits. The island, though only 4543 km² in extent, is remarkably varied in soil and topography and there is also a wide range of climate. The annual rainfall ranges from over 2500 mm to less than 1200 mm, and (which is even more significant for the vegetation) there are large variations in its seasonal distribution, some parts of the island having over 100 mm in every month of the year and others a long annual dry season. In 1946, 23% of the area was reserved forest and, owing to the population having been sparse until comparatively recently, much of this forest was primary and not greatly modified by human activities.

The forest types of Trinidad were first described by Marshall (1934) and were later reclassified by Beard (1944a, 1946b, 1955a), who added a wealth of precise data about their structure, physiognomy and floristic composition, of which a short summary is given here. In the lowlands and 'lower montane zone' (foothills up to about 760 m (2500 ft)) Beard recognized four climatic formations, each of which appeared to be a separate climax and was represented by one or more associations or consociations: (i) lower montane rain forest, (ii) evergreen seasonal forest, (iii) semi-evergreen seasonal forest, and (iv) deciduous seasonal forest (Fig. 16.3).

These four types form a series. The first (which is found only on the Northern Range from 800 to 2500 ft (ca. 245–760 m), occupies areas of ever-wet climate, the last those with the longest and most severe dry season, and the other two areas with intermediate conditions. On the neighbouring Gulf Islands and on the mainland of Venezuela, though not in Trinidad itself, there are two still more xerophilous formations of the same series, namely thorn woodland and cactus scrub. In addition to these main climatic formations there is another, littoral woodland, determined by the local climatic conditions of the sea coast (Chapter 15), and a number of edaphic formations with which we are not here concerned.

The climatic requirements of the forest formations may be stated in general terms as follows.

	Lower montane rain forest	Evergreen seasonal forest	Semi-evergreen seasonal forest	Deciduous seasonal forest
Total annual rainfall (mm)	Over 1800	Over 1800	1300–1800	800–1300
Duration of dry season	None	3 months each with under 100 but over 50 mm	5 months each with under 100 but over 25 mm	5 months each with under 100 mm; two under 25 mm

Little is known of the climatic conditions in thorn woodland and cactus scrub. The relation of these formations to rainfall is considerably modified locally by soil con-

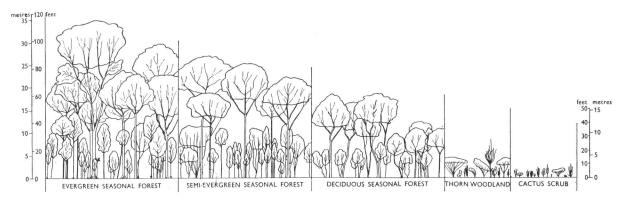

Fig. 16.3 Profile diagrams of climatic plant formations in lowlands of Trinidad and adjacent part of Venezuela. From Beard (1944a).

ditions (see p. 395); differences in the water-supplying capacity of the soil compensate to some extent for differences in rainfall.

(i) Lower montane rain forest. This closely resembles typical lowland rain forest in most of its features, but is dwarfed and somewhat modified by exposure to wind (Fig. 17.10A). Almost all the trees are evergreen. Tree species are numerous, but lianes and epiphytes are poorly developed. In Trinidad this formation is represented by only one association, the *Byrsonima–Licania* association of Beard.

(ii) Evergreen seasonal forest. This formation also differs only slightly from typical rain forest. It consists of three storeys of trees at *ca.* 30 m and upwards, 12–27 m and 3–9 m, respectively (Figs. 2.8 and 16.3), so that it is somewhat lower and less luxuriant than the best developed 'true' rain forest. Apart from the lower height of all the storeys, the highest tree storey is even more discontinuous. Individual trees may reach diameters of 3 m or more, but very large trees are not numerous. 'The general impression is of an occasional huge tree in the midst of smaller growth and the closely ranked columnar effect of rain forest is lacking' (Beard 1944a, p. 138). The larger trees branch relatively lower down than in 'true' rain forest and clean boles over 20 m in length are rare. This formation is mainly evergreen, but some large tree species are semideciduous[1], having leafless periods of a few days or weeks. In the highest storey about 17% of the species are deciduous and 17% semideciduous, but of the individuals only 3 and 0.4% are deciduous and semideciduous, respectively. The lower storeys are almost completely evergreen.

As in typical rain forest, buttressing is a prominent feature of many of the first-storey trees and the leaves of an overwhelming majority of the trees fall into Raunkiaer's 'mesophyll' size-class[2]. Lianes and epiphytes are fairly abundant. Beard recognizes four 'faciations' of the Trinidad evergreen seasonal forest; the species in them are more or less the same, but occur in different proportions. In one 'faciation' *Mora excelsa* is dominant; this community is regarded here as a distinct association (pp. 38–9).

(iii) Semi-evergreen seasonal forest. In semi-evergreen seasonal forest there are only two tree storeys, a layer at 20–26 m forming a more or less closed canopy and a lower layer at 6–14 m. Trees of large girth are rare, most of the mature trees averaging about 0.5 m in diam. Branching begins at a low level and the crowns tend to be umbrella-shaped. The upper storey consists of both evergreen and deciduous species; from 26 to 42% of the species and from 21 to 30% of the individuals, depending on the association, are regularly deciduous. Of the remaining trees in the first storey some are truly evergreen, but most are facultatively deciduous, the amount of leaf-fall varying from year to year, so that in a wet year most of the trees lose very few leaves, while in a severe dry season the whole canopy may appear leafless. The lower storey is mostly evergreen, but also includes a considerable number of deciduous species. The leaves of the upper storey are predominantly 'mesophylls', but in the lower storey there is a notable proportion of 'microphylls'.

There are few strongly buttressed trees. Thorns are not frequent, but some species in both tree stories have thorny trunks. Lianes are better developed than in any other formation in Trinidad; all large trees are heavily loaded with them. Epiphytes are relatively scarce.

There are six associations belonging to the semi-evergreen seasonal formation. All of them probably exist in habitats liable to more severe seasonal drought than the evergreen seasonal forest, but the water shortage is not wholly climatic and is due to a combination of topographic, edaphic and climatic factors.

(iv) Deciduous seasonal forest. In Trinidad itself the deciduous seasonal forest formation is represented by two communities, both of which appear to be secondary associations considerably affected by human interference. On the neighbouring island of Little Tobago, however, it is represented by what seems to be a true climax formation (Beard 1944b) and Beard's description was partly based on the latter area.

The deciduous seasonal forest is low and consists of two tree storeys, an upper open one of scattered trees up to 20 m high, and a lower layer at 3–10 m. Trunks fork or branch low down and are frequently crooked; many of the lower-storey trees tend to grow in clumps. There are few very large trees; about 0.5 m is the maximum diameter.

Over two thirds of the trees in the upper storey are deciduous and their deciduousness is obligate, taking place with unfailing regularity every dry season. The lower storey is almost entirely evergreen. About half the leaves in the upper storey are 'mesophylls', and about half 'microphylls'; in the lower storey the majority of the leaves are 'mesophylls', but there is a large proportion of 'microphylls'.

---

[1] Beard's terms. Many of the trees in the evergreen seasonal forest are probably 'leaf-exchanging' in the classification of Longman & Jeník (1987). See p. 242.

[2] Not as subdivided by Webb (1959). See pp. 93ff.

Buttressed trees are absent, but a few species have spiny or thorny trunks. Lianes are somewhat rare and epiphytes absent or scarce.

(v) Thorn woodland. This formation, which as already mentioned does not exist in Trinidad, is found in the neighbouring parts of Venezuela as a fairly open scrubby community consisting largely of hard-leaved, microphyllous, evergreen trees 3–10 m high, belonging chiefly to the Leguminosae (Mimosoideae and Caesalpinioideae); most of these trees are thorny. There is no grass on the ground (as there often is in similar types of vegetation in other parts of the world) and the only low-growing vegetation consists of a few Bromeliaceae and succulents.

(vi) Cactus scrub. This formation is met with on the island of Patos, between Trinidad and Venezuela, and on the mainland. It is an open type of vegetation dominated by columnar cacti and prickly pears (*Opuntia*) mingled with scattered gnarled micro- or leptophyllous bushes and terrestrial Bromeliaceae. There is no grass and the ground is often bare.

The six climatic formations described by Beard obviously form a continuous series, each type grading into the next. Though the dividing lines are more or less arbitrary, the divisions proposed by Beard seem on the whole to be both practically and theoretically convenient.

Taken as a whole the series shows a striking relation between the structure and physiognomy of the vegetation on the one hand and the climate on the other (Fig. 16.3). As the seasonal drought (whether due solely to climate or to a combination of climate with other factors) becomes more severe, the stature of the vegetation decreases and its structure becomes simpler, the number of layers decreasing. The deciduous habit becomes commoner, first in the upper storeys, later in the lower. That the lower storeys should remain evergreen when the top storey has become mainly deciduous is of course to be expected from their moister microclimate. There is a tendency towards a general reduction in leaf size; 'microphylls', which are rare in rain forest, become increasingly common towards the dry end of the series. Buttressing decreases (but this may be merely a function of the decreasing average size of the trees) and thorniness becomes an increasingly common feature.

The Trinidad forest types described by Beard are outliers of formations widespread (now or formerly) on the mainland of South America, as well as in the West Indian islands (Sarmiento 1972). Beard (1955a) later extended his classification to include all the forests of the American tropics. In the interior of Guyana, Fan-

shawe (1952) found all Beard's lowland forest formations except deciduous seasonal forest. In Venezuela and elsewhere in northern South America, types of seasonal forest comparable to those in Trinidad have been described by Vareschi (1980), Ellenberg (1959) and others, while thorn woodland and cactus scrub are common in dry, strongly seasonal, climates in Colombia and Venezuela as well as in Jamaica (Asprey & Robbins 1952) and other islands.

In the lowlands of Central America a similar series of plant formations occurs: many examples of these communities in Costa Rica are described in detail by Holdridge *et al.* (1971; see also Hartshorn 1983). Holdridge's 'biomes' (see pp. 164–5) are not easy to equate exactly with Beard's climatic formations. His premontane wet forest (represented at Finca La Selva; see p. 438) and premontane rain forest are within Beard's concept of tropical rain forest, while his tropical wet forest sites, and probably also his tropical moist sites, e.g. in parts of the Osa Peninsula, appear to come within the concept of evergreen seasonal forest. The dry forest of Guanacaste and elsewhere in Costa Rica is in general similar in structure and composition to the semi-evergreen seasonal forest of Trinidad.

16.2.2 Savannas in Trinidad and northern South America

Although none of the Trinidad savannas is more than a few square kilometres in extent, they are of interest because both their vegetation and soils have been carefully studied (Beard 1953). They are dominated by tussock-forming species of *Axonopus*, *Trachypogon* and other grasses, associated with numerous dicotyledonous herbs and woody plants. Small gnarled trees of characteristic savanna species such as *Curatella americana* and *Byrsonima crassifolia* are scattered a few metres apart. The well known Aripo savannas, which are in an area of high rainfall, have a different flora (Richardson 1963). Here the ground is covered mainly with Cyperaceae, associated with grasses intermixed with species of *Polygala*, *Utricularia*, *Drosera*, patches of *Sphagnum* and many other small plants.[3] There are occasional clumps of bushes and groves of tall *Mauritia* palms, but *Curatella* is absent.

The Trinidad savannas occur as pockets in what are (or were) large forests, mostly of the evergreen seasonal type. The boundary between forest and savanna, like that of the Brazilian campos (pp. 397ff.) is remarkably

---

[3] Very similar associations are found on the South American mainland, as well as in Africa and elsewhere.

sharp and cannot be determined by climatic factors. Some of the savannas are frequently burnt; according to Beard not all are subject to fire, but though he states that the Aripo savannas had never been known to burn Richardson (1963) later showed that they were subject to occasional fires. There seems to be no tendency for forest to invade the savannas. In the Northern Plain of Trinidad there are grasslands of a different kind which are known to have arisen from abandoned shifting cultivation followed by repeated burning. These are a stage in a deflected succession (see Chapter 18) and may be termed secondary savannas; the flora is quite different from that of the primary savannas discussed above. The dominant grass is *Imperata brasiliensis*, which is never abundant in primary savannas; the very similar *I. cylindrica* (alang-alang) plays a large part in secondary successions in the Palaeotropics (Chapter 18). The rich flora of the primary savannas of Trinidad, which includes a number of species of endemic and of highly discontinuous distribution, is evidence of their natural origin and great antiquity.

The lowland savannas of Trinidad are on more or less level sites with impeded drainage; in Beard's opinion they are relics of an ancient land surface which has not been forested since Pleistocene times[4]. The soils are heavy impermeable clays, sometimes overlain by a thin layer of sand. Though derived from similar parent materials to those of the forest soils and not significantly different from them in chemical properties, they are very unlike them in physical characteristics. For part of the year they are waterlogged and for the rest very dry. The 'hog-wallowed' (hummocky) surface is characteristic. Beard suggested that it is perhaps this alternation of very wet and very dry conditions which excludes forest vegetation from the savanna areas.

On the mainland of South America there are considerable areas of savanna in the Guianas and an even larger area further north in Venezuela (Sarmiento 1984).

In Guyana there are two main savanna areas, the Berbice savannas west of the Berbice river, some 100–150 km from the coast in a rain-forest climate, and the much more extensive interior savannas in the Rupununi basin and near the Pakaraima mountains. The latter are continuous with the Rio Branco savannas in Brazil, and have a relatively low rainfall with a long dry season. The Berbice savannas have been described by Martyn (1931) and Follett-Smith (1930) and are similar to the

Surinam savannas (Lindeman & Moolenaar 1959, Heyligers 1963, van Donselaar 1969). They consist of rolling plains about 24–28 m above sea-level, dominated by grasses and Cyperaceae. Here and there are islands of trees and bushes; strips of fringing forest are found in the valleys. Slightly different vegetation is found in moist depressions, termed pans, and on large ant-hills which have become colonized by woody plants. The soil is mainly a brown sand; where it changes to white sand, similar to that of the wallaba forest, the savanna flora is replaced by 'muri bush' (p. 322). The savannas are surrounded by evergreen rain forest and, as in Trinidad, the transition from savanna (or muri bush) to forest is often abrupt.

According to T.A.W. Davis (manuscript notes), the rain forest in the immediate neighbourhood of the Berbice savannas is usually dakama forest, dominated by *Dimorphandra conjugata* (p. 321); further away the *Eperua* (wallaba) association is usually met with. There is in fact a zonation of communities and a typical transect from forest to savanna might show the following sequence of zones: (i) *Eperua* association; (ii) *Eperua–Dimorphandra* association; (iii) *Dimorphandra* association; (iv) transition zone in which bania (*Swartzia* sp.), *Catostemma fragrans*, *Licania* sp. and kakarua (Sapotaceae) are prominent; (v) 'muri bush'; (vi) savanna. None of the forest types bordering the Berbice savannas is deciduous or even semideciduous; there is thus no evidence here of a transition from evergreen forest through deciduous forest to savanna.

In the interior of Venezuela and Colombia, to the north of the Guiana and Amazon rain forests, lie the Llanos, which became well known through the classical description of Humboldt (1852) (Fig. 16.8 and 16.9). They occupy an area of some 300 000 km[2], with a strongly seasonal rainfall. Modern accounts of the vegetation have been given by Beard (1953), Vareschi (1980), Sarmiento & Monasterio (1971) and Sarmiento (1984). According to Sarmiento & Monasterio (1969), the Llanos consist of four regions: (1) in the north, with a rainfall of 800–1000 mm, a deciduous forest and thorn woodland region (2) in the east with 1400–2000 mm, a semideciduous forest region; (3) a swampy region in the floodplain of the Orinoco (excluding the delta); and (4) a very extensive area in the centre, east and south, the 'dry savanna region'. This consists largely of open, park-like grasslands, with scattered groups of palms and other trees and with gallery forest along the watercourses. The annual rainfall is 100–1400 mm, with a dry season of five to six months, during which the soil dries out completely to a considerable depth. Sarmiento & Monasterio (1969) carried out association analy-

---

[4] Charter (1941) expressed a very similar view as to the origin of 'pine ridge', a type of savanna in Belize (Central America) in which *Pinus caribaea* is the chief tree; he regards the pine ridge soils as overmature (senile) and supposes that at an earlier stage in their development they carried evergreen forest.

sis of the plant communities in (4) at the Los Llanos Biological Station (8°56′N, 67°25′W).

There is thus a range of vegetation types in the Llanos comparable to those seen on a small scale in Trinidad, including evergreen seasonal, semi-evergreen and deciduous forest, thorn woodland and various types of grass-dominated savanna. The transition from the Llanos to evergreen rain forests has not been described.

16.2.3 Forest and savanna in Amazonia

Although most of the Amazon basin is (or was until recently) covered with rain forest (including much evergreen seasonal forest), it also has considerable areas of savanna. North of the equator there are the extensive Rio Branco savannas mentioned above (p. 396). Even in central Amazonia there are scattered savannas. Many of these, like some of the 'muri' and 'bana' scrub areas referred to in Chapter 12, are islands a few square kilometres in extent enclosed within the forest and no

doubt have a rain-forest climate. Eden (1974a) has described isolated patches of savanna, some only a few square kilometres in area, within the rain forest in southern Venezuela; these have a rainfall of over 2500 mm and only a brief dry season.

East of the rain forest in northeastern Brazil is the Sertão, a region of low, strongly seasonal and uncertain rainfall, where the climatic climax vegetation is chiefly deciduous forest and thorn woodland (formerly caatinga[5]). The transition from rain forest to the Sertão region is described by Aubréville (1961). On the west the Andes form a sharp boundary to the lowland rain forest, but in Peru a belt of seasonal forests lies between the rain forest and the mountains (Fig. 16.4) (Ellenberg 1959).

South of the Amazonian forest lies the vast Cerrado region, about 17×10^6 km^2 in extent. Most of it is a

[5] This term is now used mainly for types of rain forest on white sand (heath forest). See p. 322.

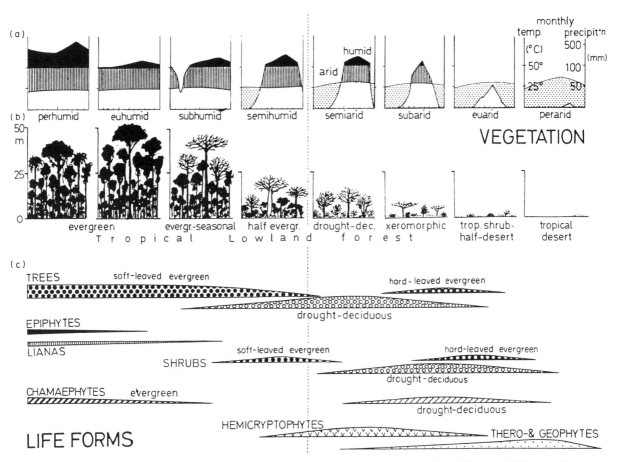

Fig. 16.4 Climate, vegetation and life-forms in the lowlands of western Peru and Ecuador. From Ellenberg (1979, Fig. 2). (a) Climate diagrams, (b) generalized profiles of 'natural' (climax) vegetation, (c) life-forms.

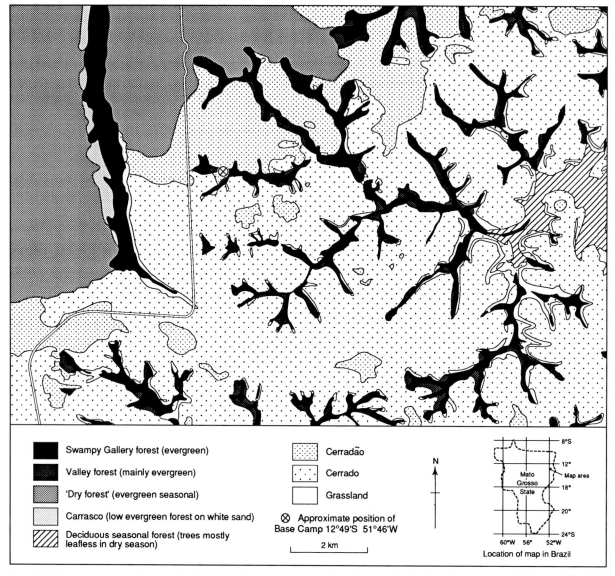

Fig. 16.5 Vegetation map of part of the Serra do Roncador region, Mato Grosso, Brazil. From Ratter *et al.* (1973), modified. The area shown is part of the research area of the Xavantina–Cachimbo Expedition to Central Brazil, 1967–69.

gently undulating plateau over 200 m above sea-level: in places there are rocky outcrops rising to 900 m or more. The annual rainfall ranges from *ca*. 750–800 to 2000 mm with a nearly rainless dry season of several months (Eiten 1972). Evaporation exceeds precipitation for more than half the year. In Brazil the word 'cerrado' is used for both the region and its most widespread type of vegetation, savanna dominated by small fire-tolerant trees and very rich in species (Fig. 16.7). This varies greatly in structure and composition: the more open types are cerrado *sensu stricto*, those with a fairly continuous tree canopy and scanty ground cover of grasses are called cerradão (cf. Eiten 1972, 1978, 1983,

Goodland 1971, Silberbauer-Gottsberger & Gottsberger 1984). There are also other kinds of vegetation in the Cerrado region, notably several kinds of forest and communities of low-growing grasses, Cyperaceae and dicotyledonous herbs and shrubs but few or no trees (campo limpo). These last are largely of natural origin, but in recent years considerable areas of cerrado and forest have been cleared and planted with *Panicum maximum* and other non-native grasses for use as cattle ranches. The whole region is intersected by rivers and streams bordered by galleries of evergreen, semideciduous and deciduous forest. The prevailing soils are sandy kaolisols, but patches of white sand occur locally.

Much of the vegetation of the Cerrado region is burnt in the dry season, when fires are started to encourage the growth of young grass for cattle (or accidentally), though not every year in the less populated areas. It is mainly the cerrado, cerradão and campo which are affected: the forests, except the deciduous forest, are burnt only on the edges and are usually not destroyed unless they are being deliberately cleared for farming or grazing.

A threefold mosaic of cerrado, campo and forest is characteristic of much of the Cerrado region and is well illustrated in an area 20 km × 20 km at 12°49'N, 51°46'W in the Serra do Roncador region (northeastern Mato Grosso) which was studied intensively in 1967–69 by the Royal Society's Xavantina–Cachimbo Expedition (Ratter *et al.* 1973, Eiten 1978). The landscape was one of low relief with a difference of 174 m between the lowest and highest points and it could be assumed that there were no significant differences in macroclimate within the area. At the time it was studied (1967–69) the area was almost uninhabited and relatively little affected by recent human activities. The distribution of the plant communities (Fig. 16.5) seemed to depend mainly on variations in topography and soil.

On about three-quarters of the 20 km × 20 km study area the vegetation was savanna (cerrado and cerradão) with forest galleries, often bordered by strips of grassland (campo) in the valleys. In the northwestern part, however, there was almost unbroken forest, most of the 'dry' type (see below) even on the interfluves. It stretched away beyond the study area for hundreds of kilometres to the north and west and was in fact the margin of the great Amazonian 'hylaea'. The boundaries between the different vegetation types were in general remarkably sharp, especially between the gallery forests and the campo and between the latter and the cerrado (Fig. 16.6).

The forest in the study area was of five types.

Swampy gallery forest. This was best developed near the headwaters of streams and in sites where the water table never fell far below the soil surface, even in the dry season. In structure, physiognomy, and to a considerable extent in composition, it resembled a swampy Amazonian rain forest, although some rain forest features such as buttressing and cauliflory were poorly developed or absent and its species richness in all synusiae was much less. The A storey trees were *ca.* 25–40 m high and mostly evergreen, although some lost part of their foliage in the dry season without becoming completely bare. On sample plots two species of *Qualea* (Vochysiaceae) together formed 58% of trees

Fig. 16.6 Aerial photograph of the vegetation near the Base Camp of the Xavantina–Cachimbo Expedition, Mato Grosso, Brazil (1968). The darker vegetation on the left is mostly swampy gallery forest; the rest is cerrado, except the paler strip in the middle, which is campo. In the wider part of the campo two zones can be seen, an outer one of low shrubs and an inner one chiefly of grasses and Cyperaceae. The white spots are bare soil exposed in soil surveys. See text, pp. 399–401.

⩾120 cm g.b.h.; both of these, as well as many of the other plants, also occur in the rain forests of Amazonia.

Valley forest. This is a type of high forest found on the gently sloping sides of valleys at a higher level than gallery forest. The soil is better drained, its surface, at least in the dry season, never becoming swampy. It is intermediate in its characteristics between swampy gallery and dry forest. The taller trees are mostly evergreen or leaf-exchanging and reach a height of 35–40 m and a diameter of *ca.* 1.5 m. The A storey is composed of four Leguminosae, *Apuleia molaris*, *Hymenaea stilbocarpa*, *Ormosia* sp. and *Copaifera langsdorfii*; all but the last of these, which also occurs in 'dry forest' and cerradão, seem to be characteristic of this association. The lower storey (*ca.* 1.5 m high) consists mainly of species also found in 'dry forest'.

Dry forest. This community occurs on deep, reddish, dystrophic sandy loams, on the interfluves as well as in valleys. It covers about a quarter of the 20 km × 20 km square and a very large area within and north of the square. Although it is classified as evergreen seasonal forest, it does not closely resemble the forest of that name in Trinidad (p. 394 above). The upper layer of trees is not more than 15–18 m high. The tree stems are slender and so crowded as to give the impression of a pole forest. Though some species become bare of leaves for part of the dry season, the majority

A

B

C

Fig. 16.7 Vegetation types near the Xavantina–Cachimbo Expedition's Base Camp. Mato Grosso, Brazil. A Aerial view of 'dry forest' (see pp. 399–401) Transamazonica road (1968). B Cerradão by road (1967). C Cerrado (top left), campo (foreground) and swampy gallery forest (right).

are evergreen or leaf-exchanging. 'Dry forest' is fairly species-rich and common trees include *Licania* spp and *Sacoglottis guianensis*; the leguminous *Copaifera langsdorfii* and *Pterodon pubescens* occur occasionally as tall emergents. In the B storey *Protium* spp. are abundant and the C storey (*ca.* 4–7 m) is dominated by *Myrciaria floribunda*. Further from the cerrado boundary the 'dry forest' becomes taller and more luxuriant (Ratter *et al.* 1973).

Carrasco. This consists of evergreen trees 5–8 m high and is similar to the Amazonian low caatinga. It occurs very locally on small patches of white sand and was not studied in detail.

Deciduous seasonal forest. This differs from the other forest types in that in the dry season all the taller trees (which are not thorny) are completely bare of leaves for many weeks. It was found only in relatively base-rich 'terra preta' soil[6] (pH *ca.* 5.6, compared with *ca.* 4.6 in other types of vegetation) in well-drained valley sites and occupied only a small area. It is the only forest type in this area which is subject to fire. The deciduous forest is very distinctive in composition; many tree species such as *Cedrela fissilis*, *Guazuma ulmifolia*, *Piptadenia macrocarpa* and *Tabebuia* sp. are not found in the other types of forest. Large lianes are common and in the undergrowth a large stemless palm (*Attalea* sp.?) and thickets of a tall bamboo (up to *ca.* 8 m high) are common.

In the 20 km × 20 km area these five types of forest exist together with cerradão, cerrado and campo; the distribution of these types of vegetation cannot be ascribed to macroclimatic factors. In places the cerrado appeared to be invading the cerradão, but there was no evidence that cerradão, cerrado or campo were invading the forests, or vice versa; as far as could be discovered the boundaries of the plant communities were stable, at least on a short time-scale. It was concluded that the cerrado–campo–forest mosaic was primarily determined by edaphic factors, especially the availability of water. Whether this would be true of the whole cerrado region is uncertain: in the study area, because it lies in a 'tension zone' between forest and cerrado, small differences in soil and water level may be more critical than further from the margin of the continuous forest.

The very thorough studies of the soils by Askew *et al.* (1970a, b, 1971) showed that the only constant difference between the forest and cerrado soils was in texture: the former were 'sandy clay loams', finer in texture (and probably more retentive of water in the dry season) than the sands and 'loamy sands' of the cerrado. Another important factor seemed to be the height of the watertable. In the gallery forests it was not far below or above the surface for most of the year, but in the cerrado and cerradão it fell several metres below the surface during the dry season. The campo–cerrado boundary was probably determined by the water level in the wet season rather than by its depth in the dry season.

All the soils in the study area were nutrient-poor and very acid (mean pH of surface horizons 4.6), except those of the deciduous forest, which have relatively high nutrient levels and a pH of 5.6. This is the only example in the area of association between a particular type of vegetation with chemical soil characteristics. It may be noted, however, that in the 'Triangulo Mineiro' (Minas Gerais), where there is a similar pattern of cerrado, cerradão and campo, Goodland & Pollard (1973) found a significant correlation between the amounts of phosphorus, nitrogen and potassium in the soil and the standing crop of trees (measured as basal area).

Fires started by farmers or wandering Indians occurred in the area during the dry season, but up to 1967–69 probably not more often than once every three or four years. There was no good evidence that fire was responsible for the cerrado–campo–forest mosaic. Fire-protection experiments elsewhere in the cerrado region do not support the idea that if fires were eliminated forest would rapidly invade the other types of vegetation.

16.2.4 Ecological status of tropical American savannas

The savannas of South and Central America have been variously regarded as climatic formations, edaphic climaxes dependent on particular soil conditions, and as biotic climaxes maintained by fire, often combined with grazing. Although some savannas are certainly secondary and derived from forest by repeated cultivation and burning, it is generally accepted that most of the South American savannas are not anthropogenic or of recent origin. The environmental factors involved in South American savanna ecosystems have been fully discussed by Beard (1953), Sarmiento & Monasterio (1975), Sarmiento (1983, 1984) and others: only some general remarks will be made here.

That savannas are most extensive in regions with long annual dry seasons is evident, but in more constantly humid tropical climates savannas and forests often exist

---

[6] This name is used for pockets of fertile, dark-coloured soil in Amazonia (Ducke & Black 1953), but these are not necessarily very similar to the 'terra preta' soils in this area.

side by side. According to Sarmiento & Monasterio (1975) 80–90% of South American savannas have 100–1200 mm annual rainfall and a dry season of five to eight months (Aw climates of Köppen). However, as was noted previously, less extensive savannas are found as islands even in rain forests where rainfall exceeds 2000 mm (or even 3000 mm) with few or no dry months. It is also of interest that in South America, though not in Africa, woody vegetation (deciduous forest and thorn woodland) rather than savanna is found where the rainfall is less than 800–1000 mm (Sarmiento & Monasterio 1975).

The soils of South American savannas are of various kinds (Chapter 10). In large savanna areas such as the Llanos and the Cerrado, red or yellowish kaolisols of sandy texture are common on the higher ground and hydromorphic soils in the valleys. In the Guyana and Rio Branco savannas the soils are mainly, but not exclusively, white sands. Both the red and white sands are very poor in plant nutrients; their infertility is partly the result of pedogenic processes, but is increased by frequent burning which tends to lead to erosion and loss of soluble material. In the cerrados of Brazil the standing crop of trees is (at least in some places) related to soil fertility, as noted above.

Some South American savannas are on more or less level, ill-drained sites which become waterlogged in the wet season. In some sites, however, there is a hard pan near the surface which prevents the percolation of water to deeper soil horizons even when the surface is wet. Beard (1953, 1955b) stressed the importance of these soil characteristics and considered that savannas in general were found particularly on very mature soils in senile landscapes. In his view the most important factor for the development of savannas is the alternation of waterlogging with long dry periods in which the soil becomes desiccated; such conditions are inimical to the growth of most forest trees. More recently Eiten (1986) has expressed similar views.

Recurrent fires are always a feature of savanna environments and even in remote uninhabited regions savannas show evidence of past fires (Myers 1936, p. 176; M. Pires, oral information). As might be expected, the frequency of burning is greater near farms and villages than in unpopulated areas. Savanna fires caused by people or by lightning have occurred for many thousands of years and are not only a recent phenomenon (Sarmiento & Monasterio 1975). As a result of selection, the woody and herbaceous flora of savannas, unlike that of rain forests, is fire-tolerant, consisting of pyrophytes adapted in various ways to survive burning.

Savannas are often said to be 'fire-maintained'; this

Fig. 16.8 Aerial photograph of savanna and gallery forest, Gran Sabana, Venezuela. (Photo. T.C. Whitmore.)

Fig. 16.9 Savanna south of R. Orinoco, road from Upata to Euasipati, Venezuela. (Photo. J.S. Beard.) The trees are *Curatella americana* and in smaller numbers, *Bowdichia virgilioides*. Ground flora of short 'bunch grasses' (*Trachypogon*, etc.). The white sandy topsoil overlies impermeable red clay. Annual rainfall *ca.* 1300 mm.

implies that if protected from fires they would become invaded by plants intolerant of burning. This has been observed in Africa (pp. 416–17), but in tropical America there are very few experiments or observations on the effects of fire protection. Kellman (1984) reports that in a savanna in Belize, where fires had been suppressed for over thirty years, colonization by both fire-sensitive and fire-tolerant woody species had taken place. However, it was noteworthy that this colonization was restricted to patches of relatively fertile soil, suggesting a synergistic relation between fires and low soil fertility.

As stated earlier, fire-protection experiments in the Brazilian cerrado (Labouriau 1966) resulted in an increase of grasses rather than an invasion by forest trees, but perhaps the duration of the experiments was too short and the distance of seed sources too great for such changes to have occurred.

From what has been said it may be concluded that climatic factors, soil and burning are all important factors in savanna ecosystems, but the latter are so diverse that their relative importance may be different in different types of savanna. Sarmiento & Monasterio (1975) divide South American savannas into three groups.

Seasonal savannas. These are found in climates with at least four successive (more or less) rainless months during which potential evapotranspiration exceeds precipitation. The soils are well drained and medium to coarse in texture. By the end of the dry season the upper layers of the soil become very dry, but water remains available to deep-rooted plants at lower levels. To this group of savannas belong much of the Llanos and cerrados.

Hyperseasonal savannas. These occur on heavy, ill-drained soils and are characterized by an alternation of dry and very wet conditions. In the dry season the vegetation dries out and is often burnt, but in the wet season the ground becomes waterlogged or flooded. Aripo and some of the other Trinidad savannas, as well as the valley campos in the Mato Grosso, are in this category.

Non-seasonal savannas. These are found in ever-wet or slightly seasonal climates, sometimes as 'islands' in rain forest. Floristically different types are found on red kaolisols and on white sands.

These non-seasonal savannas are of particular interest because of the similarity of their flora to that of the main savanna areas from which they are separated by large expanses of forest. Eden (1974a) and Sarmiento & Monasterio (1975) suggest that they are relics dating from an earlier climatic period when savanna vegetation was more extensive than in recent times. Palynological evidence suggests that this may have been in the arid phases of the Pleistocene (Chapter 1). During the subsequent period of warmer and more humid climate the forest expanded, but perhaps savanna vegetation survived on sites least favourable to forest growth and anthropogenic fires have enabled it to persist until the present.

16.3 Rain forest, monsoon forest and savanna in the eastern tropics

16.3.1 Rain forest and monsoon forest

In Southeast Asia the most typical development of lowland tropical rain forest is in the Malay Peninsula, Sumatera and Borneo. To the north and west of these areas, in Thailand, Indochina, Burma and India, as well as to the south in the Sunda islands and parts of New Guinea, there is an annual dry season of several months and the vegetation is more or less seasonal in its phenology (tropophilous in Schimper's sense). Although evergreen rain forest occurs locally, the climax forests are mainly semideciduous and deciduous; for these Schimper's term 'monsoon forest' is often used. Grasslands, with or without scattered trees and shrubs, occur widely in the Indo-Malayan region and their total area is probably considerable, but most of them seem to be anthropogenic and of fairly recent origin. It is a striking fact that there are no very extensive savanna areas in the eastern tropics comparable to the cerrado and Llanos of South America or the savannas of Africa.

The boundary between the rain forest and monsoon forest formations is determined by potential evapotranspiration and the availability of water in the soil; it is therefore more closely correlated with the length and severity of the dry season than with total annual rainfall. Owing to local climatic differences and the interaction of rainfall with drainage, geology and soil, the ecotone between the two formations is often a complex mosaic rather than a gradual transition. Even in Borneo, most of which has an ever-wet climate, there are small areas in the south and east in which rainfall is somewhat seasonal and the vegetation is monsoon forest rather than rain forest (Whitmore 1984a, p. 198). Similarly, in Vietnam, where the zonal climate is seasonal and the climax vegetation mainly monsoon forest, pockets of rain forest occur in stream valleys where the local climate and soil are favourable.

The transition from rain forest to monsoon forest is well shown in the 'Kra ecotone' in the Malay Peninsula, though now, owing to disturbance and destruction, less clearly than formerly (Whitmore 1984a, pp. 201–3). From Singapore, which has a typically ever-wet climate (lat. 1°18′N) northwards, the rainfall becomes gradually more seasonal and in Kedah (ca. 6°N) there is a dry season of about one month. In the Kra isthmus and further north in Thailand and Burma the dry season becomes longer and more severe. Parallel with the climatic gradient there is a change in flora and vegetation: according to Whitmore (1984a) the boundary between the evergreen and semi-evergreen forest lies approximately on a west–east line from Kangar (Malaysia) to Pattani (Thailand) at *ca.* 7°N. North of this line characteristic Malayan rain-forest species such as *Shorea curtisii*, *Neobalanocarpus heimii* and the undergrowth palm *Eugeissona tristis* disappear and 'monsoon-forest' species such as *Shorea guiso* begin to occur. A few species transgress the line, e.g. *Shorea leprosula*, which has isolated localities north of it, and *S. hypochra*, which though found mainly to the north also occurs in dry situations in the northern Malay States. At *ca.* 10°N in locally wet areas on the west coast there are patches of evergreen forest in the monsoon forest.

The term 'monsoon forest' covers a range of communities differing in physiognomy, structure and floristic composition. In India alone, according to Champion & Seth (1968a), there are about fourteen types of primary forest which may be regarded as monsoon forest. Like the seasonal forests of tropical America, the monsoon forest of the eastern tropics can be classified into evergreen-seasonal, semideciduous and deciduous formations, differing in structure and composition as well as in the proportions of evergreen and deciduous trees. The types of monsoon forest in regions with a not very severe dry season in Thailand, India, Indochina, etc., conform fairly well to Schimper's definition (p. 389 above), but because they are rich in economically valuable timber trees their natural structure and composition have been much modified by exploitation and silvicultural management.

An important difference between monsoon and rain forests is that the former become dry in the dry season and are frequently burnt. Bamboos and other grasses normally found in the undergrowth are favoured by fires. Some monsoon-forest trees are to some extent fire-tolerant, though probably less so than savanna trees; a well-known example is teak (*Tectona grandis*). Its natural regeneration tends to be more abundant in burnt than in unburnt forest (Champion & Seth 1968a, b) and where teak occurs in mixtures with other species it is gradually eliminated if it is completely protected from fire. A few monsoon-forest dipterocarps are fire-tolerant and differ from the rain-forest species in their very thick bark.

Floristically the monsoon forests of the eastern tropics are markedly different from rain forests; relatively few species of trees and other plants are found in both, e.g. *Carallia brachiata*, *Irvingia malayana*. Many genera and families such as the Fagaceae and Lauraceae, which are abundant in rain forests, are absent or poorly represented in monsoon forests. On the other hand, species of *Acacia*, *Albizia*, *Bombax*, *Celtis*, *Tetrameles* and many other genera are characteristic of semi-evergreen and other types of monsoon forest. The Dipterocarpaceae are often abundant as individuals in monsoon forests, forming a large percentage of the basal area, but the number of species is much less than in rain forests. Out of the 386 species of dipterocarps in Malesia only sixty-two occur in areas where one or more months have less than 100 mm rainfall; of these, twenty-four are confined to such areas (Ashton 1982). Like those of rain forests, the dipterocarps of monsoon forests are often strongly gregarious, e.g. *Shorea robusta* (sal) and *S. siamensis*.

Plants intolerant of dry atmospheric conditions, such as Hymenophyllaceae, various other ferns and epiphyllous bryophytes, are found in monsoon forests only locally in humid environments such as sheltered stream valleys.

The number of tree species in the monsoon forests is probably much less than in rain forests, but figures for a reliable comparison are not available. Stamp (1925) estimated that there were 'at least 1000' tree species in the monsoon forests of Burma; this figure can be compared with at least 2500 species in the lowland rain forests of Peninsular Malaysia.

Many monsoon forest species are very widely distributed; *Tamarindus indica* extends to West Africa as well as throughout the eastern tropics though not everywhere as a native species. Endemics of very local distribution are much fewer than in the Malesian rain forests.

The rain-forest area divides the monsoon forests into two widely separated areas, but some species that are absent from the rain forests have a disjunct distribution and occur in both Thailand and Indochina in the north and the eastern Sunda islands in the south (Whitmore 1984a, p. 200). Teak (*Tectona grandis*), which is frequent in monsoon forests in Burma, Thailand and neighbouring countries, also occurs in east Java, although there it may be an old introduction rather than a native species (van Steenis 1935a). It is interesting that teak and various other monsoon forest trees cannot be successfully cultivated in ever-wet climates such as that of southern Malaya (Burkill 1966).

16.3.2 Savannas in the eastern tropics

Throughout the Malesian rain-forest region small areas of grassland, often with occasional fire-tolerant trees and shrubs, are common. Most of these are dominated by the alang-alang grass, *Imperata cylindrica*, and are demonstrably stages of secondary or deflected successions (Chapter 18). They result from the destruction of forest, a process still proceeding at an accelerating rate, and are mostly of quite recent origin. Savannas similar to the non-seasonal and hyperseasonal savannas of South America seem to be absent from Malesia. In Indochina, India and other countries in Southeast Asia with seasonal climates there are extensive savanna woodlands (forêts claires) as well as grasslands. All these are burnt in nearly every dry season and, like the grasslands in the rain-forest regions, are probably anthropogenic.

In New Guinea at low elevations the climate is in most parts suitable for rain forest, but there are considerable areas of savanna and more or less treeless grassland. Most of these have undoubtedly resulted from the destruction of forest by 'gardening' (shifting cultivation) and burning, often dating back to precolonial times. It is possible, however, that some New Guinea grasslands are edaphically determined and not derived from forest.

In the relatively dry Port Moresby district in the southeast there are patches of savanna as well as areas of rain forest, semideciduous forest and deciduous forest. The savannas consist of communities of *Themeda* and other perennial grasses, usually associated with scattered shrubs and fire-tolerant trees such as *Eucalyptus* and *Cycas* spp. 10–15 m high (Eden 1974b). There is a gradient of climate from the coastal zone, which has a rainfall of less than 1000 mm and several successive months with less than 100 mm, to the hills and Sogeri plateau (500–1000 m) inland where the rainfall is 2200–3500 mm and there is no distinct dry season. The soils are varied and mainly clayey in texture.

Eden investigated the distribution of grassland and savanna with the help of aerial photographs and found that, although their total area decreased from the coastal to the high rainfall zones, their occurrence was not consistently correlated with any environmental factor. He concluded that most of the grasslands and savannas were due to the combined effects of cultivation and burning after forest clearance, but the possibility could not be excluded that some of them originated in a past phase of dry climate and had been maintained since then by frequent burning.

Elsewhere in New Guinea there are much larger areas of grassland, notably in the Ramu and Markham valleys in the east, while in the Middle Sepik basin in the north there are over 1500 km² of grassland (Robbins 1963). At low elevations the grassland is of two kinds, tall grasslands dominated by *Saccharum spontaneum*, *Imperata cylindrica*, *Ophiurus exaltatus* and other grasses growing to a height of up to 3 m, and short grasslands formed by species such as *Themeda australis* little more than 1 m high. Scattered fire-tolerant trees and shrubs are usually found in both tall and short

grasslands and it seems certain that most of them are derived from forest by the combined effects of cultivation and fires. In a few places where burning has ceased a succession to forest can be observed. Where burning is very frequent, tall grassland is converted to short grassland.

There is, however, some evidence that at least part of the Middle Sepik grasslands may be primarily determined more by soil conditions unfavourable to tree growth than by human activities (Reiner & Robbins 1964; Haantjens *et al.* 1965; R.G. Robbins, personal information). The rainfall here is over 1780 mm, but there is a regular annual dry period. The area is a level plain which is waterlogged for much of the year but becomes very dry and is burnt during the dry season. In both the tall and short grasslands scattered fire-tolerant trees similar to those in the Port Moresby savannas are absent and even bushes are not numerous. There are strips of forest in the stream valleys, and locally on the interfluves, but, according to Robbins, there is no evidence that all of the grassland was formerly forest-covered; he suggests that, as in the South American hyperseasonal savannas, the lack of trees is caused by the nutrient-poor soil and the alternation between very wet and very dry conditions.

16.4 Forest and savanna in Africa

In Africa, as was seen in Chapter 1, a central mass of forest occupies a large area in the Zaïre (Congo) basin, extending north into Cameroon. This is continued westwards as a belt of varying width following the coast until it eventually disappears in Sierra Leone and Guinea-Bissau at about long. 15°W and lat. 10°N. From the western frontier of Nigeria to a little west of the Volta river in Ghana the forest belt is broken by the Dahomey Gap, where savannas reach the sea and divide the forest into an eastern and a western block differing to some extent in their flora and fauna. At its margins the African 'closed forest' passes, often abruptly, into a wide zone of savannas. The total area of savannas in tropical Africa, even before the widespread forest destruction of the past two hundred years, was probably much greater than that in either tropical America or Asia.

The convenient term 'closed forest' (or 'closed-canopy forest') is equivalent to the *forêt dense* of French writers and includes all types of lowland forest in which the trees grow in closed canopy (A storey usually not less than 40 m high in mature communities), and the ground is not covered with grass. The closed forest consists mainly of evergreen and leaf-exchanging trees, but towards its periphery the proportion of deciduous and leaf-exchanging species increases and the floristic composition changes; consequently a distinction is often drawn between two main types of closed forest, evergreen forest and semideciduous forest. The latter has also been termed mixed deciduous forest, *subxerophiler Tropenwald* (Mildbraed 1922), *Forêt mésophile semi-caducifoliée* (Lebrun & Gilbert 1954) and *Forêt dense humide Semi-décidue* (C.C.T.A. 1956). Some authors, e.g. Taylor (1960), Schnell (1976–77) and Guillaumet & Adjanohoun (1971), use the term tropical rain forest (or one of its equivalents) only for the first type of closed forest, while others, e.g. Keay (1959b), use it for both types. In the UNESCO (1973) classification, most of the evergreen lowland forest in Africa is regarded as tropical evergreen seasonal forest and the more deciduous types as tropical semideciduous forest. Truly deciduous forests, like those in the tropics of Asia and America, occur in the Zambezian region and Madagascar (White 1981, 1983), but seem to be absent from Africa north of the equator. White (1983) maps all the closed forest as 'lowland rain forest', but distinguishes between 'wetter' and 'drier' types. Closed forests in which the taller tree storeys become entirely leafless in the dry season are almost unknown in West Africa.

In tropical Africa west of Cameroon there is a climatic gradient from the wet coastal belt with a dry season of about four months during which the humidity of the air usually remains high to the less humid and more strongly seasonal climates of the interior, culminating in the desert climate of the Sahara (Chapter 7). There is a corresponding ecotone in the vegetation and a number of zones running more or less parallel to the coast can be recognized.

16.4.1 Climatic ecotone in West Africa

The West African vegetation zones have been variously named and classified (see UNESCO 1973, Schnell 1976–77, White 1983, Sanford & Isichei 1986). The scheme used here is based on the well known classification of Chevalier (1900).

(1) The closed forest zone. This embraces the whole of the closed forest, as defined above, including both the wet evergreen forest near the coast and the semideciduous forest further inland. Most of the closed forests have at one time been cultivated, including most of what are generally regarded as primary. In some places over-intensive shifting cultivation has led to a deflected succession (Chapter 18), stages in the secondary succession having been invaded by grasses and converted by the effect of recurrent fires into what A.P.D. Jones (1945) and Keay (1959c) termed derived savannas. These are physiognomically slightly different from the savannas of the next zone and are usually floristically poorer. They often betray their origin by the presence of oil palms (*Elaeis guineensis*) and relics of tree species characteristic of the closed forest zone (see Sanford & Isichei 1986). Their ecological status is similar to that of the secondary savannas of Trinidad (pp. 395ff. above).

In the closed forest zone, especially from Ghana westwards, in some places near the coast there are small areas of open grassy savanna physiognomically and floristically unlike the derived and Guinea savannas (Ahn 1959, Adjanohoun 1962). These seem to be an edaphic type comparable to the coastal savannas of the Guianas (p. 396) and some of the 'esobe' of Zaïre (p. 415). They appear to depend on seasonal flooding combined with very nutrient-poor and sometimes podzolized soils. According to Miège (quoted by Schnell 1976–77, vol. 3, pp. 243–4), the coastal savannas of Côte d'Ivoire are being invaded by secondary forest trees such as *Alchornea cordifolia*.

(2) The Guinea zone. Here, except on permanently moist sites, the vegetation of uncultivated land varies from grassland with scattered fire-tolerant trees to savanna woodland with grassy undergrowth. It is usually burnt every year in the dry season. Where there is sufficient water in the soil, outliers of closed forest are found, many, but not all, of which are along water courses (gallery forests). On slopes in hilly country, as in the Mole Game Reserve in Ghana, several types of savanna, together with fringing forest, may form a catena (Lawson *et al.* 1970). Like the closed forest, much of the Guinea savanna has been affected by shifting cultivation, although some areas such as the Afram Plains in Ghana (Swaine *et al.* 1976) seem never to have been farmed.

Keay (1959b) divided the Guinea zone of Nigeria into two subdivisions:

(a) Southern Guinea subzone. The common trees here include *Vitellaria paradoxa, Daniellia oliveri, Lophira lanceolata* and *Terminalia glaucescens*. The forest outliers within this subzone resemble semideciduous forest in composition, including such species as *Khaya grandifoliola* and *Triplochiton scleroxylon*, but are usually less tall. The existing Southern Guinea savannas are mostly open, but in sparsely populated districts there are patches of savanna woodland in which not all the trees are fire-tolerant species ('transition woodland'); these may be areas where the forest is invading the savanna.

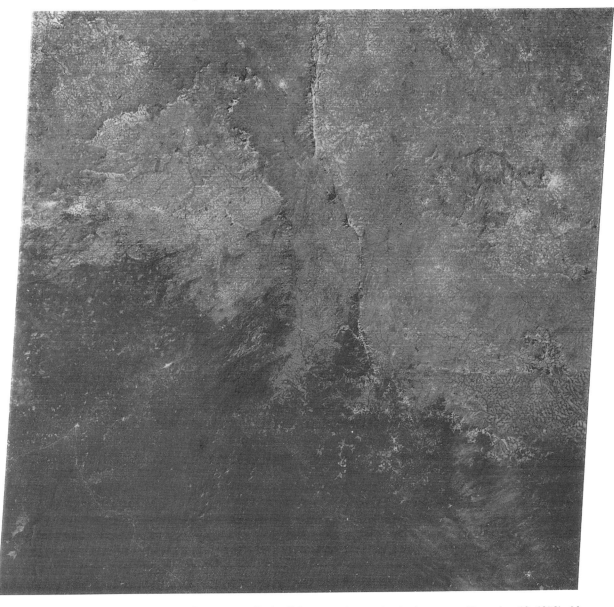

Fig. 16.10 Landsat image of southeast Guinea, near Beyla. Taken in early morning in dry season (December 18, 1973). After Hall & Swaine (1976). Darkest tones (lower centre) are closed-canopy forest; lighter tones are savanna and cultivated land. Towns appear white. In savanna at lower right a fine dendritic pattern of gallery forest can be discerned.

(b) Northern Guinea subzone. Here the common trees include species of *Combretum* and *Isoberlinia, Monotes kerstingii* and *Uapaca togoensis*. Many are identical with or closely related to species of the miombo (*Brachystegia–Isoberlinia*) woodlands of tropical Africa south of the equator. Species nearly related to those of the closed forest are less common than in the Southern Guinea subzone. Forest outliers occur chiefly as gallery forest in which the undergrowth consists mostly of closed forest plants but the trees are species common in the Southern Guinea savannas (*Anogeissus leiocarpus, Malacantha alnifolia,* etc.), together with a few found in both closed forest and savanna, e.g. *Afzelia africana* and *Diospyros mespiliformis*. The 'kurame' referred to on p. 416 are a special group of forest outliers found in this subzone. There is no clear boundary between the Southern and Northern Guinea savannas. Sanford & Isichei (1986) do not consider the two subzones worthy of recognition.

(3) The Sudanian zone. This zone is densely populated and the vegetation mostly consists of open savanna much modified by fire and cattle grazing. The trees include fine-leaved thorny acacias such as are characteristic of the Sahel zone as well as broad-leaved thornless trees which also occur in the Northern Guinea savannas. Common species include *Adansonia digitata* (baobab), *Balanites aegyptiaca*, *Combretum glutinosum*, *Lannea microcarpa* and *Sclerocarya birrea*: some of them have edible fruits and are protected or planted for this reason. Closed-forest outliers are almost entirely absent.

(4) The Sahel zone. The climax vegetation of this zone was probably originally thorn woodland dominated by *Acacia* spp. and other spiny trees. The existing woodland is generally so open that a car can drive between the trees; a ground cover of grasses or other herbaceous plants is discontinuous or absent. As in zones 2, 3 and 4 the vegetation has suffered much from over-grazing but, owing to the scanty grass cover and the openness of the vegetation, fire is not an important factor. The flora is poor in species and largely different from that of the Sudanian zone. Near water, trees common in the moister zones, such as *Adansonia*, *Anogeissus leiocarpus* and *Diospyros mespiliformis*, are found but scarcely form fringing forests.

(5) Sahara zone. In this there is only a highly specialized desert flora.

16.4.2 Gradients within the closed forest

The West African closed forest is far from homogeneous, but it has not proved easy to classify it into clearly defined phytosociological units. In the 1920s, when systematic surveys began, the forest had already been fragmented into more or less widely separated blocks; as already mentioned in Chapter 12, this has tended to obscure the fact that the variants within it could be more readily interpreted in terms of gradients than as discrete 'types' or associations. Further, most of the forests then remaining consisted of old secondary or 'depleted' forest (p. 459) which had already been exploited for mahogany and other valuable timbers; parts of some had been disturbed by 'farming' (shifting cultivation) before they became forest reserves. The reserves thus contained seral stages as well as mature forest and this makes valid comparisons between different forest areas difficult.

The earlier accounts of the closed forest, as mentioned above, distinguish between 'tropical rain forest' (tropical evergreen forest) and 'mixed deciduous' (semi-evergreen) forest. In Nigeria, Ghana and elsewhere in West Africa, the rain forests are near the coast and the mixed deciduous forests further north, bordering the Guinea savannas. As this distribution corresponds to a gradient from a high rainfall and short dry season to a drier and more seasonal climate (Chapter 8), the transition probably depends partly on climatic factors, but it was soon realised that geology and soil were also important. Mackay (1936), for instance, stated that in Nigeria the transition from rain forest to mixed deciduous forest was correlated with a change from deep soils derived from sedimentary deposits such as the Benin Sands to shallower soils overlying hard igneous rocks.

Some attempts were made to recognize different types of evergreen forest. A.P.D. Jones & R.W.J. Keay (unpublished) suggested (a proposal later withdrawn by Keay) that in Nigeria 'wet evergreen forest' could be distinguished from 'true rain forest', the latter being a type found in the high-rainfall area of the southeast. The greater part of what has been described as rain forest in West Africa, including the Nigerian forest considered in Chapters 2 and 17, is wet evergreen forest in the sense of Jones & Keay. Forests of this type have been described in detail by Aubréville (1938), Richards (1939), E.W. Jones (1955), Mangenot (1955), Guillaumet (1967) and others. When mature they are always mixed in composition and well-defined associations cannot be recognized. In most features of structure and physiognomy they resemble the rain forests of tropical Asia and America but have some characteristics which seem to reflect the comparatively severe seasonal drought of the climate (about four months with less than 100 mm rainfall). Thus the A and B storeys are more open, although it is not easy to be sure that this may not be due to past disturbance. Trees that become bare of leaves for several weeks every year are common, but less so than in semideciduous forest. Many other species are leaf-exchanging, losing all their old leaves at one time and producing a new crop more or less simultaneously. The presence in the undergrowth of geophytes with corms or tubers, some of which may lose their leaves for short periods in the dry season, at least, has been mentioned earlier (p. 123).

'True rain forest', as understood by Jones & Keay, occurs in both the eastern and western blocks of the West African rain forest. In Nigeria forests of this type exist in the Oban Hills and elsewhere in Cross River State as well as in the Korup National Park in the adjoining part of Cameroon; they may once have extended as far as the Niger delta. The Nigerian 'true rain forest' is in fact itself a westward extension of the 'forêt biafréenne' of Letouzey (1968) and Schnell (1976–77). In the western block there are forests of somewhat similar composition (but often with vicariant species) in southwestern Ghana, Côte d'Ivoire (Guillaumet 1967, Guillaumet & Adjanohoun 1971), southeastern Sierra Leone and Liberia. The most striking feature of both the eastern and western forests in this group is their species richness and the strong representation of trees of the Caesalpinioideae. Physiognomic and structural differences from the more widespread 'wet evergreen forest' (in the sense of Jones & Keay) are slight, but plants dependent on constantly humid conditions such as Hymenophyllaceae and epiphyllous bryophytes are perhaps more abundant in the undergrowth than in other types of West African lowland forest.

The semideciduous ('mixed deciduous') forests of West Africa, though mainly found along the northern fringe of the closed forest, are not in any sense transitional to savanna, although individuals of a few savanna species are sometimes found within the forest close to the savanna boundary. Even more than the rain forest, the semideciduous forest varies in its characteristics in a way difficult to define. Descriptions of its structure and composition are given by Chipp (1927), Aubréville (1938), Richards (1939), Foggie (1947), A.P.D. Jones (1948), Schnell (1952a), Taylor (1960), Lawson *et al.* (1970), Guillaumet & Adjanohoun (1971), Hall & Swaine (1976, 1981) and others. Schnell (1976–77) gives a brief account of the floristic groupings in the semideciduous forests of West Africa.

As the name implies, one of the most noticeable features of semideciduous forest is that a considerable proportion of the taller trees lose their leaves at least briefly in the dry season, but evergreen trees are also numerous and the leaf-shedding periods of the deciduous and leaf-exchanging species are not coincident: at no time does the forest look like a temperate deciduous forest in winter. The C storey trees and the smaller woody plants are almost all evergreen. Many species of trees are found in both semideciduous and evergreen forest, but the former is less rich floristically and some species common in evergreen forest are much less abundant in semideciduous forest, e.g. *Nauclea diderrichii*, *Pentaclethra macroloba* and *Strombosia* spp., or perhaps absent, e.g. *Carapa procera*, *Lophira alata*[7] and *Omphalocarpum procerum*. Other species are confined to the semideciduous forest or reach their greatest frequency there: among these may be mentioned especially *Triplochiton scleroxylon*, which is often abundant and may

form locally as much as 20% of the stand, *Celtis* spp., *Milicia* spp. and Sterculiaceae such as *Nesogordonia papavifera* and *Sterculia* spp. Some trees common in the semideciduous forest occur chiefly in clearings and secondary communities when found in evergreen forest situations. Some evergreen-forest trees extend northwards into semideciduous forest, much as closed-forest species form outliers in favourable habitats in the Guinea savanna zone. The undergrowth of the semideciduous forest, though in the wet season appearing very similar to that of the evergreen forest, tends to dry up in the dry season and includes more species that die down completely. Lianes are common, though perhaps less so than in evergreen forest, but epiphytes are usually few in species and individuals. The transition from one kind of forest to the other is usually gradual, as in the Omo and Shasha Reserves in Nigeria, but where there is a sudden change of soil it may be more abrupt.

16.4.3 Ordination studies

A less subjective view of the variations in composition of the West African closed forest than that given by the older attempts at classification is provided by statistical analysis of data from sample plots and strips. John B. Hall (1977) and Hall & Swaine (1976, 1981) have applied ordination techniques to data from Nigeria and Ghana, respectively. The results can be only briefly summarized here.

Hall's analysis deals with forty-six samples taken from enumerations made by the Nigerian Forestry Department before 1930 when the intensive exploitation and management which have greatly affected the composition of the forest reserves in later years was only beginning. The original data were enumerations of sample strips 66 ft (20 m) wide and 6600 ft (2 km) apart. The total area sampled in each reserve varied from 98 to 1450 acres (39–587 ha). Only trees ⩾2 ft (61 cm) g.b.h. were counted. In the analysis, which was on a presence–absence basis, only ninety-seven tree species were used because, although over 300 species were recorded in the data, some of the identifications were unreliable.

The sample strips were probably all old secondary forests which had been subject to man-made and natural disturbances to varying extents: different successional stages were therefore present. Samples with fewer than eleven stems (⩾2 ft g.b.h.) per 100 acres (40 ha) were rejected and it was thought that all the samples used in the analysis were at least seventy-five years old, an age when the species composition was probably no longer changing rapidly.

---

[7] Except apparently in Liberia where Kunkel (1965c, p. 12) says that *L. alata* is 'common in the northern transitional forest together with *Terminalia*, *Bombax*, *Celtis* etc.'

After standardization of the data principal components analysis was carried out. In interpreting the results two stand characteristics, species richness and sample size, were related to four environmental variables, soil type, geology, annual rainfall and altitude. The last was found not to be significant, which was not surprising, as all the samples were from below 100 m above sea-level.

Ordination involving the first component clearly separated samples from the two major soil types, the ferruginous tropical (ferrisols) and ferrallitic soils of D'Hoore (1964)[8], the former characterized by higher base saturation, cation exchange capacity and silt content than the latter. There was also a marked interaction between soil and rainfall. Below a mean annual rainfall of 1800 mm the soils were all of the ferruginous tropical type. At higher rainfalls both types of soil were found, but on the hard rocks of the basement complex, ferruginous soils were found even under rainfalls as high as 2400 mm, while on Tertiary sands (Benin Sands, etc.) ferrallitic soils occurred. At rainfalls of 2500 mm or more the soil is ferrallitic whatever the parent rock.

The ordination showed that the samples varied with soil and rainfall and could also be grouped according to their geographical location (western, central or southeastern). On the basis of the variables soil, rainfall and location the samples could be arranged in two groups with two subgroups. Group 1 is represented only in the western part of the Nigerian closed forest zone. Two forest reserves referred to in various places in this book, Omo and Okomu (pp. 36–7, 468–70, etc.) belong respectively to Groups 1b and 2a.

There is a fair amount of agreement between the grouping of the forest samples in Hall's analysis and the 'types' distinguished by earlier authors. Group 1a corresponds quite closely with semideciduous forest (Mixed deciduous forest); Groups 1b and 2a, 2b and 2c come within the definition of wet evergreen forest of Jones & Keay and of tropical evergreen seasonal forest of the UNESCO (1973) classification. Group 2c, found in the very wet southeastern part of Nigeria, is more or less equivalent to Jones & Keay's 'true rain forest' and the *forêt biafréenne* of Letouzey (p. 408–9 above).

In Ghana the forest zone differs in some respects from those in Nigeria. Except for a small area of sandy Tertiary sediments in the southwest, nearly all the Ghana closed forest is on hard, mostly igneous and metamorphic rocks of the Basement Complex. Because of the general southwesterly trend of the relief west of the Volta, both the climatic and vegetational gradients in Ghana, including the forest–savanna boundary, tend

to run northwest to southeast rather than east–west as in Nigeria. The forest soils are kaolisols of the ferruginous type (ferrisols) or ferrallitic.

The analysis of the Ghana forests by Hall & Swaine (1976, 1981) is not based on forestry enumerations like John B. Hall's Nigerian analysis but on their own sampling of selected areas, most (but not all) of which were in forest reserves. These areas all included gaps filled with climber tangles and other severely disturbed areas and some contained patches of swamp forest or savanna: such areas were rejected and all the samples were taken from 'closed-canopy forest', i.e. mature stands the composition of which might be expected to reflect the local environment rather than the recent history of the area. The plots were distributed over the whole closed-forest zone and random sampling was not attempted. Most of the plots were 25 m × 25 m (0.0625 ha) in size; a few were 50 m × 50 m (*ca.* 0.25 ha) and there were also some 'plotless' samples.

Ordination was by the reciprocal averaging technique of Hill (1973b); indicator species averaging (Hill *et al.* 1975) was also used. Some 749 species of vascular plants were used in the ordination; ground herbs, lianes and epiphytes were included and found to have as high an indicator value as tree species.

Seven major forest types were recognized and in two of these there were distinct subtypes.

(1) *Wet evergreen type.* This is found in a low-lying area of high rainfall in the extreme southwest of the forest zone. Trees seldom exceed 40 m in height and the forest has three 'more or less distinct storeys'. This is the most species-rich of all the types and includes many species not found elsewhere in Ghana, e.g. *Cynometra ananta, Pausinystalia lane-poolei* and *Scaphopetalum amoenum.* Tree species of high frequency include *Carapa procera, Dacryodes klaineana* and *Strombosia glaucescens.* Economic timber species are relatively scarce; because of this and the low soil fertility, forests of this type have been comparatively little disturbed.

(2) *Moist evergreen type.* These forests occupy a belt of mainly rather low-lying land north and east of Type 1. Most of the species are also found in other types, but some, e.g. *Diospyros gabunensis,* reach their greatest abundance here. The trees *Guarea cedrata* and *Strombosia glaucescens,* the treelet *Angylocalyx oligophyllus* and the lianes *Calycobolus africanus* and *Culcasia angolensis* are among the most frequent species. Taylor (1960) described a *Lophira–Triplochiton* association which is more or less equivalent to this type and regarded it as transitional between rain forest and semideciduous forest.

(3) *Moist semideciduous type.* This is the most extens-

---

[8] Types of kaolisol in the classification of Baillie (Chapter 10).

ive type, occupying about half the total forest area of Ghana. Most of it lies at 150–600 m above sea-level. Trees commonly reach heights of 55–60 m, greater than in any other type. Almost no species are confined to the moist semideciduous forest, but species reaching particularly high frequencies include the trees *Celtis mildbraedii* and *Nesogordonia papavifera* and the grass *Leptaspis cochleata*. The density of economically valuable trees is greater than in any other type; for this reason and because the soil is fertile and much used for food and cocoa farming, little of this forest is now undisturbed. Two subtypes, a southeastern and a northwestern, are recognizable.

(4) *Upland evergreen type.* This type is found only on two small areas above about 500 m in the Atewa Range and the Tano-Ofin Forest Reserve. In its evergreen foliage, low tree height and scarcity of economic timber species, it has much in common with the wet and moist evergreen types, but the ordination shows that it has a quite characteristic combination of species, most of which also occur in the wet evergreen, moist deciduous and other types. In the Atewa Range there are, however, numerous rare species not known elsewhere in Ghana. These were not used in the ordination; many of them are epiphytes for which the local climate is probably particularly favourable.

(5) *Dry semideciduous type.* Though not occupying as large an area as the moist semideciduous type, this one has a wider geographical range, extending from near Takoradi to the Volta region. It is also found as outliers in savannas and up to altitudes of 770 m (2500 ft). The ordination shows that its floristic composition is very varied, probably reflecting the diversity of conditions in which it occurs. Where dry semideciduous forest borders the savannas it is subject to ground fires which destroy the litter layer, though probably only at intervals of some years: in such areas there is a distinct 'fire zone subtype' in which species not normally found in closed forest occur, including the common savanna tree *Afzelia africana*, *Anogeissus* and the palm *Borassus aethiopum*. Thin-barked trees such as *Hymenostegia afzelii* are absent.

(6) *Southern marginal type.* This forms a narrow belt along the southeastern border of the closed forest where the rainfall is lower than in the dry semideciduous type, to which it has considerable floristic similarity. A pronounced feature is the tendency of some tree species to be gregarious, e.g. *Talbotiella gentii*. The region where this forest type is found has long been thickly settled and only fragments on rocky hills survive.

(7) *Southeast outliers.* Hall & Swaine used this name for the small patches of forest surrounded by grassland and thickets on the Accra Plains and the adjoining hills. The rainfall is very low, but fires are rare. In the very small area of forest now left the trees are small and few in species, yet still form a closed canopy.

Floristic characteristics such as number of species per unit area and gregariousness vary regularly and continuously; physiognomic features, for example tree height, leaf size and leaf shape, also show a similar pattern of variation. It is evident that the 'types' are not discrete natural phytosociological entities: they are based on 'areas' partitioned on a two-dimensional ordination and form parts of a continuum. It is also clear that the characteristics of the types are in some way related to environmental gradients.

Variation on axis 1 of Hall & Swaine's ordination is closely related to mean annual rainfall. This does not necessarily mean that total rainfall as such is important; in Ghana the length of the dry season is closely correlated with mean annual rainfall and is more probably the critical ecological factor. However, it seems indisputable that the characteristics of the seven forest types and their distribution are governed by moisture conditions. No doubt, as in Nigeria, soil factors and rainfall interact. Most of the soil characteristics measured were closely related to rainfall, but except for a relation to the calcium content of the soil (which mainly depends on the nature of the underlying rock) little correlation was found between forest type and soil characteristics.

The upland evergreen forest probably has a different moisture regime as well as a slightly lower mean temperature than the other types. It could be regarded as belonging to the lower montane rain forest formation (Chapter 17). 'The 'fire zone subtype' of the dry semideciduous forest also stands apart from the other types and subtypes because it is the only one in which burning is a normal feature of the environment.

It will be seen that these two ordination studies show that the gradients within the West African closed forest are much less simple than they appeared to be when it was customary to classify it into 'tropical rain forest' and 'mixed deciduous forest'. The closed forest is in fact a complex pattern of plant communities determined by climate and soil conditions. The natural pattern is of course overlaid by variations due to artificial disturbances and will become increasingly difficult to interpret in the future as more and more forest is modified by silvicultural management and exploitation, or destroyed. The nature of any particular stand depends on soil and anthropogenic factors as well as on climate, but it is evident that the most important environmental gradient is in the availability of water, especially in the dry season. This is greatest near the coast, in southwest

Ghana, southeastern Nigeria and Cameroon, and least near the borders of the savannas.

16.4.4 The closed forest – savanna boundary in West Africa

Closed forest, whether primary or secondary, and savanna are always strikingly different and the boundary between the two is usually sharp, sometimes as sharp as between woodland and cultivated land in Europe. In some places there is a 'tension zone' at the boundary formed partly by 'transition woodland', a mixture of fire-tender and fire-tolerant species in which the tree *Anogeissus leiocarpus* is often abundant, but this is usually narrow and may be no more than 50 m wide (Hopkins 1974). Although the actual transition from forest to savanna is generally sharp, in many parts of West Africa a mosaic of tongues and patches of forest and savanna occupies a wide belt between the closed forest and the Guinea savanna zones (Fig. 16.12). The complexity of these mosaics is better appreciated on aerial photographs or satellite imagery than on the ground (Fig. 16.10). The tall, straight-boled, relatively smooth-barked trees of the forest contrast strongly with the low, crooked, rugged-barked trees of the savanna, as does the shade-bearing undergrowth of the forest with the predominantly grassy ground cover of the latter (Fig. 16.11). The differences in structure, physiognomy, life-form spectrum (Fig. 16.1) and phenology (see Chapter 9) have been described by Hopkins (1962, 1968, 1970a, b), César & Menaut (1974), Sarmiento (1984), Sanford & Isichei (1986) and others. Even where they adjoin most of the forest and savanna species are different: thus in a transect at Degedege in Ghana, Swaine *et al.* (1976) found that out of 200 recorded species 60% were confined to forest or savanna (though elsewhere a few of these occurred in both). Although at the species level the two floras differ greatly, many genera are represented in each by 'vicarious' species strongly contrasting in physiognomy and phenology, e.g. *Daniellia*, *Terminalia*. In Nigeria *Khaya* is represented by *K. anthotheca* and *K. ivorensis* in the evergreen forests, by *K. grandifoliola* in the semideciduous forest and by *K. senegalensis* in the Guinea savanna. The small fire-tolerant tree *Lophira lanceolata*, often very abundant in derived and Guinea savannas, is so similar in most characters to the tall rain-forest tree *L. alata* that the two were long regarded as one species.

The contrast between closed forest and savanna vegetation is clearly dependent on their environments; the most obvious differentiating factor is certainly fire. The closed forest does not readily burn and fires penetrate

Fig. 16.11 Guinea savanna, in early dry season (December 1967). A Mole National Park, northern Ghana. Foreground recently burnt. B Open grassland, Mole National Park. C Savanna woodland on sandstone scarp, near Mpraeso, Ghana.

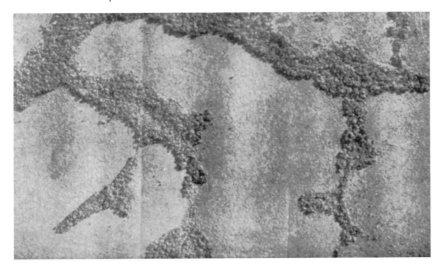

Fig. 16.12 Aerial view of gallery forest and treeless savanna, Siba area, Uganda. (Photo. Air Survey Co., London.) From Eggeling (1947, Fig. II). The savanna is dominated by *Pennisetum purpureum*.

it, if at all, only for short distances. On the other hand, most West African savannas are burned at least once a year, although, as Morgan & Moss (1965) showed from photographic evidence, the assumption that every single hectare of the West African savannas is burned every year is unfounded. Burning occurs during the dry season in scattered irregularly shaped patches. Some areas are frequently burned while others may escape, sometimes for many successive years. For this there are several reasons, an important one being that dry grass is much more inflammable than most kinds of woody vegetation, so that savannas with a considerable amount of grass are more likely to be burned than others.

Savanna fires, according to their severity, which depends partly on whether they take place early or late in the dry season, destroy most of the above-ground parts of the grasses and other herbaceous plants and other undergrowth including most young tree seedlings. The trees, though usually not killed, are often severely injured, accounting for their gnarled and deformed appearance; some species are more seriously affected than others. The taller trees shed their leaves after fires and produce flushes of young foliage later in the dry season. It is evident that the savanna flora must be subject to selection, varying in intensity from place to place and year to year: fire-tender species will be unable to compete with species resistant to burning (pyrophytes). The latter differ from rain-forest trees in their branching habit, bark (Fig. 16.2C) and many other morphological characters, as well as biological traits such as 'cryptogeal' germination (Jackson 1974).

Although savanna plants must be adapted to surviving recurrent fires, they also differ from forest plants in other respects, notably in their response to drought. Thus Okali & Doodoo (1973a, b) found that *Khaya senegalensis* is more drought-resistant than the forest species *K. ivorensis*, apparently because it is able to check transpiration when under water stress and has a greater tolerance of desiccation of the leaf tissues. Soil – vegetation relationships in forest and savanna have been compared by Sanford & Isichei (1986).

That factors besides fire and climate influence the forest – savanna boundary is shown by the work of Swaine *et al.* (1976) in west-central Ghana. Here in some districts savanna replaces forest over the whole landscape, excepting the forest galleries along streams, but in others strips of savanna penetrate into the forest along the middle slopes, forest still maintaining its hold on the hilltops as well as in the valleys. The two types of boundary are related to geology: the first ('savanna interfluve type') is mainly found on the Voltaian sandstones and the other ('forest interfluve type') on Basement Complex rocks which give rise to more clayey soils richer in nutrients. A similar situation seems to exist in Côte d'Ivoire (Spichiger & Pamard 1973).

Swaine *et al.* (1976) also showed that the forest – savanna boundary does not always coincide with the zone where annual rainfall is equal to or slightly less than annual evapotranspiration, as Davies & Robinson (1969) had found in Nigeria. In Ghana the rather inadequate data seemed to indicate that there was an interaction between soil and rainfall, so that the transition took place at different precipitation : evapotranspiration ratios on different types of soil and topography.

At Degedege, Markham & Babbedge (1979) studied the soils and vegetation on two transects from gallery forest through savanna to dry semideciduous forest on the hilltops. Principal components analysis showed that phosphorus, calcium, potassium, pH and water content were all higher in the forest topsoils than in those of the savanna, but soil depth and texture were probably the primary soil factors separating the vegetation types.

It thus seems probable, as Morgan & Moss (1965) suggested, that the forest–savanna boundary in West Africa is not, as often supposed, controlled by climate and fire alone, but by a complex of interacting factors which include soil conditions and possibly other variables.

16.4.5 Changes in the forest–savanna boundary[9]

It is often assumed that in West Africa, and tropical Africa generally, savannas are everywhere rapidly gaining ground at the expense of the forest. This is sometimes regarded as evidence for a desiccation of the climate (p. 13). Aubréville (1949) in his classic account of this process of 'savannization' says little about when the savannas originated or how fast the change was proceeding. It is certain, however, that enormous areas of Africa which are now savanna-covered were forested not more than one or two centuries ago. However, the process of 'savannization' does not necessarily imply an irreversible change and there is in fact little good evidence of major changes in the position of the boundary between the closed forest and the Guinea savanna zones in recent times. Morgan & Moss (1965) (see also Hill & Randall 1968) found that in the area of Nigeria west of Ibadan aerial photography showed no widespread increase of the savanna area between 1953–54 and 1962–63. Swaine et al. (1976) believe that in Ghana the forest–savanna boundary is at present more or less stable. It is probable that if an overall shift in the balance between forest and savanna in Africa is in fact taking place, it is very slow and could only be monitored by aerial photography or satellite imagery over a long period.

Local invasions of savannas by forest trees can be observed in many parts of Africa, e.g. in Côte d'Ivoire (Spichiger & Pamard 1973) and Nigeria, as well as in Zaïre (Germain 1945) and Cameroon. Hopkins (1962) found that in the Olokemeji Forest Reserve in Nigeria the forest had been advancing into the derived savanna at a rate of 3–4 m per year for twenty-two years. Near the forest–savanna boundary in Cameroon, Letouzey

(1968) found that thousands of hectares of 'degraded forest' detected in aerial photographs were in fact areas where forest trees had invaded the savannas; the advance of the forest could be traced in photographs taken at four-year intervals. These local expansions of the forest area seem to have followed the cessation of shifting cultivation, burning and grazing due to migrations of the human population.

In the past half century, large parts of the West African closed forest have been converted into monocultures of oil palms, *Gmelina arborea* and *Pinus* spp. Whether, if these plantations are not permanently maintained, this will lead to further extensions of derived savanna, cannot be foreseen.

Although there is no firm evidence of a climatically determined shift in the forest–savanna boundary in recent years, it is certain (see Chapter 1) that during the past million years West Africa has passed through many climatic phases, during some of which the savannas expanded and the closed forest was probably confined to small refugia. During the moist periods the forest expanded but savanna vegetation may have survived in habitats unfavourable for forest trees within the forest zone. It is thus possible that some West African savannas may, like the 'islands' of savanna in Amazonia, be relics.

Various features of present-day West African vegetation are perhaps also survivals from periods when the forest–savanna boundary was differently situated, for instance the stands of the forest trees *Heritiera utilis* and *Triplochiton* in northern Sierra Leone (Jaeger & Adam 1967) and the 'kurame' of northern Nigeria. It is also tempting to speculate that some forest species such as *Balanites wilsoniana* and *Citropsis articulata* (see p. 76), which look physiognomically 'out of place' in the closed forest, may be relics of a past dry period.

16.4.6 Forest and savanna in central Africa

In Zaïre the central block of closed forest is bordered by large areas of savanna, varying in physiognomy from open grasslands with few trees to savanna woodlands (see White 1981, 1983). The various types of savanna in Zaïre north of the equator are in general similar in structure and floristic composition to the derived, Guinea and Sudan savannas of West Africa and need not be described here. To the south and east the closed forest is bordered by the Zambezian savannas, which have similarities to the South American cerrados and are floristically richer as well as different in other respects from those of West Africa (Mullenders 1954, White 1965).

Peeters (1964) has given an account of the forest –

---

[9] See also Chapter 7, pp. 199ff.

savanna boundary north of the Zaïre river (*ca.* 2°N in the west and 1°N in the east), accompanied by maps based on aerial photographs taken in 1951–59. Three chief types of landscape are recognized:

(1) The forest region, consisting of several types of rain forest (Chapter 12) and semideciduous forest. The forest region occupies the largest part of the whole area and between *ca.* 21° and 22°E it extends to beyond 4°N.
(2) Regions mainly of savannas with scattered trees (*savanes arbustives*) with numerous, sometimes extensive, forest galleries. North of the forest this is the more widespread of the two types of savanna landscape. Locally it includes what Troll (1950) calls 'savanna with ravine forests' and termite savannas.
(3) Regions of woody or grassy savannas in which forest galleries are unimportant or quite absent. These regions are mainly found east of 29°E.

As well as these three main types of landscape there are others of less importance, such as the swamps of the lower Semliki river, the region of coffee plantations near Bambesa and recently deforested areas near roads and villages.

Direct contact between continuous forest, both evergreen and semideciduous, and the savannas can be observed, most often between forest and region 2. In the northeast, between Isiro (Paulis) and Molegbwe, semideciduous forest occupies an intermediate position between the evergreen forest and an area of savanna with galleries. The forest–savanna boundary is everywhere very indented and irregular. Within the savanna regions, islands of forest of various sizes are common, particularly in those with galleries, but they decrease in size and number northwards.

According to Peeters, the extensive savannas in northern Zaïre originated from semideciduous and deciduous forest and here the forest–savanna boundary shows no correlation with climatic, soil or other physical factors; it is probably determined by fires and other human activities. It is only where the savannas are in contact with evergreen forest that the boundary is environmental, corresponding approximately to the limits of the relatively non-seasonal rain-forest climate.

Within the forest region there are many small areas of more or less treeless grassland. All of them are subject to fires; they are probably derived savannas of fairly recent origin. In the central part of the Zaïre basin, particularly near Mbandaka (Coquilhatville) (0°3′N, 8°28′E), there are numerous islands of grassy vegetation, some only a few hectares in extent, known as 'esôbé'. Robyns (1936) believed that some of these were 'natural savannas' in which edaphic factors pre-

vented the growth of trees, though others were being invaded by the surrounding forest. However, more recent work by Deuse (1960) and Germain (1965) shows that in the esôbé the absence of trees is due to burning and not to soil conditions and succession to forest could probably take place in all of them.

Germain (1965) divides the esôbé or 'intercalary savannas' of equatorial Zaïre into two types, the 'esôbé rivulaires' on the recent alluvia of the flood-plains, and the esôbé of higher ground. Both types are usually dominated by *Hyparrhenia diplandra* and other tussock-forming grasses. On the 'esôbé rivulaires' the soils vary from white sand to pale sandy clays and are poor in nutrients. Flooding by the rivers takes place for periods depending on the level of the ground, but when the water level is low fires are frequent. On some soils associations including species of *Drosera*, *Polygala* and *Xyris* are found which are floristically similar to communities in the coastal savannas of West Africa as well as in other parts of the tropics. In permanently wet hollows patches of *Sphagnum planifolium* are found and peat is formed.

Germain found that when protected from fire the 'esôbé rivulaires' could be colonized by swamp-forest trees or secondary forest species such as *Anthocleista* spp. or *Harungana madagascariensis*, which are able to establish themselves on termite mounds and other well-drained sites. In one esôbé with a heavy soil an eight-year fire exclusion experiment showed that under such conditions colonization by trees might be very slow.

The esôbé on higher ground which, like those on the river alluvium, are frequently burnt, are typical derived savannas. The presence of relict oil palms and patches of *Imperata* often betray their origin from former cultivation. A fire-protection experiment showed that, even after only three years, invasion by *Borassus* palms and woody dicotyledons took place.

16.4.7 Climax vegetation of African savanna areas

What relation does the present vegetation of the West African savannas bear to the climatic climaxes of the areas they occupy? How far are human activities responsible for their existence or at least for their present extent? Such questions are often asked, but they can be discussed only briefly here.

The role of fire. There is no doubt that the sharp boundaries between closed forest and savanna are maintained by burning, but this does not necessarily imply that they came into existence as a result of anthropogenic fires, although the derived savannas within the forest zone certainly owe their existence to burning and cultivation. The greater part of the Guinea savannas has been frequently burnt for a very long time; as mentioned earlier, much of this zone has also been affected by shifting cultivation. Grass firing for hunting probably goes back far into African prehistory: regular burning to clear land for farming may have begun in West Africa at least 3000 years ago (Sowunmi 1986). No doubt as the population has increased fires have become more frequent and extensive. In northern Amazonia Saldarriaga & West (1986) found that fragments of charcoal were 'ubiquitous' in the soil of terra firme forests (including caatinga) and in igapó. They occurred at various depths and were radio-carbon dated to ages from 6250 yr BP to the present.

However, fires are not necessarily anthropogenic: there is good evidence that in tropical Africa, as in many other parts of the world, lightning can start fires in seasonally dry vegetation, especially if it includes a considerable amount of grass (Lebrun 1947, E.W. Jones 1963, Morton 1986, p. 253). It seems, therefore, that in the Guinea savanna zone, though not in the closed forest, fire must be regarded as a normal feature of the environment. In trying to assess the importance of fire to the past vegetation of the Guinea savanna zone, it must be remembered that in the past there have been changes of climate in West Africa, some periods having been drier than the present (see Chapter 1).

The numerous more or less isolated outliers of evergreen and semideciduous forest within savanna areas do not necessarily represent the climatic climax. Forest outliers in the Guinea savanna zone include patches of forest on interfluves as well as galleries along streams. The former, where not artificially planted, can usually be shown to occur where edaphic and/or topographic conditions are favourable; some may have been formerly connected to gallery forests by strips of forest now destroyed. All such outliers are probably dependent on special local environments, as E.W. Jones (1963) showed in his study of the kurame forests on the western foothills of the Jos Plateau in Nigeria.

Fire protection experiments. The first experimental approach to the problem of the climax in West African savannas was made by MacGregor (1937). In 1929–30 three sample plots each 142 ft × 132 ft (43 m × 40 m) and 0.43 acres (0.17 ha) in extent were laid out at Olokemeji, Nigeria, in derived savanna close to the margin of the closed forest (Fig. 16.13). At the beginning of the experiment the vegetation on all the plots was more or less similar and the whole site was burnt, coppiced and cleared of trees. Plot A was subsequently

A

B

Fig. 16.13 Fire protection experiments in derived savanna, Olokemeji, Oyo State, Nigeria (November 1977). A Secondary forest on Plot C, fire-protected since 1929. B Plot A (foreground), burnt every year in March since 1929.

burnt every year in March (end of dry season), Plot B was burnt yearly early in the dry season (December) and C was completely protected from fire. The March burning was fiercer than in December because the vegetation was then drier. These treatments have been continued up to the present. After six years the woody vegetation of the unburnt Plot C had increased considerably in height and had been invaded by eight fire-tender (forest) species, also by the exotic *Gmelina arborea* from an adjoining plantation. Plot A showed little change and on the early-burnt Plot B one fire-tender species was recorded. A survey in 1957 (Charter & Keay 1960) showed that after 28 years as many as 64% of the woody species on Plot C were fire-tender forest trees and 10% invaders from the plantations: only 17% of the species were fire-tolerant. The early-burnt Plot B now included 7% of fire-tender species and had 'assumed the features of derived savannas at a stage of development into *Anogeissus* Transition woodland' (Charter & Keay 1960, p. 20). On Plot A the trees had become 'noticeably gnarled and crooked' and fire-tender species were of course absent.

The plots as they were in 1979 are described by Sanford & Isichei (1986). The fire-protected plot C was then a 'regrowth' (young secondary) forest and its grass cover had disappeared. The vegetation of the others was derived savanna, the early burnt plot B resembling southern Guinea savanna in structure and physiognomy and the late burnt plot a northern Guinea savanna.

Fire protection experiments in derived savannas giving more or less similar results to those at Olokemeji have been carried out near Onitsha (Nigeria) (Keay 1959b) and near Bouaké (Côte d'Ivoire) (Menaut 1983). At Kpong, near the forest boundary in southeastern Ghana, a 0.5 ha plot of Guinea savanna protected from fire and grazing from 1957 became invaded by *Milicia excelsa* and other forest trees. By 1989 the largest were *Ceiba pentandra* up to 22 m high (Swaine *et al.* 1992).

It may be concluded that the Olokemeji and other fire protection experiments have demonstrated that forest can re-establish itself on derived savannas close to existing forest, but contribute little to understanding what the vegetation of the Guinea savanna zone was like before it was extensively modified by human impacts. Aubréville (1938, 1949, pp. 319–22) speculated that it might have been 'forêt claire' (open semideciduous or deciduous woodlands). Jones (1963) suggested that under primitive conditions and before the advent of shifting agriculture, forest like the present 'kurame' (outliers of semideciduous forest) were more extensive and the zone as a whole was a mosaic of forest, open woodland and savanna varying with topography and soil. Schnell

(1976–77, pp. 224–31), who described forest outliers in various parts of tropical Africa comparable to the 'kurame', held a similar opinion.

16.5 Conclusions

This very incomplete survey suffices to show that the transition from tropical rain forest to other plant formations at its climatic limits takes place in different ways in different parts of the tropics. Where soil conditions are not unfavourable for trees and artificial fires are infrequent, ecotones are found in which there is a transition from evergreen to semideciduous and deciduous forests, and as conditions become still drier, from these to thorn woodlands and other non-forest formations. These ecotones follow climatic gradients in which annual rainfall decreases and becomes more seasonal in its distribution, though in a given area the nature of the climax vegetation also depends on soil and topography. Ecotones of this kind can be observed in tropical America and the Far East, but in Africa they are uncommon. Here the closed forest generally abuts directly on savannas. In many parts of the tropics, even before the massive forest destruction of recent times, the 'tension zone' at the limit of the rain forest was often a mosaic of evergreen forest, semideciduous forest and savanna rather than a simple catena of plant formations.

Savannas are communities of very varied structure and status: their chief common feature is the dominance of grasses in the ground layer. Most, if not all of them, are subject to recurrent fires. Though savannas reach their greatest extent in climates where the rainfall is too low and too seasonal for rain forest to be the climatic climax, they also occur in ever-wet climates, sometimes as islands in rain forest.

As Sarmiento & Monasterio (1975) have emphasized, it is an over-simplification to regard all savannas as either climatically determined or fire-climaxes. The occurrence of forest or savanna vegetation is determined by competition between trees and grasses and this depends on differences in their water requirements and responses to burning (see Walter 1971, pp. 238–51). Both of these involve a complex of interacting environmental factors.

It is important to distinguish between 'derived' (or secondary) savannas which are of recent origin and the result of cultivation and frequent burning, and savannas such as the cerrados and llanos of South America which, though maintained by burning, are much more ancient and probably of natural origin. 'Derived' savannas, if protected from fire, soon revert to forest; 'natural' savannas are much more stable and do not quickly become invaded by forest trees. The flora of the 'derived' savannas is very poor in species, but that of the 'natural' savannas is species-rich with numerous endemics testifying to its age. However, the taxonomic relationship between the ancient savanna floras and that of the rain forest indicates that the former were probably derived from the latter, but perhaps as far back as the early Holocene or even earlier.

Chapter 17

The tropical rain forest at its altitudinal and latitudinal limits

17.1 Altitudinal zonation in the humid tropics

Changes in vegetation with increasing altitude are no less striking in the humid tropics than in temperate regions. As a forest-covered tropical mountain is climbed, the physiognomy of the dominant tree species as well as the structure and floristic composition of the vegetation change (Fig. 17.1, Table 17.1). The tall tropical rain forest of the lowlands gives place to other formations, also evergreen, but lower in stature, simpler in structure and less rich in tree species. The lowland tropical flora is left behind and replaced by a montane flora in which some genera, and even some species, are temperate. Tropical rain forest gives way first to lower montane rain forest; above this there is upper montane rain forest.[1] The latter, especially on exposed ridges and isolated peaks, consists of dwarf crooked trees smothered with mosses, hepatics and other epiphytes. This very characteristic type of vegetation is often called mossy forest, elfin woodland or cloud forest.[2] Lower and upper montane forests are not always easy to distinguish. The two formations are sometimes referred to as temperate rain forest, but as tropical montane climates differ fundamentally from those of temperate regions (Chapter 7),

this name should be kept for formations outside the tropics.

On the highest tropical mountains there may be a fourth forest zone, subalpine forest (pp. 425ff), extending from the upper montane forest to the climatic tree limit. Above this are grasslands, the 'woodlands' of giant senecios of the East African mountains, the paramos of the Andes and other plant formations which have no exact parallel outside the tropics. This highest vegetation zone is often termed alpine (or in Africa Afro-alpine), but is better termed the tropical-alpine or paramo zone. Troll (1955, 1959), who suggested the latter name, pointed out that its climate differs fundamentally from that of alpine zones in temperate regions in having a much greater daily than seasonal temperature range (*Tageszeitklima*), in contrast to the alternating summers and winters of higher latitudes (*Jahreszeitklima*), with quite different effects on plant and animal life.

The altitudinal zonation of vegetation on mountains in the humid tropics, which is to some extent similar in the Palaeo- and Neotropics, depends on a gradient of climate, but the relation of this gradient to the changes in vegetation is by no means simple. The variation of some of the climatic factors with altitude is considered in Chapter 7. In general temperature falls with increasing altitude, but the lapse rate (fall of temperature per 100 m increase of elevation) varies considerably and depends on cloudiness and other factors (pp. 173–4). It is important to realise that not only temperature, but every other climatic factor affecting plant growth, changes with altitude, including precipitation, atmospheric humidity, wind velocity and radiation.

---

[1] In Edn 1 these three altitudinal zones were termed Lowland, Submontane and Montane, following van Steenis (1935b) (see pp. 422–3), but in recent years the terminology used here (based on Burtt Davy 1938) has become generally adopted.

[2] The term cloud forest should perhaps be reserved for humid tropical forests on mountains in permanently or seasonally dry regions where local topographic conditions lead to frequent clouding. See pp. 439–40.

Table 17.1. *Structural and physiognomic characteristics of rain-forest formations on tropical mountains*

	Formation type			
	Lowland tropical	Lower montane	Upper montane	Subalpine
Trees				
Height (m)[a]	25–45 (67)	15–33 (37)	15–18 (26)	1.5–9 (15)
Buttresses	Frequent, often large	Infrequent; if present usually small	Usually none	None
Leaves				
Size[b]	Mesophyll	Mesophyll and notophyll	Microphyll and notophyll	Nanophyll and microphyll
Drip-tips	Common in lower storeys	Frequent in lower storeys	Usually none	None
Compound	Common	Occasional	Rare or none	Usually none
Climbers				
Large woody	Numerous	Few or none	Usually none	None
Small or herbaceous	Frequent	Sometimes abundant	Frequent; often epiphytic	Few
Epiphytes				
Vascular	Few, except on large emergent trees	Often abundant	Abundant	Abundant
Bryophytes	Rather scarce except near streams, etc.	Abundant but seldom forming thick masses	Very abundant; often forming thick blankets on trees	Very abundant

[a] The heights are those of the 'canopy' (highest more or less continuous layer of tree crowns). Heights of emergents are given in parentheses.
[b] The 'predominant' size on Raunkiaer's classification as modified by Webb (see Chapter 4).

Source: Based on Grubb & Tanner (1976).

These factors also do not vary consistently with height: thus on high tropical mountains rainfall often increases from sea-level to a zone of maximum precipitation above which it decreases. On most tropical mountains there is a distinct 'cloud belt' of persistent fog (mist), which, as will be seen later, has important effects on the vegetation.

The changes in soil characteristics with increasing altitude were briefly mentioned in Chapter 10. How far the physiognomic and other differences between the vegetation of the altitudinal zones depend directly on temperature or on indirect effects of altitude such as cloudiness, exposure to wind and soil characteristics has been much discussed (see pp. 422ff. below).

Because the rate of change of climatic factors with height above sea-level varies, the actual altitudinal limits of the vegetation zones differ considerably on different mountain ranges and often on parts of the same mountain. For example, the transition from lowland to lower montane forest may occur at 700–800 m, but on some mountains not until 1500 m or even higher. Similarly the lower limit of upper montane forest, while commonly at about 1100–1500 m, may be lower or considerably higher. In general on small isolated peaks and ridges the zones tend to be telescoped and their altitudinal limits lower than on extensive mountain ranges; usually, also, the limits are lower on coastal mountains than on those further from the sea. This is the so-called *Massenerhebung* (mass elevation) effect, which is well known in the European Alps and other temperate mountains.

In the Alps the climatic tree limit is about 700–800 m higher in the central ranges than in the fringing *Voralpen* (Schröter 1926, p. 32) and the difference seems to depend on the more oceanic climate of the latter rather than differences in mean temperature or the length of the growing season (Brockmann-Jerosch 1919, Ellenberg 1988). In the humid tropics the *Massenerhebung* effect is comparable to that on temperate mountains, but, as will be seen later, the climatic factors

Fig. 17.1 Zonation of vegetation on Mt Bellenden Ker, Queensland (17°S, 145°E). A – D photographed from cable car.
A Disturbed lowland tropical rain forest with many lianes. B Lowland or lower montane rain forest at *ca.* 300–600 m.
C Lower montane forest at *ca.* 800 m. D Transition to upper montane forest at *ca.* 1200 m. E Canopy of upper montane forest
near summit (1561 m), dominated by *Leptospermum wooroonooran* and *Cinnamomum propinquum*.

actually responsible are probably different. On some coastal mountains in the tropics, exposure to wind may be responsible for lowering the vegetation zones; Beard (1946b) found that on the Northern Range of Trinidad the zones of vegetation are depressed towards the eastern end, which faces the prevailing trade wind. On many tropical mountains, however, exposure to wind seems less important than the incidence of persistent mist and cloud.

On mountains in the humid tropics, even where there are no permanent human inhabitants, large areas of natural vegetation have been drastically modified by human activities, especially by burning. The result has been a widespread replacement of montane forests by grasslands and other secondary communities; on steep slopes this often leads to landslips and severe soil erosion. Although some montane grasslands in the humid tropics may be of natural origin and dependent on local soil conditions and lightning fires, there is little doubt that they are for the most part anthropogenic and, like lowland secondary savannas, maintained mainly by man-made fires.

17.1.1 Zonation in the mountains of Malesia

A large part of the Malesian rain-forest area is more or less mountainous; the altitudinal zonation of the vegetation has been more studied here than in any other part of the tropics. Few of the mountains in this part of the world exceed 3500 m and most of them do not reach the climatic tree limit. Only the isolated peak of Kinabalu in North Borneo and the high ranges of New Guinea are over 4000 m high; the latter alone have permanent snow. Many of the volcanoes of the Sunda Islands exceed 3000 m and are partly or completely treeless near the summit, but the absence of trees is probably due to fires or edaphic conditions; it is doubtful whether any of the Sunda Islands reach the climatic tree limit.

Van Steenis (1935b, 1961a, 1972) analysed the zonation on the mountains of Malesia on a floristic basis. Using the method adopted by Sendtner for the European Alps, he attempted to determine the delimitation of the zones objectively by tabulating the recorded upper and lower limits on the Malesian mountains of about 900 'microtherm' (montane) species (Table 17.2). He found that for a large proportion of these the limits clustered round certain levels or within narrow bands of altitude. These critical levels could be regarded as natural boundaries between the zones.

From his table van Steenis concluded that there were six altitudinal zones in Malesia (excluding the littoral zone below high tide mark); the lowest was subdivided into two subzones. The critical levels were at *ca.* 1000, 1500, 2400, 4000 and 4600 m. Though based on floristic rather than phytosociological or physiognomic criteria, the zones of van Steenis agree fairly well with those which, as stated earlier, are found to be general throughout the humid tropics.

Van Steenis' critical levels are only rough averages; the altitudinal limits of the zones vary considerably in different parts of Malesia. The variations do not depend on latitude, but chiefly on the *Massenerhebung* effect. For example, on the small isolated Gunung Santubong on the Sarawak coast the transition to mossy upper montane forest takes place at *ca.* 750 m. On another coastal mountain, G. Silam (Sabah) it occurs at 610–770 m (Proctor *et al.* 1988, Bruijnzeel *et al.* 1993). On G. Belumut in Johore at 840 m (Holttum 1924), but on the main range of the Malay Peninsula it is at *ca.* 1500 m (Whitmore 1984a) and in West Java not below *ca.* 1650 m (Seifriz 1923). The limits of the other zones vary similarly. The very marked *Massenerhebung* effects in New Guinea are discussed later.

A characteristic feature of Malesian montane vegetation is the contrast in geographical affinities between the lowland and the high mountain flora. In the lowland rain forest most of the species belong to genera of Indo-Malayan or Palaeotropical distribution and this is almost equally true of the lower montane forest, but in the upper montane forest a considerable proportion of the flora belongs to genera, such as *Leptospermum*, *Phyllocladus* and *Styphelia*, which are mainly Australasian. It is remarkable that the Australasian element does not usually increase gradually with altitude as might be expected, but increases suddenly at the boundary of the upper montane forest. In Malesia, as in tropical and subtropical Australia, the Indo-Malayan and Australasian floras tend to be ecologically separated (Richards 1943). The history of the Malesian mountain flora has been discussed at length by van Steenis (1934, 1935b, 1936, 1972).

Zonation in New Guinea. The mountains of New Guinea are probably the most favourable region in the Palaeotropics for studying the altitudinal zonation of vegetation. Since many of them are over 4000 m high and some, such as Puntjak Jaya (Carstenztop) in West Irian (5030 m), have glaciers and permanent snowfields, a complete sequence of zones from sea-level to the alpine region can be seen. The New Guinea mountains are mainly non-volcanic and the development of climax vegetation is only locally prevented by eruptions and toxic soil conditions, although landslips triggered by

Table 17.2. *Altitudinal zones in humid Malesia as defined floristically by the Sendtner method (see text)*

Forest formations as in Table 17.1.

Zone	Altitude (m)	Vegetation	Formation
Tropical	0–1000	Closed high forest	Lowland tropical rain forest and 'monsoon forest'
Lowland subzone	0–500		
Colline subzone	500–1000		
Submontane	1000–1500	Closed high forest poor in mosses	
Montane	1500–2400	Closed high forest: above 2000 m with decreasing stem diameter and increasing quantity of moss	Lower montane forest
Subalpine	2400–3600	Dense low forest with emergents, often mossy and with conifers	Upper montane and subalpine forest
	3600	FOREST LIMIT	
		Low shrubs, isolated or in clumps; conifers	
Alpine	4000–4500		
	4000	TREE LIMIT	
		Stony desert with grasses, etc.	
Nival	4600+	Permanent snow	

Source: After van Steenis (1972, p. 18), slightly modified.

earthquakes are rather frequent. A further advantage is that much of the montane forests has been little affected by people and in many places hardly at all.

Until the middle of the twentieth century, most of New Guinea had been little explored botanically. Even the presence of *Nothofagus*, now known to be one of the most important genera of trees in the montane forests, was not recognized until 1953 (van Steenis 1953). Since then the literature on New Guinea montane vegetation has become voluminous: among general accounts those of Gressit (1982), Paijmans (1976) and Hope (1980) may be mentioned. The detailed account by Grubb & Stevens (1985) of the forests in the Fatima Basin (*ca.* 6°00′S, 145°11′E) and on the adjoining Mt Kerigomna (3670 m) is particularly useful as the montane rain forests of this area seem to be fairly typical of those on the 'main spine' of the island. This paper also includes a general review of the montane and subalpine forests of Papuasia as a whole. The ecology of the subalpine and alpine vegetation of Mt Wilhelm (4510 m), some 30 km northwest of Mt Kerigomna, has been much studied (Wade & MacVean 1969).

The New Guinea montane forests are very species-rich (though less so than the lowland rain forests) and their composition shows much local variation. Although many species are endemic, the affinities of the flora are mainly Indo-Malayan, but, as might be expected, Australasian taxa are more strongly represented than in the western parts of Malesia. Southern conifers (*Araucaria, Papuacedrus, Podocarpus*, etc.) are conspicuous features of the montane forests and form an increasing proportion of the stand with increasing altitude. Of the exclusively southern-hemisphere genus *Nothofagus*, there are fourteen species in the montane forests of mainland New Guinea (Hill & Read 1987).

Although the zonation is complex and less clear-cut than on the lower mountains of Borneo and Malaya, four altitudinal zones (or formations) are now usually recognized in the primary rain forests of New Guinea: lowland, lower montane, upper montane and subalpine. In the earlier literature other classifications were used (see Brass 1941, 1956–64, Robbins 1970). These zones can be subdivided and several associations and facies can often be distinguished in each.

Lowland rain forest and freshwater swamp forest still cover a large part of New Guinea, except in the seasonally dry southeastern part (pp. 364–5), extending up to *ca.* 1500 on the main ranges, but near the coast only to *ca.* 700 m (Grubb 1977b). The lowland forest is varied in structure and composition but is in many respects similar to the lowland forests of Borneo and Malaya. Conifers such as *Agathis* and *Araucaria* occur, but are only locally frequent at low altitudes. Towards its upper limit the composition of the lowland forest changes and on high ridges stands of dipterocarps (chiefly *Anisoptera thurifera*, *Hopea* spp. and *Vatica rassak*) are characteristic (Paijmans 1976).

The second zone, the lower montane forest, is superficially rather similar to the lowland rain forest, but is less tall; large emergent dicotyledonous trees are scarce or absent. In the Fatima Basin the canopy is at *ca.* 27–33 m (Grubb & Stevens 1985). Physiognomic features typical of lowland forest, such as buttressing and cauliflory, are lacking or less developed. The leaves of the trees are predominantly mesophylls, notophylls and microphylls: their average size is probably smaller than in the lowland forests. Large woody lianes are generally absent and epiphytes, especially bryophytes, are more abundant.

Floristically the lower montane and lowland rain-forest formations differ considerably, though some species occur in both. The lower montane forest is almost certainly less rich in tree species, but data for a

valid comparison are not available; a few genera, e.g. *Elaeocarpus* and *Litsea*, have more than four species, but most have fewer. Many genera and families (including the Dipterocarpaceae) which are well represented in the lowland forests are usually absent, while others, such as the Fagaceae, are more abundant.

The composition of the lower montane forest varies locally and regionally. Three chief associations can be recognized: (1) 'oak forest', dominated by *Castanopsis acuminatissima* and/or *Lithocarpus* spp.; (2) 'beech forest', dominated by *Nothofagus*, of which some species occur locally as single dominants; and (3) 'mixed forest' in which no single genus or species forms a large proportion of the stand; the lower montane forest of the Fatima Basin and Mt Kerigomna described by Grubb & Stevens (1985) is of this last type. The distribution of these associations is complex and hard to interpret. Oak forest is generally found at *ca.* 1000–3000 m and mixed forest through the whole altitudinal range of the formation. Beech forest can probably be regarded as characteristic of the upper part of the lower montane zone; it commonly occurs at *ca.* 2100–2700 m, but Hynes (1974) found that in some places stands of *Nothofagus* occur down to 750 m and as high as 3000 m.

One of the most impressive features of New Guinea, especially seen from the air is the montane forests in which huge trees of hoop pine (*Araucaria cunninghamii*) and klinki pine (*A. hunsteinii*) over 60 m (and sometimes over 80 m) high stand out above a dense mixed canopy of broad-leaved trees as an open 'overstorey' (Fig. 17.3). *Araucaria* stands are common in the lower montane zone but not limited to it. *A. cunninghamii*, which is found throughout New Guinea, and *A. hunsteinii*, which is restricted to the eastern part of the island, have a wide altitudinal and ecological range, and are found from *ca.* 150 m to over 2000 m, and in slightly seasonal as well as ever-wet climates (Havel 1971, Gray 1973, 1975, Enright 1982). The timber of both species is much exploited and they have been extensively planted, especially in the Bulolo-Watut area.

The highest forest formations, those of the upper montane and subalpine zones, have little in common with lowland rain forest. The former, with a lower limit at *ca.* 3000 m in the Fatima area, has a characteristically compact canopy of broad-crowned trees which are seldom more than 20 m high, though there are occasional taller emergents. The leaves of the trees and shrubs are markedly smaller than in the lower montane forest and are mainly notophylls, microphylls and nanophylls, which are markedly sclerophyllous in texture. Dicotyledonous trees, especially Ericaceae and Myrtaceae, are mixed with conifers (*Dacrycarpus*, *Podocarpus* and *Papuacedrus*). The number of species, particularly

Fig. 17.2 *Nothofagus* forest, Daulo Pass (2466 m), Papua New Guinea (1975).

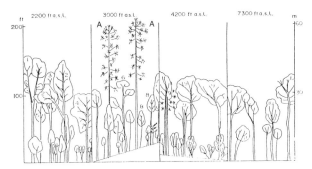

Fig. 17.3 Profile diagrams to show altitudinal gradient, Bulolo Valley, Papua New Guinea. From Havel (1971, Fig. 2). Key: A *Araucaria hunsteinii*, 1 *Garuga*, 2 *Xanthophyllum*, 3 *Dysoxylum*, 4 *Eugenia*, 5 *Mangifera*, 6 *Endiandra*, 7 *Argyrodendron*, 8 *Aglaia*, 9 *Castanopsis*, 10 *Nothofagus*, 11 *Xanthostemon*.

Fig. 17.4 Montane rain forest with conifers at *ca.* 2000 m, Arfak Mountains, western New Guinea (Irian Jaya). From van Straelen (1933, pl. xlv, Fig. 1).

of herbs, is much smaller than in lower montane forest. The abundance and luxuriance of bryophytes varies, but in typical upper montane forest the trunks and branches of the trees are covered with masses of hepatics, mosses and other epiphytes which may be many centimetres thick.

The subalpine forest extends from the upper limit of the upper montane forest to the tree limit; on Mt Kerigomna this is from *ca.* 3500 m to *ca.* 3700 m. It is often not continuous and is broken into blocks by stretches of grassland with scattered trees or shrubs.

The trees are even smaller than in the upper montane forest and are seldom over 8 m high; the foliage is predominantly nanophyll in size. Epiphytic bryophytes, though abundant, are much less luxuriant than in the upper montane forest, probably because the cloud cover is less continuous.

Owing to *Massenerhebung*, the boundaries between the altitudinal zones of vegetation are higher on the main spine of New Guinea than on mountains near the coast or on outlying islands. They are also lower on exposed ridges than in sheltered valleys. The transition between zones may be fairly abrupt or it may be gradual. For example, on Mt Wilhelm the boundary between lower and upper montane forest is rather sharp, while on the adjacent Mt Otto there is a broad transitional zone at 2700–2900 m in which species characteristic of both formations are mixed. A similar 'melange' of species from more than one zone is often found on the upper part of small mountainous off-shore islands (Grubb & Stevens 1985).

The factors controlling zonation and *Massenerhebung* in New Guinea have been discussed by Grubb (1974, 1977b) and Grubb & Stevens (1985) (see also discussion for Malesia generally by Whitmore 1984a, pp. 253–7). The altitudinal zonation of vegetation no doubt ultimately depends on the gradient of temperature, but on tropical mountains at the lower levels the effects of reduced temperature may be mainly indirect. Grubb & Stevens (1985) believe that in New Guinea the immediate controlling factor is the incidence of low cloud or fog (Fig. 17.5). They suggest that this operates by reducing the amount of radiation reaching the leaves

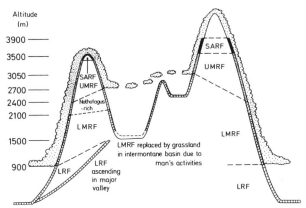

Fig. 17.5 Distribution of forest types and cloud in the Goroka region, Papua New Guinea (schematic). From Grubb & Stevens (1985, Fig. 3.3). LRF, lowland rain forest; LMRF, lower montane rain forest; UMRF, upper montane rain forest; SARF, subalpine rain forest. The density of stippling indicates the intensity and frequency of cloud cover.

and keeping leaf temperatures low. It also leads to the formation of water films over leaf surfaces, which impede the gas exchange. Both effects would tend to restrict photosynthesis, leading to slower growth rates, dwarfing of the trees and the elimination of species intolerant of such conditions. Experimental evidence supporting these suggestions would be valuable.

Grubb & Stevens also consider that the depression of the lower montane – upper montane forest boundary by much more than 1000 m is due to soil factors as well as clouding. The nature of these is not clear; possibly they are toxicity arising from high acidity, deficiency of certain nutrients or conditions arising from waterlogging.

The influence of strong winds, which is evident on some tropical mountains, e.g. in the Caribbean region, does not seem to be important in New Guinea.

Zonation in western Malesia. Mt Kinabalu in northern Borneo (4094 m) is the highest mountain in Malesia

outside New Guinea and the only one reaching the climatic tree line (here *ca.* 3500 m). The distribution of the vegetation is complicated by the geology: above *ca.* 1200 m the rock is mainly granite, but below this height there are also ultrabasic rocks (Fig. 12.12, p. 336) and Tertiary sandstones. The very rich flora of the more accessible parts of Kinabalu has been much studied, but there is no full account of the vegetation and its zonation. Meijer (1965, 1971) has written two short guides to the plant life, with references to earlier literature.

The vegetation of Gunung Dulit (1369 m), a sandstone escarpment in Sarawak (Richards 1936; see also Chapters 11 and 12), is fairly typical of the lower mountains in western Malesia. On the steep scarp slope mixed dipterocarp and secondary forest give way to lower montane forest at *ca.* 450 m. This is less tall and more irregular in structure than the primary lowland forest. Large lianes are less common, but epiphytes, especially bryophytes, more abundant. The composition of the tree

Fig. 17.6 Upper Montane forest, *ca.* 1250 m., Gunung Dulit, Sarawak. A Open facies (photo. E.F. Brünig). B Closed facies (1932).

Table 17.3. *Altitudinal zones in the Gunung Mulu National Park, Sarawak*

	Altitude (m)	Vegetation
On sandstone and shales (Mulu Formation), G. Mulu	To 800	Lowland tropical rain forest (mixed dipterocarp associations)
	800–1200	Lower montane rain forest
	1200–2371 (summit)	Upper montane rain forest (tall, short and summit facies)
On limestone (Melinau Formation), G. Api	To 800	Lowland limestone rain forest, also scree and cliff vegetation
	800–1200	Lower montane rain forest (limestone type)
	1200–summit	Upper montane rain forest (limestone type)

Source: After Anderson & Chai (1982, Table 9, p. 54), modified.

layers was not studied in detail, but clearly differs considerably from that of the forest at lower altitudes. Some dipterocarps are present, but they seemed to be relatively few in species and individuals.

On the scarp slope at 970–1100 m there is an abrupt transition to upper montane forest (montane or moss forest of Richards 1936); on the dip slope the transition from heath forest is more gradual. The most striking feature of the upper montane forest is the enormously thick covering of bryophytes and other epiphytes on the lower part of the tree-trunks and on almost every available surface in the undergrowth. The trees, most of which have small sclerophyllous leaves, are under 18 m high and have relatively thick trunks without buttresses. Although small climbers such as *Diplycosia* and *Nepenthes* spp. are common, there are no large woody lianes. Common trees and shrubs include *Quercus arbutifolia*, various Ericaceae and Myrtaceae and conifers (*Dacrydium* and *Podocarpus* spp., *Phyllocladus hypophylla*).

There are three facies of this community: closed, open and scrub. In the first, found in the more sheltered situations, there is a closed canopy of trees 15–18 m high and a sparse undergrowth (tree ferns, small palms etc.). In the open facies the trees are not more than 9–12 m high and the undergrowth denser and richer in species; much of the ground is covered by carpets of *Sphagnum*.

The montane vegetation of Gunung Mulu (4°N, 114°E), some 200 km to the northwest, has been much more fully investigated than that of G. Dulit (Martin 1977, Anderson & Chai 1982, Proctor *et al.* 1983a). It is higher (2371 m) and steeper than G. Dulit and is more varied in geology, soils and relief. The rocks are mainly shales and sandstones of Tertiary age, but on

the adjoining Gunung Api they are mainly karst limestone which in places is eroded into fantastic pinnacles up to 30 m high. The climate is ever-wet (p. 304, above; Walsh 1982a). The annual rainfall at all altitudes is over 5000 mm and even at the foot of the mountain it is seldom less than 100 mm in any month. Temperature decreases by *ca.* 5° per 1000 m (Proctor *et al.* 1983a) and at the summit may often fall below 10 °C, though probably never to freezing point. Data on cloud and mist are not available. Above *ca.* 1000 m G. Mulu (like G. Dulit) is covered with low cloud for most of the year, though there are occasional short spells of clear weather. Wind speeds, except on the exposed ridges, are usually low and cyclones do not occur.

The vegetation is divided by Anderson & Chai into zonal formations (Table 17.3). A very full account of the structure and composition of sample plots on the West Ridge at various heights from 220 to 2070 m is given by Martin (1977) (Fig. 17.7). In all zones the forest varies considerably in physiognomy and structure (as can be easily seen in aerial photographs) as well as in floristic composition. This variation evidently depends on exposure, steepness of slope and soil. It is more marked in the lower than in the higher zones. Owing to *Massenerhebung* the limits of the altitudinal zones are much higher than on mountains near the coast; thus on G. Mulu the lower boundary of the upper-montane forest is at *ca.* 1600–2177 m and at below 800 m on G. Santubong (p. 422). On the exposed ridges the zones are lower than on gentle slopes where there is more shelter.

Up to *ca.* 800 m the prevailing type of primary vegetation on rocks other than limestone is mixed dipterocarp forest similar to the lowland forest on G. Dulit. The A storey is of enormous trees, up to 55 m or more high

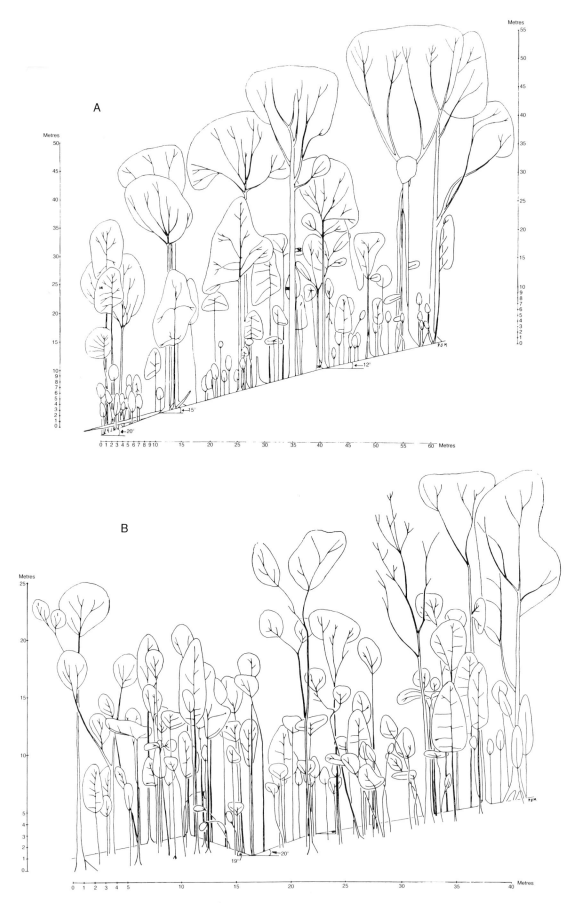

Fig. 17.7 Profile diagrams showing altitudinal zonation on West Ridge of Gunung Mulu, Sarawak. After Martin (1977), modified. A Lowland tropical rain forest at 520 m. B Lower montane rain forest at 900 m; includes various Fagaceae. C Transition, lower to upper montane rain forest at 1320 m. D Upper montane rain forest in sheltered situation at 1650 m. E Upper montane rain forest at 1860 m. F Upper montane rain forest at 2070 m. G Upper montane forest at 2070 m (second sample). H Upper montane forest at 2337 m (near summit). Depth of profiles: A 7.5 m, B–G 5.0 m, H 2.0 m.

Fig. 17.7 (*cont.*).

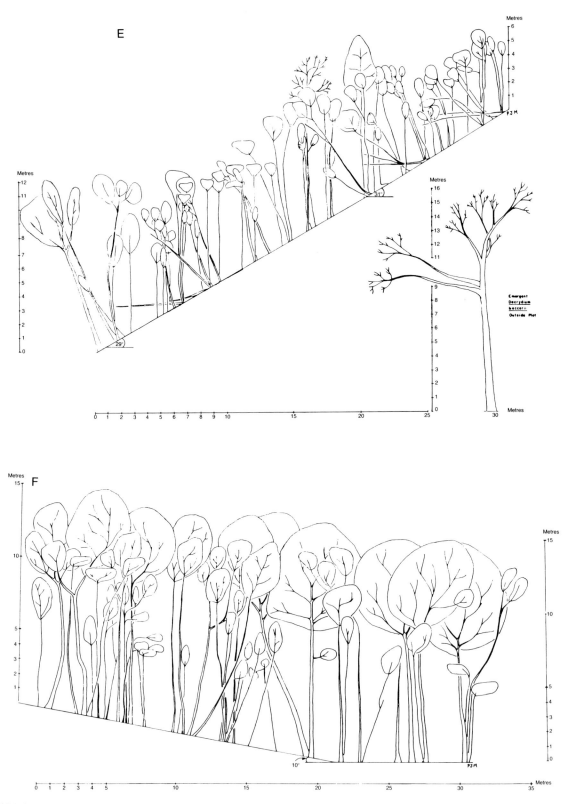

Fig. 17.7 (*cont.*). For legend see p. 428.

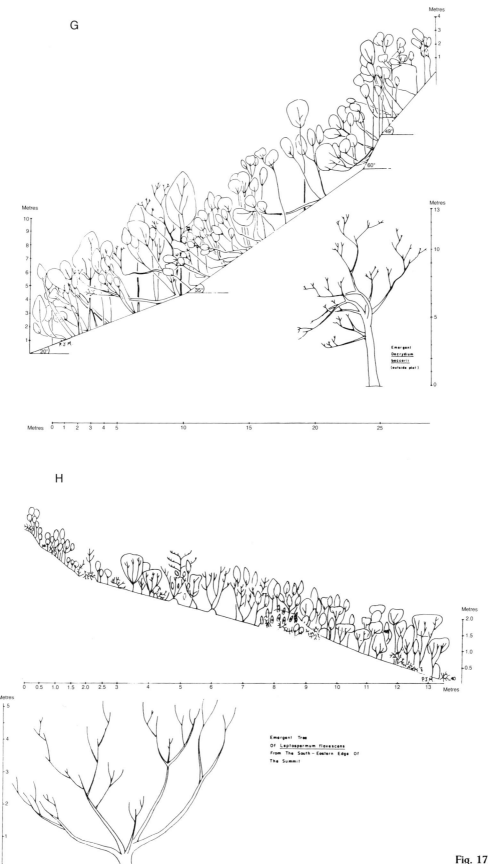

G

H

Emergent Tree
Of _Leptospermum flavescens_
From The South – Eastern Edge Of
The Summit

Fig. 17.7 (*cont.*).

and over 250 cm in girth; most of these have large buttresses. The undergrowth (D and E storeys) is not very dense and includes small palms (*Licuala* and *Pinanga* spp.). Large woody lianes and rattans are frequent. Epiphytic vascular plants and bryophytes, though abundant in the crowns of the taller trees, are not conspicuous at lower levels.

This formation is very species-rich; on three sample plots of total area 1.2 ha 284 species of trees ⩾30 cm g.b.h. were recorded. The Dipterocarpaceae are by far the most numerous family of large trees in species and individuals. On Martin's (1977) plots at 220, 500 and 700 m they formed 28.3–35.4% of the basal area and 12.7–17.8% of individual trees. The most frequent species are *Dryobalanops beccarii* and *Shorea scaberrima* at lower altitudes and *Shorea obscura* higher up. Ground herbs are few and scattered, except near streams; the sedge *Mapania cuspidata* was a widespread species.

At its upper limit the mixed dipterocarp forest is replaced by the lower montane rain forest formation which on G. Mulu covers only a relatively narrow belt. On ridges the boundary between the two formations is fairly sharp but elsewhere the transition is more gradual. The lower montane forest differs from the lowland rain forest in both structure and floristic composition. The trees are smaller, mostly not exceeding 30 m in height and 180 cm in girth: buttresses are common but usually rather small. Medium and small-sized trees are relatively more numerous than in the lowland forest (Fig. 17.7B) and stratification is less distinct. The shrub layer (D), in which *Pinanga salicifolia* and other small palms are common, is better developed and, particularly in the lower part of the zone, there is a well developed ground (E) layer of species of *Argostemma*, *Begonia* and other herbs.

Like the lowland forest, the lower montane formation consists of mixed associations without single dominants, but it is rather less species-rich; 226 tree species ⩾20 cm g.b.h. were found on three plots with a total area of 0.5 ha. The most noticeable floristic difference between the two formations is the smaller role of the Dipterocarpaceae in the lower montane forest. The number of individuals and species of dipterocarps decreases rapidly with altitude[3] and in the upper part of the lower montane zone only one species, *Shorea monticola*, occurs; on the West Ridge Martin found no dipterocarps above 1350 m. Other families, such as the Fagaceae, Guttiferae and Myrtaceae, become important. Although *Quercus subsericea* is the commonest species

in the lower part of the lower montane forest, the Fagaceae do not play such a large part as in the corresponding zone in New Guinea. Martin distinguishes two lower montane communities, one found mainly on ridges, in which the chief large tree families are the Fagaceae, Guttiferae and Myrtaceae, and another in less exposed sites where they are the Sapotaceae, Guttiferae and Myrtaceae.

The boundary between the lower montane and upper montane forest is at *ca.* 1200–1900 m and the contrast in physiognomy and floristic composition between the two formations is very marked. The upper montane forest of G. Mulu is similar in most respects to the 'Moss forest' on G. Dulit and, like the latter, has several facies. In all of their lower parts the trees have thick blankets of mosses and liverworts.

The tall facies on G. Mulu is dense and impenetrable. The trees are mostly about 15 m high, but on some slopes they are rather taller: the majority are 10–15 m in girth, a few over 100 cm. Many are bent or crooked. Buttresses are uncommon but many trees have stilt-roots concealed by a mass of bryophytes. The crowns are close together, but form a less even canopy than in the short facies. Common trees are species of *Calophyllum*, *Eugenia* and *Prunus*, also the conifers *Dacrydium beccarii* and *Phyllocladus hypophylla*. Shrubs are abundant and include two *Rhododendron* spp. and small palms. On the ground there are ferns, aroids and other herbaceous plants. The most conspicuous climbing and scrambling plants are *Nepenthes* and the bamboo *Racemobambos glabra*. On the trunks and branches there are numerous epiphytic ferns and flowering plants as well as bryophytes.

In the short facies the trees are not more than 5–9 m high; most of them have crooked trunks, leaning at various angles, but there are occasional emergent trees up to 13 m high. In the lower part of the upper montane zone the most abundant tree species in this facies are *Lithocarpus hatusimae*, *Dacrydium* and *Phyllocladus*; higher up they are *Calophyllum garcinioides*, *Eugenia kinabaluensis*, *Dacrydium* and *Phyllocladus*. The smaller woody plants of the short facies include many Ericaceae (*Rhododendron*, *Vaccinium* and *Diplycosia* spp), some of which are epiphytic. Scrambling *Nepenthes* (among them the endemic *N. muluensis*) are very abundant and there are small tree ferns as well as terrestrial orchids and other ground herbs. Tussocks of the sedge *Gahnia borneensis* occur in openings.

Near the summit of G. Mulu the upper montane forest is reduced to dense scrub 0.5–3 m high.

On the limestone of G. Api the surface consists largely of vertical cliffs and steep scree slopes on which trees

---

[3] On the altitudinal limits of Dipterocarpaceae, see Ashton (1982, p. 255).

cannot establish themselves. The plant communities are quite different from those on shale and sandstone and include many species which are probably calcicoles. At about the same altitude as on the non-limestone rocks of G. Mulu, lowland rain forest changes gradually into a distinctive type of lower montane forest. In this few trees are over 150 cm in girth and the height of the canopy is less than 25 m. The composition of this forest is very different from that of non-limestone forest at the same altitude, although the trees are mostly non-calcicolous species such as *Parishia maingayi* and *Tristaniopsis obovata*. Dipterocarps, with the exception of *Hopea cernua*, are absent.

The ground is strewn with huge boulders between which a considerable depth of litter accumulates. On this there is a luxuriant growth of herbaceous plants. On the exposed limestone rocks calcicolous plants (Gesneriaceae and Urticaceae) are found. Epiphytes are very abundant on the lower part of the tree-trunks, but there are no large woody lianes.

Above ca. 1000 m on G. Api a type of 'moss forest' is found which may be regarded as a limestone variant of upper montane forest. The broken terrain prevents the development of a tree canopy. The shrubs and small trees are thickly covered with bryophytes which also grow on the humus between the limestone outcrops. The composition of this community is similar to that of the upper montane forest on G. Mulu at the same altitude. The chief species include *Eugenia* and *Rhododendron* spp., *Leptospermum flavescens* and *Nepenthes stenophylla*, as well as *Dacrydium beccarii* and *Phyllocladus*, and are probably calcifuge. Large areas, which may be sites of fires, are dominated by species of *Pandanus*.

The zonation in the Malay Peninsula, where most of the mountains are under 2000 m, has been described by Symington (1943), Whitmore & Burnham (1969), Whitmore (1972b, 1984a), and others. The Dipterocarpaceae, the dominant tree family in the lowland forests, reach their upper limit at *ca.* 1200 m on the main ranges and at *ca.* 900 m or lower on isolated mountains. Symington recognized three subzones of dipterocarp forests: (1) lowland (0–1000 ft, 0–305 m), in which *Shorea* spp. of the 'red meranti' group are predominant; (2) hill (1000–2500 ft, 305–762 m), characterized by stands of *Shorea curtisii* (p. 330) and the abundance of other 'hill' species; and (3) upper (2500–4000 ft, 762–1219 m), of which *Shorea platyclados* and certain other species are typical.

Above the normal limit of dipterocarps Symington distinguished two zones of 'mountain forests', montane oak forests and montane ericaceous forests. The former

are tall mixed forests in which *Quercus* and *Lithocarpus* spp. are characteristic. In the montane ericaceous forests, which form the highest altitudinal zone, the trees are low and have abundant epiphytic bryophytes. The trees and shrubs include many Ericaceae (*Pieris, Rhododendron* and *Vaccinium* spp.) as well as some oaks.

Symington's three subzones of Dipterocarp forest clearly all belong to the lowland tropical rain forest formation. His montane oak forest is a type of lower montane forest and his montane ericaceous forest is similar to upper montane forests in Borneo and other parts of Malesia.

Kochummen (1982) investigated the altitudinal zonation on Gunung Jerai (1200 m) in the northwest of the Peninsula and laid out eleven sample plots of various sizes at heights from 150 m to 1140 m. On this mountain the rocks are mainly granite, schist and quartzite which weather to form white sand. Because of its isolated coastal situation the vegetation zones are markedly 'telescoped'. An interesting feature is that there is no 'montane oak forest'; at 750–780 m there is a transition from lowland Dipterocarp forest to 'montane myrtaceous forest' dominated by species of *Eugenia, Leptospermum, Tristaniopsis* and *Rhodamnia* associated with conifers (*Agathis, Dacrydium* and *Podocarpus*). Most of the trees here were less than 30 cm in diameter and in the plot at 780 m only 1.4% of the trees were over 30 m high. This forest is in fact a montane type of heath forest (kerangas) (pp. 325–7), comparable to the Ulu Koyan forest on Gunung Dulit in Borneo (pp. 305ff.).

Whitmore (1972b) remarks that on G. Benom in Pahang and other mountains in central Malaya the growth of bryophytes and other epiphytes is noticeably less luxuriant than in similar vegetation elsewhere in the tropics; this is perhaps because of local climatic conditions.

On Malesian mountains where the lowland climate is strongly seasonal, as in Java and the Philippines, the forest zonation is not very different from that in Malaya and Borneo, no doubt because at higher altitudes mist and occasional showers maintain relatively moist conditions during the dry months.

Brown's (1919) study of the vegetation of Mt Makiling (Luzon) was a pioneering attempt to relate the physiognomy and composition of the forest zones to climatic factors. It is still one of the most complete accounts of zonation on a mountain in the humid tropics.

On Mt Makiling, an isolated and long-extinct volcano 1140 m high, Brown distinguished four altitudinal zones: (1) parang (0–200 m), (2) dipterocarp forest (200–600 m), (3) mid-mountain forest (600–900 m), and (4)

Table 17.4. *Comparison of sample plots (0.25 ha), Mt. Makiling, Philippine Islands*

	Dipterocarp forest (450 m)	Mid-mountain forest (700 m)	Mossy forest (1020 m)
Height of tallest tree (m)	36	22	10
Number of tree storeys	3	2	1
Average height of storeys (m)	27, 16, 10	17, 4	6
Number of individuals of woody plants over 2 m high	353	539	610
Number of species of woody plants over 2 m high	92	70	21

Source: After Brown (1919).

mossy forest (900–1140 m). Zones (1) and (2) can be grouped as the lowland zone; both were originally forest-covered. In Brown's time the former was occupied by cultivation and secondary communities varying from grassland to closed secondary forest. Zone (2) had also been extensively cleared. In the forest which remained the most abundant tree in the A (emergent) storey was the dipterocarp *Parashorea malaanonan* and in the B storey *Diplodiscus paniculatus*. In the classification used here this association can be regarded as a type of evergreen seasonal forest. Brown's 'mid-mountain forest' is a type of lower montane forest and his 'mossy forest' is upper montane forest.

The transition from the dipterocarp to the mid-mountain forest was gradual. The latter was a mixed mainly evergreen community which at *ca.* 600–750 m consisted of what Brown called the *Quercus–Neolitsea* association and above 750 m of an *Astronia rolfei* association.

In the *Quercus–Neolitsea* association the average height of the trees diminished with altitude; at 700 m the upper (A) tree storey was *ca.* 17 m high and the lower storey *ca.* 4 m; the tallest tree noted was 21.8 m. The upper storey was thus about equal in height to the B storey in the dipterocarp forest (Table 17.4). The canopy was more open and the undergrowth much less dense than in the latter, but ground herbs (including ferns) were more abundant. Climbers were numerous and the abundance of epiphytes was a striking feature; they were less abundant than in the 'mossy forest' but considerably more so than in the dipterocarp forest and less limited to the crowns of the trees. Foliage was predominantly mesophyll.[4]

In this association species of *Quercus*[5] were the commonest trees in the A storey and *Neolitsea villosa* in

the lower storey. The number of species of woody plants was less than in the dipterocarp forest. Many species occurred in both associations.

Above *ca.* 900 m Mt Makiling was covered with dense 'mossy forest' very similar to that on other Malesian mountains. In this the single tree storey was 6–10 m high. The covering of bryophytes, ferns and other epiphytes, up to 30 cm thick, and the great development of aerial roots gave the dwarf trees a fantastic appearance. The average size of the leaves was much smaller than in the mid-mountain and dipterocarp forest; microphylls were about as numerous as meosphylls and macrophylls were absent. There was a fairly dense ground cover of herbaceous plants (including ferns); as in other 'mossy forests', the distinction between ground flora and epiphytes was vague, species usually terrestrial sometimes growing as epiphytes and vice versa. Climbing plants were plentiful, but not numerous in species.

Brown described the mossy forest as the *Cyathea-Astronia* association, the commonest trees being the tree fern *Cyathea caudata* and *Astronia lagunensis*. The composition of this community was very different from that of the mid-mountain and dipterocarp forest, although of the twenty-one species of woody plants over 2 m high on the sample plot eight occurred as understorey species in the dipterocarp forest.

Some of the differences between the forests of the three altitudinal formations on Mt Makiling are shown in Table 17.4. The most important change in structure with increasing altitude was the decrease in the height of the trees. This was accompanied by a reduction in the number of storeys, which took place by the gradual loss of first the top, then the second tree storey. The increase in the number of woody plants per unit area (which in the mossy forest is nearly double that in the dipterocarp forest) was probably dependent on the smaller size of individual trees. Apart from the disappearance of lowland forest features such as buttressing

[4] In Raunkiaer's sense (not divided into mesophylls and notophylls).
[5] Some of Brown's *Quercus* species are now placed in *Lithocarpus*.

and cauliflory, the most striking physiognomic change with increasing altitude was the reduction in leaf size.

Brown gives a full account of temperature, rainfall and other climatic factors at five elevations on the mountain and concludes that the dwarfing of the trees in the higher vegetation zones was not due to temperature alone, but to the combined effect of lower temperature and reduced illumination. The latter is much influenced by clouding; except for a short period in the dry season, clouds cover the upper part of the mountain for most of the day.

There were considerable differences between the soils in the four forest zones, those of the three lower zones being red or yellow kaolisols, while that of the mossy forest was podzolized, very acid and probably very deficient in nutrients. Brown attributed little importance to soil differences as factors responsible for the changes in vegetation with increasing altitude. It may be noted, however, that the mossy forest of Mt Makiling, like upper montane forest elsewhere, has xeromorphic features which are possibly more related to soil conditions than (as Brown implies) to climate.

The vegetation of the high volcanoes of Java has often been described (Schimper 1935, Seifriz 1923, 1924, Docters van Leeuwen 1933, van Steenis 1972) and need be mentioned only briefly here. In West Java, though most of the lowland forest vanished long ago, some montane forest still remains. The lower montane forest at Cibodas on the slopes of the dormant volcanoes Gede and Pangrango is a famous botanical site. Meijer (1959) analysed the composition of a 1 ha plot at 1450–1500 m near Cibodas; Yamada (1975–77) has described the structure, composition and litter-fall on a 1 ha plot at a slightly higher elevation, as well as plots of upper montane forest on Pangrango at 2400 and 3000 m. The Cibodas forest is unusually tall; the locally common tree *Altingia excelsa* (rasamala) grows to 40–45 m and occasionally to 60 m. Other important upper-storey trees are species of *Castanopsis* and *Lithocarpus*. No dipterocarps occur. The total number of tree species ⩾10 cm d.b.h. on the plot is about seventy-eight.

In East Java where rainfall is lower and the dry season more severe, a community dominated by *Casuarina junghuhniana* is found in the mountains at intermediate levels. This resembles savanna woodland and is probably a fire-climax.

17.1.2 Montane rain forests in tropical America

Zonation in the Andes. The Amazonian forest extends westwards to the foot of the Andes, where with increasing altitude, the lowland rain forest gradually changes, giving place to montane formations differing in physiognomy and flora. On the eastern face of the Andes, most of which has a very high annual rainfall (sometimes over 4000 mm), without a well-marked dry season, three forest zones are commonly recognized: (1) *montaña*, the basal zone of tall forest resembling that of the lowlands in its general aspect, but somewhat different in floristic composition; (2) *yungas* (or *medias yungas*), a zone of almost equally tall forest, which differs from that of the lowlands in structure and physiognomy as well as in composition and is comparable to the lower montane forests of the eastern tropics; (3) *ceja de la montaña* (brow of the mountain), a zone of low forests in which Ericaceae, Myrtaceae and other trees with small sclerophyllous leaves are abundant; the physiognomy of these forests is similar to that of the upper montane forests of the Palaeotropics.

Above the uppermost closed forests lies the mainly treeless paramo (alpine) zone, much of which is dominated by pachycaul Compositae (*Espeletia* and *Culcitium* spp.) similar in appearance to the 'giant Senecios' of the East African mountains. In the drier parts of the Andes the paramo is replaced by the more steppe-like puna (see Troll 1959).

The altitudinal limits of the zones show wide regional variations, although little exact information is available. They tend to be lower in wetter and higher in dry areas (Hermes 1955) and are also higher at the thermal equator (*ca.* 9 °N) than at the geographical equator (Vareschi 1980). Owing to *Massenerhebung*, the zones are considerably higher in the Andes than on mountains elsewhere in the American tropics which are isolated or near the coast. In the main ranges of the Andes the upper limit of lowland rain forest appears to be *ca.* 1200–1500 m and lower montane forest is found up to *ca.* 1800–2400 m (see references in Grubb & Whitmore 1966). Upper montane forest (*ceja de la montaña*) reaches its upper limit at *ca.* 3400 m in Bolivia, but ascends to 3600 m or even 3900 m in Peru (Hueck 1966).

The boundary between the *ceja* and the paramo has often been regarded as the climatic tree limit, but in fact this may not be so. Ellenberg (1979) has emphasized that the vegetation of the high Andes has been strongly affected by humans for some 10 000 years; he believes that the paramos were originally covered with forest or woodland and that deforestation was the result of grazing and the use of trees as firewood. A similar view is held by Tosi (1960) and others. Both native and introduced trees can grow on the paramos under existing conditions if planted. The small strips of open woodland of *Polylepis* spp. and other small trees found on well-drained stony soils up to 4500 m, which are not in contact with the highest closed forests (Troll 1959, Walter 1971), are perhaps relics of the former vegetation of the paramos.

There is a comprehensive general account of the montane forests of the Andes by Hueck (1966) and among regional accounts may be mentioned those of Herzog (1923) on Bolivia, Weberbauer (1911, 1945), Ellenberg (1959) and Tosi (1960) on Peru and Cuatrecasas (1958) on Colombia. Detailed studies of small areas are few: the montane forest La Carbonera, near Mérida (Venezuela) with its fine *Podocarpus* stands has been

fully described by Lamprecht (1958), Hetsch & Hoheisel (1976) and Vareschi (1980).

In the forests of the eastern Andes, as in other tropical montane forests, the increasing abundance and diversity of bryophytes with altitude is very striking (Frahm & Gradstein 1991). In a transect from 200 to 3200 m in northeastern Peru the phytomass of epiphytic bryophytes on 0.5 m samples was 5–6 g m^{-2} in lowland forest at 200–900 m and rose to a maximum of 70 g m^{-2} in subalpine forest at 3200 m, where the tree-trunks were covered by thick masses of mosses and hepatics. The number of species was also lowest at low altitudes and highest at 3000–3200 m. An attempt to determine

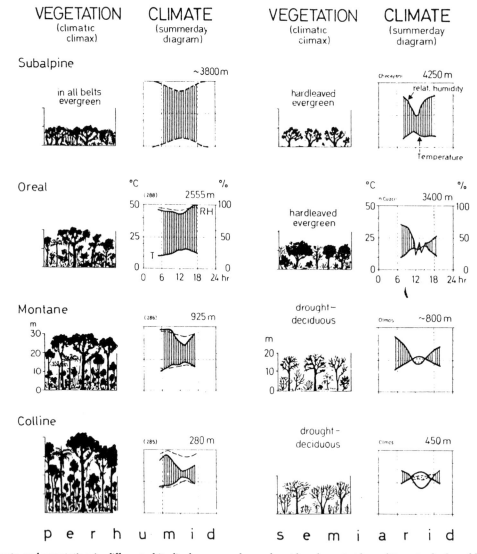

Fig. 17.8 Climate and vegetation in different altitudinal zones under perhumid and semiarid conditions in Andes of Peru and Ecuador. From Ellenberg (1979). The profile diagrams are generalized. The climate diagrams show the relative humidity and temperature at 1.5 m above bare ground (continuous line) and under the vegetation canopy (broken line) on a typical summer day.

the zonation from the upper and lower limits of bryophyte species by the Sendtner method (pp. 422ff.) (Gradstein & Frahm 1987) gave results in good agreement with Ellenberg's (1979) classification of the forest vegetation (Fig. 17.8).

The papers by Grubb *et al.* (1963) and Grubb & Whitmore (1966, 1967) on montane and lowland rain forest in Ecuador, though largely based on only two sample plots 200 ft × 25 ft (61 m × 7.6 m, *ca.* 0.5 ha) in extent, are valuable for their wealth of precise data on the structural and physiognomic differences between samples of lower montane and lowland rain forest in this part of the Andes. The areas studied were *ca.* 64 km apart, the montane plot at 1710 m and the lowland plot at 380 m.

Both forests had a fairly similar structure, with no obvious stratification, but in the montane plot the mean height of trees ≥8 in (20 cm) d.b.h. was only 21.6 m, compared with 24.7 m in the lowland plot; their mean diameter was also slightly less. The number of trees per unit area (density) was somewhat smaller in the montane plot, but probably not significantly so. More striking were the physiognomic differences, particularly in leaf size; in both forests the majority of the trees had mesophyll leaves, but in the montane forest the percentage of macrophylls was smaller and of notophylls larger, so that the average leaf size must have been significantly smaller. The trees also differed in other physiognomic features, notably in their buttresses, which were smaller in the montane forest.

No useful comparisons can be made between the floristic composition of the tree and shrub populations on the two plots, but it may be noted that twenty-seven woody species taller than 20 ft (6.1 m) occurred on the montane plot and thirty-one on the lowland plot. Only one tree species (*Carica* sp.) was found on both plots.

The climbers and epiphytes were partly responsible for the different aspect of the two forest types. Climbers, including aroids, bromeliads, orchids and various dicotyledons, were very abundant in both, especially in the undergrowth, but in the montane forest there were only five individuals reaching heights of over 18 m compared with eighteen in the lowland plot; in the former there were fewer species of climbing Araceae and of dicotyledonous climbers. Epiphytes were also abundant in both plots, but in the montane plot there were ninety-one species of vascular epiphytes (of which fifty-eight were pteridophytes) and 2555 individuals, as compared with fifty-seven species (eleven pteridophytes) and 268 individual epiphytes in the lowland plot. Epiphytic bryophytes and macrolichens were also more abundant in the montane plot.

Grubb & Whitmore (1966) compare the climate and microclimates in the lowland and montane sites and discusses the relation to climatic factors of the physiognomy and altitudinal range of the Andean forest formations generally. Only fragmentary information was available about the mean annual rainfall in the areas studied, but at the lowland site it was probably over 4000 mm and considerably more than 2000 mm at the montane site. At both sites there was a reduction in the amount of rain during the northern winter rather than a real dry season. In this part of the Andes mean temperature falls by *ca.* 0.4° per 100 m increase of elevation and monthly means vary little through the year. At the montane site mean daily temperatures were *ca.* 6° lower than in the lowland. In both sites the daily range of temperature was about 6° in the forest undergrowth and *ca.* 11° in clearings.

Grubb & Whitmore (1966) concluded from their observations that the ecologically most important climatic difference between the lowland and lower-montane forest was not mean temperature or rainfall, but the incidence of low cloud (fog) which rarely or never occurs in the lowlands but very frequently in the montane forest. At the montane site days of 'fog-bound' weather alternate with fog-free periods. During the drier months the latter may last for one or two weeks, but during the wettest part of the year fog-free periods are infrequent and fog shrouds the forest for most of the time. On foggy days the maximum temperature was 5–6° lower than on fog-free days and relative humidity averaged 94%, falling below this only for very short periods. The effects on light are considered in a further paper (Grubb & Whitmore 1967). It is evident that in foggy conditions the whole aerial environment of the

Fig. 17.9 Montane forest at *ca.* 1500–1600 m, Itatiaia (22°20′S, 44°43′W), Serra da Mantiqueira, Brazil (1967). The light-coloured crowns are *Cecropia* sp.

vegetation is very different from when it is fog-free, with important consequences for photosynthesis, respiration, transpiration and other plant functions. However, foggy periods are not continuous and are always broken by spells during which temperature, light and humidity are not very different from those in the lowland rain forest.

The climate of the upper montane forest was not investigated by Grubb & Whitmore (1966), but from general observations they suggest that its altitudinal range and characteristic physiognomy are also mainly determined by fog, which is probably even more frequent than in the lower montane zone. However, since in the upper montane forest mean temperatures are probably at least 10° lower than at their montane site and in the equatorial Andes frequent frosts occur above 3000 m (Troll 1959), it seems unlikely that low temperature is not also very important.

Lowland and montane rain forests in Costa Rica. In the humid parts of Central America there is an altitudinal zonation of rain-forest formations, comparable with that on the eastern face of the equatorial Andes. Here, because the cordillera is narrower and near the oceans on both sides, the *Massenerhebung* effect is less and the limits of the zones are probably somewhat lower. In Costa Rica (8–11°N) the mountains, some of which are dormant or intermittently active volcanoes, rise to 3820 m. According to Holdridge *et al.* (1971), the lower montane and montane 'life-zones' occupy over 7000 km, but much of the lower montane forest has been converted into cattle pastures or exploited in other ways. Considerable areas, however, remain in a natural state and some of these are now protected as forest reserves or national parks.

Owing to the great variety of topography and the wide altitudinal range in Costa Rica the plant cover is very complex (see general account by Hartshorn 1983 and the vegetation map by Tosi 1969). In a few places it is still possible to see a continuous transition from lowland to high montane rain forest, notably on the east-facing slopes of Volcán Barba (Barva) (2906 m), where a detailed study of the structure and floristics of the forest at five altitudes has been made (Heaney & Proctor 1990). None of the mountains of Costa Rica reaches the climatic tree limit, but above *ca.* 3000 m on the Cerro de la Muerte and a few other mountains there are small areas of paramo vegetation with a flora including *Puya dasylirioides* and other Andean plants (Weber 1958).

The forests of Costa Rica are usually classified according to Holdridge's system of 'life-zones' (Holdridge *et*

al. 1971) in which four tropical altitudinal belts are recognised, lowland, premontane, lower montane, and montane, above which are the subalpine, alpine and nival belts. Each altitudinal belt is subdivided into a series of life-zones based on mean annual rainfall and potential evapotranspiration. Holdridge *et al.* (1971) give a large amount of detailed information on the climate and soils, as well as the structure and composition of the vegetation, at selected sites in each altitudinal life-zone. Hartshorn's (1983) account of the forest life-zones of Costa Rica includes a description of the forest in the La Selva Biological Reserve (lowland tropical wet forest[6] and premontane wet forest) and the Monteverde 'cloud forest' Reserve, which contains premontane mist and rain forest, and lower montane and montane rain forest.

Since they are based primarily on climatic parameters, Holdridge's life-zones cannot be exactly equated with phytosociological units, though his tropical, lower montane and montane 'Altitudinal belts' are fairly closely comparable with the lowland, lower montane and upper montane forest zones of this chapter.

The forests in Holdridge's premontane belt can be regarded as transitional between what are here termed lowland and lower montane rain forests: his premontane wet and premontane rain forests are comparable to types such as the 'hill dipterocarp' forests of Malaya (pp. 304ff.) which are found towards the upper altitudinal limit of the lowland forest where rainfall is probably higher, and evapotranspiration somewhat less, than at lower elevations. The premontane wet and premontane rain forests of Costa Rica differ from the lowland forest (tropical wet of Holdridge) in various physiognomic characters as well as in floristic composition. For example, the 'canopy trees' (emergents) are less tall (30–40 m, compared with 45–55 m) and have smaller crowns and buttresses. In the lower storeys single-stemmed treelets (*Piper* spp., Rubiaceae, etc.) are more abundant than in the tropical wet forest. Tree ferns are also more frequent and small palms less so. The ground cover is denser. Woody lianes are numerous and epiphytes, both bryophytes and vascular, are very abundant.

The premontane forests are very species-rich but the wet type has rather fewer tree species ≥10 cm d.b.h. than the tropical wet (lowland) forest. Although the floristic composition of the two forest types differs considerably, many species are common to both.

Above about 1500 m lower montane forests (*sensu*

---

[6] As mentioned on p. 390 Holdridge uses 'rain forest' in a narrower sense than this book: his wet forest (and perhaps some of his moist forests) are included in the rain-forest concept used here. Lowland rain forest in Holdridge's sense does not exist in Costa Rica.

Holdridge) replace premontane forests. The former differ markedly from the latter in structure and composition. They are generally less tall, the trees being up to *ca.* 20–30 m high in the lower montane wet and lower montane rain forest, although in the Sierra de Talamanca *Quercus copeyensis* grows to 45–50 m. Lower montane forest has a more or less clearly two-storeyed structure. The trees tend to have twisted trunks with sinuous branches and are usually unbuttressed. Leaves of 'canopy trees' are *ca.* 5–10 cm long (Holdridge *et al.* 1971) and presumably mainly mesophylls.

In the undergrowth tree ferns are common. Small palms occur occasionally, though they are much fewer than in premontane forests. Woody lianes are generally scarce, but Araceae and other herbaceous climbers occur occasionally. The abundance and variety of vascular epiphytes, including shrubby species (Ericaceae, etc.) is characteristic of this type of forest. Bryophytes cover the tree-trunks in thick layers, especially in the 'elfin forest facies' found in very humid situations.

The species richness of lower montane forests is variable: probably it is usually less than that of premontane forests. Lowland tropical taxa are generally absent or rare, e.g. Annonaceae, *Licania*, *Protium*, but temperate genera such as *Cornus*, *Magnolia* and *Viburnum* are found. A characteristic feature of the lower montane life-zone in Costa Rica is the abundance of various species of *Quercus* (oaks), a genus absent from South America except in Colombia. These splendid trees have unfortunately been much reduced in number by logging and charcoal burning.

On the highest mountains, above *ca.* 2500 m, upper montane forests (the montane forest of Holdridge) are found. As in the lower montane forests the trees form two storeys: these are at *ca.* 25–30 m and 5–15 m, respectively. The trees have small rounded crowns and irregular unbuttressed trunks, bearing many twisted branches. The leaves are very sclerophyllous, the majority probably microphylls or nanophylls. A striking feature is the dense thickets of bamboos *ca.* 5 m high (*Chusquea* and probably other genera). The taller undergrowth also includes shrubs and tree ferns; palms are absent. The ground layer of tree seedlings, herbs, ferns and mosses is open. Vascular epiphytes are very abundant; most are small and some are climbers. Bryophytes form layers up to 3 cm thick on the trunks and branches of the trees.

On wind-exposed ridges the montane forest becomes reduced to thickets of dwarf trees and shrubs. On a 0.4 ha plot at 3070–3090 m on the Cerro de la Muerte (Site 6 of Holdridge *et al.* 1971) *Quercus costaricensis* was dominant (importance value 41%). Other abundant trees were *Miconia bipulifera* and *Weinmannia pinnata*. The total number of species was twenty-two.

Holdridge's life-zone system assumes that the chief factors determining the altitudinal distribution of vegetation in the tropics are temperature and humidity (rainfall and evapotranspiration). Holdridge *et al.* (1971) do not discuss the role of clouds and fog on the mountains of Costa Rica (except in relation to epiphytes). They mention the frequency of fogs in the lower montane and montane life-zones; it is hard to believe that the ecological importance of fog is less here than in other tropical mountains.

The importance of cloud and fog is shown very clearly in the Monteverde Forest Reserve in northern Costa Rica, where various types of premontane and lower montane forest cover a large area at 1200–1800 m (Lawton & Dryer 1980, Hartshorn 1983). The topography of the area and its situation astride the continental divide leads to the prevalence of cloud throughout the year and the local development within a seasonal climate of very humid montane forests which, as Hartshorn points out, are true 'cloud forests'.

Coastal 'cloud forests' in northern South America. The evergreen forests found between about 300 and 900 m on isolated coastal mountains in Columbia and Venezuela, described by Sugden (1982, 1986), are in areas of dry seasonal climate where the surrounding climax vegetation is probably deciduous forest or thorn woodland. They owe their existence to local climates dependent on topography.

The much more extensive evergreen forest in the Henri Pittier National Park (Rancho Grande) in Venezuela occupies mainly the seaward slopes of the coastal cordillera from about 850 m upwards. These face the northeast trade wind and receive rain through most of the year, and the forest can be regarded as 'cloud forest' depending on the local topography. Most of this forest is lower montane forest similar in structure and composition to that of Trinidad (see below). At higher levels this is replaced by upper montane forest. A general account, with references, is given by Vareschi (1980).

Montane rain forests in the Caribbean islands. In the West Indian islands very little lowland tropical rain forest remains; in most of them its former altitudinal limits are now difficult to trace. In Trinidad, Jamaica and the other larger islands some fairly considerable areas of montane forest still exist. In Dominica there is a high rainfall with hardly any dry season, even at low altitudes, but in most islands the lowland climate is relatively dry and the montane forests, like those in

Costa Rica and Venezuela referred to above, are 'cloud forests' dependent on frequent mist and rain induced by the local relief rather than on the regional climate. Most of the mountains do not exceed 2000 m and, excepting Trinidad which is only 11 km from the South American mainland, the islands are rather isolated, so there is no strong *Massenerhebung* effect. Perhaps it is because of the telescoping of the zonation that the montane forests in the Caribbean island differ somewhat from those in South and Central America: as will be seen later, both the lower montane and upper montane formations in the islands are somewhat untypical. Most of the islands have been affected by hurricanes, leading to the replacement of primary forest by seral communities. In several of the islands there are active volcanoes.

The montane forest formations of Trinidad and the Lesser Antilles were described in some detail by Beard (1946b, 1949). In the Northern Range of Trinidad he recognized three altitudinal zones: 'lower montane rain forest' at *ca.* 250–760 m (but above 550 m transitional to the next zone), 'montane rain forest' at *ca.* 760–880 m and 'elfin woodland' from 853 to 940 m (Fig. 17.10). The two latter are found only in the 'moist belt' on El Aripo and are small in extent; both are very poor in species.

Beard's 'lower montane forest' differs from the evergreen seasonal forest of the lowlands (regarded in this book as a type of lowland tropical rain forest) in physiognomy and structure, as well as in floristic composition. The trees are lower, forming a compact canopy at 22–30 m; below this there is an ill-defined and discontinuous middle layer. Some characteristic features of lowland rain forest are poorly developed or uncommon, e.g. buttressing and cauliflory. The leaves are mainly meso-

phyll (*sensu* Raunkiaer). Lianes and epiphytes are not particularly abundant. Palms and tree ferns occur, but are not very frequent. Beard regarded this type of forest as a single association in which *Byrsonima spicata* (*ca.* 9% of trees ⩾30 cm g.b.h.), *Licania ternatensis* (16%) and *Sterculia caribaea* (11%) are the most abundant 'canopy species'. The estimated total number of tree species (⩾30 cm g.b.h.) on 100 acres (40 ha) was eighty-seven, slightly less than in the lowland evergreen seasonal forest; many are confined, or almost so, to this association.

The 'montane rain forest' of Beard (upper montane forest) is two-storeyed. The upper storey is only 15–20 m high; no tree exceeds 23 m. Buttresses are small or absent and the leaves are predominantly mesophyll to microphyll in size. Lianes and epiphytes are abundant. Beneath the tree storeys there is a layer in which tree ferns and small palms are numerous. The 'Montane forest' is a single association in which the most abundant tree species are *Licania biglandulosa* and *Richeria grandis*. The total number of tree species on 100 acres (40 ha) is only twenty-eight.

The 'elfin woodland' is a dense thicket of small trees of which much the most abundant is the stilt-rooted *Clusia intertexta*; this forms a discontinuous upper storey 6.0–7.5 m high and constitutes over 90% of the tree layer. Below this is a very dense layer of tree ferns and small palms (two species) *ca.* 3 m high. A characteristic feature of this community is the thick covering of bryophytes and lichens on the trunks, branches, and often the leaves, of other plants.

In the Windward and Leeward islands the distribution of vegetation-types follows the patterns of climate and soils in each island (Beard 1949). On the higher islands

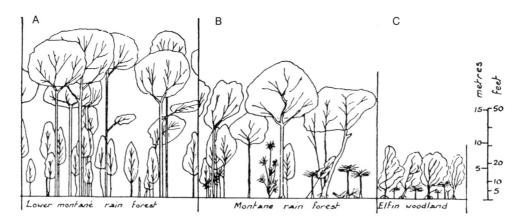

Fig. 17.10 Montane forest formations in Trinidad. From Beard (1944a). A Lower montane rain forest, 300 m, Guanapo. B Upper montane rain forest (montane of Beard) 800 m. Aripo. C Elfin woodland (facies of upper montane forest) (sketched) 1000 m, Aripo.

the zonation probably originally conformed more or less with that shown in Fig. 17.8, but as well as being complicated by topography and damaged by hurricanes and volcanic eruptions, much of the natural vegetation has been destroyed or modified by people. The largest surviving areas of forest are in Dominica (parts of which have an annual rainfall of over 7 m). In all the Lesser Antilles most of the land at low altitudes has been cultivated at one time or another and the nature of the original plant covering can usually be inferred only from climatic data and relics of secondary vegetation. Even in islands with high rainfall, rain forest may never have existed at sea-level. Relics of the *Dacryodes–Sloanea* association which Beard regards as 'true rain forest' occur on 'red earths' from about 30 m to over 1000 m in some islands, but usually this community is replaced by a *Licania ternatensis–Oxytheca* association which he classifies as lower montane forest. This resembles the similarly named forest type in Trinidad in structure and physiognomy but differs considerably from it in floristic composition. In the wettest parts of Dominica an association is found in which *Amanoa caribaea* is the commonest tree.

In the Lesser Antilles, tropical rain forest and 'lower montane forest' may extend to the tops of the mountains. There is no typical upper montane forest, but on some islands 'montane thickets', 'elfin woodlands' and 'palm brakes' are found on slopes above *ca.* 914–1067 m and on exposed ridges above *ca.* 488 m. 'Montane thickets' are dense one-storeyed communities of trees not more than 12–18 m high. Their composition is very variable and includes a few species not found in other types of vegetation. Epiphytic bryophytes and bromeliads are very abundant, though less so than in 'elfin woodland'. The latter, as in Trinidad, consists of gnarled, often wind-pruned, trees with bryophytes hanging in long streamers from their branches. Tree species are few, *Charianthus* and *Clusia* spp. and *Didymopanax attenuatum* being characteristic. 'Palm brakes', usually dominated by *Euterpe globosa*, are seral communities occurring on landslips and unstable soils.

In Puerto Rico very little forest remains except in the Luquillo mountains where rain forest over 4000 ha in area extends from *ca.* 150 m to the summit of El Yunque (1050 m). Although cut over in colonial times, this forest has remained more or less intact up to the present. Many aspects of its ecology were investigated in connection with the radiation experiments of 1965 (Odum & Pigeon 1970). Data on structure and productivity are given by Weaver & Murphy (1990).

Most of the Luquillo forest is rather similar in structure and physiognomy to the lower montane forests of

Trinidad and the Lesser Antilles. As in the latter, *Dacryodes excelsa* is the most conspicuous large emergent tree and *Sloanea berteriana* also occurs, but most of the other trees are Puerto Rican endemics or species which do not occur in the islands to the east and south. On the mountains above *ca.* 600 m, which are much exposed to the wind and receive a very high rainfall, the vegetation consists of 'palm brakes' of *Euterpe globosa* and 'elfin woodland'. The latter consists of dwarf gnarled trees and resembles the 'elfin woodland' of the Lesser Antilles in physiognomy, although it has many endemic species and is very different in floristic composition (Beard 1949, Howard 1968, 1969, Weaver *et al.* 1986).

In Jamaica there is now no lowland rain forest, although relics of what seems to have been a marshy type of rain forest existed in the western part of the island until recently (Grubb & Tanner 1976). In the mountains some tracts of forest still survive, the largest being in the Blue Mountains at *ca.* 1500–2265 m; this was the subject of a classic study by Shreve (1914a). These mountains receive *ca.* 1500–2265 mm of rain well spread through the year; at Cinchona (1500 m) no month has an average of less than 100 mm. The soils are derived from shales, limestone and hard volcanic rocks and are very varied.

Grubb & Tanner (1976), in a reconnaissance survey, distinguished nine types of montane forest, differing in physiognomy and species composition, and found in different habitats, though some were connected by continua (Fig. 17.11). Various aspects of the ecology of four of these types were later studied in considerable detail (Tanner 1977, 1980a, b, 1982, Sugden *et al.* 1985): (1) gap (or gully) forest in gullies where the soil is less acid (pH 4.0–4.9 in the A horizons) and less rich in organic matter than in the other types; (2) wet slope forest covering large areas on shallow (<30 cm deep) lithosols, chiefly on north-facing slopes; (3) mull ridge forest on ridges with moderately acid soils (pH 3.5–5.4 in A horizons) resembling a north temperate podzol or brown earth; and (4) mor ridge forest occupying small areas on knolls on ridges having an extremely acid soil (pH of A horizons 2.8–3.5) with a peat-like mor layer up to 50 cm thick.

The gap, wet slope and mull ridge types are closed-canopy forests of fairly erect trees in gap forest 12–18 m high, somewhat lower in the two other types. The mor ridge type is more open, only 5–7 m high, and the trees are leaning (but not all in one direction). The 'principal tree species'[7] in types (1), (2) and (3)

---

[7] Those with a basal area of more than 10% of the total.

A

B

C

Fig. 17.11 Types of upper montane rain forest at 1500–2265 m in the Blue Mountains, Jamaica. (Photos E.V.J. Tanner.) A Mull ridge forest. The tree fern is *Cyathea woodwardioides*. B Mor ridge forest. Note numerous epiphytic Bromeliaceae (*Tillandsia* and *Vriesia* spp.). C. Gap forest. The scale is shown by a pole marked in feet.

include *Clethra occidentalis*, *Cyrilla racemiflora*, the conifer *Podocarpus urbanii* and the tree fern *Cyathea pubescens*; in type (4) they are *Chaetocarpus globosus*, *Cyrilla racemiflora* and *Lyonia* sp. The total number of tree species (⩾10 cm g.b.h.) is greatest in the mull ridge type (47 in 0.1 ha) and lowest in mor ridge (16 in 0.8 ha).

Tanner's (1980a) estimates of the above-ground phytomass of these forests range from 40.7 kg m^{-2} in well-developed mull ridge forest to 22.9 kg m^{-2} in the mor ridge type. From comparisons with figures for lowland and montane forests in other parts of the tropics he suggests that on tropical mountains the decrease in canopy height with increasing altitude is proportionately greater than the decrease in biomass. In contrast to total phytomass, his data indicate that leaf phytomass in the Jamaican montane forests is not less than in forests at lower altitudes.

The gap and wet slope types have some 'lowland' physiognomic features, e.g. presence of large woody lianes and buttressed trees, and their foliage is predominantly notophyll in size. These were classified by Grubb & Tanner (1976) as lower montane forests, but they are untypical in their low stature and relatively small number of tree species. Mull ridge and mor ridge, which lack 'lowland' features and have smaller leaves (in the latter all the 'principal species' have microphylls), are regarded as upper montane formations. It may be noted that here (and probably in the montane forests of the Caribbean islands generally) leaf sizes tend to be larger than in comparable forests elsewhere in the tropics. Grubb & Tanner (1976, p. 347) remark that in Jamaica leaf-size spectra are not by themselves an adequate criterion for classifying montane forests. In Jamaica there is also an appreciable overlap of species between the lower and upper montane forests, perhaps comparable to the 'melange' effect on outlying mountains in New Guinea (p. 425).

It is particularly interesting that in Jamaica the distribution of these montane forest types appears to be determined more by soil and site characteristics than by climatic factors. Tanner (1977) found that mean concentrations of nutrient elements (N,P,K and Ca) in the leaves decreased from a maximum in gap forest to a minimum in mor ridge. He tentatively suggested that the distribution of mull ridge relative to gap forest is limited by the slower rate of circulation of nutrients in the former and that mor ridge is primarily limited by the very low pH of its soil. The wet slope type is limited in area relative to mull ridge by the mechanical problems of tree establishment on steep slopes (often *ca.* 30°) and the lack of support provided by its shallow

soil. Concentrations of mineral nutrients are in general lower than in the soils of comparable forests in New Guinea. These conclusions seem to suggest that the zonation of forest types on humid tropical mountains is controlled by a complex of factors and not entirely by temperature and factors closely related to temperature such as cloudiness.

17.1.3 Montane rain forests in tropical Africa

The 'Afroalpine' vegetation of Africa with its treelike senecios, lobelias and Ericaceae is very well known (Hedberg 1964), but the montane vegetation at lower levels has received much less attention.

For studying the transition from lowland rain forest to montane forest, Africa is not a favourable region. Many of the higher African mountains lie in areas where the lowland climate is strongly seasonal, so that relatively dry forest and savannas cover much of their lower slopes. Even where rain forest reaches the base of the mountains, as at the eastern margin of the Zaïre basin, areas of wet and dry climate are intricately intermingled, so that the sequence of plant formations dependent on the altitudinal gradation of climate is complicated by local variations in rainfall. Further, since much of the montane vegetation of Africa grows on young volcanic material, it may still be immature. Over large areas it has been so much changed by fire, grazing and shifting cultivation that the nature of the original vegetation is not easy to discern.

Zonation in East Africa. The zonation of vegetation at the upper limit of the African 'closed forest' on the western side of the mountains bordering the Great Rift Valley has been well described by Lebrun (1935, 1936a, 1960a,b). At one time the forest reached the mountains in many places, though even in the 1930s much of it had been destroyed and its margin was retreating westwards. At middle altitudes (*ca.* 1750–2500 m) there was a zone of evergreen montane forest in which an appreciable proportion of the flora was subtropical or temperate in its affinities. Similar forests occur elsewhere in the higher mountains of East Africa and Cameroon. According to Lebrun's map (1935, P. 1) there was actual continuity between the lowland rain forest and the upland forests in some places, e.g. northwest of Ruwenzori, east of Lake Kivu. More often, however, there was a narrow band of savanna and grassland between the lowland and the upland forest and due, according to Lebrun, to foehn winds from the mountains which produce local areas of dry climate.

Three forest zones were distinguished:

Table 17.5. *Comparison of forest formations on mountains of eastern Zaïre*

Trees ≥20 cm d.b.h.	Lowland tropical forest	Transition forest	Montane mesophilous forest
Number per ha	115	180	220
Mean height (m)	30	25	20
Mean length of boles (m)	13	12	10
Mean diameter (cm)	60	40	35
Timber volume (m³ ha⁻¹)	400–600	300	200

Source: From Lebrun (1936a).

(1) *Forêt équatoriale* (lowland tropical forest; see Chapters 12 and 16). Upper limit 1100–1300 m.

(2) *Forêt de transition* (transition forest), 1100–1300 m to 1650–1750 m.

(3) *Forêt mésophile de montagne*[8] (montane mesophilous forest), 1650–1750 to 2300–3400 m.

Immediately above the montane forest there is usually the very characteristic bamboo zone, formed by a dense association of *Arundinaria alpina* with few dicotyledonous trees which is found on most of the higher mountains in East Africa and Cameroon. At *ca.* 2600 m the bamboo zone is succeeded by the Ericaceae zone, dominated by arborescent species of *Erica* and *Philippia*, or on some mountains by dwarf *Hagenia abyssinica* woodland. Finally there is the 'Afro-alpine' (paramo) zone (3700–3800 to 4600 m), the chief home of the pachycaul senecios and lobelias, and the zone of permanent snow. The altitudinal limits given are approximate and vary on different mountains (Hedberg 1951, Lebrun 1960c,d).

Lebrun's transition forest is a distinct community which is less tall and luxuriant than lowland rain forest (Table 17.5). It is mixed in composition, rich in species, many of them endemic in the region. Some of its tree species are also found in the lowland forest, e.g. *Carapa grandiflora*, *Cynometra alexandri*, and others also in the montane forest, e.g. *Entandrophragma excelsum*, *Ocotea usambarensis*, but some are probably characteristic of the association, e.g. *Lebrunia bushiae*, *Ocotea michelsonii* (White 1983).

Where a dry foehn belt is not interposed, the transition forest at its upper limit passes into montane rain forest. This is mainly evergreen, with an upper storey of trees *ca.* 25 m high, which are mostly unbuttressed. The crowns are compact and not interwoven with lianes (which are relatively infrequent) as in the lowland forest. The shrub layer is dense, but the herb layer is open except in gaps. Epiphytes, including orchids, Loranthaceae, ferns, bryophytes and lichens, are abundant, though the growth of moss on the trunks is much less than at a higher elevation in the Ericaceae zone.

The montane forest is a mixed association differing greatly in composition from the lowland forest, although some species, e.g. *Symphonia globulifera*, occur in both. Lebrun divides this zone into three subzones ('horizons'), lower (1600–1900 m), middle (1900–2100 m) and upper (2100–2400 m). The first resembles the transition forest and the trees are taller than in the other subzones. Among the larger tree species are *Entandrophragma excelsum*, *Ficalhoa laurifolia*, *Ocotea usambarensis* and *Strombosia grandifolia*. Tree ferns are common and the herb layer includes some temperate species, e.g. *Dryopteris filix-mas*, *Sanicula europaea*. In the middle subzone the trees are less tall. The chief species in the A storey include *Entandrophragma excelsum*, *Ocotea usambarenis*, *O. viridis* and the conifers *Podocarpus latifolius* and *P. falcatus*. Tree ferns, often associated with *Ensete edule*, are common in ravines. The herb layer is very open and species of several temperate genera occur. The upper subzone has some of the aspect of elfin woodland; the trees are not more than 10–15 m high and their trunks and branches are twisted and irregular. The leaves (probably mainly notophylls and microphylls) are small and coriaceous. Epiphytes, including the lichen *Usnea*, are conspicuous. The important tree species include *Ekebergia ruppeliana*, *Olea capensis*, *Parinari mildbraedii*, *Podocarpus* spp. and *Symphonia globulifera*. The shrubby undergrowth is abundant but the herbaceous ground layer very sparse.

Relics of montane forests like Lebrun's 'forêt mesophile de montagne' are scattered through East Africa and must have been much more extensive in the past than now (Hamilton 1982); in the literature they have

---

[8] 'Forêts ombrophiles de montagne' of Lebrun & Gilbert (1954).

been variously termed subtropical, warm temperate and montane (or Afro-montane). In composition they may be mixed or have single dominants; montane forest dominated by the conifer *Juniperus procera* is widespread on the drier slopes. In later papers Lebrun (1957, 1960c, d) tried to define his zones of vegetation more precisely by dividing the flora subjectively into eight ecological groups and calculating their relative proportions in each zone. This method showed clearly the differences in the zonation on the western and eastern faces of Ruwenzori and between these and the Virunga mountains.

In Uganda surviving areas of tropical rain forest are in contact with montane rain forest in three places; Hamilton (1975) found that the number of tree species decreased linearly with altitude. He also used a quantitative method using certain physiognomic characters of selected species to test whether the composition of the tree population varies continuously with altitude or whether, as seems to be the case with the bird fauna (Moreau 1966), there are 'critical altitudes' where it changes abruptly. The study was limited to forests above 3000 ft (914 m), and to 140 tree species for which adequate data were available. It was found that leaf size, the percentage of compound leaves, deciduousness, thorniness and buttressing all decreased linearly with altitude. The relation of leaf size classes to altitude was complicated by the occurrence of macrophylls at all elevations up to over 3000 m, but if these were omitted, there was a significant linear correlation ($p > 0.99$) between leaf lamina area and altitude. The results suggested that, if Uganda were considered as a whole, the composition of the forests was a continuum with no 'critical altitudes'.

To the southeast of the mountains considered above are the isolated Usambara, Uluguru and Nguru mountains of the 'Eastern Arc' in Tanzania, which rise to over 2000 m but lack a treeless Alpine zone. They are islands of wet climate in a relatively dry region where the characteristic vegetation is deciduous forest and savanna. Although much of their primary forest was long ago replaced by cultivation, some still remains on the eastern slopes. In the lowlands the annual rainfall is 700–1000 mm and there is a dry season of four to six months, but large areas in the mountains have over 3000 mm with almost no dry period. Pócs (1974, 1976b) has given an account of the interrelations between relief, climate and vegetation in the Ulugurus (see also Hamilton & Bensted-Smith 1989). There is a preliminary report (Pócs *et al.* 1990) on the vegetation of the Nguru mountains, where there are relics of lowland rain forest and the complete zonation up to upper montane and bamboo forest can be seen.

In the humid parts of the Eastern Usambara mountains, which is one of the few areas in East Africa where lowland and montane forests are in close proximity, Moreau (1935) distinguished three forest zones, lowland (to 760 m), intermediate (760–1370 m) and highland (above 1370 m). In the parts of the Ulugurus where rainfall is high and without a dry season, Pócs (1976b) recognizes four zones: lowland semi-evergreen forest (250–500 m), submontane evergreen and semi-evergreen forest (500–1500 m), montane evergreen forest (1500–2100 m in the northern parts, slightly higher in the south), and an upper montane (or lower subalpine) zone of 'elfin forest', bamboo thickets, peat bogs and secondary grasslands.

Little detailed information on the composition of the forests in the Usambara and Uluguru mountains is available. According to Moreau (1935) the important trees in the lowland rain forest of the Eastern Usambaras are *Antiaris toxicaria*, *Ficus* spp., *Trema orientalis*, *Milicia excelsa* and species of *Sterculia* and *Albizia*; this suggests an old secondary rather than a primary community. The undergrowth consists largely of *Acacia pennata* and *Harrisonia abyssinica*, both thorny species, with thickets of the tall grass *Olyra latifolia*. Lianes are plentiful, but there are few epiphytes, except scattered plants of *Platycerium angolense*. The intermediate forest is the most luxuriant of the three types. The species of trees here are very numerous; many are the only representative of their genus in the area and are endemic. Important dominants are *Macaranga usambarica*, *Allanblackia stuhlmannii*, *Newtonia buchananii*, *Isoberlinia scheffleri* and *Parinari* spp. Tree ferns are frequent and both lianes and strangling figs are prominent. Epiphytes are numerous both as species and individuals. The general appearance of the highland forest is not unlike that of the intermediate forest, but its composition is different. The chief species here are *Podocarpus* spp. and *Ocotea usambarensis*; *Juniperus procera* is locally abundant. The 'warm temperate forest' of the Western Usambaras described by Pitt-Schenkel (1938) is similar.

The lowland forest which formerly existed in the Usambara and Uluguru mountains was probably similar to the semideciduous forest of Zaïre and West Africa (Chapter 16). The 'intermediate' forest of the Eastern Usambaras and the 'submontane' forest of the Uluguru mountains seem to be comparable to Lebrun's transition forest described above, while the 'highland' and 'montane evergreen' forests are very similar to the montane forests of eastern Zaïre and Uganda.

The Eastern Usambaras and Ulugurus are both less than 150 km from the coast. As might be expected, the altitudinal zones are considerably lower than on the higher East African mountains further inland where there is *Massenerhebung*. In addition, as Pócs (1974, 1976b) shows, they are lower on the eastern (windward) than on the western slopes. The lowering may be due in part to a higher temperature lapse rate: Moreau (1935; see also 1938) says that at 915 m in the Usambaras temperatures are no higher than at 1525 m on Mt Kenya and Kenworthy (1966) found similar differences. Pócs (1976b) found in the Ulugurus that frosts damaging to the vegetation occurred occasionally as low as 1800 m and low temperatures may well be an important limiting factor at the higher elevations. Moreau (1935) drew attention to the existence of a 'cloud belt' in the Eastern Usambaras and the relation of this to the vegetation zones does not seem to have been carefully investigated. It seems likely that, as in other parts of the tropics, the frequency of mist and low cloud is here also a significant factor in determining the altitudinal boundary between the lowland and montane forest zones.

In Malawi there is one small area of lowland rain forest and several relics of montane forest above 1525 m, but the latter are isolated and no transitions between the two types now exist (Chapman & White 1970).

Zonation in West Africa. In West Africa, owing to extensive forest clearance, there are, as in East Africa, few places where a direct transition from lowland to montane forest can be seen. Even at high altitudes, where there are no permanent human inhabitants, the natural vegetation has been much altered by anthropogenic fires, and grasslands have replaced forest over large areas. General accounts of the vegetation of the West African mountains have been given by Letouzey (1968), Schnell (1976–77) and Morton (1986).

Cameroon Mountain (4070 m), the highest peak in West Africa, is an intermittently active volcano and in some places lava flows have destroyed the vegetation. On the seaward side of the mountain the annual rainfall is very high; at Debundscha it is over 9 m. Elsewhere, especially in the grassland region, it is probably much lower. According to Fontes and Olivry (quoted by Maley 1987) the dry season lasts one month or less in the forest at 1000 m, but five to seven months in the grassland at 2500 m.

Below *ca.* 1200 m, most of the forest was long ago replaced by banana plantations and shifting cultivation, but from this height to the lower limit of the grassland which covers most of the upper part of the mountain it is more or less continuous. On the southern and western sectors of the mountain the forest–grassland boundary is mostly at *ca.* 2130 m, except where the

Fig. 17.12 Upper limit of montane forest, Cameroon Mountain, West Africa (March–April 1948). A. Forest and grassland at Mann's Spring (1438 m). The grassland, except patch on right, has been recently burnt. B Forest in small crater at *ca.* 1438 m. Foreground with fringe of *Hypericum revolutum*. Another forest outlier in crater at *ca.* 2590–2620 m in background.

forest has been destroyed by landslips or lava flows, but on the other sectors it is lower. In the grassland there are numerous small forest outliers in hollows and gullies up to 2590–2650 m (Richards 1963b) (Fig. 17.12). The grass is burned annually and at the edge of the forest there is a characteristic narrow fringe of *Hypericum revolutum* and other small fire-tolerant trees.

The lower part of the forest belt (*ca.* 1200–1800 m) is a type of lower montane rain forest. The zone of highest rainfall is probably at *ca.* 1220–2000 m and the forest belt is frequently bathed in mist. The structure and composition of this forest are not well known. It has some of the physiognomic characteristics of lowland rain forest, but the canopy is more broken and irregular.

A

B

Fig. 17.13 Interior of montane rain forest, Cameroon Mountain, West Africa (1948). A At 1890 m under typical misty conditions. Undergrowth mainly *Mimulopsis solmsii*. B At 2356 m. The large tree is *Syzygium staudtii*.

The tree flora is relatively rich and includes lowland species such as *Entandrophragma angolense* and *Turraeanthus africana*, as well as more montane species such as *Alangium chinense* (Hall 1973; see also Letouzey 1968); towards the forest's upper limit montane species become increasingly abundant. The tall tree fern *Cyathea manniana* is conspicuous in gaps, and montane species such as *Sanicula europaea* and *Viola abyssinica* are found in the herb layer. Epiphytic bryophytes are much more abundant than in the lowland forest.

From about 1800 m upwards the forest is very different; although it is untypical in some respects, it can be regarded as a type of upper montane forest (Fig. 17.13). On sample plots of 0.7 ha at 1890 and 2025 m (Richards 1963b) there was an upper layer of trees *ca.* 37–40 m high and a very discontinuous layer about 18 m high. The former consisted of *Syzygium staudtii*, *Schefflera abyssinica* and *S. mannii* only, trees which are quite unlike lowland forest trees in physiognomy. The first has a dense hemispherical crown with many thin sinuous branches radiating from a short thick trunk; it is a typical 'Kugelschirmbaum' (umbrella tree), a tree form which is characteristic of montane forests in the tropics and lowland forests in the subantarctic zone (Troll 1958). The two *Schefflera* species are rather similar to this in form when mature, but begin life as hemiepiphytes like some *Ficus* species. The lower tree layer consists mainly of *Allophylus bullatus*. The total number of tree species ⩾30 cm g.b.h. on the two plots was only nine. All were montane species, except *Canthium glabriflorum* which is common in lowland forest.

All the trees lacked buttresses and other lowland rainforest characteristics. They were evergreen (with the possible exception of *Schefflera abyssinica*, of which one individual was seen bare of leaves) and their foliage was without drip-tips and mesophyll in size, although some of the forest-marginal species had leptophyll or nanophyll leaves.

The undergrowth was considerably richer in species than the tree layers. A striking feature was the dense stands, 3–4 m high, of two semiwoody Acanthaceae, *Mimulopsis solmsii* and *Oreacanthus mannii*, which flower gregariously at intervals of several years and appear to be monocarpic (cf. *Strobilanthes*; see p. 253). The tree fern *Cyathea manniana*, which was abundant below 1800 m, was absent in both plots. Locally there were patches of grass, perhaps resulting from burning or grazing.

Climbing plants were frequent, but the only large woody liane was *Clematis simensis*. Epiphytic flowering plants, ferns and bryophytes were abundant, but the covering of bryophytes on the trunks and branches in the sample plots was not as thick as in the highest forest outliers where it rivalled that in the 'mossy forests' of Malesia and the West Indies in luxuriance.

On the volcanic islands of Bioko (Fernando Po) and S. Tomé in the Gulf of Guinea, which rise to 3007 m and 2024 m, respectively, before most of the primary forest was replaced by cultivation the zonation was probably like that on Cameroon Mountain. On S. Tomé, which has no montane grassland, the transition from lowland forest (probably similar to the existing relics of lowland rain forest on the adjacent mainland) was at about 800 m and it seems to have been 'characterized by a change in the constituents of the forest rather than in its general aspect' (Exell 1944, p. 18).

The zonation on Cameroon Mountain (and on Bioko and S. Tomé) differs in several respects from that on the high East African mountains, although in both West and East Africa the change at the upper altitudinal limit of lowland rain forest is more in species composition than in physiognomy. At higher elevations the East and West African plant formations are not closely comparable. The characteristic bamboo zone of the East African mountains is absent on Cameroon Mountain, although in the Bamenda Highlands, *ca.* 200 km to the northeast, pure stands of *Arundinaria alpina*, not forming a continuous zone, are found in the montane forest at 2000–2400 m. On Cameroon Mountain grassland replaces the ericaceous and Afro-alpine zones of the East African mountains. It has, however, been suggested that before fires were as frequent and severe as they now are, heath-like vegetation of *Hypericum revolutum*, *Philippia* and other shrubs, like that now found as a fringe round the forest outliers, may have covered the lower part of the grassland area (Richards 1963b).

In Africa west of the Dahomey Gap the vegetation

on the upper parts of the mountains is mainly grassland and montane forest; the latter is in contact with lowland rain forest in some places. A remarkable feature of the montane forests on these western mountains (Fig. 17.14) is that they are mainly single-dominant communities dominated by *Parinari excelsa*, an evergreen tree which is widely distributed in tropical Africa and occurs down to sea-level. At low elevations it is found in both semi-deciduous and rain forests, but not as a single dominant: possibly the lowland and montane populations are ecotypically distinct.

Schnell (1952b) has given a full account of the vegetation of the Nimba Mountains in southwest Guinea, which reach a height of 1752 m. The higher parts are uninhabited, but fires, usually anthropogenic, have converted much of the forest into grassland. Schnell distinguishes two zones, lowland and montane; the latter descending to *ca.* 1000 m, the height at which persistent cloud and mist become frequent. The primary lowland forest on the wetter southwestern side of the Nimba range is tropical rain forest; in the drier northern and eastern parts it is semideciduous forest like that in Ghana and Nigeria. The lowland rain forest is a mixed association (Chrysophylleto–Tarietetum utilis of Schnell) in which *Lophira alata* and *Heritiera utilis* are prominent emergent trees and the sedges *Mapania* spp. are characteristic undergrowth species.

At *ca.* 500–900 m the composition of the forest begins to change, *Heritiera* disappearing almost entirely; at about 1000 m there is an association transitional to the *Parinari*-dominated montane forest (Parinareto–Ochnetum membranaceae of Schnell). The latter has a characteristic canopy of the dense greyish-green hemispherical crowns of *Parinari excelsa*, a typical 'Kugelschirmbaum', which in ravines grows to 20–30 m, branching at 5–15 m, but on ridges is only 8–10 m tall and branches at 3–6 m.

This montane forest is considerably poorer in tree species than the lowland rain forest. The trees include various species in addition to *Parinari excelsa* which also occur in lowland forest and others, such as *Syzygium staudtii*, which are montane. Lianes and tree ferns are found. The enormous abundance of epiphytes, including species of *Begonia* and *Peperomia*, orchids, ferns, lycopods, bryophytes and lichens, is a striking difference from the lowland rain forest.

Montane forests dominated by *Parinari excelsa* similar to those of the Nimba mountains are found in other mountain ranges in Guinea and in the Loma Mountains and Tingi Hills in Sierra Leone (Morton 1986). They are somewhat anomalous and difficult to classify. They have some of the physiognomic and structural features

Fig. 17.14 Gallery forest dominated by *Parinari excelsa* in montane grassland at *ca.* 1500 m, Loma Mountains, Sierra Leone (1971). Note typical hemispherical crowns.

of the upper montane forests of Malesia and the Neo-tropics, but are very different in composition. Their flora is a mixture of lowland and montane species, but lacks conifers and temperate genera.

17.2 Tropical rain forest at its latitudinal limits

On the eastern margins of the continents, in Australia, along the Atlantic coast of Brazil[9] and in Southeast Asia, rain forests extend outside the geographical tropics. These extra-tropical rain forests resemble tropical rain forest in many of their characteristics, but can be regarded as a separate formation type, subtropical rain forest. The term temperate rain forest should be kept for the 'Subantarctic' forests of southeastern Australia, Tasmania, the South Island of New Zealand and southern Chile. These have few tropical characteristics and are dominated by evergreen trees of genera such as *Nothofagus* (southern beech), which in the tropics is found only at fairly high elevations.

As well as the relatively large areas of subtropical rain forest mentioned above, which are in more or less direct continuity with the rain forests of the tropics, there is an isolated area, once until recently quite extensive, in New Zealand, mainly in the North Island (Dawson & Sneddon 1969, Dawson 1980) and small areas exist at Knysna (Phillips 1931) and elsewhere in South Africa (Geldenhuys 1985). Much smaller outliers which, at least in their flora, recall the tropical rain forest are the 'hammock forests' of Florida (Davis 1943, Craighead 1971).

Rain forests of Australia. It is in eastern Australia that the ecological relations of tropical, subtropical and temperate rain forest have been most studied and can be best appreciated. General accounts have been given by Beadle (1981), Webb & Tracey (1981a,b), Webb *et al.* (1984) and Adam (1992).

In the Australian literature 'rainforest' is often used in a much wider sense than in this book, to include, for example, the relatively dry 'bottle-tree [*Brachychiton*] scrubs' of Queensland and the strongly seasonal relics of 'monsoon forest' in the Northern Territory, but, even when the term is thus extended, the total area of tropical, subtropical and temperate rain forests now remaining in Australia is only about 20 000 km² (Webb & Tracey 1981b). In the Tertiary period wet evergreen

forests may have covered a large part of the continent, but in recent times they have consisted of separate stands varying in extent from a few hectares to some hundreds of square kilometres, forming a narrow belt along the east coast from near Cooktown (15°28'S) in Queensland southwards to New South Wales and Victoria. The grassland and sclerophyll vegetation separating the stands varies in width from a few hundred metres to a corridor several hundred kilometres wide between Ingham (18°35'S) and Bowen (20°00'S). Tenuous connections such as strips of riverine woodland sometimes join neighbouring stands of rain forest.

Although the area occupied by rain forests is believed to be now only about a quarter of what it was 200 years ago (Webb & Tracey 1981b), its discontinuity is due as much to a complex rainfall pattern caused by the effects of coastal mountains and the varying angle of the coastline to the southeast trade winds as to clearance by people. Thus, even at the time of European settlement, and although the Australian Aborigines did not practise shifting cultivation, Australian rain forests were already, as Herbert (1967) puts it, 'an archipelago of habitats'. In tropical and subtropical Australia, forest clearance has been most complete where the land can be used for sugar-cane cultivation; as a result very little lowland forest remains. Most of the existing tropical rain forests are at altitudes of over 500 m and are not typical lowland forests.

In tropical Australia, as well as further south, the natural vegetation surrounding the rain-forest 'islands' (where it has not been replaced by grassland or other anthropogenic ecosystems) is more or less open forest or woodland dominated by species of *Eucalyptus* and other Myrtaceae, often associated with *Casuarina* spp., phyllodic acacias and other sclerophyllous trees and shrubs. In flora and physiognomy the contrast between the two formations is extreme. The location and stability of the rain-forest boundaries are determined by a complex of interacting factors (Ash 1988), among which the edaphic components are of great importance (Tracey 1969, Webb 1969). In general Australian rain-forest trees have higher mineral nutrient requirements, especially for phosphate, than those of the eucalyptus forests; rain forests are therefore mostly restricted to relatively fertile soils derived from basalt, schist or alluvium, while eucalypt forests are able to grow on nutrient-poor soils derived from sandstone and similar materials.

Where different soil types adjoin, the boundary between the two formations may be quite sharp, but there is often a transitional belt in which rain-forest and eucalypt-forest species mix. Sometimes, as in the *Eucalyptus deglupta* forest of New Guinea, this takes

---

[9] In South America the southernmost outpost of subtropical rain forest is in the Urugua-i National Park (*ca.* 25°35'S, 54°22'W) in Misiones Province, Argentina (Allen 1990).

the form of an open overstorey of tall eucalypts with an undergrowth of rain-forest species. There is often evidence that invasion of eucalypt forest by rain forest has taken place, but invasion of rain forest by eucalypts does not occur. Some species of *Eucalyptus*, e.g. *E. grandis*, *E. torelliana*, tend to grow near the rain-forest margin, but *Eucalyptus* species are intolerant of shade and are unable to regenerate under stable rain-forest conditions.

The boundary between rain forest and eucalypt forest may have been affected to some extent by cyclones, which tend to form openings littered with debris which are more readily colonized by eucalypts than by rain-forest trees. It has been considerably affected by fire. In the past the hunting fires of the aborigines, and possibly natural fires, may have eroded the rain-forest margin to some extent (Stocker in Beadle 1981). In recent times, fires have been important in restricting rain-forest stands.

Tropical rain forests of north Queensland

The finest development of the Australian tropical rain forest is in areas with over 3000 mm annual rainfall near Cape Tribulation (16°03'S), and near Tully (17°56'S) and Innisfail (18°40'S). The greatest number of tree species per unit area is probably in the Daintree river basin (*ca.* 16°S). Except in the areas of highest rainfall the climate of most of the Queensland rain forest is not ever-wet and has a relatively dry season of five to six months, each with an average of less than 100 mm. The seasonal range of temperature is greater than in typical equatorial rain forest: for example, at Innisfail there is a difference of 7.6° between the mean monthly temperatures of January and July. It is perhaps of ecological importance that the season of low rainfall is also that of low temperature, so that the vegetation may be less affected by water stress than in typical seasonal tropical climates in which temperatures are usually higher in the dry than in the wet season.

A

B

Fig. 17.15 Endemic species of Australian lowland tropical rain forest near Noah Creek, Queensland (1981). A *Bowenia spectabilis* (cycad). B Seedlings of *Ideospermum australiense*, showing four cotyledons.

In some features the tropical rain forests of Australia resemble those of the Indo-Malayan region, especially New Guinea; for example, climbing palms (*Calamus* spp., called 'lawyer vines' in Australia) are conspicuous in both, particularly in gaps whether artificial or caused by cyclones (which are very frequent). Most of the trees are evergreen, but deciduous species such as *Terminalia sericocarpa* are not uncommon. Conifers (species of *Agathis, Araucaria* and *Podocarpus*) occur, but are not abundant; erect palms (*Archontophoenix, Licuala, Linospadix*, etc.) are common.

Australian rain forests have characteristic features of their own: tree ferns (*Cyathea* spp.) are often abundant, even at low altitudes, and there is a surprising scarcity of ground herbs other than ferns, except in gaps; the lowest stratum of the forest often consists almost entirely of young trees and lianes. Cycads (*Bowenia* and other genera) are unusually common. Epiphytes, though fairly numerous in species, are not very abundant and epiphyllae, except lichens, are rather scarce.

In floristic composition as well as in structure there are similarities between the Australian tropical rain forests and those of Malesia, but although there are many families and genera (and even some species) common to the two regions, the resemblances are less close than has often been supposed. An important difference is the absence in Australia of the Dipterocarpaceae, which play such an important part in the Malesian forests and are fairly abundant even in southeastern New Guinea. A remarkable feature of the Australian forests is the large number of endemic taxa, including, as mentioned earlier, several supposedly very primitive families (Fig. 17.15). The Proteaceae are strongly represented and in north Queensland include many genera of very local distribution. Webb & Tracey (1981b) have analysed the geographical distribution of 545 genera of Australian rainforest[10] trees and lianes. Of the 442 non-endemic genera, 124 are widespread in the tropics; among the rest the largest contingent (153) is shared with Malesia and the second largest (114) with New Caledonia.

The number of tree species per unit area is often rather large. On 17 plots (0.5 ha) in northeast Queensland the number of tree species ≥10 cm diameter was 32–87 (average 58) (Stocker in Beadle 1981, p. 171). The highest numbers recorded are in the upland forests of the Daintree river (*ca.* 16°S) (Stocker). Australian tropical rain forests are usually mixed in composition but examples of local single-species dominance occur.

Thus at Woopen Creek, near Babinda (*ca.* 17°20'S) there is a tall, cathedral-like stand of *Backhousia bancroftii* (Myrtaceae) over an area of several hectares.

Besides the climax tropical rain forest on freely draining, relatively fertile soils there are edaphic types such as riverine forests and relics of swamp forest on seasonally or permanently flooded sites. A remarkable example of the latter is the swamp forest at Lacey's Creek near Tully (Queensland) where *Licuala ramsayi* and other palms are abundant. The altitudinal zonation on mountains is illustrated in Fig. 17.1.

The boundary between the tropical and subtropical rain-forest regions in Australia lies north of the geographical tropic but can be defined only rather arbitrarily.

Subtropical rain forest in Australia

Southwards from near *ca.* 20°S (between Ingham and Mackay) in Queensland to about 37°S in New South Wales, scattered areas of subtropical rain forest survive. In this zone small relict outliers of temperate rain forest dominated by *Nothofagus moorei* are found on mountains above about 1000 m (Webb & Tracey 1981a, b). Further south, in Southern New South Wales, there are only a few very small relics of subtropical rain forest, but temperate rain forest, in which *Nothofagus cunninghamii* replaces *N. moorei*, was formerly widespread and occurred down to sea-level; little of this now remains (Howard 1981).

The distribution of subtropical and temperate rain forests in Australia is related to altitude, local climates and soil conditions and there is no clear latitudinal boundary between them.

The subtropical rain forests of Australia, though perhaps rather less luxuriant in appearance, are similar in structure to the tropical rain forests further north. The trees are about as tall and there is a similar range of life-forms although, at least in the more southern subtropical rain forests, plants of characteristically temperate life-forms such as hemicryptophytes occur.

The trees are mostly evergreen, although a few species such as *Toona australis* are deciduous. An important physiognomic difference between Australian subtropical and tropical rain forests is in leaf sizes: in the former (notophyll vine forests of Webb 1959, Webb & Tracey 1981b) notophylls predominate, in the latter (complex mesophyll vine forests) mesophylls. Some trees in the subtropical forests develop very large buttresses, e.g. *Argyrodendron trifoliolatum, Sloanea woollsii*, but some features characteristic of tropical rain-forest trees, e.g. cauliflory, are rare or absent. Conifers are perhaps rather more frequent than in the tropical rain forests and

[10] Including subtropical, temperate and 'dry' rain forest.

are represented by *Agathis* spp., *Callitris macleayana*, *Podocarpus elatus, Araucaria cunninghamii* and, locally in southern Queensland, *A. bidwillii*. One type of sub-tropical rain forest (araucarian vine forest of Webb & Tracey 1981b) is characterized by an open 'overstorey' of super-emergent araucarias. Erect and climbing palms and tree ferns (*Cyathea* spp.) are generally abundant.

Large woody lianes are common and root-climbers (Araceae and ferns such as *Arthropteris* spp.) are con-spicuous, especially near gaps. Stranglers, e.g. *Ficus macrophylla*, are frequent. Epiphytes, including pendent mosses (Meteoriaceae), are numerous in species and individuals; and epiphyllous hepatics, though rather scarce, are found in sheltered gullies.

The number of plant species in Australian rain forests decreases southwards, as might be expected. According to Beadle (1981, p. 142 and Table 7.3, p. 141) the total of angiosperm species falls from 1900 in Queensland to 540 in northern New South Wales, and 227 in southern. These figures are for large undefined areas: on small sample plots there is probably a similar decrease, but few figure are available. In subtropical rain forest at Wiangarie in the north of New South Wales (*ca.* 28° 30'S) Baur (1964, p. 98) recorded 66 tree species ≥10 cm diameter on a 9.6 acre (*ca.* 4 ha) plot; this is equivalent to about 30 species per hectare, which is less than in tropical rain-forest plots in north Queensland.

Many of the trees of the subtropical forest belong to characteristically tropical families such as the Elaeocarpa-ceae, Lauraceae and Rutaceae and some to genera found in the Malesian forests or widespread in the tropics, but endemic Australian taxa are strongly represented. Some of the most abundant tree species belong to mainly southern hemisphere families such as the Cunoniaceae. According to Webb & Tracey (1981b), it is only at species level that a clear floristic distinction can be drawn between the subtropical and tropical Australian rain-forest formations.

In New South Wales, Floyd (1990) recognizes four 'alliances' of the 'subtropical rain forest subformation', differing in structure and floristic composition. In addition, he includes 'littoral rain forest' and 'dry rain forest'. The first two alliances are characterized, respect-ively, by *Argyrodendron trifoliolatum* and *A. actino-phyllum*. Both are found in southern Queensland as well as in northern New South Wales. They resemble tropical rain forest in structure more closely than the other alliances and are the richest in species.

Floyd distinguishes between 'warm temperate rain forest' and 'cool temperate Rain forest'. In two alliances of the former the characteristic tree species is *Ceratopet-alum apetalum* (coachwood) which forms a considerable proportion of the stand and is sometimes a single domi-

Fig. 17.16 Alternation of sclerophyll (*Eucalyptus*) forest on ridges with temperate (*Nothofagus moorei*) rain forest in gul-lies, seen from *ca.* 1600 m, New England National Park, New South Wales (1981).

nant, often associated with *Doryphora sassafras. Cerato-petalum* forests have some tropical characteristics, but the number of tree species is much smaller than in the *Argyrodendron* forests. Buttressed trees are rare or absent and lianes, though fairly abundant, seldom reach a large size. Large erect palms are usually absent and climbing palms are uncommon.

Scattered stands of *Ceratopetalum* forest are found from Queensland to south of Sydney (*ca.* 37°S) and, as small relics, in eastern Victoria. In some places *Cerat-opetalum* and *Argyrodendron* forests are found in close proximity, as in Dorrigo National Park, New South Wales, where the former occur on the relatively nutrient-poor soils derived from metamorphic rocks and the latter on richer soils on basalt.

In New South Wales, cool temperate rain forest is represented by two alliances, one dominated by *Notho-fagus moorei* and the other by *Eucryphia moorei*.

Another part of New South Wales where different forest formations are intermingled is the Barrington Tops (32°02'S) and Upper Williams river district, of which Fraser & Vickery (1937–39), Adam (1992) and Floyd (1990) have given detailed accounts. The area is mountainous, rising to a plateau about 1200–1300 m above sea-level. Subtropical rain forest covers the floor and sides of valleys facing east, southeast and northeast up to about 1000–1500 m. At higher elevations there are temperate rain forests. The upward extension of the subtropical forest is probably determined by low winter temperatures and the downward extension of the tem-perate forest by high summer temperatures. Various types of sclerophyll (*Eucalyptus*) forests are widespread below the lower boundary of the subtropical forest, on the crests and sides of ridges and on the plateau (Fig. 17.16).

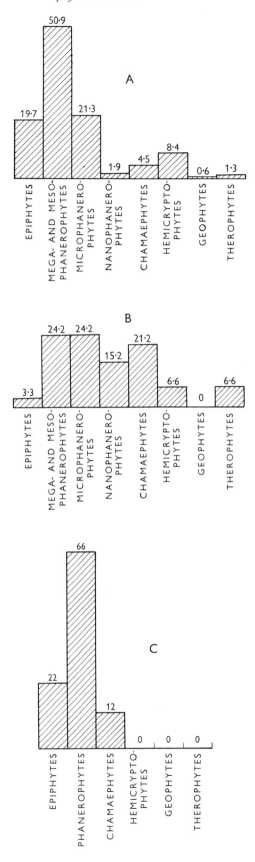

Although some species occur in more than one forest formation and mixed communities are found, the three formations remain very distinct. Some of the differences in their composition and physiognomy are well shown by their life-form spectra (Fig. 17.17). Fraser & Vickery (1937–39) found that at its margins, especially near its upper altitudinal limit, the subtropical forest is invading the sclerophyll forest, in some places quite rapidly. They regarded this expansion as part of a recovery from a retrogression due to orogenic movements in the late Tertiary; however, in the light of later work it could perhaps be due to more humid conditions following a dry period in Pleistocene or later time.

At low altitudes the subtropical forest in the eastern part of Barrington Tops consists, according to Adam (1992), of a mixture of rain-forest trees (*Sloanea woollsii*, *Cryptocarya glaucescens*, etc.) growing to a height of about 30 m, overtopped by a discontinuous layer of *Eucalyptus saligna* and *Lophostemon confertus* as tall emergents. In the western part at low levels the forest is somewhat disturbed and different in composition. Above about 1000 m the characteristic trees include *Caldcluvia paniculosa*, *Doryphora sassafras* and *Schizomeria ovata*, species typical of a 'suballiance' of Floyd's *Caldcluvia* alliance. With these are intermingled various warm temperate species but, surprisingly, *Ceratopetalum apetalum* is absent. The subtropical forest of Barrington Tops is much less rich in species than that further north. In the samples of Fraser & Vickery (total area not stated) there were 181 trees ⩾10 m (30 ft) high belonging to thirty-four species.

On the upper valley slopes and in sheltered parts of the plateau there are extensive stands of cool temperate rain forest. This is dominated by *Nothofagus moorei*, which here reaches its southern limit. In Fraser & Vickery's samples it formed 76% of trees ⩾10 m (30 ft) high.

17.3 Comparison of rain forest at its altitudinal and latitudinal limits

Humboldt (1817) was originally responsible for the generalization that the altitudinal zones of vegetation in the tropics correspond to the latitudinal zones from the

Fig. 17.17 Raunkiaer biological spectra of subtropical rain forest, temperate rain forest and tropical rain forest. A Subtropical rain forest, Upper Williams River district, New South Wales. B Temperate rain forest, Upper Williams River district, New South Wales. C Tropical rain forest, Moraballi Creek, Guyana. A and B from data of Fraser & Vickery (1938), C as in Table 1.2. Figures are for percentage of species of vascular plants in total flora, excluding pteridophytes.

equator to the poles. He regarded an increase of elevation of about 1000 m on a tropical mountain as equivalent to a difference of about 9°30′ of latitude, or about 100 m to about 1°: thus at the equator an altitude of 8800 m would in theory correspond to the poles. Although there is some general similarity between the physiognomy of the vegetation at different altitudes and latitudes with the same annual mean temperature, the correspondence is far from exact; as Lam (1945, p. 82) says, Humboldt's conception is too simple. The climate of the equivalent altitudinal and latitudinal zones is never the same, because in a given altitudinal zone at the equator, the length of day and the seasonal changes, to mention only two factors, are different from those in higher latitudes, although the mean temperature may be similar.

It is relevant to note that, as Troll (1948) pointed out, because of their small seasonal range of mean monthly temperature and the frequency of frost throughout the year at high elevations, tropical montane climates resemble those of high latitudes in the southern hemisphere more closely than north temperate climates. To this fact, he suggested, may be traced the larger representation in tropical montane floras of taxa with southern hemisphere than of northern affinities (notably in Malesia). The plant communities of tropical montane forests are also similar to those of temperate rain forests in Tasmania, New Zealand and southern Chile in physiognomic features such as the crown form and leaf-size spectrum of the trees. The resemblances between tropical montane and temperate southern climates depend on the small area of land relative to sea in high southern latitudes, which results in what Troll calls the 'asymmetry' between the vegetation zones in the two hemispheres.

Ohsawa et al. (1985) have made a detailed comparison of the forest vegetation zones on a mountain in Sumatera with the vegetation of temperate eastern Asia.

Part VI

Human impacts and the tropical rain forest

Chapter 18

Secondary and deflected successions

Throughout the tropics, enormous areas of primary rain forest have been replaced by secondary communities, varying from treeless scrub, grassland or fern brakes to tall forest not very dissimilar in appearance to the original climax vegetation. These communities, excepting those dominated by grasses and ferns (which result from deflected successions[1]) are generally called secondary forest or 'second growth'; in many countries, local names are used for them, e.g. capoeira (Brazil), low bush or mainap (Guyana), belukar (Malaysia), parang (Philippines).

Secondary vegetation derived from rain forest, whether it has the aspect of closed forest or scrub or, as is often the case, appears to be a chaotic wilderness of trees, shrubs, climbers and tall herbaceous plants, is always more or less unstable and consists of successional stages. If undisturbed by grazing, tree felling and frequent fires, this secondary vegetation is slowly invaded by primary forest trees and can eventually develop into a community similar to that which originally occupied the site. The speed of change depends on the availability of seed parents and many other factors. For a few years it is usually rapid, but later it becomes much slower and the whole process of recovery probably extends over centuries rather than decades. If the secondary vegetation is subject to recurrent fires, grazing or other disturbances, deflected ('retrogressive') successions set in, leading to apparently permanent biotic climaxes.

People have made clearings in tropical forests for various purposes and of various sizes from early prehistoric times. Pollen analysis shows that in the uplands

of Sumatera forest destruction began by 4000 yr BP, or even earlier, and in the New Guinea Highlands by 5000 yr BP, possibly long before that (Flenley 1988). In lowland Ecuadorian Amazonia, phytoliths and pollen in lake sediments show that maize cultivation was taking place at least 6000 yr BP (Bush *et al.* 1989). In recent years (since about 1950) many thousands of square kilometres of rain forest in South America (Jari river, Colombian Chocó) and Malesia have been logged by highly mechanized methods for pulpwood and other industrial uses. In Amazonia in recent decades much forest has been converted into cattle pastures and in both Amazonia and Malaya considerable areas of forest have been destroyed by opencast mining. In Vietnam over a million hectares of forest, mainly monsoon forest and mangrove, but some of it rain forest, were partly or totally destroyed by herbicide spraying and other military actions in the 1961–75 war (US National Academy of Sciences 1974, Richards 1984b, Westing 1984). However, today, as in the past, shifting cultivation is probably the most important cause of deforestation. Shifting subsistence agriculture ('slash-and-burn', swidden or ladang farming) is practised everywhere in the humid tropics and subtropics, except in Australia and a few regions (Java, parts of India) where local conditions allowed the early development of sedentary agriculture.

The economic and sociological aspects of shifting agriculture have been much studied (e.g. Conklin 1957, 1961, 1963 Meggers 1971, Watters 1975, Dove 1985, Hong 1987, Jordan 1986). The methods of shifting cultivation in the humid tropics vary in detail in different regions and cultures, but the general procedure is always

---

[1] In Godwin's (1929) sense.

[457]

similar. A patch of forest is felled, usually by a family or small group, the size of the clearing depending on the amount of labour available and the productivity of the site; usually it is between about one and fifteen hectares. Often, for superstitious or other reasons, some of the larger trees are spared. The brushwood and fallen timber are usually, but not always, burnt when dry enough. Crops are then planted, the commonest being cassava (*Manihot esculenta*), maize, 'hill' (unirrigated) rice, yams (*Dioscorea*), taro (*Colocasia esculenta*), bananas (*Musa*) and sweet potatoes (*Ipomoea batatas*) or mixtures of these. Smaller quantities of fruit trees and other edible or useful plants are also often grown.

Because humus is rapidly destroyed by exposure to the sun and the ground receives little cultivation and no manure, the soil soon becomes impoverished and infertile. Weeds become increasingly difficult to control and crop yields decrease. After about three years, the period depending on the extent of leaching and erosion

as well as the inherent fertility of the soil (Chapter 10), the field is abandoned. During the secondary succession which follows, soil fertility is gradually restored and after a fallow period, usually of about ten or more years, the forest can be cleared and cultivated again.

Whether the farmers make their new fields in primary forest or on forest land which has been previously cultivated depends on the availability of suitable land and the density of population. In sparsely inhabited regions such as remote parts of Amazonia virgin forest may be used, but under modern conditions the growth of populations usually makes it necessary to use land which has been previously farmed; there is also pressure to shorten the fallow periods. Even early in this century Lane-Poole (1925) noted that in New Guinea cultivation cycles had become shorter after the colonial governments had established peaceful conditions, so that secondary

Fig. 18.2 'Farm' with maize crop, in forest, Omo Forest Reserve, Nigeria (1935).

Fig. 18.1 Shifting agriculture in Amazonia. A Conocos (fields) in rain forest, Upper Rio Negro (Venezuelan–Colombian border). (Photo. R. Herrera.) In the field on the right the felled trees are being burnt. B A village with fields, dry season, R. Xingú, Mato Grosso, Brazil (August 1968). The white areas have been recently burnt and are covered with wood ash.

Fig. 18.3 Clearing a forest for irrigated rice, Ulu Tinjar, Sarawak (1932).

forests were being cleared before fertility had been fully restored. A similar change can now be seen in most parts of the tropics; soil deterioration, combined with other effects of repeated burning of secondary vegetation, is leading to the establishment of enormous areas of the grass *Imperata* (alang-alang), ferns such as *Pteridium aquilinum* and *Dicranopteris* (*Gleichenia*), or other weed species.

It has been estimated that in the humid tropics shifting agriculture alone cannot support a population greater than about 7 per km², unlike sedentary systems such as irrigated rice in tropical deltas which can support populations several hundred times as large (Whitmore 1984a; see also Watters 1975). Although in sparsely inhabited forest regions shifting systems can be ecologically stable, at high population densities they must inevitably give way to more productive land-use systems or lead to soil deterioration (see Bayliss-Smith 1982).

Plantation agriculture in the tropics has often proved little more permanent than shifting cultivation. In most tropical countries large areas of secondary vegetation occupy the sites of plantations abandoned because of soil impoverishment, animal pests, plant diseases or other reasons. Individuals of long-lived crop plants, such as oil palms and fruit trees, often survive in secondary vegetation as relics of past cultivation.

Timber exploitation in tropical rain forests has greatly changed over the last hundred years in scale, methods and the range of species utilized. In the early years of this century felling was generally highly selective; often only large logs of the most valuable trees such as greenheart (*Ocotea rodiei*) and mahoganies (Meliaceae) were extracted, using teams of men along 'hauling roads' (Fig. 18.4). The 'depleted forest' which remained had a very irregular structure, with numerous large gaps filled with climber tangles or dense stands of young trees. In the 1930s there were large areas of such forest in Guyana and the coastal region of Nigeria (Richards 1939). Ng (1983) found that in Malaya trees regenerating after timber exploitation have shorter boles and are 25–50% less tall than those in the original forest because of the different surroundings in which the young trees develop.

In modern mechanized forest exploitation a wide range of species are extracted; when the timber is required for 'wood chips' (pulpwood) almost all the trees are usually removed (Fig. 18.5). Even when logging is selective large numbers of trees are damaged or destroyed incidentally. In dipterocarp forests in Malaya, for example, about 10% of the trees ≥10 cm d.b.h. may be felled for use as timber, 55% damaged or destroyed and only some 35% left undamaged (Burgess 1971). In Para State (Amazonia) Uhl & Vieira (1989) found that

Fig. 18.4 Old timber hauling road, Moraballi Creek, Guyana (1929).

Fig. 18.5 Forest clearance, Gogol Project, Papua New Guinea (1975).

in the sampled area 26% of the 'pre-harvest trees' were killed or damaged and nearly half the canopy cover was removed. Data on the effects of selective logging in Sarawak on the growth rates of surviving trees of Moraceae are given by Primack *et al.* (1985). Johns (1988) has discussed some of the general ecological effects, especially on birds, of selective logging in Malaysian dipterocarp forest.

Although in historic times human activities, particularly shifting cultivation and timber felling, have been the main causes of rain-forest destruction, considerable areas have also been destroyed by natural causes such as wind, landslips, volcanic eruptions and fire. Windthrows a few hectares in extent can be seen in most rain forests. An example on a much larger scale is the old secondary forest ('storm forest') which has replaced hundreds of square kilometres of mixed dipterocarp forest in Kelantan (Malaya) after an exceptional storm in 1880 (pp. 196–9). The 'hurricane forests' of the West Indies and the 'cyclone forests' in the Solomon Islands (Whitmore 1974, 1989) and Queensland (p. 196) are the result of recurrent forest destruction by wind. Landslips, often triggered by road building and other human activities, are common in hilly rain forest in some areas, e.g. New Guinea.

Fires, mostly anthropogenic, are now one of the commonest causes of rain-forest destruction. In undisturbed rain forest single trees or small groups are often burnt by lightning and extensive fires occur only in periods of exceptional drought. The most extensive rain-forest fires ever recorded were those which destroyed vast areas of forest in eastern Borneo during the drought associated with the El Niño perturbation of 1982–83.[2] In general these fires swept through the undergrowth and after one year many large trees were still living (S. Riswan).

From what has been said, it will be evident that deforested sites invaded by secondary vegetation differ greatly from each other in their previous history as well as in local climatic and soil conditions. On some sites the pre-existing vegetation has been almost completely destroyed, on others only partly. The smallest clearings are comparable with natural gaps formed by the death of large trees (pp. 52–3). As might be expected, the secondary succession depends in part on the size of the denuded areas and their distance from intact forest. Where the land has been cultivated the early successional stages will be determined by such factors as whether the previous vegetation was burnt after clearance, how many crops were harvested, and the extent to which the

soil was eroded or otherwise impoverished. Similarly, in forests that have been logged, the structure and composition of the secondary vegetation will be affected by the method of timber extraction and the amount of damage to vegetation and soil, as well as by the proportion of trees felled. It is not surprising that secondary successions in rain-forest areas following disturbance are very varied.

18.1 General features of secondary rain forests

Before considering the successions themselves some of the more obvious characteristics of 'typical' secondary forests, i.e. young forests on areas which have been cultivated or exploited for timber, but not subsequently burnt or grazed, will be noted briefly. Such forests have many distinctive features and are easily recognized. After about fifty to a hundred years they begin to resemble primary forest and in the absence of artifacts such as buried fragments of pottery or other historical evidence, very old secondary forests may be difficult or impossible to recognize as such.

One distinguishing feature of 'typical' secondary forests is that they contain relatively few very large trees and are less tall than primary forests on similar soils. Most early seral (pioneer) trees seldom grow to more than 20–30 m in height and about 90 cm in diameter, while primary forest trees of the A storey commonly grow to 50 m or more and to a diameter of over a metre. The canopy of secondary forest is generally much more even than that of primary forest. From G. Dulit (Sarawak) the young secondary forest fallows in the valley 1200 m below looked like a smooth lawn, while in the darker coloured primary forest the huge crowns of the emergent trees stood out distinctly (Richards 1936).

The structure of secondary forest varies greatly. When very young the abundance of saplings and climbers generally give it a dense tangled appearance and make it difficult to penetrate. After a few years, particularly on abandoned cultivated ground, even-aged stands of trees of remarkably regular structure often grow up (Fig. 18.6). These may consist of a single fast-growing species such as *Cecropia* (American tropics), *Musanga cecropioides* (Africa) and species of *Macaranga, Mallotus, Adinandra* or other genera in Malesia. All these trees are light-demanding and short-lived; they are dominant for only a single generation and are unable to regenerate in their own shade. When they die they are replaced by a mixture of less fast-growing, more

---

[2] Much of the burnt forest had been logged earlier (T.C. Whitmore).

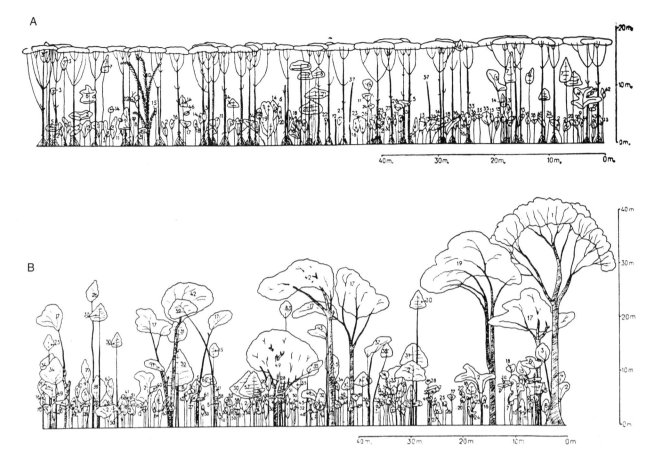

Fig. 18.6 A Profile diagram of six-year-old secondary (pioneer) forest, near Yangambi, Zaïre. After Louis (1947b). The upper storey consists only of *Musanga cecropioides*. In the lower storey there are forty-five tree species, of which the most abundant are *Macaranga spinosa*, *Ricinodendron heudelotii* and *Tetrorchidium didymostemon*. B Mature (primary or old secondary) forest at Isalowe (in same region). After Louis (1947b). In this type of forest the commonest large tree is *Scorodophloeus zenkeri*. There are seventy-one tree species in this diagram. In both A and B the profile strips are 10 m wide. Lianes and herbaceous plants are omitted.

shade-tolerant and longer-lived trees, some of which are seral and some primary forest species. After logging, in contrast to seres following cultivation, the secondary vegetation, at least during its earlier years, is very irregular in structure, with damaged survivors from the former forest scattered among climber tangles, patches of 'razor grass' (*Scleria* spp.) and dense stands of saplings; such vegetation grades into the 'depleted forest' mentioned earlier.

At a later stage, a structure similar to that of primary forest develops. In post-cultivation successions in Panama and Costa Rica, Budowski (1970) (p. 477 below) found that in the early seral stages (1–15 years) there were one or two well-differentiated strata, but in the 'late secondary stage' (20–50 years) there were three strata, which became increasingly difficult to discern as the forest became older.

Secondary forests, even when fairly old, differ in floristic composition from primary forests, although secondary species (including early pioneers) often occur, usually in small numbers, in natural gaps in the latter. In the American tropics the abundance of such trees as *Goupia glabra*, *Vismia guianensis*, species of *Inga*, *Miconia*, etc. is generally an indication that a forest is secondary. In Africa *Zanthoxylum (Fagara) macrophyllum*, *Harungana madagascariensis*, *Pycnanthus angolensis*, etc., and in Malesia species of *Alstonia*, *Ficus*, *Glochidion*, *Litsea* and many other genera, have a similar significance. Conversely, a scarcity of mature individuals of common primary forest species, e.g. *Mora* in Guyana, Dipterocarpaceae in Malesia, also suggests secondary status. Burkill's (1919) analysis of thirty-year-old secondary forest at Singapore showed that the representation of families was very different from that in primary forest. The families with the largest number of individuals were Lauraceae, Myrtaceae, Rhizophoraceae and Urticaceae; only the first two are among the most abundant families in Malayan primary forest.

It is often supposed that secondary forests always have fewer species per unit area than primary forest, especially of trees, but this is not necessarily so; their species richness probably depends on age and other factors; data for valid comparisons are scarce[3]. In the early stages of secondary successions the number of species per hectare on small plots rises rapidly, but after a few years the increase is slower. In old secondary forests the number of tree species on plots of 1–2 ha may be as great as in comparable samples of mature forest in the same region, but in very extensive areas of secondary forest the number of species is probably less than in primary forest because secondary forest species usually have wider geographical ranges than primary forest species (pp. 462–4 below).

It is noteworthy that semideciduous and deciduous tree species characteristic of seasonal climates are often common in secondary rain forest. In West Africa, for example, *Milicia excelsa* and other mixed semi-deciduous forest trees often occur in clearings and young secondary evergreen seasonal rain forest. In Central America, deciduous species common in seasonal forest formations, e.g. various Bombacaceae, are also often found in old and middle-aged secondary forests (Budowski 1970).

Data on the biomass and productivity of secondary rain forests have been compiled by Brown & Lugo (1990).

18.2 Characteristics of secondary rain-forest trees

As mentioned in earlier chapters, the trees that colonize deforested areas and dominate the early and middle

---

[3] Uhl (1987) (see p. 474 below) gives data on species diversity indices during the first five years of succession on a sample plot at S. Carlos (Venezuela).

Fig. 18.7 Old secondary forest, Omo Forest Reserve, Nigeria (1935).

Fig. 18.8 Secondary forest, forty-five years old, Florencia, Turrialba, Costa Rica (1965). The large trees are all *Goethalsia meiantha*.

Fig. 18.9 Early seral (pioneer) trees. A Young *Musanga cecropioides* by forest road, Atewa Range, Ghana (1969). B *Cecropia sciadophylla*, Masura, Guyana 1992. (Photo T.C. Whitmore). C Young trees of *Ochroma lagopus* (balsa) by road, Osa Peninsula, Costa Rica (1965). D Three-year-old regrowth on experimental destruction plot, Pasoh Forest Reserve, Malaysia (1974), showing young trees of *Macaranga hypoleuca, M. triloba* and early seral species.

stages of secondary successions (often called pioneers) differ greatly from those of mature rain forest in ecological requirements, vegetative features and reproductive strategies. They share a syndrome of characters indepen-dent of their taxonomic affinities. Van Steenis (1941, 1958c) referred to them as 'biological nomads' because, in contrast to the 'dryads' (or climax species) of the primary forest, they are necessarily migratory. 'Nomads' and 'dryads' are not sharply defined groups; some species are intermediate in their characters (see Alvarez-Buylla & Martinez-Ramos 1992).

The most important feature of pioneer trees is that they are light-demanding and intolerant of shade, especially when young. Their seeds, unlike those of typical primary forest trees, germinate only in open sites and juvenile plants cannot establish themselves under a closed canopy[4]. In clearings and large forest gaps they grow very fast, but they are eventually replaced by more shade-tolerant species. Because they cannot regenerate in their own shade, typical pioneers are unable to occupy any site permanently. At Barro Colorado Island (Panama), the size of gap required for establishment differs in different pioneer species (Brokaw 1987).

Two other characters essential to the biological equipment of pioneers are their reproductive strategies and their capacity for rapid growth, especially in height, when young. The former enables them to colonize open ground as soon as it becomes available and the latter allows them to reach maturity and reproduce before being suppressed by the slower-growing and more shade-tolerant trees of the next successional stage.

In their reproductive strategies, rain-forest pioneers are typical R-selected plants, as was seen in Chapter 5. They begin to flower at an early age and their seed production is prolific. The common Malayan seral tree *Adinandra dumosa*, for instance, begins to flower at a height of 6 ft (2 m) when it is two to three years old and continues daily for some hundred years (Corner 1988). Other common rain-forest pioneers such as *Cecropia* spp., *Musanga* and *Macaranga* spp. reproduce equally precociously and probably as continuously. Opler *et al.* (1980a) found that at La Selva (Costa Rica) the mean flowering period of plants in one-year seral vegetation was 301 days, compared with 243 days in the two-year-old stage and 107 days in primary forest. The early and more or less continuous seed production of early seral trees contrasts with that of primary forest species such as dipterocarps which under forest conditions take many years to reach the flowering stage and fruit only at long intervals (see pp. 250–1).

In addition to copious seed production, some early seral trees[5] reproduce vegetatively by suckers from the roots or at the base of the stem, e.g. *Cecropia, Goupia glabra, Trema.*

The seeds of secondary forest trees, though they vary considerably in size (see Putz & Appanah 1987), are generally smaller and lighter than those of primary forest trees (p. 111). In most species they seem adapted for transport by frugivorous birds or bats, or by wind or explosive mechanisms. Many have berries, drupes or pulpy fruits, e.g. *Cecropia, Musanga*, various Myrtaceae, *Trema, Vismia*. Winged or plumed seeds or fruits are found in *Goethalsia, Ochroma, Terminalia* and many other seral trees. In young regrowth in Ghana, Swaine & Hall (1983) found that 64% of the tree species and 41% of the individuals were probably bird- or bat-dispersed; 17% of the species (46% of individuals) were wind-dispersed. Uhl *et al.* (1981) found that at a site in Amazonia the four commonest woody pioneers were dispersed by birds or bats, but many of the forbs and grasses were wind-dispersed. Guevara *et al.* (1986) have pointed out the importance in succession of surviving unfelled trees on which birds and bats perch and drop or excrete seeds. This may account for the clumped distribution often seen in young secondary forest trees. Heavy fruits eaten and dispersed by rodents and other ground-living mammals are absent (or rare) among secondary forest trees.

An important feature of pioneer tree seeds, discussed in Chapter 5, is that, unlike those of typical shade-tolerant primary forest trees, they can retain their viability for a long time and remain dormant in the soil as a seed bank: when brought to the surface they germinate, though not necessarily immediately. The abundance of viable seeds in the top 50 cm or so of rain-forest soils varies greatly; in general the density seems always to be low in little disturbed forest (see Chapter 5, p. 116). Most of the buried viable seeds are those of pioneer species, mainly trees and lianes; in primary forests they are generally of species absent or rare in the vegetation growing above them (Saulei & Swaine 1988).

When forest is cleared a large proportion of the seeds in the seed bank germinate. As Uhl *et al.* (1981) and Riswan (1982) have shown, burning the vegetation reduces but does not altogether eliminate the stock of viable seeds. Saulei & Swaine (1988) suggest that the density of seeds in the soil at a given moment depends on the time that has elapsed since the last major disturbance as well as the nearness of the sampling site to seed parents.

The seeds of *Cecropia* spp, *Musanga* and many other early seral species require light for germination, although some germinate equally well in the dark and in *Ochroma* heat triggers germination (Whitmore 1990).

The even-aged stands of *Cecropia* and *Macaranga* spp. and other pioneers in clearings and abandoned fields

---

[4] Swaine & Whitmore (1988) define rain-forest pioneers as trees whose seeds germinate and establish only under gaps in the forest canopy which receive 'full sunlight' for at least part of the day.

[5] Especially on white sand soils.

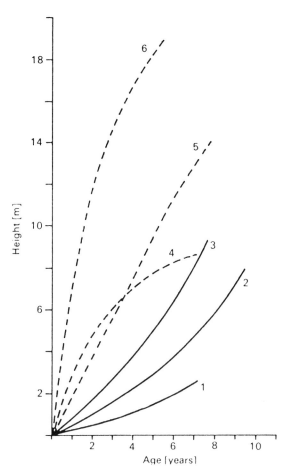

Fig. 18.10 Height growth under natural conditions of young trees in Zaïre. After Lebrun & Gilbert (1954). Continuous lines, primary forest species: 1 *Scorodophloeus zenkeri*, 2 *Oxystigma oxyphyllum*, 3 *Gilbertiodendron dewevrei*. Broken lines, seral (pioneer) species: 4 *Caloncoba welwitschii*, 5 *Terminalia superba*, 6 *Musanga cecropioides*.

no doubt usually originate from buried seeds, but many plants common in early seral vegetation, particularly *Imperata* and other grasses, probably arrive in the seed rain.

The fast growth of secondary rain-forest trees has often been commented on and is one of their most important characteristics. The growth rate of young pioneer trees is especially remarkable, as a few examples will show. Balsa (*Ochroma lagopus*) reaches a height of 18 m in five years (Hueck 1966) and in plantations grows at a rate of 5.5 m per year (Longman & Jeník 1987). In forest in Ghana *Musanga cecropioides* grows to 11 m in three years and to 24 m in nine years, reaching its maximum height in about ten years; *Terminalia ivorensis*, a taller and longer-lived tree, grows to 14 m in four years (Swaine & Hall 1983). In Malaya

eight individuals of the common pioneer *Macaranga tanarius* had a mean height of 11 m after three years (Symington 1933). In Borneo *Trema cannabina* reached a height of 7.77 m in 1.5 years (Riswan 1982).

The difference in growth rates between secondary and primary forest trees is less marked as they get older. Fig. 18.10 shows the average height growth of six tree species under natural forest conditions in Zaïre. It will be seen that the three secondary forest species, *Caloncoba welwitschii*, *Musanga cecropioides* and *Terminalia superba*, grow rapidly at first but after about five years their growth slows down. At that age the growth rate of the primary forest species (then at the sapling stage) begins to accelerate; earlier it is much slower than that of the seral species.

Growth in stem diameter is also faster in secondary than in primary forest trees. Brown (1919) found that at Mt Makiling (Philippines) the annual increase in diameter of young parang (second-growth) trees in the 5–10 cm d.b.h. class was 0.44–3.27 cm a^{-1} in species not reaching 20 cm d.b.h. and 1.43–4.03 cm a^{-1} in those growing to larger diameters. In the neighbouring primary forest the average increase was less than 1.0 cm a^{-1}. Similar data are given by Whitmore (1984a, Chapter 9).

The physiological basis for the very rapid growth of young rain-forest pioneer trees has only recently become clear. Stephens & Waggoner (1970) and Koyama (1978) found that, as might be expected, they have higher net photosynthetic rates than large primary forest trees such as dipterocarps, but growth analysis of *Trema orientalis* (Coombe 1960) and *Musanga* (Coombe & Hadfield 1962) seemed to show that high rates of dry matter production could not be the only factor responsible for their fast growth in height. Even when young their unit leaf rates (net assimilation) were lower, under similar conditions, than that of the sunflower (*Helianthus annuus*), a fast-growing temperate herbaceous plant. This is true of trees generally, according to Evans (1972). Riddoch *et al.* (1991) found that in seedlings of nine species in the semideciduous forest of Nigeria the pioneer trees, e.g. *Ricinodendron heudelotii* and *Ceiba pentandra*, and herbaceous weeds, the light compensation point, quantum efficiency and other photosynthetic characteristics indicated a greater capacity to utilize the high light intensities found in gaps and clearings than was shown by seedlings of two 'climax' tree species.

Later work, however, has shown conclusively that the unit leaf rate of some young tropical pioneers can equal that of *Helianthus*. Okali (1971) made growth analyses of seedlings of various trees in Nigeria under plantation conditions and found that one species, the long-lived pioneer *Ceiba pentandra*, had a rate not significantly less than that of *Helianthus*. Experiments in Costa Rica (Oberbauer & Strain 1984, Oberbauer & Donnelly 1986; see also Whitmore 1990, pp. 114–16) suggest that not only is the fast growth of early pioneers dependent on high photosynthetic efficiency, but the unit leaf rate is closely related to the seral status of the species. Seedlings of six species were grown in containers in an open site. Two of them, *Ochroma lagopus* and *Heliocarpus appendiculatus*, were common early pioneers of the region; two, *Cordia alliodora* and *Terminalia oblonga*, were light-demanding trees of mature forest which germinate only in gaps; and two, *Brosimum alicastrum* and *Pentaclethra macroloba*, were shade-tolerant rain-forest trees, The results show that the unit leaf rate of *Ochroma* was not significantly less than that of *Helianthus*, while that of *Heliocarpus* was only slightly lower. The two 'gap-dependent' species had rates significantly less than that of the pioneers; those of the two shade-tolerant species were less than half that of *Helianthus*. These results need to be extended to a wider range of species.

Although the fast growth of rain-forest pioneers thus seems to be primarily dependent on high rates of dry matter production, other factors such as their shoot architecture and their capacity for uninterrupted growth in height may also be of some importance. The comparatively small height and girth attained by pioneers such as *Trema* spp., *Musanga* and *Cecropia* spp. result from their early maturity and short life. Although little exact information is available, the average life-span of most secondary forest trees is certainly much less than that of shade-tolerant primary forest trees, which may live for 100–400 years or more (Chapter 3).

It is convenient to divide secondary forest trees into two, not sharply separated, groups, early and late seral species (or ephemeral and persistent pioneers). The former, e.g. *Cecropia* spp., many *Macaranga* species, *Trema*, *Ochroma*, grow very rapidly and seldom live for more than 20–30 years, some for considerably less. Brown (1919, p. 138) says that at Mt Makiling 'average secondary forest trees' (i.e. early seral species) attain a diameter of 30–35 cm and then, or soon after, begin to lose vitality. The exceptionally fast-growing *Melochia umbellata* probably dies within three to four years. In West Africa *Trema orientalis* dies at less than 10 years (Taylor 1960) and *Musanga* usually at 15–20 years (p. 469). It is doubtful if a sharp distinction can be made between long-lived seral trees and primary forest species which can regenerate only in gaps.

Late seral species grow less fast and reach ages of 50–100 years or more. Whitmore (1984a, p. 110) men-

tions stands of *Albizia falcataria* and *Octomeles sumatrana* in New Guinea which were dying when they were 84 years old and says that *Anthocephalus chinensis* lives to at least 60 years. *Ceiba pentandra*, a frequent late seral species in West Africa, lives to a much greater age and reaches very large dimensions.

The wood of early seral trees is of soft texture and low density (Saldarriaga 1986); usually it is paler in colour than that of the slower-growing 'tropical hardwoods' of primary forests. Mostly it is of little economic value, though that of balsa (*Ochroma lagopus*), which is lighter than cork, has many commercial uses. Late seral species have wood which is denser and more durable; some such as *Anthocephalus chinensis*, *Terminalia ivorensis* and *T. superba* provide useful timber. *Aucoumea klaineana* (okoumé), a pioneer which grows in extensive stands on abandoned farmland in Gabon, is particularly valuable and is much used for veneers. Gregarious pioneers such as *Cecropia* and *Musanga* could be used as pulpwood if their wood fibres were longer. Pines such as *Pinus caribaea*, which are more suitable for paper-making, are being increasingly planted on rain-forest sites.

The root systems of secondary forest trees have been little studied. Early seral species usually have superficial roots and lack tap-roots. Uhl (1987, p. 389) found that at S. Carlos they developed loose networks of roots within the upper 15 cm of soil; these roots often grew outwards for several metres beyond the crown of the tree. Early seral species allocated large amounts of energy to root production and had high root : shoot ratios. In *Cecropia ficifolia* the ratio was 0.29 : 1, in longer-lived species it was from 0.14 : 1 to 0.20 : 1. In Zaïre most of the fine roots of *Musanga* in young regrowth are within 25 cm of the soil surface, although when the tree is cultivated in isolation it is able to produce a tap-root (Lebrun & Gilbert 1954). Many small and medium-sized secondary forest trees have stilt-roots; these probably give stability to trees in which the need for rapid growth in height precludes early increase of girth or the early development of plank buttresses.

Most early seral trees probably have vesicular–arbuscular mycorrhizas, but ectomycorrhizas seem to occur only in primary forest trees. Janos (1980, 1983) says that cultivation reduces the population of mycorrhizal fungi in the soil; during the early stages of secondary succession there is an increase in both the vesicular–arbuscular fungi and in the number of obligate mycorrhizal plants.

The leaves of secondary forest trees, like those of primary forest trees, are predominantly mesophyll in

Fig. 18.11 A *Barteria fistulosa* 'circle' in secondary rain forest, Sakpoba, Benin State, Nigeria (1978). (Photo. S.R. Edwards.) The tree of the myrmecophyte *Barteria* is at the centre of a circular patch of relatively low vegetation, which has been browsed by *Pachysima* ants occupying the *Barteria* stems (see Janzen 1972).

size: they seem to be less uniform in shape, and perhaps a greater percentage of them are larger or smaller than the mesophyll class (see Plate 1 in Janzen 1975), but quantitative information is lacking. They tend also to be softer in texture and (as Spruce 1908, vol. 2, p. 355, noted) of a paler green. They suffer more damage from herbivores than those of primary forest trees (Coley 1983) and there are indications that they tend to have less chemical defence than primary forest leaves (Janzen 1970). Perhaps for this reason myrmecophytes, e.g. *Cecropia, Barteria* (Fig. 18.11), some *Macaranga* spp., are common in secondary communities.

Geographical range and origin of the secondary forest flora. In general secondary forest taxa are more widely distributed than those of primary forests; unlike the latter (see Chapter 12) very few of them have narrowly localized or disjunct areas. Some genera of secondary forest trees, such as *Trema* and *Ficus*, are pantropical or range over nearly the whole of the humid Palaeotropics or Neotropics, e.g. *Alstonia, Bridelia, Macaranga* and many others. In Trinidad, Greig-Smith (1952) found that there was a higher percentage of wide-ranging species of trees and shrubs and a smaller percentage of endemics in the secondary and degraded forests than in undisturbed forest. Closely related genera sometimes play a similar part in the seral vegetation of different continents: a striking example is *Cecropia*, which has some 100 species in tropical America, mostly pioneers of secondary forest and bare ground, and the monotypic genus *Musanga*, found on old fields and roadsides throughout the moister parts of the African 'closed

forest' (Janzen 1977c). The lianes and herbaceous plants of secondary communities are also wide-ranging: *Bauhinia, Ipomoea* and other genera of lianes are pantropical, as are many herbaceous genera and species.

The wide range of second-growth plants is no doubt chiefly due to their efficient dispersal, copious reproduction and tolerance of a wide range of environments. The biological equipment which enabled them to colonize gaps and other temporarily available habitats before the extensive destruction of tropical forests by humans has thus in recent centuries allowed them to expand their geographical range enormously. Van Steenis (1937) suggested that Malesian 'anthropogenic' plant communities consist of species which existed earlier in much smaller numbers in suitable habitats within the primary forest; large-scale clearing and burning has selected in favour of species able to take advantage of the new conditions. This interpretation is probably correct for the majority of secondary forest plants, but some, especially the herbs, may be invaders from other plant formations. In Africa and Asia the gaps available for pioneers in the primitive forests no doubt included openings made by elephants and other 'bulldozer herbivores' (Kortlandt 1982).

Pioneer trees such as species of *Alstonia, Cecropia* and *Bridelia* often occur in swamp forests with a rather open canopy; some may well have evolved in such habitats. River alluvium is another natural habitat where pioneers may have originated. A large-scale example of seral forest developing on alluvium is on the interfluves between the numerous small tributaries of the Peruvian Amazon, described in Chapter 14.

Whitmore's work (1969, 1984a) on *Macaranga* in Malaya shows that the evolution of the secondary forest flora is still continuing today. Of the twenty-six Malayan species eighteen grow gregariously on roadsides and in recent clearings. In openings in high forest, particularly on landslides, which Whitmore believes was their original habitat, they often grow to a height of 24 m and a girth of 1.8 m, but in anthropogenic habitats they are small bushy trees not more than 6–7 m high, which flower at an early age. Their migration has thus been accompanied by a reduction in size and a tendency to precocious reproduction. On Gunung Benom, where logging began in 1966, the invasion of a new habitat could actually be witnessed. *M. constricta*, until then a rare species of forest gaps, had an opportunity to colonize the extraction roads and behave like other gregarious early seral species.

Weeds introduced from other parts of the tropics, as well as cultivated plants which have run wild, are often very abundant in young secondary vegetation. Thus in

Malaya three of the commonest species in second-growth are the shrub *Lantana camara* and the herbs *Ageratum conyzoides* and *Chromolaena (Eupatorium) odorata*, all weeds of tropical American origin. The last named has spread explosively in Africa; in Nigeria it began to be common about 1948 and by 1978 dominated recent clearings and roadsides throughout the closed forest belt. In Fiji and other Pacific islands large areas of former forest have been replaced by thickets of *Lantana, Melastoma malabathricum, Psidium guajava* and other aliens. In southwestern Cameroon *Cecropia palmata* was introduced about 1900, probably from the Caribbean region. Since then it has spread widely along roadsides and in clearings; in some places it is replacing the native *Musanga* (McKey 1988). It has also appeared in Zaïre and in Ghana.

18.3 Observations on secondary successions

Although many accounts of the early stages of secondary successions in tropical forests have appeared in recent years, most observations are over periods not longer than about five years and little is known about the later successional stages. Much of the available information derives from 'chronosequences' of seral vegetation of known age and history growing under similar soil and other conditions. Only in a few places, as at Kepong (Malaya) (see below) have permanent plots been kept under observation for many years.

18.3.1 Secondary and deflected successions in Africa

Successions in Nigeria. Ross (1954)[6] described the early stages of a typical post-cultivation succession near Akilla in the Omo (formerly part of the Shasha) Forest Reserve, a low-lying area where at that time (1935) there was a considerable area of high forest. Within the reserve there were scattered villages dating from before it was constituted; each was surrounded by an 'enclave' in which farming was permitted. Near the houses there were small plantations of cola and other fruit trees; the rest of the enclaves was used for 'bush-fallow' cultivation. As available land was limited, new fields had always been previously farmed fairly recently; all the uncultivated land in the enclaves was seral vegetation of various ages. When forest was cleared for farming a few large trees, mostly hard-wooded species such as *Lophira alata*, were left unfelled, but often some of

---

[6] The summary given here includes some information added by the present author.

these were subsequently killed by fire. Most of the debris except the larger trees trunks was removed by burning (Fig. 18.2). The chief crops were maize, yams, cassava, bananas, plantains and coco yams (*Colocasia*). After a few years the ground became unprofitable, a new field was made and secondary succession began on the old one.

By studying sample plots on comparable sites where it was known at what date cultivation had been abandoned, Ross was able to reconstruct the early stages of succession. Plots 200 ft × 100 ft (60 m × 30 m), which had been fallow for 5.5, 14.5 and 17.5 years, respectively, were examined, and one of unknown, probably greater, age.

When cultivation was abandoned the ground was nearly bare, except for a few surviving tall trees and perennial crop plants such as bananas, which persisted for a while, but within a few weeks it became covered with a dense mass of low vegetation. Within a few years this was replaced by secondary forest about 16 m high. In the short period after the last crop was harvested young plants of a large variety of herbs and woody species grew up. At that time, it should be remembered, conditions were very unlike those in high forest: competition from other plants was minimal, but the ground was freely exposed to sun, wind and rain; illumination and the daily range of temperature and humidity were consequently much greater. In addition, during the period of cultivation the soil will have been directly and indirectly modified in various ways. These environmental changes were considered in Chapters 8 and 10.

The first invaders fell into three groups: (*a*) herbaceous weeds of cultivated land, e.g. *Phyllanthus* and *Solanum* spp; (*b*) species, mainly woody, characteristic of secondary vegetation, but also found in small openings in high forest, e.g. *Musanga cecropioides, Zanthoxylum, Trema orientalis, Vernonia conferta* and *V. frondosa*, which are typical early pioneers with the characteristics described above; (*c*) light-demanding high-forest trees whose seedlings are able to establish themselves under open conditions, e.g. *Erythrophleum ivorense, Khaya ivorensis, Lophira alata*.

Group (*a*), the herbaceous weeds, may form a closed cover very rapidly, but their dominance is transient and the part they play in the succession seems to be a minor one. It should be noted also that woody lianes, which are conspicuous in the early stages of many secondary rain-forest successions, were not important in this one. The group (*b*) species, though not dominant from the first, soon assumed dominance and maintained it for fifteen to twenty years; they could do this because they

Fig. 18.12 Secondary vegetation after 5.5 years on abandoned 'farm', Omo Forest Reserve, Nigeria (1935). In foreground, *Zanthoxylum macrophyllum* (left, pinnate leaves), *Vernonia conferta* (left, large leaves).

were tall enough to suppress the herbaceous weeds and grew much faster when young than the high-forest species in group (*c*). The two *Vernonia* spp., Compositae with large leaves, were most commonly the first woody dominants, *V. conferta* forming pure stands locally. In other places *Trema orientalis*, and perhaps other species, were the first woody dominants; one small area cleared of all vegetation in March 1935 by May had become covered with a dense growth of young *Trema*, mixed with a few smaller seedlings of *Albizia adianthifolia*, *Carica papaya* and *Solanum* sp. Which species attains dominance during the invasion phase seemed to be to some extent a matter of chance, depending on the propagules available at the time when the site became available for colonization and what plants survived (as stumps, roots or seeds) the burning that preceded cultivation. An unknown, but probably considerable, proportion of the pioneers may have sprung from the 'seed bank' dormant in the soil. As soon as a closed plant community had formed, chance factors presumably decreased in importance and the composition began to be determined mainly by competition.

In each sample plot *Musanga cecropioides* had become dominant within about three years. This tree, which seems to have high moisture requirements when young,

was apparently unable to establish itself on bare ground,[7] but once a cover of vegetation had been formed its rapid growth gave it a great advantage, in spite of its comparatively open canopy (Fig. 18.12). On the 5.5 and 14.5 year plots the subdominant tree was *Macaranga barteri*, a group (*b*) species which grew less fast than *Musanga*. Beneath the tree layer there was a dense lower story of shrubs and young trees, including *Tabernaemontana penduliflora*, *Discoglypremna caloneura*, *Rinorea* spp. and *Rauwolfia vomitoria*. After five years this second storey was already 4–5 m high. During the early years of the Musanga phase small herbs (*Geophila* spp., *Leea guineensis*, Commelinaceae, etc.) were much more abundant than in the high forest and formed a closed carpet. Later, as shade increased, the ground vegetation became more sparse.

After about fifteen to twenty years the *Musanga* trees died, and on the 17.5-year-old plot the number of individuals was already less than on the 14.5-year-old plot. The cause of death seemed to be merely senescence; just before they died the trees showed a scaling of the bark on the stilt-roots, but no other obvious symptoms of disease. Dead and dying trees were commonly blown over during the 'tornado season' (March–May). The dead trees had no successors of their own species; *Musanga* seems unable to regenerate beneath its own shade and, like other early pioneers, forms a single-generation community.

By this time, however, young trees of various other slower-growing species had reached a height of 20–25 m; among these were *Albizia adianthifolia*, *Anthocleista vogelii*, *Diospyros suaveolens*, *Discoglypremna caloneura* and *Nauclea diderrichii*. Some of these are group (*b*) species and some are normal constituents of high forest.

The secondary succession thus consisted of three stages: (1) a short invasion stage, in which a cover of low-growing vegetation (including many non-woody species) was established; (2) a *Musanga* stage, during which a single pioneer tree species was dominant, but for one generation only; and (3) a stage in which fast-growing, soft-wooded pioneer trees are gradually replaced by slower-growing, more hard-wooded species, characteristic of the old secondary forest of the region, which may be presumed to have been components of the hypothetical 'original climax forest'. These hardwood

[7] In the Shagamu-Ijebu-Ode area, *ca.* 50 km northwest of Akilla, in the semideciduous forest belt (Chapter 16), *Musanga* plays no part in the early seral stages (Aweto 1981), presumably because of the lower rainfall (1300–1600 mm) and somewhat more severe dry season. It is also of interest that Aubréville (1947) found that in Côte d'Ivoire attempts to establish plantations of *Musanga* on bare ground failed completely.

species were probably present as juveniles or dormant seeds from the beginning of the succession. Each stage of the succession seemed to prepare the way for the next: just as the ground cover formed in the first stage provided an environment in which *Musanga* could establish itself, the *Musanga* canopy provided conditions in which the high forest trees could survive and eventually replace it.

The seral communities, it should be noted, were dominated by woody plants almost from the start; the herb-dominated stage was brief. Shade-tolerant grasses, such as *Leptaspis cochleata*, were present in the undergrowth, but did not play an important part in the succession. *Imperata cylindrica* and other robust perennial grasses, which invade secondary vegetation which is burnt or grazed, were absent.

Ross (1954) was unable to follow the succession beyond the beginning of the third stage[8], but the work of E.W. Jones (1955–56) on the Okomu Reserve (*ca.* 150 km west of Akilla) shows that the later stages may be complex as well as long-continued.

The Okomu Forest Reserve lies on the Benin Sands on a plateau less than 100 m above sea-level. The climate is more or less similar to that of the Omo Reserve, but the soil is mainly a brown sand of considerable depth. In 1947–48 the vegetation was a mosaic of patches of tall 'high forest', some only a few hectares in extent, interspersed with 'broken high forest' and low forest of irregular structure which Jones termed 'scrub'. In some types of the latter the trees were not more than 3–5 m high, draped with lianes and often separated by thickets of Scitamineae, grasses, etc.

In most of the reserve there was no evidence of recent cultivation, except for some small areas occupied by young secondary forest which were known to have been farmed about twenty years earlier. Timber felling had been in progress for some forty years, but until a few years previous to Jones' work exploitation had been relatively light. In the 1940s it was generally believed that the high forest patches were relics of true primary forest. Jones found, however, that throughout the reserve, even in the high forest, there was evidence of human occupation such as buried fragments of pottery and charcoal, and soil mounds, indicating that most of the reserve may have been cultivated at some time in the past. Redhead (1992) later came to a similar conclusion. He suggested from historical evidence that much of the Benin forests, including Okomu, had been farmed

in the fifteenth and sixteenth centuries when the Benin Kingdom was at the height of its power, and that the forest had grown up later when large areas of cultivation were abandoned. If this is so, the Okomu high forest was probably about 300–400 years old.

From a study of the 'phases' of the forest mosaic Jones tentatively suggested that the 'broken high forest' and the various types of 'scrub' had developed in windfall gaps and were normal components of the forest, though they had been much extended by timber exploitation. He also found evidence that elephants, which were frequent in the reserve up to at least the 1950s, enlarged the gaps by damaging the trees and browsing on the vegetation (Fig. 18.13B). Jones' view of the relationships between the 'phases' is shown in Fig. 18.14. His interpretation rests partly on the assumption that the composition of the juvenile tree population is a reliable indication of future canopy composition: this has been criticized by Swaine & Hall (1988). From the early 1950s there has been much logging and other kinds of disturbance in the Okomu Reserve. In 1985 part of it was constituted a wildlife sanctuary; a report on this was made by Dawson (1989).

Secondary successions in Ghana. Swaine & Hall's (1983) account of early succession on a clearing of two hectares in the Atewa Forest Reserve is particularly valuable because their sample area was kept under observation for five years. The site, which unlike Ross' plots in Nigeria had been cleared but not recently cultivated or burnt, was at an altitude of 750 m in a 'climatic island' where frequent fogs mitigated the effects of the dry season. The adjacent mature forest was classified by Hall & Swaine (1976, 1981) as upland evergreen forest and had for a long time been more or less undisturbed.

A permanent transect 80 m × 10 m was marked out soon after the clearing was made. All trees, including seedlings and saplings, were counted and, with a few exceptions, identified: enumerations were made at intervals from 1975 to 1980. The species were classified as secondary or primary; the former were defined as requiring gaps for germination and establishment, e.g. *Ceiba pentandra*, *Musanga*, the second group as 'capable of germinating and establishing under at least a light shade and persisting in mature forest' (Swaine & Hall 1983, p. 603), e.g. *Albizia zygia*, *Hymenostegia afzelii*.

At the first enumeration (Feb. 1975) there were 290 trees, of which 252 were secondary and 28 primary. After one year there were 2007 trees and their density had reached a maximum (2.5 m^{-2}); 95% were secondary species. After five years (Feb. 1980) the total number

---

[8] After about 1950 much of the Omo forest was logged and modified by silvicultural treatments. The regeneration and successional patterns in relation to long-term changes in the tree population have been investigated by Okali & Ola-Adams (1987).

Fig. 18.13 Secondary forest, Okomu Forest Reserve, Nigeria (1948). A Old secondary forest, mature phase. B Gap in which grazing by elephants has led to invasion by grasses and mutilation of the trees.

of trees was 752 and the number of secondary trees was 482 (64%). The number of primary trees (which had risen temporarily to 1064 in February 1979, owing to a massive immigration of *Albizia zygia*) was 270.

The changes in the density of some individual species are shown in Fig. 18.15. The most numerous large secondary tree, *Terminalia ivorensis*, was not present at the first enumeration, but appeared later in the same year; it reached a peak (536 individuals) early in the following year, but declined to 147 by 1980. *Musanga* behaved much as in Ross' plots in Nigeria: the number of trees was 22 in 1975, rising to 481 after a year and falling to 28 in 1980. If its mortality rates are extrapolated, it would probably disappear in twenty years.[9]

The total number of secondary species on the transect rose to 35 after about 2.5 years; afterwards some of them disappeared. The primary species continued to

increase and there were 94 at the end of the fifth year. In both groups most new arrivals appeared in the first year. Most species prominent in the later stages were already present in the early years.

Secondary successions elsewhere in West Africa. Secondary successions similar to those in Nigeria and Ghana can be seen throughout the West African rain-forest belt. Aubréville (1947) gave an account of second-growth vegetation in Côte d'Ivoire, Cameroon and Gabon which includes lists of species and profile diagrams of seral communities of various ages. Secondary vegetation in Côte d'Ivoire has been discussed by Alexandre *et al.* (1978), Alexandre (1989) and Dosso *et al.* (1981). Schnell (1976–77) gives a general account for the whole Guinean region.

Deflected successions in West Africa. In deflected successions, which are very common in West Africa, the early seral communities are invaded by *Imperata cylindrica* var. *africana* and other tall perennial grasses, or on sandy or very degraded soils, as in parts of southeastern

[9] The transect was reassessed in 1990 (Swaine *et al.* 1990). By then some of the early pioneers had disappeared, e.g. *Trema*. Others including *Musanga* had been reduced to a few survivors. Some shade-tolerant trees, e.g. *Albizia* spp. *Bussea occidentalis*, had become 'prominent'.

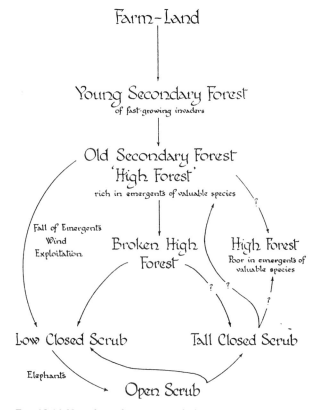

Fig. 18.14 Hypothetical successional phases in secondary forest, Okomu Forest Reserve, Nigeria. From Jones (1955–56, Fig. 8).

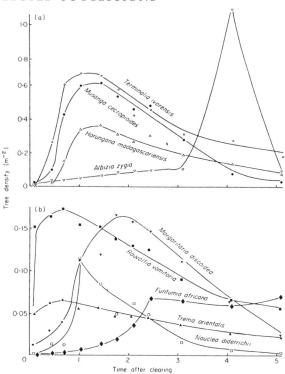

Fig. 18.15 Density changes in some of the commoner primary and secondary tree species following clearance in Atewa Range Forest Reserve, Ghana. After Swaine & Hall (1983).

Nigeria, by the bracken fern *Pteridium aquilinum*. These plants are fire-tolerant, their rhizomes surviving beneath the soil surface. This type of succession is found when the fallow periods between cultivation are very short and when the seral vegetation is subject to burning, grazing, gathering of fuelwood, etc. On poor soils in parts of southern Sierra Leone, *Imperata* appears as a weed in fields of 'hill' rice which becomes dominant when cultivation is abandoned. Deflected successions lead to the replacement of forest by 'derived' (secondary) savannas, tracts of grassland with or without scattered trees and bushes. The development of these communities in West Africa has not been described in detail, but their method of origin probably differs little from that of similar biotic climaxes in tropical America and Malesia. Like other deflected successions, that from forest to savanna is not irreversible; as was seen in Chapter 16, forest may eventually re-establish itself if the grassland is protected from fire.

Secondary and deflected successions in Zaïre. In the Zaïre rain forest the post-cultivation successions are very like those in West Africa and many of the species

involved are the same. Vermoesen (quoted by Lebrun 1936a) distinguished two divergent series of successions, 'progressive' and 'regressive'. The former, found when the fallow period between successive cultivations was long, led to the re-establishment of forest: the latter, the deflected succession found when fallow periods are short, led to the dominance of grasses or *Pteridium*.

In the progressive successions the abandoned cultivated ground is invaded by fast-growing herbs which complete their life-cycle in a few weeks. These are followed by a community of perennial herbs, climbers and young trees. After some months the trees begin to form a canopy. Vermoesen recognized three tree phases, the first dominated by *Musanga*, *Trema orientalis*, *Harungana madagascariensis* and *Pycnanthus angolensis* and the second by species of *Alstonia*, *Ricinodendron*, *Sterculia*, *Trilepisium*, *Zanthoxylum*, etc. During the third phase, primary forest trees gradually return. Vermoesen believed that the latter did not become dominant until 60–100 years after the abandonment of cultivation, but Lebrun considered this an overestimate.

In the Ubangi region the typical course of events leading to regressive succession is as follows. After an area of forest has been cleared and burnt, several harvests of maize are gathered. When the yield has become insufficient the field is abandoned, but after the subsequent succession has reached the first phase of the tree stage (in this region usually dominated by *Trema orientalis*), the land is again cleared, burnt and cultivated. During the herbaceous stage of the succession after the first cultivation *Imperata* appears as isolated tufts which do not flower. In the succeeding tree stage these are suppressed but not killed and they benefit from the second period of cultivation. When the field is abandoned for the second time *Imperata* very soon becomes dominant, forming nearly pure stands; these are frequently burnt and persist indefinitely. On moist clayey soils *Pennisetum purpureum* takes the place of *Imperata* and on sandy soils secondary grasslands are replaced by extensive stands of *Pteridium*. On the upper Uelé, according to Gérard (1960), similar communities of *Imperata* and *Pteridium* develop after clearing and burning *Gilbertiodendron dewevrei* forests and are permanent as long as they are periodically burnt.

In 1956 parts of the Uelé region were becoming depopulated and the present author was shown grasslands where burning had ceased about 3–5 years previously. These grasslands were being colonized by *Musanga*, *Harungana*, *Vitex doniana* and other pioneer trees; 'mesophytic forest' was evidently re-establishing itself.

Lebrun & Gilbert (1954) give a general account of the composition and ecology of secondary forests in Zaïre. In the young second-growth in the rain-forest region *Musanga* is usually (but not invariably) dominant. In the older secondary forests Lebrun & Gilbert distinguish two 'alliances', one characterised by *Pycnanthus angolensis* and *Zanthoxylum* (*Fagara*) spp. (Pycnantho-Fagarion), the other by *Triplochiton scleroxylon* and *Terminalia superba* (Triplochito-Terminalion). The former is found in rain-forest areas and the latter in the more seasonal semideciduous forest region.

18.3.2 Secondary and deflected successions in tropical America

The vegetation of clearings and old fields in the rain-forest regions of South and Central America is similar in structure and appearance to that in other parts of the humid tropics, but different in floristic composition.

Most of the trees and shrubs common in secondary communities in tropical America belong to genera confined to the New World, e.g. *Inga*, *Ochroma* and many Melastomataceae, or which are predominantly Neotropical but with a few species in the Palaeotropics, e.g. *Vismia*. Pantropical genera such as *Sterculia* and *Ter-*

minalia, which include many secondary forest trees, are represented by different species in the New and Old World tropics. Throughout the tropical American rain forest species of *Cecropia*, much resembling the African *Musanga* in foliage, growth habit and biological characteristics, are conspicuous in young second-growth and along river banks. Stands of *Heliconia* spp. with brightly coloured scarlet inflorescences are characteristic of clearings and disturbed sites; this genus is wholly Neotropical, except for one (or perhaps more) species in Melanesia.

In deflected successions the grass *Imperata brasiliensis* replaces the Palaeotropical *I. cylindrica* (of which it can be regarded as a subspecies).

Successions in clearings in Surinam. Schulz (1960) described secondary successions in clearings in northern Surinam which had not been burned or cultivated. Openings of not more than 0.01–0.10 ha were soon filled by saplings from the existing regeneration of primary forest trees; only a few light-demanding pioneers such as *Cecropia* and *Goupia glabra* appeared. In larger openings where the illumination was 50–100% of full daylight, surviving primary forest species were soon suppressed by secondary vegetation. The first successional stage was generally the development of a closed cover of herbs such as Melastomataceae, *Costus* and *Heliconia* spp.; where the top soil had been removed by erosion or logging *Axonopus*, *Panicum* and other grasses were the chief early colonists. This stage was transient. Pioneer trees, e.g. *Cecropia* spp., *Goupia glabra*, *Palicourea guianensis*, *Trema micrantha* and species of *Inga* and *Vismia*, usually soon became dominant; *Cecropia* reached a height of over 10 m in two years. Here and there climbers, e.g. Cucurbitaceae, *Bauhinia* spp. and razor grass (*Scleria* spp.), made impenetrable tangles.

Rapid establishment of *Cecropia* was possible because its light-sensitive seeds were abundant everywhere in the soil. *Goupia glabra* may also have been recruited from the seed bank. Many of its seedlings were suppressed by larger-leaved plants such as *Cecropia*; however, in old secondary and disturbed forests it forms large populations.

In eastern Surinam, Boerboom (1975) cleared three rain-forest plots to show the effects of burning and soil differences on the succession. In Plot 1, on well-drained sandy loam, the debris was not burnt but was removed. After eight months herbaceous weeds, shrubs and young trees had covered 10–30% of the surface, but at eighteen months *Cecropia obtusa* had formed an open canopy at 8–11 m, with a dense lower layer of *Goupia glabra*, species of *Inga*, *Palicourea* and other young trees. *Virola surinamensis* appeared only after four years, its late arrival perhaps due to its large heavy seeds. After seven years *Cecropia* had reached a height of 11–14 m and was beginning to decrease. The lower layers included representatives of all the strata in the original forest and the *Cecropia* canopy was beginning to be penetrated by other trees.

In Plot 2, on sticky loam, the debris was also removed. After a period of flooding due to heavy rain, savanna-like vegetation developed: only after seven years was this slowly invaded by trees spreading from small elevations on the site.

On the third plot the felling debris was burnt. The wood-ash became overgrown by algae, delaying the succession. Even after seven years part of the plot was covered only by weeds and grasses. A report on this experiment after eleven years (with profile diagrams) has been given by Zwetsloot (1981).

Successions on kaolisols in Amazonia. On clearings and abandoned fields near San Carlos on the Upper Rio Negro, common early colonists include *Eupatorium cerasifolium*, *Phyllanthus* sp., various Melastomataceae and grasses such as *Andropogon bicornis*, *Panicum* spp. and *Paspalum* spp. Among important pioneer trees are species of *Cecropia*, *Miconia* and *Vismia* and *Palicourea guianensis*.

At three sites 4 km east of San Carlos, Uhl *et al.*

(1981) followed the succession for twenty-two months on 0.25 ha plots which were cleared and burnt about three months later but not cultivated. Uhl (1987; see also Uhl *et al.* 1982a) also studied the changes in vegetation and soil over five years on a 0.25 ha plot in the same locality which was cleared and burnt without removing any timber; cassava (*Manihot esculenta*) was then cultivated for three years. All these plots were in primary mixed forest on nutrient-poor oxisols or ultisols (kaolisols) derived from Precambrian rocks.

On all the uncultivated plots, sprouts from stumps dominated the vegetation three months after clearance; on one plot they formed 87% of the individual plants, the remainder being seedlings. Nearly all the sprouts were killed by the burning. After the burning, which took place at the beginning of a period of relatively dry weather, seedlings did not appear for a few weeks, but at four months the plant density was 0.6 m^{-2}, though the tallest was only 0.25 m high; 41% of the individuals were species common in undisturbed forest. By ten months the vegetation had increased considerably in height and density (Fig. 18.16) and the primary forest plants were outnumbered by secondary forest trees and herbs, which now formed 74% of the individuals. By sixteen months the herbs began to decrease in density, but the pioneer trees had increased further in numbers. At twenty-two months *Cecropia ficifolia* had formed a 'loose canopy' and constituted *ca.* 78% of the living phytomass. The total number of species (on 40.5 m^2) had risen to 55, of which 23 were represented by single individuals. Sixteen were primary, and fourteen secondary, forest trees.

Experiments in which the number of seeds germinating in soil samples from one of the cleared and burnt sites was compared with that in soil from adjacent unburnt forest showed that burning much reduced the size of the seed bank: in the burnt soil the density of viable seeds was 157 m^{-2}, compared with 752 m^{-2} in forest samples. To evaluate the relative contributions to the plant cover from the seed bank and from immigration after burning, plots were treated with methyl bromide (which kills all seeds) and after four weeks compared with untreated plots. From these experiments it was concluded that *Cecropia* and other woody pioneers originated from buried seeds, but the grasses, and probably also the forbs, in the plots came mainly from immigrant seeds.

Small-scale differences in surface conditions had a considerable effect on seedling establishment. Three types of surface were available, bare mineral soil, root-mat and charred wood, each of which might or might not be overlain with slash (making six microhabitats in

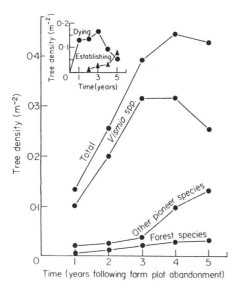

Fig. 18.16 Changes in the density, establishment and mortality of trees ⩾2 m tall on 0.15 ha plot during the first five years after the abandonment of shifting cultivation near San Carlos de Rio Negro, Venezuela. From Uhl (1987).

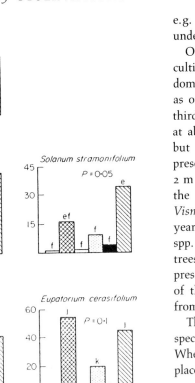

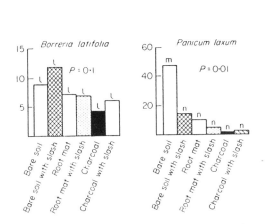

Fig. 18.17 **Mean number of plants of seven common early successional species established in each of six microhabitats on cleared oxisol sites near San Carlos de Rio Negro, Venezuela. From Uhl et al. (1981). Letters indicate differences significant at the value of P shown.**

all). Different pioneer species showed significant preferences (Fig. 18.17); for example, *Cecropia* and *Vismia lauriformis* established better on charcoal or slash-covered soil than in other microhabitats, while *Panicum laxum* did best on bare soil. These preferences may be partly dependent on seed mass; heavy seeds with long hypocotyls seemed to be at an advantage on exposed surfaces, but those with light seeds and short hypocotyls,

e.g. *Clidemia sericea* and *Eupatorium cerasifolium*, under slash.

On the plot which had been abandoned after being cultivated for three years (Uhl 1987) the vegetation was dominated by forbs and grasses during the first year, as on the sites discussed above. During the second and third years pioneer trees formed a partly closed canopy at about 8 m; these were chiefly *Vismia* (two species), but *Cecropia* spp., *Miconia* spp. and others were also present. A dense layer of shrubs (Melastomataceae) *ca.* 2 m high had replaced the herbaceous vegetation. By the fifth year the canopy had reached 12 m. The two *Vismia* spp., which had formed 90.6% of the trees in year 3, were now decreasing and giving way to *Miconia* spp. and other longer-lived pioneers. Primary forest trees, e.g. *Caryocar* sp. and *Eschweilera collina*, were present, but in small numbers (3.2 per 100 m²); 79% of these had grown from seeds, the rest were sprouts from stumps.

The slow rate of establishment of primary forest species seemed to be in part due to seed predation. When seeds of primary forest trees were artificially placed in the formerly cultivated area a large proportion was removed, presumably by animals. For example, out of 192 seeds of *Ocotea costulata* 54% disappeared within four days and all but one of 192 seeds of the palm *Jessenia bataua*. Losses of larger seeds, such as those of *Aldina*, *Licania* and *Ormosia* spp., were less. Removal of seeds was consistently greater on the farmed plot than in forest gaps.

The number of tree species had risen from 17 in year 1 to 35 in year 5, but it was still far below the number in an adjacent area of mature forest. During this five-year period the trees on the plot had grown *ca.* 1–2 m in height and *ca.* 1–2 cm in diameter each year; these growth rates, though higher than those of trees in mature rain forests, are not as high as those of some young secondary forest trees. At year 5 the basal area was 7.0 m² ha⁻¹; the total above-ground biomass was 3386 g m⁻² and root biomass 426 (± 110) g m⁻². The average above-ground productivity over the five years was 1258 g m⁻², similar to values recorded for mature rain forest, but the total live and dead plant mass (including litter and soil organic matter) was only 35–40% of that in the pre-existing mature forest.

When cultivation was abandoned soil nutrient concentrations had fallen to about the level of those in mature forest, but scattered undecomposed slash still provided a source of additional nutrients. The chief cause of declining crop yields had probably been nutrient limitation, yet the secondary vegetation on the same site grew rapidly. Thus the seral species appeared to be better adapted to obtaining scarce nutrients than the crop (cassava). Uhl also suggests that other reasons for the relative success of the seral species may have been that they allocated more energy to root production; and because their leaves were longer-lived, with lower nutrient concentrations and more resistance to herbivores, they probably used the soil nutrients more efficiently.

Saldarriaga (1986) and Saldarriaga *et al.* (1988) followed later stages of secondary successions in the San Carlos area by means of a 'chronosequence' of twenty-three stands of secondary vegetation of different ages on sites where the approximate date when cultivation was abandoned was known from oral or other evidence. The stands, which were all on terra firme within a radius of 30 km from San Carlos, were grouped into age-classes of 9–14, 20–23, 30–40, 55–60 and 75–85 years. Four stands not known to have been cultivated were also studied: these were regarded as 'mature' (primary) forest because of their structural characteristics and because of the absence of buried charcoal with a radio-carbon date of less than 200 years. The soils varied from sandy oxisols on hill tops to ultisols of heavier texture elsewhere; all were very poor in nutrients. In each stand vegetation and soil were sampled on three 10 m × 10 m plots.

During the first ten to twenty years of succession early seral trees (chiefly *Cecropia* and *Vismia* spp.) were dominant; a large proportion of these died within twenty years. At thirty to forty years longer-lived, less fast-growing trees such as *Alchornea* sp., *Vochysia* sp. and *Jacaranda copaia* replaced the early seral species and remained dominant for about fifty years. Between about forty and eighty years gaps were produced by the death of old trees; in these, saplings of primary forest trees grew up. In the larger gaps *Cecropia* sp. and other pioneers also occurred.

The average number of species (⩾10 cm d.b.h.) at eighty years was 23.5, only slightly less than in the mature forest plots. However, the floristic composition of the eighty-year-old and mature forest plots was very different: of the twenty-two species with highest importance value, only seven were found in both.

The changes in the structure of the vegetation with age are as striking as the changes in its composition. Basal area and biomass had increased greatly, though not as rapidly from nine to eighty years as in the early years of succession. At eighty years the basal area is still much less than that of mature forest and there are no trees of over 60 cm d.b.h. Both the living and the total above-ground biomass are only about two thirds of the average for the mature forest plots.

The increase in biomass was matched by the increase in the nutrients potassium, calcium and phosphorus in leaves, stems and roots (Saldarriaga 1986). At about 50–60 years the rate of increase fell off slightly. Changes in nutrient stocks in the soil (0–30 cm) were less marked than in the biomass.

It is evident that an ecosystem more or less similar to the existing mature forest could eventually develop on these sites of former cultivation, but after eighty years recovery in both vegetation and soil was still incomplete. Saldarriaga *et al.* (1988) estimated by extrapolation from their data that about 190 years would be required for vegetation to develop comparable to mature rain forest in basal area and biomass.

Successions on Amazonian podzols. The successions following forest clearance on South American white sands (podzols) have been little studied: they have features in common with those on similar soils in Malesia. The climax on these soils, which are extremely infertile and cannot be successfully cultivated, is Amazonian caatinga (heath forest) (Chapters 11 and 12).

Anderson (1981) says that when white-sand forests are destroyed by fire the first colonists are plants such as *Cecropia*, *Byrsonima*, *Vismia* spp. and bracken (*Pteridium*). At first these grow rapidly, but later much more slowly than on oxisols (latosols); for example, Uhl & Jordan (1984) found that on white sand ('spodosols') at San Carlos *Cecropia ficifolia* took about six years to reach a height of 12 m as compared with two to three years on oxisols and its leaves were small and yellowish green in colour.

Uhl *et al.* (1982b) compared the effects of clear-felling, with and without burning, and of removing the topsoil by bulldozer on caatinga forest near San Carlos. After three years the results of these treatments were strikingly different. Sites cleared but not burnt became covered with a dense stand of trees most of which were resprouts from the stumps of the original trees. These formed 73% of the above-ground phytomass, the remainder being pioneer trees (*Cecropia*, Melastomataceae, *Palicourea*) and herbs. On sites burnt after clearing most of the tree-stumps were killed and pioneer trees (*Cecropia* etc.) formed a loose canopy: herbs (*Eupatorium* spp., *Pityrogramma*, etc.) were abundant (15.4 individuals m^{-2}). Where the topsoil had been removed by bulldozer the forest was replaced by grasses (*Andropogon*) and herbs such as *Xyris* and Araceae.

The above-ground phytomass was 1291 g m^{-2} on the cleared unburnt site, 870 g m^{-2} on the cleared and burnt site and 77 g m^{-2} on the bulldozed site. After three years soil nutrient contents were at first higher on the cleared sites, whether burnt or not, than in the undisturbed forest, but later they probably fell owing to leaching of the sandy soil. On the bulldozed sites nutrient contents were much lower than in the undisturbed forest.

It was estimated that where the forest had been felled, whether burnt or not, recovery to a biomass equal to that of mature caatinga forest might take a hundred years, but where it had been bulldozed recovery might take over a thousand years.

Deflected successions in Amazonia. Successions in which rain forest is converted into grassland or bracken (*Pteridium*) communities can be seen in many parts of Amazonia. Usually they are the result of burning or shortened cycles of shifting cultivation, but the details of the change have seldom been recorded.

Scott (1986) describes the development of secondary forest, grassland and bracken communities at 400–2000 m in eastern Peru. The Campo Indians of the Gran Pajonal make shifting 'gardens' of *Manihot esculenta*. Land suitable for cultivation is scarce and the fallows are often cleared and replanted after only a few years. The second-growth grows to a height of 15 m in about fifteen years and it is often burnt to prevent the development of impenetrable forest. Burning encourages the spread of *Imperata brasiliensis*, a common weed in the 'gardens'. If the fallows are not burnt, forest may re-establish itself; aerial photographs show that some parts of the region which were grassland in 1958 had become forest by 1976. More often burning results in a deflected succession with the following stages:

(1) Forest (primary or secondary) is cut, burnt and planted with *Manihot*.
(2) The 'garden' is cultivated for one to four years, the chief weeds being *Baccharis floribunda*, *Imperata* and *Pteridium*.
(3) Cultivation ceases and burning takes place yearly in the dry season. *Pteridium* becomes dominant and some surface soil erosion becomes evident.
(4) Grasses, chiefly *Imperata* and *Andropogon* spp., replace *Pteridium* and erosion continues.
(5) Grassland dominated by *Andropogon* spp. and small Cyperaceae is established. This is maintained by frequent fires as a biotic climax.

Very similar deflected successions leading to communities of grasses, ferns and fire-tolerant woody plants occur in many parts of the New and Old World tropics. The artificially planted grasslands that are now widespread in Amazonia differ from fire-induced anthropogenic grasslands such as those of the Gran Pajonal in being composed mainly of introduced species such as *Panicum maximum* rather than native grasses, which

are of little value for cattle grazing (Buschbacher *et al.* 1986). The succession on abandoned pastures in eastern Amazonia has been described by Uhl *et al.* (1988a). In general the abandoned pastures revert to forest, but if they are used too intensively or too long, or if the second-growth is repeatedly burnt, the succession is deflected and the re-establishment of forest is prevented.

Secondary successions in Central America and the West Indies. Although secondary successions in Central America are in general similar to those in South America, the composition of the seral communities is somewhat different. For example, *Ochroma lagopus* (balsa), which is absent or uncommon in the Guianas and Amazonia, is an abundant and conspicuous pioneer in the humid parts of Central America. *Goethalsia meiantha*, a large and relatively long-lived pioneer tree, is common in Costa Rica (Fig. 18.8), but is apparently absent in South America, except in Colombia.

Budowski (1965, 1970) gave a general account of secondary successions in the rain forests of Costa Rica and Panama based on data from sixteen sample plots of known history. He distinguished four successional stages: (i) pioneer (1–3 years old), (ii) early secondary (5–15 years), (iii) late secondary (20–50 years) and (iv) climax (over 100 years). In the pioneer stage there were not more than five woody species per plot, the 'dominants' being *Cecropia* spp., *Ochroma*, *Trema micrantha* and various Euphorbiaceae. In stage (ii) there were up to ten woody species and *Heliocarpus appendiculatus* was an additional 'dominant'. Stage (iii) was mixed forest 20–30 m high and the number of woody species, which included various Bombacaceae, Meliaceae and Tiliaceae, had increased to 50–60 per plot. Many of these trees were deciduous, unlike those of the climax forest. As in other secondary successions dominance shifted from shade-intolerant pioneers to trees that, at least when young, were more tolerant of shade.

At Barro Colorado (Panama), although the climate is moderately seasonal, the secondary succession is not very different from that in non-seasonal rain forests. Kenoyer (1929) found that in old clearings after two years *Cecropia* spp., *Ochroma*, *Trema* and other pioneers began to replace a mixed community of herbs, lianes and young trees. At about fifteen years this had become a dense, relatively species-rich forest, including species of *Ficus*, *Inga*, *Miconia*, *Protium* and other genera; lianes, including the ribbon-like *Bauhinia excisa*, were numerous.

The later course of secondary succession at Barro Colorado is largely conjectural. According to Foster & Brokaw (1983) the whole area (before it became an island) had probably been cultivated; parts of it had been affected by 'blow-downs' and other disturbances. The present vegetation is a mosaic of various ages, some parts possibly several centuries old, some much younger. Some of the existing 'old forest' is similar in structure and composition to primary forest. A phytosociological analysis of the forest by ordination methods was made by Knight (1975), who discussed its dynamics.

Various aspects of secondary successions in Central America were considered at a symposium in 1977 (Ewel 1980). Data on the stand characteristics of twelve-year-old secondary forest at La Selva (Costa Rica) are given by Hartshorn (1983).

In the West Indian islands there is much secondary vegetation of all ages, but most of it has been frequently disturbed and the successions are not easy to trace. In the Lesser Antilles stands of tree ferns are common in secondary forest even at low altitudes (Beard 1949). Seral vegetation in forest clearings in Trinidad was described by Marshall (1934); Greig-Smith (1952) has given an account of degraded and secondary forest at five sites in Trinidad where the original vegetation had probably been evergreen seasonal or semievergreen, seasonal forest. Various types of secondary communities at low and medium altitudes in Guadeloupe were described by Stehlé (1935).

18.3.3 Secondary and deflected successions in Malesia

Much of the rain forest that once covered most of the lowlands of Malesia has now been replaced by secondary vegetation and biotic climaxes resulting from burning,

grazing and shifting cultivation. The structure and floristic composition of the seral communities are remarkably varied, depending on local soil and climatic conditions as well as on the age and previous treatment of the cleared areas. It is thus characteristic of Malesia that young second-growth (belukar) is less uniform in composition, and probably richer in species, than in Africa or tropical America. Many pioneer trees and shrubs such as *Adinandra dumosa*, *Melastoma malabathricum*, *Trema orientalis* and various Euphorbiaceae (especially *Macaranga* and *Mallotus* spp.) are widespread and very abundant, often forming nearly pure even-aged stands, but few genera or species are as omnipresent as *Musanga* in Africa or *Cecropia* spp. in tropical America. The large part played by introduced plants in the earlier stages of secondary successions in Malesia was mentioned earlier (pp. 462–4).

Much former forest land in the Malayan region is now occupied by stands of the fire-resistant grass *Imperata cylindrica* (alang-alang or 'lalang' in Malaya, cogon in the Philippines) which soon becomes dominant if the secondary vegetation is burnt (Fig. 18.18). If protected from fire and grazing, forest can re-establish itself in *Imperata* grassland.

For a fuller account of secondary successions and human influences on Malesian rain forests than can be given here reference should be made to Whitmore (1984a).

Secondary successions in Malaya. The secondary vegetation near Kepong (Selangor) has been studied for many years. Symington (1933) found that on abandoned

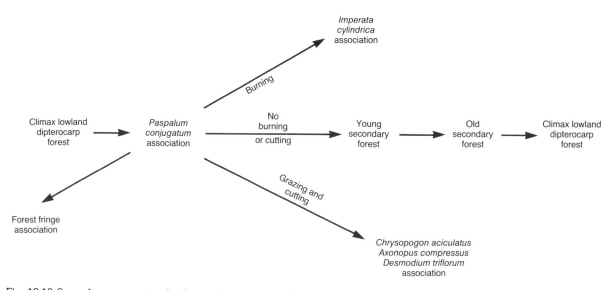

Fig. 18.18 Secondary successions leading to four main grassland communities in tropical rain forest, Pahang, Malaysia. After Verboom (1968), modified.

vegetable gardens, which had probably been under high forest a few years previously, there were seral communities in various stages of development. Most of them had been subject to burning, grass-cutting and other disturbances. The first successional stage was colonization by herbaceous plants, including climbers and grasses. This was soon followed by the dominance of shrubs, such as *Melastoma polyanthum* and *Lantana camara*, and later by the development of typical belukar of *Macaranga* spp. and other pioneer trees. Symington supposed that this community would gradually change until after many years mixed dipterocarp forest would be re-established. Although the general course of the succession was fairly easy to reconstruct, several problems remained to be solved, particularly why lalang grass (*Imperata*) became dominant in some places, arresting the succession for long periods. Plots, 1 chain square (0.04 ha), were marked out and kept under observation for three years.

Two plots, on which cultivation had been abandoned shortly before, had a flora of about fifty-two species, including *Bridelia tomentosa*, *Macaranga tanarius*, *Trema* sp. and other young trees, shrubs, wild banana (*Musa malaccensis*), many climbing and herbaceous species, and a few surviving cultivated plants. Patches of the grass *Brachiaria distachya* and some small colonies of *Imperata* were present. During the first few months of observation, generations of herbs succeeded one another rapidly and the climbers and grasses spread vegetatively. By the end of one year most of the herbaceous weeds had declined in abundance, owing partly to the growth of woody plants and *Musa* and partly to the increase of *Imperata*, which had spread by means of its rhizomes.

After three years trees had formed a canopy over part of these plots, with an understorey of *Lantana* in some places but practically no herbaceous undergrowth. Over the remainder *Imperata* was almost exclusively dominant, though one patch of *Brachiaria* resisted invasion by *Imperata* for a long time. Thus on these two plots *Imperata* had established itself as a stage in succession but not over the whole area.

Another plot was chosen as the purest *Imperata* stand in the neighbourhood; its previous history was unknown. Although at first sight *Imperata* seemed to exclude everything else, scattered shrubs, herbs and creepers were present. The chief change during the three years the plot was under observation was that *Imperata* became distinctly less vigorous and over part of the area it was replaced by the fern *Nephrolepis biserrata*. Bushes of *Melastoma* reached a height of 3 m and now cast sufficient shade greatly to reduce the vigour of the

Imperata growing beneath them. In this plot there was no evidence that woody species could establish themselves among the *Imperata* after it had become a closed community; the shifting of dominance from *Imperata* to woody species which appeared to be taking place was brought about by the increase of species which had established themselves earlier in the succession.

One more of Symington's plots deserves mention; this was laid out in a small clearing on the fringe of the primary forest. This area had never been cultivated and it was chosen in order to see whether the succession on rich forest soil was different from that on the impoverished soil of abandoned cultivated land. The vegetation before clearing consisted of primary forest species with a small admixture of belukar species; the latter may have been invaders from neighbouring secondary vegetation or may have arisen from dormant seeds in the soil. In the three years the plot was studied a typical secondary forest dominated by *Macaranga tanarius* established itself. The first stage of the succession was dominated by the herbs *Physalis minima* and *Amaranthus viridis*, beneath which masses of seedlings of *Trema*, *Musa* and other belukar species had germinated. *Trema* later became dominant, but eventually it was overtopped by the *Macaranga*. It was noteworthy that no *Imperata* appeared on this plot.

Symington's work was ended by World War II, but in 1947 a new successional plot of 0.9 acres (0.36 ha) was established at Kepong. This has been kept under observation up to the present time and the results for the first thirty years were reported by Kochummen & Ng (1977). The site had been farmed for several periods before cultivation was finally abandoned in 1945. At that time it was surrounded by plantations of young forest trees and was some 800 m distant from the nearest primary forest. It was protected from fire and counts of woody stems over 3 ft (0.9 m) high were made on subplots totalling 0.2 ha, but not every year (Table 18.1).

In 1947, when regular observations began, most of the plot was covered with pioneer shrubs and trees. *Melastoma malabathricum* was ca. 1.5–1.8 m high and formed 79.3% of the stems. There were twenty other woody species on the sampled area, the most numerous being *Mallotus paniculatus*, *Rhodamnia cinerea* and *Vitex pubescens*. A few trees overtopped the *Melastoma*, *Grewia tomentosa* reaching the 14–15 ft (4.3–4.6 m) height-class.

In 1949, four years after cultivation had ended, the number of woody species had risen to twenty-six. Several of them had become less abundant, especially *Melastoma*, the number of stems of which had fallen from

Table 18.1. *Secondary succession at Kepong (W. Malaysia), 1947–76*

Number of woody stems on samples (total 0.2 ha) in 0.9 acre (0.36 ha) plot (see text).

	1947	1949	1960	1976
Number of stems				
3–5 ft (0.91–1.52 m)	1670	254	123	578
6–9 ft (1.82–2.74 m)	1605	654	37	115
⩾10 ft (3.04 m)	69	644	27	12
Total (⩾3 ft) (0.91 m)	3344	1553	274	876
Number of woody species (⩾3 ft)	21	26	25	51
Most abundant species[a]				
Clerodendrum villosum	64	31	1	2
Eugenia polyantha	3	—	2	50
E. syzygioides	6	10	13	139
Grewia paniculata	81	77	37	73
Macaranga triloba	—	1	1	97
Mallotus paniculatus	150	38	—	—
Melastoma malabathricum	2652	1075	2	1
Rhodamnia cinerea	88	90	95	111
Trema sp.	66	—	—	—
Vitex pubescens	124	109	52	10

[a] Species with 50 or more individuals in any of the years of sampling.

Source: Data of Kochummen & Ng (1977).

2652 to 1075 on 0.2 ha. The tallest trees were now *ca.* 7 m high.

Five years later (1954) there were scattered trees *ca.* 9 m high. *Melastoma* had almost disappeared and been replaced by the fern *Dicranopteris linearis*, which had begun to spread in 1951 and now formed a dense tangle 2 m high. Nearly all the woody plants were survivors of those which had originally colonized the plot. By 1960 the tallest trees were about 15 m high and the number of stems was reduced to 274. The number of species was twenty-five, almost the same as in 1949.

No enumerations were made between 1960 and 1976. By July 1976 a closed canopy of trees 6–9 m high had developed with emergents up to 18 m high. The number of woody stems was now 876, still fewer than normal in a primary forest. Of these 12 were trees ⩾10 ft (3.04 m) high and 33 had a girth of ⩾15 in (38 cm). The total basal area was 12 548 square feet (1165.74 m²). Lianes, of which eleven species were present, formed an important part of the canopy. The total number of woody species, which had increased very little up to 1960, had risen to 51. *Dicranopteris* remained only here and there in unshaded gaps; it was evident that the dominance first of *Melastoma* and later of *Dicranopteris* had prevented the immigration of many species earlier in the succession. Some early colonists, e.g. several

Macaranga spp., had disappeared. An interesting new record in 1976 was two stems of the dipterocarp *Shorea leprosula*, a primary forest tree. Yet after thirty years, the composition of the plot was still characteristically that of a seral community.

Kochummen & Ng (1977) contrast the relatively slow succession on the Kepong plot with that on one at Sungei Kroh (Wyatt-Smith 1955, Kochummen 1966), which had been incompletely cleared and then abandoned after carrying a single crop of rice. At Sungei Kroh after four years *Macaranga gigantea* became dominant and remained so until at seventeen years it was 12–15 m high and becoming senescent. The number of woody species was much greater than on the Kepong plot. The much slower recovery of the latter was no doubt due to unfavourable soil conditions and a seed bank depleted by several periods of cultivation, as well as to its distance from seed-parents of forest species.

Secondary successions in Borneo. On clayey and loamy soils (kaolisols) in the lowlands of Borneo the climax is mixed dipterocarp forest and the successions in clearings and old fields are similar to those in Malaya, described above, although not all the species involved are the same. On white sands and podzols, however, where the climax is heath forest (kerangas) and which are seldom cultivated, the second-growth vegetation differs greatly from that on kaolisols in both physiognomy and floristic composition.

Riswan (1982) and Riswan & Kartawinata (1988a,b) have given a detailed account of the vegetation and soil conditions of primary, secondary and experimentally cleared lowland rain forest at Lempake near Samarinda (0°30′S, 117°09′E) and 10 km from Samboja, *ca.* 57 km northeast of Balikpapan (1°15′S, 116°50′E) in eastern Borneo (Kalimantan). At Lempake the soils are kaolisols and the primary vegetation is very species-rich mixed dipterocarp forest. The emergent trees in this are mainly Dipterocarpaceae (over 60% of basal area), although the most numerous family of trees ⩾10 cm d.b.h. in species and individuals is the Euphorbiaceae. At Samboja the soil is white sand and the primary vegetation is a type of heath forest more than usually poor in species: the common trees include *Tristaniopsis obovata*, *Brackenridgea hookeri* and the dipterocarp *Cotylelobium burckii*. In both localities the climate is ever-wet with only a brief relatively dry season.

At Lempake a plot of 1.6 ha was marked out in primary mixed dipterocarp forest and one of 0.8 ha in secondary forest on the site of a pepper plantation abandoned thirty-five years earlier. In addition, two experimental plots of primary forest each 0.5 ha in area were cleared: one was not burnt and the cut vegetation was removed, the other was burnt after cutting. On the heath-forest site two plots 0.5 ha in area were laid out in primary forest and one in secondary forest thirty-five years old; there were also unburnt and burnt clear-cut experimental plots of the same size.

On the four experimental plots the changes in the vegetation were recorded at intervals for 78 weeks (1.5 years) and data were collected on nutrient concentrations in the foliage and soil conditions. A wealth of interesting information was obtained to which it is impossible to do justice here.

On both mixed dipterocarp experimental plots a mass of young woody and herbaceous plants grew up during the first weeks after clearance. These consisted of 'seedlings' (defined as plants with stems ⩾2 cm d.b.h.) and resprouts from stumps and roots. There were many differences between the two plots.

At six weeks on the unburnt plot there were more species of resprouts than of seedlings, but the cover and frequency of the seedlings were much greater. Most of the resprouts were from tree-stumps, e.g. *Eusideroxlon zwageri*, *Hopea rudiformis*, *Pentace laxiflora*, but they also included some from herbs and lianes. Of the forty-two seedling species, twenty-nine were trees and nineteen of these were pioneers such as *Anthocephalus chi-*

nensis, *Endospermum diadenum* and *Macaranga* spp. Many of the seedlings of pioneer trees grew in dense clumps, which probably originated from seeds in the seed bank. The commonest herbs were *Curculigo glabrescens* and the grass *Paspalum conjugatum*.

In the burnt plot both resprouts and tree seedlings were at first much fewer than in the unburnt plot at this stage: the fire had evidently killed many seeds lying on or beneath the soil surface, as well as many tree-stumps. There were seedlings of twenty-nine primary forest and pioneer tree species, but the seedlings of the latter were much fewer than in the unburnt plot; for example, there were only ninety *Macaranga pruinosa* seedlings, compared with 3466 in the unburnt plot. No liane seedlings were recorded and the most abundant herbs were *Blumea balsamifera* and *Costus speciosus*. It was noticeable that in this plot seedlings did not appear in large numbers simultaneously, as they did in the unburnt plot, probably because most of the seeds came in the seed rain and not from the seed bank in the soil.

On both plots the time from six to eight weeks seemed to be a critical period in the succession when changes in the soil and vegetation were most rapid. At one year the total number of seedlings was still increasing but the number of tree seedlings declined sharply. Many lianes and herbs also died, probably because of the increased shade from the pioneer trees. The grass *Imperata* had appeared on the unburnt plot by twelve weeks, and on the burnt plot by twenty-four weeks, but failed to spread, probably also because of shading.

At seventy-eight weeks (1.5 years), when the observations on these plots ended, the mean height of the seedlings was 3.9 m on the unburnt and 4.1 m on the burnt plot; growth rates in diameter as well as height were higher on the burnt plot. The stem/wood biomass was 46.29 kg per 10 m^2 on the burnt plot compared with 26.91 per 10 m^2 on the unburnt plot. Throughout the succession resprouts remained less important than seedlings and failed to attain as high a frequency or cover value as the latter.

Although the two plots were contiguous, their floristic composition at seventy-eight weeks differed considerably. On the unburnt plot the total number of species had reached 105, of which twenty-six were primary forest trees and twenty-eight pioneer trees, while on the burnt plot there were eighty-one species, twenty-three of which were primary and thirty-two pioneer trees. The commonest trees on the unburnt plot included four species of *Macaranga*, *Endospermum diadenum*, *Ficus arfakensis*, *Callicarpa pentandra* and the dipterocarp *Hopea rudiformis*. Two other dipterocarps were also present, *Shorea leprosula* and *S. parvifolia*. Although dipterocarps are in general primary forest trees[10], these three species, as noted by Wyatt-Smith (1958) have some pioneer characteristics, e.g. fast growth and ability to regenerate in unshaded situations. Young trees of *Anthocephalus*, though common up to twenty-four weeks, had all died by seventy-eight weeks.

On the burnt plot the commonest trees were *Anthocephalus* and *Endospermum diadenum*; *Macaranga* spp. were less abundant. The only dipterocarp was a single seedling of *Shorea parvifolia*. *Eusideroxylon zwageri*, which is the commonest non-dipterocarp tree in the primary forest, though at maturity slow-growing, had grown rapidly from resprouts on both plots.

The vegetation of the 35-year-old plot, which was on a site similar to the two experimentally cleared plots, was dense secondary forest with a canopy about half as tall as the primary forest. It had 576 trees (≥10 cm d.b.h.) per hectare, with a total basal area of 21.94 m² ha⁻¹, compared with 445 trees and 33.74 m² ha⁻¹ in the primary forest plot. The emergent trees were 25–30 m high, the largest being a *Macaranga conifera* with a girth of 265 cm. Below the emergents two tree strata could be recognized, at 11–20 and 5–10 m, respectively. Most of the emergents, and all the trees over 50 cm d.b.h., were pioneers, e.g. *Macaranga* spp. and *Octomeles sumatrana*, which were assumed to be about thirty-five years old. As in the experimental plots, the trees often had markedly clumped distribution patterns. Many of the saplings and seedlings were of pioneer species, but immature primary forest trees were also numerous. There were 121 plant species in the plot, compared with 209 in the primary forest plot (1.6 ha): about 70% were primary forest species.

In one group of species on the 35-year-old plot seedlings and saplings were absent, e.g. *Anthocephalus chinensis*, *Macaranga conifera*, and in another group saplings were present but no seedlings. These groups were pioneer species which would probably disappear when their mature trees died. A third group had a 'normal' age structure, with many saplings and seedlings as well as older trees: these were mostly primary forest species, e.g. *Payena lucida* and *Fordia gibbsiae*. It is remarkable, however, that *Eusideroxylon zwageri*, the commonest large tree in the primary forest, was represented by only eight small individuals (10–15 cm d.b.h.). The only dipterocarps in this plot were two young trees of *Hopea rudiformis* with no seedlings or saplings. The poor representation of dipterocarps is surprising, because mature trees and seedlings of twelve species occurred in primary forest

only 250 m distant. This may have been due to the well-known inefficient dispersal and intermittent reproduction of this family (Chapters 5 and 9), but the plot was possibly in some way unsuitable for their establishment.

It was evident that, even after thirty-five years, the composition of the secondary forest was still changing and the succession still at an early stage. Although the pioneer flora was being replaced by primary forest species, this was taking place very slowly. From his data Riswan estimated that it would be at least another 152 years before this plot would become similar in composition to the existing primary forest. Riswan *et al.* (1986) estimated from various lines of evidence that in eastern Borneo the 'stabilization' of secondary forest took *ca.* 60–70 years and the re-establishment of forest similar to the original primary dipterocarp forest *ca.* 200–250 years.

On the white sand at Samboja the early successional stages were quite different from those in the mixed dipterocarp forest at Lempake: the secondary kerangas (heath forest) was also very unlike secondary mixed dipterocarp forest. The most striking feature of the young second-growth at Samboja was that it consisted mainly of resprouts from the stumps of the former forest trees: seedlings and fast-growing pioneers such as *Trema* and *Macaranga* spp. played only a very small part in the succession. On both the unburnt and the burnt experimental plots, seedlings were extremely few; those present were mostly not field weeds but plants characteristic of 'padang' vegetation (p. 326), e.g. *Nepenthes reinwardtiana*, *Scleria* sp. and the fern *Schizaea dichotoma*. No species was recorded both as a seedling and as a resprout.

At six weeks there were resprouts of fourteen species in the hundred samples (each 1 m²) on the unburnt plot, the commonest being *Tristaniopsis obovata*, the dominant tree in the primary forest. Only five seedlings were found, including one of *Trema cannabina*, two of the palm *Oncosperma* and two of herbaceous plants. The total cover of resprouts and seedlings was 6%, and 6% of the samples were bare of vegetation. On the burnt plot there were resprouts of eleven species and two seedlings (*Scleria* sp. and *Trema cannabina*). The total cover was 0.73% and 59% of the samples were bare.

At one year the number of resprout species had risen to twenty-six on the unburnt plot and nineteen on the burnt plot. The number of seedling species was seven on the unburnt and eight on the burnt plot. Cover increased to 16.28% on the unburnt plot, only one sample remaining bare; on the burnt plot cover rose to 30.24% and ten samples remained bare.

After seventy-eight weeks the cover was 58.6% on the unburnt plot and 38.5% on the burnt plot; one sample was still bare in the former and ten were in the latter.

---

[10] In New Guinea and the Philippines the dipterocarp *Anisoptera thurifera* regenerates profusely in secondary forest and reinvades abandoned cultivated land (Ashton 1982).

Some resprouts had died; the number of species had fallen to twenty-three. The density of seedlings on the unburnt plot, which had been 25 per 100 m² at 52 weeks, was now 19 per 100 m². The number of seedling species was still seven; five had died and others, including *Melastoma affine*, had appeared; seedlings of the semi-parasitic shrub *Dendrotrophe varians* and the climber *Hoya multiflora* were common. *Imperata*, which often invades kerangas after burning, did not appear on either plot. There were no significant differences between the plots in floristic composition.

The mean growth rate of the resprouts was also similar on both plots, although some species grew faster on one than on the other; for example, *Barringtonia reticulata*, which was perhaps favoured by the nutrients in the wood-ash, grew to a height of 1.55 m on the burnt plot but only to 0.36 m on the unburnt one.

Little can be said about the 0.5 ha plot of 35-year-old secondary kerangas. It had only eight tree species and no large trees of the common primary kerangas species *Tristaniopsis obovata* and *Cotylelobium burckii*; both are used by local people for stakes and piles and may have been removed after the original clearance. The distribution of the trees was patchy with bare spaces between. The total number of trees was only 106 ha⁻¹ (basal area 1.15 m² ha⁻¹), compared with 750 ha⁻¹ (b.a. 16.90 m² ha⁻¹) on one primary kerangas plot and 554 (b.a. 6.42 m² ha⁻¹) on the other. Though the floristic composition of the experimental plots differed little from that of the primary kerangas, the similarity coefficients (Sørensen, modified by Bray & Curtis 1951) between the 35-year plot and the two primary plots were only 12.00% and 9.06%.

The predominance of resprouts rather than seedlings at Samboja may be a general feature of regrowth in Bornean and other heath forests. Their ability to regenerate from stumps depends on the readiness of heath forest trees to produce coppice shoots, which seems to be greater than that of most trees in mixed dipterocarp and other mixed rain forests on kaolisols. It is unlikely that the scarcity of seedlings is due to the small number of seeds in the seed bank and seed rain. As Riswan points out, the seedlings grew very slowly and their survival rate was poor. It is thus more likely that the scarcity of seedlings is due to the difficulty of establishment on a very porous infertile soil fully exposed to the sun than to a shortage of viable seeds. In mature kerangas forests, it should be noted, seedlings are generally abundant (Brünig 1974) and the tree populations have a normal age structure.

Regrowth chiefly from resprouts is also seen in the Amazonian caatinga forests on nutrient-poor white sands. In the latter also, as in the kerangas, fast-growing

pioneer species play a much less important part in regrowth after clearance than in secondary successions on kandisols. In regenerating after clearance by vegetative growth rather than by seed, kerangas (and other types of heath forest) resembles savanna communities such as the Brazilian cerrado (pp. 397–9).

Ewel *et al.* (1985) studied young secondary forests in Sabal Reserve (Sarawak) and found that their biomass and floristic composition were both related to the soil fertility level. Three sites were examined where the forest had been logged and then one crop of rice had been grown. On Sites 1 and 2 sampling was done four and a half years after cultivation had been abandoned, on Site 3 nine and a half years after. Site 1 was on level ground and the soil was relatively fertile recent alluvium; Sites 2 and 3 were on slopes on 'red-yellow podsolic' soils (kaolisols). The soil of Site 2 was of low fertility and probably much poorer in nutrients than that of Site 1. The Site 3 soil was intermediate in quality between that of the other sites. The plots on Site 1 had the largest mean biomass of trees (4662 g m⁻²) but they also had the largest biomass of lianes (603 g m⁻²) and there is no consistent correlation on the three sites between high tree and low liane biomass.

On all three sites the trees were 11–12 m high and were mainly fast-growing pioneer species. The most abundant were *Callicarpa pentandra* on Site 1, *Callicarpa* and *Macaranga conifera* on Site 2 and *Macaranga hypoleuca* on Site 3. On Site 2 young trees of two species of *Shorea* were present.

The mean number of trees per hectare was *ca.* 3900 on Sites 1 and 2 (b.a. 16.3 m² ha⁻¹), while on Site 3 it was *ca.* 2200 (b.a. 12.7 m² ha⁻¹). The leaf biomass was not significantly different on the three plots. The total above-ground phytomass on Site 1, which had the most fertile soil, was 5.4 kg m⁻², comparable with a young tree plantation, while on Sites 2 and 3 it was 2.1 and 3.9 kg m⁻², respectively. The smaller number of trees in the oldest stand (3) was probably due to competition, which might also account for the smaller number of species.

Secondary successions in the Philippines. Whitford (1906) and Brown (1919) described the secondary (parang) vegetation at low altitudes on Mt Makiling (Luzon). The climate here is somewhat seasonal and the now almost vanished climax vegetation was evergreen seasonal dipterocarp forest (chapter 12). Clearings soon became covered with a 'dense heterogeneous tangle' of many kinds of small trees, lianes and other plants. The trees included *Ficus* spp., *Litsea glutinosa*, *Macaranga tanarius*, *Melochia umbellata* and other pioneers, as well as some taller species such as *Bischofia javanica*:

the relatively rapid growth rates of these trees were referred to earlier. If left undisturbed this vegetation developed into high forest, but often the succession was deflected by repeated burning and secondary grasslands dominated by *Imperata* or *Saccharum spontaneum* developed. Much former forest land was occupied by secondary savannas with scattered individuals or clumps of fire-tolerant trees such as *Antidesma ghaesembilla*, *Bauhinia malabarica* and *Acacia farnesiana*. When fires were excluded the grasslands were invaded by secondary forest.

Kellman (1970) has given a detailed account of secondary succession on Mt Apo in southeastern Mindanao. Here the annual rainfall is over 4000 mm and there are no regularly recurrent dry periods. The study area was at 870–1235 m in the 'upper fringe of lowland dipterocarp forest', of which small relics still remained; its most obvious montane characteristic was the abundance of tree ferns (*Cyathea* spp.) in some seral communities. The soil was a 'brown forest soil' derived from relatively nutrient-rich volcanic materials. Most of the area had been selectively logged and much of it had been used at various times for shifting cultivation or plantations of manila hemp (*Musa textilis*) and other cash crops.

Of the eighteen stands chosen for analysis, most were secondary vegetation from one to twenty-seven years old on land which had been burnt and cultivated one or more times; one had been cleared a year previously, but not burnt or cultivated, one was forest which had been selectively logged and one was a relic of old forest containing mature trees of *Shorea* spp. and other primary forest species. Some stands consisted mainly of herbs, including grasses, *Pteridium* and other ferns; on one the tall grass *Panicum maximum* formed 75% of the phytomass; in two tree ferns were dominant and in one (in addition to the forest relic) there were many 'hardwood' (primary forest) trees. The rest were dominated mainly by 'softwoods' (pioneers such as species of *Ficus*, *Homalanthus* and *Trema*). The degree of past disturbance had been very varied, even in different parts of the same stand, making it difficult to reconstruct the successional changes.

From the quadrat data and other observations made during the twenty months of field work Kellman concluded that on lightly cropped land where most of the stumps and seeds had been destroyed during clearing and cultivation, the phases of succession would have been as follows.

(i) During cultivation a weed flora of small herbs, Zingiberaceae, tree ferns and 'softwoods' develops.

(ii) In the first year after cultivation ended the weeds persist and many spread vegetatively.

(iii) After a year the 'softwoods' begin to overtop and eliminate the herb layer. The tree ferns continue to grow slowly in their shade.

(iv) If the pioneer trees are small, e.g. *Homalanthus* spp., they form a dense even canopy and die when ten to fifteen years old; if taller and longer-lived (thirty years or more), e.g. *Mallotus paniculatus*, *Trema orientalis*, the tree density is less and the canopy more uneven. In the first case tree ferns become dominant and remain so until they are eventually overtopped by 'hardwoods'; in the latter case dominance passes gradually from 'softwoods' to 'hardwoods', the tree ferns remaining suppressed in the undergrowth. Unlike the 'softwoods', the 'hardwoods' are not able to establish themselves during phase (i) when the pioneers have created a favourable environment.

This account of the 'typical' succession is very generalized: there were many complicating factors and the composition of a stand could not be predicted solely from its age. Kellman tested statistically the relative importance of various factors affecting the number of 'regrowth' (secondary) and 'forest' (primary) species. The number of 'regrowth' species showed a strong negative correlation with the number of 'cleanings' (weedings) during the previous cultivation period: the partial correlation coefficient (PCC) was -0.453. However, there was little correlation with the other factors tested. The number of 'forest' species, however, had strong positive correlations (PCC $+0.500$) with stand age and altitude and strong negative correlations with burning (PCC -0.634) and number of 'cleanings' (PCC -0.491), but little correlation with distance from existing forest.

Burning during forest clearance reduces the number of seeds at or near the surface of the soil and kills most of the tree-stumps. On areas which had been cleared but not burnt, such as selectively logged forest, the succession consequently differs from that usually following cultivation, though the structural phases are similar. Hardwood seedlings are present from the start and there are fewer field weeds in the herb phase.

Kellman investigated microclimatic and soil conditions in the stands. The main changes in microclimate with age of stand (decreased radiation near the ground and lower daily maximum temperature) depended on the increasingly complete cover of foliage. Soil changes with age, such as increase in soil moisture content, carbon content and cation exchange capacity, also resulted from, rather than caused, changes in the structure and composition of the vegetation and seemed to have little effect on the course of the succession.

Even Kellman's oldest secondary stands represent only

relatively early stages in a very long process. One stand, cleared twenty-seven years earlier, seemed at first sight to have a typical forest structure, but in fact it lacked large emergent trees (in this area mostly dipterocarps) characteristic of the neighbouring primary forest. Its above-ground phytomass was only 41.92 kg m^{-2}, compared with 1120.46 kg m^{-2} in the latter. From Brown's (1919) data on the growth rates of dipterocarps under forest conditions at Mt Makiling, Kellman estimated that these emergents were several centuries old and concluded that the re-establishment of mature forest would take a similar length of time.

18.4 General features of secondary successions in lowland rain forest

The observations summarized above show that secondary successions at low altitudes throughout the humid tropics have many features in common. There is, however, considerable variation depending on local differences in climate, soil and other environmental conditions, as well as on the history of the site between the destruction of the original forest and the beginning of succession. If the seral vegetation is left undisturbed, succession leads eventually to the restoration of forest similar to the climatic climax. This process is very slow and probably takes several centuries, even when the cleared area is only a short distance from intact forest.

Under present-day conditions burning, grazing, fuel-gathering and other human disturbances often deflect the succession and result in the development of biotic climaxes dominated by grasses, ferns or fire-tolerant trees. These changes are not irreversible and if the vegetation is protected from fire and other disturbances the forest can gradually re-establish itself.

In typical post-cultivation successions on clay or loam soils (kaolisols) four fairly well marked stages can be recognized: (i) invasion, (ii) early seral, (iii) late seral, and (iv) mature (climax). During (i) the vegetation is a mixture of herbs, including grasses, climbers, shrubs, sprouts from stumps and young trees. This stage is transient, seldom lasting more than a few months. In stage (ii) a canopy is formed by fast-growing pioneer trees, sometimes as nearly pure single-species stands. These early seral trees have a life-span of not more than twenty to thirty years (sometimes much less); they are shade-intolerant and unable to regenerate in situ. In stage (iii) the forest becomes more mixed in composition and age structure, as well as increasing in biomass. Some of the trees are now late seral (persistent pioneer) species, which have some of the characteristics of early seral species but are longer-lived and slower-growing, and some are primary forest species. The latter form an increasing proportion of the stand until eventually they become dominant.

Some of the components of stages (i) and (ii) originate partly from the seed bank in the soil and some from seeds dispersed by birds and other animals or by wind, as well as from stumps and roots surviving from the pre-existing vegetation. Pioneer trees and other plants which are intolerant of shade when young can establish themselves only at the beginning of the succession or in gaps formed accidentally later on, but more tolerant species can do so at any stage, some probably being present from the early stages and surviving as suppressed individuals through stages (ii) and (iii).

Differences in site characteristics, such as slope, drainage, soil texture and fertility, affect the floristic composition of the seral stages, but not very markedly, probably because, as noted earlier, most secondary forest species seem to tolerate a wide range of environmental conditions. Only on porous, nutrient-deficient white sands does the course of secondary successions appear to be very different from the typical successions described above.

In general the history of a site, particularly whether the previous vegetation was burnt and whether the soil has become eroded or degraded, seems to have more obvious effects than differences in intrinsic site characteristics. Post-cultivation successions differ, at least in their early stages, from those in which forest clearance has not been followed by cultivation. Burning and weeding reduce but do not eliminate the seed bank: burning temporarily enriches the soil with nutrients from the ash and kills most tree-stumps. On sites that are cleared but not cultivated many trees from the pre-existing forest may survive as stumps or juveniles and early seral species are less important. There is little evidence to show whether the effects of site treatment persist into the later stages of the secondary succession.

Forest clearance, as well as drastically changing the microclimate (Chapter 8), has important effects on the soil (Chapter 10). The soil surface temperature rises and exposure to sun and rain quickly changes the physical and chemical properties of its superficial horizons. Much of its reserves of organic matter are destroyed and large amounts of nutrients are released in soluble forms, especially when clearance is followed by burning. During the succession the organic matter content of the soil increases and the stock of nutrients is slowly replaced, mainly from the deeper layers of the soil. The original nutrient cycle is thus eventually restored. These processes take many years and for this reason the soil becomes impoverished when the fallow periods between shifting cultivation are shortened. The ground is then invaded by plants tolerant of degraded soils, such as ferns and *Imperata*, especially if the seral vegetation is burnt.

Tropical post-cultivation successions have usually been regarded as 'obligatory' or 'Clementsian' in the sense of Horn (1976), i.e. successions in which the early stages are necessary precursors of those that follow. Although this view has been questioned (Swaine & Hall 1983), there is a certain amount of evidence to support it. For example, in Africa *Musanga* seems to establish itself more readily in a pioneer herb community than on bare ground (pp. 468–9). Lebrun & Gilbert (1954, pp. 57–8) stress the role of *Musanga* and other early pioneer trees in creating favourable conditions for their successors by increasing the organic content of the soil and restoring the nutrient cycle. However, some secondary successions, e.g. where forest clearance has not been followed by cultivation, more resemble what Horn (1976) terms a 'competitive hierarchy', i.e. a succession in which many species invade the site simultaneously and only the longer-lived persist into the later stages.

The seral vegetation on artificial clearings is comparable to the building phase in primary forest gaps made by windfalls or the death of old trees, but not in all respects (Chapters 3, 9 and 10). As mentioned earlier, pioneer trees and other secondary forest plants are often found in natural gaps, but the microclimate, soil and flora of clearings, even when not burnt or cultivated, are not exactly like those in gaps. Herwitz (1981a) compared the tree flora of abandoned clearings in the Corcovado National Park (Costa Rica) with that of twenty-five natural gaps 100–1200 m² in area. *Cecropia*, *Trema* and other pioneers, which were abundant in the clearings, were found in the gaps in only small numbers; the chief part in filling the latter was played by various primary forest tree species capable of germinating and establishing themselves under gap conditions. As Alexandre (1989) says, the differences between the successions in clearings and those in natural forest gaps are quantitative rather than qualitative.

Comparison of secondary successions with primary xeroseres on bare volcanic materials and river alluvium (Chapter 13) shows that the two are essentially similar; the same pioneer species often play a large part in the early stages of both kinds of succession. However, primary successions begin on immature mineral soils generally lacking in organic matter and they have no seed bank, so colonization is entirely by propagules carried by wind, water or animals. For these reasons primary xeroseres tend to be more prolonged than secondary successions, although the sequence of stages is similar.

There are of course general resemblances between secondary successions in all climates where the climax is broad-leaved forest. On old fields in Europe and North America, the stages dominated by herbs, shrubs and trees are comparable to those in the humid tropics. The birches (*Betula*), elders (*Sambucus*), poplars (*Populus*) and other seral trees of temperate regions are shade-intolerant and have many of the other characteristics of rain-forest pioneers, such as short life, rapid growth and small seeds dispersed by birds or wind; they also often spring up in large numbers from seed banks in the soil. The deflected successions leading to the replacement of temperate forests by grasslands, heaths and *Pteridium* communities are also quite closely analogous to those by which secondary savannas and other anthropogenic communities develop from tropical rain forests.

Chapter 19

Postscript: the future of the tropical rain forest

The history of the tropical rain forest, as was seen in Chapter 1, is still very imperfectly known, but it is now clear that forests probably very like modern rain forests existed in the late Cretaceous period some 70 million years ago. Since then climatic changes have caused their boundaries sometimes to retreat and sometimes to advance. During the dry, and probably relatively cool, phases of the Quaternary, tropical rain forests seem to have been restricted to comparatively small refuges and to have had a much smaller total area than during later, more humid phases. Until the most recent stages in its history, humans have had only relatively small effects on the tropical rain forest: most of it probably long remained uninhabited or inhabited only by very small nomadic populations. These early forest dwellers, as were until very recently their modern representatives such as the Sakai and Penan of Malesia and the African pygmies, were hunter–gatherers, probably with hardly more influence on the ecosystem than any of its other animal inhabitants.

After surviving for millions of years, apparently with little change in its general features, the rain forest was rudely disturbed, first by invasions of agricultural peoples and, much later, by the spread of modern technology into the humid tropics. The former seems to have begun in the palaeotropics at least 10 000 yr BP and about the same time or somewhat later in the neotropics, the latter not until recent centuries. Until some two hundred years ago, the impact of indigenous agriculture on the tropical rain forest was on a comparatively small scale because, except on sites where soil and other conditions were particularly favourable, such as river deltas and on volcanic soils, cultivation was

shifting and unable to support dense populations. The effects of European colonization were far more drastic, leading eventually to the clearance of large areas of virgin forest for growing sugar cane, coffee, rubber, oil palms and other export crops, as well as to the selective logging of valuable hardwoods such as mahoganies in the remaining forests. At the same time, in some countries the rapid increase in human populations led to a huge expansion of shifting cultivation and the replacement of much primary forest by secondary forest and savannas.

Until modern times, most of the tropical rain forest was comparable to the forests of Europe in the Mesolithic period when human habitations were mainly confined to the forest fringes and sites accessible by river, as they were in Malaya until the mid-nineteenth century. The widespread clearance of tropical forests in modern times is comparable to the clearance of European forests in the Neolithic, Bronze and Iron Ages, except that the former has been accomplished in a few decades while the latter extended over several thousand years.

The destruction of the tropical rain forest, already proceeding apace in the early nineteenth century, accelerated enormously after World War II. For this there were many reasons, among them the rapid development of air transport and road building in the tropics and the introduction of power saws, bulldozers and other forest-clearing machinery, but above all it was due to the inexorable growth of land-hungry populations and the insatiable demand from developed countries for timber and other tropical products. Between 1950 and the early 1970s, world consumption of tropical hardwoods quadrupled (Pringle 1976). Vast areas of forest have now been cleared and planted with oil palms, pines and *Gmelina arborea*. In tropical America (though not in Africa, because of the prevalence there of tsetse-borne diseases) large areas of rain forest have been converted to cattle ranches.

The decrease in the forest area in Costa Rica and Malaya is shown in Figs. 19.1 and 19.2. Maps for various other countries, based on recent information, are given by Prance & Campbell (1988). A full account of 'forest conversion' in Malaysia is given by Aiken & Leigh (1992). Fifty years ago Heske (1938–39) estimated that tropical forests (including rain forests) occupied about 1507 million hectares and formed half the world's forest area. More recent maps for Asia and the Pacific, with accompanying text, are given in the *Conservation Atlas* by Collins *et al.* (1991) and for Africa by Sayer *et al.* (1992). The present area of intact rain forest is certainly only a small fraction of what it was before World War II, but estimates of the exact area are subject to large errors and must be treated with reserve. A careful assessment of the world's resources of 'tropical moist forests' and their rate of destruction in 1980 was made by FAO (1981) and updated in 1988. The results have been reviewed by Lanly (1982) and by Sayer & Whitmore (1991). The 1980 assessment was based on vegetation maps, land-use reports and similar published

information, but for Brazil and some other countries information from satellite imagery later became available (see Chapter 1). The term 'tropical moist forests' includes various types of seasonal tropical forests as well as tropical rain forest in the sense of this book. Global surveys do not distinguish between primary and secondary forests or different degrees of forest disturbance and degradation.

The FAO assessment (1981) forecast that some 7.5 million hectares per year of tropical moist forest would be cleared during 1981–85 and a further 4 million hectares logged but not completely destroyed. In the event the loss of forest was considerably greater than predicted, in part because of the extensive fires in Borneo in 1982–83 referred to on pp. 191–3 and the extensive clearing in Brazil, especially in Rondonia. Sayer & Whitmore (1991) believe that the rate of destruction was still accelerating in 1990. The 'conversion' of forest is proceeding less fast in some countries than in others and it is possible that in the Guiana region, parts of Amazonia, Gabon, and perhaps Zaïre, some considerable areas of lowland rain forest may remain relatively undisturbed into the twenty-first century (Myers 1984).

The prediction in the first edition of this book (Richards 1952, p. 405) is already near to fulfilment: 'Unless determined efforts are made to halt the destruc-

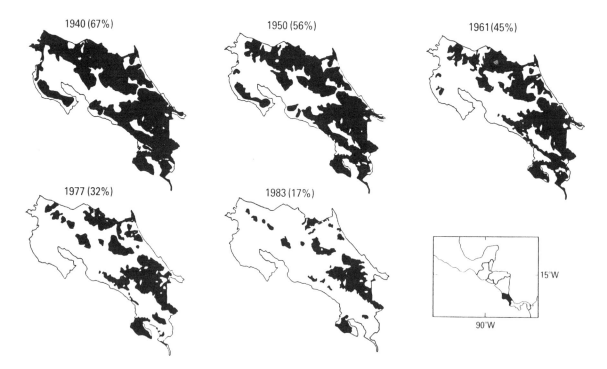

Fig. 19.1 Loss of primary forest in Costa Rica, 1940–83. After Sader & Joyce (1988, Fig. 1), redrawn by Whitmore (1990, Fig. 10.10). Based on aerial photography and Landsat images.

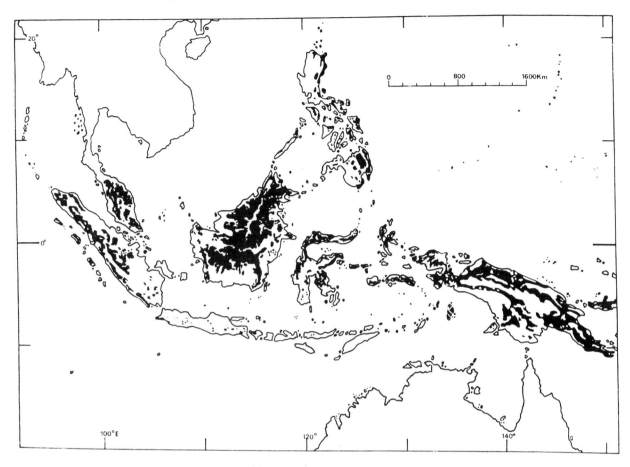

Fig. 19.2 Remaining lowland tropical forest in Malesia in the early 1980s. Based on Whitmore (1984b). The forest area shown is mostly evergreen rain forest, but a small part consists of seasonal forest. In the first half of the nineteenth century the greater part of the Malay Peninsula, Sumatera and Borneo, as well as New Guinea, was covered with tropical rain forest.

tion, the whole of the Tropical Rain forest may disappear in the lifetime of those now living, except for a few inaccessible areas and small 'forest reserves' artificially maintained mainly for the production of timber.' At best only a small fraction of the former forests will survive. Thus most of a plant formation which for millions of years covered a large part of the earth's surface is doomed to disappear almost entirely and for ever.

The virtual disappearance of the world's tropical rain forests will inevitably have profound consequences. Some of the possible climatic consequences, due to effects on rainfall both local and world-wide, and on atmospheric carbon dioxide, were discussed in Chapters 7 and 8. Biologically the most far-reaching result of rain-forest destruction will be the probable extinction of vast numbers of plant and animal species. Published estimates of the number and speed of these extinctions vary very widely and can be no more at best than informed guesses (see discussion in Sayer & Whitmore 1991 and Whitmore & Sayer 1992). This is partly because knowledge of the number of species, particularly of invertebrate animals, in existing forests is still very inadequate (pp. 289–92). It is also impossible to forecast accurately how many species will be able to survive in the secondary forests and fragments of primary rain forest which may still remain.

A large number of rain-forest species have probably already become extinct in the past fifty years and many are listed as endangered, but it is remarkable that very few (mostly birds and mammals) can be regarded as extinct with the same certainty as the moa, the dodo and the North American tree *Franklinia alatamaha* (which was last seen wild in 1803). Very few rain-forest trees can yet be said to be definitely extinct, although many species, chiefly local endemics, are so rare that it is unlikely that they will survive very long in the wild state. Gentry (1979b) found that various trees in the rain forests of Ecuador were close to extinction. *Persea theobromifolia*, formerly the most important timber tree in its region, had already been reduced to fewer than a dozen mature individuals. Of the lianoid *Dicliptera dodsonii*, only a single plant remained.

There are many reasons why the loss of a large number of plant and animal species is a matter for grave concern. On a geological time-scale it may have important effects on the future course of evolution.

Species of very restricted distribution are of course particularly endangered. In Perak (Malaya) various tree species, e.g. *Burkilliodendron album*, have been recorded only once and as their habitats have become rubber plantations they are presumably extinct (T.C. Whitmore). The rain-forest flora with its immense wealth of species belonging to scores of families and thousands of genera has in the past acted as a reservoir of genetic diversity and potential variability for both plants and animals. During at least the more recent epochs of the earth's history it has been a centre of evolutionary activity from which much of the flora of other regions of the world has originated. There have been other foci of plant evolution, but as was mentioned in Chapter 1, there is evidence that a large part of the flora of the temperate zones is derived directly or indirectly from the humid tropics; other centres such as South Africa and Australia have had much less influence on the rest of the world. Tropical rain forests have thus played a part in plant evolution greater than that of any other of the major plant formations, partly no doubt because in them genetic changes could be continuous and not interrupted by cold winters or seasonal droughts although, as we now know, they have not been unaffected by secular climatic changes. The consequences of the disappearance of the rain forests for animal evolution will certainly be at least equally profound.

The consequences for the world economy of tropical deforestation are already becoming evident. The destruction of tropical rain forests removes what is still the only source of a wide range of plant and animal products. Mahoganies, merantis and many other tropical hardwoods used as saw-timber and veneers on an ever-increasing scale all over the world are at present obtained almost entirely from trees in natural rain and semi-deciduous monsoon forests. In some countries, such as Nigeria, stocks of native hardwoods are already so depleted that export of all the more valuable species is prohibited. Peninsular Malaysia is already a net importer of timber. As mentioned earlier, in most tropical countries the highly selective felling of a few valuable species of earlier years has been replaced by the utilization of a much wider range of species. Where 'wood chip' factories have been set up, as in Papua New Guinea (Webb 1977, Lamb 1990) (Fig. 18.5) the forest is clear-felled and practically the whole woody biomass is converted into paper pulp, fibreboards and other reconstituted wood products.

But it is mistaken to regard tropical rain forests as no more than producers of timber and pulpwood. They are also a source of nuts, fruits, fibres, rattans, gums, resins, rubber, drugs and pesticides, as well as of animals useful for food and other purposes. Some of these non-timber forest products (or 'minor forest products', as they used to be called in official reports) are of considerable importance as exports, as well as being sold in local markets.

Rainforest plants produce a large variety of secondary

Fig. 19.3 Amerindian collecting edible fruit of *Pourouma cecropifolia* in secondary forest, upper Rio Napo, Ecuador. (Photo. J. Friedman.)

compounds, particularly alkaloids, of actual or possible value as drugs or pesticides. Some drugs have long been obtained from wild rain-forest plants. The curares, traditionally used as arrow poisons by the Amerindians of Amazonia and the Guianas, are extracted from *Chondrodendron tomentosum*, *Strychnos* spp. and other lianes. L-dopa, used to control Parkinson's disease, is obtained from the seeds of tropical American *Mucuna* species. *Dioscorea* spp., collected in the Mexican rain forest, are the chief source of diosgenin and other sterols used in the manufacture of various drugs, including cortisone and contraceptives; after timber they are Mexico's most valuable forest product. Some Flacourtiaceae of the West African rain forest contain 'marketable molecules' which are promising as sources of antibacterial and anticancer drugs (P. Waterman) and tropical American Simaroubaceae seem to be promising sources of antimalarial compounds. (O'Neill *et al.* 1988).

So far only a very small proportion of the rain-forest flora has been investigated for compounds of potential value, either in themselves or as templates[1] for synthesizing other products. Many of the secondary compounds in rain-forest plants may have evolved as defences against herbivores; as Allen (1975) says, 'tropical trees, thanks to their constant battle with herbivores, are source books of chemical invention of which man has scarcely turned the pages'. As Lowry (1971), Schultes (1988a, b), Olaniyi (1988), Schultes & Raffauf (1990) and others have pointed out, the extensive and fast vanishing knowledge of the medical virtues of rain-forest plants possessed by some primitive peoples may provide useful clues for pharmacological research.

Forest-living peoples in fact find uses for a very wide range of forest plants. Prance *et al.* (1987) give quantitative data on the utilization of forest trees by four Amerindian tribes of the Amazonian terra firme, the Ka'apor and Tembé of Brazil, the Venezuelan Panare and the Bolivian Chácobo. All trees ⩾10 cm d.b.m. were identified on a one hectare plot of rain forest in the territory of each tribe: it was found that the percentage of tree species used varied from 48.6% by the Panare to 78.7% by the Chácobo. The chief uses were for construction, food and medicine; other uses included *Carapa guianensis* as an insect repellent and *Tetragastris altissima* as a catalyst in the fermentation of beer. Some products were collected for sale.

In more advanced societies the range of forest products used is generally much less; nevertheless, the total value, actual and potential of 'minor forest products' is considerably greater than is commonly realized. Peters *et al.* (1989) calculated that on a hectare of species-rich rain forest in Peru, fruits and rubber latex collected and sold in a neighbouring market could give an annual revenue of US $422; using a standard economic model, these products could be valued at US $6330 per hectare. If all the trees on the plot were felled and sold (which would probably eliminate future production of fruits and latex) there would be an immediate profit of US $1000, while selective, presumably sustainable, logging of timber might yield US $490 per hectare. Added to the value of the fruits and latex, this gives a value of US $6820 per hectare. Similar calculations for samples of rain forest in other parts of the tropics would be of interest.

Apart from their timber and other marketable produce, rain forests are of potential economic importance as gene pools. They are the original home of species from which cultivars of cocoa, bananas, rubber and other important crop plants are derived. Wild populations of these plants still exist and may be of value in breeding programmes. In the rain forest of Ulu Kelantan (Malaya), Whitmore (1984a) found wild types of fourteen species of fruit tree, including species of *Artocarpus*, *Durio* and *Nephelium*, also three other tree species of actual or potential pharmacological importance. Among underexploited tropical plants with promising economic value, a considerable number are rain-forest species (National Academy of Sciences 1975).

The significance of tropical rain forest for humans extends far beyond its influence on the environment and the economic value of its products. Tropical rain forests are uniquely rich communities of plants and animals which include many of the most beautiful and bizarre forms of life. Only in them are found butterflies such as the brilliant blue morphos of tropical America and the giant birdwings (*Troides*) of Malesia, many species of parrots, macaws and toucans and countless other strange and lovely creatures, as well as magnificent trees and orchids. A large proportion of living primates occur only in tropical rain forests and are likely to become extinct if their habitat is lost. It has been estimated that about a third of the world's flowering plants are found in tropical rain forests and over 10% in Amazonia alone (Prance & Lovejoy 1985, p. xiii).

Yet scientific knowledge of tropical biota is still so incomplete that a large proportion of rain-forest plant species and the majority of rain-forest invertebrates are unnamed and undescribed. Even among described species the biological characteristics of only a very few are more or less adequately known. As the author wrote in 1973

---

[1] An example of the latter is the extensively used drug atrocurarine, which could not have been synthesized without knowledge of the molecular structure of the alkaloid tubocurarine, a natural constituent of curare (J.D. Phillipson).

'If we believe that all living creatures should be a source of wonder, enjoyment and instruction to man, a vast realm of potential human experience may disappear before there is even a bare record of its existence' (Richards 1973b, p. 67).

Because of its immense species diversity and long history, intricate and subtle relations have evolved in rain forests between different organisms and between organisms and their environment, making the tropical rain forest a field for biological research which has no substitute. In it must lie the key to a vast amount of scientific knowledge which cannot be obtained elsewhere. It was from their experiences in the rain forests of tropical America and the Far East that the contributions to science of Humboldt, Wallace, Bates, Darwin and other great naturalists of the past were derived.

Human attitudes to rain-forest conservation vary greatly. For the last remaining hunter-gatherers the forest is the source of their livelihood and their own survival largely depends on it (Denslow & Padoch 1988), but for the Brazilian caboclos, making a poor living from scanty crops of cassava and maize, it is 'green hell'.

Similarly, the Mbuti pygmies of the Ituri forest in Zaïre regard the forest as their 'mother' because it provides the natural products on which they depend, but the Bantu people living alongside who fell it to grow their crops fear and despise it. The attitudes to the rain forest of the tribal peoples in the Baram area of Sarawak have been discussed by Lian (1988) and are similar to those of hunter-gatherers. They had a 'folk ecology' with a concern for the future of the forest and they had evolved methods of exploiting its timber and other resources which are sustainable, in contrast to the destructive methods practised in the same region today. Many politicians regard the rain forest as a 'desert of trees' which it is their duty to remove (Schultes 1988b). Most entrepreneurs and many governments have made the 'Faustian bargain' (Richardson 1978): for them a tropical forest is a resource to be 'mined' and converted into immediate wealth, regardless of future consequences for themselves or for humanity in general.

Fortunately, an awareness that the rich biota of tropical rain forests are a precious human heritage, to be conserved for posterity rather than consumed for immediate gain, is spreading fast in the developing as well as the economically advanced nations. One result is that in nearly all countries in which tropical rain forest exists considerable areas have been set aside as national parks or biological reserves. Some of these are more than fifty years old. In Zaïre, national parks were established early in this century; Barro Colorado Island in the Panama Canal became a nature reserve in 1923 and since 1946 has been under the management of the Smithsonian Institution. The Henri Pittier National Park (Ranch Grande) in Venezuela was established in 1937; it consists mainly of rain forest (largely montane; see Chapter 17).

Some tropical national parks are very large. Taman Negara in Malaysia is 434 000 ha in extent and forms *ca.* 4% of the land area of the Peninsula: in 1982 it included nearly a fifth of its remaining rain forests. Large national parks have also been made in Brazil, including parts of Amazonia, and in India and other tropical countries. Costa Rica has an excellent system of national parks, some of which contain rain forest, in addition to the rain-forest reserve of the Organization of Tropical Studies at La Selva (McDade *et al.* 1994). Among the national parks in the Far East the Gunung Mulu National Park in Sarawak, constituted in 1975 is of outstanding importance because of the great variety of primary rain forest included in it, ranging from types of lowland forest at altitudes down to 60 m to montane forest and scrub up to 2377 m (Anderson *et al.* 1982).

In 1988 the 'Wet Tropics' of Queensland, an area including all the remaining tropical rain forest in Australia, became a World Heritage site. Parts of these forests were already protected as national parks or scientific reserves; management plans in which all logging will be prohibited are being drawn up for the rest.

Many existing national parks in the humid tropics have been designated primarily for their scenic attractions, or to conserve their larger fauna, and consist of montane vegetation or savanna; relatively few wholly or mainly of lowland tropical rain forest. They vary considerably in the effectiveness of their protection from encroachment, illegal logging and other disturbances, as well as in size.

Isolated fragments of forest a few hectares in extent may survive for many years, but they gradually degenerate because of wind damage, fires, weed invasions and probably also from loss of ecosystem components such as birds and other animals essential for the reproduction of the trees and other plants. An example of such a fragment is the 'Gardens jungle' at Singapore, a relic of 11 acres (4.45 ha) of dipterocarp forest left unfelled when the Botanic Garden site was cleared in 1859 (Purseglove 1968). In recent years, although some of the original flora survives, the forest has been seriously damaged by invading lianes and other weeds (in contrast to the not far distant Bukit Timah forest, which is 71 ha in area and structurally still relatively well preserved, with a flora of 787 angiosperm species (Corlett 1990, Turner 1994). According to Ng (1983) many undisturbed forest reserves in Malaya, including the well-known Sungei Menyala Reserve are too small to be self-sustaining without 'remedial management'.

The size of the smallest area of rain forest that can be expected to maintain itself permanently cannot be easily predicted. It is the resultant of many factors such as the reproductive strategies of its component tree species and the size of the foraging areas of its larger animals, as well as climatic and other 'edge effects' (Soulé 1987). Research is in progress in which changes in isolated rain-forest stands of various sizes within a large cleared area in Amazonia are being monitored (Lovejoy et al. 1986, Kapos 1989). Natural rain forests, as we have seen, normally consist of mature, gap and building phases; in a viable stand all three need to be represented.

Terborgh (1992) has drawn attention to the changes in the fauna and flora of Barro Colorado Island and the nearby islets which became isolated when the Panama Canal was constructed in 1914. The islets are too small to support large vertebrates and the absence of large seed-eaters is probably why large-seeded palms and other trees are more abundant on them than on the adjacent mainland. On Barro Colorado itself, ocelots are the only large carnivores remaining; this has led to an abnormal abundance of seed-eating mammals. Effects on plants have not been recorded.

More (and larger) national parks and inviolate forest reserves in the humid tropics are urgently needed, particularly in lowland rain forest, but in an already over-populated world this will be hard to achieve. Most remaining rain forests have some human inhabitants; if the forest is to be conserved, provision has to be made for these inhabitants to support themselves without damage to the ecosystem, or they have to be relocated. Such a policy is being attempted in the recently created Korup (Oban) National Park on the frontier of Cameroon and Nigeria.

What will be the fate of the remaining forests not protected as national parks or nature reserves? As Poore et al. (1989) have pointed out, forests not managed for some economic purpose nearly always disappear as soon as they become accessible. This has been the fate of large areas of the Amazonian forest since the construction of the Trans-Amazonian highways. It has often been proposed that as much as possible of the world's remaining rain forests should be managed under sustainable systems in which selective logging should take place under strictly controlled conditions and on a previously determined rotation. Such systems, which imitate the regeneration cycles of natural forests, are theoretically feasible. But for them to succeed, according to Poore et al.

(1989), four conditions are necessary; governments must provide long-term protection; timber extraction must be carried out with minimum damage to seedlings and immature trees, soil and water courses; reliable information must also be available on the composition of the forest and its rate of regeneration. Lastly there must be marketing arrangements which bring a fair share of the returns to the people living in and near the forest.

At present only a very small proportion of the world's rain forests are managed on genuinely sustainable systems: according to Poore et al. (1989), the only examples are in Malaya and Trinidad. Experiments in sustainable forestry have also been made in Brazil (Rankin 1985), Venezuela and elsewhere. In Surinam sustained-yield logging was abandoned after twenty years for political reasons. In Nigeria the Shelterwood management system established by the former colonial government has also been abandoned, partly because of the increasing demand for land for other purposes.

Up to now sustained-yield management has not been widely practised in the tropics mainly because it does not give such quick financial returns as clear-felling. It also demands strict control and skilled oversight. Sustained-yield forestry might avert the prospect of still more forest land on poor infertile soils being invaded by savanna and scrub, but doubts remain as to whether logging rotations of forty or fifty years, as have been proposed, will really allow the survival of forests dominated by trees which naturally live to two hundred years or more and usually do not begin to reproduce at an early age. It seems more likely that there will be a gradual replacement of long-lived primary forest species by shorter-lived seral trees. It may also be unrealistic to expect that the careful management necessary for sustained-yield forestry can be maintained indefinitely.

For assuring future supplies of the more valuable tropical hardwoods, plantations may be a more viable option (Evans 1992). For these, land already logged or deforested could be used and much natural forest might be saved as a result. Plantations also have obvious technical advantages in harvesting. Some species of rain-forest tree may not be amenable to plantation cultivation, but some in fact grow faster in plantations than in natural forests (Whitmore 1984a, Rankin 1985). However, both economic and ecological problems have hitherto restricted the large-scale development of tropical hardwood plantations. In the short term it is more expensive to establish and maintain plantations than to extract timber from natural forests. There are also problems of pest and disease control, as well as possible environmental damage caused by the effects of the long continued plantation silviculture on soil fertility and microclimate.

Trees grown in monoculture are more subject to disease than in their native forests. A well-known example is rubber (*Hevea brasiliensis*) which cannot be grown successfully in plantations in Amazonia where it is a widespread native species. Its success in plantations in the Old World depends largely on the absence there of the leaf-blight fungus, *Microcyclus ulei*.

The ecological problems of tree plantations in the tropics are probably not insoluble. It seems likely that the difficulties would be less severe in mixed plantations approaching natural forests in composition and structure. The relatively high cost of plantations will become less important as the available supply of tropical hardwoods from natural forests decreases.

Managed forests and tree plantations could make an important contribution to the survival of some, but not all, components of the rain-forest ecosystem. For others, botanic gardens and zoos may provide 'Noah's arks'. But if all natural rain forests are destroyed an enormous number of plant and animal species will probably be lost, and 'extinction is forever'. Co-ordinated international efforts are urgently needed if as much as possible of the tropical rain forests still left are to be preserved.

Appendix 1

Tree recognition in the field and the use of vernacular names

One of the most perplexing problems in studying the floristic composition of tropical rain forests is the identification of trees in the field. The majority of them are tall and their flowers, fruit and foliage are overhead and far out of sight. A pair of binoculars is a useful aid, but a sweet scent, fallen petals or the buzzing of insects is often the first indication that a tree is in flower. Herbarium specimens can be obtained by climbing or felling the tree, but for obvious reasons felling should be avoided except where forest clearance is actually in progress. Mitchell (1982, 1986) and Moffett (1993) have given a useful account of methods for climbing tall trees and exploring the forest canopy. Catapults (Hyland 1972), shotguns and even trained monkeys (E.J.H. Corner) have been used for obtaining specimens from tall trees, but these are not methods which can be extensively used.

The flowering seasons of different trees tend to be spread over most of the year and some species flower only at intervals of several years (Chapter 9), so to collect herbarium material with flowers and fruits of every species in even a 1 ha plot is a time-consuming, and often an impossible, task. For this and other reasons, a small proportion of the trees in most sample plots usually remains unidentified, or doubtfully identified, as mentioned in Chapter 11.

For field recognition of rain-forest trees, much use can be made of vegetative characters. Fallen leaves (when it is certain from which tree they have come) can be very helpful. In Brunei, Ashton (1964) found it possible to recognize all the Dipterocarpaceae (137 species) from fallen leaves, but often trees of other families without flowers could not be precisely identified.

The trunk and bark of rain-forest trees, as well as the leaves, provide many useful diagnostic characters which may surprise those accustomed only to the meagre tree floras of temperate regions. The form of the trunk is sometimes distinctive, curved or straight, smoothly rounded or, as for example in *Aspidosperma* (tropical America) (Fig. 4.7), conspicuously fluted. When buttresses are present, their shape, size and number may be distinctive (pp. 78–84).

Surface features of the bark, its colour, whether it is furrowed, flaking or peels in thin strips, are often helpful, but much more important is the 'slash' (blaze). This shows the macroscopic appearance of the wood, the distinctive characters of the inner bark, and the presence of latex or other secretions. Some trees are at once recognizable by the colour of the slash, e.g. *Enantia chlorantha* (Africa) in which it is bright yellow. Sometimes the inner bark and wood have a characteristic smell or taste (a smell of burnt feathers in *Meliosma*, a sweet taste in the Amazonian *Glycoxylon*). Some trees emit a hissing sound when slashed, e.g. Malesian species of *Dillenia* and *Alangium*.

Bark and wood characters often occur in specific combinations and together enable a practised eye to identify a large proportion of the tree flora. Using these and other field characters, guides to the trees of various rain-forest areas are available, e.g. Gentry (1993, northwest South America), Den Outer (1972, Côte d'Ivoire), Meijer (1974, West Malesia), Newman *et al.* (1995, dipterocarps of SE Asia). Trunk and bark characters usually run parallel with those normally used by taxonomists, but sometimes species of the same genus differ in bark and slash, while unrelated species may be much alike. The full appreciation of these characters requires a sharp eye and considerable experience; visiting workers seldom become as expert in using them as do local people who have spent their whole lives in the forest (treefinders, mateiros). For them tree recognition is part of daily life.

Most tropical trees have vernacular names, which can be

very useful for tree recognition. Among the Arawak people of Guyana, for example, hundreds of tree names are in common use. Many of these are applied to only a single botanical species, but others refer to related groups; for example, baromalli refers to two related species of *Catostemma*, and warakusa to *Inga* spp., excluding a few outstandingly distinct species. Only a few names refer to members of several genera or families; for example, kokeritiballi means species of various genera of Sapotaceae, kulashiri refers to various Meliaceae, Flacourtiaceae and Sapindaceae. The Malays, who also have numerous names for native trees, attach many of these names to one particular species, but have collective names for groups of species which have some obvious characteristic in common, e.g. medang (soft-wooded trees), balek angin (trees with white undersides to their leaves). Many other forest-dwelling peoples have extensive vocabularies of tree and plant names. The Kelabits of the Sarawak highlands have few names for plants, perhaps because they are relatively recent immigrants to the area (T.H. Harrisson). Plants to which vernacular names are given are mostly used for timber, fibre, food, drugs, poisons, etc., or have some superstitious significance. Their names form what Conklin (1957) called 'folk taxonomy'.

It should be emphasized that, although any forest-dwelling person may know the names of some trees, only a few experts will know the names of a large number. In tropical countries with well-organized forestry services some of these experts have been trained by European foresters. By constant practice they acquire a knowledge of the trees of their district which is astonishing in its extent and accuracy. Expert tree finders exist in Guyana, Nigeria, Borneo, Brazil and doubtless in most other tropical countries.

Enumerations of trees on sample areas by their vernacular names can be very useful where the tree flora is inadequately known, but it will be necessary to translate as many of these names as possible into their botanical equivalents. This involves much careful checking and comparison. The use of each name must be repeatedly checked in the field to find out to what species or group of species it corresponds. As many collections as possible of herbarium material of the same 'vernacular species' should be made for subsequent study by specialists. It should be remembered that the same local name may be used for different species in different districts, so an equivalent which has been carefully checked in one area cannot be used without

further investigation in another. When a local name is applied to a group of similar species, it may happen that in a given district, or in a particular sample plot, only one of these species is present; within the plot or limited district the name can therefore be safely treated as the equivalent of one species. Approximate information should not be despised; it is often useful to know that a certain genus is present, even if it is impossible to decide which species or how many.

When listing the species on a sample plot or transect both the vernacular names and their botanical equivalents should usually be given, even when the equivalent of every name has not been worked out. The evidence for the correctness of the equivalents will probably vary in completeness. Thus if herbarium material of a certain 'vernacular species' has been collected only once, the evidence that the name in question is the equivalent of a particular taxonomic species is less complete than if ten collections have been made. An equivalent based on material collected actually on the plot or transect under consideration is more reliable than when it is based on material collected in another locality in the same district or in another district. A scheme of conventions to indicate the 'degree of reliability' of the names used may be adopted (e.g. Richards, 1939, p. 53 and Tables 6–9). Whenever an identification is based on a collected specimen a serial number should be quoted and the herbarium where the specimen is deposited should be indicated.

Systematists accustomed to working only with temperate floras will perhaps doubt the value of methods of studying floristic composition which involve the use of native 'tree-finders' and vernacular names; yet such methods can give results of great scientific value provided sufficient precautions are taken. In any ecological investigation involving the listing of species that cannot be immediately identified in the field there are two stages at which errors are liable to arise; one is in the determination of the specimens collected and the other in deciding which individuals are in fact specifically identical with the specimens collected. The possible error at the second stage depends on the taxonomic ability and experience of the field worker and can never be entirely eliminated. When a 'tree-finder' is employed the possibilities of error are necessarily somewhat increased but, if care is taken, the lists of species, though inevitably less complete and accurate than for a well investigated temperate flora, may have a high degree of reliability.

Appendix 2

Application of numerical methods in rain forest

P. Greig-Smith

Introduction

Numerical methods of analysing data on the floristic composition of rain forests can contribute to understanding of their ecology at several levels. A major objective is commonly to seek correlations between composition and features of the environment, history of management or disturbance, as a basis of hypotheses about the factors determining composition. This has long been an important approach in community ecology; the contribution of numerical methods is to extract major trends in composition from a mass of data too complex to interpret by direct inspection. Similar difficulties of interpretation of complex data apply to the detection of mosaics of cyclically related gap, building, mature and degenerate phases within one environmentally determined forest type (Chapter 3).

At a more detailed level, it is useful to consider the pattern of distribution shown by species within a stretch of apparently uniform forest. 'Pattern' in this context refers to any departure from randomness of arrangement by individuals either being more regularly arranged than random expectation or, more commonly, being to some degree clustered or clumped. The objective is again to detect correlation between the occurrence or performance of species and influencing factors which may, at this scale, be not only slight environmental variation, but also the presence of other species nearby or the reproductive characteristics of the species under investigation (Greig-Smith 1979). At this level examination of correlations is rarely possible without prior simplification of the data by appropriate numerical methods.

The use of numerical methods in these ways is essentially for data exploration; they clarify features of the available data in a way that permits erection of plausible hypotheses, but they cannot provide a test of hypotheses. In principle, a test requires experiment, which in practice may be difficult or impossible, although additional field observations may serve to provide additional support for a hypothesis.

Characteristics of rain forest affecting the use of numerical methods

Although some other types of vegetation approach or possibly exceed rain forest in number of species per unit area (Chapter 11), only in rain forest is such species richness combined with a physiognomy that dwarfs the observer (Greig-Smith 1971a). There are thus greater difficulties in identification of species than in other types of vegetation, a difficulty still accentuated in many areas by inadequate taxonomic knowledge. In any field data there is some 'background noise', trivial or irrelevant information resulting from errors in data collection and chance effects. The difficulties of species identification in rain forest increase the amount of background noise.

A high degree of species richness in itself also contributes to background noise. Assessment of variation in composition depends on the tendency of species to occur together. Even numerical methods that are not explicitly based on measures of association between pairs of species in fact mostly depend on association between species. In very species-rich vegetation most of the species will necessarily be rare and joint occurrence of rare species in a plot may be due to chance. The effect is clear from Austin & Greig-Smith's (1968) analysis of an enumeration of all trees of $\geqslant 12$ in (30 cm) g.b.h in a block of fifty 0.04 ha plots in dipterocarp forest in Sabah. Of the 198 tree species, 60 were represented by a single individual only. With fifty plots there are necessarily ten joint occurrences of such species, giving perfect correlation between them, and seventeen[1] plots with two or more species represented by single individuals may be expected by chance. Some joint occurrences of rare species in a particular plot may be meaningful, in that they are responding to a peculiarity of that plot, but it is not possible to distinguish between meaningful and chance joint occurrences. Chance joint occurrences are more likely to be numerous in sets of data which are relatively homogeneous.

The confusing effect of background noise is illustrated by an ordination of these Sabah data based on fifty species selected randomly from the total 198. This ordination was uninterpretable; analysis of the correlation matrix between species showed that the first axis of the ordination was determined primarily by three species, all of which occurred once only, but all in the same plot.

Since rare species are a principal source of background noise, it is helpful, particularly with relatively homogeneous data sets, to eliminate such species from the data before analysis. With the Sabah data a satisfactory ordination was obtained using only the twenty-five most abundant species (ca. 12.5% of the total) and there was little improvement on including further species.

When there is a large number of species in relation to number of stands (plots), the stands are necessarily over-defined; fewer species than the total present are sufficient to summarize the differences between stands. Although the capacity of computers to handle large bodies of data speedily continues to increase, computation has a financial cost; elimination of redundant information is therefore desirable. Webb *et al.* (1967) examined this problem in relation to complete enumerations of all vascular plants in eighteen 0.1 ha stands spread over 7° of latitude in tropical Australia and including 818 species. Classification by association analysis using the 269 'big trees' (species capable of reaching the canopy) produced a hierarchy identical with that based on all species. Using a two-way table comparing stand and species classifications to eliminate species largely confined to one stand group, they reduced the number of species required to produce a fully satisfactory classification to 65 big tree species.

These results must be treated with some reservation. The stands were deliberately chosen to represent a very wide range of physiognomic–structural types, some of which are doubtfully rain forest as understood in this book. With such a wide range of composition there may be more redundant information than in less heterogeneous data sets. John B. Hall (1977) examined tree enumerations of forty-six samples of Nigerian closed forest, which were of varying composition but much more structurally uniform than the Australian samples. He reduced species number in a similar way, eliminating less common species and, effectively, those confined to a single stand group, and obtained a satisfactory ordination.

[1] Not thirty-five, as quoted by Austin & Greig-Smith (1968).

Ashton (1964) analysed data from dipterocarp forest in Brunei which certainly have a narrower range of composition than the geographically widespread samples of Nigerian forest examined by Hall. He used the fifty most abundant species that were certainly identifiable out of 472 tree species in his Andalau data and seventy-nine out of 420 species in his Kuala Belalong data to obtain satisfactory ordinations. The Sabah data quoted above, being drawn from a compact block of forest, are probably still more homogeneous.

It is clear that satisfactory simplification by classification or ordination can be made by using many fewer species than the total. Unfortunately, it is not normally possible, at least in an initial survey, to distinguish in advance which are the appropriate species to use. If further stands within the same range of composition are to be placed into an existing classification or ordination, as may be needed in a management context, then field recording may be limited to the species list used in the initial analysis.

Data collection

Three decisions are required: the layout and number of stands, the type of data to be gathered, and the size of plot to be used as a stand. These are not independent of one another and are further dependent on the objectives of the investigation and the precise procedure of analysis to be used.

The considerations involved are not peculiar to rain forest (see Greig-Smith 1983, Chapter 10). The optimal placing of stands depends on objectives. If inventory of resources, the relative amounts of different vegetation types (or of different species), is the main objective, stands are being treated as samples of the whole area under investigation and must be placed in a way that gives unbiased estimates. This can generally be attained by a systematic arrangement of stands unless the sampling grid happens to coincide with periodic variation in the vegetation, e.g. where there is regular alternation of ridge and valley. Only if confidence limits of estimates are required need the stands be randomly placed; in rain forest this is likely to be very time-consuming.

A more usual objective is to examine correlation between composition and possible influencing factors. As far as possible all variants of vegetation should be included, with equal representation of the variants in the data set. This is difficult to achieve completely (Greig-Smith 1983). There is no objection to the selection of stand positions on the basis of observed differences in vegetation or environment; to do so may, however, result in overemphasis of distinctiveness between groups in a subsequent classification.

With the general species richness and consequent over-definition of rain forest, it is to be expected that relatively small stands will be adequate to quantify the degree of similarity between stands. In practice this has proved to be correct; even recording larger trees only, stands as small as 20 m × 20 m (0.04 ha) have given satisfactory interpretations of variation in mixed dipterocarp forest (Austin & Greig-Smith 1968) and heath forest (Brünig 1968) in Borneo. In the much less species-rich forests of Ghana, Hall & Swaine (1976), recording all angiosperm species, successfully used stands 25 m × 25 m (0.0625 ha) in an informative survey covering all forest types.

The use of defined plots as stands is well established. An alternative, which eliminates the time-consuming demarcation of plots, is a form of plotless sampling (Newbery & Proctor 1984, Williams et al. 1969b). This involves taking a point or an individual tree as the centre of a stand and recording the n nearest trees, n being fixed to give an adequate stand. It has the further advantage of equalizing the amount of information from different stands even if density varies widely within the area being studied. Hall (1991) investigated the accuracy of such n nearest-tree sampling in montane forest in Tanzania. He examined samples of 5, 10, 15 and 20 nearest trees and concluded that in this forest a sample of the 15 nearest trees gave acceptable results not only for species composition but also for proportions of different tree sizes.

If, as a result of gap regeneration or localized disturbance, there is a marked mosaic pattern not related to environmental differences, the use of small stands may emphasize the mosaic differences at the cost of less efficiency in detecting environmentally related variation. If large stands are thought to be necessary, the labour of complete enumeration can be reduced by appropriate sampling within a stand. For example, in Guyana Ogden (1966) used twenty-five systematically arranged points within each of his 1 ha stands; at each point four trees were recorded by the 'point centred quarter' method of Cottam & Curtis (1956), in which the nearest tree is recorded in each of four quadrants around the point (the orientation of the quadrants being fixed in advance). On a larger scale Hall (1977), in an analysis of forest composition in relation to climate and soil type in Nigeria, used forestry enumerations based on large blocks (39–587 ha) of pre-exploitation forest.

The usual practice in enumerating forest plots is to record species and girth or diameter of all trees above a stated minimum size, allowing estimates of density and biomass (as total basal area) of each species. With species-rich vegetation, however, the presence or absence of species often provides sufficient information to quantify differences in composition between stands. That this should be so, if at first surprising, is clear. Consider the simplified situation of a single environmental gradient along which species show a unimodal (though not necessarily symmetric) response curve (Fig. A2.1). The amount of a particular species present in a stand will give an indication of the position of that stand on the gradient (species 1 and 11 in this example) or of two alternative positions (species 2–10). Any particular combination of species present, regardless of the amount in which they occur, will, however, also indicate

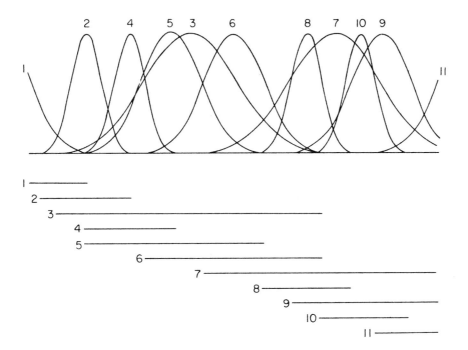

Fig. A.2.1 Unimodal response curves of species along an environmental gradient, with corresponding species presence shown below.

the position on the gradient within narrow enough limits to be satisfactory in sorting out the relationships between stands.

If subsequent classification or ordination is the only objective, a list of the species occurring in each stand is thus an adequate basis. Moreover, growth forms such as lianes, rhizomatous herbs, etc., which are difficult to assess quantitatively, can be included if presence only is recorded. The additional information thus available allows smaller stands to be used (see, for example, Hall & Swaine 1976). If a survey is to serve other objectives, e.g. description of vegetation structure, pattern of regeneration, or potential yield of economically valuable species, much more extensive data will be needed. This raises the question of whether to allocate the available time in the field to gathering more data from fewer stands or fewer data from more stands. A further consideration is the time required for travelling to and demarcating stands, which in difficult terrain may be so considerable that the gain in recording only minimal information from each stand is slight. In planning a survey these considerations should be borne in mind (see Greig-Smith 1983 (Chapter 10), Williams 1971); Noy-Meir (1971) has an informative discussion of a particular case, though in a very different type of vegetation.

Analysis of survey data

Since the introduction of the first numerical techniques a very wide range of procedures has been developed (Greig-Smith 1980, Noy-Meir & Whittaker 1977). It is not appropriate to discuss these in detail here; general accounts have been given by Digby & Kempton (1987), Gauch (1982), Greig-Smith (1983), Orloci (1978) and Williams (1976).

Two groups of techniques, classification and ordination, are available. Classification, assigning stands to groups such that differences within groups are minimized and differences between groups are maximized, is a familiar and well-established approach. The advantage of numerical methods is that they allow the data themselves to indicate the optimal classification, rather than relying on preconceptions about which species give the most efficient classification (dominants, constants, etc.) or on subjective assessment of the degree of difference between stands. Almost all numerical techniques produce hierarchical classifications. At each division in a hierarchy the environmental characteristics of the two groups formed may be compared. Marked differences between them indicate correlation between forest composition and environmental factors, providing a basis for hypotheses about the factors controlling composition.

When used in this way, classification is a tool for the exploration of a particular data set. It is the successive divisions in the hierarchy rather than the final groups which are of interest; classification of a different set of stands from the same range will not necessarily give exactly the same hierarchy. However, if the data include enough stands to give representation of all variants, the final groups can provide a classification of general validity for use in management, vegetation mapping, etc.

Ordination is an alternative to classification as a means of simplifying large bodies of field data. Often described as arranging a set of stands in relation to axes of variation, it may also be regarded as a reduction in the dimensionality of the data set. The composition of a stand can be related to orthogonal species axes. With many species, the resulting configuration of a set of stands has a very large number of dimensions (up to as many dimensions as there are species or one less than the number of stands, whichever is the less). Ordination aims to project this configuration into fewer dimensions efficiently, i.e. to retain the relative positions of stands as nearly as possible. If values of an environmental variable plotted on the arrangement of stands produced by an ordination show a degree of order, that factor is to some extent correlated with the composition of the vegetation. With very heterogeneous data including a wide range of forest composition, it may be helpful to classify the stands first to produce a set of more homogeneous groups. Environmental correlations can then be examined both by an ordination of the classes (based on their mean composition) and by separate ordinations of the members of each class (Greig-Smith et al. 1967). If an ordination covers a wide range of composition, it can assist in delimiting broadly defined forest types, e.g. Hall & Swaine's (1976, 1981) account of Ghanaian forests. Ordination may also be used to place further stands in a previously erected classification (Swaine & Hall 1976). It is often worthwhile both to classify and to ordinate a set of data.

Although we are not concerned here with details of techniques, the standardization of data before analysis should be mentioned because the investigator must make decisions on this based on biological criteria (Greig-Smith 1971b). The effect of standardization is to alter the weighting given to different aspects of the information in the data. Thus with quantitative data, standardization by species standard deviate (division of the observed amount of a species in a stand by the standard deviation of that species in the total data) will equalize the weighting of rare and common species. With presence/absence data it will emphasize the importance of presence of rare species and absence of common species. Standardization of quantitative data by stand total will eliminate the effect of differences in standing crop, concentrating attention on differences in the proportions of different species. With presence/absence data it will remove the effects of differences in species richness. Noy-Meir et al. (1975) give a valuable discussion of a wide range of possible standardizations. See also Greig-

Smith (1983) and, for a discussion in relation to rain forest, Austin & Greig-Smith (1968).

Classification and ordination of stands in terms of their composition, described above, is referred to as 'normal analysis'. The initial data matrix can also be regarded as describing species behaviour in terms of the stands in which they occur and, for quantitative data, their abundance in those stands; in this case species, rather than stands, are classified or ordinated. Species in the same group in a classification, or occurring close together in an ordination have similar habitat requirements. Such 'inverse analysis', though potentially valuable, particularly in comparing the ecological behaviour of taxonomically closely related species, has been relatively little used with rain-forest data (but see, for example, Poore 1968). Analyses of stands and of species, being based on the same data matrix, are inter-related and some techniques provide simultaneous normal and inverse analyses, e.g. reciprocal averaging (Hill 1973b) or detrended correspondence analysis (DECORANA) (Hill & Gauch 1980).

If a number of environmental variables have been recorded, e.g. various soil characteristics, it is useful to simplify them by an appropriate ordination technique, generally principal component analysis, to facilitate comparison with floristic composition. For examples dealing with rain forest, see Austin et al. (1972) and Newbery & Proctor (1984). If adequate environmental data are available for all stands, ter Braak's (1986) canonical correspondence analysis, which extracts axes maximising the indications of correlation between composition and environmental variables, is more efficient and economical of computing (see Palmer 1993).

Vegetational change

Techniques for analysing change in forest composition are less developed than those for analysing differences between contemporaneous stands. Two scales of change may be of interest: relatively major changes in total composition, such as those involved in succession after clearance or disturbance, and changes in a small area such as those resulting from the death of an individual tree and its replacement by a tree of another species. At one extreme the latter may represent part of a major unidirectional change and at the other the maintenance of a stable but dynamic mosaic structure.

Two numerical approaches are available. If successive records have been made for the same set of stands, an ordination treating each record as a separate 'stand' will allow the movement of a stand in the ordination space with time to be plotted (Austin 1977). In this way changes in a number of stands at different stages in a succession can be combined to obtain an indication of longer-term

trends or convergence in composition. With the requirement for repeated recording, which in species-rich forests needs to be at relatively long intervals for change to be evident, this is not generally helpful. For a particular species, however, tree diameter may sometimes be used as an approximate measure of age and records of stands at different successional stages can be used to elucidate the course of change in terms of the appearance and disappearance of species (Goff & Zedler 1972). This approach has apparently not been attempted for rain-forest data, but it is potentially valuable.

A further disadvantage of direct ordination of repeated records of stand composition is that the effects of changing composition are confounded with stand differences other than those due to stage in succession, e.g. slight environmental differences. Swaine & Greig-Smith (1980) suggested a modification of principal component analysis, applicable to a reasonably homogeneous set of stands, which takes account only of change. Swaine (1973) used this method to analyse records of sub-plots within a 1 ha plot of hill forest in Malaya, extending over twenty-two years, but failed to detect consistent trends, perhaps because of the relatively high minimum girth limit (31.5 cm) used in recording.

An alternative approach to the analysis of repeated records uses transition matrices tabulating the probability (estimated from the data) of one 'state' changing to another after a time interval. The 'states' may be either total composition of a stand, previously assigned to a type by numerical classification, or the species occupying a point. The transition probabilities not only elucidate the course of change, but may also be used to predict future composition by repeated multiplication of the initial composition by the appropriate probabilities. The method has severe limitations, especially in assuming that the transition probabilities remain constant with time (Usher 1981), but Williams et al. (1969a) successfully used it to analyse the early stages of succession after clearing in an Australian rain forest (see also Webb et al. 1972). Enright & Ogden (1979) found it effective in an examination of species replacement in lower montane forest in Papua New Guinea.

Pattern

Spatial variation in forest composition occurs at scales varying from that corresponding to major environmental difference, explored by the techniques of classification and ordination discussed above, to patchiness or point-to-point variation (pattern) within what in such analyses would be considered as a single stand.

If the occurrence of a species in small plots systematically arranged, generally in transects, is recorded, nested-block analysis of variance, 'pattern analysis', allows the scale of patchiness to be assessed (Greig-Smith 1961, 1983, Hill 1973a). Interest is generally not so much in the absolute scale of patchiness as in comparison between the patterns of different species or between the pattern of species and that of environmental factors, which can be analysed in the same way. Similar analysis of covariance between species, or between a species and an environmental factor, allows the relationship to be examined with some precision, e.g. Pemadasa & Gunatilleke's (1981) study of the pattern of some tree species in Sri Lanka in relation to levels of soil macronutrients. Greig-Smith (1991) has used analysis of the pattern of 'total covariance' (sum of all covariances between species) on data for savanna trees and found evidence of some degree of small-scale regularity of distribution, attributed to interference between individual trees. This approach is potentially useful in forest also. Application of these techniques to rain forest is limited to some

extent by the large number of species and consequent low density of most of them. Large samples are then necessary for the patterns of many species to be detected.

An alternative, which high species richness permits, is to classify or ordinate small plots or multiple-nearest-neighbour samples and relate the resulting classes or ordinational axes to concomitant environmental measures, i.e. the same approach as for whole stands. For studies at this small scale the time-consuming complete mapping of the position of individual trees, allowing variable numbers of nearest neighbours to be sampled, is valuable (see, for example, Newbery & Proctor 1984). Another possible treatment of data from small plots is by 'community pattern analysis' in which ordination axis values, rather than representation of individual species, are subjected to pattern analysis (Gibson & Greig-Smith 1986).

References

Absy, M.L. (1985). Palynology of Amazonia. Pp. 72–82 in *Key Environments: Amazonia*, Prance, G.T. & Lovejoy, T.E. (Eds), Pergamon Press, Oxford.

Adam, P. (1992). *Australian Rainforests*. Clarendon Press, Oxford.

Adams W.W., III, Terashima, I., Brugnøli, E. & Demmig, B. (1988). Comparisons of photosynthesis and photoinhibition in the CAM vine *Hoya australis* and several C₃ vines growing on the coast of eastern Australia. *Pl. Cell Environ.* 11, 173–81.

Adamson, R.S. (1910). Notes on the roots of *Terminalia arjuna* Bedd. *New Phytol.* 9, 150–6.

Adedeji, F.O. (1984). Nutrient cycles and successional changes following shifting cultivation in moist semi-deciduous forests in Nigeria. *For. Ecol. Mgmt* 9, 87–9.

Adjanohoun, E.J. (1962). Etude phytosociologique des savanes de la Basse Côte d'Ivoire (savanes lagunaires). *Vegetatio* 11, 1–39.

Ahmad, N. & Jackson, P. (1965). Studies of nutrient levels of Valencia oranges in North Trinidad and associated soil characteristics. *Soil Sci.* 100, 428–32.

Ahmad, N., Jones, R.L. & Beavers, A.H. (1968). Genesis, mineralogy and related properties of West Indian soils. I Montserrat series, derived from glauconitic sandstone, Central Trinidad. *J. Soil Sci.* 19, 1–8.

Ahn, P. (1959). The savanna patches of Nzima, south-western Ghana. *J. West Afr. Sci. Assn* 5, 10–25.

Ahn, P.M. (1979). Microaggregation in tropical soils: its measurement and effects on the maintenance of soil productivity. Pp. 75–85 in *Soil physical Conditions and Crop Production in the Tropics*, Lal, R. & Greenland, D.J. (Eds), J. Wiley & Son, Chichester.

Ahn, P.M. (1993). *Tropical Soils and Fertiliser Use*. Longman, Harlow.

Aiken, S.R. & Leigh, C.R. (1992). *Vanishing Rainforests: the Ecological Transition in Malaysia*. Clarendon Press, Oxford.

Aiyar, T.V.V. (1932). The Sholas of the Palghat Division. Parts I and II. *Indian For.* 58, 414–32, 473–586.

Alegre, J.C., Cassel, D.K. & Bandy, D.E. (1986). Effect of land clearing and subsequent management on physical properties. *Soil Sci. Soc. Amer. Jl* 50, 1379–84.

Alexander, E.B. (1973). A comparison of forest and savanna soils in northeastern Nicaragua. *Turrialba* 23, 181–91.

Alexander, E.B., Wildman, W.E. & Lynn, W.C. (1985). Ultramafic (serpinitic) mineralogy class. Pp. 135–46 in *Mineral Classification of Soils*, Kittrick, J.A. (Ed.), Soil Sci. Soc. Amer. & Amer. Soc. Agron., Madison, Wisconsin.

Alexander, I. (1989). Mycorrhizas in tropical forests. Pp. 169–88 in *Mineral Nutrients in Tropical Forest and Savanna Ecosystems*, Proctor, J. (Ed.), Br. Ecol. Soc. Spec. Pubn. 9, Blackwell Scientific, Oxford.

Alexandre, D.Y. (1989). Dynamique de la régénération naturelle en forêt dense de Côte d'Ivoire. *Etudes et Thèses*, ORSTOM, Paris.

Alexandre, D.Y., Guillaumet, J.J., Kahn, F., Namur, C. de & Moreau, R. (1978). Observations sur les premières stades de la reconstitution de la forêt dense humide (Sud-Ouest de la Côte d'Ivoire). *Cahiers ORSTOM, ser. Biol.* 8, 189–270.

Allee, W.C. (1926). Measurement of environmental factors in the tropical rain forest of Panama. *Ecology* 7, 273–302.

Allen, R. (1975). Interdependence, the trend we cannot buck. *IUCN Bulletin*, N.S. 6, 9–10.

Allen, S. (1990). Subtropical rain forest at risk. *Tropinet* 4 (*Biotropica* Supplement 27), 2.

Alvarez-Buylla, E.R. & Martinez-Ramos, M. (1992). Demography and allometry of *Cecropia obtusifolia*, a neotropical pioneer tree – an evaluation of the climax-pioneer paradigm for tropical rain forests. *J. Ecol.* 80, 275–90.

Alvim, P. de T. & Alvim, R. (1978). Relation of climate to growth periodicity of tropical trees. Pp. 445–64 in *Tropical Trees as Living Systems*, Tomlinson, P.B. & Zimmermann, M.H. (Eds), Cambridge University Press.

Amobi, C.C. (1973). Observation on bud break and elongation growth in the lowland rainforest and the Southern Guinea savanna in Nigeria. *Nigerian J. Forestry* 3, 64–73.

Anderson, A.B. (1981). White-sand vegetation of Brazilian Amazonia. *Biotropica* 13, 199–210.

Anderson, A.B., Prance, G.T. & Albuquerque, B.W.P.de (1975). A vegetacão lenhosa da campina da Reserva Biológica INPA-SUFRAMA (Manaus-Caracarai, Km. 62). *Acta Amazon.* 5, 225–46.

Anderson, J.A.R. (1961a) The destruction of *Shorea albida* forest by an unidentified insect. *Emp. For. Rev.* 40, 19–29.

Anderson, J.A.R. (1961b). *The Ecology and Forest Types of the Peat Swamp Forests of Sarawak and Brunei in Relation to their Silviculture*. Ph.D. Thesis, Edinburgh University.

Anderson, J.A.R. (1963). The flora of the peat swamp forests of Sarawak and Brunei including a catalogue of all recorded species of flowering plants, ferns and fern allies. *Gdns Bull. Singapore* 20, 131–228.

Anderson, J.A.R. (1964). The structure and development of the peat swamps of Sarawak and Brunei. *J. Trop. Geogr.* 18, 7–16.

Anderson, J.A.R. (1965). Limestone habitat in Sarawak. Pp. 49–57 in *Proceedings of the Symposium on Ecological Research in Humid Tropics Vegetation*, Kostermans, A.J.G.H. & Fosberg, F.R. (Eds), Science Cooperation Office for S.E.Asia, Bangkok.

Anderson, J.A.R. (1966). A note on two tree fires caused by lightning in Sarawak. *Malay. For.* 29, 18–20.

Anderson, J.A.R. & Chai, P.P.K. (1982). Vegetation formations. Pp. 53–84 in *Gunung Mulu National Park, a Management and Development Plan*, Anderson, J.A.R., Jermy, A.C. & Cranbrook, Earl of (Eds), Roy. Geogr. Soc., London.

Anderson, J.A.R., Jermy, A.C. & Cranbrook, Earl of (Eds) (1982). *Gunung Mulu National Park, a Management and Development Plan*. Roy. Geogr. Soc., London.

Anderson, J.M., Proctor, J. & Vallack, H. (1983). Ecological studies in four contrasting lowland rain forests in Gunung Mulu National Park, Sarawak. III Decomposition processes and nutrient losses from leaf litter. *J. Ecol.* 71, 503–27.

Anderson, J.M. & Spencer, T. (1991). Carbon, nutrient and water balances of tropical rain forest ecosystems subject to disturbance: management implications and research proposals. *Man and Biosphere Digest* 7, UNESCO, Paris.

Anderson, J.M. & Swift, M.J. (1983). Decomposition in tropical rain forests. Pp. 287–309 in *Tropical Rain Forest Ecology and Management*, Sutton, S.L., Whitmore, T.C. & Chadwick, A.C. (Eds), Br. Ecol. Soc. Spec. Pubn 2, Blackwell Scientific, Oxford.

Andriesse, J.P. (1972). *The Soils of West Sarawak (East Malaysia)*. Dept. Agric., Kuching.

Andriesse, J.P. (1977). Nutrient level changes during a 20-year shifting cultivation cycle in Sarawak (Malaysia). Pp. 479–91 in *Proceedings CLIMATROPS*. Kuala Lumpur.

Andriesse, J.P. (1987). *Monitoring Project of Nutrient Cycling in Soils Used for Shifting Cultivation in Various Climatic Conditions in Asia. Final Report*. Royal Tropical Inst., Amsterdam.

Anon. (1971). Men seek to save 160-foot tall elm. *New York Times* 23 Nov., 1971.

Aoki, M., Yabuki, K. & Koyama, H. (1978). Micrometeorology of Pasoh forest. *Malay. Nat. J.* 30, 149–59.

Appanah, S. (1980). Pollination in Malaysian primary forests. Pp. 177–82 in *Tropical Ecology and Development*, Furtado, J.I. (Ed.), International Society of Tropical Ecology, Kuala Lumpur.

Appanah, S. (1982). Pollination of androdioecious *Xerospermum intermedium* Radlk. (Sapindaceae) in a rain forest. *Biol. J. Linn. Soc. Lond.* 18, 11–34.

Appanah, S. & Chan, H.T. (1981). Thrips: the pollinators of some dipterocarps. *Malay. For.* 44, 234–52.

Argent, G. (1979). New Guinea bananas – a problematic genetic resource. Pp. 57–63 in *Biological Aspects of Plant Genetic Resource Conservation in Southeast Asia*, Jong, K. (Ed.), Univ. Hull, Dept. Geogr., Misc. Ser. 21.

Armesto, J.J., Mitchell, J.D. & Villagram, C. (1986). A comparison of spatial patterns of trees in tropical and temperate forests. *Biotropica* 18, 1–11.

Ash, J. (1988). The location and stability of rainforest boundaries in northeastern Queensland, Australia. *J. Biogeogr.* 15, 619–30.

Ashton, P.S. (1964). Ecological studies in the mixed dipterocarp forests of Brunei State. *Oxford For. Mem.* 25.

Ashton, P.S. (1969). Speciation among tropical forest trees: some deductions in the light of recent evidence. *Biol. J. Linn. Soc. Lond.* 1, 155–96.

Ashton, P.S. (1976). Mixed dipterocarp forest and its variation with habitat in the Malayan lowlands: a re-evaluation at Pasoh. *Malay. For.* 39, 56–72.

Ashton, P.S. (1977). A contribution of rain forest research to evolutionary theory. *Ann. Missouri Bot. Gdn.* 64, 694–705.

Ashton, P.S. (1982). Dipterocarpaceae. Pp. 237–552 in *Flora Malesiana* Ser. I, 9, Steenis, C.G.G.J. van (Ed.), M. Nijhoff, The Hague.

Ashton, P.S. (1988). Dipterocarp biology as a window to the understanding of tropical forest structure. *Ann. Rev. Ecol. Syst.* 19, 347–70.

Ashton, P.S. (1989). Dipterocarp reproduction biology. Pp. 219–40 in *Tropical Rain Forest Ecosystems: Biogeographical Systems and Ecological Studies*, Lieth, H. & Werger, M.J.A. (Eds), Elsevier, Amsterdam.

Ashton, P.S., Givnish, T.J. & Appanah, S. (1988). Staggered flowering in the Dipterocarpaceae: new insights into floral induction and the evolution of mast fruiting in the aseasonal tropics. *Amer. Nat.* 132, 44–66.

Ashton, P.S. & Hall, P. (1992). Comparison of structure among Mixed Dipterocarp forests of north-western Borneo. *J. Ecol.* 80, 459–81.

Ashton, P.S. Soepadmo, E. & Yap, S.K. (1977). Current research into the breeding systems of rain forest trees and its implications. Pp. 176–81 in *Trans. Intern. MAB IUFRO Worksh. Trop. Rainf. Ecosystem Res.*, Brünig, E.F. (Ed.), Reinbek, Hamburg.

Askew, G.P., Moffatt, D.J., Montgomery, R.F. & Searl, P.L. (1970a). Soil landscapes in north eastern Mato Grosso. *Geogr. J.* 136, 211–27.

Askew, G.P., Moffatt, D.J., Montgomery, R.F. & Searl, P.L. (1970b). Interrelationships of soils and vegetation in the savanna-forest boundary zone of north eastern Mato Grosso. *Geogr. J.* 136, 370–6.

Askew, G.P., Moffatt, D.J., Montgomery, R.F. & Searl, P.L. (1971). Soils and soil moisture as factors influencing the distribution of the vegetation of the Serra do Roncador, Mato Grosso. Pp. 150–60 in *III Simposio Sôbre O Cerrado*, Universidade de São Paulo.

Asprey, G.F. & Robbins, R.G. (1953). The vegetation of Jamaica. *Ecol. Monogr.* 23, 359–412.

Aubréville, A. (1933). La forêt de la Côte d'Ivoire. *Bull. Com. Afr. Occid. Franç.* 15, 205–61.

Aubréville, A. (1938). La forêt coloniale; les forêts de l'Afrique occidentale française. *Ann. Acad. Sci. Coloniale* 9.

Aubréville, A. (1947). Les brousses secondaires en Afrique équatoriale, Côte d'Ivoire, Caméroun. *Bois For. Trop.* 1, 24–49.

Aubréville, A. (1949). *Climats, Forêts et Désertification de l'Afrique Tropicale*. Soc. d'Edit. Géogr., Marit. Colon., Paris.

Aubréville, A. (1950–51). La concept d'association dans la forêt dense équatoriale de la Basse Côte d'Ivoire. *Mém. Soc. Bot. France* 98, 145–56.

Aubréville, A. (1961). Etude écologique des principales formations végétales du Brésil et contribution à la connaissance des forêts de l'Amazonie brésilienne. Centre Technique For. Tropical, Nogent-sur-Marne.

Aubréville, A. (1963). Classification des formes biologiques des plantes vasculaires en milieu tropical. *Adansonia* 3, 221–6.

Aubréville, A. (1965). Principes d'une systématique des formations végétales tropicales. *Adansonia*, Sér. 2, 5, 153–97.

Aubréville, A. (1969). Essais sur la distribution et l'histoire des angiospermes tropicales dans le monde. *Adansonia*, N.S. 9, 189–247.

Aubréville, A. (1970). Legumineuses, Cesalpinioidées. In *Flore du Caméroun*, Aubréville, A. & Leroy, J.F. (Eds), Mus. d'Hist. Nat., Paris.

Austin, M.P. (1977). Use of ordination and other multivariate methods to study succession. *Vegetatio* 35, 165–75.

Austin, M.P., Ashton, P.S. & Greig-Smith, P. (1972). The application of quantitative methods to vegetation survey III. A re-examination of rain forest data from Brunei. *J. Ecol.* 60, 305–24.

Austin, M.P. & Greig-Smith, P. (1968). The application of quantitative methods to vegetation survey II. Some methodological problems of data from rain forest. *J. Ecol.* 56, 827–44.

Aweto, A.O. (1981). Secondary succession and soil fertility restoration in south-western Nigeria. I. Succession. II. Soil fertility restoration. III. Soil and vegetation interrelationships. *J. Ecol.* 69, 601–7, 609–14, 957–63.

Axelrod, D.I. (1952). A theory of angiosperm evolution. *Evolution* 6, 29–59.

Axelrod, D.I. (1959). Poleward migration of the early angiosperms. *Science* 130, 203–7.

Axelrod, D.I. (1966). Origin of deciduous and evergreen habits in Temperate forests. *Evolution* 20, 1–15.

Axelrod, D.I. (1972). Ocean-floor spreading in relation to ecosystematic problems. Pp. 15–76 in *A Symposium on Ecosystematics*, Allen, R.T. & James, F.C. (Eds), *Univ. Arkansas Mus. Occ. Papers* 4.

Backer, C.A. (1909). De flora van het eiland Krakatau. *Jaarversl. Topogr. Dienst Ned.-Ind. 1909*, Batavia (Java).

Backer, C.A. (1929). *The Problem of Krakatao as seen by a Botanist*. Weltevreden (Java) and The Hague.

Bailey, I.W. & Sinnott, E.W. (1916). The climatic distribution of certain types of angiosperm leaves. *Amer. J. Bot.* 3, 24–39.

Baillie, I.C. (1970). *Report on the Detailed Soil Survey of the Experimental Afforestation Sites.* Report F3, Research Section, Forest Dept., Kuching, Sarawak.

Baillie, I.C. (1971). *Semi-detailed Survey of the Bakam Road Silvicultural Reserve, 4th Division.* Report F4, Research Section, Forest Dept., Kuching, Sarawak.

Baillie, I.C. (1972). *Further Studies on the Occurrence of Drought in Sarawak.* Soil Survey Report F7. Forest Dept., Kuching, Sarawak.

Baillie, I.C. (1975). Piping as an erosion process in the uplands of Sarawak. *J. Trop. Geogr.* 41, 9–15.

Baillie, I.C. (1976). Further studies on drought in Sarawak, East Malaysia. *J. Trop. Geogr.* 43, 20–9.

Baillie, I.C. (1989). Soil characteristics and classification in relation to the mineral nutrition of tropical wooded ecosystems. Pp. 15–26 in *Mineral Nutrients in Tropical Forest and Savanna Ecosystems* Proctor, J. (Ed.), Blackwell Scientific Publications, Oxford.

Baillie, I.C. & Ahmad, M.I. (1984). The variability of Red Yellow Podzolic soils under mixed dipterocarp forest in Sarawak. *Malay. J. Trop. Geogr.* 9, 1–13.

Baillie, I.C. & Ashton, P.S. (1983). Some aspects of the nutrient cycle of mixed Dipterocarp forest in Sarawak. Pp. 347–56 in *Tropical Rain Forest: Ecology and Management*, Sutton, S.L., Whitmore, T.C. & Chadwick, A.C. (Eds), Blackwell, Oxford.

Baillie, I.C., Ashton, P.S., Court, M.N., Anderson, J.A.R., Fitzpatrick, E.A. & Tinsley, J. (1987). Site characteristics and the distribution of tree species in mixed Dipterocarp forest on Tertiary sediments in central Sarawak. *J. Trop. Ecol.* 3, 201–20.

Baillie, I.C., Carr, J.P., Gibson, G.A. & Wright, A.C.S. (1991). Throughflow in fine textured soils in the coastal lowlands of Southern Belize. *Caribbean Geogr.* 3, 94–106.

Baillie, I.C. & Mamit, J.D. (1983). Observations on rooting in mixed dipterocarp forest, Central Sarawak. *Malay. For.* 46, 369–74.

Baillie, I.C. & Wright, A.C.S. (1988). *Soil and Citrus Suitability of Topco National Land, Toledo, Belize.* British Development Division (Caribbean), Bridgetown, Barbados.

Baker, H.G. (1965). The evolution of the cultivated kapok tree: a probable West African product. Pp. 185–215 in *Ecology and Economic Development in Tropical Africa*, Brokesha, D. (Ed.), Inst. for Internat. Studies, Univ. California, Berkeley, 9.

Baker, H.G. (1970). Evolution in the tropics. *Biotropica* 2, 101–11.

Baker, H.G. (1973). Evolutionary relationships between flowering plants and animals in American and African tropical forests. Pp. 145–59 in *Tropical Forest Ecosystems in Africa and South America: A Comparative Review*, Meggers, B.J., Ayensu, E.S. & Duckworth, W.D. (Eds), Smithsonian Inst. Press, Washington, D.C.

Baker, H.G. (1978). Chemical aspects of the pollination biology of woody plants in the tropics. Pp. 57–82 in *Tropical Trees as Living Systems*, Tomlinson, P.B. & Zimmermann, M.H. (Eds), Cambridge University Press.

Baker, H.G. & Baker I. (1981). Floral nectar constituents in relation to pollinator type. Pp. 119–41 in *Handbook of Experimental Pollination Biology*, Jones, C.E. & Little, R.J. (Eds), Van Nostrand-Reinhold, New York.

Baker, H.G., Bawa, K.S., Frankie, G.W. & Opler, P.A. (1983). Reproductive biology of plants in tropical forests. Chap. 12 in *Tropical Rain Forest Ecosystems, Structure and Function. Ecosystems of the World*, Golley, F.B. (Ed.), 14a. Elsevier, Amsterdam.

Baker, H.G. & Harris, B.J. (1957). The pollination of *Parkia* by bats and its attendant evolutionary problems. *Evolution* 11, 449–60.

Baker, J.R. & Baker, I. (1936). The seasons in a tropical rain-forest (New Hebrides). Part 2. Botany. *J. Linn. Soc. Lond. (Zool.)*, 39, 507–19.

Baker, N.R. & Hardwick, K. (1973). Biochemical and physiological aspects of leaf development in cocoa (*Theobroma cacao*). I Development of chlorophyll and photosynthetic activity. *New Phytol.* 72, 1315–24.

Bakshi, B.K. & Singh, S. (1970). Heart rot in trees. *Int. Rev. For. Res.* 3, 197–251.

Balée, W. & Campbell, D.G. (1990). Evidence for the successional status of liana forest (Xingu river basin, Amazonian Brazil). *Biotropica* 22, 36–47.

Baren, J. van (1931). Properties and constitution of a volcanic soil, built in 50 years in the East-Indian Archipelago. *Med. Landbouwhoogesch. Wageningen* 35 (Verh.6), 1–29.

Barkman, J.J. (1958). *Phytosociology and Ecology of Cryptogamic Epiphytes.* Assen, Netherlands.

Barry, R.G. & Chorley, R.J. (1982). *Atmosphere, Weather and Climate.* Fourth edition, Methuen, London.

Barton, A.M., Fetcher, N. & Redhead, S. (1989). The relationship between treefall gap size and light flux in a Neotropical rain forest in Costa Rica. *J. Trop. Ecol.* 5, 437–9.

Baseden, S.C. & Southern, P.J. (1959). Evidence of potassium deficiency in coconut palms on coral-derived soils in New Ireland from analysis of nut waters, husks, fronds and soils. *Papua New Guinea Agric. J.* 11, 101–15.

Baur, G.N. (1964). *The Ecological Basis of Rain Forest Management.* Ministry for Conservation N.S.W., Sydney.

Bawa, K.S. & Crisp, J.E. (1980). Wind-pollination in the understorey of a rain forest in Costa Rica. *J. Ecol.* 68, 871–6.

Bawa, K. & Hadley, M. (1990). *Reproductive Ecology of Tropical Forest Plants. (UNESCO Man and the Biosphere Series 7.)* Parthenon, Carnforth, Lancashire.

Bawa, K.S. & Opler, P.A. (1975). Dioecism in tropical trees. *Evolution* 29, 167–79.

Bayliss-Smith, T.P. (1982). *The Ecology of Agricultural Systems.* Cambridge University Press.

Bazzaz, F.A. (1984). Dynamics of wet tropical forests and their species strategies. Pp. 233–44 in *Physiological Ecology of Plants of the Wet Tropics*, Medina, E., Mooney, H.A. & Vasques-Yanes, C. (Eds), Junk, The Hague.

Beadle, N.C.W. (1981). *The Vegetation of Australia.* Cambridge University Press.

Beaman, R.S., Beaman, J.H., Marsh, C.W. & Wood, P.V. (1985). Drought and forest fires in Sabah in 1983. *Sabah Soc. J.* 8, 10–30.

Beard, J.S. (1944a). Climax vegetation in tropical America. *Ecology* 25, 127–58.

Beard, J.S. (1944b). The natural vegetation of the island of Tobago, British West Indies. *Ecol. Monogr.* 14, 135–63.

Beard, J.S. (1945). The progress of plant succession on the Soufriere of St Vincent. *J. Ecol.* 33, 173–92.

Beard, J.S. (1946a). The Mora forests of Trinidad, British West Indies. *J. Ecol.* 33, 173–92.

Beard, J.S. (1946b). The natural vegetation of Trinidad. *Oxford For. Mem.* 20.

Beard, J.S. (1949). The natural vegetation of the Windward and Leeward Islands. *Oxford For. Mem.* 21.

Beard, J.S. (1953). The savanna vegetation of northern tropical America. *Ecol. Monogr.* 23, 149–215.

Beard, J.S. (1955a). The classification of Tropical American vegetation types. *Ecology* 36, 89–100.

Beard, J.S. (1955b). Note on gallery forests. *Ecology* 36, 339–40.

Beccari, O. (1904). *Wanderings in the Great Forests of Borneo*, transl. E.H. Gigliolo, revised and ed. F.H.H. Guillemard. Constable, London. Repr. with introduction by the Earl of Cranbrook, Oxford University Press, 1987.

Becking, R. (1967). The ecology of the coastal redwood forest and the impact of the 1964 floods upon redwood vegetation. Final Report to National Science Foundation, Grant NSF GB 3468. (Unpubl.)

Beirnaert, A. (1941). La technique culturale sous l'équateur. I. Influence de la culture sur les réserves en humus et en azote des terres équatoriales. *Publ. Inst. Nat. Etude Agron. Congo Belge (I.N.E.A.C.)*, série technique 26.

Bell, T.I.W. (1969). *An Investigation into Some Aspects of Management in the Mora (Mora excelsa Benth.) Forests of Trinidad with Special Reference to the Matura Forest Reserve.* Ph.D. Thesis, University of the West Indies.

Bell, T.I.W. (1971). *Management of the Trinidad Mora Forests with Special Reference to the Matura Forest Reserve.* Government Printery, Trinidad.

Bellingham, P.J. (1991). Landforms influence patterns of hurricane damage: evidence from Jamaican montane forests. *Biotropica* 23, 427–33.

Belt, A. (1874). *The Naturalist in Nicaragua.* John Murray, London. Repr. Edward Bumpus, London 1888, Univ. of Chicago Press, Chicago & London 1985.

Bentley, B.L. (1979). Longevity of individual leaves in a tropical rainforest understorey. *Ann. Bot.* 43, 119–21.

Bentley, B.L. & Carpenter, E.J. (1984). Direct transfer of newly-fixed nitrogen from free-living epiphyllous micro-organisms to their host plant. *Oecologia* 63, 52–6.

Benzing, D.H. (1970). An investigation on two bromeliad myrmecophytes: *Tillandsia butzii* Mez, *T. caputmedusae* E. Morren and their ants. *Bull. Torrey Bot. Cl.* 97, 109–15.

Benzing, D.H. (1973). The Monocotyledons: their evolution and comparative biology, I. Mineral nutrition and related phenomena in Bromeliaceae and Orchidaceae. *Quart. Rev. Biol.* 48, 277–90.

Benzing, D.H. (1978). The life history profile of *Tillandsia circinnata* (Bromeliaceae) and the rarity of extreme epiphytism among the angiosperms. *Selbyana* 2, 325–37.

Benzing, D.H. (1990). *Vascular Epiphytes: General Biology and Related Biota.* Cambridge, University Press.

Benzing, D.H. & Davidson, E.H. (1979). Oligotrophic *Tillandsia circinnata* Schlecht. (Bromeliaceae): an assessment of its patterns of mineral allocation and reproduction. *Amer. J. Bot.* 66, 386–97.

Benzing, D.H. & Renfrow, A. (1971a). The significance of photosynthetic efficiency to habitat preference and phylogeny among tillandsoid Bromeliads. *Bot. Gaz.* 132, 19–30.

Benzing, D.H. & Renfrow, A. (1971b). Significance of the patterns of CO_2 exchange to the ecology and phylogeny of the Tillandsioideae (Bromeliaceae). *Bull. Torrey Bot. Cl.* 98, 322–7.

Benzing, D.H. & Renfrow, A. (1974). The nutritional status of *Encyclia tampense* and *Tillandsia circinnata* on *Taxodium ascendens* and the availability of nutrients to epiphytes on this host in South Florida. *Bull. Torrey Bot. Cl.* 101, 191–7.

Berger, W. (1931). Das Wasserleitungssystem von krautigen Pflanzen und Lianen im quantitativer Betrachtung. *Beih. Bot. Centralbl.* 48, 363–90.

Bernard, E. (1945). Le climat écologique de la Cuvette Centrale Congolaise. *Publ. Inst. Nat. Etude Agron. Congo Belge (I.N.E.A.C.)*, Brussels.

Berrie, G.K. & Eze, J.M.O. (1975). The relationship between an epiphyllous liverwort and host leaves. *Ann. Bot.* 39, 955–63.

Beveridge, A.E. (1953). The Menchali forest reserve. *Malay. For.* 16, 87–93.

Bews, J.W. (1927). Studies in the ecological evolution of the angiosperms. *New Phytol.* 26, 1–21, 65–84, 129–48, 209–48, 273–94.

Birch, H.F. & Friend, M.T. (1956). The organic matter and nitrogen status of East African soils. *J. Soil Sci.* 7, 156–67.

Black, G.A., Dobzhansky, T. & Pavan, C. (1950). Some attempts to estimate species diversity and population density of trees in Amazonian forests. *Bot. Gaz.* 111, 413–25.

Blackie, J.R. (1965). A comparison of methods of estimating evaporation in East Africa. Proc. Third Specialist Meeting on Applied Meteorology in E. Africa, EAAFRO, Muguga, Kenya.

Blanford, H.R. (1929). Regeneration of evergreen forests in Malaya. *Indian For.* 55, 333–9, 383–95.

Bleeker, P. (1983). *Soils of Papua New Guinea.* Australian National University Press, Canberra.

Bodley, J.H. & Benson, F.C. (1980). Stilt-root walking by an Iriarteoid palm in the Peruvian Amazon. *Biotropica* 12, 67–71.

Boerboom, J.H.A. (1975). Succession studies in the humid tropical lowlands of Surinam. *Proc. First Internat. Congr. of Ecol. (Structure, Functioning and Management of Ecosystems).* The Hague, Netherlands, Sept. 8–14, 1974, 1, 343–7. PUDOC, Wageningen.

Bollen, W., Chen, W., Lu, K. & Tarrant, R. (1967). Effect of stemflow precipitation on chemical and microbiological soil properties beneath a single alder tree. Pp. 149–56 in *Biology of Alder*, Trappe, J.M., Franklin, J.F., Tarrant, R. & Hansen, G. (Eds), Pacific Northwest Forest and Range Experiment Station, Portland.

Bonell, M., Cassells, D.S. & Gilmour, D.A. (1983). Vertical soil water movement in a tropical rainforest catchment in northeast Queensland. *Earth Surf. Proc. Landf.* 8, 253–73.

Bonell, M. & Gilmour, D.A. (1978). The development of overland flow in a tropical rainforest catchment. *J. Hydrol.* 39, 365–82.

Bonell, M. with Balek, J. (1993). Recent scientific developments and research needs in hydrological processes of the humid tropics. Pp 167–260 in *Hydrology and Water Management in the Humid Tropics*, Bonell, M., Hufschmidt, M.M. & Gladwell, J.S. (Eds), Cambridge University Press.

Bongers, F., Engelen, D. & Klinge, H. (1985). Phytomass structure of natural plant communities on spodosols in southern Venezuela: the Bana woodland. *Vegetatio* 63, 13–34.

Bookers (1981). *Put Put Land Resource Study.* Bookers Agriculture International, London.

Borchert, R. (1978). Feedback control and age-related changes of shoot growth in seasonal and nonseasonal climates. Pp. 497–513 in *Tropical Trees as Living Systems*, Tomlinson, P.B. & Zimmermann, M.H. (Eds), Cambridge University Press.

Borchert, R. (1983). Phenology and control of flowering in tropical trees. *Biotropica* 15, 81–9.

Borssum-Waalkes, J. van (1960). Botanical observations on the Krakatau Islands in 1951 and 1952. *Ann. Bogorienses* 4, 5–64.

Boucher, D.H. (1990). Growing back after hurricanes; catastrophes may be critical to rain forest dynamics. *Bioscience* 40, 163–6.

Boudot, J.-P., Hadj Brahim, A.B. & Chone, T. (1988). Dependence of carbon and nitrogen mineralisation rates upon amorphous metallic constituents and allophanes in highland soils. *Geoderma* 42, 245–60.

Boughey, A.S. (1957). Ecological studies of tropical coastlines. I. The Gold Coast. *J. Ecol.* 45, 665–87.

Bouillenne, R. (1930). Un voyage botanique dans le Bas-Amazone. Pp. 1–185 in *Une Mission Biologique Belge au Brésil (Août 1922 – Mai 1923), à la mémoire de Jean Massart* 2, Bouillenne, R., Ledoux, P., Brien, P. & Navez, A. (Eds), Imprimerie Méd. et Sci., Brussels.

Boulet, R., Brugière, J.M. & Humbel, F.X. (1979). Relations entre organisation des sols et dynamique de l'eau en Guyane Française: conséquences agronomiques d'une évolution déterminée par un déséquilibre d'origine principalement tectonique. *Science Sol.* 1, 3–17.

Bourlière, F. (1983). Species-richness in tropical forest vertebrates. Biology Internat. Special Issue No. 6 (Report I.U.B.S. Working Group on Species Diversity/Decade of the Tropics Programme 1983, Paris). Pp. 49–60.

Bourlière, F. & Hadley, M. (1983). Present-day savannas: an overview. Pp. 1–16 in *Tropical Savannas*, Bourlière, F. (Ed.), Elsevier, Amsterdam.

Bowen, G.D. (1980). Mycorrhizal roles in tropical plants and ecosystems. Pp. 165–90 in *Tropical Mycorrhizal Research*, Mikola, P. (Ed.), Oxford University Press, New York.

Boyer, Y. (1964). Contribution a l'étude de l'écophysiologie de deux fougères épiphytes, *Platycerium stemaria* (Beauv.) Desv. et *P. angolense* (Welch.). *Ann. Sci. Nat. (Bot.)*, sér. 5, 12, 87–228.

Braak, C. (1931). Klimakunde von Hinterindien und Insulinde. In *Handbuch der Klimatologie* 4, Part R, Köppen, W. & Geiger, R. (Eds), Gebrüder Bornträger, Berlin.

Brandt, J. (1988). The transformation of rainfall energy by a tropical rain forest canopy in relation to soil erosion. *J. Biogeogr.* 15, 41–8.

Brass, L.J. (1941). The 1938–39 expedition to the Snow Mountains, Netherlands New Guinea. *J. Arnold Arboretum* 22, 271–95, 297–342.

Brass, L.J. (1956–64). Results of the Archbold Expeditions. Summaries of the fourth, fifth and sixth expeditions to New Guinea. *Bull. Amer. Mus. Nat. Hist.* 111, 80–152 (1956); 118, 1–70 (1959); 127, 145–215 (1964).

Brassell, H.M. & Gilmour, D. (1980). The cation composition of precipitation at four sites in far north Queensland. *Australian J. Ecol.* 5, 397–405.

Braun, E.L. (1964). *Deciduous Forests of Eastern North America*. (Facsimile of Ed. of 1950.) Hafner, New York.

Bray, J.R. & Curtis, J.T. (1957). An ordination of the upland forest communities of southern Wisconsin. *Ecol. Monogr.* 27, 325–49.

Brenan, J.P.M. (1978). Some aspects of the phytogeography of tropical Africa. *Ann. Miss. Bot. Gard.* 65, 437–78.

Brenner, W. (1902). Klima und Blatt bei der Gattung *Quercus*. *Flora* 90, 114–60.

Brinkmann, W.L.F. (1983). Nutrient balance of a central Amazonian rainforest: comparison of natural and man-managed systems. *Assoc. Int. Hydrol. Sci. Publ.* 140, 153–63.

Brockmann-Jerosch, H. (1913). Der Einfluss des Klimacharakters auf die Verbreitung der Pflanzen und Pflanzengesellschaften. *Bot. Jahrb. (Beibl.)* 49, 19–43.

Brockmann-Jerosch, H. (1919). Baumgrenze und Klimacharakter. *Beitr.z. Geobot. Landesaufnahme der Schweiz. Herausgegeben von der Pflanzengeogr. Kommission der Schweizer Naturforsch. Gesellschaft* 48.

Brokaw, N.V.L. (1983). Treefalls: frequency, timing and consequences. Pp. 101–8 in *The Ecology of a Tropical Forest*, Leigh, C.H., Rand, A.S. & Windsor, D.M. (Eds), Oxford University Press (for Smithsonian Inst., Washington, D.C.).

Brokaw, N.V.L. (1987). Gap-phase regeneration of three pioneer tree species in a tropical forest. *J. Ecol.* 75, 9–19.

Brokaw, N.V.L. & Walker, L.R. (1991). Summary of the effects of Caribbean hurricanes on vegetation. *Biotropica* 23, 442–7.

Brown, K.S. Jr. (1987). Soils. Pp. 19–28 in *Biogeography and Quaternary History in Tropical America*, Whitmore, T.C. & Prance, G.T. (Eds), Clarendon Press, Oxford.

Brown, N.D. (1990). *Dipterocarp Regeneration in Tropical Rain Forest Gaps of Different Sizes*. D.Phil. Thesis, Oxford University.

Brown, N.D. & Whitmore, T.C. (1992). Do dipterocarp seedlings really partition tropical rain forest gaps? *Phil. Trans. R. Soc. Lond.* B 335, 369–78.

Brown, S. & Lugo, A.E. (1990). Tropical secondary forests. *J. Trop. Ecol.* 6, 1–32.

Brown, W.H. (1919). *Vegetation of the Philippine Mountains*. Bureau of Science (Publ. 13), Manila.

Brown, W.H. & Matthews, D.M. (1914). Philippine Dipterocarp forests. *Philipp. J. Sci.* 9 (sect.A), 413–561.

Bruijnzeel, L.A. (1983). Hydrological and biogeochemical aspects of man-made forests in south-central Java, Indonesia. Seraya Valley Project Final Report, 9, Amsterdam (Ph.D. Thesis, Inst. of Earth Sciences, Free University, Amsterdam).

Bruijnzeel, L.A. (1989). Nutrient cycling in moist tropical forests: the hydrological framework. Pp. 383–415 in *Mineral Nutrients in Tropical Forest and Savanna Ecosystems*, Proctor, J. (Ed.), Br. Ecol. Soc. Spec. Pubn. 9, Blackwell Scientific, Oxford.

Bruijnzeel, L.A. (1990). *Hydrology of Moist Tropical Forests and Effects of Conversion: a State of Knowledge Review*. UNESCO, Paris.

Bruijnzeel, L.A., Waterloo, M.J., Proctor, J., Kuiters, A.T. & Kotterlink, B. (1993). Hydrological observations in montane rain forests on Gunung Silam, Sabah, Malaysia, with special reference to the 'Massenerhebung' effect. *J. Ecol.* 81, 145–67.

Brünig, E.F. (1961). An introduction to the vegetation of Bako National Park. Pp. 13–35 in *Report of the Trustees for National Parks, 1959–1960*, Kuching, Sarawak.

Brünig, E.F. (1964). A study of damage attributed to lightning in two areas of *Shorea albida* forest in Sarawak. *Emp. For. Rev.* 43, 134–44.

Brünig, E.F. (1965). Guide and introduction to the vegetation of the kerangas forests and the padangs of the Bako National Park. Pp. 289–318 in *Proc. Symp. on Humid Tropics Vegetation*. UNESCO. (Science Cooperation Office for S.E. Asia).

Brünig, E.F. (1968). Der Heidewald von Sarawak und Brunei. *Mitt. der Bundesforschungsanstalt Forst- u. Holzwirtsch. Reinbek* (Hamburg) 68, Band 1 & 2.

Brünig, E.F. (1969). On the seasonality of drought in the lowlands of Sarawak (Borneo). *Erdkunde* 2, 127–33.

Brünig, E.F. (1970). Stand structure, physiognomy and environmental factors in some lowland forests in Sarawak. *Tropical Ecology* 11, 26–43.

Brünig, E.F. (1971). On the ecological significance of drought in the equatorial wet evergreen (rain) forest of Sarawak (Borneo). Pp. 66–9 in *The Water Relations of Malesian Forests*, Flenley, J.R. (Ed.), Univ. Hull, Dept. Geogr., Misc. Ser. 11.

Brünig, E.F. (1973a). Some further evidence on the amount of damage attributed to lightning and wind-throw in *Shorea albida* Forest in Sarawak. *Commonw. For. Rev.* 52, 260–5.

Brünig, E.F. (1973b). Species richness and stand diversity in relation to size and succession of forests in Sarawak and Brunei (Borneo). *Amazoniana* 4, 293–320.

Brünig, E.F. (1974). *Ecological Studies in the Kerangas Forests of Sarawak and Brunei*. Borneo Literature Bureau for Sarawak Forest Dept., Kuching, Sarawak.

Brünig, E.F., Alder, D. & Smith, J.P. (1979). The International MAB Amazon Rainforest Pilot Project at San Carlos de Rio Negro: vegetation and structure. Trans. Second Internat. MAB-IUFRO Workshop of Tropical Rainforest Ecosystem Research. Chair of World Forestry Hamburg-Reinbek, Special Report No. 2.

Brünig, E.F. & Huang, Y.-W. (1989). Patterns of tree species diversity and canopy structure in humid tropical evergreen forest in Borneo and China. Pp. 77–88 in *Tropical Forests: Botanical Dynamics, Speciation and Diversity*, Holm-Nielsen, L.B., Nielsen, I.C. & Balslev, H. (Eds), Academic Press, London.

Buchmann, S.L. & Buchmann, M.D. (1981). Autecology of *Mouriri myrtilloides* (Melastomataceae: Memcycleae), an oil flower in Panama. *Biotropica* 13 (suppl.), 7–24.

Budowski, G. (1961). *Studies on Forest Succession in Costa Rica and Panama*. Ph.D. thesis, Yale University, New Haven, Connecticut.

Budowski, G. (1965). Distribution of tropical American rain forest species in the light of successional processes. *Turrialba* 15, 40–2.

Budowski, G. (1970). The distinction between old secondary and climax species in tropical Central American lowland forests. *Tropical Ecology* 11, 44–8.

Bullock, S.H. (1980). Demography of an undergrowth palm in littoral Cameroon. *Biotropica* 12, 247–55.

Bullock, S.H. & Bawa, K.S. (1981). Sexual dimorphism and the annual flowering pattern in *Jacartia dolichaula* (D. Smith) Woodson (Caricaceae) in a Costa Rican rain forest. *Ecology* 62, 1494–1504.

Bultot, F. & Griffiths, J.F. (1972). The equatorial wet zone. Pp. 259–311 in *Climates of Africa*, Griffiths, J.F. (Ed.), Elsevier, Amsterdam.

Bünning, E. (1947). *In den Wäldern Nord-Sumatras*. F. Dümmler, Bonn.

Buol, S.W. (1985). Mineralogy classes in soil families with low activity clays. Pp. 169–78 in *Mineral Classification of Soils*, Kittrick, J.A. (Ed.), Special Pubn. 16, Soil Sci. Soc. Amer. & Amer. Soc. Agron., Madison, Wisconsin.

Burgess, P.F. (1971). Effect of logging on hill dipterocarp forest. *Malay. Nat. J.* 24, 231–7.

Burgess, P.F. (1972). Studies on the regeneration of the hill forests of the Malay Peninsula – The phenology of dipterocarps. *Malay. For.* 35, 103–23.

Burgess, P.F. (1975). Silviculture in the hill forests of the Malay Peninsula. *Malay. For. Dept. Res. Pamph.* 66, Kepong, Malaysia.

Burkill, I.H. (1919). The composition of a piece of well-drained secondary jungle thirty years old. *Gdns Bull. Singapore* 2, 145–57.

Burkill, I.H. (1928). The main features of the vegetation of Pahang. *Malay. Nat. J.* 2, 11–21.

Burkill, I.H. (repr. 1966). *A Dictionary of the Economic Products of the Malay Peninsula*. 2 vols. Ministry of Agriculture and Co-operatives, Kuala Lumpur.

Burnham, C.P. (1989). Pedological processes in temperate and tropical soils. Pp. 27–41 in *Mineral Nutrients in Tropical Forest and Savanna Ecosystems*, Proctor, J. (Ed.), Br. Ecol. Soc. Spec. Pubn 9, Blackwell Scientific, Oxford.

Burtt, B.L. (1977). Notes on rain-forest herbs. *Gdns Bull. Singapore* 29, 73–80.

Burtt Davy, J. (1938). *The Classification of Tropical Woody Vegetation Types*. Imp. For. Inst. Oxford, Paper 13.

Buschbacher, R.J. (1987). Cattle productivity and nutrient fluxes on an Amazon pasture. *Biotropica* 19, 200–7.

Buschbacher, R., Uhl, C. & Serrão, E.A.S. (1986). Pasture management and environmental effects near Paragominas, Pará. Pp. 90–9 in *Amazonian Rain Forests: Ecosystem Disturbance and Recovery*, Jordan, C.F. (Ed.), Springer, New York.

Buschbacher, R., Uhl, C. & Serrão, E.A.S. (1988). Abandoned pastures in eastern Amazonia. II. Nutrient stocks in the soil and vegetation. *J. Ecol.* 76, 682–99.

Büsgen, M. (1903). Einige Wachstumsbeobachtungen aus den Tropen. *Ber. dtsch. bot. Ges.* 21, 435–40.

Büsgen, M. & Münch, E. (1929). *The Structure and Life of Forest Trees*. Transl. T. Thomson. Chapman & Hall, London.

Bush, M.B., Piperno, D.R. & Colinvaux, P.A. (1989). A 6,000 year history of Amazonian maize cultivation. *Nature* 340, 303–5.

Bush, M.B., Piperno, D.R., Colinvaux, P.A., De Oliveira, P.E., Krissek, L.A., Miller, M.C. & Rowe, W.L. (1992). A 14,300-yr paleoecological profile of a lowland tropical lake in Panama. *Ecol. Monogr.* 62, 251–75.

Buurman, P. (1978). Red soils in Indonesia, a state of knowledge report. Pp. 1–12 in *Proc. First National Soil Classification Workshop*, Buurman, P. (Ed.), Soils Research Inst., Bogor, Indonesia.

Buxbaum, F. (1969). Die Entwicklungs wege der Kakteen in Südamerika. Pp. 583–623 in *Biogeography and Ecology in South America*, Fittkau, E.J., Illies, J., Klinge, H., Schwabe, G.H. & Sioli, H. (Eds), W. Junk, The Hague.

Cachan, P. (1963). Signification écologiques des variations microclimatiques verticales dans la forêt sempervirente de Basse Côte d'Ivoire. *Ann. Fac. Sci. Univ. Dakar* 8, 89–155.

Cachan, P. & Duval, J. (1963). Variations microclimatiques verticales et saisonières dans la forêt sempervirente de Basse Côte d'Ivoire. *Ann. Fac. Sci. Univ. Dakar* 8, 5–87.

Cain, S.A., Castro, G.M. de O., Pires, J.M. & Silva, N.T. da (1956). Application of some phytosociological techniques to Brazilian rain forest. *Amer. J. Bot.* 43, 911–41.

Cain, S.A., & Castro, G.M. de O. (1959). *Manual of Vegetation Analysis*. Harper & Row, New York.

Calder, I.R., Wright, I.L. & Murdiyarso, D. (1986). A study of evaporation from tropical rain forest – West Java. *J. Hydrol.* 89, 13–31.

Campbell, D.G., Daly, D.C., Prance, G.T. & Maciel, U.N. (1986). Quantitative ecological inventory of terra firme and várzea tropical forest on the Rio Xingú, Brazilian Amazon. *Brittonia (New York)* 38, 369–93.

Campbell, D.G., Stone, J.L. & Rosas, A.Jr. (1992) A comparison of the phytosociology and dynamics of three floodplain (Várzea) forests of known ages, Rio Juruá, western Brazilian Amazon. *J. Linn. Soc. Lond. (Bot.)* 108, 213–37.

Capon, M. (1947). Observations sur la phénologie des éssences de la forêt de Yangambi. Pp. 849–62 in Comptes Rendus de la Semaine Agricole de Yangambi 1947, 2. *Publ. d' Inst. Nat. pour l'Etude agron. du Congo Belge* (Hors sér.).

Carter, G.S. (1934). Reports of the Cambridge Expedition to British Guiana, 1933. The fresh waters of the rain-forest areas of British Guiana. *J. Linn. Soc. (Zool.)* 39, 147–86.

Castro Soares, L. de (1953). Limites meridionais e orientais da área de ocorrência da floresta amazônica em território brasileiro. *Rev. Bras. Geogr.* 15, 1, 1–122.

C.C.T.A. (1956). C.S.A. specialist meeting on phytogeography, Yangambi (28th July-8th August, 1956). *Sci. Council for Africa South of the Sahara (C.S.A.)*, Pubn. 22.

César, J. & Menaut, J.C. (1974). Analyse d'un écosysteme tropicale humide: la savane de Lamto (Côte d'Ivoire). II. Peuplement végétale. *Bull. Liaison Chercheurs de Lamto, Numéro Spéc.* 2. Station d'écol. trop. de Lamto (Côte d'Ivoire).

Chalk, L. & Akpala, J.D. (1963). Possible relation between the anatomy of wood and buttressing. *Commonw. For. Rev. Lond.* 42, 53–8.

Champion, H.G. & Seth, S.K. (1968a). *A Revised Survey of the Forest types of India*. Government of India Press Nasik, Delhi.

Champion, H.G. & Seth, S.K. (1968b). *General Sylviculture for India*. Government of India Press Nasik, Delhi.

Chan, H.T. (1980). Reproductive biology of some Malaysian dipterocarps. Pp. 169–75 in *Tropical Ecology and Development*, Furtado, J.I. (Ed.), International Society of Tropical Ecology, Kuala Lumpur.

Chandler, G. (1985). Mineralisation and nitrification in three Malaysian forest soils. *Soil Biol. Biochem.* 17, 347–53.

Chang, C.P. (1970). Westward propagating cloud patterns in the tropical Pacific as seen from time-composite satellite photographs. *J. Atmos. Sci.* 27, 133.

Chapman, J.D. & White, F. (1970). *The evergreen forests of Malawi*. Commonw. For. Inst., Oxford.

Chapman, V.J. (1976). *Mangrove Vegetation*. J. Cramer, Vaduz.

Chapman, V.J. (Ed.) (1977). *Wet Coastal Ecosystems. (Ecosystems of the world, I.)* Elsevier, Amsterdam.

Charter, C.F. (1941). *A Reconnaissance Survey of the Soils of British Honduras*. Govt. of British Honduras, Trinidad.

Charter, J.R. & Keay, R.W.J. (1960). Assessment of the Olokemeji fire-control experiment (Investigation 254) 28 years after institution. *Niger. For. Inf. Bull.*, N.S. 3, 11–32.

Chatterjee, K. (1989). Surface wash in rainforest of Singapore. *Singapore J. Trop. Geogr.* 10, 95–109.

Chazdon, R.L. & Fetcher, N. (1984). Photosynthetic light environments in a lowland tropical rain forest in Costa Rica. *J. Ecol.* 72, 553–64.

Chevalier, A. (1900). Les zones et les provinces botaniques de l'Afrique Occidentale française. *C.R. Acad. Sci., Paris* 130, 458–61.

Chevalier, A. (1909). L'extension et la régression de la forêt vierge de l'Afrique tropicale. *C.R. Acad. Sci., Paris* 149, 458–61.

Chiariello, N.R., Field, C.B. & Mooney, M.A. (1987). Midday wilting in a tropical pioneer tree. *Functional Ecol.* 1, 3–4.

Chin, S.C. (1977). The limestone hill flora of Malaya. Part 1. *Gdns. Bull. Singapore* 30, 165–219.

Chipp, T.F. (1922). Buttresses as an aid to identification. *Kew Bull.* (1922), 265–8.

Chipp, T.F. (1927). The Gold Coast forest. A study in synecology. *Oxford For. Mem.* 7.

Chujo, H. (1982). Ecological observations of the tropical rain forest in Papua New Guinea with particular reference to the *Eucalyptus deglupta* forest with a single dominant species. *Bull. Biol. Sci. Hiroshima Univ.* 48, 2–6.

Church, A.H. (1919). Thalassiophyta and the subaerial transmigration. *Oxford Bot. Mem.* 3.

Cintron, G. (1970). Variation in size and frequency of stomata with altitude in the Luquillo Mountains. Pp. H133-5 in *A Tropical Rain Forest*, Odum, H.T. & Pigeon, R.F. (Eds), Divn. Tech. Inf., U.S. Atomic Energy Commission, Oak Ridge, Tennessee.

Clark, D.B. & Clark, D.A. (1990). Distribution and effects on tree growth of lianes and woody hemiepiphytes in a Costa Rican tropical wet forest. *J. Trop. Ecol.* 6, 321–31.

Clayton, H.H. (1927). World Weather Records. *Smithsonian Misc. Collns.* 79.

Clayton, H.H. (1944). World Weather Records. *Smithsonian Misc. Collns.* 90.

Clayton, H.H. & Clayton, F.L. (1947). World Weather Records. *Smithsonian Misc. Collns.* 105.

Clements, R.G. & Colon, J.A. (1975). The rainfall interception process and mineral cycling in a montane rainforest in Puerto Rico. Pp. 813–23 in *Mineral Cycling in Southeastern Ecosystems*, Howell, H.G., Gentry, J.B. & Smith, M.H. (Eds), United States Energy Research and Development Administration, Oak Ridge, Tennessee.

Coley, P.D. (1983). Herbivory and defensive characteristics of tree species in a lowland tropical forest. *Ecol. Monogr.* 53, 209–33.

Collins, N.M. (1983). Termite populations and their role in litter removal in Malaysian rain forests. Pp. 311–25 in *Tropical Rain Forest: Ecology and Management*, Sutton, S.L., Whitmore, T.C. & Chadwick, A.C. (Eds), Br. Ecol. Soc. Spec. Pubn. 2, Blackwell, Oxford.

Collins, N.M., Harcourt, C.S. & Sayer, J.A. (1992). *The Conservation Atlas of Tropical Forests: Africa*. Macmillan, Basingstoke, UK.

Collins, N.M., Sayer, J.A. & Whitmore, T.C. (1991). *The Conservation Atlas of Tropical Forests: Asia and the Pacific*. Macmillan, Basingstoke, U.K.

Collinson, M.E. (1983). *Fossil Plants of the London Clay*. Palaeontological Society, London.

Collinson, M.E. & Hooker, J.J. (1987). Vegetational and mammalian faunal changes in the Early Tertiary of southern England. Pp. 259–304 in *The Origins of Angiosperms and their Biological Consequences*, Friis, E.M., Chaloner, W.G. & Crane, P.R. (Eds), Cambridge University Press.

Conklin, H.C. (1957). Hanunóo agriculture. *FAO Forestry Development Paper* 12. Rome (Repr. 1988, Elliots Books, Northford, Conn. U.S.A.).

Conklin, H.C. (1961). The study of shifting cultivation. *Current Anthropology* 2, 27–61.

Conklin, H.C. (1963). The study of shifting cultivation. *Panamerican Union Studies & Monogr.* 6.

Connell, J.H. & Lowman, M.D. (1989). Low-diversity tropical rain forests: some possible mechanisms for their existence. *Amer. Nat.* 134, 88–109.

Connor, E.F. (1986). The role of Pleistocene forest refugia in the evolution and biogeography of tropical biota. *Trends in Ecol. Evolution* 1, 165–8.

Conway, D. & Alexander, I. (1992). Soil conditions under monodominant *Gilbertiodendron dewevrei* and mixed forest in Ituri Forest Reserve, Zaïre. *Tropical Biology Newsletter*, Aberdeen Univ., no. 62, June 1992.

Coombe, D.E. (1960). An analysis of the growth of *Trema guineensis*. *J. Ecol.* 48, 219–31.

Coombe, D.E. & Hadfield, W. (1962). An analysis of the growth of *Musanga cecropioides*. *J. Ecol.* 50, 221–34.

Cooper, A. (1979). Muri and white sand savannah in Guyana and French Guiana. Pp. 471–81 in *Ecosystems of the World 19 A*, Specht, R.L. (Ed.), Elsevier, Amsterdam.

Copeland, E.B. (1907). Comparative ecology of the San Ramon Polypodiaceae. *Philipp. J. Sci.* (sect. C) 2, 1–76.

Corlett, R.T. (1990). Flora and reproductive phenology of the rain forest at Bukit Timah, Singapore. *J. Trop. Ecol.* 6, 55–63.

Corner, E.J.H. (1933). A revision of the Malayan species of *Ficus*: Covellia and Neomorphe. *J. Malayan Branch Roy. Asiatic Soc.* 11, 1–65.

Corner, E.J.H. (1935). The seasonal fruiting of agarics in Malaya. *Gdns Bull. Singapore* 9, 79–88.

Corner, E.J.H. (1938). The systematic value of the colour of withering leaves. *Chronica Bot.* 4, 119–21.

Corner, E.J.H. (1939). A revision of *Ficus*, subgenus Synoecia. *Gdns Bull. Singapore* 10, 82–161.

Corner, E.J.H. (1949). The Durian Theory or the origin of the modern tree. *Ann. Bot.*, N.S. 13, 367–414.

Corner, E.J.H. (1953). The Durian Theory extended – I. *Phytomorphology* 3, 465–76.

Corner, E.J.H. (1954a). The Durian Theory extended – II. The arillate fruit and the compound leaf. *Phytomorphology* 4, 152–65.

Corner, E.J.H. (1954b). The Durian Theory extended – III. Pachycauly and Megaspermy – Conclusion. *Phytomorphology* 4, 263–74.

Corner, E.J.H. (1964). *The Life of Plants*. Weidenfeld & Nicholson, London.

Corner, E.J.H. (1966). *The Natural History of Palms*. Weidenfeld and Nicholson, London.

Corner, E.J.H. (1978). The freshwater swamp-forest of south Johore and Singapore. *Gdns Bull. Singapore* Suppl. 1.

Corner, E.J.H. (1988). *Wayside Trees of Malaya*. Third edition, 2 vols. Malay. Nature Soc., Kuala Lumpur, Malaysia.

Corner, E.J.H. (1990). On *Trigobalanus* (Fagaceae). *Bot. J. Linn. Soc. Lond.* 102, 219–33.

Coster, C. (1923). Lauberneuerung und andere periodische Lebensprozesse in dem trockenem Monsun-gebiet Ost-Javas. *Ann. Jard. Bot. Buitenz.* 33, 117–89.

Coster, C. (1926a). Periodische Blüteerscheinungen in den Tropen. *Ann. Jard. Bot. Buitenz.* 35, 125–62.

Coster, C. (1926b). Die Buche auf dem Gipfel des Pangerango. *Ann. Jard. Bot. Buitenz.* 35, 105–19.

Coster, C. (1927–28). Zur Anatomie und Physiologie der Zuwachszonen und Jahresringbildung in den Tropen. *Ann. Jard. Bot. Buitenz.* 37, 49–160; 38, 1–114.

Coster, C. (1932a). Wortelstudiën in de Tropen. I. De jeugdontwikkeling van het wortelstelsel van een zeventigtal boomen en groenbemesters. *Tectona* 25, 828–72.

Coster, C. (1932b). Die Geschwindigkeit des Transpirationsstromes. *Planta (Berl.)* 15, 540–66.

Coster, C. (1933). Wortelstudiën in de Tropen. III. De zuurstof behoefte van het wortelstelsel. *Tectona* 26, 450–97.

Cottam, G. & Curtis, J.T. (1956). The use of distance measures in phytosociological sampling. *Ecology* 37, 451–60.

Cousens, J.E. (1951). Some notes on the composition of lowland tropical rain forest in Rengam Forest Reserve, Johore. *Malay. For.* 14, 131–9.

Cousens, J.E. (1965). Some reflections on the nature of Malayan lowland rain forest. *Malay. For.* 28, 122–8.

Coutinho, L.M. (1964). Untersuchungen über die Lage des Lichtkompensationspunkts zu verschiedenen Tageszeiten mit besonderer Berücksichtigung des 'de-Saussure-Effektes' bei Sukkulenten. *Beitr. Phytologie (Walter Festschr.)*, 101–8.

Cowling, R.M., Holmes, P.M. & Rebelo, A.G. (1992). Plant diversity and endemism. Pp. 62–112 in *The Ecology of Fynbos*, R. Cowling (Ed.), Oxford University Press.

Craighead, F.C. (1971). *The Trees of South Florida*, 1. *The Natural Environments and Their Succession*. Univ. Miami Press, Coral Gables, Florida.

Crane, P.R. (1987). Vegetational consequences of angiosperm diversification. Pp. 107–44 in *The Origin of Angiosperms and their Biological Consequences*, Friis, E.M., Chaloner, W.G. & Crane, P.R. (Eds), Cambridge University Press.

Crawford, R.M.M. & Wishart, D. (1967). A rapid multivariate method for the detection and classification of groups of ecologically related species. *J. Ecol.* 55, 505–24.

Cremers, G. (1973). Architecture de quelques lianes d'Afrique Tropicale, 1. *Candollea* 28, 249–80.

Cremers, G. (1974). Architecture de quelques lianes d'Afrique Tropicale, 2. *Candollea* 29, 57–110.

Croat, T.B. (1969). Seasonal flowering behavior in Central Panama. *Ann. Missouri Bot. Gdn.* 56, 295–307.

Croat, T.B. (1975). Phenological behavior on Barro Colorado Island. *Biotropica* 7, 270–7.

Croat, T.B. (1978). *Flora of Barro Colorado Island*. Stanford Univ. Press, Stanford, California.

Croat, T.B. (1979). The sexuality of the Barro Colorado Island flora (Panama). *Phytologia* 42, 319–48.

Cromer, D.A.N. & Pryor, L.D. (1942). A contribution to rain forest ecology. *Proc. Linn. Soc. N.S.W.* 67, 249–68.

Crow, T.R. (1980). A rain forest chronicle: a 30 year record of change in structure and composition at El Verde, Puerto Rico. *Biotropica* 12, 42–55.

Crowther, J. (1986). Karst environments and ecosystems in Peninsular Malaysia. *Malay. Nat. J.* 39, 231–57.

Crowther, J. (1987). Ecological observations in tropical karst terrain, West Malaysia. II. Rainfall interception, litterfall and nutrient cycling. *J. Biogeogr.* 14, 145–55.

Crozier, C.R. & Boerner, R.E.J. (1984). Correlations of understory herb distribution patterns with microhabitats under different tree species in a mixed mesophytic forest. *Oecologia* 62, 337–43.

Crutcher, H.L. & Quayle, R.G. (1974). *Mariners' Worldwide Climatic Guide to Tropical Storms at Sea*. U.S. Dept. Commerce, NOAA/EDS, National Climate Center, Asheville, Tennessee.

Crutzen, P.J. (1987). Role of the tropics in atmospheric chemistry. Pp. 107–30 in *The Geophysiology of Amazonia: Vegetation and Climate Interactions*, Dickinson, R.E. (Ed.), Wiley, Chichester.

Cuatrecasas, J. (1958). Aspectos de la vegetación natural de Colombia. *Rev. Acad. Colomb. Ciencias Nat.* 10, No. 40.

Cuatrecasas, J. (1964). Cacao and its allies: a taxonomic revision of the genus *Theobroma*. *Contrib. U.S. Nat. Herb.* 35, 379–614.

Cuevas, E. & Medina, E. (1988). Nutrient dynamics within Amazonian forests. II. Fine root growth, nutrient availability and leaf litter decomposition. *Oecologia* 76, 222–35.

Curi, N. & Franzmeier, D.P. (1987). Effects of parent rocks on chemical and mineralogical properties of some oxisols in Brazil. *Soil Sci. Soc. Amer. J.* 51, 153–8.

Cypert, E. (1972). Plant succession on burned areas in Okefenokee Swamp following the fires of 1954 and 1955. *Proc. Ann. Tall Timbers Fire Ecology Conference* 12, 199–217.

Czapek, F. (1909). Ueber die Blattentfaltung der Amherstien. *S.B. Akad. Wiss. Wien* 118, 201–30.

Dabin, F. (1957). Note sur le fonctionnement des parcelles experimentales pour l'étude de l'érosion à la station d'Adiopodoumé (Côte d'Ivoire). *Décret. Permanent Bureau Sols AOF, Dakar* (mimeographed).

Dabral, B.G., Pant, S.P. & Pharasi, S.C. (1984). Microsite characteristics vis-a-vis rooting behaviour in sal (*Shorea robusta*). *Indian For.* 110, 997–1013.

Dale, W.L. (1959). The rainfall of Malaya. *J. Trop. Geogr.* 13, 23–37.

Daly, D.C. & Prance, G.T. (1988). Brazilian Amazon. Pp. 401–26 in *Floristic Inventory of Tropical Countries*, Campbell, D.G. & Hammond, D. (Eds), N.Y. Bot. Gdn, New York.

Damuth, J.E. & Fairbridge, R.W. (1970). Equatorial Atlantic deep-sea arkosic sands and ice-age aridity in tropical South America. *Geol. Soc. Amer. Bull.* 81, 189–206.

Dansereau, P. & Lems, K. (1957). The grading of dispersal types in plant communities and their ecological significance. *Contrib. Inst. Bot. Univ. Montreal* 71.

Daubenmire, R. (1972). Phenology and other characteristics of tropical semi-deciduous forest in north-western Costa Rica. *J. Ecol.* 60, 147–70.

Davies, J.A. & Robinson, P.J. (1969). A simple energy balance approach to the moisture balance climatology of Africa. Pp. 23–56 in *Environment and Land Use in Africa*, Thomas, H.F. & Whittington, G.W. (Eds), Methuen, London.

Davis, J.H. (1940). The ecology and geologic role of mangroves in Florida. (Papers from Tortugas Lab. no. 22). *Pubn. Carnegie Inst.* 517, 303–412.

Davis, J.H. (1943). The natural features of southern Florida, especially the vegetation of the Everglades. *State of Florida, Dep. of Conservation, Geol. Bull.* 25.

Davis, T.A.W. (1929). Some observations on the forest of the North West District. *Agric. J. Brit. Guiana* 2, 157–66.

Davis, T.A.W. (1941). On the island origin of the endemic trees of the British Guiana peneplain. *J. Ecol.* 29, 1–13.

Davis, T.A.W. & Richards, P.W. (1933–34). The vegetation of Moraballi Creek, British Guiana: an ecological study of a limited area of Tropical Rain Forest. Parts I and II. *J. Ecol.* 21, 350–84; 22, 106–55.

Dawkins, H.C. (1963). Crown diameters, their relation to bole diameter in tropical forest trees. *Commonw. For. Rev.* 42, 318–37.

Dawkins, H.C. (1965). The time dimension of tropical forest trees. *J. Ecol.* 53, 837–8.

Dawkins, H.C. (1966). Dimensional behaviour of tropical high forest trees as related to stratification. *J. Ecol.* 54, 281–2.

Dawson, C. (1989). Nigeria Expedition 1988. Tree enumeration project. (Typescript.) Dept. Biol., Univ. College, London.

Dawson, J.W. (1980). Middle-latitude rainforests in the Southern Hemisphere. *Biotropica* 12, 159–60.

Dawson, J.W. & Sneddon, B.V. (1969). The New Zealand rain forest: a comparison with tropical rain forest. *Pacific Science* 23, 131–47.

Delamare-Deboutteville, C. (1951). Les dépendances du sol et les sols suspendus. *Ann. Biol.* 27, 107–118.

Denny, C.S. & Goodlett, J.C. (1968). Tree-thrown origin of patterned ground on beaches of the ancient Champlain sea, near Plattsburgh, New York. Professional Paper 600B, U.S. Geol. Survey, Washington, D.C.

Den Outer, R.W. (1972). Tentative determination key to 600 trees, shrubs and climbers from the Ivory Coast, Africa, mainly based on characters of the living bark, besides the rhytidome and the leaf, ((i) Large trees, (ii) Small trees). *Meded. Landbouwhoogesch. Wageningen* 72, 18 & 19.

Denslow, J.S. (1980). Gap partitioning among tropical rain forest trees. *Biotropica* 12 (Suppl.), 47–55.

Denslow, J.S. (1987). Tropical rain forest gaps and tree species diversity. *Ann. Rev. Ecol. Syst.* 18, 431–51.

Denslow, J.S. & Padoch, C. (Eds) (1988). *People of the tropical rain forest.* University of California Press, Berkeley.

Deuse, P. (1960). Etude écologique et phytosociologique de la végétation des ''Esobe'' de la région est du lac Tumba (Congo Belge). *Acad. Roy. des Sci. d'Outre-Mer (Brussels), Cl. Sci. nat. et méd. Mém. in-8° 11, fasc.* 3, 1–115.

D'Hoore, J.L. (1964). *Soil map of Africa (Scale 1 : 5,000,000).* Explanatory monograph. CCTA, Lagos.

Dickinson, R.E. & Virji, H. (1987). Climate change in the humid tropics, especially Amazonia, over the last twenty thousand years. Pp. 91–101 in *The Geophysiology of Amazonia: Vegetation and Climate Interactions*, Dickinson, R.E. (Ed.), Wiley, Chichester.

Dickinson, T.A. & Tanner, E.V.J. (1978). Exploitation of hollow trunks by tropical trees. *Biotropica* 10, 231–3.

Diels, L. (1918). Das Verhältnis von Rhythmik und Verbreitung bei der Perennen der europäischen Sommerwaldes. *Ber. dtsch. bot. Ges.* 36, 337–51.

Diels, L. & Hackenberg, C. (1926). Beiträge zur Vegetationskunde und Floristik von Süd-Borneo. *Bot. Jahrb.* 60, 293–316.

Dietrich, W.E., Windsor, D.M. & Dunne, T. (1983). Geology, climate and hydrology of Barro Colorado Island. Pp. 21–46 in *The Ecology of a Tropical Forest: Seasonal Rhythms and Long-Term Changes*, Leigh, E.G., Rand, A.S. Windsor, D.M. (Eds), Oxford University Press.

Digby, P.G.N. & Kempton, R.A. (1987). *Multivariate Analysis of Ecological Communities.* Chapman & Hall, London.

Dijkerman, J.C. & Miedma, R. (1988). An Ustult-Aquult-Tropept catena in Sierra Leone. I. Characteristics, genesis and classification. *Geoderma* 42, (1), 1–27.

Dingler, H. (1911a). Versuche über die Periodizität einiger Holzgewächse in den Tropen. *Sitzber. bayer. Akad. Wiss. München, math. naturw. Kl.* 1911, 127–43.

Dingler, H. (1911b). Ueber Periodizität sommergrüner Bäume Mitteleuropas in Gebirgsklima Ceylons. *Sitzber. bayer. Akad. Wiss. München, math. naturw. Kl.* 1911, 217–47.

Dobzhansky, T. (1950). Evolution in the tropics. *Amer. Sci.* 38, 209–21.

Docters van Leeuwen, W.M. (1929). Kurze Mitteilung über Ameisen Epiphyten aus Java. *Ber. dtsch. bot. Ges.* 47, 90–7.

Docters van Leeuwen, W.M. (1933). Biology of plants and animals occurring in the higher parts of Mount Pangrango-Gedeh in West Java. *Verh. Akad. Wet., Amsterdam* (sect. 2) 31, 1–270.

Docters van Leeuwen, W.M. (1936). Krakatau, 1883 to 1933. A. Botany. *Ann. Jard. bot. Buitenz.* 46–7, 1–506.

Donselaar, J. van (1969). Observations on savanna vegetation-types in the Guianas. *Vegetatio* 17, 271–312.

Dosso, H., Guillaumet, J.L. & Hadley, M. (1981). The Tai project: land use problems in a tropical forest. *Ambio* 10, 120–5.

Douglas, I. (1968). Erosion in the Sungei Gombak catchment, Selangor, Malaysia. *J. Trop. Geogr.* 26, 1–16.

Douglas, I. (1969). The efficiency of humid tropical denudation systems. *Trans. Inst. Br. Geogr.* 46, 1–16.

Douglas, I & Spencer, T. (1985). Present-day processes as a key to the effects of environmental change. Pp. 39–73 in *Environmental Change and Tropical Geomorphology*, Douglas, I. & Spencer, T. (Eds), Allen & Unwin, London.

Douglas, I., Spencer, T., Greer, T., Bidin, K., Sinun, W. & Wong, W.M. (1992). The impact of selective commercial logging on stream hydrology, chemistry and sediment loads in the Ulu Segama rain forest, Sabah. *Phil. Trans. R. Soc. Lond.* B 335, 397–406.

Dove, M.R. (1985). *Swidden Agriculture in Indonesia: the Subsistence Strategies of the Kalimantan Kantu.* Mouton, Berlin.

Doyle, J.A. (1978). Fossil evidence on the evolutionary origin of tropical trees and forests. Pp. 3–30 in *Tropical Trees as Living Systems*, Tomlinson, P.B. & Zimmermann, M.H. (Eds), Cambridge University Press.

Dransfield, J. (1978). Growth forms of rain forest palms. Pp. 247–68 in *Tropical Trees as Living Systems*, Tomlinson, P.B. & Zimmermann, M.H. (Eds), Cambridge University Press.

Duchaufour, P. (1982). *Pedology.* (transl. T.R. Paton). Allen & Unwin, London.

Ducke, A. & Black, G.A. (1953). Phytogeographical notes on the Brazilian Amazon. *An. Acad. Brasil. Ciencias* 25, 1–46.

Dudgeon, W. (1923). Succession of epiphytes in the *Quercus incana* forest at Landour, western Himalayas. *J. Indian Bot. Soc.* 3, 270–2.

Du Rietz, G.E. (1931). Life-forms of terrestrial flowering plants. *Acta Phytogeogr. suec.* 3 (1).

Dycus, A.N.M. & Knudson, L. (1957). The role of the velamen of the aerial roots of orchids. *Bot. Gaz.* 119, 78–87.

Edelman, C.H. & van der Voorde, P.K.J. (1963). Important characteristics of alluvial soils in the tropics. *Soil Sci.* 95, 258–63.

Eden, M.J. (1974a). Palaeoclimatic influences and the development of savanna in southern Venezuela. *J. Biogeogr.* 1, 95–109.

Eden, M.J. (1974b). The origin and status of savanna and grassland in southern Papua. *Trans. Inst. Brit. Geographers* Nov. 1974 (Publ. No. 63), 97–110.

Eden, M.J. (1985). Forest cultivation and derived savanna and grassland: comparative studies from southern Papua and south west Guyana. Pp. 260–4 in *Ecology and Management of the World's Savannas*, Tothill, J.C. & Mott, J.J. (Eds), Australian Academy of Sciences, Canberra.

Eden, M.J. (1990). *Ecology and Land Management in Amazonia.* Belhaven Press, London.

Eden, M.J. & McGregor, D.F.M. (1992). Dynamics of the forest-savanna boundary in the Rio Brianco – Rapunun region of northern Amazonia. Pp. 77–89 in *Nature and Dynamics of Forest/Savanna Boundaries*, Furley, P.A., Proctor, J. & Ratten, J.A. (Eds), Chapman & Hall, London.

Edmisten, J. (1970). Preliminary studies of the nitrogen budget of a tropical rain forest. Pp. 211–15 in *A Tropical Rain Forest: a Study of Irradiation and Ecology at El Verde, Puerto Rico*, Odum, H.T. & Pigeon, R.F. (Eds), Div. Tech. Inf., U.S. Atomic Energy Commission, Oak Ridge, Tennessee.

Edwards, K.A. & Blackie, J.R. (1981). Results of the East African Catchment Experiments 1958–74. Pp. 168–88 in *Tropical Agricultural Hydrology*, Lal, R. & Russell, E.W. (Eds), Wiley, Chichester.

Edwards, P.J. & Grubb, P.J. (1977). Studies of mineral cycling in a montane rain forest in New Guinea. I The distribution of organic matter in the vegetation and soil. *J. Ecol.* 65, 943–69.

Eggeling, W.J. (1947). Observations on the ecology of Budongo rain forest, Uganda. *J. Ecol.* 34, 20–87.

Egler, F.E. (1952). Southeast saline everglades vegetation, Florida and its management. *Vegetatio* 3, 213–65.

Ehringer, J.R., Ullmann, I., Lange, O.L., Farquahar, G.D., Cowan, I.R., Shulze, E.-D. & Ziegler, H. (1986). Mistletoes: a hypothesis concerning morphological and chemical avoidance of herbivory. *Oecologia* (Berlin) 70, 234–7.

Eilers, R.G. & Looi, K.S. (1982). *The Soils of Northern Interior Sarawak (East Malaysia).* Memoir 3, Soil Surv. Div., Res. Branch, Dept. Agric., Kuching.

Eiten, G. (1972). The cerrado vegetation of Brazil. *Bot. Rev.* 38, 201–41.

Eiten, G. (1975). The vegetation of the Serra do Roncador. *Biotropica* 7, 112–35.

Eiten, G. (1978). Delimitation of the cerrado concept. *Vegetatio* 36, 169–78.

Eiten, G. (1983). *Classificacão da Vegetacão do Brasil.* Conselho Nac. de Desenvolv. Cientif. e Technol., Brasilia.

Eiten, G. (1984). Vegetation of Brasília. *Phytocoenologia* 12, 271–92.

Eiten, G. (1986). The use of the term 'savanna'. *Trop. Ecol.* 27, 10–23.

Ellenberg, H. (1959). Typen tropischer Urwälder in Peru. *Schweiz. Zeitschr. Forstwirtsch.* 110, 167–87.

Ellenberg, H. (1979). Man's influence on tropical mountain ecosystems in South America. *J. Ecol.* 67, 401–16.

Ellenberg, H. (1985). Unter welchen Bedingungen haben Blätter sogenannte 'Träufelspitzen'? *Flora* 176, 169–88.

Ellenberg, H. (1988). *Vegetation Ecology of Central Europe.* Cambridge University Press.

Ellenberg, H. & Mueller-Dombois, D. (1967). A key to Raunkiaer plant life forms with revised subdivisions. *Ber. geobot. Inst. Rübel (Zurich)* 37, 56–73.

Ellison, J.C. (1991). The Pacific palaeography of *Rhizophora mangle* L. (Rhizophoraceae). *J. Linn. Soc. Lond. (Bot.)* 105, 271–84.

Elsenbeer, H. & Cassel, D. K. (1990) Surficial processes in the rainforest of western Amazonia. Pp. 289–97 in *Research Needs and Applications to Reduce Erosion and Sedimentation in Tropical Steeplands,* Zimmer, R.R., O' Loughlin, C.L. & Hamilton, L.S. (Eds), Proc. Fiji Symposium June 1990, Pubn 192, International Assocn Hydrol. Sciences, Geneva.

Elsenbeer, H. & Cassel, D.K. (1991). The mechanisms of overland flow generation in a small catchment in Western Amazonia. Pp. 213–22 in *Water Management of the Amazon Basin,* Braga, B.P.F.Jr. & Fernandez-Jauregui, C.A. (Eds), UNESCO, Montevideo.

Elwes, H.J. & Henry, A. (1906–7). *The Trees of Great Britain and Ireland.* 2 vols. Privately printed, Edinburgh.

Emberger, L., Mangenot, G. & Miège, J. (1950). Caractères analytiques et synthétiques des associations de la forêt équatoriale de Côte d'Ivoire. *C. R. Acad. Sci. Paris* 231, 812–14.

Endress, P.K. (1994). *Diversity and Evolutionary Biology of Tropical Flowers.* Cambridge University Press.

Enright, N.J. (1982). The *Araucaria* forests of New Guinea. In *Biogeography and Ecology of New Guinea,* Monographiae Biologicae 42, Gressitt, J.L. (Ed.), Junk, The Hague.

Enright, N. & Ogden, J. (1979). Applications of transition matrix models in forest dynamics: *Araucaria* in Papua New Guinea and *Nothofagus* in New Zealand. *Australian J. Ecol.* 4, 3–23.

Ernst, A. (1908). *The New Flora of the Volcanic Island of Krakatau.* (Transl. A.C. Seward). Cambridge University Press.

Ernst, A., Bernard, C. *et al.* (1910–14). Beiträge zur Kenntnis der Saprophyten Javas. *Ann. Jard. Bot. Buitenz.* 23, 20–61; 24, 55–97; 25, 161–88; 26, 219–57; 28, 99–124.

Evans, G.C. (1939). Ecological studies on the rain forest of Southern Nigeria. II. The atmospheric environmental conditions. *J. Ecol.* 27, 436–82.

Evans, G.C. (1956). An area survey method of investigating the distribution of light intensity in woodlands, with particular reference to sunflecks. *J. Ecol.* 44, 391–428.

Evans, G.C. (1966). Model and measurement in the study of woodland light climates. Pp. 53–76 in *Light as an Ecological Factor,* Bainbridge, R., Evans, G.C. & Rackham, O. (Eds), Blackwell, Oxford.

Evans, G.C. (1972). *The Quantitative Analysis of Plant Growth.* Blackwell, Oxford.

Evans, G.C., Whitmore, T.C. & Wong, Y.K. (1960). The distribution of light reaching the ground vegetation in a tropical rain forest. *J. Ecol.* 48, 193–204.

Evans, J. (1992). *Plantation Forestry in the Tropics.* Second edition, Clarendon Press, Oxford.

Evans, P.G.H. (1986). Dominica Multiple Land Use Project. *Ambio* 15, (2), 82–9.

Evrard, C. (1968). Recherches écologiques sur le peuplement forestier des sols hydromorphes de la Cuvette centrale congolaise. *Publ. Inst. Agron. Congo Belge (INEAC),* Ser. Sci. 110.

Ewel, J. (Ed.) (1980). Tropical succession. *Biotropica* 12 (Supplement), 1–95.

Ewel, J., Benedict, F., Berish, C. & Brown, B. (1981). Slash and burn impacts on a Costa Rican wet forest site. *Ecology* 62, 816–29.

Ewel, J.J., Chai, P. & Lim, M.T. (1985). Biomass and floristics of three young second-growth forests in Sarawak. *Malay. For.* 46 (1983), 347–64.

Ewers, F.W., Fisher, J.B. & Fichtner, K. (1991). Water flux and xylem structure in vines. Pp. 127–60 in *Biology of Vines,* Putz, F.E. & Mooney, H.A. (Eds), Cambridge University Press.

Exell, A.W. (1944). *Catalogue of the Vascular Plants of S. Tomé (with Principe and Annobon).* Brit. Mus. (Nat. Hist.), London.

Eyre, L.A. & Gray, C. (1990). Utilization of satellite imagery in the assessment of the effect of global warming on the frequency and distribution of tropical cyclonic storms in the Caribbean, East Pacific and Australian regions. Pp. 365–75 in *Proc. of the Twenty-third International Symposium on Remote Sensing of Environment, Thailand, April 18–25 1990.*

Faber-Langendoen, D. & Gentry, A.H. (1991). Structure and diversity of rain forests at Bajo Calima, Chocó region, western Colombia. *Biotropica* 23, 2–11.

Faegri, K. & Pijl, L. van der (1982). *The Principles of Pollination Biology.* Fourth edition, Pergamon Press, London.

Fanshawe, D.B. (1952). The vegetation of British Guyana. A preliminary review. Imp. For. Inst., Oxford Univ., Inst. Paper No. 29.

FAO (1963). *World Forest Inventory.* FAO, Rome.

FAO (1976). *Attempt at a Global Appraisal of the Tropical Moist Forest.* Multigr. FAO, Rome.

FAO (1981). *Tropical Forest Resources Assessment Project* (4 vols). FAO, Rome.

FAO (1988). *An Interim Report on the State of Forest Resources in the Developing Countries.* Forest Resources Division, For. Dept. FO:MISC/88/7. FAO, Rome.

FAO (1993). *Forest Resources Assessment 1990. Tropical Countries.* (FAO Forestry Paper 112.) FAO, Rome.

FAO/UNESCO (1974). *Soil Map of the World 1 : 5 000 000. I. Legend.* UNESCO, Paris.

FAO/UNESCO (1988). *Soil Map of the World. Revised Legend. World Soil Resources Report* 60. FAO, Rome.

Fassi, B. (1963). Die Verteilung der ektotrophen Mykorrhiza in der Streu und in der oberen Bodenschicht der *Gilbertiodendron dewevrei* (Caesalpiniaceae-Wälder) im Kongo. Pp. 297–302 in *Mykorrhiza. Proc. Internat. Mykorrhiza-Symposium, Weimar 1960,* Rawald, W. & Lyr, H. (Eds), G. Fischer, Jena.

Fearnside, P.M. (1980). The effects of cattle pasture on soil fertility in the Brazilian Amazon: consequences for beef production sustainability. *Trop. Ecol.* 21, 125–37.

Fedorov, A.A. (1966). The structure of the tropical rain forest and speciation in the humid tropics. *J. Ecol.* 54, 1–11.

Ferri, M. (1960). Contribution to the knowledge of the ecology of the 'Rio Negro caatinga' (Amazon). *Bull. Res. Council Israel* 8D (Botany), 195–208.

Fetcher, N., Oberbauer, S.F. & Strain, B.R. (1985). Vegetation effects on microclimate in lowland tropical forest in Costa Rica. *Int. J. Biomet.* 29, 145–55.

Fisher, J.B. (1976). Adaptive value of rotten tree cores. *Biotropica* 8, 264.

Fisher, J.B. (1982). A survey of buttresses and aerial roots of tropical trees for the presence of reaction wood. *Biotropica* 14, 56–61.

Fisher, J.B. & Ewers, F.W. (1992). Xylem pathways in liana stems with variant secondary growth. *Bot. J. Linn. Soc. Lond.* 108, 181–202.

Fitting, H. (1910). Über die Beziehungen zwischen den epiphyllen Flechten und den von ihnen bewohnten Blättern. *Ann. Jard. Bot. Buitenz.* (suppl. 3), 505–18.

Fittkau, E.J. (1973). Arten mannigfältigkeit amazonischer Lebensräume aus ökologischer Sicht. *Amazoniana* 4, 321–40.

Fittkau, E.J. & Klinge, H. (1973). On biomass and trophic structure of the central Amazonian Rain forest ecosystem. *Biotropica* 5, 2–14.

Fleming, T.H. & Heithaus, E.R. (1981). Frugivorous bats, seed shadows and the structure of tropical forests. *Biotropica* 13, (suppl.), 45–53.

Flenley, J.R. (1979). *The Equatorial Rain Forest: A Geological History*. Butterworth, London & Boston.

Flenley, J.R. (1985). Relevance of Quaternary palynology to geomorphology in the tropics and subtropics. Pp. 153–64 in *Environmental Change and Tropical Geomorphology*, Douglas, I. & Spencer, T. (Eds), Allen & Unwin, London.

Flenley, J.R. (1988). Palynological evidence for land use changes in South-east Asia. *J. Biogeogr.* 15, 185–97.

Floyd, A.G. (1990). *Australian Rainforests of New South Wales*. Surrey Beatty (in association with National Parks & Wildlife Service of New South Wales), Chipping Norton, N.S.W.

Foggie, A. (1947). Some ecological observations on a tropical forest type in the Gold Coast. *J. Ecol.* 34, 88–106.

Follett-Smith, R.R. (1930). Report of an investigation of the soils and of the mineral content of pasture grasses occurring at Waranama Ranch, Berbice River. *Agric. J. Brit. Guiana* 3, 142–59.

Forget, P.-M. (1989). La regénération naturelle d'une espèce autochore de la forêt guyanaise: *Eperua falcata* Aublet (Caesalpiniaceae). *Biotropica* 21, 115–25.

Forget, P.-M. (1992). Regeneration ecology of *Eperua grandiflora* (Caesalpiniaceae), a large-seeded tree in French Guiana. *Biotropica* 24, 146–56.

Forti, M.C. & Moreira-Nordemann, L.M. (1991). An ion budget for a 'Terra firme' rainforest in Central Amazonia. Pp. 13–18 in *Water Management of the Amazon Basin*, Braga, B.P.F. Jr. & Fernandez-Jauregui, C.A. (Eds), UNESCO, Montevideo.

Fosberg, F.R., Garnier, B.J. & Küchler, A.W. (1961). Delimitation of the Humid Tropics. *Geogr. Rev.* 51, 333–47.

Foster, R.B. (1977). *Tachygalia versicolor* is a suicidal neotropical tree. *Nature (Lond.)* 268, 624–6.

Foster, R.B. (1983). The seasonal rhythm of fruitfall on Barro Colorado island. Pp. 151–85 in *The Ecology of a Tropical Forest: Seasonal Rhythms and Long-term Changes*, Leigh, E.G., Rand, A.S. & Windsor, D.M. (Eds), Oxford University Press.

Foster, R.B. & Brokaw, N.V.L. (1983). The structure and history of the vegetation of Barro Colorado island. Pp. 67–81 in *The Ecology of a Tropical Forest: Seasonal Rhythms and Long-term Changes*, Leigh, E.G., Rand, A.S. & Windsor, D.M. (Eds), Oxford University Press.

Fournier, F. (1967). Research on soil erosion and soil conservation in Africa. *African Soils* 12, 53–96.

Fox, J.E.D. (1968). *Didelotia idae* in the Gola Forest, Sierra Leone. *Economic Bot.* 22, 338–46.

Fox, J.E.D. (1969). Climbers in the Lowland Dipterocarp forest. *Commonw. For. Rev.* 48, 196–8.

Fox, J.E.D. (1972). *The Natural Vegetation of Sabah and Natural Regeneration of the Dipterocarp Forests*. Ph.D. Thesis, University of Wales.

Fox, J.E.D. (1973). Dipterocarp seedling behaviour in Sabah. *Malay. For.* 36, 205–14.

Fox, J.E.D. (1976). Constraints on the natural regeneration of Tropical Moist forest. *For. Ecol. Mgmt* 1, 37–65.

Fox, J.E.D. & Tan, H. (1971). Soils and forest on an ultrabasic hill north east of Ranau, Sabah. *J. Trop. Geogr.* 32, 38–48.

Foxworthy, F.W. (1927). Commercial timber trees of the Malay Peninsula. *Malay. For. Rec.* 3.

Frahm, J.-P. & Gradstein, S.R. (1991). An altitudinal zonation of tropical rain forests using bryophytes. *J. Biogeogr.* 18, 669–78.

Francis, W.D. (1924). The development of buttresses in Queensland trees. *Proc. Roy. Soc. Queensland* 36, 21–37.

Francis, W.D. (1951). *Australian Rain-Forest Trees*. Second edition, Angus & Robertson, Sydney.

Franco, W. & Dezzeo, N. (1991). Soils and soil water regime in the terra firme-Caatinga forest complex near San Carlos de Rio Negro, Amazon Territory, Venezuela. Unpublished.

Franken, N.A.P. & Roos, M.C. (1981). *Studies in Lowland Equatorial Forest in Jambi Province, Central Sumatra*. Biotrop, Bogor.

Franken, W., Leopoldo, P.R., Matsui, E. & Ribeiro, M.N.G. (1982). Interceptação das precipitacoes em floresta Amazonica de terra firme. *Acta Amazon.* 12, 15–22.

Frankie, G.W., Baker, H.G. & Opler, P.A. (1974). Comparative phenological studies of trees in tropical Wet and Dry forests in the lowlands of Costa Rica. *J. Ecol.* 62, 881–919.

Fränzle, O. (1977). Biophysical aspects of species diversity in Tropical Rain forest ecosystems. Pp. 69–83 in *Ecosystem Research in South America*, Müller, P. (Ed.), W. Junk, The Hague.

Fraser, L. & Vickery, J.W. (1937–39). The ecology of the Upper Williams river and Barrington Tops districts, I, II & III. *Proc. Linn. Soc. N.S.W.* 62, 269–83; 63, 139–84; 64, 1–33.

Frei, J.K. & Dodson, C.H. (1972). The chemical effects of certain bark substances on the germination and early growth of epiphytic orchids. *Bull. Torrey Bot. Cl.* 99, 301–7.

Freise, F. (1936). Das Binnenklima von Urwäldern im subtropischen Brasilien. *Peterm. Geogr. Mitt.* 82, 301–4, 346–8.

Fritz-Sheridan, R.P. (1987). Nitrogen fixation on a tropical volcano, La Soufriere. II. Nitrogen fixation by *Scytonema* sp. and *Stereocaulon virgatum* Ach. during colonization of phreatic material. *Biotropica* 19, 297–300.

Funke, G.L. (1929). On the biology and anatomy of some tropical leaf joints. I. *Ann. Jard. Bot. Buitenz.* 40, 45–74.

Funke, G.L. (1931). On the biology and anatomy of some tropical leaf joints. II. *Ann. Jard. Bot. Buitenz.* 41, 33–64.

Furch, K. (1984). Amazon water chemistry. Pp. 167–99 in *The Amazon. Limnology and Landscape Ecology of a Mighty Tropical River and its Basin*, Sioli, H. (Ed.), Junk, The Hague.

Furch, K. & Klinge H. (1989). Chemical relationships between vegetation, soil and water in contrasting innundation areas of Amazonia. Pp. 189–204 in *Mineral Nutrients in Tropical Forest and Savanna Ecosystems*, Proctor, J. (Ed.) Blackwell Scientific Publications, Oxford.

Furtado, J.L. & Mori, S. (Eds) (1982). *Tasek Bera. The Ecology of a Freshwater Swamp*. Junk, The Hague.

Galil, J. & Eisikowitch, D. (1971). Studies on mutualistic symbiosis between syconia and sycophilous wasps in monoecious figs. *New. Phytol.* 70, 773–87.

Gams, H. (1918). Prinzipienfragen der Vegetationsforschung. *Vierteljahrschr. Naturf. Ges. Zürich* 63, 293–493.

Gan, Y.-Y., Robertson, F.W. & Soepadmo, E. (1981). Isozyme variation in some rain forest trees. *Biotropica* 13, 20–8.

Gartlan, J.S. (1983). Ecology of Central African rain forests. Unpublished.

Gartlan, J.S., Newberry, D. McC., Thomas, D.W. & Waterman, P.G. (1986). The influence of topography and soil phosphorus on the vegetation of Korup Forest Reserve, Cameroun. *Vegetatio* 65, 131–48.

Garwood, N.C. (1983). Seed germination in a seasonal tropical forest in Panama: a community study. *Ecol. Monogr.* 53, 159–81.

Garwood, N.C. (1989). Tropical soil seed banks: a review. Pp. 149–209 in *Ecology of Soil Seed Banks*, Leck, M.A., Parker, V.T. & Simpson, R.A. (Eds), Academic Press, San Diego.

Garwood, N.C., Janos, D.P. & Brokaw, N. (1979). Earthquake-caused landslides: a major disturbance to tropical forests. *Science* 205, 997–9.

Gauch, H.G. (1982). *Multivariate Analysis in Community Ecology.* Cambridge University Press.

Gaussen, H. (1955). Expression des milieux par des formules écologiques; leur représentation cartographique. *Coll. Int. Cent. Nat. Rech. Sci.* 59, 257–69.

Gaussen, H. (1959). The vegetation maps. *Trav. Sect. Sci. & Tech. Inst. Français de Pondichéry* 1, fasc. 4, 155–79.

Gay, H. (1993). Animal-fed plants: an investigation into the uptake of ant-derived nutrients by the far-eastern epiphytic fern *Lecanopteris* Reinw. (Polypodiaceae). *Biol. J. Linn. Soc. Lond.* 50, 221–33.

Geel, B. van, & Hammen, T. van der (1973). Upper Quaternary vegetational and climatic sequence of the Fuquene area (Eastern Cordillera, Colombia). *Paleogeogr. Paleoclim. Palaeoecol.* 14, 9–92.

Geldenhuys, C.J. (1985). Annotated bibliography of South African indigenous Evergreen forest ecology. *S. African Nat. Sci. Prog. Rep. no. 107*, Foundation for Research Development, Council for Sci. Industr. Res. Pretoria.

Gentner, G. (1909). Ueber den Blauglanz auf Blättern und Früchten. *Flora* 99, 337–54.

Gentry, A.H. (1974). Flowering phenology and diversity in tropical Bignoniaceae. *Biotropica* 6, 64–8.

Gentry, A.H. (1979a). Distribution patterns of neotropical Bignoniaceae: some phytogeographic implications. Pp. 339–54 in *Tropical Botany*, Larson, K. & Holm-Nielson, L.B. (Eds), Academic Press, London.

Gentry, A.H. (1979b). Extinction and conservation of plant species in tropical America: a phytogeographical perspective. Pp. 110–26 in *Systematic Botany, Plant Utilization and Biosphere Conservation*, Hedberg, I. (Ed.), Almquist & Wiksell, Stockholm.

Gentry, A.H. (1982a). Patterns of neotropical plant species diversity. *Evolutionary Ecology* 15, 1–84.

Gentry, A.H. (1982b). Neotropical floristic diversity: phytogeographical connections between Central and South America, Pleistocene climatic fluctuations or an accident of Andean orogeny? *Ann. Missouri Bot. Gdn.* 69, 557–93.

Gentry, A.H. (1988a). Tree species richness of upper Amazonian forests. *Proc. Nat. Acad. Sci. USA* 85, 156–9.

Gentry, A.H. (1988b). Changes in plant community diversity and floristic composition on environmental and geographical gradients. *Ann. Missouri Bot. Gard.* 75, 1–34.

Gentry, A.H. (1992). Tropical forest biodiversity: distribution patterns. *Oikos* 63, 19–28.

Gentry, A.H. (1993). *A field guide to the families and genera of woody plants of northwest South America (Colombia, Ecuador, Peru) with supplementary notes on herbaceous taxa.* Conservation International, Washington, DC

Gentry, A.H. & Dodson, C. (1987). Contribution of non-trees to species-richness of a tropical forest. *Biotropica* 19, 149–56.

Gentry, A.H. & Emmons, L.H. (1987). Geographical variation in fertility, phenology and composition of the understorey in Neotropical forests. *Biotropica* 19, 216–27.

Gérard, P. (1960). Etude écologique de la forêt dense à *Gilbertiodendron dewevrei* dans la région de l'Uelé. *Publ. INEAC.*, Sér. Sci. 87.

Gerasimov, I.P. (1973). A genetic approach to the subdivision of tropical soils, regolith and keir products of redeposition. *Soviet Geography* (Review and translation) 14, 165–77.

Germain, R. (1945). Note sur les premiers stades de la reforestation naturelle des savanes du Bas-Congo. *Bull. Agr. Congo Belge* 36, 16–25.

Germain, R. (1957). Un essai d'inventaire de la flore et des formes biologiques en forêt équatoriale congolaise. *Bull. Jard. Bot. de l'Etat Bruxelles* 27 (Vol. Jubilaire Walter Robyns), 563–76.

Germain, R. (1965). Les biotopes alluvionnaires herbeux et les savanes intercalaires du Congo équatorial. *Mém. Acad. Roy. des Sci. D'Outre-Mer*, N.S. XV-4, Bruxelles.

Germain, R. & Evrard, C. (1956). Etude écologique et phytosociologique de la forêt à *Brachystegia laurentii*. *Publ. Inst. Agron. Congo belge (INEAC)*, Sér. Sci. 67.

Germeraad, J.H., Hopping, C.A. & Muller, J. (1968). Palynology of Tertiary sediments from tropical areas. *Rev. Palaeobot. Palynol.* 6, 189–348.

Gersper, P.L. & Holowaychuk, N. (1971). Some effects of stem flow from forest canopy trees on chemical properties of soils. *Ecology* 52, 691–702.

Gessner, F. (1956a). Die Wasseraufnahme durch Blätter und Samen. Pp. 222–3 in *Encyclopaedia of Plant Physiology* 3, Socker, O. (Ed.), Springer, Berlin.

Gessner, F. (1956b). Der Wasserhaushalt der Epiphyten und Lianen. Pp. 915–42 in *Handbuch der Pflanzenphysiologie* Bd. 3, *Pflanzen und Wasser*, Ruhland, W. (Ed.), Springer, Berlin.

Ghesquière, J. (1925). Note sur les racines tabulaires ou accottements ailés de quelques arbres congolais. *Rev. Zool. Afr.* 13 (suppl. 2), 1–2.

Ghuman, B.S. & Lal, R. (1987). Effects of deforestation on soil properties and microclimate of a high Rain Forest in Southern Nigeria. Pp. 225–44 in *The Geophysiology of Amazonia: Vegetation and Climate Interactions*, Dickinson, R.E. (Ed.), Wiley, Chichester.

Gibbs, R.J. (1967). The geochemistry of the Amazon river system. Part 1. The factors that control the salinity and the composition and concentration of the suspended solids. *Bull. Geol. Soc. Amer.* 78, 1203–32.

Gibson, D.J. & Grieg-Smith, P. (1986). Community pattern analysis: a method for quantifying community mosaic structure. *Vegetatio* 66, 41–7.

Gilbert, L.E. (1972). Pollen feeding and reproductive biology of *Heliconius* butterflies. *Proc. Nat. Acad. Sci. (Washington)* 69, 1403–7.

Gilbert, L.E. (1975). Ecological consequences of a coevolved mutualism between butterflies and plants. Pp. 210–40 in *Coevolution of animals and plants*, Gilbert, L.E., Raven, P.H. (Eds), Univ. Texas Press, Austin.

Gill, A.M. & Tomlinson, P.B. (1969). Studies on the growth of Red Mangrove (*Rhizophora mangle* L.). I. Habit and general morphology. *Biotropica* 1, 1–9.

Gill, A.M. & Tomlinson, P.B. (1971a). Studies on the growth of Red Mangrove (*Rhizophora mangle* L.). II. Growth and differentiation of aerial roots. *Biotropica* 3, 63–77.

Gill, A.M. & Tomlinson, P.B. (1971b). Studies on the growth of Red Mangrove (*Rhizophora mangle* L.). 3. Phenology of the shoot. *Biotropica* 3, 109–24.

Gilmour, D.A. (1977). Effect of rainforest logging and clearing on water yield and quality in a high rainfall zone of north-east Queensland. *Institution of Engineers, Australia, Hydrology Symp. 1977 Brisbane, Nat. Conf. Publ.* 77/5, 156–60.

Givnish, T.J. (1978a). In discussion on Jeník, J., 'Roots and root systems'. P. 344, in *Tropical trees as living systems*, Tomlinson, P.B. & Zimmermann, M.H. (Eds), Cambridge University Press.

Givnish, T.J. (1978b). On the adaptive significance of compound leaves, with particular reference to tropical trees. Pp. 351–80 in *Tropical trees as living systems*, Tomlinson, P.B. & Zimmermann, M.H. (Eds), Cambridge University Press.

Givnish, T.J. (1984). Leaf and canopy adaptations in tropical forests. Pp. 51–84 in *Physiological ecology of plants in the wet tropics*, Medina, E., Mooney, H.A. & Vázquez-Yánes, C. (Eds), Junk, Dordrecht.

Gleason, H.A. & Cook, M.T. (1926). Plant ecology of Porto Rico. *Scientific Survey of Porto Rico and the Virgin Islands* 7, Part 1. New York Acad. Sci., New York.

Gledhill, D. (1963). The ecology of the Aberdeen Creek mangrove swamp. *J. Ecol.* 51, 693–703.

Godwin, H. (1929). The subclimax and deflected succession. *J. Ecol.* 17, 144–47.

Goebel, K. von (1888). Morphologische und biologische Studien 1. Ueber epiphytische Farne und Muscineen. *Ann. Jard. bot. Buitenz.* 7, 1–73.

Goebel, K. von (1889). *Pflanzenbiologische Schilderungen*, vol. III *Epiphyten*. Marburg, N.G. Elvert'sche Verlagsbuchhandlung.

Goebel, K. von (1922). Erdwurzeln mit Velamen. *Flora* (N.F.) 15, 1–26.

Goebel, K. von (1930). Organographie der Pflanzen. Teil I. *Allgemeine Organographie* (ed. 3). G. Fischer, Jena.

Goedert, W.J. (1983). Management of the cerrado soils of Brazil: a review. *J. Soil Sci.* 34, 405–28.

Goff, F.G. & Zedler, P.H. (1972). Derivation of species succession vectors. *Am. Midl. Nat.* 87, 397–412.

Golley, F.B. (1986). Chemical plant-soil relationships. *J. Trop. Ecol.* 2, 219–29.

Golley, F.B., McGinnis, J.T., Clements, R.G., Child, G.I. & Duever, M.J. (1975). *Mineral Cycling in a Tropical Moist Forest Ecosystem*. Univ. Georgia Press, Athens, Georgia.

Golley, F.B., Yantko, J., Richardson, T. & Klinge, H. (1980). Biochemistry of tropical forests. I. The frequency distribution and mean concentration of selected elements in a forest near Manaus, Brazil. *Trop. Ecol.* 21, 59–70.

Gomez-Pompa, A. & Vázquez-Yanes, C. (1974). Studies on the secondary succession of tropical lowlands: the life cycle of secondary species. *Proc. 1st Internat. Congr. of Ecol., The Hague, Netherlands 1974*, pp 336–42.

Goodland, R. (1971). A physiognomic analysis of the 'cerrado' vegetation of central Brazil. *J. Ecol.* 59, 411–19.

Goodland, R. & Pollard, R. (1973). The Brazilian cerrado vegetation: a fertility gradient. *J. Ecol.* 61, 219–24.

Goodland, R.J.A. & Irwin, H.S. (1975). *Amazon Jungle; Green Hell to Red Desert?* Elsevier, Amsterdam.

Gonggrijp, L. (1941). De verdamping van het gebergte bosch in West Java op 1750–2000 m zeehoogte. *Tectona* 34, 437–47.

Goosem, S. & Lamb, D. (1986). Measurement of phyllosphere nitrogen fixation in a tropical and two subtropical rain forests. *J. Trop. Ecol.* 2, 373–6.

Gottsberger, G. (1970). Beiträge zur Biologie von Annonaceen-Blüten. *Oesterr. Bot. Zeitschr.* 118, 237–79.

Gottsberger, G. (1974). The structure and function of the primitive angiosperm flower – a discussion. *Acta Bot. Néerl.* 23, 401–71.

Gottsberger, G. (1978). Seed dispersal by fish in the innundated regions of Humaitá, Amazonia. *Biotropica* 10, 170–83.

Gould, S.J. & Lewontin, R.C. (1979). The spandrels of San Marco and the Panglossian paradigm: a critique of the adaptationist programme. *Proc. R. Soc. Lond.* B 205, 581–98.

Goulding, M. (1980). *The Fishes and the Forest. Explorations in Amazonian natural history*. Univ. California Press, Berkeley.

Gradstein, S.R. & Frahm, J.-P. (1987). Die floristische Höhengliederung der Moose entlang des Bryotrop-Transektes in NO-Peru. Pp. 105–13 in *Moosflora Und-vegetation in Regenwäldern NO-Peru: Ergebnisse der Bryotrop-Expedition nach Peru 1982*, Frey, W. (Ed.), Beiheft z. *Nova Hedwigia* 88.

Gray, B. (1973). Distribution of *Araucaria* in Papua New Guinea. *Dept. Forests, Papua New Guinea, Res. Bull.* 1.

Gray, B. (1975). Size-composition and regeneration of *Araucaria* stands in New Guinea. *J. Ecol.* 63, 273–89.

Gray, W.M. & Jacobson, R.W. (1977). Diurnal variation of deep cumulus convection. *Mon. Wea. Rev.* 105, 1171–88.

Greathouse, D.C., Laetsch, W.M. & Phinney, B.O. (1971). The shoot growth rhythm of a tropical tree, *Theobroma cacao*. *Amer. J. Bot.* 58, 281–6.

Green, S., Green, T. & Heslop-Harrison, Y. (1979). Seasonal heterophylly and leaf gland features in *Triphyophyllum* (Dioncophyllaceae) a new carnivorous plant genus. *Bot. J. Linn. Soc. Lond.* 78, 99–116.

Greenland, D.J. (1977). Soil damage by intensive arable cultivation: temporary or permanent? *Phil. Trans. R. Soc. Lond.* B 281, 193–208.

Greig-Smith, P. (1952). Ecological observations on degraded and secondary forest in Trinidad, British West Indies. I. General features of the vegetation. *J. Ecol.* 40, 283–315.

Greig-Smith, P. (1961). Data on pattern within plant communities. I. The analysis of pattern. *J. Ecol.* 49, 695–702.

Greig-Smith, P. (1971a). Application of numerical methods to tropical forests. Pp. 195–206 in *Statistical Ecology* 3, Patil, G.P., Pielou, E.C. & Waters, W.E. (Eds), Pennsylvania State Univ. Press.

Greig-Smith, P. (1971b). Some problems of analytical data. *Statistician* 21, 215–19.

Greig-Smith, P. (1979). Pattern in vegetation. *J. Ecol.* 67, 755–79.

Greig-Smith, P. (1980). The development of numerical classification and ordination. *Vegetatio* 42, 1–9.

Greig-Smith, P. (1983). *Quantitative Plant Ecology*. Third edition, Blackwell Scientific, London.

Greig-Smith, P. (1991). Pattern in a derived savanna in Nigeria. *J. Trop. Ecol.* 7, 491–502.

Greig-Smith, P., Austin, M.P. & Whitmore, T.C. (1967). The application of quantitative methods of vegetation survey. I. Association analysis and principal component ordination of rain forest. *J. Ecol.* 55, 483–503.

Grenand, F. & Grenand, P. (1993). Histoire du peuplement de la várzea en Amazonas. *Amazoniana* 12, 509–26.

Gresser, E. (1919). Resumeerend rapport over het voorkomen van ijzerhout op de olieterreinen Djambi. I. *Tectona* 12, 283–304.

Gressit, J.L. (Ed.) (1982). *Biogeography and Ecology of New Guinea. Monographicae Biologicae* 42. Junk, The Hague.

Gressit, J.L., Sedlacek, J. & Szent-Ivany, J.J.H. (1965). Flora and fauna on backs of large Papuan moss-forest weevils. *Science* 150, 1833–5.

Gressit, J.L., Samuelson, G.A. & Vitt, D.H. (1968). Moss growing on living Papuan moss-forest weevils. *Nature (Lond.)* 217, 765–7.

Griffiths, H. & Smith, J.A.C. (1983). Photosynthetic pathways in the Bromeliaceae, relations between the life-forms habitat preference and the occurrence of CAM. *Oecologia* 60, 176–84.

Griffiths, J.F. (1972). Nigeria. Pp. 167–92 in *Climates of Africa*, Griffiths, J.F. (Ed.), Elsevier, Amsterdam.

Grimm, U. & Fassbender, H.W. (1981a). Ciclos bioquimicos en un ecosistema forestal de los Andes Occidentales de Venezuela. I. Inventario de las reservas organicos y minerales (N, P, K, Ca, Mg, Mn, Fe, Al, Na). *Turrialba* 31, 27–37.

Grimm, U. & Fassbender, H.W. (1981b). Ciclos bioquimicos en un ecosistema forestal de los Andes Occidentales de Venezuela. III. Ciclo hidrologico y translocacion de elements quimicos con el agua. *Turrialba* 31, 89–99.

Grubb, P.J. (1974). Factors controlling the distribution of forest types on tropical mountains – new facts and a new perspective. Pp. 13–46 in *Altitudinal Zonation of Forest in Malesia*, Flenley, J.R. (Ed.), *Univ. Hull, Dept. Geogr. Misc. Ser.* 16.

Grubb, P.J. (1977a). The maintenance of species richness in plant communities: the importance of the regeneration niche. *Biol. Rev.* 52, 107–45.

Grubb, P.J. (1977b). Control of forest growth and distribution on wet tropical mountains with special reference to mineral nutrition. *Ann. Rev. Ecol. Syst.* 8, 83–107.

Grubb, P.J., Lloyd, J.R., Pennington, T.D. & Whitmore, T.C. (1963). A comparison of montane and lowland rain forest in Ecuador. I The forest structure, physiognomy and floristics. *J. Ecol.* 51, 567–601.

Grubb, P.J. & Stevens, P.F. (1985). *The Forests of the Fatima Basin and Mt. Kerigomna, Papua New Guinea*. Res. Sch. Pacific Studies, Austr. Nat. Univ., Canberra.

Grubb, P.J. & Tanner, E.V.J. (1976). The montane forests and soils of Jamaica: a reassessment. *J. Arnold Arboretum* 57, 313–68.

Grubb, P.J. & Whitmore, T.C. (1966). A comparison of montane and lowland forest in Ecuador. II. The climate and its effects on the distribution and physiognomy of the forests. *J. Ecol.* 54, 303–33.

Grubb, P.J. & Whitmore, T.C. (1967) A comparison of montane and lowland forest in Ecuador. III. The light reaching the ground vegetation. *J. Ecol.* 55, 33–57.

Guariguata, M.R. (1990). Landslide disturbance and forest regeneration in the Upper Luquillo Mountains, Puerto Rico. *J. Ecol.* 78, 814–32.

Guerrant, E.O. & Fiedler, P.L. (1981). Flower defenses against nectar-pilferage by ants. *Biotropica* 13 (suppl.), 25–33.

Guevara, S. & Gomez-Pompa, A. (1972). Seeds from surface soils in a tropical region of Veracruz, Mexico. *J. Arnold Arboretum* 53, 312–35.

Guevara, S., Purata, S.E. & van der Maarel, E. (1986). The role of remnant forest trees in Tropical Secondary Successions. *Vegetatio* 66, 77–84.

Guillaumet, J.-L. (1967). *Recherches sur la Végétation et la Flore de la Région du Bas-Cavally (Côte d'Ivoire)*. ORSTOM, Paris.

Guillaumet, J.-L. (1984). The vegetation: an extraordinary diversity. Pp. 27–74 in *Key Environments: Madagascar*, Jolly, A., Oberlé, P. & Albignac, R. (Eds), Pergamon Press, London.

Guillaumet, J.-L. & Adjanohoun, E. (1971). La végétation de la Côte d'Ivoire. *Mem. Off. Sci. Récherches Techn. Outre-Mer* 50, 157–263.

Ha, C.O., Sands, V.E., Soepadmo, E. & Jong, K. (1988a). Reproductive patterns of selected understorey trees in the Malaysian rain forest: the sexual species. *Bot. J. Linn. Soc. Lond.* 97, 295–316.

Ha, C.O., Sands, V.E., Soepadmo, E. & Jong, K. (1988b). Reproductive patterns of selected understorey trees in the Malaysian rain forest: the apomictic species. *Bot. J. Linn. Soc. Lond.* 97, 317–31.

Haantjens, H.A., Mabbutt, J.A. & Pullen, R. (1965). Anthropogenic grasslands in the Sepik Plains, New Guinea. *Pacific Viewpoint* 6, 215–19.

Haberlandt, G. (1926). *Eine Botanische Tropenreise*. Third edition, Englemann, Leipzig.

Haddow, A.J., Gillett, J.D. & Highton, R.B. (1947). The mosquitoes of Bwamba County, Uganda. V. The vertical distribution and biting-cycle of mosquitoes in rain-forest, with further observations on microclimate. *Bull. Ent. Res.* 37, 301–30.

Haffer, J. (1969). Speciation in Amazonian forest birds. *Science* 165, 131–7.

Hall, John B. (1973). Vegetational zones on the southern slopes of Mount Cameroun. *Vegetatio* 27, 49–69.

Hall, John B. (1977). Forest-types in Nigeria: an analysis of pre-exploitation forest enumeration data. *J. Ecol.* 65, 187–99.

Hall, John B. (1991). Multiple-nearest-tree sampling in an ecological survey of Afromontane catchment forest. *For. Ecol. Mgmt.* 42, 245–66.

Hall, J.B. & Swaine, M.D. (1976). Classification and ecology of closed canopy forest in Ghana. *J. Ecol.* 64, 913–51.

Hall, J.B. & Swaine, M.D. (1980). Seed stocks in Ghanaian forest soils. *Biotropica* 12, 256–63.

Hall, J.B. & Swaine, M.D. (1981). *The Distribution and Ecology of Vascular Plants in a Tropical Rain Forest: Forest Vegetation in Ghana*. Junk, The Hague.

Hallé, F. & Martin, R. (1968). Etude de la croissance rythmique chez l'Hévéa (*Hevea brasiliensis* Müll. Arg. Euphorbiacées-Crotonioidées). *Adansonia*, N.S. 8, 475–503.

Hallé, F. & Ng, F.S.P. (1981). Crown construction in mature dipterocarp trees. *Malay. For.* 44, 222–33.

Hallé, F. & Oldeman, R.A.A. (1970). *Essai sur l'Architecture et la Dynamique de Croissance des Arbres Tropicaux*. Masson, Paris.

Hallé, F, Oldeman, R.A.A. & Tomlinson, P.B. (1978). *Tropical Trees and Forests. An Architectural Analysis*. Springer-Verlag, Berlin.

Hambler, D.J. (1964). The vegetation of granitic outcrops in western Nigeria. *J. Ecol.* 52, 573–94.

Hamilton, A. (1974). The history of the vegetation. Pp. 188–209 in *East African Vegetation*, Lind, E.M. & Morrison, M.E.S. (Eds), Longman, London.

Hamilton, A.C. (1975). A quantitative analysis of altitudinal zonation in Uganda forests. *Vegetatio* 30, 99–106.

Hamilton, A.C. (1976). Significance of patterns of distribution shown by forest plants and animals in tropical Africa for the reconstructions of Upper Pleistocene palaeoenvironments: a review. *Palaeoecology of Africa* 9, 63–97.

Hamilton, A.C. (1982). *Environmental History of East Africa. A Study of the Quaternary*. Academic Press, London.

Hamilton, A.C. & Bensted-Smith, R. (1989). Forest conservation in the East Usambara mountains, Tanzania. IUCN & Forest Division, Min. Lands, Nat. Resources & Tourism, Tanzania.

Hamilton, L.S. (1985). Overcoming myths about soil and water impacts of tropical land uses. Pp. 680–90 in *Soil Erosion and Conservation*, El-Swaify, S.A., Modenhauer, W.C. & Lo, A. (Eds), Soil conservation Society of America, Ankeny, Ohio.

Hammen, T. van der (1974). The Pleistocene changes in vegetation and climate in tropical South America. *J. Biogeogr.* 1, 3–26.

Harcombe, P.A. (1980). Soil nutrient loss as a factor in early tropical secondary succession. *Biotropica* 12 (Suppl.), 8–15.

Hardon, H.J. (1937). Padang soil, an example of podsol in the Tropical Lowlands. *Verh. Akad. Wetensch. Amsterdam* 40, 530–8.

Hardy, F. (1928). An index of soil texture. *J. Agric. Sci.* 18, 252–6.

Haridasan, M. (1982). Aluminium accumulation of some native species of central Brazil. *Plant Soil* 65, 265–73.

Harper, J.L. (1977). *Population Biology of Plants*. Academic Press, London.

Harper, J.L., Lovell, P.H. & Moore, K.G. (1970). The shapes and sizes of seeds. *Ann. Rev. Ecol. Syst.* 1, 327–56.

Hart, J.W. (1988). *Light and Plant Growth*. Unwin Hyman, London.

Hart, T.B., Hart, J.A. & Murphy, P.G. (1989). Monodominant forests. *Amer. Nat.* 133, 613–33.

Hartshorn, G.S. (1972). *The Ecological Life History and Population Dynamics of* Pentaclethra macroloba, *a Tropical Wet forest dominant, and* Stryphnodendron excelsum, *an occasional associate*. Ph.D. Thesis, Univ. Wash., Seattle, U.S.A.

Hartshorn, G.S. (1978). Tree falls and tropical forest dynamics. Pp. 617–38 in *Tropical Trees as Living Systems*, Tomlinson, P.B. & Zimmermann, M.H. (Eds), Cambridge University Press.

Hartshorn, G.S. (1983). Plants – introduction. Pp. 118–57 in *Costa Rican Natural History*, Janzen, D.A. (Ed.), Univ. Chicago Press, Chicago & London.

Hastenrath, S. (1985). *Climate and Circulation of the Tropics*. D. Reidel, Dordrecht.

Havel, J.J. (1971). The *Araucaria* forests of New Guinea and their regenerative capacity. *J. Ecol.* 59, 203–14.

Heaney, A. & Proctor, J. (1990). Preliminary studies on forest structure and floristics on Volcán Bárva, Costa Rica. *J. Trop. Ecol.* 6, 307–20.

Hecht, S.B. (1983). Cattle ranching in the eastern Amazon: environmental and social implications. In *The dilemma of Amazonian development*, Moran, E.F. (Ed.), Westview, Boulder, Colorado.

Hedberg, O. (1951). Vegetation belts of the East African mountains. *Svensk Bot. Tidskr.* 45, 140–202.

Hedberg, O. (1964). Features of Afroalpine plant ecology. *Acta Phytogeogr. Suecica* 49, 1–139.

Hegarty, E.L. & Caballé, G. (1991). Distribution and abundance of vines in forest communities. Pp. 313–35 in *Biology of Vines*, Putz, F.E. & Mooney, H.A. (Eds), Cambridge University Press.

Heinsdijk, D. (1965). (quoted by Pires & Prance 1977). A distribuçao dos diametros nas florestas brasileiras. *Min. Agric. Dept. Rec. Nat. Renov., Boletim* 11, 1–56.

Henderson, M.R. (1931). The 'Padang' flora of Jemaja in the Anambas islands, N.E.I. *Gdns Bull. Straits Settl.* 5, 234–40.

Henderson-Sellers, A. (1987). Effect of change in land use on climate in the humid tropics. Pp. 463–93 in *The Geophysiology of Amazonia: Vegetation and Climate Interactions*, Dickinson, R.E. (Ed.), Wiley, Chichester.

Henderson-Sellers, A. & Gornitz, V. (1984). Possible climatic impacts of land cover transformations, with particular emphasis on tropical deforestation. *Climatic Change* 6, 231–58.

Henwood, K. (1973). A structural model of forces in buttressed tropical rain-forest trees. (With an appendix by H.G. Baker.) *Biotropica* 5, 83–93.

Herbert, D.A. (1929). The major factors in the present distribution of the genus *Eucalyptus*. *Proc. Roy. Soc. Queensland* 40, 165–93.

Herbert, D.A. (1935). The climatic sifting of Australian vegetation. *Rep. Melbourne Meeting, Aust. & N.Z. Ass. Adv. Sci.*, pp. 349–70.

Herbert, D.A. (1967). Ecological segregation and Australian phytogeographic elements. *Proc. Roy. Soc. Queensland* 78, 110–1.

Hermes, K. (1955). Die Lage der oberen Waldgrenze in den Gebirge der Erde und ihr Abstand von der Schneegrenze. *Geogr. Arbeiten* 5, 1–277.

Herrera, R. (1979). *Nutrient distribution and cycling in an Amazon caatinga forest on spodosol in southern Venezuela*. Ph.D. Thesis, University of Reading, U.K.

Herrera, R. & Jordan, C.F. (1981). Nitrogen cycle in a tropical Amazonian rain forest: the caatinga of low mineral status. Pp. 493–505 in *Terrestrial nitrogen cycles*, Clark, F.E. & Rosswall, T. (Eds), *Ecol. Bull.* (Stockholm) 33.

Herrera, R., Merida, T., Stark, N. & Jordan, C.F. (1978). Direct phosphorus transfer from leaf litter to roots. *Naturwissenschaften* 65, 208–9.

Herwitz, S.R. (1981a). Regeneration of selected tropical tree species in Corcovado National Park, Costa Rica. *Univ. California Pubn. in Geography* 24, 1981, Berkeley, California.

Herwitz, S.R. (1981b). Landforms under tropical wet forest cover on the Osa Peninsula, Costa Rica. *Zeitschrift für Geomorphologie* 25, 259–70.

Herwitz, S.R. (1985). Interception storage capacities of tropical rainforest canopy trees. *J. Hydrol.* 77, 237–52.

Herwitz, S.R. (1986a). Episodic stemflow inputs of magnesium and potassium to a tropical rainforest floor during heavy rainfall events. *Oecologia* 70, 423–5.

Herwitz, S.R. (1986b). Infiltration-excess caused by stemflow in a cyclone-prone tropical rainforest. *Earth Surf. Processes Landf.* 11, 401–12.

Herwitz, S.R. (1987a). Rainfall totals in relation to solute inputs along an exceptionally wet altitudinal transect. *Catena* 14, 25–30.

Herwitz, S.R. (1987b). Raindrop impact and water flow on the vegetative surfaces of trees and the effects on stemflow and throughfall generation. *Earth Surf. Processes Landf.* 12, 425–32.

Herzog, T. (1923). Die Pflanzenwelt der bolivischen Anden. *Veget. Erde* 15.

Heske, F. (1938–39). Der tropische Wald als Rohstoffquelle. *Zeitschr. Weltforstwirtsch.* 6, 413–85.

Hetsch, W. & Hoheisel, H. (1976). Standorts- und Vegetationsgliederung in einem tropischen Nebelwald. *Allgemeine Forst- und Jagdzeitung* (Frankfurt am Main) 147, 200–9.

Heyligers, P.C. (1963). Vegetation and soil of a white-sand savanna in Suriname. *Verh. Akad. Wetensch.*, 2e Reeks, 54 (3), 148.

Hill, M.O. (1973a). The intensity of spatial pattern in plant communities. *J. Ecol.* 61, 225–35.

Hill, M.O. (1973b). Reciprocal averaging: an eigenvector method of ordination. *J. Ecol.* 61, 237–49.

Hill, M.O. (1979). *TWINSPAN – a FORTRAN Program for Arranging Multivariate Data in an Ordered Two-way Table by Classification of Individuals and Attributes.* Ecology and Systematics, Cornell University, Ithaca, New York.

Hill, M.O., Bunce, R.G. & Shaw, M.W. (1975). Indicator species analysis, a divisive polythetic method of classification and its application to a survey of native pinewoods in Scotland. *J. Ecol.* 63, 597–613.

Hill, M.O. & Gauch, H.G. (1980). Detrended correspondence analysis: an improved ordination technique. *Vegetatio* 42, 47–58.

Hill, R.S. & Read, J. (1987). Endemism in Tasmanian cool temperate rainforest: alternative hypotheses. *Bot. J. Linn. Soc. Lond.* 95, 113–24.

Hills, T.L. & Randall, R.E. (Eds) (1968). The ecology of the forest/savanna boundary. Savanna Res. Ser. No. 13. Dept. of Geogr., McGill Univ., Montreal.

Hilty, S.L. (1980). Flowering and fruiting periodicity in a Premontane rain forest in Pacific Colombia. *Biotropica* 12, 292–306.

Hladik, A. (1974). Importance des lianes dans la production foliare de la forêt du nord-est du Gabon. *C.R. Acad. Sci. Paris* 278, 2527–30.

Hladik, A. (1978). Phenology of leaf production in rain forest of Gabon: distribution and composition of food for folivores. Pp. 51–71 in *The Ecology of Arboreal Folivores*, Montgomery, G.G. (Ed.), Smithsonian Inst. Press, Washington, D.C.

Hladik, C. & Hladik, A.M. (1969). Rapports trophiques entre végétation et primates dans la forêt de Barro Colorado (Panama). *Terre Vie* 116, 25–117.

Ho, C.C., Newbery, D.McC. & Poore, M.E.D. (1987). Forest composition and inferred dynamics in Jengka Forest Reserve, Malaysia. *J. Trop. Ecol.* 3, 25–56.

Högberg, P. (1989). Root symbioses of trees in savannas. Pp. 121–36 in *Mineral Nutrients in Tropical Forest and Savanna Ecosystems*, Proctor, J. (Ed.), Br. Ecol. Soc. Spec. Pubn. 9, Blackwell Scientific, Oxford.

Högberg, P. & Alexander, I.J. (1995). Roles of root symbioses in African woodland and forest: evidence from ^{15}N abundance and foliar analysis. *J. Ecol.* 83, 217–24.

Holdridge, L.R. (1947). Determination of world plant formation from Simple climatic data. *Science* 105, 367–8.

Holdridge, L.R., Grenke, W.C., Hatheway, W.H., Liang, T. & Tosi, J.A. Jr. (1971). *Forest Environments in Tropical Life Zones: a Pilot Study*. Pergamon Press, Oxford.

Holdsworth, M. (1963). Intermittent growth of the mango tree. *J. W. Afr. Sci. Ass.* 7, 163–71.

Holtermann, C. (1907). *Die Einfluss des Klimas auf den Bau der Pflanzengewebe. Anatomisch-physiologische Untersuchungen in den Tropen.* W. Englemann, Leipzig.

Holthuijzen, A.M.A. & Boerboom, J.H.A. (1982). The *Cecropia* seedbank in the Surinam lowland rain forest. *Biotropica* 14, 62–8.

Holttum, R.E. (1924). The vegetation of Gunong Belumut in Johore. *Gdns Bull. Singapore* 3, 245–57.

Holttum, R.E. (1931). On periodic leaf-change and flowering of trees in Singapore. *Gdns Bull. Singapore* 5, 173–206.

Holttum, R.E. (1935). The flowering of Tembusu trees (*Fagraea fragrans* Roxb.) in Singapore 1928–35. *Gdns Bull. Singapore* 9, 73–8.

Holttum, R.E. (1938a). The ecology of tropical pteridophytes. Pp. 420–50 in *Manual of Pteridology*, Verdoorn, F. (Ed.), M. Nijhoff, The Hague.

Holttum, R.E. (1938b). Leaf-fall in a non-seasonal climate (Singapore). *Proc. Linn. Soc. Lond.*, 150th session 1937–8, pp. 78–81.

Holttum, R.E. (1940). Periodic leaf-change and flowering of trees in Singapore. II. *Gdns Bull. Singapore* 11, 295–302.

Holttum, R.E. (1953). Evolutionary trends in an equatorial climate. *Symp. Soc. Exp. Biol.* 7 (*Evolution*), 159–73.

Holttum, R.E. (1954). *Plant Life in Malaya*. Longman, Green, London.

Holttum, R.E. (1957). *A Revised Flora of Malaya*, Vol. 1. *Orchids of Malaya*. Second edition, Govt. Print. Off., Singapore.

Hong, E. (1987). *Natives of Sarawak; Survival in Borneo's Vanishing Forests*. Institut Masyarakat, Kuching, Sarawak.

Hong, L.T., Thillainathan, F. & Omar, A. (1984). Observations on the fruiting and growth of some agarics in a dipterocarp stand. *Malay. Nat. J.* 38, 81–8.

Hope, G.S. (1976). The vegetational history of Mt. Wilhelm, Papua New Guinea. *J. Ecol.* 64, 627–64.

Hope, G.S. (1980). New Guinea mountain vegetation communities. *The Alpine Flora of New Guinea* I (*General part*), P. van Royen (Ed.). Cramer, Vauduz.

Hopkins, B. (1962). Vegetation of the Olokemeji Forest Reserve, Nigeria. I General Features of the reserve and the research sites. (Appendix IV by R.N. Jenkins.) *J. Ecol.* 50, 559–98.

Hopkins, B. (1968). Vegetation of the Olokemeji Forest Reserve, Nigeria. V The vegetation of the savanna site with special reference to its seasonal changes. *J. Ecol.* 56, 97–115.

Hopkins, B. (1970a). Vegetation of the Olokemeji Forest Reserve, Nigeria. VI The plants on the forest site with special reference to their seasonal growth. *J. Ecol.* 58, 765–93.

Hopkins, B. (1970b). Vegetation of the Olokemeji Forest Reserve Nigeria. VII The plants on the savanna site with special reference to their seasonal growth. *J. Ecol.* 58, 795–825.

Hopkins, B. (1974). *Forest Ecology*. Second edition, Heinemann Educational Books, Kingswood, Surrey.

Hopkins, H.C. (1984). Floral biology and pollination ecology of the neotropical species of *Parkia*. *J. Ecol.* 72, 1–23.

Hopkins, H.C.F. (1986). *Parkia* (Leguminosae: Mimosoideae). *Flora Neotropica*, Monograph 43. New York Botanical Garden, New York.

Hopkins, M.S. & Graham, A.W. (1983). The species composition of soil seed banks beneath lowland tropical rainforests in north Queensland, Australia. *Biotropica* 15, 90–9.

Horn, H.S. (1971). *The Adaptive Geometry of Trees*. Princeton University Press, Princeton, New Jersey.

Horn, H.S. (1976). Succession. Pp. 252–71 in *Theoretical Ecology*, May, R.M. (Ed.), Blackwell, Oxford.

Hosokawa, T. (1943). Studies on the life-forms of vascular epiphytes and the epiphyte flora of Ponape, Micronesia. *Trans. Nat. Hist. Soc. Taiwan* 33, 35–55; 71–89; 113–41.

Hosokawa, T. (1950). Epiphyte-quotient. *Bot. Mag. Tokyo* 63, 18–20.

Hosokawa, T. (1957). Outline of the mangrove and strand forests of the Micronesian islands. *Mem. Fac. Sci. Kyushu Univ.*, Ser. E (Biol.) 2, 102–8.

Hosokawa, T. (1968). Ecological studies of tropical epiphytes in forest ecosystems. Pp. 482–501 in *Proc. Symp. Recent. Adv. Trop Ecol. (Varanasi) 1968*.

Hou, H.Y. (Ed.) (1979). *Vegetation Map of China*. Chinese Academy of Sciences. Inst. Botany. Lab. Plant Ecol. and Geobot., Publisher of People's Republic of China, Peking.

Howard, R.A. (1968). The ecology of an elfin forest in Puerto Rico, 1. Introduction and composition studies. *J. Arnold Arboretum* 49, 381–418.

Howard, R.A. (1969). The ecology of an elfin forest in Puerto Rico, 8. Studies of stem growth and form, and of leaf structure. *J. Arnold Arboretum* 50, 225–67.

Howard, R.A. (1970). The summit of Pico de Oeste, Puerto Rico. Pp. B325–8 in *A Tropical Rain Forest*, Odum, H.T. & Pigeon, R.F. (Eds), Divn. Tech. Inf., U.S. Atomic Energy Commission, Oak Ridge, Tennessee.

Howard, T.M. (1981). Southern closed forests. Ch. 5 in *Australian Vegetation*, Groves, R.H. (Ed.), Cambridge University Press.

Howe, H.F. (1977). Bird activity and seed dispersal of a tropical wet forest tree. *Ecology* 58, 539–50.

Howe, H.F. & Primack, R.B. (1975). Differential seed dispersal by birds of the tree *Casearia nitida* (Flacourtiaceae). *Biotropica* 7, 278–83.

Hubbell, S.P. (1979). Tree dispersion, abundance and diversity in a tropical dry forest. *Science* 203, 1299–309.

Hubbell, S.P. & Foster, R.B. (1986). Canopy gaps and the dynamics of a neotropical forest. Pp. 77–96 in *Plant Ecology*, Crawley, M.J. (Ed.), Blackwell Scientific Publications, Oxford.

Huber, J. (1902). Contribuição geographia physica dos Furos de Breves e da parte occidental de Marajó. *Bol. Mus. Paraense Hist. Nat.* 3, 447. (*Abstr. Bot. Zentralbl. 1903*, 93, 235–7.)

Hudson, N.W. (1971). *Soil Conservation*. Batsford, London.

Hueck, K. (1966). *Die Wälder Südamerikas*. Fischer, Stuttgart.

Hueck, K. & Seibert, P. (1981). Vegetationskarte von Südamerika (*Vegetationsmonographien der einzelnen Grossräume* 2a). Second edition, Fischer, Stuttgart.

Hughes, N.F. (1976). *Palaeobiology of Angiosperm Origins*. Cambridge University Press.

Hughes, N.F. (1991). *The Enigma of Angiosperm Origins*. Cambridge University Press.

Humboldt, F.H.A. von (1808). *Ansichten der Natur mit wissenschaftlichen Erläuterungen*. Stuttgart & Augsburg.

Humboldt, F.H.A. von (1817). *De Distributione Geographica Plantarum Secundum Coeli Temperiem et Altitudinem Montium Prolegomena*. Paris.

Humboldt, F.H.A. von (1852). *Personal Narrative of Travels to the Equinoctial Regions of America*, 2. Transl. T. Ross, London.

Huston, M. (1980). Soil nutrients and tree species in Costa Rican forests. *J. Biogeogr.* 7, 147–57.

Hutchinson, J. & Dalziel, J.M. (1954–72). *Flora of West Tropical Africa*. Second edition, 3 Vols. (Vol. 1, R.W.J. Keay (Ed.); vols. 2 & 3, F.N. Hepper (Ed.).) Crown Agents for Overseas Governments, London.

Huttel, C. (1975). Recherches sur l'écosystème de la forêt subéquatoriale de basse Côte d'Ivoire. IV. Estimation du bilan hydrique. 29, 192–202.

Huttel, C. & Bernhard-Reversat, F. (1975). Recherches sur l'écosystème de la forêt subéquatoriale de basse Côte d'Ivoire. V. Biomasse végétale et productivité primaire. Cycle de la matière organique. *Terre Vie* 29, 203–28.

Huxley, C. (1978). The ant-plants *Hydnophytum* and *Myrmecodia* (Rubiaceae) and the relationships between their morphology, ant occupants, physiology and ecology. *New Phytol.* 80, 231–68.

Huxley, C. (1980). Symbiosis between ants and epiphytes. *Biol. Rev.* 55, 321–40.

Huxley, P.A. & Eck, W.A. van (1974). Seasonal changes in growth and development of some woody perennials near Kampala, Uganda. *J. Ecol.* 62, 579–92.

Hyland, B.P.M. (1972). A technique for collecting botanical specimens in rain forest. *Flora Malesiana Bull.* 26, 2038–40.

Hynes, R.A. (1974). Altitudinal zonation in New Guinea *Nothofagus* forest. Pp. 75–109 in *Altitudinal Zonation of*

Forests in Malesia, Flenley, J.R. (Ed.), *Univ. Hull, Geogr. Dept. Misc. Ser.* 16.

Ihering, H. von (1923). Die periodische Blattwechsel der Bäume in tropischen und subtropischen Südamerika. *Bot. Jahrb.* 58, 524–98.

Jackson, G. (1974). Cryptogeal germination and other seedling adaptations to the burning of vegetation in savanna regions: the origin of the pyrophytic habit. *New Phytol.* 73, 771–80.

Jackson, I.J. (1971). Problems of throughfall and interception measurement under tropical forest. *J. Hydrol.* 12, 234–54.

Jackson, I.J. (1975). Relationships between rainfall parameters and interception by tropical forest. *J. Hydrol.* 24, 215–38.

Jackson, I.J. (1989). *Climate, Water and Agriculture in the Tropics.* Second edition, Longman, London.

Jackson, J.F. (1978). Seasonality in a Brazilian forest. *Biotropica* 10, 38–42.

Jacobs, M. (1974). Botanical panorama of the Malesian Archipelago (Vascular Plants). In *Natural Resources of Humid Tropical Asia. Natural Resources Research* 12, 263–94. UNESCO, Paris.

Jacobs, M. (1976). The study of lianas. *Flora Malesiana Bull.* 29, 2610–8.

Jaeger, P. & Adams, J.G. (1967). Sur la présence en Piedmont Ouest des Monts Loma (Sierra Leone) d'un groupement forestier relictuel à *Tarrietia utilis* Sprague (Sterculiacées). *C.R. Acad.Sci. Paris* 265, 1627–9.

Jane, C. (1930). *Select Documents Illustrating the Four Voyages of Columbus.* Translated and edited by Cecil Jane. Vol. I (*First and second voyages*). Hakluyt Society, London.

Janos, D.P. (1980). Mycorrhizae influence tropical succession. *Biotropica* 12 (suppl.), 56–64.

Janos, D.P. (1983). Tropical mycorrhizas, nutrient cycles and plant growth. Pp. 327–45 in *Tropical Rain Forest Ecology and Management*, Sutton, S.L., Whitmore, T.C. & Chadwick, A.C. (Eds), Br. Ecol. Soc. Spec. Pubn 2, Blackwell, Oxford.

Janson, C.H., Terborgh, J. & Emmons, L.H. (1981). Non-flying mammals as pollinating agents in the Amazonian forest. *Biotropica* 13 (suppl.), 1–6.

Janzen, D.H. (1966). Coevolution of mutualism between ants and acacias in Central America. *Evolution* 20, 249–75.

Janzen, D.H. (1967). Interaction of the bull's-horn acacia (*Acacia cornigera* L.) with an ant inhabitant (*Pseudomyrmex ferruginea* F. Smith) in eastern Mexico. *Univ. Kansas Sci. Bull.* 47, 315–558.

Janzen, D.H. (1970). Herbivores and the number of tree species in Tropical Forests. *Amer. Nat.* 104, 501–28.

Janzen, D.H. (1971). Euglossine bees as long-distance pollinators of tropical plants. *Science* 171, 203–5.

Janzen, D.H. (1972). Protection of *Barteria* (Passifloraceae) by *Pachysima* ants (Pseudomyrmecinae) in a Nigerian rain forest. *Ecology* 53, 885–92.

Janzen, D.H. (1974a). Tropical blackwater rivers, animals and mast fruiting by the Dipterocarpaceae. *Biotropica* 6, 69–103.

Janzen, D.H. (1974b). Epiphytic myrmecophytes in Sarawak: mutualism through feeding of plants by ants. *Biotropica* 6, 237–59.

Janzen, D.H. (1975). *Ecology of Plants in the Tropics.* Inst. of Biol. Studies in Biology 58. E. Arnold, London.

Janzen, D.H. (1976a). Why bamboos wait so long to flower. *Ann. Rev. Ecol. Syst.* 7, 347–91.

Janzen, D.H. (1976b). Why tropical trees have rotten cores. *Biotropica* 8, 110.

Janzen, D.H. (1977a). Why don't ants visit flowers? *Biotropica* 9, 252.

Janzen, D.H. (1977b). Promising direction of study in tropical animal-plant interactions. *Ann. Missouri Bot. Gdn.* 64, 700–36.

Janzen, D.H. (1977c). *Musanga cecropioides* is a *Cecropia*. *Biotropica* 9, 57.

Janzen, D.H. (1978). Seeding patterns of tropical trees. Pp. 83–128 in *Tropical Trees as Living Systems*, Tomlinson, P.B. & Zimmermann, M.H. (Eds), Cambridge University Press.

Janzen, D.H. (1979). How to be a fig. *Ann. Rev. Ecol. Syst.* 10, 13–51.

Janzen, D.H. (1981). Digestive seed predation by a Costa Rican Baird's tapir. *Biotropica* 13 (suppl.), 59–63.

Janzen, D.H. (1983). (Ed.) *Costa Rican Natural History.* Chicago Univ. Press.

Jeník, J. (1967). Root adaptations in West African trees. *Bot. J. Linn. Soc. Lond.* 60, 25–9.

Jeník, J. (1969). The life-form of *Scaphopetalum amoenum* A. Chev. *Preslia* (Praha) 41, 109–12.

Jeník, J. (1970a). Root system of tropical trees 4. The stilted peg-roots of *Xylopia staudtii* Engl. et Diels. *Preslia* (Praha) 42, 25–32.

Jeník, J. (1970b). Root system of tropical trees 5. The peg-roots and the pneumathodes of *Laguncularia racemosa* Gaertn. *Preslia* (Praha) 42, 105–13.

Jeník, J. (1971a). Root structure and underground biomass in equatorial forests. Pp. 323-31 in *Productivity of Forest Ecosystems, Proc. Brussels Symp. UNESCO 1969*, UNESCO, Paris.

Jeník, J. (1971b). Root system of tropical trees 6. The aerial roots of *Entandrophragma angolense* (Welw.) C. DC. *Preslia* (Praha) 43, 1–4.

Jeník, J. (1971c). Root system of tropical trees 7. The facultative peg-roots of *Anthocleista nobilis* G. Don. *Preslia* (Praha) 43, 97–104.

Jeník, J. (1973). Root system of tropical trees 8. Stilt-roots and allied adaptations. *Preslia* (Praha) 45, 250–64.

Jeník, J. (1978). Roots and root systems in tropical trees: morphologic and ecologic aspects. Ch. 14 in *Tropical Trees As Living Systems*, Tomlinson, P.B. & Zimmermann, M.H. (Eds), Cambridge University Press.

Jeník, J. & Harris, B.J. (1969). Root-spines and spine-roots in dicotyledonous trees of tropical Africa. *Osterr. Bot. Z.* 117, 128–38.

Jeník, J. & Kubikova, J. (1969). Root system of tropical trees 3. The heterorhizis of *Aeschynomene elaphroxylon* (Guill. et Perr.) Taub. *Preslia* (Praha) 41, 220–6.

Jeník, J. & Lawson, G.W. (1968). Zonation of microclimate and vegetation on a tropical shore in Ghana. *Oikos* 19, 198–205.

Jiménez, J.A. (1984). A hypothesis to explain the reduced distribution of the mangrove *Pelliciera rhizophorae* Tr. & Pl. *Biotropica* 16, 304–8.

Johansson, D. (1974). Ecology of vascular epiphytes in West African rain forest. *Acta Phytogeogr. Suecica* 59, 1–129.

Johns, A.D. (1988). Effects of 'selective' timber extraction on rain forest structure and composition and some consequences for frugivores and folivores. *Biotropica* 20, 31–7.

Johns, R.J. (1986). The instability of the tropical ecosystem in Papua New Guinea. *Blumea* 31, 341–71.

Johnson, A.M. (1976). The climate of Peru, Bolivia and Ecuador. Pp. 147–218 in *Climates of Central and South America*, Schwerdtfeger, W. (Ed.), Elsevier, Amsterdam.

Johnson, N.E. (1976). Biological opportunities and risks associated with fast-growing plantations in the tropics. *J. Forestry* 74, 206–11.

Johnson, P.L. & Atwood, D.M. (1970). Aerial sensing and photographic study of the El Verde Rain forest. Pp. B63–78 in *A Tropical Rain Forest: A Study of Irradiation and Ecology at El Verde, Puerto Rico*, Odum, H.T. & Pigeon, R.F. (Eds), Divn. Tech. Inf., U.S. Atomic Energy Commission, Oak Ridge, Tennessee.

Johnston, M.H. (1992). Soil-vegetation relationships in a tabonuco forest community in the Luquillo mountains of Puerto Rico. *J. Trop. Ecol.* 8, 253–63.

Johnstone, I.M. (1978). The mangroves and mangrove ecosystems of Papua New Guinea. Paper presented at 2nd Internat. Congr. Ecol., Jerusalem, 1978.

Johow, F. (1885). Die chlorophyllfreien Humusbewohner West-Indiens. *Jahrb. wiss. Bot.* 16, 415–49.

Johow, F. (1889). Die chlorophyllfreien Humuspflanzen nach ihrer biologischen und anatomisch-entwicklungsgeschichtlichen Verhältnissen. *Jahrb. wiss. Bot.* 20, 475–525.

Jones, A.P.D. (1945). Notes on terms for use in vegetation description in southern Nigeria. *Farm & Forest (Nigeria)* 6, 130–6.

Jones, A.P.D. (1947). Botanica nigerica. *Farm & Forest (Nigeria)* 8, 10–16.

Jones, A.P.D. (1948). *The Natural Forest Inviolate Plot.* Nigerian Forest Dept., Ibadan.

Jones, E.W. (1950). Some aspects of natural regeneration in the Benin rain forest. *Emp. For. Rev.* 29, 108–24.

Jones, E.W. (1955). Ecological studies on the rain forest of southern Nigeria. IV. The plateau forest of the Okomu Forest Reserve, Part I, The environment, the vegetation types of the forest and the horizontal distribution of species. *J. Ecol.* 43, 564–94.

Jones, E.W. (1956). Ecological studies on the rain forest of southern Nigeria. IV. The plateau forest of the Okomu Forest Reserve, Part II, The reproduction and history of the forest. *J. Ecol.* 44, 83–117.

Jones, E.W. (1963). The forest outliers in the Guinea zone of northern Nigeria. *J. Ecol.* 51, 415–34.

Jordan, C., Golley, F., Hall, J. & Hall, J. (1980). Nutrient scavenging of rainfall by the canopy of the Amazonian rain forest. *Biotropica* 12, 61–6.

Jordan, C.F. (Ed.) (1970). A progress report on studies of mineral cycles at El Verde. Pp. H217–19 in *A Tropical Rain Forest: A Study of Irradiation and Ecology at El Verde, Puerto Rico,* Odum, H.T. & Pigeon, R.F. (Eds), Divn. Tech. Inf., U.S. Atomic Energy Commission, Oak Ridge, Tennessee.

Jordan, C.F. (1977). Distribution of elements in tropical montane rain forest. *Trop. Ecol.* 18, 124–30.

Jordan, C.F. (1978). Stem flow and nutrient transfer in a tropical rain forest. *Oikos* 31, 257–63.

Jordan, C.F. (1982). The nutrient balance of an Amazonian rain forest. *Ecology* 63, 647–54.

Jordan, C.F. (1985). *Nutrient Cycling in Tropical Forest Ecosystems.* John Wiley, Chichester.

Jordan, C.F. (Ed.) (1986). *Amazonian Rain Forests: Ecosystems Disturbance and Recovery (Ecological Studies* 60). Springer, New York.

Jordan, C.F. (1989). An Amazonian rain forest. The structure and function of a nutrient stressed ecosystem and the impact of slash and burn agriculture. MAB Series 2, Parthenon, Carnforth, Lancs., for UNESCO.

Jordan, C.F. & Herrera, R. (1981). Tropical rainforests: are nutrients really critical? *Amer. Nat.* 117, 167–80.

Jordan, C.F. & Kline, J.R. (1977). Transpiration of trees in a tropical rainforest. *J. Appl. Ecol.* 14, 853–60.

Junghuhn, F. (1852). *Java, seine Gestalt, Pflanzendecke und innere Bauart.* Transl. Hasskarl. Leipzig.

Jungner, J.R. (1891). Anpassungen der Pflanzen an das Klima in den Gegenden der regenreichen Kamerungebirge. *Bot. Zbl.* 47, 353–60.

Junk, W. (1970). Investigations on the ecology and production-biology of the 'Floating Meadows', Part 1, the floating vegetation and its ecology. *Amazoniana* 2, 449–95.

Junk, W.J. (1989) Flood tolerance and tree distribution in the central Amazonian floodplain. Pp. 47–64 in *Tropical Forests: Botanical Dynamics, Speciation and Diversity,* Holm-Nielsen, L., Balslev, H. & Nelson, I. (Eds) (*Proc. Conf. Tropical Forests,* University of Aarhus, Denmark, August 1988), Academic Press, London.

Juo, A.S.R. (1981). Chemical characteristics. Pp. 51–79 in *Characterisation of Soils,* Greenland, D.J. (Ed.), Oxford University Press.

Juo, A.S.R. & Kang, B.T. (1989). Nutrient effects of modification of shifting cultivation in West Africa. Pp. 289–309 in *Mineral Nutrients in Tropical Forest and Savanna Ecosystems,* Proctor, J. (Ed.), Br. Ecol. Soc. Spec. Pubn 9, Blackwell Scientific, Oxford.

Kadomura, H. (1984). Problems of past and recent environmental changes in the humid areas of Cameroon. Pp. 7–20 in *Natural and Man-Induced Environmental Changes in Tropical Africa: Case Studies in Cameroon and Kenya,* Kadomura, H. (Ed.), Hokkaido Univ., Sapporo.

Kadomura, H. (1989). Savannization in Tropical Africa. Pp. 3–16 in *Savannization Processes in Tropical Africa* I, Kadomura, H. (Ed.), Dept. of Geography, Tokyo Metropolitan Univ. and Zambia Geographical Association.

Kahn, F. & de Castro, A. (1985). The palm community in a forest of central Amazonia, Brazil. *Biotropica* 17, 210–16.

Kalliola, R., Puhakka, M., Salo, J., Tuomisto, H. & Ruokolainen, K. (1991a). The dynamics, distribution and classification of swamp vegetation in Peruvian Amazonia. *Ann. Bot. Fennici* 28, 225–39.

Kalliola, R., Salo, J. Puhakka, M. & Rajasilta, M. (1991b). New site formation and colonizing vegetation in primary succession on the western Amazon floodplain. *J. Ecol.* 79, 877–901.

Kang, B.T. & Moormann, F.R. (1977). Effects of some biological factors on soil variability in the tropics. I Effect of pre-clearing vegetation. *Plant and Soil* 47, 441–9.

Kapos, V. (1989). Effects of isolation on the water status of forest patches in the Brazilian Amazon. *J. Trop. Ecol.* 5, 173–85.

Kartawinata, K. (1978). The 'Kerangas' Heath forest in Indonesia. In *Glimpses of Ecology (Professor R.R. Misra, Commemoration vol.),* Singh, J.S. & Gopal, B. (Eds). Internat. Scientif. Publs, Jaipur, India.

Kartawinata, K. (1980). A note on kerangas (Heath) forest at Sebulu, East Kalimantan. *Reinwardtia* 9, 429–47.

Kartawinata, K., Abdulhadi, R. & Partomihardjo, T. (1981). Composition and structure of a lowland dipterocarp forest at Wanariset, East Kalimantan. *Malay. For.* 44, 397–406.

Kato, R., Tadaki, Y. & Ogawa, H. (1978). Plant biomass and growth increment studies in Pasoh Forest. *Malay. Nat. J.* 30, 211–24.

Kaur, A., Ha, C.O., Jong, K., Sands, V.E., Chan, H.T., Soepadmo, E. & Ashton, P.S. (1978). Apomixis may be widespread among trees of the climax rain forest. *Nature, Lond.* 271, 440–2.

Kaur, A., Jong, K., Sands, V.E. & Soepadmo, E. (1986). Cytoembryology of some Malaysian dipterocarps, with some evidence of apomixis. *Bot. J. Linn. Soc. Lond.* 92, 75–88.

Keay, R.W.J. (1953). *Rhizophora* in West Africa. *Kew Bull.* 1953, 121–7.

Keay, R.W.J. (1957). Wind-dispersed species in a Nigerian forest. *J. Ecol.* 45, 471–8.

Keay, R.W.J. (1959a). Lowland vegetation on the 1922 lava flow, Cameroons Mountain. *J. Ecol.* 47, 25–9.

Keay, R.W.J. (1959b). *An Outline of Nigerian Vegetation.* Third edition, Govt. Printer, Nigeria.

Keay, R.W.J. (1959c). Derived savanna – derived from what? *Bull. IFAN* 21, sér. A, 28–438.

Keay, R.W.J. (1960). Seeds in forest soils. *Niger. For. Inf. Bull.,* N.S. 4, 1–12.

Keel, S.H.K. & Prance, G.T. (1979). Studies of the vegetation of a white-sand black-water igapó (Rio Negro, Brazil). *Acta Amazon.* 9, 645–55.

Kellman, M.C. (1969). Some environmental components of shifting cultivation in upland Mindanao. *J. Trop. Geogr.* 28, 40–56.

Kellman, M.C. (1970). *Secondary Plant Succession in Tropical*

Montane Mindanao. Research School for Pacific Studies Pubn. BG2, Australian Nat. Univ., Canberra.

Kellman, M.C. (1984). Synergistic relationships between fire and low soil fertility in neotropical savannas: a hypothesis. *Biotropica* 16, 158–60.

Kellman, M.C. (1989). Mineral nutrient dynamics during savanna-forest transformation in Central America. Pp. 137–51 in *Mineral Nutrients in Tropical Forest and Savanna Ecosystems,* Proctor, J. (Ed.), Br. Ecol. Soc. Spec. Pubn 9, Blackwell Scientific, Oxford.

Kelly, D.L., Tanner, E.V.J., Kapos, V., Dickinson, T.A., Goodfriend, G.A. & Fairbairn, P. (1988). Jamaican limestone forests: floristics, structure and enironment of three examples along a rainfall gradient. *J. Trop. Ecol.* 4, 121–36.

Kendall, R.L. (1969). An ecological history of the Lake Victoria basin. *Ecol. Monogr.* 39, 121–76.

Kennedy, J.D. (1936). *The Forest Flora of Southern Nigeria.* Government Printer, Lagos.

Kenoyer, L.A. (1929). General and successional ecology of the Lower Tropical Rain-Forest at Barro Colorado Island, Panama. *Ecology* 10, 201–22.

Kenworthy, J.B. (1971). Water and nutrient cycling in a tropical rain forest. Pp. 49–65 in *The Water Relations of Malesian Forests,* Flenley, J.R. (Ed.), *Univ. Hull, Dept. Geogr. Misc. Ser.* 11.

Kenworthy, J.J. (1966). Temperature conditions in the tropical highland climates of East Africa. *East Afr. Geogr. Rev.* 4, 1–11.

Kesel, R.H. (1977). Slope runoff and denudation in the Rupununi Savanna, Guyana. *J. Trop. Geogr.* 44, 33–42.

Kiew, R. (1982). Observations on leaf color, epiphyll cover, and damage on Malayan *Iguanura wallichiana. Principes* 26, 200–4.

Kiew, R. (1986). Phenological studies on some rain forest herbs in Peninsular Malaysia. *Kew Bull.* 41, 733–46.

King, R.B., Baillie, I.C., Bissett, P.G., Grimble, R.J., Johnson, M.S. and Silva, G.L. (1986). *Land Resource Survey of Toledo District, Belize.* Report P 177. Land Resources Development Centre, Tolworth.

King, R.B., Baillie, I.C., Dusmore, J.R., Grimble, R.J., Johnson, M.S., Williams, J.B. & Wright A.C.S. (1989). *Land resource assessment of Stann Creek district, Belize.* Bulletin 19. Overseas Development Nat. Res. Inst., Chatham.

Kira, T. (1978). Community architecture and organic matter dynamics in tropical lowland rain forests of southeast Asia with special reference to Pasoh forest, West Malaysia. Pp. 561–90 in *Tropical Trees as Living Systems,* Tomlinson, P.B. & Zimmermann, M.H. (Eds), Cambridge University Press.

Kira, T. (1991). Forest ecosystems of East and Southeast Asia in a global perspective. *Ecol. Research (Ecol. Soc. of Japan)* 6, 185–200.

Kira, T. & Ogawa, H. (1971). Assessment of primary production in tropical and equatorial forests. Pp. 309–21 in *Proc. Brussels Symposium, 1969. Productivity of Forest Ecosystems,* Duvigneaud, P. (Ed.), UNESCO, Paris.

Kira, T., Ogawa, H. & Yoda, K. (1962). Some unsolved problems in tropical forest ecology. *Proc. Ninth Pacific Sci. Congress 1957* 4, 124–34.

Kira, T.C., Shinozaki, K. & Hozumi, K. (1969). Structure of forest canopies as related to their primary productivity. *Plant Cell Physiol.* 10, 129–42.

Klebs, G. (1911). Ueber die Rhythmik in der Entwicklung der Pflanzen. *Sitz. ber. heidelberg. Akad. Wiss., math.-nat. Kl.,* Abhandl. 23.

Klebs, G. (1926). Ueber periodisch wachsende tropischer Baumarten. *Sitz. ber. heidelberg. Akad. Wiss., math.-nat. Kl.,* Abhandl. 2.

Kline, J.R. & Jordan, C.F. (1968). Tritium movement in soil of tropical rain forest. *Science* 160, 550–1.

Kline, J.R., Martin, J.R., Jordan, C.F. & Koranda, J.J. (1970). Measurement of transpiration in tropical trees with tritiated water. *Ecology* 51, 1068–73.

Klinge, H. (1962). Ueber Epiphyten humus aus El Salvador, Zentralamerika. I. Die Standorte und ihre gesamt-ökologische Situation. *Pedobiologia* 2, 1–8.

Klinge, H. (1963). Ueber Epiphyten aus El Salvador, Zentralamerika. II. Kennzeichnung des Humus durch analytische Merkmale. *Pedobiologia* 2, 102–7.

Klinge, H. (1966). Humus im Kronenraum tropischer Wälder. *Die Umschau in Wissensch. u. Technik* (Frankfurt a. M.) 1966 (hft. 4).

Klinge, H. (1967). Podzol soils: a source of blackwater rivers. Pp. 117–25 in *Atas do Simposio sôbre a Biota Amazônica* 3 (*Limnologia*).

Klinge, H. (1973). Struktur und Artenreichtum des zentralamazonischen Regenwaldes. *Amazoniana* (Kiel) 4, 283–92.

Klinge, H. (1978). Studies on the ecology of Amazonian caatinga forest in southern Venezuela. I. Biomass dominance of selected tree species in the Amazon caatinga near San Carlos de Rio Negro. *Acta Cient. Venezolana* 29, 258–62.

Klinge, H., Furch, K., Harms, E. & Revilla, J. (1983). Foliar nutrient levels of native tree species from Central Amazonia. I. Inundation forests. *Amazoniana* 8, 19–45.

Klinge, H. & Herrera, R. (1983). Phytomass structure of natural plant communities in southern Venezuela: the tall Amazon caatinga forest. *Vegetatio* 53, 65–84.

Klinge, H. & Medina, E. (1979). Rio Negro caatingas and campinas, Amazonas States of Venezuela and Brazil. Ch. 22 in *Heathlands and Related Shrublands of the World. A. Descriptive Studies,* Specht, R.L. (Ed.) (Ecosystems of the World 9B), Elsevier, Amsterdam).

Klinge, H. & Rodrigues, W.A. (1974). Phytomass estimation in a central Amazonian rain forest. Pp. 339–49 in *IUFRO Biomass Studies,* Young, H.E. (Ed.), Univ. Press, Orono, Maine.

Klinge, H., Rodrigues, W.A., Brünig, E.F. & Fittkau, E.J. (1975). Biomass and structure in a central Amazonian forest. Pp. 115–22 in *Tropical Ecological Systems: Trends in Terrestrial and Aquatic Research,* Golley, F.B. & Medina, E. (Eds), Springer, New York.

Klink, C.A. & Joly, C.A. (1989). Identification and distribution of C3 and C4 grasses in open and shaded habitats in São Paulo State, Brazil. *Biotropica* 21, 30–4.

Knight, D.H. (1975). A phytosociological analysis of species-rich tropical forest on Barro Colorado Island, Panama. *Ecol. Monogr.* 45, 259–84.

Knoch, K. (1930). Klimakunde von Sudamerika. Vol. 2, Part G. In *Handbuch der Klimatologie,* Köppen, W. & Geiger, R. (Eds), Gebrüder Bornträger, Berlin.

Koch, M.S. & Mendelssohn, I.A. (1989). Sulphide as a phytotoxin: differential responses in two marsh species. *J. Ecol.* 77, 565–78.

Kochummen, K.M. (1966). Natural plant succession after farming in Sg. Kroh. *Malay. For.* 29, 170–81.

Kochummen, K.M. (1982). Effects of elevation on vegetation on Gunung Jerai, Kedah. *For. Res. Inst. Kepong (Peninsular Malaysia) Res. Pamph.* 87.

Kochummen, K.M., LaFrankie, J.V. Jr & Manokaran, N. (1990). Floristic composition of Pasoh Forest Reserve, a lowland rain forest in Peninsular Malaysia. *J. Trop. For. Sci.* 3, 1–15.

Kochummen, K.M., LaFrankie, J.V. Jr & Manokaran, N. (1992). Diversity of trees and shrubs in Malaya at regional and local level. *Malayan Nature J.* 45, 545–54.

Kochummen, K.M. & Ng, F.S.P. (1977). Natural plant succession after farming in Kepong. *Malay. For.* 40, 61–78.

Koelmeyer, K.O. (1959, 1960). The periodicity of leaf change and flowering in the principal forest communities of Ceylon. *Ceylon For.* 4, 157–89, 308–64.

Koernicke, M. (1910). Biologische Studien an Loranthaceen. *Ann. Jard. Bot. Buitenz.*, Suppl. 3, 665–97.

Koopman, M.J.F. & Verhoef, L. (1938). *Eusideroxylon zwageri* T. & B., het ijzerhout van Borneo en Sumatra. *Tectona* 31, 381–99.

Koppel, C. van der (1926). Winning van copal in het Gouvernement Celebes en onderhoorigheden, de uitvoer uit Makassar, en eenige details over het gebruik van copal. *Tectona* 19, 525–74.

Köppen, W. (1918). Klassifikation der Klimate nach Temperatur, Niederschlag und Jahreslauf. *Peterm. Geogr. Mitt.* 64, 193–203, 243–8.

Köppen, W. (1936). Das geographische System der Klimate. Vol. 1, Part C. In *Handbuch der Klimatologie*, Köppen, W. & Geiger, R. (Eds), Gebrüder Bornträger, Berlin.

Koriba, K. (1958). On the periodicity of tree-growth in the tropics. *Gdns Bull. Singapore* 17, 11–81.

Korning, J. & Balslev, H. (1994). Growth and mortality of trees in Amazonian tropical rain forest in Ecuador. *J. Veg. Sci.* 4, 77–86.

Kortlandt, A. (1982). Vegetation research and the 'bulldozer herbivores' in tropical Africa. (Unpubl.).

Kousky, V.E. (1980). Diurnal rainfall variation in Northeast Brazil. *Mon. Wea. Rev.* 108, 488–98.

Koyama, H. (1978). Photosynthesis studies in Pasoh forest. *Malay. Nat. J.* 30, 253–8.

Kramer, F. (1933). De natuurlikke verjonging in het Goenoeng-Gedehcomplex. *Tectona* 26, 156–85.

Kramer, P.J. (1983). *Water Relations of Plants*. Academic Press, New York.

Kress, W.J. (1986). A symposium: the biology of tropical epiphytes. *Selbyana* 9, 1–22.

Kubitzki, K. & Ziburski, A. (1994). Seed dispersal in flood plain forests of Amazonia. *Biotropica* 26, 30–43.

Kuijt, J. (1969). *The Biology of Parasitic Flowering Plants*. Univ. California Press, Berkeley.

Kunkel, G. (1965a). Der Standort: Kompetenzfaktor in der Stelzwurzelbildung. *Biol. Zentralbl.* 84, 641–51.

Kunkel, G. (1965b). Ueber die Struktur und sukzession der Mangrove Liberias und deren Randformationen. *Ber. schweiz. bot. Ges.* 75, 20–40.

Kunkel, G. (1965c). *The Trees of Liberia* (Report no. 3, German Forestry Mission to Liberia). Bayer, Landwirtschafteverlag, München.

Kursar, T.A. & Coley, P.D. (1992). Delayed greening in tropical leaves: an antiherbivore defense? *Biotropica* 24 (suppl.), 256–62.

Laan, E. van der (1926). Analyse der bosschen in de onderafdeeling Pleihari van der afdeeling Banjermasin der Zuider- an Oostafdeeling van Borneo. *Tectona* 19, 103–23.

Labouriau, L.G. (1966). Revisão da situação da ecologia vegetal nos cerrados. *Anais Acad. Braz. Ciencias* 38, Supplemento (Segundo Simposio sôbre o Cerrado), 5–38.

Laessle, A.M. (1961). A micro-limnological study of Jamaican bromeliads. *Ecology* 42, 499–517.

Lal, R. (1986). *Tropical Ecology and Physical Edaphology*. J. Wiley, Chichester.

Lam, H.J. (1945). Fragmenta Papuana. (Observations of a naturalist in Netherlands New Guinea). Transl. L.M. Perry. *Sargentia* 5.

Lamb, D. (1990). *Exploiting the Tropical Rain Forest: an account of pulpwood logging in Papua New Guinea*. (MAB Series 3). Parthenon, Carnforth, Lancs (for UNESCO).

Lamb, H.H. (1966). Climate in the 1960s. *Geogr. J.* 132, 183–212.

Lambert, J.D., Arnason, J.T. & Gale, J. (1984). Mineral cycling in a tropical palm forest. *Plant Soil* 79, 211–25.

Lamprecht, H. (1958). Der Gebirgs-Nebelwald der Venezolanischen Anden. *Schweiz. Zeitschr. f. Forstwesen* 2.

Landsberg, J.J. (1984). Physical aspects of the water regime of wet tropical vegetation. Pp. 13–25 in *Physiological Ecology of Plants of the Wet Tropics*, Medina, E., Mooney, H.A. & Vasques-Yanes, C. (Eds), Junk, The Hague.

Lane-Poole, C.E. (1925). The forests of Papua and New Guinea. *Emp. For. J.* 4, 206–34.

Langdale-Brown, I., Osmaston, H.A. & Wilson, J.G. (1964). *The Vegetation of Uganda and its bearing on land-use*. Govt of Uganda, Entebbe.

Lanjouw, J. (1936). Studies of the Surinam savannahs and swamps. *Nederl. Kruidk. Arch.* 46, 823–51.

Lanly, J.P. (1982). Tropical forest resources. *FAO For. Pap.* 30.

Lausberg, T. (1935). Quantitative Untersuchungen über die kutikulare Excretion des Laubblattes. *Jahrb. Wiss. Bot.* 81, 769–806.

Lawson, G.W. (1986). Coastal vegetation. Pp. 195–213 in *Plant Ecology in West Africa*, Lawson, G.W. (Ed.), J. Wiley, Chichester.

Lawson, G.W., Armstrong-Mensah, K.O. & Hall, J.B. (1970). A catena in tropical moist semi-deciduous forest near Kade, Ghana. *J. Ecol.* 58, 371–98.

Lawton, R. & Dryer, V. (1980). The vegetation of the Monteverde Cloud Forest Reserve. *Brenesia* 18, 101–16.

Leakey, R.J.G. & Proctor, J. (1987). Invertebrates in the litter and soil at a range of altitudes on Gunung Silam, a small ultrabasic mountain in Sabah. *J. Trop. Ecol.* 3, 119–29.

Lebrun, J. (1935). Les essences forestières des régions montagneuses du Congo oriental. *Publ. Inst. Agron. Congo Belge Sér. Sci.* 1.

Lebrun, J. (1936a). Répartition de la forêt équatoriale et des formations végétales limitrophes. Ministr. Colon., Direct. Gén. Agr. Elev., Brussels.

Lebrun, J. (1936b). Le forêt équatoriale congolaise. *Bull. Agric. Congo Belge* 27, 163–92.

Lebrun, J. (1936c). Observations sur la morphologie et l'écologie des contreforts du *Cynometra alexandri* au Congo Belge. *Inst. Roy. Colon. Belge, Bull. des Séances* 7, 573–84.

Lebrun, J. (1947). La végétation de la plaine alluviale au sud du Lac Edouard. *Exploration du Parc National Albert, Mission J. Lebrun (1937–8)*, Fasc. 1. Inst. Parcs Nat. Congo Belge, Brussels.

Lebrun, J. (1957). Sur les éléments et groupes phytogéographiques de la flore de Ruwenzori (versant occidental). *Bull. Jard. Bot. Etat. Bruxelles* 27, 453–78.

Lebrun, J. (1960a). Etudes sur la flore et la végétation des champs de lave au nord du Lac Kivu (Congo Belge). *Exploration du Parc National Albert, Mission J. Lebrun (1937–8)*, Fasc. 2. Inst. Parcs Nat. Congo Belge, Brussels.

Lebrun, J. (1960b). Sur la richesse de la flore de divers territoires africains. *Bull. Acad. Roy. Sci. d'Outre-Mer (Bruxelles)* N.S. 6, 669–90.

Lebrun, J. (1960c). Sur une méthode de délimitation des horizons et étages de végétation des montagnes du Congo oriental. *Bull. Jard. Bot. Etat. Bruxelles* 30, 75–94.

Lebrun, J. (1960d). Sur les horizons et étages de végétation de divers volcans du masse des Virunga (Kivu-Congo). *Bull. Jard. Bot. Etat. Bruxelles* 30, 255–77.

Lebrun, J. (1961). Le concept de 'synusie' en écologie végétale. *Bull. Acad. Roy. Belgique*, Sci. Sér. 5, 47, 169–78.

Lebrun, J. (1966). Les formes biologiques dans les végétations tropicales. *Bull. Soc. Bot. France, Mémoires* (1966), 164–75.

Lebrun, J. (1968). A propos du rythme végétatif de l'*Acacia albida* Del. *Collectanea Botanica* 7 (in memoriam Dr P. Font Quer), 625–36.

Lebrun, J. (1969). La végétation psammophile du littoral congolais. *Acad. Roy. Sciences d'Outre-Mer, (Brussels)*, Cl. Sci. nat. méd. N.S. 18 (1), 1–166.

Lebrun, J. & Gilbert, G. (1954). Une classification écologique des forêts du Congo. *Publ. INEAC, Sér. sci. no. 63.*

Ledger, D.C. (1975). The water balance of an exceptionally wet catchment area in West Africa. *J. Hydrol.* 24, 207–14.

Lee, D.W. (1986). Unusual strategies of light absorption in rain-forest herbs. Pp. 105–31 in *On the Economy of Plant Form and Function*, Givnish, T.J. (Ed.), Cambridge University Press.

Lee, D.W., Brammeir, S. & Smith, A.P. (1987). The selective advantages of anthocyanins in developing leaves of mango and cacao. *Biotropica* 19, 40–9.

Lee, D.W. & Graham, R. (1986). Leaf optical properties of rainforest of sun and extreme shade plants. *Amer. J. Bot.* 73, 1100–8.

Lee, D.W. & Lowry, J.B. (1975). Physical basis and ecological significance of iridescence in blue plants. *Nature, Lond.* 254, 50–1.

Lee, D.W. & Lowry, J.B. (1980). Young-leaf anthocyanin and solar ultraviolet. *Biotropica* 12, 75–6.

Lee, D.W., Lowry, J.B. & Stone, B.G. (1979). Abaxial anthocyanin layer in leaves of tropical rain forest plants: enhancer of light capture in deep shade. *Biotropica* 11, 70–7.

Lee, P.C. (1967). *Ecological Studies of* Dryobalanops aromatica *Gaertn. F.* Ph.D. Thesis, University of Malaya, Kuala Lumpur.

Legris, P. & Viart, M. (1959). Documentation and method proposed for vegetation mapping at one millionth scale. *Trav. Sect. Sci. Tech. Inst. Français Pondichéry* 1, fasc. 4, 197–206.

Leigh, C.H. (1978a). Slope hydrology and denudation in the Pasoh Forest Reserve. I – Surface wash: experimental techniques and some preliminary results. *Malay. Nat. J.* 30 (2), 179–97.

Leigh, C.H. (1978b). Slope hydrology and denudation in the Pasoh Forest Reserve. II – Throughflow: experimental techniques and some preliminary results. *Malay. Nat. J.* 30 (2), 199–210.

Leigh, E.G., Jr. (1983). Introduction: why are there so many kinds of tropical trees? Pp. 63–6 in *The Ecology of a Tropical Forest: Seasonal Rhythms and Long-Term Changes*, Leigh, E.G. Jr., Rand, A.S. & Windsor, D.M. (Eds), Oxford University Press.

Leigh, E.G., Jr. & Windsor, D.M. (1983). Forest production and regulation of primary consumers on Barro Colorado Island. Pp. 111–22 in *The Ecology of a Tropical Forest: Seasonal Rhythms and Long-Term Changes*, Leigh, E.G., Jr., Rand, A.S. & Windsor, D.M. (Eds), Oxford University Press.

Leighton, M. & Leighton, D.R. (1983). Vertebrate responses to fruiting seasonality within a Bornean rain forest. Pp. 181–96 in *Tropical Rain Forest Ecology and Management*, Sutton, S.L., Whitmore, T.C. & Chadwick, A.C. (Eds), Br. Ecol. Soc. Spec. Pubn 2, Blackwell, Oxford.

Leighton, M. & Wirawan, N. (1986). Catastrophic drought and fire in Borneo rain forests associated with the 1982–3 El Nino southern oscillation event. In *Tropical rain forests and the world atmosphere*, Prance, G.T. (Ed.), Westview, Boulder, Colorado.

Lemon, E., Allen, L.H. & Muller, L. (1970). Carbon dioxide exchange of a tropical rain forest. Part II. *BioScience* 20, 1054–9.

Leopoldo, P.R., Franken, W., Matsui, E. & Salati, E. (1982). Estimava da evapotranspiração de floresta Amazonica de terra firme. *Acta Amazon.* 12, 23–8.

Letouzey, R. (1968). *Etude phytogéographique du Caméroun.* (*Encycl. Biologie* no. 69). Lechevalier, Paris.

Lettau, H.H. (1968). Small- to large-scale features of the boundary layer structure over mountain slopes. Colorado State University, Department of Atmospheric Science Paper 122, 1–74.

Lettau, H., Lettau, K. & Molion, L.C.B. (1979). Amazonia's hydrologic cycle and the role of atmospheric recycling in assessing deforestation effects. *Mon. Wea. Rev.* 107, 227–38.

Lewis, L.A. (1976). Soil movement in the tropics – a general model. *Z. Geomorph.*, N.F., Suppl. -Bd. 25, 132–44.

Lezine, A.-M. (1988). Les variations de la couverture forestière mésophile d'Afrique occidentale au cours de l'Holocène. *C. R. Acad. Sci., Paris* 307, sér. 2, 439–45.

Li, X. & Walker, D. (1986). The plant geography of Yunnan Province, southwest China. *J. Biogeogr.* 13, 367–97.

Li Jin, Zhou Min Zong & Zhou Huizhen (1988). *Soil map of China.* Science Press, Beijing.

Lian, F.J. (1988). The economics and ecology of the production of the tropical rainforest resources by tribal groups of Sarawak, Borneo. Pp. 103–25 in *Changing Tropical Forests: Historical Perspectives in Asia, Australasia and Oceania*, J. Dargavel, K. Dixon & N. Semple (Eds), Centre for Resources and Environmental Studies, Australian National University, Canberra.

Lieberman, D. & Lieberman, M. (1987). Forest tree growth and dynamics at La Selva, Costa Rica (1969–82). *J. Trop. Ecol.* 3, 347–58.

Lieberman, M., Lieberman, D., Hartshorn, G.S. & Peralta, R. (1985a). Small-scale altitudinal variation in lowland wet tropical vegetation. *J. Ecol.* 73, 505–16.

Lieberman, D., Lieberman, M., Hartshorn, G.S. & Peralta, R. (1985b). Growth rates and age-size relationships of wet forest trees in Costa Rica. *J. Trop. Ecol.* 1, 97–109.

Lieberman, D., Lieberman, M., Peralta, R. & Hartshorn, G.S. (1985c). Mortality patterns and stand turnover rates in a wet tropical forest in Costa Rica. *J. Ecol.* 73, 915–24.

Lieske, R. (1915). Beiträge zur Kenntnis der Ernährungsphysiologie extrem atmosphärischer Epiphyten. *Jarhb. wiss. Bot.* 56 (Pfeffer Festschr.) 112–22.

Liew, T.C. (1973). Occurrence of seeds in virgin forest topsoil with particular reference to secondary species in Sabah. *Malay. For.* 36, 185–93.

Liew, T.C. (1974). A note on soil erosion at Tawau Hills Forest Reserve. *Malay. Nat. J.* 27, 20–6.

Lind, E.M. & Morrison, M.E.S. (1974). *East African vegetation.* Longman, London.

Lindeman, J.C. (1953). The vegetation of the coastal region of Suriname. *The Vegetation of Suriname* 1, part 1, Hulster, I.A. de & Lanjouw, J. (Eds), Van Eedenfonds, Amsterdam, for Bot. Mus. & Herb, Rijksuniversitet, Utrecht.

Lindeman, J.C. & Moolenaar, S.P. (1959). Preliminary survey of the vegetation types of northern Suriname. *The vegetation of Suriname* 1, part 2, Hulster, I.A. de & Lanjouw, J. (Eds), Van Eedenfonds, Amsterdam, for Bot. Mus. & Herb, Rijksuniversitet, Utrecht.

Lindmann, C.A.M. (1900). *Vegetationen i Rio Grande do Sul (Sydbrasilien).* Nordin & Josephson, Stockholm.

Livingstone, D.A. (1975). Late Quaternary climatic change in Africa. *Ann. Rev. Ecol. Syst.* 6, 249–80.

Livingstone, D.A. (1976). The Nile – palaeolimnology of headwaters. Pp. 21–30 in *The Nile, biology of an ancient river*, Rzoska, J. (Ed.), W. Junk, The Hague.

Livingstone, D.A. & Clayton, W.D. (1980). An altitudinal cline in tropical African grass floras and its paleoecological significance. *Quaternary Research* 13, 392–402.

Lloyd, C.R., Gash, J.H.C., Shuttleworth, W.J. & Marques Filho, A. de O. (1988). The measurement and modelling of rainfall interception by Amazonian rain forest. *Agric. For. Meteorol.* 43, 277–94.

Lloyd, C.R. & Marques Filho, A. de O. (1988). Spatial variability of throughfall and stemflow measurements in Amazonian rain forest. *Agric. For. Meteorol.* 42, 63–73.

Lockwood, J.G. (1974). *World Climatology: an Environmental Approach.* Arnold, London.

Longman, K.A. (1969). The dormancy and survival of plants in the humid tropics. *Sym. Soc. Exp. Biol.* 23, 471–88.

Longman, K.A. & Jeník, J. (1987). *Tropical Forest and its Environment.* Second edition, Longman Scientific & Technical, Harlow, England.

Lötschert, W. (1969). *Pflanzen an Grenzstandorten*. G. Fischer, Stuttgart.

Louis, J. (1939). Rapport de la Division de Botanique. Pp. 17–36 in *Rapport Annuel Pour l'Exercice 1938*, Inst. Nat. Etude Agron. Congo belge. (INEAC), hors série, Brussels.

Louis, J. (1947a). Contribution à l'étude des forêts équatoriales congolaises. Pp. 902–15 in *Comptes Rendus de la Semaine Agricole de Yangambi (26 Fév.–5 Mars 1947)*, Deuxième Partie. INEAC (Hors Sér.), Bruxelles.

Louis, J. (1947b). La phytosociologie et le problème des jachères au Congo. Pp. 916–23 in *Comptes Rendus de la Semaine Agricole de Yangambi (26 Fév.–5 Mars 1947)*, Deuxième Partie. INEAC (Hors Sér.), Bruxelles.

Louis, J. (1947c). L'origine et la végétation des îles du fleuve de la région de Yangambi. Pp. 924–33 in *Comptes Rendus de la Semaine Agricole de Yangambi (26 Fév.–5 Mars 1947)*, Deuxième Partie. INEAC (Hors Sér.), Bruxelles.

Louis, J. & Fouarge, J. (1949). Essences forestières et bois du Congo. Fasc. 6, *Macrolobium dewevrei*. INEAC, Bruxelles.

Lovejoy, T.E. *et al.* (1986). Edge and other effects of isolation on Amazon forest fragments. Ch. 12 in *Conservation Biology*, Soulé, M.E. (Ed.), Sinauer, Sunderland, Massachusetts.

Low, K.S. (1974). Interception loss of precipitation in Lowland Dipterocarp forest. Paper presented at the IBP Synthesis Meeting, Kuala Lumpur, Aug. 1974.

Low, K.S. & Goh, K.C. (1972). The water balance of five catchments in Selangor, West Malaysia. *J. Trop. Geogr. 35*, 60–6.

Lowry, J.B. (1971). Conserving the forest, a phytochemical view. *Malay. Nat. J. 24*, 225–30.

Lu, X.-H. (1989). Characteristics and utilization of tropical laterites in China. Pp. 47–57 in *Soils and their management: a Sino-European Perspective*, Maltby, E. & Wollersen, T. (Eds), Elsevier, Barking, Essex.

Lugo, A.E. (1980). Mangrove ecosystems: successional or steady state? *Biotropica 12* (suppl.), 65–72.

Lugo, A.E., Applefield, M., Pool, D.J. & McDonald, R.B. (1983). The impact of Hurricane David on the forests of Dominica. *Canadian J. For. Res. 13*, 201–13.

Lugo, A.E. & Snedaker, S.C. (1974). The ecology of mangroves. *Ann. Rev. Ecol. Syst. 5*, 39–64.

Lütge, U. (Ed.) (1989). *Vascular Plants as Epiphytes*. Springer, Berlin.

Maas, P.J.M. (1971). Floristic observations on forest types in western Suriname. I. *Koninkl. Nederl. Akad. van Wetenschappen-Amsterdam, Proc. Ser. C 74*, 269–302.

Maas, P.J.M. & collaborators (1986). *Flora Neotropica Monographs 40, 41 and 42. Saprophytes*, pro parte. New York Botanical Garden, New York.

Mabberley, D.J. (1979). The species of *Chisocheton* (Meliaceae). *Bull. Brit. Mus. (Nat. Hist.) Bot. Ser. 6*, 301–86.

MacArthur, R.H. (1965). Patterns of species diversity. *Biol. Reviews* (Cambridge) *40*, 510–33.

MacArthur, R.H. (1972). *Geographical Ecology: Patterns in the Distribution of Species*. Harper & Row, New York.

McCarthy, J. (1962). The form and development of knee roots in *Mitragyna stipulosa*. *Phytomorphology 12*, 20–30.

McClure, H.E. (1966). Flowering, fruiting and animals in the canopy of a tropical rain forest. *Malay. For. 29*, 192–203.

McColl, J.G. (1970). Properties of some natural waters in a tropical wet forest of Costa Rica. *BioScience 20*, 1096–100.

McDade, L.A., Bawa, K.S., Hespenheide, H.A. & Hartshorn, G.S. (1994). *La Selva: Ecology and Natural History of a Neotropical Rain Forest*. University of Chicago Press, Chicago and London. [Review by T.C. Whitmore: *J. Biogeogr. 2.1.94*.]

MacFarlane, M.J. (1976). *Laterite and Landscape*. Academic Press, London.

MacGregor, W.D. (1934). Silviculture of the Mixed Deciduous forests of Nigeria. *Oxford. For. Mem. 18*.

MacGregor, W.D. (1937). Forest type and succession in Nigeria. *Emp. For. J. 16*, 234–42.

Mackay, J.H. (1936). Problems of ecology in Nigeria. *Emp. For. J. 15*, 190–200.

McKey, D. (1988). *Cecropia peltata*, an introduced neotropical pioneer tree, is replacing *Musanga cecropioides* in southwestern Cameroon. *Biotropica 20*, 262–4.

McLean, R.C. (1919). Studies in the ecology of Tropical Rain-forest: with special reference to the forests of south Brazil. *J. Ecol. 7*, 15–54, 121–72.

MacNae, W. (1968). A general account of the fauna and flora of mangrove swamps and forests in the Indo-West Pacific region. *Adv. Marine Biol. 6*, 73–270.

McNaughton, K.G. & Jarvis, P.G. (1983). Predicting effects of vegetation changes on transpiration and evoporation. In *Water Deficits and Plant Growth*, Vol. 7, Kozolowski, T.T. (Ed.), Academic Press, New York.

Madge, D.S. (1969). Litter disappearance in forest and savanna. *Pedobiologia 9*, 288–99.

Madison, M. (1977). Vascular epiphytes: their systematic occurrence and salient features. *Selbyana 2*, 1–13.

Maguire, B., Ashton, P.S., Zeeuw, C. de, Giannasi, D.E. & Niklas, K.J. (1977). Pakaraimoideae, Dipterocarpaceae of the Western Hemisphere. *Taxon 26*, 341–85.

Maksymowych, R. (1973). *Analysis of Leaf Development*. Cambridge University Press.

Maley, J. (1987). Fragmentation de la forêt dense humide africaine et extension des biotopes montagnards au Quaternaire Récent: nouvelles données polliniques et chronologiques; implications paléoclimatiques et biogéographiques. Pp. 307–32 in *Palaeoecology of Africa and the Surrounding Islands 18*, Balkema, Rotterdam.

Maley, J. (1990). L'histoire recente de la forêt dense humide africaine: essai sur le dynamisme de quelques formations forestières. Conclusions de la quatrieme partie, Synthèse sur le domaine forestier africain au Quaternaire récent. Pp. 367–82 and 383–89 in *Paysages de l'Afrique Centrale Atlantique*, R. Lanfranchi & D. Schwartz (Eds), ORSTOM, Paris.

Maley, J. & Livingstone, D.A. (1983). Extension d'un élément montagnard dans le sud du Ghana (Afrique de l'Ouest) au Pléistocène supérieur et à l'Holocène inférieur: premières données polliniques. *Comptes Rendus des Séances de l'Académie des Sciences 296*, Sér. II, 1287–92.

Malloch, D.W., Pirozynski, K.A. & Raven, P.H. (1980). Ecological and evolutionary significance of mycorrhizal symbioses in vascular plants. A review. *Proc. Nat. Acad. Sci. USA 77*, 2113–18.

Malmer, A. (1990). Stream suspended sediment load after clear-felling and different forestry treatments in tropical rain forest, Sabah, Malaysia. *Int. Ass. Sci. Hyd. Publ. 192*, 62–71.

Mangenot, G. (1950). Essai sur les forêts denses de la Côte d'Ivoire. *Bull. Soc. Bot. France 97*, 159–62.

Mangenot, G. (1955). Etude sur les forêts des plaines et plateaux de la Côte d'Ivoire. *Etudes Eburnéennes (Inst. Français d'Afrique Noire, Adiopodoumé) 4*, 1–83.

Manokaran, N. (1979). Stemflow, throughfall and rainfall interception in a lowland tropical rain forest in peninsular Malaysia. *Malay. For. 42*, 174–201.

Manokaran, N. (1980). The nutrient contents of precipitation, throughfall and stemflow in a lowland tropical rain forest in Peninsular Malaysia. *Malay. For. 43*, 266–89.

Manokaran, N. & Kochummen, K.M. (1987). Recruitment, growth and mortality of tree species in a lowland dipterocarp forest in Peninsular Malaysia. *J. Trop. Ecol. 3*, 315–30.

Markham, R.H. & Babbedge, A.J. (1979). Soil and vegetation catenas on the forest-savanna boundary in Ghana. *Biotropica 11*, 224–34.

Marshall, R.C. (1934). The physiography and vegetation of Trinidad and Tobago. *Oxford For. Mem. 17*.

Martin, C. (1991). *The Rainforests of West Africa. Ecology, Threats, Conservation.* Transl. L. Tarsadakas. Birkhäuser, Basle.

Martin, P.J. (1977). *The Altitudinal Zonation of Forests along the West Ridge of Gunong Mulu.* Report, For. Dept., Sarawak. (Unpublished.)

Martinez-Ramos, M., Alvarez-Buyulla, E., Sarukhan, J. & Pinero, D. (1988). Treefall age determination and gap dynamics in a tropical forest. *J. Ecol.* 76, 700–16.

Martyn, E.B. (1931). A botanical survey of the Rupununi Development Company's ranch at Waranama, Berbice River. *Agric. J. Brit. Guiana* 4, 18–25.

Matsumoto, T. & Abe, T. (1979). The role of termites in an equatorial rain forest ecosystem of West Malaysia. II. Leaf litter consumption on the forest floor. *Oecologia* 38, 261–74.

Medina, E. & Cuevas, E. (1989). Patterns of nutrient accumulation and release in Amazonian forests of the upper Rio Negro basin. Pp. 217–40 in *Mineral Nutrients in Tropical Forest and Savanna Ecosystems*, Proctor, J. (Ed.), Br. Ecol. Soc. Spec. Pubn 9, Blackwell, Oxford.

Medina, E., Garcia, V. & Cuevas, E. (1990). Sclerophylly and oligotrophic environments: relationships between leaf structure, mineral nutrient content and drought resistance in tropical rain forests of the Upper Rio Negro. *Biotropica* 22, 51–64.

Medina, E. Montes, G., Cuevas, E. & Rokzandic, Z. (1986). Profiles of CO_2 concentration and $d^{13}C$ values in tropical rain forests of the upper Rio Negro Basin, Venezuela. *J. Trop. Ecol.* 2, 207–17.

Medina, E., Mooney, H.A. & Vazquez-Yanes, C. (Eds) (1984). *Physiological Ecology of plants of the Wet Tropics.* Tasks for Vegetation Science 12, Junk, Dordrecht.

Medway, Lord. (1972). Phenology of a tropical rain forest in Malaya. *Biol. J. Linn. Soc. Lond.* 4, 117–46.

Meggers, B.J. (1971). *Amazonia: man and culture in a counterfeit paradise.* Aldine--Atherton, Chicago.

Meggers, B.J. (1987). The early history of man in Amazonia. Pp. 151–74 in *Biogeography and Quaternary history in tropical America*, Whitmore, T.C. & Prance, G.T. (Eds), Clarendon Press, Oxford.

Meijer, W. (1959). Plantsociological analysis of montane rainforest near Tjibodas, West Java. *Acta. Bot. Néerlandica* 8, 277–91.

Meijer, W. (1965). A botanical guide to the flora of Mount Kinabalu. Pp. 325–66 in *Symposium on ecological research in Humid Tropical Vegetation, Kuching, Sarawak 1963*, Kostermans, A.J.G.H. & Fosberg, F.R. (Eds), Science and Cooperation Office for S.E. Asia, Bangkok.

Meijer, W. (1971). Plant life in Kinabalu National Park. *Malay. Nat. J.* 24, 184–9.

Meijer, W. (1974). *Field guide to trees of West Malesia.* Univ. Kentucky, Lexington, Kentucky.

Meijer, W. & Wood, G.H.S. (1964). Dipterocarps of Sabah (North Borneo). *Sabah For. Rec.* 5.

Menaut, J.C. (1983). The vegetation of African savannas. In *Tropical savannas*, Boulière, F. (Ed.), Elsevier, Amsterdam.

Mensah, K.O.A. & Jenik, J. (1968). Root system of tropical trees 2. Features of the root system of Iroko (*Chlorophora excelsa* Benth. et Hook). *Preslia (Praha)* 40, 21–7.

Mervart, J. (1972). Growth and mortality rates in the natural high forest of Western Nigeria. *Nigeria For. Inf. Bull.* N.S. 22 (Fed. Dept. For., Ibadan).

Mervart, J. (1974). Appendix to paper on growth and mortality rates in the natural high forest of western Nigeria. *Niger. For. Inf. Bull.* N.S. 28 (Fed. Dept. For., Ibadan).

Meteorological Office (1958). *Tables of temperature, relative humidity and precipitation for the world.* Part II *Central and South America, the West Indies and Bermuda.* HMSO, London.

Meteorological Office (1967). *Tables of temperature, relative humidity and precipitation for the world.* Part V *Asia.* HMSO, London.

Michaloud, G. & Michaloud-Pelletier, S. (1987). *Ficus* hemi-épiphytes (Moraceae) et arbres supports. *Biotropica* 19, 125–36.

Miehe, H. (1911). Javanische Studien 1. Zur Frage der mikrobiologischen Vorgänge im Humus einiger humussammelnder Epiphyten. *Abh. sachs. Ges. (Akad.) Wiss., math.-phys. Kl.* 32, 376–98.

Mikola, P. (Ed.) (1980). *Tropical mycorrhiza research.* Clarendon Press, Oxford.

Mildbraed, J. (1914). Die Vegetationsverhältnisse im Sammelgebiet der Expedition. *Ergeb. d. deutsch. Zentral-Afrika-Expedition 1907–8* 2, 603–91. Leipzig.

Mildbraed, J. (1922). *Wissenschaftliche Ergebnisse der zweiten deutschen Zentral-Afrika-Expedition 1910–1 unter Führung Adolf Friedrichs, Herzogs zu Mecklenburg.* Klinkhardt & Biermann, Leipzig.

Milton, K. (1991). Leaf change and fruit production in six neotropical Moraceae species. *J. Ecol.* 79, 1–26.

Milton, K., Laca, E.A. & Demment, M.W. (1994). Successional patterns of mortality and growth of large trees in a Panamanian lowland forest. *J. Ecol.* 82, 79–87.

Mitchell, A.W. (1982). *Reaching the rain forest roof. A handbook on techniques of access and study in the canopy.* Leeds Lit. & Phil. Soc., Leeds.

Mitchell, A.W. (1986). *The enchanted canopy.* Collins, Glasgow.

Moeyersons, J. (1988). The complex nature of creep movements on steeply sloping ground in Southern Rwanda. *Earth Surface Processes and Landforms* 13, 511–24.

Moffett, M.W. (1993). *The high frontier.* Harvard University Press, Cambridge, Massachusetts.

Mohr, E.C.J. & Van Baren, F.A. (1954). *Tropical soils.* Interscience, London.

Mohr, E.C.J., Van Baren, F.A. & Van Schuylenborgh, J. (1972). *Tropical soils: a comprehensive study of their genesis.* Third edition, Mouton – Ichtiar Baru – Van Hoeve, The Hague – Paris – Djakarta.

Molion, L.C. (1987). Micrometeorology of an Amazonian Rain Forest. Pp. 255–70 in *The geophysiology of Amazonia: vegetation and climate interactions*, Dickinson, R.E. (Ed.), Wiley, Chichester.

Monk, C.D. (1965). Southern mixed hardwood forest of north central Florida. *Ecol. Monogr.* 35, 335–54.

Monk, C.D. (1966). An ecological significance of evergreenness. *Ecology,* 47, 504–5.

Moraes, V.H.F. & Pires, J.M. (1967). Estudos sôbre regeneração natural na mata Amazônica. In *Area de pesquisas ecológicas do Guamá (Belém, Pará, Brasil), Relatorio Anual Abril – Dezembro 1966.* (Unpublished.)

Moreau, R.E. (1935). A synecological study of Usambara, Tanganyika Territory, with particular reference to birds. *J. Ecol.* 23, 1–43.

Moreau, R.E. (1938). Climatic classification from the standpoint of East African biology. *J. Ecol.* 26, 466–96.

Moreau, R.E. (1966). *The bird faunas of Africa and its islands.* Academic Press, London.

Morgan, W.B. & Moss, R.P. (1965). Savanna and forest in western Nigeria. *Africa* 35, 286–94.

Morley, R.J. & Flenley, J.R. (1987). Late Cenozoic vegetational and environmental changes in the Malay archipelago. Pp. 50–9 in *Biogeographical evolution of the Malay archipelago*, Whitmore, T.C. (Ed.), Clarendon Press, Oxford.

Morton, J.K. (1986). Montane vegetation. Pp. 247–71 in *Plant ecology in West Africa*, Lawson, G.W. (Ed.), Wiley, Chichester

Mueller, O.P. & Cline, M.G. (1959). Effects of mechanical soil barriers and soil wetness on rooting of trees and soil mixing by blow-down in central New York. *Soil Sci.* 88, 107–11.

Mullenders, W. (1954). La végétation de Kaniama. *Publ. INEAC, sér. sci.* 61.

Müller, D. & Nielsen, J. (1965). Production brute, pertes par respiration et production nette dans la forêt ombrophile tropicale. *Forst. Forsøgsv. Danmark* 29, 69–160.

Muller, J. (1964). A palynological contribution to the history of mangrove vegetation in Borneo. Pp. 33–42 in *Ancient Pacific Floras*, Cranwell, L. (Ed.), University of Hawaii Press, Honolulu.

Muller, J. (1972). Palynological evidence for change in geomorphology, climate and vegetation in the Mio-Pliocene of Malesia. Pp. 6–34 in *The Quaternary era in Malesia*, Ashton, P. & Ashton, M. (Eds), *Univ. Hull, Dept. Geogr. Misc. Ser* 13.

Muller, J. (1981). Fossil pollen records of extant angiosperms. *Bot. Rev.* 47,

Mullette, K.J., Hannon, N.J. & Elliott, A.G.L. (1974). Insoluble phosphorus usage by Eucalyptus. *Plant Soil* 41, 199–205.

Muul, I.M. & Lim, B.L. (1970). Vertical zonation in a Tropical Rain forest in Malaysia: a method of study. *Science* 169, 788–9.

Myers, J.G. (1935). Zonation of vegetation along river courses. *J. Ecol.* 23, 256–360.

Myers, J.G. (1936). Savannah and forest vegetation of the interior Guiana Plateau. *J. Ecol.* 24, 162–84.

Myers, N. (1980). *Conversion of Tropical Moist Forests.* (Committee on Research Priorities in Tropical Biology of the Nat. Res. Council.) Nat. Acad. Sciences, Washington, D.C.

Myers, N. (1984). *The Primary Source: Tropical Forests and Our Future.* W.W. Norton, New York & London.

Nakano, K. & Syahbuddin (1989). Nutrient dynamics in forest fallows in South-East Asia. Pp. 325–36 in *Mineral Nutrients in Tropical Forest and Savanna Ecosystems*, Proctor, J. (Ed.), Br. Ecol. Soc. Spec. Pubn. 9, Blackwell, Oxford.

National Academy of Sciences (1975). Underexploited tropical plants with promising economic value. Report of an ad hoc Panel of the Advisory Committee on Technology Innovation, Board on Science and Technology for International Development, Comm. Internat. Rel. Nat. Acad. Sci., Washington, D.C.

Naveh, Z. & Whittaker, R.H. (1980). Structural and floristic diversity of shrublands and woodlands in northern Israel and other Mediterranean areas. *Vegetatio* 41, 171–90.

Navez, A. (1930). On the distribution of tabular roots in *Ceiba* (Bombacaceae). *Proc. Nat. Acad. Sci., Wash.* 16, 339–44.

Nevling, L.I. Jr. (1971). The ecology of an elfin forest in Puerto Rico, 16. The flowering cycle and an interpretation of its seasonality. *J. Arnold Arboretum* 52, 586–613.

Newbery, D.McC., Alexander, I.J., Thomas, D.W. & Gartlan, J.S. (1988). Ectomycorrhizal rain-forest legumes and soil phosphorus in Korup National Park, Cameroon. *New Phytol.* 109, 433–50.

Newbery, D.McC., Gartlan, J.S., McKay, D.B. & Waterman, P.G. (1986). The influence of drainage and soil phosphorus on the vegetation of Douala-Edea forest reserve, Cameroun. *Vegetatio* 65, 149–62.

Newbery, D.McC. & Proctor, J. (1984). Ecological studies in four contrasting lowland rain forests in Gunung Mulu National Park, Sarawak IV. Associations between tree distribution and soil factors. *J. Ecol.* 72, 475–93.

Newman, M.S., Burgess, P.S. & Whitmore, T.C. (1995). *Manual of Dipterocarpaceae for Foresters.* Royal Botanic Garden, Edinburgh.

Ng, F.S.P. (1966). Age at first flowering in dipterocarps. *Malay. For.* 29, 290–5.

Ng, F.S.P. (1977a). Gregarious flowering of dipterocarps in Kepong, 1976. *Malay. For.* 40, 126–37.

Ng, F.S.P. (1977b). Shyness in trees. *Nature Malaysiana* 2, 35–7.

Ng, F.S.P. (1978). Strategies of establishment in Malaysian forest trees. Pp. 129–62 in *Tropical Trees as Living Systems*,

Tomlinson, P.B. & Zimmermann, M.H. (Eds), Cambridge University Press.

Ng, F.S.P. (1980a). Germination ecology of Malaysian woody plants. *Malay. For.* 43, 406–37.

Ng, F.S.P. (1980b). The phenology of the Yellow Flame Tree (*Peltophorum pterocarpum*). *Malayan Nat. J.* 33, 201–8.

Ng, F.S.P. (1983). Ecological principles of tropical lowland rain forest conservation. Pp. 359–75 in *Tropical Rain Forest Ecology and Management*, Sutton, S.L., Whitmore, T.C. & Chadwick, A.C. (Eds), Br. Ecol. Soc. Spec. Pubn 2, Blackwell, Oxford.

Ng, F.S.P. (1984). Plant phenology in the humid tropics. *For. Dept., Res. Pamphlet* 96, Kepong, Malaysia.

Ng, F.S.P. & Low, C.M. (1982). *Check List of Endemic Trees of the Malay Peninsula.* For. Dept., Res. Pamphlet 88, Kepong, Malaysia.

Nicholaides, J. (Ed.) (1978). *Soil Diversity in the Tropics.* Spec. Pubn. 34. Amer. Soc. Agronomy, Madison, Wisconsin.

Nicholaides, J.J. II, Bandy, D.E., Sanchez, P.A., Villachia, J.M., Coutu, A.J. & Valverde, C.S. (1984). From migratory to continuous agriculture. *FAO Soils Bulletin* 53, 141–68.

Nicholson, D.I. (1965). A study of virgin forest near Sandakan, North Borneo. Pp. 67–87 in *Proc. Symposium on Ecological Research in Humid Tropics Vegetation*, UNESCO (1963), Kostermans, A.J.G.H. & Fosberg, F.R. (Eds), Science Cooperation Office for S.E. Asia, Bangkok.

Nieuwolt, S. (1968a). Diurnal variation of rainfall in Malaya. *Annls Ass. Am. Geog.* 58, 313–26.

Nieuwolt, S. (1968b). Uniformity and variation in an equatorial climate. *J. Trop. Geogr.* 27, 23–39.

Nieuwolt, S. (1977). *Tropical Climatology: an Introduction to the Climates of the Low Latitudes.* Wiley, London.

Njoku, E. (1963). Seasonal periodicity in the growth and development of some forest trees in Nigeria. I. Observations on mature trees. *J. Ecol.* 51, 617–24.

Njoku, E. (1964). Seasonal periodicity in the growth and development of some forest trees in Nigeria. II. Observations on seedlings. *J. Ecol.* 52, 19–26.

Nortcliff, S., Ross, S.M. & Thornes, J.B. (1990). Soil moisture, runoff and sediment yield from differentially cleared tropical rain forest plots. Pp. 419–35 in *Vegetation and Erosion*, Thornes, J.B. (Ed.), Wiley, Chichester.

Nortcliff, S. & Thornes, J.B. (1978). Water and carbon movement in a tropical rainforest environment. I. Objectives, experimental methods and preliminary results. *Acta Amazon.* 8, 245–58.

Nortcliff, S. & Thornes, J.B. (1981). Seasonal variations in the hydrology of a small forested catchment near Manaus, Amazonas, and the implications for its management. Pp. 37–57 in *Tropical Agricultural Hydrology*, Lal, R. & Russell, E.W. (Eds), Wiley, Chichester.

Nortcliff, S. & Thornes, J.B. (1988). The dynamics of tropical floodplain environment with reference to forest ecology. *J. Biogeogr.* 15, 49–59.

Nortcliff, S. & Thornes, J.B. (1989). Variations in soil nutrients in relation to soil moisture status in a humid tropical forested ecosystem. Pp. 43–54 in *Mineral Nutrients in Tropical Forest and Savanna Ecosystems* Proctor, J. (Ed.), Br. Ecol. Soc. Spec. Pubn. 9, Blackwell, Oxford.

Noy-Meir, I. (1971). Multivariate analysis of the semi-arid vegetation in south-eastern Australia: nodal ordination by component analysis. *Proc. Ecol. Soc. Aust.* 6, 159–93.

Noy-Meir, I., Walker, D. & Williams, W.T. (1975). Data transformation in ecological ordination. II. On the meaning of data standardisation. *J. Ecol.* 63, 779–800.

Noy-Meir, I. & Whittaker, R.H. (1977). Continuous mutivariate methods in community analysis: some problems and developments. *Vegetatio* 33, 79–98.

Nye, P.H. (1961). Organic matter and nutrient cycles under moist tropical forest. *Plant and Soil* 13, 333–46.

Obaton, M. (1960). Les lianes ligneuses à structure anormale des forêts denses d'Afrique occidentale. *Ann. Sci. Nat.* (Sér. bot.) 12, 1–220.

Oberbauer, S.F. & Donnelly, M.A. (1986). Growth analysis and successional status of Costa Rican rain forest trees. *New Phytol.* 104, 517–21.

Oberbauer, S.F. & Strain, B.S. (1984). Photosynthesis and successional status of Costa Rican rain forest trees. *Photosynthesis Res.* 5, 227–32.

O'Connell, A.M. (1988). Nutrient dynamics in decomposing litter in Karri (*Eucalyptus diversicolor* F. Muell.) forests in southwestern Australia. *J. Ecol.* 76, 1186–203.

Odum, H.T. (1970a). Summary: an emerging view of the ecological system at El Verde. Ch. 1–10 in *A Tropical Rain Forest. A Study of Irradiation and Ecology at El Verde, Puerto Rico*, Odum, H.T. & Pigeon, R.F. (Eds), Divn. Tech. Inf., U.S. Atomic Energy Commission, Oak Ridge, Tennessee.

Odum, H.T. (1970b). Rain forest structure and mineral cycling homeostasis. Ch. H-1 in *A Tropical Rain Forest. A Study of Irradiation and Ecology at El Verde, Puerto Rico*, Odum, H.T. & Pigeon, R.F. (Eds.), Divn. Tech. Inf., U.S. Atomic Energy Commission, Oak Ridge, Tennessee.

Odum, H.T. (1970c). Holes in leaves and the grazing control mechanism. Ch. I–6 in *A Tropical Rain Forest. A Study of Irradiation and Ecology at El Verde, Puerto Rico*, Odum, H.T. & Pigeon, R.F. (Eds.), Divn. Tech. Inf., U.S. Atomic Energy Commission, Oak Ridge, Tennessee.

Odum, H.T., Copeland, B.J. & Brown, R.Z. (1963). Direct and optical assay of leaf mass of the Lower Montane Rain forest of Puerto Rico. *Proc. Nat. Acad. Sci. USA* 49, 429–34.

Odum, H.T., Drewry, G. & Kline, J.R. (1970a). Climate at El Verde, 1963–1966. Pp. B347–418 in *A Tropical Rain Forest. A Study of Irradiation and Ecology at El Verde, Puerto Rico*, Odum, H.T. & Pigeon, R.F. (Eds), Divn. Tech. Inf., U.S. Atomic Energy Commission, Oak Ridge, Tennessee.

Odum, H.T., Lugo, A., Cintron, G. & Jordan, C.F. (1970b). Metabolism and evapotranspiration of some rain forest plants and soil. Pp. I103–64 in *A Tropical Rain Forest. A Study of Irradiation and Ecology at El Verde, Puerto Rico*, Odum, H.T. & Pigeon, R.F. (Eds), Divn. Tech. Inf., U.S. Atomic Energy Commission, Oak Ridge, Tennessee.

Odum, H.T. & Pigeon, R.F. (Eds) (1970). *A Tropical Rain Forest. A Study of Irradiation and Ecology at El Verde, Puerto Rico.* Divn. Tech. Inf., U.S. Atomic Energy Commission, Oak Ridge, Tennessee.

Ogawa, H. Yoda, K., Kira, T., Ogino, K., Shidei, T., Ratanawongse, D. & Apasutaya, C. (1965). Comparative ecological study on three main types of forest vegetation in Thailand. I. Structure and floristic composition. *Nature and Life in Southeast Asia* 4, 13–48.

Ogden, J. (1966). *Ordination Studies on a Small Area of Tropical Rain Forest.* M.Sc. Thesis, University of Wales.

Ogden, J. (1981). Dendrochronological studies and the determination of tree ages in the Australian tropics. *J. Biogeogr.* 8, 405–20.

Ohsawa, M., Nainggolan, P.H.J., Tanaka, N. & Anwar, C. (1985). Altitudinal zonation of forest vegetation in Mount Kerinci, Sumatra, with comparisons to vegetation in the temperate region of east Asia. *J. Trop. Ecol.* 1, 193–214.

Ojanuga, A.G. & Lee, G.B. (1973). Characteristics, distribution and genesis of nodules and concretions in the soils of the southwestern uplands of Nigeria. *Soil Sci.* 116, 282–91.

Okali, D.U.U. (1971). Rates of dry matter production in some tropical forest tree seedlings. *Ann. Bot.* 35, 87–97.

Okali, D.U.U. & Doodoo, G. (1973a). Leaf water relations of seedlings of two West African mahogany species. *Ghana J. Sci.* 13, 174–8.

Okali, D.U.U. & Doodoo, G. (1973b). Seedling growth and transpiration of two West African mahogany species in relation to water stress in the root medium. *J. Ecol.* 61, 421–38.

Okali, D.U.U. & Ola-Adams, B.A. (1987). Tree population changes in treated forest at Omo Forest Reserve, southwestern Nigeria. *J. Trop. Ecol.* 3, 291–313.

Okigbo, B.N. (1984). Improved permanent production systems as an alternative to shifting intermittent culture. *FAO Soils Bull.* 53, 1–100.

Olaniyi, A.A. (1988). Chemical and biological investigations on medicinal plants of West Africa. *Acta Amazonica* 18 (suppl.), 381–92.

Oldeman, R.A.A. (1972). L'architecture de la végétation ripicole forestière des fleuves et criques guyanais. *Adansonia* sér. 2, 12, 253–65.

Oldeman, R.A.A. (1974). L'architecture de la forêt guyanaise. *Mém. ORSTOM* 73.

Oldeman, R.A.A. (1983). Tropical rain forest, architecture, sylvigenesis and diversity. Pp. 139–50 in *Tropical Rain Forest Ecology and Management*, Sutton, S.L., Whitmore, T.C. & Chadwick, A.C. (Eds), Br. Ecol. Soc. Spec. Pubn 2, Blackwell, Oxford.

Olsen, C. (1917). Studies on the succession and ecology of epiphytic bryophytes on the bark of common trees in Denmark. *Bot. Tidskr.* 34, 313–42.

Omotosho, J.B. (1985). The separate contributions of line squalls, thunderstorms and the monsoon to the total rainfall in Nigeria. *J. Clim.* 5, 543–52.

O'Neill, M.J., Bray, D.H., Boardman, P., Wright, C.W., Phillipson, J.D., Warhurst, D.C., Gupta, M.P., Correya, M. & Solis, P. (1988). Plants as sources of antimalarial drugs. Part 6: Activities of *Simarouba Amara* fruits. *J. Ethnopharmacol.* 22, 183–90.

Opler, P.A., Baker, H.G. & Frankie, G.W. (1975). Reproductive biology of some Costa Rican *Cordia* species (Boraginaceae). *Biotropica* 7, 234–47.

Opler, P.A., Baker, H.G. & Frankie, G.W. (1980a). Plant reproductive characteristics during secondary succession in neotropical lowland forest ecosystems. *Biotropica* 12 (suppl.), 40–6.

Opler, P.A., Frankie, G.W. & Baker, H.G. (1980b). Comparative studies of treelet and shrub species in wet and dry forests in the lowlands of Costa Rica. *J. Ecol.* 68, 167–88.

Orloci, L. (1978). *Multivariate Analysis in Vegetation Research.* Second edition, Junk, The Hague.

Paijmans, K. (1970). An analysis of four tropical rain forest sites in New Guinea. *J. Ecol.* 58, 77–101.

Paijmans, K. (1976). Vegetation. Pp. 23–105 in *New Guinea Vegetation*, part 2, Paijmans, K. (Ed.), Elsevier, Amsterdam.

Palmer, M.W. (1993). Putting things in even better order: the advantages of canonical correspondence analysis. *Ecology* 74, 2215–30.

Parfitt, L. (1985). The nature of andic and vitric materials. Pp. 413–35 in *Volcanic Soils Catena*, Caldas, E.F. & Yaalon, D.H. (Eds) Supplement 7.

Parkhurst, D.F. & Loucks, O.L. (1972). Optimal leaf size in relation to environment. *J. Ecol.* 60, 505–37.

Parkin, J. (1953). The durian theory – a criticism. *Phytomorphology* 3, 80–8.

Parsons, R.F. & Cameron, D.G. (1974). Maximum plant species diversity in terrestrial communities. *Biotropica* 6, 202–3.

Pascal, J.-P. (1984). Les forêts denses humides sempervirentes des Ghâts Occidentaux de l'Inde. *Travaux Sect. Sci. Techn.* 20. Institut français de Pondichéry, Pondicherry, India.

Peace, W.J.H. & MacDonald, F.D. (1981). An investigation of the leaf anatomy, foliar mineral levels and water relations of trees of a Sarawak forest. *Biotropica* 13, 100–9.

Peeters, L. (1964). Les limites forêt-savane dans le nord du Congo en relation avec le milieu géographique. *Rev. Belge de Géogr.* 88, 239–73.

Pemadasa, M.A. & Gunatilleke, C.V.S. (1981). Pattern in a rain forest in Sri Lanka. *J. Ecol.* 69, 117–24.

Peñalosa, J. (1984). Basal branching and vegetative spread in two tropical rain forest lianas. *Biotropica* 16, 1–9.

Penman, H.L. (1948). Natural evaporation from open water, bare soil and grass. *Proc. R. Soc. Lond.* 193, 120–45.

Penman, H.L. (1963). Vegetation and hydrology. *Tech. Commun.* 53. Commonwealth Bureau of Soil Science, Harpenden, UK.

Penzig, O. (1902). Die Fortschritte der Flora des Krakatau. *Ann. Jard. Bot. Buitenzorg* 18, 82–113.

Percival, M. & Womersley, J.S. (1975). Floristics and ecology of the mangrove vegetation of Papua New Guinea. *Bot. Bull.* 8. Papau New Guinea Nat. Herb., Lae, Papua New Guinea.

Perry, D.R. & Starrett, A. (1980). The pollination ecology and blooming strategy of a neotropical emergent tree, *Dipteryx panamensis. Biotropica* 12, 307–13.

Persson, R. (1974). World forest resources. Review of world's forest resources in the early 1970s. *Stockholm Institutionen för Skogstatzering Pubn.* 17.

Petch, T. (1924). Gregarious flowering. *Ann. Roy. Bot. Gdns, Peradeniya* 9, 101–17.

Petch, T. (1930). Buttress roots. *Ann. Roy. Bot. Gdns, Peradeniya* 11, 277–85.

Peters, C.M., Gentry, A.H. & Mendelsohn, R.O. (1989). Valuation of an Amazonian rain forest. *Nature* 339, 655–6.

Petters, S.W. (1981). Caves and tower karst near Calabar, Nigeria. *Nigerian Field* 46, 9–20.

Phillips, J.F.V. (1931). Forest succession and ecology in the Knysna region. *Bot. Surv. S. Africa Mem.* 14, Govt. Printer, Pretoria.

Phillips, O.L. & Gentry, A.H. (1994). Increasing turnover through time in tropical forests. *Science* 263, 954–8.

Pianka, E.R. (1966). Latitudinal gradients in species diversity: a review of concepts. *Amer. Nat.* 100, 33–46.

Picado, C. (1912). Les maires aeriennes de la forêt vierge américaines: les Bromeliacées. *Biologica* 11, 110–5.

Picado, C. (1913). Les Bromeliacées épiphytes, considerées comme milieu biologique. *Bull. Sci. Fr. Belg., Sér.* 7, 47, 215–360.

Pickett, S.T.A. (1983). Differential adaptation of tropical tree species to canopy gaps and its role in community dynamics. *Trop. Ecol.* 24, 68–84.

Pijl, L. van der (1934). Die Mycorrhiza von *Burmannia* and *Epirrhizanthes* und die Fortpflanzung ihres Endophyten. *Rec. Trav. bot. néerland.* 31, 761–779.

Pijl, L. van der (1941). Flagelliflory and cauliflory as adaptations to bats in *Mucuna* and other plants. *Ann. Jard. bot. Buitenz.* 51, 83–93.

Pijl, L. van der (1952). Absciss-joints in the stems and leaves of tropical plants. *Proc. Akad. Wet., Amst.* 30, Ser. C, 55, 574–86.

Pijl, L. van der (1954). *Xylocopa* and flowers in the tropics, I. The bees as pollinators, lists of flowers visited. *Koninkl. Nederl. Akad. Wetensch. Amsterdam Proc.*, Ser. C 57, 413–551.

Pijl, L. van der (1955). Some remarks on myrmecophytes. *Phytomorphology* 5, 190–200.

Pijl, L. van der (1956). Remarks on pollination by bats in the genera *Freycinetia, Duabanga* and *Haplophragma*, and on chiropterophily in general. *Acta bot. néerland.* 5, 135–44.

Pijl, L. van der (1969). Evolutionary action of tropical animals on the reproduction of plants. *J. Linn. Soc. Lond. (Biol.)* 1, 85–96.

Pijl, L. van der (1982). *The Principles of Dispersal in Higher Plants.* Third edition, Springer, Berlin.

Piñero, D. & Sarukhan, J. (1982). Reproductive behaviour and its variability in a tropical palm, *Astrocaryum mexicanum. J. Ecol.* 70, 461–72.

Piñero, D., Sarukhán, J. & Gonzalez, E. (1977). Estudios demograficos en plantas. *Astrocaryum mexicanum* Liebm. 1. Estrutura de las poblaciones. *Bol Soc. Bot. Mexico* 37, 67–118.

Pires, J.M. (1973). Tipos de vegetação da Amazonia. *Publ. Aculsas Museu Goeldi,* Belém, Brazil. 20, 179–202.

Pires, J.M. (1984). The Amazonian forest. Pp. 581–602 in *The Amazon: limnology and landscape ecology of a mighty tropical river and its basin,* Sioli, H. (Ed.), W. Junk, Dordrecht.

Pires, J.M. & Moraes, V.H. (1966). Composicão e estrutura da mata da Reserva Mocambo. In *Area de Pesquisas Ecológicas do Guamá.* Relatoria Anual, Abril-Dezembro 1966. Ministério da Agricultura IPEA, Belém, Brazil.

Pires, J.M. & Prance, G.T. (1977). The Amazon forest: a natural heritage to be preserved. Pp. 158–94 in *Extinction is Forever,* Prance, G.T. & Elias, T.S. (Eds.), New York Bot. Garden, New York.

Pires, J.M. & Prance, G.T. (1985). The vegetation types of the Brazilian Amazon. Pp. 109–45 in *Key environments: Amazonia,* Prance, G.T. & Lovejoy, T.E. (Eds), Pergamon Press, Oxford.

Pittendrigh, C.S. (1948). The bromeliad-*Anopheles*-malaria complex in Trinidad. 1 – The bromeliad flora. *Evolution* 2, 58–89.

Pitt-Schenkel, C.J.W. (1938). Some important communities of Warm Temperate Rain forest at Magamba, West Usambara, Tanganyika Territory. *J. Ecol.* 26, 50–81.

Ploey, J. de (1965). Position géomorphologique, génèse et chronologie de certains dépôts superficiels au Congo occidental. *Quaternaria* 7, 131–54.

Pócs, T. (1974). Bioclimatic studies in the Uluguru Mountains (Tanzania, East Africa), I. *Acta Bot. Acad. Sci. Hungaricae* 20, 115–35.

Pócs, T. (1976a). The role of the epiphytic vegetation in the water balance and humus production of the rain forests of the Uluguru mountains, East Africa. *Comptes Rendus VIIIe Réunion AETFAT, Boissiera* 24, 499–503.

Pócs, T. (1976b). Bioclimatic studies in the Uluguru Mountains (Tanzania, East Africa), II. Correlations between orography, climate and vegetation. *Acta Bot. Acad. Sci. Hungaricae* 22, 163–83.

Pócs, T. (1980). The epiphytic biomass and its effect on the water balance of two rain forest types in the Uluguru mountains (Tanzania, East Africa). *Acta Bot. Acad. Sci. Hungaricae* 26, 143–67.

Pócs, T., Temu, R.P.C. & Minja, T.R.A. (1990). Short survey of the natural vegetation and flora of the Nguru mountains. *Proc. Workshop on Usambara Integrated Rain Forest Project,* Hedberg, I. (Ed.).

Poels, R.H.L. (1987). *Soils, water and nutrients in a forested ecosystem in Suriname.* Agricultural University, Wageningen.

Polak, B. (1933a). Een tocht in het zandsteen gebied bij Mandor (West Borneo). *Trop. Natuur* 22, 23–8.

Polak, E. (1933b). Ueber Torf und Moor in Niederlandisch Indien. *Verh. Akad. Wet., Amsterdam* 30, 1–85.

Poore, D., Burgess, P., Palmer, J., Rietbergen, J. & Synott, T. (1989). *No timber without trees: sustainability in the tropical forest. A study for ITO.* Earthscan Publications, London.

Poore, M.E.D. (1968). Studies in Malaysian rain forest. I. The forest on Triassic sediments in Jengka forest reserve. *J. Ecol.* 56, 143–96.

Porter, D.M. (1971). Buttressing in a tropical xerophyte. *Biotropica* 3, 142–4.

Portig, W.H. (1976). The climate of Central America. Pp. 405–78 in *Climates of Central and South America,* Schwedtfeger, W. (Ed.), Elsevier, Amsterdam.

Potter, G.L., Elsaesser, H.W., MacCracken, M.C. & Luther, F.M. (1975). Possible climatic impact of tropical deforestation. *Nature* 258, 697–8.

Potter, M.C. (1891). Observations on the protection of buds in the tropics. *J. Linn. Soc. Lond. (Bot.)* 28, 343–52.

Prance, G.T. (1976). The pollination and androphore structure of some Amazonian Lecythidaceae. *Biotropica* 8, 235–41.

Prance, G.T. (1977). The phytogeographic subdivisions of Amazonia and their influence on the selection of biological reserves. Pp. 195–213 in *Extinction is Forever*, Prance, G.T. & Elias, T.S. (Eds.), New York Botanical Garden, New York.

Prance, G.T. (1979). Notes on the vegetation of Amazonia. III. The terminology of Amazonian forest types subject to inundation. *Brittonia* 31, 26–38.

Prance, G.T. (1985). The changing forests. Pp. 146–65 in *Key Environments: Amazonia*, Prance, G.T. & Lovejoy, T.E. (Eds), Pergamon Press, Oxford.

Prance, G.T. (1987). Vegetation. Pp. 46–65 in *Biogeography and Quaternary History in Tropical America*, Whitmore, T.C. & Prance, G.T. (Eds), Clarendon Press, Oxford.

Prance, G.T., Balee, W. & Boom, B.M. (1987). Quantitative ethnobotany and the case for conservation in Amazonia. *Conservation Biol.* 1, 296–309.

Prance, G.T. & Campbell, D.G. (1988). The present state of tropical floristics. 37, 519–48.

Prance, G.T. & Lovejoy, T.E. (Eds) (1985). *Key environments: Amazonia*. Pergamon Press, Oxford.

Prance, G.T. & Mori, S.A. (1979). Lecythidaceae – Part 1. The Actinomorphic-flowered New World Lecythidaceae. *Flora Neotropica, Monograph* 21. New York Botanical Garden, New York.

Prance, G.T., Rodrigues, W.A. & da Silva, M.F. (1976). Inventario florestal de um hectare de mata de terra firme km 30 da estrada Manaus – Itacoatiara. *Acta Amazonica* 6, 9–35.

Priestley, C.H.B. & Taylor, R.J. (1972). On the assessment of surface heat flux and evaporation using large scale parameters. *Mon. Wea. Rev.* 100, 81–92.

Primack, R.B., Ashton, P.S., Chai, P. & Lee, H.S. (1985). Growth rates and population structure of Moraceae trees in Sarawak, East Malaysia. *Ecology* 66, 577–88.

Pringle, S.L. (1976). Tropical moist forests in world demand, supply and trade. *Unasylva* 28, 106–18.

Proctor, J. (1987). Nutrient cycling in primary and old secondary rain forests *Applied Geogr.* 7, 135–52.

Proctor, J. (Ed.) (1989). *Mineral Nutrients in Tropical Forest and Savanna Ecosystems*. Br. Ecol. Soc. Spec. Pubn 9, Blackwell Scientific, Oxford.

Proctor, J., Anderson, J.M., Chai, P. & Vallack, H.W. (1983a). Ecological studies in four contrasting rain forests in Gunung Mulu National Park, Sarawak. I. Forest environment: structure and floristics. *J. Ecol.* 71 237–60.

Proctor, J., Anderson, J.M., Fogden, S.C. & Vallack, H.W. (1983b). Ecological studies in four contrasting rain forests in Gunung Mulu National Park, Sarawak. II. Litterfall, litter standing crop and preliminary observations on herbivory. *J. Ecol.* 71, 261–83.

Proctor, J., Lee, Y.F., Langley, A.M., Munro, W.R.C. & Nelson, T. (1988). Ecological studies on Gunung Silam, a small ultrabasic mountain in Sabah, Malaysia. I. Environment, forest structure and floristics. *J. Ecol.* 76, 320–40.

Proctor, J., Phillipps, C., Duff, G.K., Heaney, A. & Robertson, F.M. (1989). Ecological studies on Gunung Silam, a small ultrabasic mountain in Sabah, Malaysia. II. Some forest processes. *J. Ecol.* 77, 317–31.

Proctor, J. & Woodell, S.R.J. (1975). The ecology of serpentine soils. *Adv. Ecol. Res.* 9, 255–366.

Projeto RADAM (1973–75). *Levantamento de Recursos Naturais* 1–7. Ministério de Minas e Energia, Rio de Janeiro.

Projeto RADAMBRASIL (1975–83). *Levantamento de Recursos Naturais* 8–32. Ministério de Minas e Energia, Rio de Janeiro.

Pundir, Y.P.S. (1981). Cauliflory in *Ficus hispida* L. *J. Ind. Bot. Soc.* 60, 33–50.

Purseglove, J.W. (1968). *Tropical Crops, Dicotyledons*. 2 vols. Longmans Green, London and Harlow.

Putz, F.E. (1983a). Developmental morphology of *Desmoncus isthmius*, a climbing colonial cocosoid palm. *Principes* 27, 38–47.

Putz, F.E. (1983b). Liana biomass and leaf area of a 'Tierra firme' forest in the Rio Negro basin, Venezuela. *Biotropica* 15, 185–9.

Putz, F.E. (1984). How trees avoid and shed lianas. *Biotropica* 16, 19–23.

Putz, F.E. (1990). Liana stem diameter growth and mortality rates on Barro Colorado Island, Panama. *Biotropica* 22, 103–5.

Putz, F.E. & Appanah, S. (1987). Buried seeds, newly dispersed seeds, and the dynamics of a lowland forest in Malaysia. *Biotropica* 19, 326–33.

Putz, F.E. & Chai, P. (1987). Ecological studies of lianas in Lambir National Park, Sarawak, Malaysia. *J. Ecol.* 75, 523–31.

Putz, F.E. & Chan, H.T. (1986). Tree growth, dynamics and productivity in a mature mangrove forest in Malaysia. *Forest Ecol. & Management* 17, 211–30.

Putz, F.E. & Holbrook, N.M. (1986). Notes on the natural history of hemiepiphytes. *Selbyana* 9, 61–9.

Putz, F.E., Lee, H.S. & Goh, R. (1987). Effects of post-felling silvicultural treatments of woody vines in Sarawak. *Malay. For.* 47, 214–26.

Putz, F.E. & Mooney, H.A. (1991). *The Biology of Vines.* Cambridge University Press.

Pynaert, L. (1933). La mangrove congolaise. *Bull. Agric. Congo Belge* 23, 184–207.

Raich, J.W. (1989). Seasonal and spatial variation in the light environment in a tropical dipterocarp forest and gaps. *Biotropica* 21, 299–302.

Ramage, C.S. (1964). Diurnal variation of summer rainfall of Malaya. *J. Trop. Geogr.* 19, 62–8.

Ramirez, W. (1970). Taxonomic and biological studies of neotropical fig wasps (Hymenoptera: Agaonidae). *Univ. Kansas Sci. Bull.* 49, 1–44.

Ramsay, J.M. & Rose Innes, R. (1963). Some quantitative observations on the effects of fire on the Guinea Savanna vegetation of northern Ghana over a period of eleven years. *African Soils* 8, 41–85.

Rankin, J.M. (1985). Forestry in the Brazilian Amazon. Pp. 369–92 in *Key Environments: Amazonia*, Prance, G.T. & Lovejoy, T.E. (Eds), Pergamon Press, Oxford.

Ratisbona, L.R. (1976). The climate of the World. Pp. 219–93 in *Climates of Central and South America*, Schwerdtfeger, W. (Ed.), Elsevier, Amsterdam.

Ratter, J.A., Richards, P.W., Argent, G. & Gifford, D.R. (1973). Observations on the vegetation of northeastern Mato Grosso. I. The woody vegetation types of the Xavantina-Cachimbo Expedition area. *Phil. Trans. R. Soc. Lond.* B 266, 449–92.

Raunkiaer, C. (1934). *The Life-Forms of Plants and Statistical Plant Geography.* Transl. H. Gilbert-Carter. Oxford University Press.

Raven, P.H. & Axelrod, D.I. (1974). Angiosperm biogeography and past continental movements. *Ann. Missouri Bot. Gdn.* 61, 539–673.

Reading, A.J. (1986). Landslides, heavy rainfalls and hurricanes in Dominica, West Indies. Ph.D. Thesis, Univ. of Wales.

Redhead, J.F. (1992). The forest kingdom of Benin. *Nigerian Field* 57, 113–18.

Reed, R.J. (1970). Structure and characteristics of easterly waves in the equatorial western Pacific during July-August 1967. *Proc. Symp. Trop. Meteorol., Amer. Met. Soc., World Met. Organization, Hawaiian Inst. Geophysics, Honolulu.*

Rees, A.R. (1964a). Some observations on the flowering behaviour of *Coffea rupestris* in Southern Nigeria. *J. Ecol.* 52, 1–7.

Rees, A.R. (1964b). The flowering behaviour of *Clerodendron incisum* in Southern Nigeria. *J. Ecol.* 52, 9–17.

Reich, P.B. & Borchert, R. (1982). Phenology and ecophysiology of the tropical tree, *Tabebuia neochrysantha* (Bignoniaceae). *Ecology* 63, 294–99.

Reich, P.B. & Borchert, R. (1984). Water stress and tree

phenology in a tropical dry forest in the lowlands of Costa Rica. *J. Ecol.* 72, 61–74.

Reid, E.M. & Chandler, M.E.J. (1933). *The London Clay Flora*. Brit. Mus. (Nat. Hist.), London.

Reiner, E.J. & Robbins, R.G. (1964). The Middle Sepik Plains, New Guinea. A physiographic study. *Geogr. Rev.* 54, 20–44.

Resvoll, T.R. (1925). Beschuppte Laubknospen in den immerfeuchten Tropenwäldern. *Flora* (N.F.) 18–19 (*Goebel Festschr.*), 409–20.

Rice, B. & Westoby, M. (1983). Plant species richness at the 0.1 hectare scale in Australian vegetation compared to other continents. *Vegetation* 52, 129–40.

Richards, A.J. (1990). Studies in *Garcinia*, dioecious tropical forest trees: agamospermy. *Bot. J. Linn. Soc. Lond.* 103, 233–50.

Richards, P. & Williamson, G.B. (1975). Treefalls and patterns of understorey species in a wet lowland tropical forest. *Ecology*, 56, 1226–9.

Richards, P.W. (1936). Ecological observations on the rain forest of Mount Dulit, Sarawak. I, II. *J. Ecol.* 24, 1–37, 340–60.

Richards, P.W. (1939). Ecological studies on the rain forest of Southern Nigeria. I. The structure and floristic composition of the primary forest. *J. Ecol.* 27, 1–61.

Richards, P.W. (1941). Lowland Tropical Podsols and their vegetation. *Nature* (Lond.) 148, 129–31.

Richards, P.W. (1943). The biogeographic division of the Indo-Australian Archipelago. 6. The ecological segregation of the Indo-Malayan and Australian elements in the vegetation of Borneo. *Proc. Linn. Soc. Lond.* (Sess. 154, 1941–42), 154–6.

Richards, P.W. (1952). *The Tropical Rain Forest*. First edition, Cambridge University Press.

Richards, P.W. (1957). Ecological notes on West African vegetation. I. The plant communities of the Idanre Hills, Nigeria. *J. Ecol.* 45, 563–77.

Richards, P.W. (1961). The types of vegetation of the humid tropics in relation to the soil. Pp. 15–23 in *Tropical Soils and Vegetation*, Proc. Abidjan Symposium. UNESCO, Paris.

Richards, P.W. (1963a). Ecological notes on West African vegetation. II. Lowland forest of the Southern Bakundu Forest Reserve. *J. Ecol.* 51, 123–49.

Richards, P.W. (1963b). Ecological notes on West African vegetation. III. The upland forests of Cameroons Mountain. *J. Ecol.* 51, 529–54.

Richards, P.W. (1965). Soil conditions in some Bornean lowland plant communities. Pp. 198–205 in *Proc. Symposium on Ecological Research in Humid Tropics Vegetation*, Kuching, Sarawak 1963. UNESCO Science Cooperation Office for S.E. Asia, Bangkok.

Richards, P.W. (1969). Speciation in the tropical rain forest and the concept of the niche. *Biol. J. Linn. Soc. Lond.* 1, 149–53.

Richards, P.W. (1973a). Africa, the 'Odd man out'. Ch. 3 in *Tropical Forest Ecosystems in Africa and South America: a Comparative Review*, Meggers, B.J., Ayensu, E.S. & Duckworth, W.D. (Eds), Smithsonian Inst. Press, Washington, D.C.

Richards, P.W. (1973b). The tropical rain forest. *Scientific American* 229 (Dec. 1973), 58–67.

Richards, P.W. (1977). Tropical forests and woodlands: an overview. *Agro-Ecosystems* 3, 225–38.

Richards, P.W. (1983). The three-dimensional structure of tropical rain forest. Pp. 3–10 in *Tropical Rain Forest Ecology and Management*, Sutton, S.L., Whitmore, T.C. & Chadwick, A.C. (Eds), Br. Ecol. Soc. Spec. Pubn 2, Blackwell, Oxford.

Richards, P.W. (1984a). The ecology of tropical forest bryophytes. Pp. 1233–70 in *New Manual of Bryology*, Schuster, R.M. (Ed.), Hattori Botanical Laboratory, Nichinan, Miyazaki, Japan.

Richards, P.W. (1984b). The forests of South Viet Nam in 1971–72: a personal account. *Environmental Conserv.* 11, 147–53.

Richardson, J.H. (1982). Some implications of tropical forest

replacement in Jamaica. *Zeitschrift Für Geomorphologie* Suppl. 44, 107–18.

Richardson, S.D. (1978). Foresters and the Faustian bargain. In *Papers for Conference on Improved Utilization of Tropical Forests, 21–28 May 1978, Madison, Wisconsin.* U.S. For. Ser. Products Lab., Washington, D.C.

Richardson, W.D. (1963). Observations on the vegetation and ecology of the Aripo savannas, Trinidad. *J. Ecol.* 51, 295–314.

Ricklefs, R.E. (1973). *Ecology*. Nelson, London.

Riddoch, I, Grace, J., Fasehun, F.E., Riddoch, B. & Ladipo, D.O. (1991). Photosynthesis and successional status of seedlings in a tropical semideciduous rain forest in Nigeria. *J. Ecol.* 79, 491–503.

Ridley, H.N. (1905). On the dispersal of seeds by the wind. *Ann. Bot.* 19, 351–63.

Ridley, H.N. (1922). *Flora of the Malay Peninsula*, vol. 1. Lovell Reeve, London.

Ridley, H.N. (1930). *The Dispersal of Plants Throughout the World*. Lovell Reeve, Ashford, Kent.

Riehl, H. (1954). *Tropical Meteorology*. McGraw-Hill, New York.

Riehl, H. (1979). *Climate and Weather in the Tropics*. Academic Press, London.

Rikli, M. (1943). *Das Pflanzenkleid der Mittelmeerländer*, vol. 1. Huber, Bern.

Riou, C. (1984). Experimental study of potential evapotranspiration (PET) in central Africa. *J. Hydrol.* 72, 275–88.

Riswan, S. (1982). *Ecological Studies on Primary, Secondary and Experimentally Cleared Mixed Dipterocarp Forest and Kerangas Forest in East Kalimantan, Indonesia*. Ph.D. Thesis, Aberdeen University.

Riswan, S. & Kartawinata, K. (1988a). A lowland dipterocarp forest 35 years after pepper plantation in East Kalimantan, Indonesia. Pp. 1–39 in *Some ecological aspects of Tropical Forest of East Kalimantan*, Soemodihardjo, S. (Ed.), Indonesian Inst. Sciences (LIPI), Jakarta, Indonesia.

Riswan, S. & Kartawinata, K. (1988b). Regeneration after disturbance in a Kerangas (Heath) forest in East Kalimantan, Indonesia. Pp. 61–85 in *Some Ecological Aspects of Tropical Forest of East Kalimantan*, Soemodihardjo, S. (Ed.), Indonesian Inst. Sciences (LIPI), Jakarta, Indonesia.

Riswan, S., Kenworthy, J.B. & Kartawinata, K. (1986). The estimation of temporal processes in tropical rain forests: a study of Mixed Dipterocarp forest in Indonesia. *J. Trop. Ecol.* 2, 171–82.

Robbins, R.G. (1959). The use of the profile diagram in rain forest ecology. *J. Biol. Sci.* (CSIRO) 2, 53–63.

Robbins, R.G. (1961). The vegetation of New Guinea. Pp. 1–12 in *Australian Territories* 1.6, Govt. Printer, Canberra.

Robbins, R.G. (1963). The anthropogenic grasslands of Papua and New Guinea. Pp. 313–29 in *Symposium on the Effect Impact of Man on Humid Tropics Vegetation, Goroka, Papua New Guinea (Sept. 1960)*. UNESCO Science Coop. Office for S.E. Asia, Jakarta, Indonesia.

Robbins, R.G. (1970). Vegetation of the Goroka-Mount Hagen area. Pp. 104–17 in *Lands of the Goroka-Mount Hagen Area, Papua New Guinea*. Lands Research Series, 27, 104–17. CSIRO, Australia.

Robertson, C. (1904). The structure of the flowers and the mode of pollination of the primitive angiosperms. *Bot. Gaz.* 37, 294–8.

Robertson, G.P. (1989). Nitrification and denitrification in humid tropical ecosystems: potential controls on nitrogen retention. Pp. 55–69 in *Mineral Nutrients in Tropical Forest and Savanna Ecosystems*, Proctor, J. (Ed.), Br. Ecol. Soc. Spec. Pubn 9, Blackwell, Oxford.

Robinson, G.W. (1932). *Laterite and Lateritic Soils*. Tech. Comm. 24, Imp. Bureau Soil Sci., Rothamsted.

Robyns, W. (1936). Contribution à l'étude des formations herbeuses du District Forestier Central du Congo Belge. *Mém. Inst. Roy. Colon. Belge* 5.

Roche, E. & Van Grunderbeek, M.C. (1985). Apports de la palynologie à l'étude du Quaternaire supérieur au Rwanda. In IXe Symposium *Palynologie et Milieux Tropicaux*, Montpellier 1–3 Octobre 1985, Association de Palynologie de Langue Française.

Rodda, J. (1987). Comments on 'The forest and the hydrological cycle'. Pp. 273–96 in *The Geophysiology of Amazonia: Vegetation and Climate Interactions*, Dickinson, R.E. (Ed.), Wiley, Chichester.

Rodrigues, W.A. (1961a). Aspectos fitosociologicos das catingas do Rio Negro. *Bol. Mus. Paraense Emílio Goeldi, N.S.* 15 (Bot.), Belém, Brazil.

Rodrigues, W.A. (1961b). *Estudo Preliminar de Mata de Várzea Alta de Uma Ilha do Baixo Rio Negro de Sólo Argiloso e úmido.* Inst. Nac. Pesquis. Amazonia, Publ. 10 (Bot.), Manaus, Brazil.

Rodrigues, W.A. (1962). *Arvore hapaxanta na Flora Amazônica.* Inst. Nac. Pesquis. Amazonia, Publ. 14, Manaus, Brazil.

Rodrigues, W.A. (1963). Estudo de 2.6 hectares de mata de terra firme da Serra do Navio. Território de Amapá. *Bol. Mus. Paraense Emílio Goeldi N.S.* 19 (Bot.), Belém, Brazil.

Rogstad, S.H. (1990). The biosystematics and evolution of the *Polyalthia hypoleuca* species complex (Annonaceae) of Malesia. II. Comparative distribution ecology. *J. Trop. Ecol.* 6, 387–408.

Rollet, B. (1974). *L'architecture des Forêts Denses Humides Sempervirentes de Plaine.* Centre Technique Forestier Tropical, Nogent-sur-Marne, France.

Rollet, B. (1981). *Bibliography on Mangrove Research, 1600–1975.* UNESCO, Paris.

Room, P.M. (1972). The fauna of the mistletoe *Tapinanthus bangwensis* (Engl. & K. Krause) growing on cocoa in Ghana: relationships between fauna and mistletoe. *J. Anim. Ecol.* 41, 611–21.

Room, P.M. (1973). Ecology of the mistletoe *Tapinanthus bangwensis* growing on cocoa trees in Ghana. *J. Ecol.* 61, 729–42.

Roosmalen, M.G.M. van (1985). *Fruits of the Guianan Flora.* Inst. Syst. Bot., Utrecht.

Rorison, I.H. (1973). The effect of extreme soil acidity on the nutrient uptake and physiology of plants. Pp. 223–74 in *Acid Sulphate Soils*, Dost, H. (Ed.), Internat. Inst. Land Reclamation and Irrigation, Wageningen, Holland.

Ross, R. (1954). Ecological studies on the rain forest of Southern Nigeria. III. Secondary succession in the Shasha Forest Reserve. *J. Ecol.* 42, 259–82.

Ross, S.M., Thornes, J.B. & Nortcliff, S. (1990). Soil hydrology, nutrient and erosional response to the clearance of terra firme forest, Maraca Island, Roraima, northern Brazil. *Geogr. J.* 156, 267–82.

Rougerie, G. (1960). Le façonnement actuel des modèles en Côte d'Ivoire forestière. *Mem. Inst. Française Afrique Noire* 58, Dakar.

Rouse, W.C., Reading, A.J. & Walsh, R.P.D. (1986). Volcanic soil properties in Dominica, West Indies. *Engin. Geol.* 23, 1–28.

Ruinen, J. (1953). Epiphytosis. A second view on epiphytism. *Ann. Bogor.* 1, 101–58.

Ruinen, J. (1956). Occurrence of *Beijerinckia* species in the 'phyllosphere'. *Nature (Lond.)* 177, 220–1.

Ruinen, J. (1974). Nitrogen fixation in the phyllosphere. Pp. 121–67 in *The Biology of Nitrogen Fixation*, Quispel, A. (Ed.), North-Holland Publ. Co., Amsterdam, Holland.

Ruinen, J. (1975). Nitrogen fixation in the phyllosphere. Pp. 85–100 in *Nitrogen Fixation by Free-Living Micro-Organisms*, Stewart, W.D.P. (Ed.), Cambridge University Press.

Ruxton, B.P. (1967). Slopewash under primary rainforest in Northern Papua. Pp. 85–94 in *Landform Studies from Australia*

and New Guinea, Jennings, J.N. & Mabbutt, J.A. (Eds), Cambridge University Press.

Sader, S.A. & Joyce, A.T. (1988). Deforestation rates and trends in Costa Rica 1940–1983. *Biotropica* 20, 11–19.

Saenger, P., Specht, M.M., Specht, R.L. & Chapman, V.J. (1977). Mangal and coastal salt-marsh communities in Australasia. Pp. 293–345 in *Ecosystems of the World 1, Wet Coastal Ecosystems*, Chapman, V.J. (Ed.), Elsevier, Amsterdam.

Salati, E. (1987). The Forest and the Hydrological Cycle. Pp. 273–96 in *The Geophysiology of Amazonia: Vegetation and Climate Interactions*, Dickinson, R.E. (Ed.), Wiley, Chichester.

Saldarriaga, J.G. (1986). Recovery following shifting cultivation. A century of succession in the upper Rio Negro. Pp. 24–33 in *Amazonian Rain Forests: Ecosystem Disturbance and Recovery*, Jordan, C.F. (Ed.), Springer, New York.

Saldarriaga, J.G. & West, D.C. (1986). Holocene fires in the northern Amazon basin. *Quaternary Res.* 26, 358–66.

Saldarriaga, J.G., West, D.C., Tharp, M.L. & Uhl, C. (1988). Long-term chronosequence of forest succession in the upper Rio Negro of Colombia and Venezuela. *J. Ecol.* 76, 938–58.

Salick, J., Herrera, R. & Jordan, C.F. (1983). Termitaria: nutrient patchiness in nutrient-deficient rain forests. *Biotropica* 15, 1–7.

Salisbury, E.J. (1925). The structure of woodlands. Pp. 334–54 in *Veröff. Geobot. Inst. Rübel*, Hft 3. Festschr. C. Schroeter. Geobot. Inst. Rübel, Zürich.

Salisbury, E.J. (1929). The biological equipment of species in relation to competition. *J. Ecol.* 17, 197–222.

Salisbury, E.J. (1942). *The Reproductive Capacity of Plants.* G. Bell, London.

Sanchez, P.A. (1976). *Properties and Management of Soils in the Tropics.* J. Wiley, New York.

Sandwith, N.Y. (1931). Contributions to the flora of tropical America, VI. New and noteworthy species from British Guyana, Dilleniaceae-Connaraceae. *Kew Bull. 1931* 4, 1–18.

Sanford, W.W. (1974). The ecology of orchids. Pp. 1–98 in *The Orchids: Scientific Studies*, Withner, C.L. (Ed.), Wiley, New York.

Sanford, W.W. & Adanlawo, I. (1973). Velamen and exodermis characters of West African epiphytic orchids in relation to taxonomic grouping and habitat tolerance. *Bot. J. Linn. Soc. Lond.* 66, 307–21.

Sanford, W.W. & Isichei, A.O. (1986). Savanna. Pp. 95–149 in *Plant Ecology in West Africa*, Lawson, G.W. (Ed.), John Wiley, Chichester.

Sapper, K. (1932). Klimakunde von Mittelamerika. In *Handbuch der Klimatologie* Vol. 2, Part H, Köppen, W. & Geiger, R. (Eds), Gebrüder Bornträger, Berlin.

Sarmiento, G. (1972). Ecological and floristic convergence between seasonal plant formations of tropical and subtropical South America. *J. Ecol.* 60, 367–410.

Sarmiento, G. (1983). The savannas of tropical America. Pp. 245–288 in *Ecosystems of the World 13, Tropical Savannas*, Bourliere, F. (Ed.), Elsevier, Amsterdam.

Sarmiento, G. (1984). *The Ecology of Neotropical Savannas.* Transl. O. Solbrig. Harvard University Press, Cambridge, Massachusetts.

Sarmiento, G. & Monasterio, M. (1969). Studies on the savanna vegetation of the Venezuelan Llanos. I. The use of association analysis. *J. Ecol.* 57, 579–98.

Sarmiento, G. & Monasterio, M. (1971). Ecologio de las sabanas de America Tropical (Analisis macrosociologico de los Llanos de Calabozo, Venezuela). *Cuadernos Geograficos* 4. Facultad de Ciencias Forestales, Universidad de los Andes, Mérida, Venezuela.

Sarmiento, G. & Monasterio, M. (1975). A critical consideration of the environmental conditions associated with the occurrence of savanna ecosystems in Tropical America. Pp. 223–50 in *Tropical Ecological Systems*, Golley, F.B. & Medina, E. (Eds), Springer, New York.

Sarukhán, J. (1978). Studies on the demography of tropical trees. Pp. 163–84 in *Tropical Trees as Living Systems*, Tomlinson, P.B. & Zimmerman, M.H. (Eds), Cambridge University Press.

Sastre, C., Baudoin, R. & Portecop, J. (1983). Evolution de la végétation de la Soufrière de Guadeloupe depuis les éruptions de 1976–77 par l'étude de la répartition d'espèces indicatrices. *Adansonia* 5, 63–92.

Saulei, S.M. & Swaine, M.D. (1988). Rain forest seed dynamics during succession at Gogol, Papua New Guinea. *J. Ecol.* 76, 1133–52.

Saxton, W.T. (1924). Phases of vegetation under monsoon conditions. *J. Ecol.* 12, 1–38.

Sayer, J.A., Harcourt, C.S. & Collins, N.M. (Eds) (1992). *The Conservation Atlas of Tropical Forests: Africa*. Macmillan, Basingstoke, U.K.

Sayer, J.A. & Whitmore, T.C. (1991). Tropical moist forests: destruction and species extinction. *Biol. Conservation* 55, 199–213.

Schatz, G.E., Williamson, G.B. & Stam, A.C. (1985). Stilt roots and growth of arboreal palms. *Biotropica* 17, 206–9.

Scheffler, G. (1901). Ueber die Beschaffenheit des Usambara-Urwaldes und über den Laubwechsel an Bäumen derselben. *Notizbl. bot. Gart. Berlin* 3, 139–66.

Schenck, H. (1892–93). Beiträge zur Biologie und Anatomie der Lianen. I. Beiträge zur Biologie der Lianen. *Bot. Mitt. Trop.* 4 (1892). II. Beiträge zur Anatomie der Lianen. *Ibid.* 5 (1893).

Schimper, A.F.W. (1888). Die epiphytische Vegetation Amerikas. *Bot. Mitt. Trop.* 2.

Schimper, A.F.W. (1891). Die indo-malayische Strandflora. *Bot. Mitt. Trop.* 3.

Schimper, A.F.W. (1898). *Pflanzengeographie auf Physiologischer Grundlage*. Second edition, G. Fischer, Jena.

Schimper, A.F.W. (1903). *Plant-geography upon a Physiological Basis* (transl. Fisher, W.R.; ed. Groom, P. & Balfour, I.B.). Oxford University Press.

Schimper, A.F.W. (1935). *Pflanzengeographie auf Physiologischer Grundlage*. Third edition (revised Faber, F.C. von), G. Fischer, Jena.

Schlesinger, W.H. & Marks, P.L. (1977). Mineral cycling and the niche of Spanish Moss, *Tillandsia usneoides* L. *Amer. J. Bot.* 64, 1254–62.

Schmidt, F.H. & Ferguson, J.H.A. (1951). Rainfall types based on wet and dry period ratios for Indonesia with western New Guinea. *Verhandel. Djawatan Meteorol. dan Geofis, Djakarta*, 42.

Schnell, R. (1952a). Contribution à l'étude phytosociologique et phytogéographique de l'Afrique occidentale: les groupements et les unités géobotaniques de la Région guinéene. *Mém. Inst. français d'Afrique Noire* 18, 43–234.

Schnell, R. (1952b). Végétation et flore de la région montagneuse du Nimba. *Mém. Inst. français d'Afrique Noire* 22, Dakar.

Schnell, R. (1965). Aperçu préliminaire sur la phytogéographie de la Guyane. *Adansonia* 5, 309–55.

Schnell, R. (1970–71). *Introduction à la Phytogéographie des Pays Tropicaux*. 2 vols. Gauthier-Villars, Paris.

Schnell, R. (1976–77). *La Flore et la Végétation de l'Afrique Tropicale*. 2 vols. Gauthier-Villars, Paris.

Scholander, P.F. (1958). The rise of sap in lianas. Pp. 3–17 in *The Physiology of Forest Trees (Symposium Held at Harvard Forest, April 1957)*, Thimann, K.V. (Ed.), Ronald Press, New York.

Scholander, P.F. (1968). How mangroves desalinate water. *Physiologia Plantarum* 21, 251–61.

Scholander, P.F., Bund, B. & Leivestad, H. (1957). The rise of sap in a tropical liane. *Plant Physiol.* 32, 1–6.

Scholander, P.F., Van Dam, L. & Scholander, S.I. (1955). Gas exchange in the roots of mangroves. *Amer. J. Bot.* 42, 92–8.

Scholten, J.J. & Andriesse, W. (1986). Morphology, genesis and classification of three soils over limestone, Jamaica. *Geoderma* 39, 1–40.

Schott, G. (1938). Klimakunde der Südsee-Inseln. Vol. 4, Part T in *Handbuch der Klimatologie*, Köppen, W. & Geiger, R. (Eds), Gebrüder Bornträger, Berlin.

Schröter, C. (1926). *Das Pflanzenleben den Alpen*. Albert Raustein, Zürich.

Schroo, H. (1963). A study of highly phosphatic soils in a karst region of the humid tropics. *Neth. J. Agric. Sci.* Vol. II, 209–31.

Schultes, R.E. (1988a). Ethnopharmacological conservation: a key to progress in medicine. *Acta Amazon.* 18 (suppl.), 393–406.

Schultes, R.E. (1988b). *Where the Gods Reign. Plants and peoples of the Colombian Amazon*. Synergetic Press, Oracle, Arizona.

Schultes, R.E. & Raffauf, R.F. (1990). *The Healing Forest: Medicinal and Toxic Plants of Northwest Amazonia*. Timber Press, Portland, Oregon.

Schulz, J.P. (1960). Ecological studies on the rain forest of northern Suriname. *The Vegetation of Suriname*, 2, De Hulster, I.A. & Lanjouw, J. (Eds), Van Eedenfonds, Amsterdam for Bot. Mus. & Herb., Rijksuniversitet Utrecht.

Schwabe, G.H. (1962). Aus Böden von El Salvador kultivierte Blau-algen. *Nova Hedwigia* 4, 495–545.

Schwartz, O. (1988). Some podzols on Bateke Sands and their origins, People's Republic of Congo. *Geoderma* 43, 229–47.

Schwertman, U. & Latham, M. (1986). Properties of iron oxides in some New Caledonian oxisols. *Geoderma* 39, 105–23.

Scott, G.A.J. (1986). Campa Indian agriculture in the Gran Pajonal of Peru. Pp. 34–45 in *Amazonian Rain Forests: Ecosystem Disturbance and Recovery*, Jordan, C.F. (Ed.), Springer, New York.

Scott, I.M. (1985). *The Soils of Central Sarawak Lowlands, East Malaysia*. Nat. Printing Dept., Kuching, Malaysia.

Seifriz, W. (1923). The altitudinal distribution of plants on Mt. Gedeh, Java. *Bull. Torrey Bot. Cl.* 50, 283–305.

Seifriz, W. (1924). The altitudinal distribution of lichens and mosses on Mt. Gedeh, Java. *J. Ecol.* 12, 307–13.

Seiler, W. & Conrad, R. (1987). Contributions of tropical ecosystems to the global budgets of trace gases, especially CH_4, H_2, CO, and N_2O. Pp. 133–60 in *The Geophysiology of Amazonia: Vegetation and Climate Interactions*, Dickinson, R.E. (Ed.), Wiley, Chichester.

Sengele, N. (1981). Estimating potential evapotranspiration from a watershed in the Loweo region of Zaïre. Pp. 83–95 in *Tropical Agricultural Hydrology: Watershed Management and Land Use*, Lal, R. & Russell, E.W. (Eds), Wiley, Chichester.

Senn, G. (1913). Der osmotische Druck einiger Epiphyten und Parasiten. *Verh. Naturf. Ges. Basel* 24, 179–83.

Senn, G. (1923). Ueber die Ursachen der Brettwurzelbildung bei der Pyramiden-Pappel. Pp. 405–35 in *Verh. Naturf. Ges. Basel* 359, (Festbd. H. Christ).

Sewandono, M. (1937). Inventarisatie en inrichting van de veenmoerasbosschen in het panglonggebied van Sumatra's Oostkust. *Tectona* 30, 660–75.

Seybold, A. (1957). Träufelspitzen. *Beitr. Biol. Pflanz.* 33, 257–64.

Shackleton, N.J. and other CLIMAP members. (1981). Seasonal reconstruction of the Earth's surface at the last glacial maximum. *The Geological Society of America Map and Chart Series* (R. Cline, Ed.), MC36. The Geological Society of America, Boulder, Colorado.

Shreve, F. (1914a). *A Montane Rain-Forest. A Contribution to the Physiological Plant Geography of Jamaica*. Pubn 199, Carnegie Instn, Washington, D.C.

Shreve, F. (1914b). The direct effects of rainfall on hygrophilous vegetation. *J. Ecol.* 2, 82–98.

Shukla, R.P. & Ramakrishnan, P.S. (1986). Architecture and growth strategies of tropical trees in relation to successional status. *J. Ecol.* 74, 33–46.

Shuttleworth, W.J.C. & Calder, I.R. (1979). Has the Priestley-Taylor equation any relevance to forest evaporation? *J. Appl. Met.* 18, 639–46.

Shuttleworth, W.J. et al. (1984a). Eddy correlation measurements of energy partition for Amazonian forest. *Quart. J. R. Met. Soc.* 110, 1143–62.

Shuttleworth, W.J. et al. (1984b). Observations of radiation exchange above and below Amazonian forest. *Quart. J. R. Met. Soc.* 110, 1163–9.

Shuttleworth, W.J. et al. (1985). Daily variations of temperature and humidity within and above Amazonian forest. *Weather* 40, 102–8.

Silberbauer-Gottsberger, I. & Eiten, G. (1987). A hectare of cerrado. I. General aspects of the trees and thick-stemmed shrubs. *Phyton (Austria)* 27, 55–91.

Silberbauer-Gottsberger, I. & Gottsberger, G. (1984). Cerrado-Cerradão. A comparison with respect to number of species and growth forms. *Phytocoenologia* 12, 293–303.

Simmonds, N.W. (1949). Notes on the biology of the Araceae of Trinidad. *J. Ecol.* 38, 277–91.

Simmonds, N.W. (1980). Monocarpy, calendars and flowering cycles in angiosperms. *Kew Bull.* 35, 235–45.

Simon, S.V. (1914). Studien über die Periodizität der Lebensprozesse der in dauernd feuchten Tropengebieten heimischen Bäume. *Jahrb. Wiss. Bot.* 54, 71–187.

Singer, R. (1984). Role of fungi in Amazonian forests and in reforestation. Ch. 23 in *The Amazon: Limnology and Landscape Ecology of a Mighty Tropical River and its Basin*, Sioli, H. (Ed.), Junk, Dordrecht.

Singer, R. & Silva Araujo, J. (1979). Litter decomposition and ectomycorrhiza in Amazonian forests. *Acta Amazon.* 9, 25–41.

Sinia, H.R. (1938). Zur Phylogenie der Fiederblätter der Burseraceen und verwandten Familien. *Ann. Jard. Bot. Buitenz.* 48, 69–102.

Sinun, W., Wong, W.M., Douglas, I. & Spencer, T. (1992). Throughfall, stemflow, overland flow and throughflow in the Ulu Segama rain forest, Sabah, Malaysia. *Phil. Trans. R. Soc. Lond.* B. 335, 389–95.

Sioli, H. (1964). General features of the limnology of Amazonia. *Verh. Internat. Ver. Limnol.* 15, 1053–8.

Sioli, H. (1983). *Amazonien. Grundlagen der Okologie des Grössten Tropischen Waldlandes.* Wissenschaftl. Verlagsgesellsch. MBH, Stuttgart.

Sioli, H. (Ed.) (1984). *The Amazon: Limnology and Landscape Ecology of a Mighty Tropical River and its Basin.* Junk, Dordrecht.

Skutch, A.F. (1946). A compound leaf with annual increments of growth. *Bull. Torrey Bot. Cl.* 73, 542–6.

Smith, A.P. (1972). Buttressing of tropical trees: a descriptive model and new hypothesis. *Amer. Nat.* 106, 32–46.

Smith, A.P. (1979). Buttressing of tropical trees in relation to bark thickness in Dominica, B.W.I. *Biotropica* 11, 159–60.

Smith, B.N. & Brown, W.V. (1973). The Kranz syndrome in the Gramineae as indicated by carbon isotopic ratios. *Amer. J. Bot.* 60, 505–13.

Smith, J.J. (1923). Periodische Laubfall bei *Breynia cernua* Muell. Arg. *Ann. Jard. Bot. Buitenz.* 32, 97–102.

Smits, W.T.M. (1983). Dipterocarps and mycorrhiza. An ecological adaptation and a factor in forest regeneration. *Flora Malesiana Bull.* 38, 3926–37.

Snedaker, S.C. (1978). Mangroves: their value and perpetuation. *Nature & Resources (UNESCO)* 14, 6–13.

Snow, D.W. (1966). A possible selective factor in the evolution of flowering seasons in tropical forest. *Oikos* 15, 274–81.

Snow, D.W. (1981). Tropical frugivorous birds and their food plants: a world survey. *Biotropica* 13, 1–14.

Snow, J.W. (1976). The climate of northern South America. Pp. 295–403 in *Climates of Central and South America*, Schwerdtfeger, W. (Ed.), Elsevier, Amsterdam.

Soderstrom, T.R. & Calderón, C.E. (1971). Insect pollination in Tropical Rain-forest grasses. *Biotropica* 3, 1–16.

Soepadmo, E. (1972). Fagaceae. Pp. 265–75 in *Flora Malesiana* ser. 7 (2), Steenis, C.G.G.van (Ed.), M. Nihoff, The Hague.

Soepadmo, E. & Eow, B.K. (1976). The reproductive biology of *Durio zibethinus* Murr. *Gdns Bull Singapore* 29, 25–33.

Soil Management Support Services. (1990). *Keys To Soil Taxonomy.* Fourth edition, Virginia Polytech. Inst. and State Univ., Blacksburg, Virginia.

Soil Survey Staff. (1975). *Soil Taxonomy: A Basic System of Soil Classification for Making and Interpreting Soil Surveys.* Handbook 436. U.S. Dept. Agric., Washington, D.C.

Sollins, P. (1989). Factors affecting nutrient cycling in tropical soils. Pp. 85–95 in *Mineral Nutrients in Tropical Forest and Savanna Ecosystems*, Proctor, J. (Ed.), Br. Ecol. Soc. Spec. Pubn 9, Blackwell, Oxford.

Sombroek, W.G. (1966). *Amazon Soils; a Reconnaissance of the Soils of the Brazilian Amazon Region.* Centre for Agricultural Publications and Documentation, Wageningen.

Soong, N.K. (1980). Influence of soil organic matter on aggregation of soils in Peninsular Malaysia. *J. Rubber Res. Inst. Malaya* 28, 32–46.

Soulé, M.E. (Ed.) (1987). *Viable Populations for Conservation.* Cambridge University Press.

Southard, R.J. & Buol, S.W. (1988). Subsoil blocky structure formation in some North Carolina Piedmont Palendults and Paleaquults. *Soil Sci. Soc. Amer. J.* 52, 1069–76.

Sowunmi, M.A. (1986). Change of vegetation with time. Ch. 11 in *Plant Ecology in West Africa*, Lawson, G.W. (Ed.), John Wiley, Chichester.

Spanner, L. (1939). Untersuchungen über den Wärme- und Wasserhaushalt von *Myrmecodia* und *Hydnophytum.* *Jahrb. Wiss. Bot.* 88, 243–83.

Spichiger, R. & Pamard, C. (1973). Recherches sur le contact forêt-savane en Côte d'Ivoire: étude du recru forestier sur parcelles cultivées en lisière d'un îlot forestier dans le sud du pays baoulé. *Candollea* 28, 21–37.

Sporne, K.R. (1970). The advancement index and tropical rain-forest. *New Phytol.* 69, 1161–6.

Sporne, K.R. (1973). The survival of archaic dicotyledons in tropical rain-forests. *New Phytol.* 72, 1175–84.

Sporne, K.R. (1980). A re-investigation of character correlations among dicotyledons. *New Phytol.* 85, 419–49.

Spruce, R. (1908). *Notes of a Botanist on the Amazon and Andes*, 2 vols., Wallace, A.R.(Ed.), Macmillan, London.

Stahl, E. (1893). Regenfall und Blattgestalt. *Ann. Jard. Bot. Buitenz.* 11, 98–182.

Stahl, E. (1896). Ueber bunte Laubblätter. Ein Beitrag zur Pflanzenbiologie. *Ann. Jard. Bot. Buitenz.* 13, 137–216.

Stamp, L.D. (1925). *The Vegetation of Burma from an Ecological Standpoint.* Thacker, Spink, Calcutta.

Stanhill, G. (1963). The accuracy of meteorological estimates of evapotranspiration in Nigeria. *Nigerian Meteorol. Serv., Tech. Note* 31.

Stark, J., Hill, H., Rutherford, G.H. & Jones, T.A. (1959). *British Guiana*, Part II, *The Bartica Triangle*. Soil and Land-use Surveys No 5, Reg. Res. Centre, I.C.T.A., Trinidad.

Start, A.N. & Marshall, A.G. (1976). Nectarivorous bats as pollinators of trees in west Malaysia. Pp. 141–59 in *Tropical Trees, Variation, Breeding and Conservation (Linn. Soc. Symp. Ser.* 2), Burley, J. & Styles, B.T. (Eds), Academic Press, London.

Stebbing, E.P. (1937). *The Forests of West Africa and the Sahara.* Chambers, Edinburgh.

Stebbins, G.L. (1974). *Flowering Plants; Evolution above the Species Level.* Arnold, London.

Steenis, C.G.G.J. van (1934). On the origin of the Malaysian mountain flora. I. Facts and Statement of the problem. *Bull. Jard. Bot. Buitenz. (Ser.* 3) 13, 135–262.

Steenis, C.G.G.J. van (1935a). Maleische Vegetatieschetsen. *Tidschr. Ned. Aardrijksk. Genoot. Reeks* 2, 52.

Steenis, C.G.G.J. van (1935b). On the origin of the Malaysian mountain flora. Part 2. Altitudinal zones, general considerations and renewed statement of the problem. *Bull. Jard. Bot. Buitenz. (Ser.* 3) 13, 289–417.

Steenis, C.G.G.J. van (1936). On the origin of the Malaysian mountain flora. Part 3. Analysis of floristical relationships (1st instalment). *Bull. Jard. Bot. Buitenz. (Ser.* 3) 14, 56–72.

Steenis, C.G.G.J. van (1937). De invloed van den mensch op het bosch. *Tectona* 30, 634–52.

Steenis, C.G.G.J. van (1941). Oekologische eigenschappen van pionierplanten. Pp. 195–205 in *28ste Vergadering van de Vereeniging van Proefstation-Personeel, Maart 1941* (Buitenzorg).

Steenis, C.G.G.J. van (1953). Results of the Archbold expeditions: Papuan *Nothofagus. J. Arnold Arboretum* 34, 301–73.

Steenis, C.G.G.J. van (1954). Vegetatie en flora. Pp. 218–75 in *Nieuw Guinea*, Vol. 2, Klein, W.C. (Ed.), Staatsdrukerij, The Hague.

Steenis, C.G.G.J. van (1957). Outline of vegetation types in Indonesia and some adjacent regions. *Proc. Eighth Pacific Science Congress* (Manila) 4, 61–97.

Steenis, C.G.G.J. van (1958a). *Vegetation Map of Malaysia.* UNESCO, Paris.

Steenis, C.G.G.J. van (1958b). Rhizophoraceae (introduction on ecology). Pp. 431–44 in Ding Hou, *Rhizophoraceae (Flora Malesiana* 1 (5)), M. Nijhoff, The Hague.

Steenis, C.G.G.J. van (1958c). Rejuvenation as a factor for judging the status of vegetation types: the biological nomad theory. Pp. 212–5 in *Study of Tropical Vegetation* (Proc. Kandy Symposium), UNESCO, Paris.

Steenis, C.G.G.J. van (1961a). An attempt towards an explanation of the effect of mountain mass elevation. *Proc. Koninkl. Ned. Akad. Wetensch., Amsterdam*, Ser.C, 64, 435–42.

Steenis, C.G.G.J. van (1961b). Discrimination of tropical shore formations. Pp. 215–17 in *Proc. Symp. Humid Tropics Vegetation Tjiawi* (1958), UNESCO, Tijawi.

Steenis, C.G.G.J. van (1962a). The distribution of mangrove plant genera and its significance for palaeogeography. *Proc. Koninkl. Ned. Akad. Wetensch., Amsterdam*, ser. C, 65, 164–9.

Steenis, C.G.G.J. van (1962b). The Land-bridge Theory in botany. *Blumea* 11, 235–42.

Steenis, C.G.G.J. van (1969). Plant speciation in Malesia, with special reference to the theory of non-adaptive saltatory evolution. *Biol. J. Linn. Soc. Lond.* 1, 97–133.

Steenis, C.G.G.J. van (1972). *The Mountain Flora of Java.* E.J. Brill, Leiden.

Steenis, C.G.G.J. van (1980). The pollination syndrome. *Flora Malesiana Bull.* 33, 3437–9.

Steenis, C.G.G.J. van (1981). *Rheophytes of the World.* Sijthoff & Noordhoff, Alphen a. d. Rijn.

Steenis, C.G.G.J. van (1987). Rheophytes of the world: supplement. *Allertonia (Pacific Trop. Bot. Gard.)* 4, 267–330.

Stehlé, H. (1935). Essai d'écologie et géographie botanique. *Flore de la Guadeloupe et Dépendances*, 1. Imprimerie Catholique, Basse Terre (Guadeloupe) & Lechevalier, Paris.

Stehlé, H. (1945). Forest types of the Caribbean Islands. Part I. *Caribb. For.* 6, 273–408.

Steinhardt, U. (1979). Untersuchungen über den Wasser- und Nahrstoffhaushalt eines andinen Wolkenwaldes in Venezuela. *Göttinger Bodenkundliche Berichte* 56, 1–185.

Stephens, G.R. & Waggoner, P.E. (1970). Carbon dioxide exchange of a tropical rain forest. Part I. *BioScience* 20, 1050–9.

Stiles, F.G. (1978). Temporal organization of flowering among the hummingbird foodplants of a Tropical Wet forest. *Biotropica* 10, 194–210.

Stocker, O. (1935). Transpiration und Wasserhaushalt in verschiedenen Klimazonen. III. Ein Beitrag zur Transpirationsgrosse in javanischen Urwald. *Jahrb. Wiss. Bot.* 81, 464–96.

Stoddart, D.R. (1971). Coral reefs and islands and catastrophic storms. Pp. 155–97 in *Applied Coastal Geomorphology*, Steers, J.A. (Ed.), Macmillan, London.

Stoddart, D.R. & Walsh, R.P.D. (1979). Long-term climatic change in the western Indian Ocean. *Phil. Trans. R. Soc. Lond.* B286, 11–23.

Stoddart, D.R. & Walsh, R.P.D. (1992). Environmental variability and environmental extremes as factors in the island ecosystem. *Atoll Res. Bull.* 356.

Stopp, K. (1956). Botanische Analyse des Driftgutes vom Mittellauf des Kongoflusses, mit kritischen Bemerkungen über die Bedeutung fluviatiler Hydatochorie. *Beitr. Biol. Pflanzen* 32, 427–49.

Straelen, V. van (1933). Résultats scientifiques du voyage aux Indes Néerlandaises de LL.AA.RR. le Prince et la Princesse Léopold de Belgique, 1. Introd. *Mém. Mus. Hist. Nat. Belgique*, hors sér., Brussels.

Street-Perrott, F.A. & Roberts, N. (1983). Fluctuations in closed basin lakes as an indicator of past atmospheric cirulation patterns. In *Variations in the global water budget*, Street-Perrott, F.A., Meran, M. & Ratcliffe, R.A.S. (Eds), Reidel, Dordrecht.

Street-Perrott, F.A. & Roberts, N. & Metcalfe, S. (1985). Geomorphic implications of late Quaternary hydrological and climatic changes in the Northern Hemisphere tropics. Pp. 167–83 in *Environmental Change and Tropical Geomorphology*, Douglas, I. & Spencer, T. (Eds), Allen & Unwin, London.

Stubblebine, W., Langenheim, J. & Lincoln, D. (1978). Vegetative response to photoperiod in the tropical leguminous tree, *Hymenaea courbaril. Biotropica* 10, 18–29.

Sugden, A.M. (1982). The vegetation of the Serrania de Macuira, Guajira, Colombia: a contrast of arid lowlands and isolated Cloud forest. *J. Arnold Arboretum* 63, 1–30.

Sugden, A.M. (1986). The montane vegetation and flora of Margarita Island, Venezuela. *J. Arnold Arboretum* 67, 187–232.

Sugden, A.M., Tanner, E.V.J. & Kapos, V. (1985). Regeneration following clearing in a Jamaican montane forest: results of a ten-year study. *J. Trop. Ecol.* 1, 329–51.

Sussman, R.W. & Raven, P.H. (1978). Pollination of flowering plants by lemurs and marsupials: a surviving archaic coevolutionary system. *Science (N.Y.)* 200, 731–6.

Sutcliffe, J. (1979). *Plants and water.* Second edition, Arnold, London.

Swaine, M.D. (1973). *Some aspects of vegetation change.* Ph.D. Thesis, University of Wales.

Swaine, M.D. & Beer, T. (1977). Explosive seed dispersal in *Hura crepitans* L (Euphorbiaceae). *New Phytol.* 78, 695–708.

Swaine, M.D. & Greig-Smith, P. (1980). An application of principal components analysis to vegetation change in permanent plots. *J. Ecol.* 68, 33–41.

Swaine, M.D. & Hall, J.B. (1976). An application of ordination to the identification of forest types. *Vegetatio* 32, 83–6.

Swaine, M.D. & Hall, J.B. (1983). Early succession on cleared forest land in Ghana. *J. Ecol.* 71, 601–27.

Swaine, M.D. & Hall, J.B. (1986). Forest structure and dynamics. Pp. 47–93 in *Plant Ecology in West Africa* Lawson, G.W., (Ed.), John Wiley & Sons, Chichester.

Swaine, M.D. & Hall, J.B. (1988). The mosaic theory of forest regeneration and the determination of forest composition in Ghana. *J. Trop. Ecol.* 4, 253–69.

Swaine, M.D., Hall, J.B. & Lock, J.M. (1976). The forest-savanna boundary in west-central Ghana. *Ghana J. Sci.* 16, 35–52.

Swaine, M.D., Hawthorne, W.D. & Orgle, T.K. (1992). The effects of fire exclusion on savanna vegetation at Kpong, Ghana. *Biotropica* 24, 166–72.

Swaine, M.D., Hawthorne, W.D., Orgle, T.K. & Agyemang, V.K. (1990). Fifteen years of forest succession in Ghana. *Univ. Aberdeen Trop. Biol. Newsltr.* 58, 1–2.

Swaine, M.D. & Whitmore, T.C. (1988). On the definition of ecological species groups in tropical rain forests. *Vegetatio* 75, 81–6.

Symington, C.F. (1933). The study of secondary growth on rain forest sites. *Malay. For.* 2, 107–17.

Symington, C.F. (1934). Notes on Malayan Dipterocarpaceae – II. *Gdns Bull. Singapore* 8, 1–40.

Symington, C.F. (1943). Foresters' Manual of Dipterocarps. *Malayan For. Rec.* 16 (repr. with plates and introduction, Penerbit Universiti Malaya, Kuala Lumpur, 1974).

Sys, C. (1960). The principles of soil classification in the Belgian Congo. *Trans. Seventh Internat. Congr. Soil Science* 5, 112–18.

Tagawa, T., Suzuki, E., Partomihardjo, T. & Suriadarma, A. (1985). Vegetation and succession on the Krakatau islands, Indonesia. *Vegetatio* 60, 131–45.

Takeuchi, M. (1961a). The structure of the Amazonian vegetation, III. Campina forest in the Rio Negro region. *J. Fac. Sci. Univ. Tokyo,* Sect. III, Bot. 8, 27–37.

Takeuchi, M. (1961b). The structure of the Amazonian vegetation, IV. High campina forest in the Rio Negro region. *J. Fac. Sci. Univ. Tokyo,* Sect. III, Bot. 8, 279–88.

Takeuchi, M. (1962). The structure of the Amazonian vegetation, VI. Igapó. *J. Fac. Sci. Univ. Tokyo,* Sect. III, Bot. 8, 297–304.

Takhtajan, A. (1969). *Flowering plants, origin and dispersal.* Transl. C. Jeffrey. Oliver & Boyd, Edinburgh.

Talbot, M.R. & Delibrias, G. (1980). A new Pleistocene-Holocene water-level curve for Lake Bosumtwi, Ghana. *Earth Planet. Sci. Lett.* 47, 336–44.

Talbot, M.R., Livingstone, D.A., Palmer, P.G., Maley, J., Melack, J.M., Delibrias, G. & Guliksen, S. (1984). Preliminary results from sediment cores from Lake Bosumtwi, Ghana. *Palaeocol. Afr.* 16, 173–92.

Tammes, P.M.L. (1938). Movement of water in tropical lianas. *Ann. Jard. bot. Buitenz.* 48, 1–9.

Tanner, E.V.J. (1977). Four montane forests of Jamaica: a quantitative characterization of the floristics, the soils and the foliar mineral levels, and a discussion of the interrelations. *J. Ecol.* 65, 883–918.

Tanner, E.V.J. (1980a). Studies on the biomass and productivity in a series of montane rain forests in Jamaica. *J. Ecol.* 68, 573–88.

Tanner, E.V.J. (1980b). Litterfall in montane rain forests in Jamaica and its relation to climate. *J. Ecol.* 68, 833–48.

Tanner, E.V.J. (1982). Species diversity and reproductive mechanisms in Jamaican trees. *Biol. J. Linn. Soc. Lond.* 18, 263–78.

Tanner, E.V.J., Kapos, V. & Healey, J.R. (1991). Hurricane effects on forest ecosystems in the Caribbean. *Biotropica* 23, 513–21.

Tansley, A.G. & Fritsch, F.E. (1905). Sketches of vegetation at home and abroad. I. The flora of the Ceylon littoral. *New Phytol.* 4, 1–17, 27–55.

Tardy, Y. & Nahon, D. (1985). Geochemistry of laterites: stability of Al-goethite, Al-hematite, and Fe^{3+}-kaolinite in bauxites and ferricretes: an approach to the mechanism of concretion formation. *Amer. J. Sci.* 285, 865–903.

Taylor, B.W. (1957). Plant succession on recent volcanoes in Papua. *J. Ecol.* 45, 233–43.

Taylor, B.W. (1959). The classification of lowland swamp communities in northeastern Papua. *Ecology* 40, 703–11.

Taylor, C.J. (1960). *Synecology and Silviculture in Ghana.* T. Nelson, Edinburgh.

Taylor, F.J. (1970–71). Some aspects of the development of mango (*Mangifera indica*) leaves. I. Leaf area, dry weight and water content. *New Phytol.* 69 (1970), 377–94. II. Methanol-soluble and cell wall fractions, *Ibid.* 70 (1971), 567–79. III. A mechanical analysis, *Ibid.* 70 (1971), 911–22.

Teas, H.J. (Ed.) (1983). *Biology and Ecology of Mangroves.* Junk, The Hague.

Terborgh, J. (1992). Maintenance of diversity in tropical forests. *Biotropica* 24 (Suppl.), 283–92.

Terborgh, J. & Petren, K. (1989). Development of habitat structure through succession in an Amazonian floodplain forest. (Unpublished.)

ter Braak, C.J.F. (1986). Canonical correspondence analysis: a new eigenvector technique for multivariate direct gradient analysis. *Ecology* 67, 1167–79.

ter Steege, J. & Cornelissen, J.H.C. (1989). Distribution and ecology of vascular epiphytes in lowland rain forest of Guyana. *Biotropica* 21, 331–9.

Thien, L. (1980). Patterns of pollination in the primitive angiosperms. *Biotropica* 12, 1–13.

Thom, A.S. & Oliver, H.R. (1977). On Penman's equation for estimating regional evaporation. *Quart. J. R. Met. Soc.* 103, 345–57.

Thomas, M. (1994). *Geomorphology in the Tropics.* Wiley, Chichester.

Thomas, M.F. & Thorp, M.B. (1980). Some aspects of the geomorphological interpretation of Quaternary alluvial sediments in Sierra Leone. *Z. Geomorph. N.F.,* suppl. 36, 140–61.

Thompson, J.McL. (1943). A modern study of cauliflory. *Nature (Lond.)* 151, 481–2.

Thompson, J.McL. (1946). The study of plant-behaviourism: a common meeting-ground for those engaged in botanical enquiry. *Proc. Linn. Soc. Lond., Sess.* 157, 72–91.

Thompson, J.McL. (1951). A further contribution to our knowledge of cauliflorous plants (with special reference to *Swartzia pinnata* Willd.). *Proc. Linn. Soc. Lond., Sess.* 162, 212–22.

Thornthwaite, C.W. (1948). An approach towards a rational classification of climate. *Geogr. Rev.* 38, 55–94.

Thornton, I.W.B. (Ed.) (1987). 1986 Zoological Expedition to the Krakataus, Preliminary Report. *La Trobe University (Melbourne), Misc. Ser.* 3.

Thornton, I.W.B., New, T.R., Zann, R.A. & Rawlinson, P.A. (1990). Colonization of the Krakatau Islands by animals: a perspective from the 1980s. *Phil. Trans. R. Soc. Lond.* B328, 131–65.

Thorp, J. (1936). *Geography of Soils of China.* Nat. Geol. Surv. China, Nanking.

Tie, Y.L. (1990). *Studies of Peat Swamps in Sarawak with Particular Reference to Soil-Forest Relationships and Development of Dome-Shaped Structures.* Ph.D. Thesis, Polytechnic of North London.

Tie, Y.L., Baillie, I.C., Phang, C.M.S. & Lim, C.P. (1979). *Soils of the Gunong Mulu National Park.* Soil Survey Div., Dept. Agric., Kuching, Sarawak.

Tiemann, H.D. (1935). What are the largest trees in the world? *J. For.* 33, 903–15.

Tietze, M. (1906). Physiologische Bromeliaceenstudien. II. Die Entwicklung der wasseraufnehmenden Bromeliaceen Trichome. *Zeitschr. f. Naturwissenschaften (Halle)* 78, 1–49.

Todzia, C. (1986). Growth habits, host tree species and density of hemiepiphytes at Barro Colorado Island, Panama. *Biotropica* 18, 22–7.

Tomin, N.M. (1991). Profile and tree morphology of a Heath forest in Peninsular Malaysia. Pp. 205–20 in *L'arbre Biologie et Developpment,* Edelin, C. (Ed.), Naturalia Monspeliensis, Montpellier.

Tomlinson, P.B. (1980). *The Biology of Trees Native to Tropical Florida*. Harvard Univ. Printing Office, Allston, Massachusetts.

Tomlinson, P.B. (1986). *The Botany of Mangroves*. Cambridge University Press.

Tomlinson, P.B. & Longman, K.A. (1981). Growth phenology of tropical trees in relation to cambial activity. Pp. 7–19 in *Age and Growth Rate of Tropical Trees*, Bormann, F.H. & Berlyn, G. (Eds), Yale Univ. Sch. of For. and Env. Stud., 94.

Torquebiau, E.F. (1986). Mosaic patterns in dipterocarp rain forest in Indonesia, and their implications for practical forestry. *J. Trop. Ecol.* 2, 301–25.

Tosi, J.A. Jr. (1960). Zonas de vida natural en el Peru: memoria explicativa sobre el mapa ecológico del Peru. Bol. tecnico No. 5, Inst. Interamericano de Ciencias Agricolas de la O.E.A., Programa de Cooperacion de Tecnica, Zona Andina, Lima, Peru.

Tosi, J.A. Jr. (1969). *Mapa Ecológico, República de Costa Rica. Según la Clasificación de Zonas de Vida del Mundo de L.R. Holdridge*. Centro Científico Tropical, San José, Costa Rica.

Tracey, J.G. (1969). Edaphic differentiation of some forest types in eastern Australia, I. Soil physical factors. *J. Ecol.* 57, 805–16.

Tracey, J.G. (1982). *The Vegetation of the Humid Tropical Region of North Queensland*. CSIRO, Canberra.

Trenbath, B.R. (1989). The use of mathematical models in the development of shifting cultivation systems. Pp. 353–389 in *Mineral Nutrients in Tropical Forest and Savanna Ecosystems*, Proctor, J. (Ed.), Br. Ecol. Soc. Spec. Pubn 9, Blackwell, Oxford.

Treub, M. (1883). Observations sur les plantes grimpantes du Jardin Botanique de Buitenzorg. *Ann. Jard. bot. Buitenz.* 3, 160–83.

Treub, M. (1888). Notice sur la nouvelle flore de Krakatau. *Ann. Jard. bot. Buitenz.* 7, 213–23.

Tricart, J. (1985). Evidence of Upper Pleistocene dry climates in northern South America. Pp. 197–217 in *Environmental Change and Tropical Geomorphology*, Douglas, I. & Spencer, T. (Eds), Allen & Unwin, London.

Troll, C. (1948). Der asymmetrische Aufbau der Vegetationszonen und Vegetationsstufen auf der Nord- und Südhalbkugel. *Ber. Geobot. Inst. Rübel f.*, 1947, 46–83.

Troll, C. (1950). Savannentypen und das Problem der Primärsavannen. *Proc. 7th Int. Bot. Congr. Stockholm*, 670–4.

Troll, C. (1955). Der jahreszeitliche Ablauf des Naturgeschehens in den verschiedenen Klimagürteln der Erde. *Studium Generale* 8, 713–33.

Troll, C. (1958). Zur Physiognomik der Tropengewächse. *Jahresber. d. Ges. d. Freunde u. Förderer der Universität Bonn*, 1958, pp. 3–15.

Troll, C. (1959). Die tropischen Gebirge. Ihre dreidimensionale klimatische and pflanzengeographische Zonierung. *Bonner Geogr. Abhandl.* 25, 1–93.

Troll, W. (1930) Ueber die sogennanten Atemwurzeln von Mangroven. *Ber. Dtsch. Bot. Ges.* 48 (Generalversammlungs-Heft), 81–99.

Troll, W. (1938–43). *Vergleichende Morphologie der Pflanzen*. 1. Bornträger, Berlin.

Troll, W. & Dragendorf, O. (1931). Ueber die Luftwurzeln von *Sonneratia* Linn f. und ihre biologische Bedeutung. *Planta* 13, 311–473.

Tschinkel, H.M. (1966). Annual growth rings in *Cordia alliodora*. *Turrialba* (Costa Rica) 16, 73–80.

Turner, B.L. (1986). Mystery of the Maya revealed: the agricultural base of a tropical civilisation. *Amer. Geogr. Soc. Focus* 36 (2), 2–7.

Turner, I.M. (1994). The taxonomy and ecology of the vascular plant flora of Singapore: a statistical analysis. *Bot. J. Linn. Soc. Lond.* 114, 215–27.

Tutin, C.E.G., Williamson, E.A., Rogers, M.E. & Fernandez, M. (1991). A case study of a plant-animal relationship: *Cola lisae*

and lowland gorillas in the Lope Reserve, Gabon. *J. Trop. Ecol.* 7, 181–99.

Tüxen, R. (1929). Ueber einige nordwestdeutsche Waldassoziationen von regionaler Verbreitung. *Jahrb. Geogr. Ges. Hannover* 1929, 55–116.

Uhe, G. (1974). The composition of the plant communities inhabiting the recent lava flows of Savaii, Western Samoa. *J. Trop. Ecol.* 15, 140–51.

Uhl, C. (1987). Factors controlling succession following slash-and-burn agriculture in Amazonia. *J. Ecol.* 75, 377–407.

Uhl, C., Buschbacher, R. & Serrão, E.A.S. (1988a). Abandoned pastures in eastern Amazonia. I. Patterns of plant succession. *J. Ecol.* 76, 663–81.

Uhl, C., Clark, H., Clark, K. & Maquirino, P. (1982a). Successional patterns associated with slash-and-burn agriculture in the Upper Rio Negro region of the Amazon basin. *Biotropica* 14, 249–54.

Uhl, C., Clark, K., Clark, H. & Murphy, P. (1981). Early plant succession after cutting and burning in the Upper Rio Negro region of the Amazon basin. *J. Ecol.* 69, 631–50.

Uhl, C., Clark, K., Dezzeo, N. & Maquirino, P. (1988b). Vegetation dynamics in Amazonian treefall gaps. *Ecology* 69, 751–63.

Uhl, C. & Jordan, C.F. (1984). Succession and nutrient dynamics following cutting and burning in Amazonia. *Ecology* 65, 1474–90.

Uhl, C., Jordan, C., Clark, K., Clark, H. & Herrera, R. (1982b). Ecosystem recovery in Amazon Caatinga forest after cutting, cutting and burning and bulldozer clearing treatment. *Oikos* 38, 313–20.

Uhl, C. & Vieira, I.C.G. (1989). Ecological impact of selective logging in the Brazilian Amazon: a case study from the Paragominas region of the State of Pará. *Biotropica* 21, 98–106.

Uhl, N.W. & Dransfield, J. (1987). *Genera Palmarum: a Classification of Palms Based on the Work of Harold E. Moore, Jr.* For L.H. Bailey Hortorium & Internat. Palm Soc., Allen Press, Lawrence, Kansas.

Ulbrich, E. (1928). *Biologie der Früchte und Samen*. Springer, Berlin.

Ule, E. (1905). Epiphyten des Amazonasgebietes. *Vegetationsbilder*, Reihe 2, 1.

Ule, E. (1906). Ameisenpflanzen. *Bot. Jahrb.* 37, 335–52.

Ule, E. (1908). Die Pflanzenformationen des Amazonas-Gebietes, *Bot. Jahrb.* 40, 114–72.

UNESCO (1973). *International Classification and Mapping of Vegetation* (Ecology and Conservation 6). UNESCO, Paris.

UNESCO (1978). Tropical forest ecosystems: a state-of-knowledge report prepared by UNESCO/UNEP/FAO (Nat. Resources Res. 14). UNESCO, Paris.

Upadhyay, V.P. & Singh, J.S. (1989). Patterns of nutrient immobilisation and release in decomposing forest litters in Central Himalaya, India. *J. Ecol.* 77, 127–46.

Upchurch, G.R. Jr. & Wolfe, J.A. (1987). Mid-Cretaceous to Early Tertiary vegetation and climate: evidence from fossil leaves and woods. Pp. 75–105 in *The Origins of Angiosperms and their Biological Consequences*, Friis, E.M., Chaloner, W.G. & Crane, P.R. (Eds) Cambridge University Press.

U.S. National Academy of Sciences (1974). The effects of herbicides in south Vietnam. Part A. Summary and conclusions. Nat. Acad. Sci., Washington, D.C.

Usher, M.B. (1981). Modelling ecological succession, with particular reference to Markovian models. *Vegetatio* 46, 11–18.

Van der Weert, R. (1974). The influence of mechanical forest clearing on soil conditions and resulting effects on root growth. *Trop. Agric.* 51, 325–31.

Vareschi, V. (1980). *Vegetationsökologie der Tropen*. E. Ulmer, Stuttgart.

Vaughan, R.E. & Wiehe, P.O. (1937). Studies on the vegetation of Mauritius. I. A preliminary survey of the plant community. *J. Ecol.* 25, 289–342.

Vaughan, R.E. & Wiehe, P.O. (1941). Studies on the vegetation of Mauritius. III. The structure and development of the Upland Climax forest. *J. Ecol.* 29, 127–60.

Vaughan, R.E. & Wiehe, P.O. (1947). Studies on the vegetation of Mauritius. IV. Some notes on the internal climate of the Upland Climax forest. *J. Ecol.* 34, 126–36.

Vaupel, F. (1910). Die Vegetation der Samoa-Inseln. *Bot. Jahrb.* 44, Beibl. 102, 47–58.

Vazquez-Yanes, C. (1974). Studies on the germination of seeds of *Ochroma lagopus* Swartz. *Turrialba* 24, 176–9.

Vazquez-Yanes, C. (1976). Seed dormancy and germination in secondary vegetation tropical plants: the role of light. *Comp. Physiol. Ecol.* 1, 30–4.

Vazquez-Yanes, C. & Smith, H. (1982). Phytochrome control of seed germination in the tropical rain forest pioneer trees *Cecropia obtusifolia* and *Piper auritum* and its ecological significance. *New Phytol.* 92, 477–86.

Verboom, W.C. (1968). Grassland successions and associations in Pahang, central Malaya. *Trop. Agric. (Trinidad)* 45, 47–59.

Vermoesen, C. (1931). *Les essences forestières du Congo Belge. I. Manuel des essences de la région equatoriale et du Mayombe.* Brussels.

Vis, M. (1986). Interception, drop-size distributions and rainfall kinetic energy in four Colombian forest ecosystems. *Earth Surf. Proc. Landf.* 11, 591–603.

Vitousek, P.M. (1982). Nutrient cycling and nutrient use efficiency. *Amer. Nat.* 119, 553–72.

Vitousek, P.M. (1984). Litterfall and nutrient limitation in tropical forests. *Ecology* 65, 285–98.

Vitousek, P.M., Cleve, K. van, Balakrishnan, N. & Mueller-Dombois, M. (1983). Soil development and nitrogen turnover in montane rainforest soil in Hawaii. *Biotropica* 15, 268–74.

Vitousek, P.M. & Denslow, J.S. (1986). Nitrogen and phosphorus availability in treefall gaps of a lowland tropical rain forest. *J. Ecol.* 74, 1167–78.

Vitousek, P.M., Gosz, J.R., Grier, C.C., Melillo, J.M. & Reiners, W.A. (1982). A comparative analysis of potential nitrification and nitrate mobility in forest ecosystems. *Ecol. Monogr.* 52, 155–77.

Vitousek, P.M. & Sanford, R.L. (1986). Nutrient cycling in moist tropical forest. *Ann. Rev. Ecol. Syst.* 17, 137–67.

Vogel, S. (1974). *Olblumen und Olsammelnde Bienen.* Steiner, Wiesbaden.

Volkens, G. (1912). *Laubfall und Lauberneuerung in den Tropen.* Bornträger, Berlin.

Voorhoeve, A.G. (1964). Some notes on the Tropical Rain forest of the Yoma-Gola National Forest near Bomi Hills, Liberia. *Commonw. For. Rev.* 43, 17–24.

Wada, K. & Kakuto, Y. (1985). Embryonic halloysite in Ecuadorian soils derived from volcanic ash. *Soil Sci. Soc. Amer. J.* 49, 1309–18.

Wade, L.K. & MacVean, D.N. (1969). *Mt. Wilhelm Studies. 1. The Alpine and Subalpine Vegetation.* Pubn. BG/1, Dept. Biogeogr. & Geomorph., Australian Nat. Univ., Canberra.

Wadsworth, F.H. (1947). Growth in the Lower Montane Rain forest of Puerto Rico. *Caribbean For.* 8, 27–35.

Walker, D. (1966). Vegetation of the lake Ipea region, New Guinea highlands. I. Forest, grassland and 'garden'. *J. Ecol.* 54, 503–33.

Walker, D. (1986). Tropical rainforests. *Sci. Progr. (Oxford)* 70, 461–72.

Walker, D. & Chen, Y. (1987). Palynological light on tropical rainforest dynamics. *Quaternary Sci. Rev.* 6 77–92.

Walker, D. & Flenley, J.R. (1979). Late Quaternary vegetational history of the Enga Province of Upland Papua New Guinea. *Phil. Trans. R. Soc. Lond.* B286, 265–344.

Walker, D. & Singh, G. (1981). Vegetation history. Pp. 26–43 in *Australian Vegetation,* Groves, R.H. (Ed.), Cambridge University Press.

Wallace, A.R. (1878). *Tropical Nature and Other Essays.* Macmillan, London.

Wallace, A.R. (1890). *The Malay Archipelago.* Third edition, Macmillan, London.

Wallace, B.J. (1981). *The Australian Vascular Epiphytes: Flora and Ecology.* Ph.D. Thesis, University of New England, Armidale, N.S.W., Australia.

Walsh, R.P.D. (1980a). *Drainage Density and Hydrological Processes in a Humid Tropical Environment: the Windward Islands.* Ph.D. Thesis, University of Cambridge.

Walsh, R.P.D. (1980b). Runoff processes and models in the humid tropics. *Z. Geomorph. N.F.,* 36, 176–201.

Walsh, R.P.D. (1982a). Climate. Pp. 29–67 in *Gunung Mulu National Park, Sarawak. An Account of its Environment and Biota being the Results of The Royal Geographical Society/Sarawak Government Expedition and Survey 1977–78,* Part 1, Jermy, A.C. & Kavanagh, K.P. (Eds), Sarawak Museum Journal 30, No. 51.

Walsh, R.P.D. (1982b). A provisional survey of the effects of Hurricanes David and Frederic on the terrestrial environment of Dominica. *Swansea Geogr.* 19, 28–34.

Walsh, R.P.D. (1982c). Hydrology and water chemistry. Pp. 121–81 in *Gunung Mulu National Park, Sarawak. An Account of its Environment and Biota being the Results of The Royal Geographical Society/Sarawak Government Expedition and Survey 1977–78,* Part 1, Jermy, A.C. & Kavanagh, K.P. (Eds), Sarawak Museum Journal 30, No. 51.

Walsh, R.P.D. (1985). The influence of climate, lithology and time on drainage density and relief development in the tropical volcanic terrain of the Windward Islands. Pp. 93–122 in Douglas, I. & Spencer, T. (Eds), *Environmental Change and Tropical Geomorphology,* Allen & Unwin, London.

Walsh, R.P.D. (1987). Interception and stemflow in tropical rain forest environments. *Swansea Geogr.* 24.

Walsh, R.P.D. (1992). Representation and classification of tropical climates for ecological purposes using the perhumidity index. *Swansea Geogr.* 24, 109–29.

Walsh, R.P.D. (1993). Problems of the climatic geomorphological approach with reference to drainage density, chemical denudation and slopewash in the humid tropics. *Würzburger Geographische Arbeiten* 87, 221–39.

Walsh, R.P.D. & Howells, K.A. (1988). Soil pipes and their role in runoff generation and chemical denudation in a humid tropical catchment in Dominica. *Earth Surf. Processes Landf.* 13, 9–17.

Walsh, R.P.D. & Reading, A.J. (1991). Historical changes in tropical cyclone frequency within the Caribbean since 1500. *Würzburger Geographische Arbeiten* 80, 199–240.

Walter, H. (1964). *Die Vegetation der Erde in Oeko-Physiologischer Betrachtung,* Vol. 1: *Die Tropischen und Subtropischen Zonen.* Second edition, G. Fischer, Jena.

Walter, H. (1971). *Ecology of Tropical and Subtropical Vegetation.* Transl. D. Mueller-Dombois; Burnett, J.H. (Ed.), Oliver & Boyd, Edinburgh.

Walter, H. & Lieth, H. (1960). *Klimadiagramm-Weltatlas.* G. Fischer, Jena.

Wambeke, A. van (1989). Tropical soils and soil classification updates. *Adv. Soil Sci.* 10, 171–93.

Wambeke, A. van (1992). *Soils of the Tropics.* McGraw-Hill, New York.

Wang, C. (1961). *The Forests of China.* Maria Moors Cabot Foundation (Harvard Univ.), Pubn. 5.

Warming, E. (1909). *Oecology of Plants.* Transl. P. Groom & I.B. Balfour. Clarendon Press, Oxford.

Waterman, P.G. & Mole, S. (1989). Soil nutrients and plant secondary compounds. Pp. 241–53 in *Mineral Nutrients in*

Tropical Forest and Savanna Ecosystems, Proctor, J. (Ed.), Br. Ecol. Soc. Spec. Pubn 9, Blackwell, Oxford.

Watson, J.G. (1928). The mangrove swamps of the Malay Peninsula. *Malay. For. Rec.* 6.

Watson, J.G. (1937). Age-class representation in virgin forest. *Malay. For.* 6, 146–7.

Watt, A.S. (1947). Pattern and process in the plant community. *J. Ecol.* 35, 1–22.

Watters, R.F. (1975). Shifting agriculture – its past, present and future. Pp. 77–88 in *The Use of Ecological Guidelines for Development in the American Humid Tropics (Proc. Internat. Meeting, Caracas, Venezuela, 20–22 Feb. 1974)*, IUCN Publ. N.S. 31.

Watts, I.E.M. (1969). Climates of China and Korea. Pp. 1–117 in *Climates of Northern and Eastern Asia*, Arakawa, H. (Ed.), Elsevier, Amsterdam.

Weaver, P.L., Birdsley, R.A. & Lugo, A.E. (1987). Soil organic matter in secondary forests in Puerto Rico. *Biotropica* 19, 17–23.

Weaver, P.L., Medina, E., Pool, K., Gonzales-Libby, J. & Cuevas, E. (1986). Ecological observations in the Dwarf Cloud forest of the Luquillo mountains in Puerto Rico. *Biotropica* 18, 79–85.

Weaver, P.L. & Murphy, P.G. (1990). Forest structure and productivity in Puerto Rico's Luquillo mountains. *Biotropica* 22, 69–82.

Webb, L.J. (1958). Cyclones as an ecological factor in tropical lowland rain forest, north Queensland. *Aust. J. Bot.* 6, 220–8.

Webb, L.J. (1959). A physiognomic classification of Australian rain forests. *J. Ecol.* 47, 551–70.

Webb, L.J. (1969). Edaphic differentiation of some forest types in eastern Australia, II. Soil chemical factors. *J. Ecol.* 57, 817–30.

Webb, L.J. (1977). *Ecological Considerations and Safeguards in the Modern Use of Tropical Lowland Rain Forests as a Source of Pulpwood: Example, the Madang Area, Papua New Guinea*. Dept. of Envir. & Nat. Resources, PNG, Waigani, Papua New Guinea.

Webb, L.J. & Tracey, J.G. (1981a). The rainforests of northern Australia. Ch. 4 in *Australian Vegetation*, Groves, R.H. (Ed.) Cambridge University Press.

Webb, L.J. & Tracey, J.G. (1981b). Australian rain forests: pattern and change. Ch. 22 in *Ecological Biogeography of Australia*, Keast, A. (Ed.), Junk, The Hague.

Webb, L.J., Tracey, J.G. & Jessup, L.W. (1986). Recent evidence for autochthony of Australian tropical and subtropical rainforest floristic elements. *Telopea* 2, 575–89.

Webb, L.J., Tracey, J.G. & Williams, W.T. (1972). Regeneration and pattern in the Subtropical rain forest. *J. Ecol.* 60, 675–96.

Webb, L.J., Tracey, J.G. & Williams, W.T. (1984). A floristic framework of Australian rainforests. *Austr. J. Ecol.* 9, 169–98.

Webb, L.J., Tracey, J.G., Williams, W.T. & Lance, G.N. (1967). Studies in the numerical analysis of complex rain forest communities. II. The problem of species sampling. *J. Ecol.* 55, 525–38.

Webber, M.L. (1934). Fruit dispersal. *Malay. For.* 3, 18–19.

Webber, M.L. (1954). The mangrove ancestry of a fresh water swamp forest suggested by its diatom flora. *Malay. For.* 17, 25–6.

Weber, H. (1958). Die Paramos von Costa Rica. *Abhandl. Akad. Wiss. u. Litt., Wiesbaden, Math.-Naturwiss. Kl. Jg.* 1958, 3.

Weberbauer, A. (1911). Die Pflanzenwelt der peruanishcen Anden. *Veget. Erde* 12.

Weberbauer, A. (1945). *El Mundo Vegetal de Los Andes Peruanos*. Minist. Agric. Lima.

Welch, I.A., Sampson, O.R. & Bell, G.S. (1972). Vegetation types of Guyana. Photo-interpretation key. *For. Bull.* 4 (N.S.), Min. Mines & Forests, Guyana.

Went, F.W. (1940). Soziologie der Epiphyten eines tropischen Urwaldes. *Ann. Jard. bot. Buitenz.* 50, 1–98.

Wernstedt, F.L. (1972). *World climatic data*. Climatic Data Press, Lemont, Penn., U.S.A.

West, R.G. (1956). Mangrove swamps of the Pacific coast of Colombia. *Ann. Ass. Amer. Geogr.* 46, 98–121.

Westing, A.H. (Ed.) (1984). *Herbicides in War: The Long-term Ecological and Human Consequences*. Taylor & Francis (for SIPRI), London.

White, F. (1965). The savanna woodlands of the Zambezian and Sudanian domains. *Webbia* 19, 651–71.

White, F. (1981, 1983). *Vegetation Map of Africa with Descriptive Memoir* (map 1981, Memoir 1983). UNESCO, Paris.

White, K.G. (1975). The effect of natural phenomena on the forest environment. Pres. Addr. Papua New Guinea Scientific Soc., March 1975. Dept. of Forests, Port Moresby, Papua New Guinea.

Whitford, H.N. (1906). The vegetation of the Lamao Forest Reserve. *Philipp. J. Sci.* 1, 373–431.

Whitmore, T.C. (1962a, b). Studies in systematic bark morphology. I. Bark morphology in Dipterocarpaceae. II. General features of bark construction in Dipterocarpaceae. *New Phytol.* 61, 191–220.

Whitmore, T.C. (1963). Studies in systematic bark morphology, III. Bark taxonomy in Dipterocarpaceae. *Gdns Bull. Singapore* 19, 321–71.

Whitmore, T.C. (1969). First thoughts on species evolution in Malayan *Macaranga* (Studies in *Macaranga* III). *Biol. J. Linn. Soc. Lond.* 1, 223–31.

Whitmore, T.C. (1972a). (Ed.) *Tree flora of Malaya* 1. Longman, Kuala Lumpur & London.

Whitmore, T.C. (1972b). The Gunong Benom Expedition. 2. An outline description of the forest zones on north east Gunong Benom. *Bull. Brit. Mus. (Nat. Hist.) Zool. Ser.* 23, 11–15.

Whitmore, T.C. (1974). Change with time and the role of cyclones in tropical rain forest on Kolombangara, Solomon Islands. *Commonw. For. Inst. (Oxford), Paper* 46.

Whitmore, T.C. (1978). Gaps in the forest canopy. Pp. 639–55 in *Tropical Trees as Living Systems*, Tomlinson, P.B. & Zimmermann, M.H. (Eds), Cambridge University Press.

Whitmore, T.C. (1982). On pattern and process in forests. Pp. 45–59 in *The Plant Community as a Working Mechanism*, Newman, E.I. (Ed.), Brit. Ecol. Soc. Special Pubn 1, Blackwell, Oxford.

Whitmore, T.C. (1984a). *Tropical Rain Forests of the Far East*. Second edition, Clarendon Press, Oxford.

Whitmore, T.C. (1984b). A new vegetation map of Malesia at scale 1 : 5 million with commentary by T.C. Whitmore. *J. Biogeogr.* 11, 461–71.

Whitmore, T.C. (1989). Changes over 21 years in the Kolombangara rain forests. *J. Ecol.* 77, 469–83.

Whitmore, T.C. (1990). *An Introduction to Tropical Rain Forests*. Clarendon Press, Oxford.

Whitmore, T.C. & Burnham, C.P. (1969). The altitudinal sequence of forests and soils on granite near Kuala Lumpur. *Malay. Nat. J.* 22, 99–118.

Whitmore, T.C. & Gong, W.-K. (1983). Growth analysis of the seedlings of balsa (*Ochroma lagopus*). *New Phytol.* 95, 305–11.

Whitmore, T.C., Peralta, R. & Brown, K. (1986). Total species count in a Costa Rican tropical rain forest. *J. Trop. Ecol.* 1, 375–8.

Whitmore, T.C. & Prance, G.T. (Eds) (1987). *Biogeography and Quaternary History in Tropical America*. Clarendon Press, Oxford.

Whitmore, T.C. & Sayer, J.A. (Eds) (1992). *Tropical Deforestation and Species Extinction*. Chapman & Hall, London.

Whitmore, T.C. & Sidiyasa, K. (1986). Composition and structure of a lowland rain forest at Toraut, northern Sulawesi. *Kew Bull.* 41, 747–56.

Whittaker, R.H., Levin, S.A. & Root, R.B. (1973). Niche, habitat and ecotope. *Amer. Nat.* 107, 321–8.

Whittaker, R.H. & Likens, G.E. (1975). The biosphere and man. Pp. 305–28 in *Primary Productivity of the Biosphere,* Springer, New York.

Whittaker, R.J., Bush, M.B. & Richards, K. (1989). Plant recolonization and vegetation succession on the Krakatau islands, Indonesia. *Ecol. Monogr.* 59, 59–123.

Whitten, A.J., Damanik, S.J., Anwar, J. & Hisyam, N. (1984). *The Ecology of Sumatra.* Second edition, Gadjah Mada Univ. Press, Yogyakarta, Indonesia.

Wiebes, J.T. (1982). Fig wasps (Hymenoptera). Pp. 735–55 in *Biogeography and Ecology of New Guinea,* Gressit, J.L. (Ed.), Junk, The Hague.

Wierda, A., Veen, A.W.L. & Hutjes, R.W.A. (1989). Infiltration at the Tai Rain Forest (Côte d'Ivoire): measurements and modelling. *Hydrol. Processes* 3, 371–82.

Wiersum, K. (1985). Effects of various vegetation layers in an *Acacia auriculiformis* forest plantation on surface erosion in Java, Indonesia. Pp. 79–89 in *Soil Erosion and Conservation,* Swaify, E.I., Mohlenhauer, W.C. & Lo, A. (Eds), Soil and Water Conserv. Soc. America, Ankery, Iowa.

Wiesner, J. (1895). Untersuchungen über den Lichtgenuss der Pflanzen mit Rücksicht auf die Vegetation von Wien, Kairo und Buitenzorg (Java). *S. B. Akad. wiss. Wien,* Math.-naturw. Cl. 104, 605–711.

Wiesner, J. (1907). *Der Lichtgenuss der Pflanzen.* Leipzig.

Wilkinson, G. (1939). Root competition and sylviculture. *Malay. For.* 8, 11–15.

Williams, C.B. (1947). The logarithmic series and the comparison of island floras. *Proc. Linn. Soc. Lond.* 158, 104–8.

Williams, M.A.J. (1985). Pleistocene aridity in tropical Africa, Australia and Asia. Pp. 219–33 in *Environmental Change and Tropical Geomorphology,* Douglas, I. & Spencer, T. (Eds), Allen & Unwin, London.

Williams, W.T. (1971). Strategy and tactics in the acquisition of ecological data. *Proc. Ecol. Soc. Aust.* 6, 57–62.

Williams, W.T. (Ed.) (1976). *Pattern Analysis in Agricultural Science.* CSIRO, Melbourne; Elsevier, Amsterdam.

Williams, W.T., Lance, G.N., Webb, L.J., Tracey, J.G. & Connell, J.H. (1969a). Studies in the numerical analysis of complex rain forest communities. IV. A method for the elucidation of small-scale forest pattern. *J. Ecol.* 57, 635–54.

Williams, W.T., Lance, G.N., Webb, L.J., Tracey, J.G. & Dale, M.B. (1969b). Studies in the numerical analysis of complex rain forest communities. III. The analysis of successional data. *J. Ecol.* 57, 515–35.

Wilson, E.O. & Peter, F.M. (Eds) (1988). *Biodiversity.* Nat. Acad. Press, Washington, DC.

Wilson, M.F. (1984). *Construction and Use of Land Surface Information in a General Circulation Climate Model.* Ph.D. Thesis, University of Liverpool.

Winkler, H. (1914). Die Pflanzendecke Südost-Borneos. *Bot. Jahrb.* 50 (Festbd), 188–208.

Winkler, S. (1967). Die epiphyllen Moose der Nebelwälder von San Salvador (C.A.). *Rev. Bryol. Lichénol.* 35, 303–69.

Witkamp, H. (1925). De ijzerhout als geologische indicator. *Trop. Natuur* 14, 97–103.

Witkamp, M. (1970). Mineral retention by epiphyllic organisms. Ch. H-14 in *A Tropical Rain Forest. A Study of Irradiation and Ecology at El Verde, Puerto Rico,* Odum, H.T. & Pigeon, R.F. (Eds), Divn. Tech. Inf., US Atomic Energy Commission, Oak Ridge, Tennessee.

WMO/UNEP/ICSU (1986). Report of the International Conference on the Assessment of the Role of Carbon Dioxide and Other Greenhouse Gases in climate Variations and Associated Impacts. World Meteorol. Organ. Pubn. 661, Geneva, Switzerland.

Wolfe, J.A. (1972). An interpretation of Alaskan Tertiary floras. Pp. 201–33 in *Floristics and Palaeofloristics of Asia and Eastern North America* (Symp. of XI Int. Bot. Congr. Seattle and Japan-US Coop. Sci. Progr. Corvallis, 1969). Elsevier, Amsterdam

Wong, Y.K. & Whitmore, T.C. (1970). On the influence of soil properties on species distribution in a Malayan lowland dipterocarp rain forest. *Malay. For.* 33, 42–54.

Wood, D. (1970). The tropical forest and Sporne's advancement index. *New Phytol.* 69, 113–15.

Wood, G.H.S. (1956). The dipterocarp flowering season in North Borneo in 1955. *Malay. For.* 19, 193–214.

Worbes, M. (1986). Lebensbedingungen und Holzwachstum in zentralamazonischen Ueberschwemmungswäldern. *Scripta Geobotanica (Göttingen)* 17, 1–112.

Wright, A.C.S., Romney, D.H., Arbuckle, R.H. & Vial, V.E. (1959). *Land in British Honduras.* HMSO, London.

Wyatt-Smith, J. (1954). Storm forest in Kelantan. *Malay. For.* 17, 5–11.

Wyatt-Smith, J. (1955). Changes in composition in early natural plant succession. *Malay. For.* 18, 44–9.

Wyatt-Smith, J. (1958). Seedling/sapling survival of *Shorea leprosula, Shorea parvifolia* and *Koompassia malaccensis. Malay. For.* 21, 185–93.

Wyatt-Smith, J. (1959). Peat swamp forests in Malaya. *Malay. For.* 23, 5–32.

Wyatt-Smith, J. (1960). Stems per acre and topography. *Malay. For.* 23, 57–8.

Wyatt-Smith, J. (1963). Manual of Malayan silviculture for inland forests (2 vols.). *Malay. For. Rec.* 23.

Wyatt-Smith, J. (1966). *Ecological Studies on Malayan Forests.* I. Composition of and dynamic studies in Lowland Evergreen Rain-Forest in two 5-acre plots in Bukit Lagong and Sungei Menyala Forest Reserve and in two half-acre plots in Sungei Menyala Forest Reserve, 1947–59. *Res. Pamph.* 52, For. Res. Inst. Kepong, Malaya.

Wycherley, P.R. (1963). Variation in the performance of *Hevea* in Malaya. *J. Trop. Geogr.* 17, 144–71.

Wycherley, P.R. (1973). The phenology of plants in the Humid Tropics. *Micronesia* 9, 75–96.

Yabuki, K. & Aoki, M. (1978). Micrometeorological assessment of primary production rate of Pasoh Forest. *Malay. Nat. J.* 30 (2), 281–9.

Yamada, I. (1975–77). Forest ecological studies of the montane forest of Mt. Pangrango, W. Java, I–IV. *Tonan Ajia Kenkyu (S.E. Asian Studies)* 13, 402–26 (1975), 513–34 (1976); 14, 194–229 (1976); 15, 226–54 (1977).

Yampolsky, C. (1924). The pneumathodes on the roots of the Oil Palm (*Elaeis guineensis* Jacq.). *Amer. J. Bot.* 11, 502–12.

Yap, S.K. (1980). Phenological behaviour of some fruit tree species in a lowland Dipterocarp forest of West Malaysia. Pp. 161–7 in *Tropical Ecology and Development* 1, Furtado, J. (Ed.), Internat. Soc. Trop. Ecol., Kuala Lumpur, West Malaysia.

Yew, F.K. (1979). Potassium-supplying power of seven soils under rubber. *Planters Bull.* 158, 28–9.

Yih, K., Boucher, D.H., Vandermeer, J.H. & Zamorra, Z. (1991). Recovery of the rain forest of southeastern Nicaragua after destruction by Hurricane Joan. *Biotropica* 23, 106–13.

Yoda, K. (1978). Three-dimensional distribution of light intensity in a tropical rain forest of West Malaysia. *Malay. Nat. J.* 30 (2), 161–77.

Young, A. (1976) *Tropical Soils and Soil Survey.* Cambridge University Press.

Young, K.R. (1985). Deeply buried seeds in a tropical forest in Costa Rica. *Biotropica* 17, 336–8.

Young, T.P. & Perkocha, V. (1994). Treefalls, crown asymmetry and buttresses. *J. Ecol.* 82, 319–24.

Zimmermann, M.H. & Brown, C.L. (1971). *Trees: Structure and Function*. Springer, Berlin.

Zon, P. van (1915). Mededeelingen omtrent den kamferboom (*Dryobalanops aromatica*). *Tectona* 8, 220–4.

Zulkifli, Y., Anhar, S. & Baharuddin, K. (1990). Postlogging effects on suspended solids and turbidity – five years observation. Paper presented at the Workshop on Watershed Development & Management, 19–23 February 1990, Kuala Lumpur.

Zwetsloot, H. (1981). Forest succession on a deforested area in Suriname. *Turrialba* 31, 369–79.

Index of plant names

Acacia, Leguminosae (Mimosoideae) 404, 408
  A. cornigera (L.) Willd. 133
  A. farnesiana (L.) Willd. 484
  A. pennata (L.) Willd. 445
Acanthaceae 85, 122
Acanthus, Acanthaceae 384
  A. ilicifolius L. 375
  A. montanus T. Anders 123
Acer, Aceraceae 243
  A. rubrum L. 3
Aceraceae 243
Acrostichum, Adiantaceae
  A. aureum L. 343, 373, 378, 381
  A. danaeifolium Langsd. & Fisch. 353
Adansonia, Bombaceae 408
  A. digitata L. (Baobab) 408
Adiantaceae
Adiantum, Adiantaceae 323
Adinandra, Theaceae 460
  A. dumosa Jack 250, 464, 478
Aechmea, Bromeliaceae 141, 149
Aegiceras, Myrsinaceae 375
Aeschynanthus, Gesneriaceae 142, 144
Aeschynomene, Leguminosae (Papilionoideae)
  A. elaphroxylon (Guill. & Perr.) Taub. 77
Aesculus, Hippocastanaceae
  A. hippocastanum L. 2
Afrotrilepis, Cyperaceae
  A. pilosa (Boeck.) J. Raynal 350
Afzelia, Leguminosae (Caesalpinioideae)
  A. africana Sm. ex Pers. 407, 411
Agathis, Araucariaceae 19, 321, 423, 433, 452
  A. alba Foxworthy 4
  A. australis Steud. 4
  A. borneensis Warb. 42, 64, 96, 307, 325
  A. microstachys J.F. Bailey & C.T. White 50
Agelaea, Connaraceae
  A. trifolia (Lam.) Gilg 131

Ageratum, Compositae
  A. conyzoides L. 362, 468
Aglaia, Meliaceae 425
Agropyron, Gramineae 383
Agrostistachys, Euphorbiaceae 72
Alang-alang (Lalang), *see Imperata cylindrica*
Alangium, Alangiaceae 14
  A. chinense (Lour.) Harms 447
Albizia, Leguminosae (Mimosoideae) 99, 404, 445
  A. adianthifolia (Schum.) W.F. Wight 469
  A. falcataria (L.) Fosb. (*A. falcata* (L.) Back. ex Merrill) 350, 466
  A. saman (Jacq.) F. Muell. (*Samanea saman* (Jacq.) Merrill) 148, 255
  A. zygia (DC.) J.F. Macbride 470, 471
Alchornea, Euphorbiaceae 476
  A. castaneifolia (Willd.) A. Juss. 357, 358, 362
  A. cordifolia (Schum. & Thonn.) Muell. Arg. 361, 362, 363, 406
  A. laxiflora (Benth.) Pax & K. Hoffm. (*Lepidoturus laxiflorus* Benth.) 334
Alder (*Alnus*), Betulaceae 85
Aldina, Leguminosae (Papilionoideae) 40, 135, 136, 475
  A. discolor Spruce ex Benth. 323, 324
  A. heterophylla Benth. 324
  A. latifolia Benth. 244, 358
Allagoptera, Palmae
  A. arenaria (Gomez) O. Ktze. 386
Allanblackia, Guttiferae
  A. stuhlmannii (Engl.) Engl. 445
Allophyllus, Sapindaceae
  A. bullatus Radlk. 447
Alphitonia, Rhamnaceae
  A. zizyphoides A. Gray 237
Alstonia, Apocynaceae 461, 467, 472
  A. angustiloba Miq. 78
  A. boonei De Wild. 63, 77, 89, 350
Alternanthera, Amaranthaceae
  A. maritima (Mart.) St.-Hil. 383, 385, 386
Altingia, Hammamelidaceae
  A. excelsa Noronha 147, 435

[541]

Ficus, Moraceae (*cont.*)
   F. lepicarpa Bl. 118
   F. leprieurii Miq. 77
   F. macrophylla Desf. ex Pers. 452
   F. mucuso Ficalho 362
   F. obpyramidata King 118
   F. pubinervis Bl. 346
   F. religiosa L. 98
   F. retusa Wall. 381
   F. sumatrana Miq. 238
   F. tinctoria Forst. f. subsp. *gibbosa* (Bl.) Corner 346
   F. variegata Bl. 241, 242
Fimbristylis, Cyperaceae 124
Fissidens, Fissidentaceae (Musci) 34, 150
Fissidentaceae (Musci) 34, 150
Flacourtiaceae 491
Floribundaria, Meteoriaceae (Musci) 152, 360
Fordia, Leguminosae (Papilionoideae)
   F. gibbsiae Dunn. & E.G. Baker 482
Forrestia, Commelinaceae 125
Forsythia, Oleaceae 119
Franklinia, Theaceae
   F. alatamaha Marshall 489
Fraxinus, Oleaceae 243
   F. excelsior L. 111
Freycinetia, Pandanaceae 128
Frullania, Frullaniaceae (Hepaticae) 150
Fuchsia, Onagraceae
   F. excorticata L. f. 119

Gaertnera, Rubiaceae
   G. paniculata Benth. 362
Gahnia, Cyperaceae
   G. borneensis Benl. 432
Galeola, Orchidaceae 153
Garcinia, Guttiferae 96, 238
   G. mangostana L. 103
   G. parvifolia (Miq.) Miq. 103
Garuga, Burseraceae 425
   G. floribunda Dcne. 248, 327
Gaylussacia, Ericaceae
   G. amazonica Huber 324
Geissospermum, Apocynaceae
   G. sericeum (Sagot) Benth. & Hook. 60, 61
Gentianaceae 153
Geonoma, Palmae 75
Geophila, Rubiaceae 122, 331, 469
   G. herbacea (Jacq.) K. Schum. 248
Gesneriaceae 122, 138
Gilbertiodendron, Leguminosae (Caesalpinioideae)
   G. dewevrei (De Wild.) J. Léonard 40, 41, 54, 63, 68, 82, 251, 252, 329, 331–2, 473
Gleichenia, see *Dicranopteris*
Glochidion, Euphorbiaceae 461
Gluta, Anacardiaceae 14, 50, 67
   G. renghas L. 238
Glycoxylon, Sapotaceae 495
   G. inophyllum (Mart. ex Miq.) Ducke 324
Gmelina, Verbenaceae
   G. arborea Roxb. 414, 417, 488
Gnetaceae 128, 130
Gnetum, Gnetaceae 128, 130

Goethalsia, Tiliaceae 464
   G. meiantha (D. Sm.) Burret 64, 462, 477
Goniothalamus, Annonaceae
   G. malayanus Hook. f. & Thoms. 124
Gonystylus, Thymelaeaceae 326
   G. bancanus (Miq.) Kurz. 366, 371
Goupia, Celastraceae
   G. glabra Aubl. 60, 3124, 315–16, 461, 464, 473
Gossweilerodendron, Leguminosae 110
Greenheart, see *Ocotea rodiei*
Grewia, Tiliaceae
   G. coriacea Mast. 78, 101
   G. paniculate Roxb. 480
   G. tomentosa Juss. 479
Grias, Lecythidaceae
   G. cauliflora L. 118
Groutiella, Orthotrichaceae (Musci) 150
Guaduella, Gramineae
   G. humilis W.D. Clayton 123
Guarea, Meliaceae
   G. cedrata (A. Chev.) Pellegr. 50, 63
   G. glabra Vahl 39
   G. thompsonii Sprague & Hutch. 63
Guazuma, Sterculiaceae
   G. ulmifolia Lam. 327, 401
Guibourtia, Leguminosae (Caesalpinioideae)
   G. demeusii (Harms.) J. Léonard 87, 110, 244, 363
Guttiferae 432
Gymnosiphon, Burmanniaceae 153
Gymnostoma nobile, see *Casuarina nobilis*
Gynerium, Gramineae
   G. sagittatum (Aubl.) P. Beauv. 353

Haemanthus, Amaryllidaceae
   H. cinnabarinus Decne. 123
Hagenia, Rosaceae
   H. abyssinica (Bruce) Gmel. 444
Harrisonia, Simaroubaceae
   H. abyssinica Oliv. 445
Harungana, Guttiferae 473
   H. madagascariensis Lam. ex Poiret 350, 415, 461, 472
   H. robynsii Spirlet 363
Haumania, Marantaceae
   H. liebrechtsiana (De Wild. & Th. Dur.) J. Léonard 8, 333
Hedychium, Zingiberaceae
   H. longicornis Bak. 149
Helianthus, Compositae
   H. annuus L. 465, 466
Heliconia, Heliconiaceae 473
Heliocarpus, Tiliaceae
   H. appendiculatus Turcz. 466
Heliocostylis Moraceae
   H. sp. cf. asperifolia Ducke 324
Helosis, Balanophoraceae
   H. cayenensis (Sw.) Spreng. 155
Heritiera, Sterculiaceae
   H. littoralis Aiton 382
   H. macrophylla Wall. ex Voigt 241
   H. utilis (Sprague) Sprague (*Tarrietia utilis* (Sprague) Sprague) 77, 85, 414, 448
   H. nymphaeifolia (Presl.) Kubitzki (*H. peltata* Meissn.) 342

General Index

Note: page numbers in *italics* refer to figures and tables.

[559]